Dynamic Routing in Telecommunications Networks

Other McGraw-Hill Communications Books of Interest

Dynamic Routing in Telecommunications Networks

Gerald R. Ash

McGraw-Hill

New York   San Francisco   Washington, D.C.   Auckland   Bogotá
Caracas   Lisbon   London   Madrid   Mexico City   Milan
Montreal   New Delhi   San Juan   Singapore
Sydney   Tokyo   Toronto

Library of Congress Cataloging-in-Publication Data

Ash, Gerald R.
    Dynamic routing in telecommunications networks / Gerald R. Ash,
        p.   cm.
    Includes index.
    ISBN 0-07-006414-8
      1. Telecommunication–Switching systems.   2. Telecommunication–
Traffic.   I. Title.
    TK5103.8.A84   1997
    621.382ʹ16–dc21   97-42105
                                                   CIP

McGraw-Hill

A Division of The **McGraw·Hill** *Companies*

1 2 3 4 5 6 7 8 9 0 DOC/DOC 9 0 2 1 0 9 8 7

ISBN 0-07-006414-8

*The sponsoring editor for this book was Steve Chapman and the
production supervisor was Tina Cameron. It was set in Century
Schoolbook by Publication Services, Inc.*

Printed and bound by R. R. Donnelley & Sons Company.

 This book is printed on recycled, acid-free paper containing a
minimum of 50% recycled, de-inked fiber.

McGraw-Hill books are available at special quantity discounts to use
as premiums and sales promotions, or for use in corporate training
programs. For more information, please write to the Director of Special
Sales, McGraw-Hill, 11 West 19th Street, New York, NY 10011. Or
contact your local bookstore.

For Lynsie

Contents

Preface

This book is about the major evolution of telecommunications networks that has taken place over the past two decades. In 1976, when electronic switching (the 4ESS™ switch) and common-channel interoffice signaling (CCIS) were first introduced to the AT&T network, the author took the job of leading a newly formed group, the routing studies group, to examine the possibilities of introducing dynamic routing into the network. The time was ripe for such an undertaking, because stored program control in the 4ESS switch and the greatly expanded capability to exchange data between switches with CCIS enable far more elaborate routing strategies than the hierarchical routing plan in place since the earliest days of the network. This, of course, did not mark the first time that dynamic routing was considered, having been the subject of many studies over a long period of time, but it was the first time that a practical implementation technology was being introduced to the network.

Over a period of several years, the group reviewed earlier and ongoing work on dynamic routing and made several fundamental advances in new routing algorithms, network design methods, network management methods, network impact analysis, cost/benefit analysis, and implementation plans. From 1976 to the present, the author has led the evolution of dynamic routing in all facets of the AT&T worldwide intelligent network. Through visionary leadership of many people including Vernon Mummert, who was the Network Management Department Head at Bell Laboratories, and Billy Oliver, who was Vice President of Technology at Long Lines, the compelling results of studies performed by the group led to the implementation of dynamic nonhierarchical routing (DNHR), beginning in July 1984. Over a period of seven years, through October 1987, DNHR was implemented throughout the AT&T network and was the first national dynamic routing network implemented in the world. DNHR was followed by a major upgrade to real-time network routing (RTNR) in 1991 and to real-time internetwork routing (RINR) in the global international network in 1995.

This DNHR, RTNR, and RINR deployment has revolutionized the switching, configuration, management, and design of the entire global AT&T network. This book gives the technical details of the wide variety of routing methods studied and ultimately implemented, network design and management methods that were developed and used, network impact studies and cost/benefit analysis, network implementation plan details, and many examples of the actual performance of dynamic routing networks. This is a practical engineering book describing alternatives studied and methods successfully used to implement dynamic routing networks. There are no books as yet devoted to this subject, although many papers have appeared on the implementation of dynamic routing and its clear success in improving network performance while reducing its cost. These papers are referenced in the book, as are books that in part discuss dynamic routing networks.

The successful deployment of dynamic routing and its proven benefits, together with many related studies discussed and referenced in the book, has inspired many other administrations to plan and implement dynamic routing networks. Dynamic routing has made major strides in the past few years and is now deployed in ten networks that include AT&T USA, FTS-2000 USA, MCI USA, Sprint USA, Bell Canada, Stentor Canada, NTT Japan, British Telecom United Kingdom, Norwegian Telecom Norway, and the worldwide international network (WIN) consortium of carriers. In all these applications dynamic routing has provided considerable benefits in improved performance quality and reduced costs. Dynamic routing in telecommunications networks therefore has become the subject of worldwide study and interest. Service providers, equipment providers, and academic institutions throughout the world are now engaged in active research and development in this area. This book provides helpful guidance in supporting studies and network evolution plans for administrations considering dynamic routing evolution and provides state-of-the-art technological details of interest to researchers engaged in studies of dynamic routing techniques. This book addresses many dynamic routing configurations, technologies, and topics of active research. Dynamic routing network configurations discussed in the book include metropolitan area networks, national intercity networks, and global international networks. Network technologies discussed include circuit-switched networks, packet-switched networks, and broadband networks. Research topics discussed in the book include multiservice integrated dynamic routing networks, dynamic internetwork routing, dynamically rearrangeable transport routing, and advanced dynamic routing network design methods.

The book describes and compares all categories of dynamic routing networks in operation today and many others that are proposed, across various criteria including network design efficiency, implementation cost, and performance. The dynamic routing categories treated include dynamic nonhierarchical routing (DNHR), deployed in the AT&T network from 1984 to 1991 and in the FTS-2000 network in 1987; real-time network routing (RTNR), deployed in the AT&T network in 1991; real-time internetwork routing (RINR), deployed in the AT&T network in 1995; dynamically controlled routing (DCR), deployed in the Stentor Canada long-distance network in 1991, in the Bell Canada Toronto metropolitan area network in 1991, in the Sprint network in 1994, and in the MCI network in 1995; dynamic alternative routing (DAR), deployed in the British Telecom network in 1996 and the Norwegian Telecom Network in 1996; state- and time-dependent routing (STR), deployed in the Nippon Telephone and Telegraph network in 1992; system to test adaptive routing (STAR), developed and tested by France Telecom in the Paris network; DR-5, developed and tested by Bellcore for application to metropolitan Local Exchange Carrier networks; and worldwide international network (WIN) dynamic routing, deployed in 1994 by a consortium of several countries. DNHR is a preplanned time-variable approach, RTNR/RINR is a distributed real-time state-dependent approach, DCR is a centralized real-time periodic state-dependent approach, DAR is a distributed real-time event-dependent approach, STR is a hybrid distributed real-time event-dependent and preplanned time-variable approach, and WIN is a distributed real-time periodic state-dependent approach.

The book is intended as a definitive work on dynamic routing and a practical engineering guide to teach and enable the reader to analyze routing models, design dynamic routing networks, and operate and manage a dynamic routing network. Theoretical material that is presented gives the reader a foundation in the basic concepts of traffic theory and network design. Extensive material is presented that illustrates practical methods and results for all classes of routing networks, as summarized in Table 1.1 in Chapter 1. Within each class of routing method there are often several, sometimes many, variations. Within preplanned dynamic routing, there are multilink path routing, two-link path routing, cyclic path routing, CGH path routing, and many other methods; in the book we study DNHR as the representative method for preplanned dynamic routing. Within real-time state-dependent routing, there are WIN, STAR, DR-5, DCR, TSMR, and RTNR. Within these, the periodic update real-time state-dependent routing systems include WIN, STAR, DR-5, DCR, and TSMR, and in the book we study TSMR as a representative method within this class. Real-time event-dependent routing systems include DAR, STR, LRR, and STT. We study LRR and STT as representative methods of this class. Comparisons of generic classes of routing methods are given in terms of performance, network cost, implementation complexity, development cost, network impacts, and other measures. We do not attempt to study in detail each implemented routing technique, because (1) the exact details of implemented routing methods, such as DCR, DAR, or STR, are not publicly available and (2) there are small differences between routing methods within the same class, such as between TSMR with 10-second updates and DCR or between LRR and DAR. Rather, the book compares generic classes of routing methods by developing detailed examples within each class and illustrating how these routing methods are designed, developed, and evaluated. This development of the specific examples will then enable readers to develop their own variations of routing methods within each class.

A brief outline is given of the 18 chapters of the book, followed by acknowledgments of the many people who collaborated in the work, most of whom have worked in the author's group over the past 20 years. These people, and their collaborations with the author in advancing the state of the art so dramatically in dynamic networking, form the heart and soul of the book.

Chapter 1 covers an overview of dynamic routing networks and discusses various network configurations, routing methods, and network management and design methods. It addresses the design, analysis, operation, implementation, and performance of dynamic routing networks and discusses basic concepts, examples of current applications, and topics of ongoing research. It gives a tutorial overview of the principles of dynamic routing networks and factors driving their evolution, which include performance quality improvements, introduction of new services, and technological evolution. We discuss in some detail network architectures including metropolitan area networks, national intercity networks, and global international networks. The roles of network elements in providing various services such as long-distance service, software-defined network service, 800 service, and international long-distance service are described. The chapter discusses three stages of network routing evolution, which include hierarchical traffic routing, deployed in most networks in the world today; dynamic traffic routing, illustrated by AT&T's DNHR network and currently deployed RTNR network; and dynamic traffic and transport routing, illustrated by AT&T's fast automatic restoration (FASTAR™) capability. Performance examples are given of operating dynamic routing networks that include high-day loads on the Monday after Thanksgiving, peak-day loads on Thanksgiving and Christmas, a focused overload caused by Hurricane Bob, an international network overload during the Chinese New Year holiday, and a cable cut near Austin, Indiana. The elements of network design are discussed for the various network load variations, including minute-to-minute, hour-to-hour, day-to-day, and week-to-week variations. Three multihour dynamic network design models are described and compared, which include erlang flow optimization models, transport flow optimization models, and discrete event flow optimization models. Using the design models described in the chapter, comparisons are made across different routing systems including DNHR, RTNR, DCR-type routing methods, DAR-type routing methods, and WIN-type routing methods. These comparisons are made for metropolitan area network designs, national intercity network designs, and global international network designs. The comparisons include network design efficiency, implementation complexity, and performance across a range of network scenarios including various network overloads and element failures.

Chapter 2 addresses hierarchical networks, which, except for the ten dynamic routing networks mentioned above, are ubiquitous throughout the world today in all network applications. In this chapter we discuss the elements of networking, including the fundamentals of switching, trunk terminations and hunting, numbering plans and digit translations, signaling, and billing. The chapter reviews the routing rules applied in hierarchical network design, which are perhaps more complex than may be realized and lead to many different variations of hierarchical network implementation. We review the fundamentals of network design traffic engineering, including basic traffic models, the Erlang B formula, the Wilkinson equivalent random method for peaked traffic, hierarchical network design principles including Truitt's ECCS technique for sizing high-usage links, cluster busy-hour sizing techniques, Neal-Wilkinson methods for sizing final links, and hierarchical network design principles used in the comparisons presented in Chapter 1.

Chapters 3 to 6 discuss preplanned routing methods that have been implemented or are representative of existing applications. Chapters 4 to 6 develop a useful network design methodology for a very wide class of networks. The path/route erlang flow optimization method (also known as the unified algorithm) is widely recognized as a breakthrough network design technique that has broad, deep, and lasting importance. It enables the design of multilink path routing networks (Chapter 4), progressive routing networks (Chapter 5), and two-link path routing networks (Chapter 6). The method is comprehensively taught to the reader through the examples in these chapters, and the reader can then apply the method to a large class of other network designs. Chapters 3, 4, 5, and 6 discuss various dynamic routing methods and network designs that include preplanned dynamic hierarchical routing networks in Chapter 3, preplanned dynamic multilink path routing networks in Chapter 4, preplanned dynamic progressive routing networks in Chapter 5, and preplanned dynamic two-link path routing networks in Chapter 6.

Preplanned dynamic hierarchical routing, described in Chapter 3, uses an underlying hierarchical network structure but allows link capacity to be dynamically rearranged by formation of triad sublink paths. A linear programming model is described that provides the design of the triad sublink path routing patterns and the capacity requirements of the network. Such networks are shown to have good design efficiency but practical limitations for implementation. Dynamic hierarchical networking was at one time tested in the AT&T network and could well be considered in other network applications. The design methods developed are unique and useful for other network applications.

In Chapter 4 we present preplanned dynamic multilink path routing networks and a network design methodology called the route-erlang flow optimization model, which provide a fundamentally new approach for evaluating a wide variety of dynamic routing methods not possible prior to this advance. In particular, preplanned path routing with cyclic path selection and ordered path selection are examined for multilink, as well as two-link, routing design. It is shown that preplanned two-link path routing networks show promise for design efficiency and implementation simplicity.

This route-erlang flow optimization model is applied again in Chapter 5 to examine preplanned dynamic progressive routing networks in which the routing patterns are again preplanned time-variable, but in which calls route progressively from the originating switch to the destination switch rather than selecting a complete path from the originating switch to the destination switch, as in path routing. Progressive routing networks are implemented in the AUTOVON and other defense networks, and path routing networks are implemented in the FTS-2000 network, and both are candidates for future dynamic networking applications.

In Chapter 6, studies of alternative dynamic routing networks zero in on preplanned two-link dynamic routing networks and introduce the path-erlang flow optimization model, which is an important advance in designing preplanned two-link path networks. In this chapter various path-selection methods are examined and compared, including CGH, skip-one-path, and sequential path selection. It is shown that the simple preplanned two-link sequential path routing method achieves nearly the same design efficiencies as far more complex methods such as CGH but has significant implementation advantages. This method, then, forms the basis for DNHR as introduced in 1984.

Chapters 7 to 10 further develop essential elements of dynamic routing networks in relation to dynamic routing design for a four-switch network example (Chapter 7), real-time routing design (Chapter 8), dynamic routing design under forecast uncertainty (Chapter 9), and dynamic routing design for multiservice integrated networks (Chapter 10). Although we use preplanned dynamic two-link path routing for illustration in these chapters, the principles developed are generally applicable to all types of dynamic routing networks.

Chapter 7 gives a detailed example of the design of a four-switch network using the path-erlang flow optimization model and illustrates the detailed workings of that design method implemented as part of the DNHR introduction in 1984. Because the model is illustrated in detail for a small network, the step-by-step implementation and design principles applied are made very clear. These principles include shortest-path selection, multihour dynamic routing design, multihour capacity optimization, and link flow optimization, which are generally applicable to the design of dynamic routing networks of all types and are applied throughout the book.

In Chapter 8 we study dynamic routing design under random load variations, in which real-time routing methods are combined with preplanned dynamic routing and where the design principles discussed apply in general to all dynamic routing networks. Real-time dynamic routing can lead to more efficient and better-performing networks under random day-to-day variations, week-to-week variations, and deviations from design loads caused by forecast errors. This chapter develops in-depth analytical and simulation models for real-time dynamic routing techniques that include trunk reservation, call gapping, and real-time path selection. We develop the notion of service protection through trunk reservation, which is used in all dynamic routing networks to ensure their stability and efficient use of network capacity. Call-by-call simulation models are described in some detail and demonstrated to be powerful tools able to go far beyond the ability of analytical models to represent complex network behavior. Such models and techniques are applied and extended in later chapters to real-time state-dependent, real-time event-dependent, and dynamic transport routing networks.

Chapter 9 discusses dynamic routing design under forecast uncertainty and shows how ongoing routing design based on current traffic load levels and patterns leads to reduction in network reserve capacity through more efficient utilization of capacity available in the network. The chapter discusses the trade-offs between reserve capacity and the amount of short-term capacity adjustment and illustrates the effect of dynamic routing adaptation on this trade-off. Examples are given for dynamic routing table updates with either preplanned or real-time dynamic routing, and in the latter case the routing table updates are made automatically in the network according to the real-time dynamic routing method.

Chapter 10 describes dynamic routing design principles employed in multiservice integrated networks, which include bandwidth allocation strategies, dynamic routing call setup, network management procedures, and integrated network design models. These principles are illustrated for preplanned and real-time dynamic routing networks and apply in general to all dynamic routing networks. This routing and design methodology was initially applied to the extension of DNHR to ISDN services, including switched 64-kbps data services, and the desire to extend dynamic routing to new services such as switched 64-kbps data services and other emerging services led ultimately to the upgrade of DNHR to multiservice integrated dynamic routing (with RTNR) in 1991. In the next three chapters we elaborate on the application of these principles to multiservice integrated networks with real-time dynamic routing.

Chapters 11, 12, and 13 provide details of the design and implementation of real-time state-dependent routing. In Chapter 11 we study trunk status map routing (TSMR) as a representative example of centralized real-time dynamic routing methods, and in Chapter 12 we study RTNR as a representative example of distributed real-time dynamic routing methods. The methods and principles applied to the design and evaluation of these examples also apply to the many possible variations of each method. In applying these methods and principles, the reader is able to evaluate any of the possible variations.

Chapter 11 examines centralized real-time dynamic routing networks, particularly TSMR. TSMR envisions the use of a central status database maintained in real time for computation of optimum routing tables based on the instantaneously available capacity and combines two-link sequential path design with real-time path selection. Studies described in Chapter 11 illustrate the advantages of real-time dynamic routing in improving the design efficiency and performance in comparison with preplanned dynamic routing. The protective and expansive mechanisms of TSMR, such as trunk reservation, extended routing logic, and other mechanisms, are discussed in detail in the chapter. All good routing techniques, in fact, must employ such service protection mechanisms. Many of the TSMR protective and expansive routing mechanisms studied in the chapter are now implemented in RTNR.

Much of the routing and congestion control methodology developed in Chapter 11 is appled in Chapter 12, which discusses distributed real-time dynamic routing for multiservice integrated networks. The illustrative routing method studied in Chapter 12 is RTNR. We present details of class-of-service routing for multiservice integrated networks, RTNR, multiple ingress/egress routing, and RINR. Simulation modeling results are given related to the design of many of the routing parameters and methods and which illustrate RTNR and RINR performance across many different network scenarios. The phenomenon of network instability is illustrated through these simulation studies, and the effectiveness of trunk reservation in eliminating instability is demonstrated. Real-time event-dependent routing methods are further developed in Chapter 12.

Chapter 13 discusses network design for real-time dynamic routing networks and gives design methods and illustrations of path-erlang flow optimization with alternative day-to-day variation models, fixed-point erlang flow optimization models, transport flow optimization models, and discrete event flow optimization models. In particular, the latter two models provide practical techniques for design of real-time dynamic routing networks and therefore have been applied in RTNR network operation. Methods are also illustrated in Chapter 13 for selection of RTNR paths based on traffic pattern analysis, in which it is shown that limiting path selection to a subset of preferred paths, such as the path selection used in preplanned dynamic routing, is also potentially beneficial for real-time dynamic routing for maximum network utilization efficiency.

It is shown in Chapters 14 and 15 that joint consideration of traffic and transport dynamic routing is important within network design to meeting various performance objectives under normal conditions and under overload and failure conditions. As such, these chapters illustrate dynamic routing evolution to dynamic transport networking, discussing the role of dynamic routing as networks evolve to asynchronous transfer mode (ATM) high-speed packet broadband networks. Chapter 14 discusses methods for reliable traffic and transport network design, and in particular shows how link diversity routing for transport reliability, dynamic transport restoration routing for transport reliability, and multiple homing for switching reliability lead to improved network performance that meets network design objectives under failure.

Chapter 15 discusses dynamic traffic and transport routing networks, and in particular illustrates rearrangeable transport routing mesh network designs and real-time transport routing ring network design. A hybrid circuit-switched and high-speed packet-switched transport network design is presented, which incorporates elements of both the rearrangeable network and real-time ring network design. It is shown that there are considerable benefits in performance improvement and reduction in reserve capacity for these dynamic transport network alternatives.

Chapters 16, 17, and 18 give implementation feasibility study results, implementation requirements, economic studies, network management and design experience, and performance experience for dynamic routing networks. Chapter 16 discusses dynamic routing feasibility, economic analysis, and implementation requirements. Feasibility studies include real-time traffic management, transmission performance, processing load on the ESS/CCS network, traffic data collection, capacity management, network design, and large-scale optimization. Implementation requirements are discussed for ESS/CCS network development requirements and network management and design system development requirements. Study results reported in the chapter show that dynamic routing implementation is both technically feasible and economically justifiable. Business justification analysis shows that dynamic routing benefits far outweigh the costs, where costs include development, network impact, and transition costs. Benefits include (1) capital savings, (2) customer revenue retention through improved network reliability, (3) additional revenue through improved call completions, (4) new service revenue based on priority routing technology, (5) service revenues through integration of voice and ISDN data services allowing competitive price reduction of ISDN data services, (6) operational expense savings, (7) switch development cost avoidance and fast feature introduction through standardized class-of-service routing functions, and (8) network cost avoidance through class-of-service network integration versus building dedicated overlay networks. Standardized information exchange is necessary so that switching equipment from different vendors interacts to implement dynamic routing methods in a coordinated fashion, and thereby service providers who have multivendor networks can reap the efficiency and performance benefits of dynamic routing in their networks. This information exchange is illustrated with the use of standardized common-channel signaling messages for various dynamic routing methods.

Chapter 17 discusses network management and design implementation for dynamic routing networks and uses the 16-switch initial DNHR network implemented in 1984 to illustrate advances in real-time traffic management, capacity management forecasting, capacity management performance monitoring, and capacity management short-term network adjustment. Actual network experience and reports are used to illustrate the workings of dynamic routing network management and design, which revolutionized network operational methods to more centralized and automated operations.

Chapter 18 illustrates dynamic routing implementation experience through numerous examples drawn from the DNHR and RTNR network operation, including peak-day performance on Thanksgiving and Christmas, high-day performance on the Monday after Thanksgiving, network overload experience under focused overload during natural disasters, and network performance under various failure conditions. Chapters 16 to 18 reinforce the practicality and considerable economic and performance benefits of dynamic routing implementation.

The author has been privileged to work with a large number of highly talented, gifted individuals over the course of the work covered in this book. Vernon Mummert promoted him into the position of Supervisor of the Routing Studies Group, Bell Laboratories, in June 1976, at which time the first 4ESS switch and CCIS were being deployed in the AT&T network. Richard Cardwell, Pamela Turner, and Alan Westreich were among the charter members of the group and contributed fundamentally to the early studies. Ms. Turner did the dynamic hierarchical routing modeling discussed in Chapter 3, Mr. Cardwell did the development and analysis of the route-erlang flow optimization model and the path-erlang flow optimization model (also known as the unified algorithm) described in Chapters 4 and 6, and Mr. Westreich applied the route-erlang flow optimization model to progressive routing networks, as described in Chapter 5. Ronald Graham, Fan Chung, and Frank Hwang of the Mathematics of Networks research department collaborated in the development of the path-erlang flow optimization model, and in particular developed the CGH routing technique discussed in Chapter 6. Richard Wong and Robert Murray performed the fundamental work on large-scale optimization methods and developed the heuristic solution method within the path-erlang flow optimization model discussed in Chapter 6. Luc Nguyen and Hemant Thapar were instrumental in developing traffic analysis methods and helped in overall development of the path-erlang flow optimization model.

The four-switch example of the path-erlang flow optimization model given in Chapter 7 was formulated by Frank Field, who also performed studies of dynamic routing applied to the Chicago metropolitan area network model described in Chapter 1 and performed design and economic analysis of DNHR networks discussed in Chapter 16. Alan Kafker collaborated with the author in the development of the real-time routing design models described in Chapter 8 and also did studies of the ESS and CCS impacts discussed in Chapter 16.

K. R. Krishnan performed studies of dynamic routing design under forecast uncertainties, as discussed in Chapter 9, and also analyzed real-time dynamic routing methods. Bruce Blake collaborated with the author on the extension of the path-erlang flow optimization model to multiservice integrated networks, as discussed in Chapter 10. Many people in 4ESS switch development contributed very significantly to DNHR implementation, including Elliott Baral, James Beck, Paul Carestia, James Carroll, and Mel DiCarlo Cottone.

Jin-Shi Chen, BaoSheng Huang, and Fu Chang made fundamental contributions to the development of the RTNR strategies described in Chapter 12. The concept of exchanging bit maps to encode network status was conceived by Alan Frey. Many other people in 4ESS switch development contributed very significantly to RTNR/RINR implementation, including James Beck, Dana Garoutte, Robert Gerritson, Kathy Laskowski, Clement Liu, and Andrew Peck. David McGuigan made fundamental contributions to the design of class-of-service routing, and Matthew Liotine and Gail MacDonald to the design of multiple ingress/egress routing discussed in Chapter 12. Jiayu Chen, Saul Fishman, Chin Lee, BaoSheng Huang, and David Zerling were deeply involved in the design of real-time internetwork routing discussed in Chapter 12.

Lindsay Hiebert and Fu Chang collaborated on the transport flow optimization models, and BaoSheng Huang contributed to the fixed-point erlang flow optimization models discussed in Chapter 13. Fu Chang also performed the network analysis of the network bandwidth allocation techniques developed in Chapter 13 and applied the Karmarkar algorithm to solving the large-scale linear programming optimization models described in Chapter 13. Fu Chang, Joshua Dayanim, and Deepankar Medhi worked with the author in developing the reliable network design techniques examined in Chapter 14, and Kenneth Chan, Steven Dodd, John Labourdette, and Steven Schwartz all did basic studies of dynamic transport networks examined in Chapter 15. Frank Field, Detlev Haenschke, Alan Kafker, Richard Jessup, and Robert Melville were all involved in various aspects of the feasibility, economic, and implementation studies discussed in Chapter 16. Joseph Alfred and Arik Kashper worked with the author on the dynamic routing interworking methods discussed in Chapter 16. Interviews with Terry Brown, John Dudash, and Gerald McCurdy formed the basis for the material presented in Chapter 17 on network management and design experience in dynamic routing networks. Finally, the examples provided in Chapter 18 on DNHR and RTNR network experience were the result of collaborations with Chin Lee and Eric Oberer.

A large community of people outside of AT&T has provided many stimulating exchanges of ideas over the years, including Hugh Cameron of Bell Northern Research Canada, Prosper Chemouil of France Telecom France, Lars Engvall of the International Telecommunications Union Switzerland, Januz Filipiak of the University of Krakow Poland, David Garbin of the Defense Communications Agency USA, Richard Gibbens of Cambridge University UK, Joseph Hui of Rutgers University USA, Akiya Inoue of NTT Japan, Narendra Karmarkar of Lucent Technologies USA, Konosuke Kawashima of NTT Japan, Frank Kelly of Cambridge University UK, Anis Khalil of MCI USA, Edward Knepley of the Defense

Communications Agency USA, Xiong-Jian Liang of Beijing University of Posts and Telecommunications China, Kinichi Mase of NTT Japan, Debasis Mitra of Lucent Technologies USA, K.S. Narendra of Yale University USA, Toshikane Oda of KDD Japan, Baron Peterssen of Telkom South Africa, Michal Pioro of Warsaw University of Technology Poland, K.G. Ramakrishnan of Lucent Technologies USA, Jean Regnier of Bell Northern Research Canada, Martin Reiman of Lucent Technologies USA, James Roberts of France Telecom France, Keith Ross of the University of Pennsylvania USA, Gyula Sallai of the Communications Authority Hungary, Mischa Schwartz of Columbia University USA, and Yu Wantanabe of KDD Japan.

The author is deeply indebted to all of these people for their pioneering, creative work over the past 20 years, and it has been the author's privilege to work with these talented individuals. However, all the great work ever done would never see the light of day without executives who have the vision to see the worth of these ideas and convert them to plans, to forge ahead with them in spite of the normal resistance to change usually encountered in such large-scale migration to radical new ideas. Among these visionaries I would count Vernon Mummert, Billy Oliver, Richard Ambler, John Healy, William Roach, Phillip Timm, and Alan Wright in the implementation of DNHR; Douglas Rom and Lynne Schaeffer in the implementation of RTNR; and Edgar Grijalva, Ralph Hohmann, and Helen McGrath in the implementation of RINR. To all these people, my heartfelt thanks—it's been quite a ride.

Gerald R. Ash
g.r.ash@att.com
October 1997

About the Author

Gerald R. Ash is the Manager of Routing Evolution Planning in the Network Technology Development division of AT&T Labs, where dynamic routing was pioneered. He has worked in this area since 1976 and is widely regarded as a world expert on the subject of routing. His research has been quoted extensively in textbooks and articles on telecommunications networks. Mr. Ash lives and works in Monmouth County, New Jersey.

Architectures for Dynamic Routing Networks

1.1 Introduction

Routing is an indispensable telecommunications network function that connects a call from origin to destination, and is at the heart of the architecture, design, and operation of any network. The term *network* denotes an arrangement of switching systems interconnected by transmission links. The rapid deployment of stored program control networks, consisting of electronic switching systems (ESSs) interconnected by common channel signaling (CCS) links, provides an opportunity to extend network routing rules beyond conventional fixed hierarchical routing to dynamic nonhierarchical routing. The term *dynamic* describes routing methods that are time-sensitive, or possibly real-time state-dependent, as opposed to present-day hierarchical routing rules, which are time-fixed. The introduction of dynamic routing into several telecommunications networks has resulted in a marked improvement in network connection availability while simultaneously reducing network costs.

Factors driving network routing evolution are performance quality improvements, new services introduction, and technological evolution. Performance quality improvement is achieved when network routing increases real-time adaptivity and robustness to load shifts and failures. Introduction of new services is aided when network routing extends dynamic routing and flexible bandwidth allocation to new services within an integrated network. Technological evolution is supported when network routing capitalizes on new transmission, switching, and network management/design technologies to achieve simpler, more automated, and more efficient networks than previously possible. We discuss in the book three stages of network routing evolution to achieve these goals:

1. Fixed traffic routing—in a fixed traffic routing network, such as fixed hierarchical routing networks deployed throughout the world today, there is minimal flexibility to share bandwidth among switch pairs.

2. Dynamic traffic routing—here the traffic routing allocates bandwidth among switch pairs and services in an efficient manner, perhaps with preplanned hourly routing tables, as in a dynamic nonhierarchical routing (DNHR) network, or perhaps in real time, as in a real-time network routing (RTNR) network for integrated classes-of-service.

3. Dynamic traffic and transport routing—-here dynamic traffic routing combines with dynamic transport routing, in which the latter allocates transport bandwidth rapidly among switch pairs and services, to provide efficient transport capacity utilization, automatic transport provisioning, diverse link routing, and rapid transport restoration, as with the fast automatic restoration (FASTAR™) capability.

Figure 1.1 illustrates a model for network routing and for network management/design. The central box represents the network, which can have various architectures and configurations, and the traffic routing tables and transport routing tables within the network. Network configurations include metropolitan area networks, national intercity networks, and global international networks, which support both hierarchical and nonhierarchical structures and combinations of the two. Routing tables describe the route choices from an originating switch to a terminating switch, for a connection request for a particular service. Hierarchical and nonhierarchical traffic routing tables are possible, as are fixed routing tables and dynamic routing tables. Routing tables are used for a multiplicity of traffic and transport services on the telecommunications network. Terminology used in the book, as illustrated in Figure 1.2, is that a *link* connects two switches, a

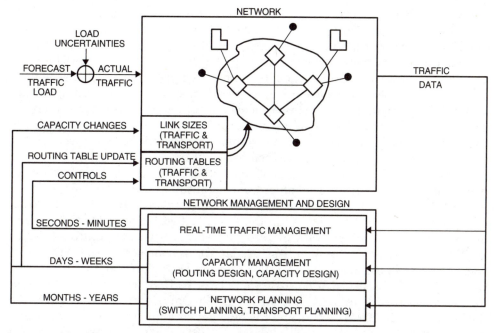

Figure 1.1 Network routing, management, and design

path is a sequence of links connecting an origin and destination switch, and a *route* is the set of different paths between the origin and destination that a call might be routed on within a particular routing discipline. Various implementations of routing tables are discussed in Chapters 2–15.

Dynamic routing networks have been in operation since 1984, when DNHR cut over into operation in the AT&T network, and now many such networks are in the planning or deployment stage. Dynamic routing is now deployed in 10 networks

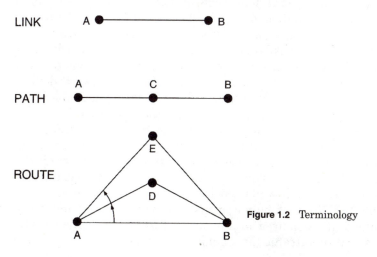

Figure 1.2 Terminology

(AT&T Switched Network, AT&T FTS-2000 Network, Bell Canada Ontario Network, British Telecom National Network, MCI National Network, Norwegian Telecom National Network, NTT Japan National Network, Sprint National Network, Stentor Canada National Network, and the Worldwide International Network) and has provided considerable benefits in improved performance quality and reduced cost. Such benefits have motivated the extension of dynamic routing to integrated networks with multiple classes-of-service. Dynamic routing techniques are applicable to all networks, including metropolitan area networks, national intercity networks, global international networks, future broadband networks, packet switched networks, and circuit switched networks.

A number of dynamic routing methods have been proposed and/or implemented including (1) dynamic nonhierarchical routing (DNHR) [ACM81, AKK81, AKK83, Ash87, Ash90, Ash95, AsM84, AsO89, CDK83, CGH81, DaF83, Fie83, HKO83, HSS87, Wol90], deployed in the AT&T network from 1984 to 1991 and in the AT&T FTS-2000 network in 1987; (2) real-time network routing (RTNR) [ACF91, ACF92, AsC93, Ash95, AsH93, MGH91], deployed in the AT&T network in 1991; (3) dynamically controlled routing (DCR) [BNR86, Cam81, Car88, CGG80, HSS87, LaR91, RBC83, RBC95, SzB79, WaD88], deployed in the Stentor Canada network in 1991, the Bell Canada Ontario network in 1992, the Sprint network in 1994 [Usr95], and the MCI network in 1995 [Kha97]; (4) state- and time-dependent routing (STR) [KaI95, YMI91], deployed in the NTT Japan network in 1992; (5) Worldwide International Network (WIN) dynamic routing [ACK89, AKA94, AsH94, KAK88, KaW95, WaM87], with 5-minute information exchange implemented in 1993 by a group of international carriers, with the goal of worldwide implementation of dynamic routing; (6) dynamic alternative routing (DAR) [Gib86, Mee86, StS87], deployed in the British Telecom network in 1996 and the Norway national network in 1996; (7) dynamic routing/5-minute (DR-5) [CKP91, KrO88], under development by Bellcore for application to Regional Bell Operating Company networks; and (8) planned dynamic routing implementations announced by France, Germany [Dre95], and Spain [Gar97]. Some packet-switched network implementations such as the ARPANET have used link-state and distance vector dynamic routing methods [MRS80, Ste95]. Current packet-switched data network routing standards, such as the Private Network–Network Interface (PNNI) from the ATM Forum [ATM96], have specified dynamic routing protocols for future implementation in cell-based asynchronous transfer model (ATM) networks. With the exception of these dynamic network deployments and initiatives, fixed hierarchical routing is in use throughout the world for essentially all other telecommunications network implementations.

Network management and design functions include real-time traffic management, capacity management, and network planning. Real-time traffic management ensures that network performance is maximized under all conditions including load shifts and failures. Capacity management ensures that the network is designed and provisioned to meet performance objectives for network demands at minimum cost. Network planning ensures that switching and transport capacity is planned and deployed in advance of forecasted traffic growth. Figure 1.1 illustrates real-time traffic management, capacity management, and network planning as three interacting feedback loops around the network. The input driving the network ("system") is a noisy traffic load ("signal"), consisting of predictable average demand components added to unknown forecast error and load variation components. The load variation components have different time constants ranging from instantaneous variations, hour-to-hour variations, day-to-day variations, and week-to-week or seasonal variations. Accordingly, the time constants of the feedback controls are matched to the load variations, and function to regulate the service provided by the network through capacity and routing adjustments. Real-time traffic management provides monitoring of network performance through collection and display of real-time traffic and performance data, and allows traffic management controls, such as code blocks, call gapping, and reroute controls, to be inserted when circumstances warrant.

Capacity management plans, schedules, and provisions needed capacity over a time horizon of several months to one year or more. Under exceptional circumstances, capacity can be added on a shorter-term basis, perhaps one to several weeks, to alleviate service problems. Network design embedded in capacity management encompasses both routing design and capacity design. Routing design takes account of the capacity provided by capacity management, and on a weekly or possibly real-time basis adjusts traffic routing tables and transport routing tables as necessary to correct service problems. The updated routing tables are sent to the switching systems either directly or via an automated routing update system. Network planning includes switch planning and transport planning, operates over a multiyear forecast interval, and drives network capacity expansion over a multiyear period based on network forecasts. Network management and design implementation experience for dynamic routing networks is discussed in Chapter 17, for the DNHR network implementation in 1984, and RTNR implementation in 1991.

Dynamic routing brings benefits to customers in terms of new service flexibility and improved service quality and reliability, at reduced cost. Dynamic routing, as shown through examples with DNHR, RTNR, and FASTAR, achieves the following improvements over fixed hierarchical traffic and transport routing: (1) essentially zero blocking performance on normal business days and weekends, (2) improved performance under network overloads and failures, (3) increased revenue through improved call completion, (4) lower capital costs through improved network design and integration, (5) lower expense costs through centralization and automation of network management and design functions, and (6) overall higher quality service for customers.

Several performance illustrations are given in Chapter 18, and here we give highlights. These examples are drawn from AT&T experience. RTNR network performance for the highest business day load on the network up to that date—the Monday after Thanksgiving on December 2, 1991, which is normally the highest business calling day of the year—set a record for the number of calls on the network: 157.5 million, of which only 228 calls were blocked on AT&T facilities. This provided a completion rate on AT&T facilities of 99.999% on the first try, or about 1 out of every 1 million calls blocked. This illustrates that the dynamic traffic routing network under normal load conditions is virtually

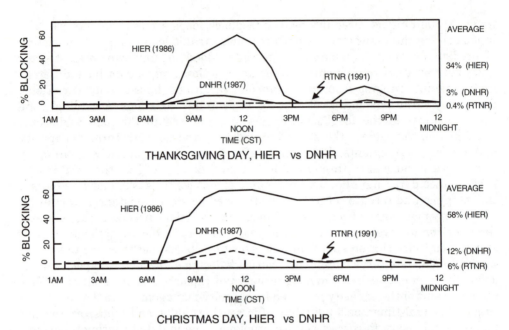

Figure 1.3 Dynamic network blocking performance with peak day traffic loads

nonblocking, in comparison to hierarchical routing with about 1 out of every 100 or more calls blocked, especially on the busiest day of the year. Peak days, such as Christmas, Mother's Day, and Thanksgiving, in addition to such unpredictable events as earthquakes, hurricanes, and switch failures, are a very great test for dynamic routing, because the traffic loads and patterns of traffic deviate severely from normal business day traffic for which the network is designed. Peak day blocking results show that the performance improvements of the DNHR network over the hierarchical (HIER) network, and the RTNR network over the DNHR network, are dramatic. As illustrated in Figure 1.3, in the case of Thanksgiving, average network blocking, which in 1986 is about 34% for HIER, is down to 3% in 1987 with DNHR, and then down to 0.4% in 1991 with RTNR, which is nearly a factor of 100 improvement. The average blocking for Christmas day is down from 58% for the HIER network, to 12% for the DNHR network, to 6% for the RTNR network, which is almost a factor of 10 improvement. Also, in 1991 versus 1986, there is less network capacity relative to demand, because of the efficiencies of dynamic traffic network design, and there is more traffic load.

Dynamic routing provides a self-healing network capability to ensure a networkwide path selection and immediate adaptation to failure. As illustrated in Figure 1.4, a fiber cut near Nashville severed 67,000 trunks in the switched network, and after FASTAR dynamic transport restoration a total of 30,000 trunks were still out of service in the switched network. FASTAR implements centralized automatic control of DCS3/3 transport switching devices to quickly restore service following a transport failure, such as caused by a cable cut. Over

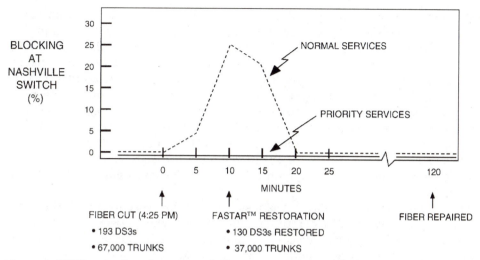

Figure 1.4 RTNR network performance with fiber cut in Austin, IN (9/25/91)

the duration of this event, more than 12,000 calls were blocked in the switched network, almost all of them originating or terminating at the Nashville, TN switch, and it is noteworthy that the blocking in the network returned to zero after the 37,000 trunks were restored in the first 11 minutes, even though there were 30,000 trunks still out of service. RTNR was able to find paths on which to complete traffic even though there was far less link capacity than normal even after the FASTAR restoration. Hence RTNR in combination with FASTAR transport restoration provides a self-healing network capability, and even though the cable was repaired two hours after the cable cut, degradation of service was minimal. RTNR provides priority routing for selected customers and services, which permits priority calls to be routed in preference to other calls, and blocking of the priority services is essentially zero throughout the whole event.

This improved network performance provides additional service revenues as formerly blocked calls are completed, and it improves service quality to the customer. As discussed in Section 1.5, dynamic routing lowers capital costs through improved network design, in the range of 10 to 30% of network cost over a range of network configurations and applications. Results show that voice and data services integration leads to more efficient and robust network performance with independent traffic control for each voice or data class-of-service, provides efficient sharing of integrated transport network capacity, and implements an integrated class-of-service routing feature for extending dynamic routing to emerging services.

As discussed in Chapters 16–18, dynamic routing lowers operations expense through centralized and automated network management and design. A number of dynamic routing feasibility and economic issues are discussed in Chapter 16, as are implementation and standardization issues. Dynamic routing and originating call control are achieved by permitting information exchange between switches through standardized CCS messages. These messages include via switch indicator messages to control route choices at via switches, crankback messages to return call control to originating switches from via switches or terminating switches, query messages to request link status from a switch, and status messages to contain link status information. Standardized information exchange is necessary so that switching equipment from different vendors interacts to implement dynamic routing methods in a coordinated fashion, which is discussed in Chapter 16. Network management and design implementation experience is discussed in Chapter 17. Finally, the performance experience of dynamic routing implementations is described in Chapter 18, in which highlights are cited above. Chapters 16–18 reinforce the significant economic and service benefits provided by dynamic routing implementation, in all network applications.

The model shown in Figure 1.1 is also a model for presentation in the book, in which the network block, routing tables block, and network management and design blocks are each discussed; that is, for each routing method under discussion, we first present the network architecture and configuration impacts within the network block, next we describe the routing disciplines employed within the routing tables block, then we present models for network management and design employed within the network management and design block, and finally we present modeling results. This is the presentation outline for this chapter and for Chapters 2 to 15.

1.2   Network Architectures and Configurations

Network architecture encompasses the distribution of functions within the network elements to implement the services provided by the network. Such functions include many hardware-related functions, such as switching and transmission, and software-based functions, such as call processing, routing, billing, and network management and design. In this section we discuss network configurations that include metropolitan area network configurations, national intercity network configurations, and global international network configurations for fixed hierarchical routing and dynamic routing. In many ways these configurations are similar in that they support simultaneously both hierarchical and nonhierarchical structures, as well as both dynamic routing and fixed routing subnetworks which interwork with each other. First we illustrate network architectural functions, and then discuss various network configurations.

Networks consist of various elements to provide network services. Figure 1.5, for example, illustrates the AT&T network elements, management, and interconnection to other networks. It shows that (1) approximately one million trunks interconnect 135 4ESS™ switches, (2) there are approximately 110 network control points providing advanced database features and service processing, (3) there are 24 signal transfer points which switch common channel signaling (CCS) network messages among the traffic switches and network control points, (4) real-time traffic management is provided by the Network Operations Center in Bedminister, NJ, (5) the network interconnects directly to customer equipment, such as private branch exchanges or customer networks, (6) the network interconnects

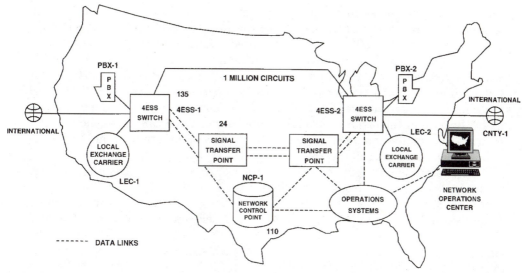

Figure 1.5 AT&T network

to seven regional Bell Operating Companies and many other local exchange carriers, and (7) the network interconnects to 265 countries throughout the world.

Services provided by these elements are now illustrated. An ordinary long-distance service voice call might originate, for example, from local exchange carrier LEC-1 destined for LEC-2, when a customer in LEC-1 dials the 10-digit North American Numbering Plan (NANP) number identifying another customer in LEC-2. The NANP number is translated in the 4ESS switch 4ESS-1 to determine the destination 4ESS switch, which is 4ESS-2, the class-of-service, and the specific routing parameters for routing the call. The call is routed either on the direct link or, if not available, via a two-link path through another one of the 133 4ESS switches from 4ESS-1 to 4ESS-2. For the call setup, initial address messages, call completion messages, and other messages are sent between the 4ESS switches via the CCS network. The CCS network consists of the 24 signal transfer point switches, interconnected by signaling links, which switch the CCS signaling messages between 4ESS switches, and also between the switched network and connecting networks, such as LEC-1, LEC-2, and international destinations. Once the call reaches the destination 4ESS switch 4ESS-2, 4ESS-2 determines the routing to the destination local exchange carrier LEC-2 and routes the call accordingly.

An 800 services voice call might originate, for example, from local exchange carrier LEC-1 destined for an 800 service provider at customer private branch exchange switch PBX-2, when a customer in LEC-1 dials the 10-digit 800 number identifying the 800 service provider at PBX-2. 4ESS-1 determines that the 800 number needs to be translated in a database, for example, one of the network control points, such as NCP-1. The 4ESS switch communicates with NCP-1 through CCS messages to and from the database. At NCP-1, the 800 number is translated to a routing number, which specifies the destination 4ESS switch, 4ESS-2, to which the 800 call is to be routed. 4ESS-1 receives the destination 4ESS switch from NCP-1, and also determines the class-of-service and the specific routing parameters for routing the call. The call is routed either on the direct link or, if not available, via a two-link path through another one of the 133 4ESS switches from 4ESS-1 to 4ESS-2, and CCS messages are used in the call setup as in the case of the ordinary long-distance service voice call. Once the 800 call reaches the destination 4ESS switch 4ESS-2, 4ESS-2 determines the routing to PBX-2 and routes the call accordingly.

A software-defined network (SDN) services voice call might originate, for example, from private branch exchange PBX-1 customer location for another SDN customer at PBX-2. The customer at PBX-1 dials a number corresponding to an internal SDN numbering plan. 4ESS-1 determines that the SDN number needs to be translated in a database, for example NCP-1. At NCP-1, the SDN number is translated to a routing number, which specifies the destination 4ESS switch, 4ESS-2, to which the SDN call is to be routed. 4ESS-1 receives the destination 4ESS switch from NCP-1, and also determines the class-of-service, and the specific routing parameters for routing the call. The call is routed either on the direct link or, if not available, via a two-link path through another one of the 133 4ESS switches from 4ESS-1 to 4ESS-2, and CCS messages are used in the call setup as in the case of the ordinary long-distance service voice call. Once the SDN call reaches the destination switch 4ESS-2, 4ESS-2 determines the routing to PBX-2.

An international long-distance services voice call might originate, for example, from local exchange carrier LEC-1 destined for country CNTY-1, when a customer in LEC-1 dials the 12 (or more)-digit International Standard E.164 number identifying the country code, city code, and national number of another customer in CNTY-1. The E.164 number is translated in switch 4ESS-1 to determine the destination 4ESS switch—which is an international switching center serving CNTY-1—4ESS-2, the class-of-service, and the specific routing parameters for routing the call. The call is routed either on the direct link or, if not available, via a two-link path through another one of the 133 4ESS switches from 4ESS-1 to 4ESS-2. The country CNTY-1 could be served by more than one international switching center, in which case multiple ingress/egress routing is used, described further in Section 1.3.2 and Chapter 12. With multiple ingress/egress routing, if the international long-distance service call cannot be completed on circuits connecting 4ESS-2 to CNTY-1, then the call can return to 4ESS-1 through use of a CCS crankback message, for possible further routing to another 4ESS international switching center (not shown), which also has circuits to CNTY-1. For the call setup between 4ESS-1 and 4ESS-2, initial address messages, call completion messages, and other messages are sent between the 4ESS switches via the CCS network. Once the call reaches the international switching center 4ESS-2, 4ESS-2 determines the routing to the destination country CNTY-1 and routes the call accordingly. Signaling messages are sent between switches in CNTY-1 and 4ESS-2 in setting up the call in CNTY-1. In completing the call to CNTY-1, 4ESS-2 can use real-time internetwork routing to dynamically select a direct link, a multiple link path through an alternate switch in CNTY-1, or perhaps a multiple link path through an alternate switch in another country CNTY-2 (not shown). Real-time internetwork routing is discussed in Section 1.3.2.

We now discuss various network configurations including metropolitan area networks, national intercity networks, and global international networks.

1.2.1 Metropolitan area network configuration

Many metropolitan networks are two-level hierarchical networks and use multi-alternate routing as shown in Figure 1.6 [Fie83]. Final links, sized to 1% blocking, connect each end to its home tandem and interconnect all of the tandems in the metropolitan network. Economic CCS engineering techniques are used to size the high-usage links in the network [Tru54]. Depending on the development of high-usage links, this network configuration provides at most four paths between any pair of end-offices. Figure 1.6 shows two nondynamic routing end-offices, E_1 and E_6. These end-offices might use electromechanical switches or ESS switches without a dynamic routing capability. Calls between two nondynamic routing end-offices, or a nondynamic routing end-office and a dynamic routing end-office, are routed hierarchically. For example, a call between E_1 and E_2 would first attempt the primary high-usage link E_1–E_2 with overflow offered to the path

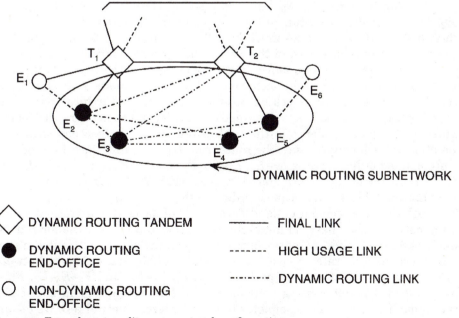

Figure 1.6 Example metropolitan area network configuration

through T_1. Calls that overflow the first path are offered to path E_1–T_1–E_2. All tandem switching is done by the dedicated tandems T_1 and T_2. If the network uses combined "metropolitan/intercity" links, then the end-office-to-tandem final links also carry traffic destined for or completing from distant cities. In order to ensure that such calls would continue to receive the same blocking probability grade-of-service as they would experience in a fully hierarchical network, the links that connect dynamic routing end-offices to their home tandems would be sized to 1% blocking. Note that for such *hierarchically* routed calls, the tandems T_1 and T_2 have hierarchical relationships to the dynamic routing end-offices.

A metropolitan network with dynamic routing is also shown in Figure 1.6. End-offices E_2 through E_5 and tandems T_1 and T_2 form a nonhierarchical dynamic routing subnetwork. Each of these switches is assumed to be an ESS switch equipped with software that implements dynamic routing. Intrametropolitan calls originate and terminate at the end-offices. Dynamic routing is used to carry calls between the dynamic routing end-offices. Dynamic routing end-offices originate, terminate, and tandem switch dynamically routed intrametropolitan calls. Many dynamic routing methods are possible, as described in the next section; here we give an example. The originating end-office retains control of call routing through the use of the CCS network. A nonhierarchical route consists of an ordered set of one- and two-link paths. For example, in Figure 1.6 a call from end-office E_2 to E_3 is offered to the first path in the route, the one-link path E_2–E_3. If all trunks in the E_2–E_3 link are busy, E_2 route advances the call to path E_2–E_4–E_3. If there is a free trunk in the E_2–E_4 link, E_2 seizes a trunk and sends called number information to E_4, which attempts to route the call directly to E_3. If all trunks in the E_4–E_3 link are busy, E_4 returns control of the call to E_2, which advances the call to the next path in the route. This sequential path-to-path overflow continues until the call is set up or until the call is blocked after overflowing the last path in the route. The dynamic routing subnetwork is sized to a *switch-to-switch* (end-office-to-end-office) *blocking objective*. A sufficient number of paths is included in each route so that the combined blocking of all paths does not exceed the blocking objective. The nonhierarchical routes are changed by time of day to take advantage of capacity in the network that is idle because of the noncoincidence of the peak traffic loads between end-offices. At some other time during the day, the route from E_2 to E_3 in Figure 1.6 might be the preplanned ordered paths E_2–E_3, E_2–T_2–E_3, E_2–E_4–E_3, and E_2–T_1–E_3. All of these preplanned routes are stored in the switches. The switches select the correct routing table at call setup according to the time of day.

Although the pure tandems (T_1 and T_2) are part of the nonhierarchical dynamic routing network, they have no hierarchical relationship to the dynamic routing end-offices for the purpose of routing intrametropolitan dynamically routed traffic. For instance, as illustrated above, a permissible nonhierarchical route between E_3 and E_5 is the set of ordered paths E_3–E_5, E_3–T_2–E_3, and E_3–E_4–E_5. Final links between dynamic routing end-offices and their home tandem might also carry traffic that accesses or completes from the intercity network(s). We ensure a satisfactory blocking probability grade-of-service for these calls by requiring that the dynamic routing "final" links be sized to 1% blocking. Although it might be possible with certain transmission and switching network elements and transmission loss plans to dynamically route these intercity accessing/completing calls in the metropolitan network, we restrict these calls to route only on the one-link path from the end-offices directly to the tandems.

Combining metropolitan area and intercity traffic flows in the metropolitan network leads to multiple ways of interfacing metropolitan and national intercity networks. Each approach has different economic implications for the national intercity network and also for the metropolitan area network. If intercity traffic could be dynamically routed on metropolitan area links, design efficiency can be increased in comparison with a separate design for metropolitan area and intercity traffic.

A model used to study metropolitan networks is based on the Chicago metropolitan network, which incorporates actual traffic load patterns from the 312 Number Plan Area (NPA) in the North American network. The 312 NPA is the Chicago metropolitan area and includes Chicago itself along with the surrounding urban and suburban areas within a radius of about 10 to 50 miles from downtown Chicago. End-office-to-end-office traffic data were collected among all of the end-offices in the 312 NPA during the first 10 business days of December, 1977. The data, which include both link measurements and detailed billed call accounting data, are processed to form hourly average switch-to-switch loads for each of the 15 hours, 8 A.M. through 11 P.M. of the business day. A set of 38 end-offices and four tandems is selected to model the entire metropolitan Chicago traffic load. Figure 1.7 is a sketch of the locations of the 42 switches. Figure 1.8 compares the daily variation of the total intrametropolitan traffic in the model and in the entire 312 NPA. In the aggregate, the hourly variation of traffic in the model matches that in all of Chicago very closely. An effort is made in the model to represent the entire geographical area, to reflect business-residence sectors, to reflect the small-end-office–large-end-office split, and to represent the various socioeconomic regions.

The metropolitan area model is used to address unique problems of dynamic routing within the metropolitan network. These include differences in the transmission constraints and also special considerations with respect to combined metropolitan area and intercity traffic. Metropolitan networks must be subdivided into manageable areas which can be separately designed and managed. In some areas, such as the New York/Long Island and Tri-State metropolitan areas, boundaries are more unclear than in other areas, such as Atlanta. A wide variation of network environments may be encountered for different metropolitan areas. For example, the Chicago metropolitan network and Los Angeles differ markedly in demographic makeup and traffic patterns, but models show similar benefits for dynamic network design.

1.2.2 National intercity network configuration

Figure 1.9 illustrates a network configuration that incorporates dynamic routing into a national intercity network, which is based on the intercity network

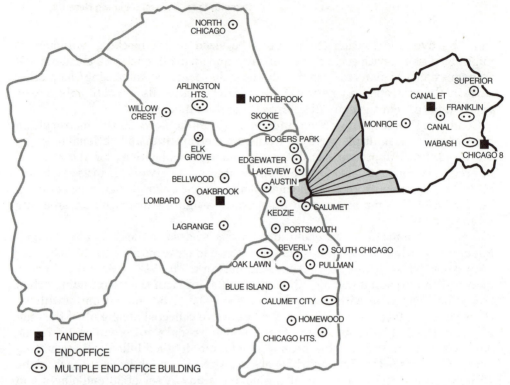

Figure 1.7 Chicago metropolitan area model

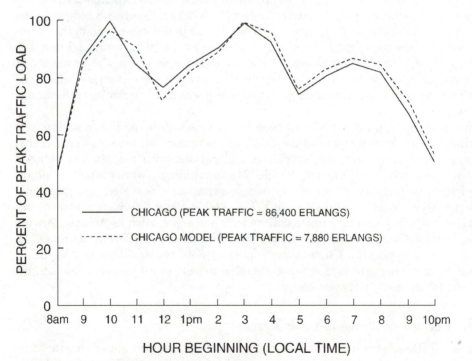

Figure 1.8 Daily variation of intrametropolitan traffic

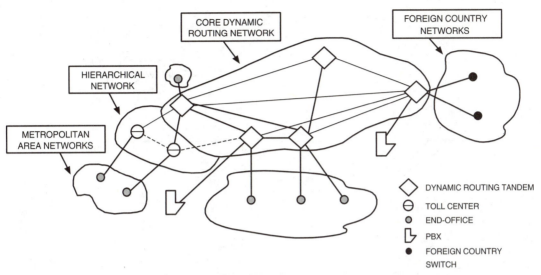

Figure 1.9 Example national intercity network configuration

configuration [ACM81, AKK81, AKK83]. As shown in Figure 1.9 the dynamic routing subnetwork has only one class of switching system, and end-offices continue to home on tandem switches in the dynamic routing environment as they would in a fixed hierarchical network. The dynamic routing portion of the network consists of ESS switches with dynamic routing capability interconnected by CCS. Dynamic nonhierarchical routing rules are used only between dynamic routing switches, and fixed hierarchical routing rules are used between all other pairs of switches. In addition, there is a large number of smaller metropolitan and rural switches employing only the fixed hierarchical rules. Many of these smaller switches home directly on dynamic routing switches and, as illustrated by tandem T_1 in Figure 1.10, traffic is also concentrated in the hierarchical network through the use of hierarchical homing patterns. From the viewpoint of the intercity hierarchical switches, the intercity dynamic routing switches appear as a network of the top-level hierarchical (e.g., regional) switching centers, and hierarchical routing patterns are therefore defined by this hybrid structure.

The dynamic routing network handles three kinds of load. The first is the load that originates and terminates at end-offices homed directly on dynamic routing switches; for example, E_2 to E_4 traffic illustrated in Figure 1.10. The dynamic routing network must also route overflow and through-switched loads from the hierarchy. Overflow load from the E_1 to E_3 traffic and through-switched load from E_1 to E_5 are examples of these latter two categories of load, which are also regarded as T_1-to-T_2 loads within the dynamic routing network. A dynamic routing originating switch is able to identify a call first entering the dynamic routing network as one of the three components of load and also determines the terminating switch within the dynamic routing network. Once a call is identified from the routing translations as being a dynamic routing

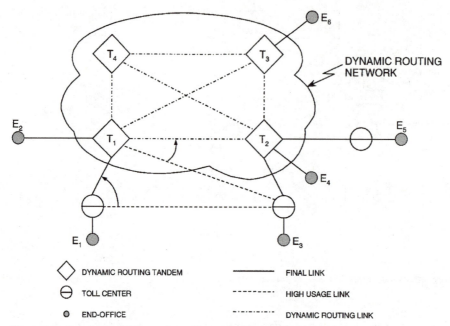

Figure 1.10 Interaction of dynamic routing network and hierarchical network

call, dynamic routing rules are employed to complete the call on at most two dynamic routing links between the dynamic routing originating switch and the dynamic routing terminating switch.

The dynamic routing method could be one of many methods discussed in the next section, but for purposes of illustration we describe an example method that uses preplanned two-link dynamic routing, which is very similar to the dynamic routing example given for metropolitan area networks. The preplanned dynamic, or time-varying, nature of the dynamic routing method is achieved by introducing several route choices, which consist of different sequences of paths, and each path has one or at most two links in tandem. In Figure 1.10, the originating switch at T_1, for example, retains control over a dynamically routed call from dynamic routing tandem T_1 to T_4 until it is either completed to its destination at T_4 or blocked. Switch T_1 may first try the direct path T_1–T_4, and if blocked overflow to the T_1–T_2–T_4 path. If the call overflows the second link (T_2–T_4) of the two-link T_1–T_2–T_4 path it is returned to the originating switch T_1 for possible further alternate routing, such as to path T_1–T_3–T_4. Control is returned by sending a CCS crankback signal from the via switch T_2 to the originating switch T_1.

A 135-switch full-scale national intercity network model, along with a 28-switch and 10-switch subset, are used for several studies throughout the book and are illustrated in Figures 1.11 and 1.12. A weekend and average business day load profile for the national intercity network, with 72 separate hours of traffic load data, is illustrated in Figure 1.13 for 135-switch, 28-switch, and 10-switch networks for March, August, and October, 1995.

Figure 1.11 135-switch national intercity network model

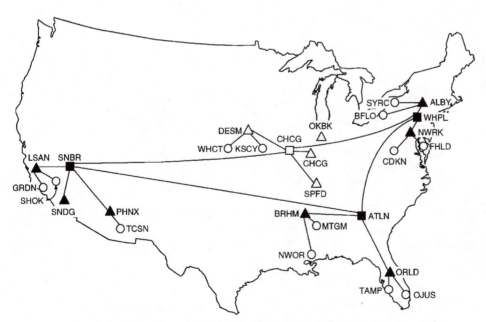

Figure 1.12 28-switch national intercity model (shaded switches indicate 10-switch model)

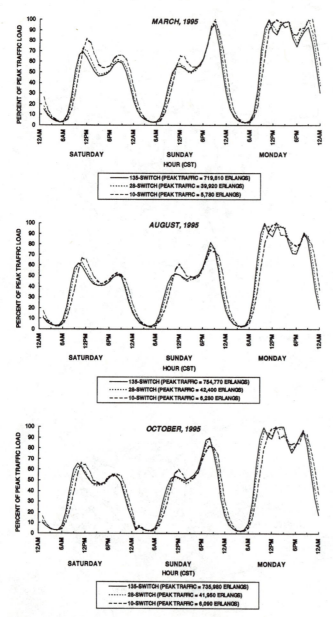

Figure 1.13 March, August, and October, 1995, traffic load profiles for 135-switch, 28-switch, and 10-switch national intercity network

1.2.3 Global international network configuration

The international dynamic routing configuration illustrated in Figure 1.14 envisions two tiers: an upper tier of international switches and a subtending tier of national intercity switches. Here there is a direct parallel to the national intercity network configuration in which the subtending national switches can employ fixed hierarchical routing rules to interface to the international dynamic routing switches, which employ dynamic routing rules. A difference is that in traversing the national network, there can be multiple ingress or egress points to that network, for example S_1, S_2, and S_3 in Country A, Figure 1.14. An S_4–S_5 dynamic routing call from Country B to Country C, for example, originates at S_4 and may try first to route on S_4 to S_5, and if blocked, then route S_4–S_2–S_5, and if blocked, then S_4–S_1–Country C. In this third choice path, the call control could be taken over temporarily by S_1, which could try to route the call to S_2 or S_3 and then S_5 using the dynamic routing rules within Country A. If this is not successful, that is, Country A cannot reach Country C through any route choice, then the call may be sent back to Country B S_4 with a CCS crankback message for possible further alternate routing attempts by S_4 through other countries in the international dynamic routing network.

In the global international network there is minimal centralized network management and design, and distributed design and control are preferred. Also, each participating country can use its preferred dynamic routing method, that is, not all countries need use the same dynamic routing method, which leads to a mixed dynamic routing (MXDR) network, which is discussed further in the next section.

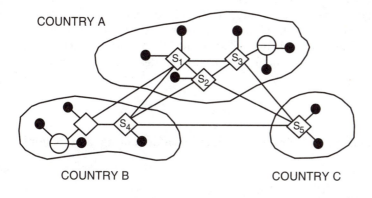

Figure 1.14 Example of global international network configuration

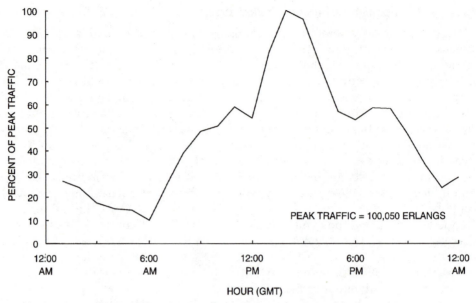

Figure 1.15   Hourly traffic load profile for 13-switch global international network

A model used to study global international networks for various dynamic routing methods is a small global model of a 13-switch (13-country) network [ScW94]. The 13 countries modeled in the small global model include Australia, Canada, France, Germany, Italy, Japan, Netherlands, Norway, Spain, Switzerland, Taiwan, United Kingdom, and United States. Standard CCITT traffic profiles are used and the maximum network traffic load is 100,050 erlangs. Figure 1.15 shows the 24-hour load pattern for this network model. Average switch termination and transport costs are assumed for purposes of designing both network models.

1.3   Routing Methods

Below we elaborate on fixed traffic routing, dynamic traffic routing, and dynamic transport routing.

1.3.1   Fixed traffic routing

Fixed hierarchical traffic routing is discussed in this section and is descriptive of almost all current networks, including most metropolitan area networks, national intercity networks, and global international networks. Nonhierarchical, fixed routing is exemplified by the existing AUTOVON network deployed by the United States Department of Defense. Fixed traffic routing networks are the subject of Chapter 2.

Network routing plans have always been closely aligned with technological capabilities. Such is the case with hierarchical traffic routing when in 1930 a hierarchy of switching centers was established by the General Toll Switching Plan [Osb30]. This plan limited the number of switches in a "long-distance" call to six in any connection including the switching centers to which the metropolitan area switches are connected. As illustrated in Figure 1.16, routing through the network was usually handled manually by operators at specialized switchboards following rather involved operating procedures. During the late 1940s, planning for nationwide customer dialing required that the 1930 network plan be revised [Pil52]. The new plan established ten regions and a five-level hierarchy of switching systems centering on the regional switches, as illustrated in Figure 1.17. By 1951, this five-level hierarchy of switching centers had evolved, and direct distance dialing was introduced concurrently with automatic alternate routing. The technologies which enabled these improvements included the No. 4A crossbar system, automatic message accounting, and efficient network design techniques for alternate routing networks. The planning for direct distance dialing and the five-level hierarchical network was of such quality that only minor refinements were necessary until it was replaced by DNHR in the 1980s.

Figure 1.16 Manual routing

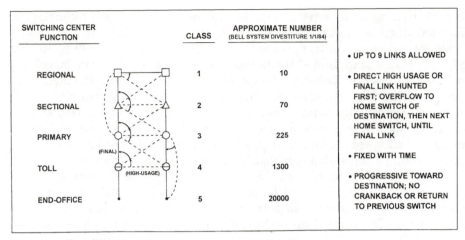

Figure 1.17 Hierarchical routing

Figure 1.17 illustrates hierarchical traffic routing which serves most networks today. The hierarchical plan establishes regions and a hierarchy of switching systems centering on the regional switches. The switching hierarchy used in the Bell System network prior to divestiture in 1984 had five levels, as illustrated in Figure 1.17, in which successively higher level switches (also called *classes*) concentrate traffic from increasingly larger geographical areas. At the highest level of switching were 10 regional switching centers, or Class 1 switches, shown as squares in Figure 1.17. At the next level of switching were the sectional switching centers, then primary switching centers, then toll switching centers, and at the lowest level end-offices, or Class 5 end-offices, which are part of the metropolitan area switching network. The approximate numbers of these switching centers at the time of divestiture in 1984 are given in Figure 1.17.

Two types of links connect these switches. *High-usage links* (dashed lines) connect any two switches that have sufficient traffic between them to make a direct route economical. *Final links* are the links between each switch and its immediate superior in the hierarchy, together with the final links interconnecting all the regional centers (solid lines). A switch connected by a final link to a higher-class switch is said to "home" on that switch. High-usage links are sized to handle only a portion of the traffic directed to them, and the switching systems redirect traffic by automatic alternate routing to a different link when all circuits of a high-usage link are busy. At each stage, the alternate routing plan shifts overflow calls from the more direct route toward the final route. Final links are designed to handle their own direct traffic, plus overflow traffic from high-usage links, with low blocking, such as an average loss of one call in a hundred. The hierarchical routing tables preclude switching or looping calls back on themselves or using an excessive number of links on a call.

When the hierarchical network was developed, computers were in their infancy and design rules had to be kept simple to allow manual network design. It also was necessary that routing decisions of the early electromechanical switches be quite limited. Hierarchical traffic routing served for more than 50 years prior to the first conversion to dynamic routing starting in 1984, and is still in widespread use throughout the world today. In fact, it is the only standard routing method that permits interworking among switching products from multiple vendors. Hierarchical traffic routing is, however, less efficient and beneficial than dynamic traffic routing methods, since the ability to share available bandwidth among switch pairs is minimal.

1.3.2 Dynamic traffic routing

Dynamic traffic routing in telecommunications networks has been deployed in ten networks to date, and is the subject of worldwide study and interest. Dynamic routing capitalizes on technological advances to improve the efficiency and performance of all networks. See the reference list for a comprehensive bibliography of work in this area. During the 1970s network technology evolved rapidly toward stored program control capabilities. In the AT&T network, for example, the network evolved toward handling considerably higher traffic volumes with fewer switches because the 4ESS switch [BST77, BST81], a very large switching system with more than three times the capacity of the No. 4A crossbar, became the primary switching vehicle in the network and enabled consolidation of many electromechanical switches into a single large ESS switch. ESS switches provide extensive stored program control capabilities that enable sophisticated routing instructions to be executed. In addition, the CCS network, a high-speed, high-capacity signaling system for linking ESS switches, has been deployed. CCS enables far more routing information to be sent between switches in call setup messages than is possible with in-band signaling, and allows ESS switches to make full use of their stored program control capabilities. Finally, computerized network management/design has made major advances in automating complex network functions. These systems can process vast amounts of raw data, implement very complex network design procedures, and automate the flow of routing instructions and other network management/design information into and out of ESS switches. Therefore the simple manual design rules for hierarchical networks could be made more efficient and sophisticated with computerized implementation.

Dynamic traffic routing allows ESS routing tables to be changed dynamically, either in a preplanned time-varying manner, or on-line in real time. With preplanned (time-sensitive) dynamic traffic routing methods, routing patterns contained in routing tables might change every hour or at least several times a day to respond to known shifts in traffic loads, and in general preplanned dynamic routing tables change with a time constant normally much greater than a call holding time. A typical preplanned dynamic routing method may change routing tables every hour, which is longer than a typical call holding time of a few minutes. Various implementations of preplanned dynamic routing are illustrated in Figure 1.18, which shows dynamic hierarchical routing, discussed in Chapter 3,

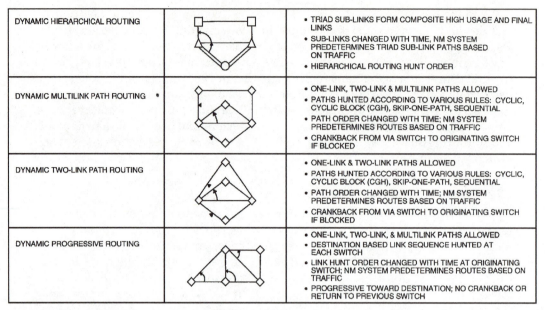

DYNAMIC HIERARCHICAL ROUTING		• TRIAD SUB-LINKS FORM COMPOSITE HIGH USAGE AND FINAL LINKS • SUB-LINKS CHANGED WITH TIME, NM SYSTEM PREDETERMINES TRIAD SUB-LINK PATHS BASED ON TRAFFIC • HIERARCHICAL ROUTING HUNT ORDER
DYNAMIC MULTILINK PATH ROUTING		• ONE-LINK, TWO-LINK & MULTILINK PATHS ALLOWED • PATHS HUNTED ACCORDING TO VARIOUS RULES: CYCLIC, CYCLIC BLOCK (CGH), SKIP-ONE-PATH, SEQUENTIAL • PATH ORDER CHANGED WITH TIME; NM SYSTEM PREDETERMINES ROUTES BASED ON TRAFFIC • CRANKBACK FROM VIA SWITCH TO ORIGINATING SWITCH IF BLOCKED
DYNAMIC TWO-LINK PATH ROUTING		• ONE-LINK & TWO-LINK PATHS ALLOWED • PATHS HUNTED ACCORDING TO VARIOUS RULES: CYCLIC, CYCLIC BLOCK (CGH), SKIP-ONE-PATH, SEQUENTIAL • PATH ORDER CHANGED WITH TIME; NM SYSTEM PREDETERMINES ROUTES BASED ON TRAFFIC • CRANKBACK FROM VIA SWITCH TO ORIGINATING SWITCH IF BLOCKED
DYNAMIC PROGRESSIVE ROUTING		• ONE-LINK, TWO-LINK, & MULTILINK PATHS ALLOWED • DESTINATION BASED LINK SEQUENCE HUNTED AT EACH SWITCH • LINK HUNT ORDER CHANGED WITH TIME AT ORIGINATING SWITCH; NM SYSTEM PREDETERMINES ROUTES BASED ON TRAFFIC • PROGRESSIVE TOWARD DESTINATION; NO CRANKBACK OR RETURN TO PREVIOUS SWITCH

Figure 1.18   Preplanned dynamic routing methods

dynamic multilink path routing, discussed in Chapter 4, dynamic progressive routing, discussed in Chapter 5, and dynamic two-link path routing, discussed in Chapters 6–10. These routing tables are preplanned, preprogrammed, and recalculated perhaps each week within the capacity management network design function. In Section 3.2.1 we discuss an example implementation of preplanned two-link sequential dynamic routing networks, which is the DNHR network deployed in 1984. By using the CCS "crankback" signal, DNHR limits tandem connections to at most two links, and such two-link preplanned sequential routing allows nearly as much network utilization and performance improvement as the most flexible preplanned dynamic routing possible. Preplanned dynamic routing networks are the subject of Chapters 3–10, including preplanned dynamic hierarchical routing in Chapter 3, preplanned dynamic multilink path routing in Chapter 4, preplanned dynamic progressive routing in Chapter 5, preplanned dynamic two-link path routing in Chapter 6, a four-switch design example for a preplanned dynamic two-link path routing network in Chapter 7, combined preplanned and real-time dynamic routing networks in Chapter 8, preplanned dynamic routing under forecast uncertainty in Chapter 9, and preplanned dynamic routing for multiservice integrated networks in Chapter 10. In Chapters 6–10, we focus on the special case of preplanned two-link sequential path routing networks, which include the DNHR network deployed in 1984.

Real-time dynamic traffic routing is the most sophisticated method for traffic routing. Real-time dynamic routing does not depend on precalculated routing tables. Rather, the switching system or network management system senses

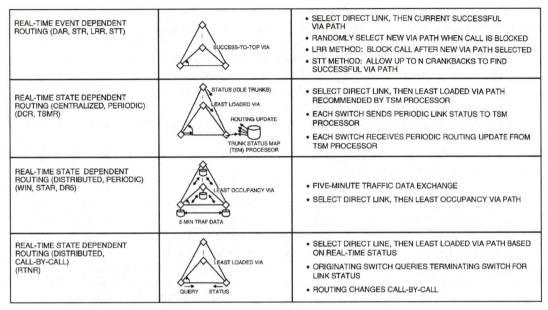

REAL-TIME EVENT DEPENDENT ROUTING (DAR, STR, LRR, STT)	SUCCESS-TO-TOP VIA	• SELECT DIRECT LINK, THEN CURRENT SUCCESSFUL VIA PATH • RANDOMLY SELECT NEW VIA PATH WHEN CALL IS BLOCKED • LRR METHOD: BLOCK CALL AFTER NEW VIA PATH SELECTED • STT METHOD: ALLOW UP TO N CRANKBACKS TO FIND SUCCESSFUL VIA PATH
REAL-TIME STATE DEPENDENT ROUTING (CENTRALIZED, PERIODIC) (DCR, TSMR)	STATUS (IDLE TRUNKS) LEAST LOADED VIA ROUTING UPDATE TRUNK STATUS MAP (TSM) PROCESSOR	• SELECT DIRECT LINK, THEN LEAST LOADED VIA PATH RECOMMENDED BY TSM PROCESSOR • EACH SWITCH SENDS PERIODIC LINK STATUS TO TSM PROCESSOR • EACH SWITCH RECEIVES PERIODIC ROUTING UPDATE FROM TSM PROCESSOR
REAL-TIME STATE DEPENDENT ROUTING (DISTRIBUTED, PERIODIC) (WIN, STAR, DR5)	LEAST OCCUPANCY VIA 5-MIN TRAF DATA	• FIVE-MINUTE TRAFFIC DATA EXCHANGE • SELECT DIRECT LINK, THEN LEAST OCCUPANCY VIA PATH
REAL-TIME STATE DEPENDENT ROUTING (DISTRIBUTED, CALL-BY-CALL) (RTNR)	LEAST LOADED VIA QUERY STATUS	• SELECT DIRECT LINE, THEN LEAST LOADED VIA PATH BASED ON REAL-TIME STATUS • ORIGINATING SWITCH QUERIES TERMINATING SWITCH FOR LINK STATUS • ROUTING CHANGES CALL-BY-CALL

Figure 1.19 Real-time dynamic routing methods

the immediate traffic load and if necessary searches out new paths through the network possibly, as in RTNR, on a call-by-call basis. With real-time dynamic routing methods, routing tables change with a time constant on the order of or less than a call holding time. As illustrated in Figure 1.19, real-time dynamic routing methods can be event-dependent or state-dependent. Examples of event-dependent real-time routing are the DAR method used in the British Telecom network and the STR method deployed in NTT Japan. In the DAR and STR learning approaches, the path last tried, which is also successful, is tried again until blocked, at which time another path is selected at random and tried on the next call. STR path choices are changed with time in accordance with changes in traffic load patterns.

Examples of state-dependent real-time routing networks include RTNR and DCR. RTNR uses real-time exchange of network status information, with CCS query and status messages, to determine an optimal route from a very large number of possible choices. With RTNR, the originating switch first tries the direct path and if it is not available finds an optimal two-link path by querying the terminating switch through the CCS network for the busy-idle status of all links connected to the terminating switch. The originating switch compares its own link busy-idle status to that received from the terminating switch, and finds the least loaded two-link path to route the call. DCR is a centralized routing method with 10-second updates of network status. The DCR centralized network management system provides 10-second routing updates to all switches in the dynamic routing network, in which routing tables are determined from analysis of the link status data using the DCR dynamic routing method. A fixed trunk reservation technique is used, and path choices are changed with time in accordance with changes in traffic load patterns. State-dependent real-time routing methods may change routing tables up to every few minutes based on traffic load estimates, for example, through decentralized exchange of 5-minute traffic data, as illustrated in Figure 1.19. Real-time dynamic routing is the subject of Chapters 11–13, including centralized real-time dynamic routing networks in Chapter 11, distributed real-time dynamic routing networks in Chapter 12, and network management and design for real-time dynamic routing networks in Chapter 13.

We now describe several dynamic traffic-routing methods to be discussed in more detail in the book.

1.3.2.1 Preplanned dynamic hierarchical routing.

Preplanned dynamic hierarchical routing is the subject of Chapter 3. As illustrated in Figures 1.18 and 1.20, dynamic hierarchical routing uses triad sublinks to satisfy the hourly demands, in which the triads augment high-usage or final links with trunks borrowed from other links in the network. The number of trunks in each triad sublink is allowed to vary dynamically from hour to hour within the underlying fixed hierarchical routing structure, and by means of triad sublinks the actual number of trunks in each high-usage and final link is allowed to vary hour by hour in accordance with the demand for circuits between switches in the network. Network utilization efficiencies with dynamic hierarchical routing can be attributed exclusively to the effects of load noncoincidence and do not include the advantages of nonhierarchical optimum path selection.

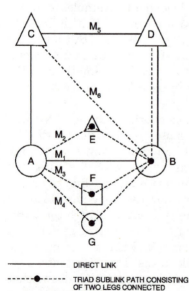

- HIGH USAGE AB IS COMPOSITE OF:
  - M_1 DIRECT TRUNKS
  - M_2 TRIAD TRUNKS THROUGH SWITCH E
  - M_3 TRIAD TRUNKS THROUGH SWITCH F
  - M_4 TRIAD TRUNKS THROUGH
  - SWITCH G
- FINAL LINK CD IS COMPOSITE OF:
  - M_5 DIRECT TRUNKS
  - M_6 TRIAD TRUNKS THROUGH SWITCH B
- TRIAD CONFIGURATION CHANGE FROM HOUR TO HOUR

———— DIRECT LINK

----●---- TRIAD SUBLINK PATH CONSISTING OF TWO LEGS CONNECTED TOGETHER AT VIA SWITCH

Figure 1.20 Example of preplanned dynamic hierarchical routing with triad sublink routing

1.3.2.2 Preplanned dynamic multilink path routing. Preplanned dynamic (and fixed) multilink path routing is the subject of Chapter 4. As illustrated in Figures 1.19, 1.21, and 1.22, path routing implies selection of an entire path between originating and terminating switches before a connection is actually attempted on that path. If a connection on one link in a path is blocked, the call then attempts another complete path. Implementation of such a routing method can be done through control from the originating switch, plus a multiple-link crankback capability to allow paths of two, three, or more links to be used. As discussed earlier, crankback is a CCS message capability that allows a call blocked on a link in a path to return to the originating switch for further alternate routing on other paths. Path-to-path routing is nonhierarchical and allows the choice of the most economical paths rather than being restricted to hierarchical paths. One method of implementing dynamic multilink path routing is to allocate fractions of the traffic to routes and to allow the fractions to vary as a function of time. One approach is *cyclic routing,* illustrated in Figure 1.21, which has as its first route $(1, 2, \ldots, M)$, where the notation (i, j, k) means that all traffic is offered first to path i, which overflows to path j, which overflows to path k. The second route of a cyclic route choice is a cyclic permutation of the first route: $(2, 3, \ldots, M, 1)$. The third route is likewise $(3, 4, \ldots, M, 1, 2)$, and so on. This approach has computational advantages because its cyclic structure requires considerably fewer calculations in the design model than does a general collection of paths. The route blockings of cyclic routes are identical; what varies from route to route is the proportion of flow on the various links.

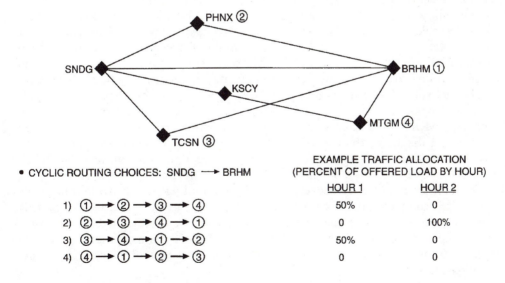

Figure 1.21 Preplanned dynamic multilink cyclic path routing

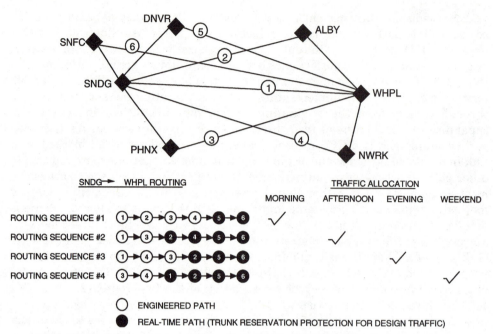

Figure 1.22   Preplanned dynamic two-link sequential path routing (DNHR)

1.3.2.3   Preplanned dynamic progressive routing.   Preplanned dynamic (and fixed) progressive routing is the subject of Chapter 5. As illustrated in Figure 1.18, in this method, a call is sent from one switch to another switch, and control of the call is also passed to the next switch. No return to a previous switch, or crankback, is allowed, but the call must continue toward its destination at each switch or be blocked. A familiar example of progressive routing is fixed hierarchical routing. The main difficulty with progressive routing is to avoid looping. In hierarchical routing this is prevented automatically by the structure of the network. In nonhierarchical progressive routing, one approach to prevent looping is to carry the call history in the CCS call-setup message. In this way each switch knows the switches to which the call has already been routed and disallows them as the next outlet choice. Besides preventing looping, this route control method can be used to limit the maximum path length of a given call and thereby prohibit excessive alternate routing. Eliminating paths with too many links prevents calls from "stealing" trunks from other calls that might complete on one or two links. Otherwise, excessive alternate routing can take place, having a cascading effect that causes inefficient link use, with fewer call completions than otherwise possible. In dynamic progressive routing, traffic is allocated to the most economical next-switch choices on a time-varying basis.

1.3.2.4   Preplanned dynamic two-link path routing.   Preplanned dynamic (and fixed) two-link path routing is the subject of Chapters 6–10. In the design of multilink path networks, about 98 percent of the traffic is routed on one- and two-link paths, even though paths of greater length are allowed. Because of switching costs, paths with one or two links are usually less expensive than paths with more links. Therefore, as illustrated in Figure 1.18, two-link path routing uses the simplifying restriction that paths can have only one or two links, which requires only single-link crankback to implement and uses no common links as is possible with multilink path routing. Alternative two-link routing methods include the cyclic routing method described above, CGH routing, and sequential routing. *CGH routing,* named after Chung, Graham, and Hwang [CGH81], who developed it, is composed of cyclic blocks. For example, suppose there are seven paths. One possible cyclic block realization of the seven paths is (1)(2 3 4)(5 6)(7). This notation means that all the offered load to this route is first offered to path 1. The overflow from path 1 is then offered to a cyclic block composed of paths 2, 3, and 4. The term *cyclic block* means that a proportion β_i^k of the total load offered to the kth block is offered to cyclic permutation i, where the cyclic permutation i is selected so that the ordering within the block is preserved but a different path appears first. In the cyclic block under consideration, a proportion β_1^2 of the input traffic will be offered to the paths in the order (2, 3, 4) and proportion β_2^2 to (3, 4, 2), and so on. Routing traffic in this manner may be accomplished by generating a random number when a call is offered to the cyclic block. Note that all calls see the same blocking probability within the cyclic block, since all paths are searched.

In *sequential routing,* all traffic in a given hour is offered to a single route, and the first path is allowed to overflow to the second path, which overflows to the third path, and so on. This is illustrated in Figures 1.17 and 1.22. Thus, traffic is routed sequentially from path to path with no probabilistic methods being used to influence the realized flows. The reason that sequential routing works well is that permuting path order provides sufficient flexibility to achieve desired flows without the need for probabilistic routing. Both fixed and dynamic versions of two-link path routing are compared, and the results are discussed below. In fixed two-link sequential path routing, a fixed sequential route is used in all hours. In preplanned dynamic two-link sequential path routing, which is the method used in dynamic nonhierarchical routing (DNHR) implementation, the sequential route is allowed to change from hour to hour. Figure 1.23 compares some of the key differences between hierarchical routing and DNHR.

The preplanned dynamic, or time-varying, nature of the dynamic routing method is achieved by introducing several route choices, which consist of different sequences of paths, and each path has one or, at most, two links in tandem. In Figure 1.22, the originating switch at SNDG, for example, retains control over a dynamically routed call from dynamic routing tandem SNDG to WHPL until it is either completed to its destination at WHPL or blocked. Switch SNDG may first try the direct path SNDG–WHPL and, if blocked, overflow to the SNDG–ALBY–WHPL path. If the call overflows the second link (ALBY–WHPL) of the SNDG–ALBY–WHPL path, it is returned to the originating switch SNDG for possible further alternate routing, such as to path SNDG–PHNX–WHPL. Control is returned by sending a CCS crankback signal from the via switch ALBY to the

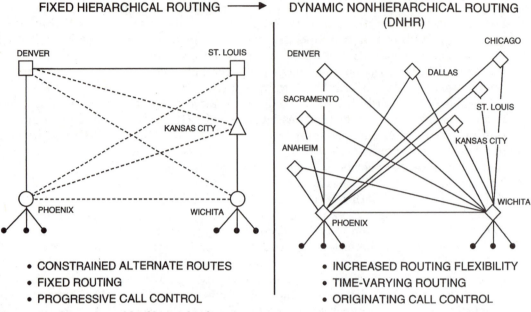

Figure 1.23 Comparison of fixed hierarchical routing and dynamic nonhierarchical routing (DNHR)

originating switch SNDG. The real-time, traffic-sensitive aspect of the routing method involves the use of additional "real-time" paths for possible completion of calls that overflow the preplanned routing sequences. The preplanned, or "engineered," paths are designed to provide the objective blocking probability grade-of-service. The candidate real-time paths designed for this dynamic routing method provide additional possible completion of calls that would otherwise be blocked, subject to trunk reservation restrictions. Trunk reservation requires that one more than a specified number of trunks—the "reservation level"—be free on each link before a real-time connection is allowed. This prevents calls that normally use a link from being swamped by real-time–routed calls. Each of the time-varying routing sequences, including the engineered and real-time paths, uses a subset of the available paths in a different order, and the paths used in various time periods need not be the same.

Each routing sequence results in a different allocation of link flows (for example, first choice paths normally carry the maximum flow of a switch-to-switch load), but all satisfy a switch-to-switch blocking requirement. Allocating traffic to the optimum route choice during each time period leads to design benefits due to the noncoincidence of loads. Since many intercity traffic demands change with time in a reasonably predictable manner, the routing also changes with time to achieve maximum link utilization and minimum network cost. Several dynamic routing time periods are used to divide up the hours on an average business day and weekend into contiguous routing intervals. The network design is performed in an off-line, centralized computation that determines the optimal routing tables from a very large number of possible alternatives in order to minimize the network cost. In this dynamic routing example, rather than search for optimal routing tables in real time, we perform this search off-line using a centralized design system that employs the dynamic network design model described in Section 1.4.4. The effectiveness of the design depends on how accurately we can estimate the load on the network. Forecast errors are corrected in the short-term capacity management process, which allows routing table updates to replace link augments whenever possible. The only decisions necessary in real time involve network conditions that become known in real time, such as actual (not forecasted) loads, failures, and overloads.

1.3.2.5 Real-time event-dependent routing. A number of real-time dynamic traffic routing methods have been studied. Real-time dynamic routing methods can be event-dependent or state-dependent. Event-dependent real-time routing methods may also use learning models, such as learning with random routing (LRR); success-to-the-top (STT) routing; DAR, developed by British Telecom [Gib86, Mee86, StS87]; and the STR method implemented by NTT Japan [KaI95,YMI91]. As illustrated in Figure 1.19, LRR is a decentralized call-by-call method with update based on random routing. LRR uses a simplified decentralized learning method to achieve flexible adaptive routing. The direct link is used first if available, and a fixed alternate path is used until it is blocked. In this case a new alternate path is selected at random as the alternate route choice for the next call overflow from the direct link. No crankback is used at a via switch or an egress switch in LRR, so a call blocked at a via or egress switch will be lost. Dynamically activated trunk reservation is used under call-blocking conditions. STT routing is an extension of the LRR method, in which crankback is allowed when a via path is blocked at the via switch, and the call advances to a new random path choice. In a limiting case of STT, all possible one- and two-link path choices can be tried by a given call before the call is blocked. In the DAR and STR learning approaches, as in LRR and STT, the direct path is used first and then the path last tried, which is also successful, is tried again until blocked, at which time another path is selected at random and tried on the next call, if needed. A fixed trunk reservation technique is used. In the STR enhancement of DAR, path choices are changed with time in accordance with changes in traffic load patterns. Event-dependent real-time routing methods perform well but are not as efficient as state-dependent real-time routing methods.

1.3.2.6 Real-time state-dependent routing. State-dependent real-time routing methods may change routing tables (1) every few minutes, as in the system to test adaptive routing (STAR) method tested by French Telecom [GCK87], the DR-5 method tested by Bellcore [CKP91, KrO88], Worldwide International Network (WIN) routing [ACK89, AKA94, AsH94, KAK88, KaW95, WaM87], and learning automata models [KuN79, McN78, NaT78, NaT80, NWM77]; (2) every few seconds, as in the DCR method [BNR86, Cam81, Car88, CGG80, HSS87, LaR91, RBC83, RBC95, SzB79, WaD88] or trunk status map routing (TSMR) [Ash85]; or (3) on every call, as in the RTNR method

[ACF91, ACF92, AsC93, Ash95, AsH93, MGH91]. Real-time dynamic routing methods may change routing tables every few minutes based on traffic load estimates. One method estimates the traffic load a few minutes in advance based on sequential estimation of recent traffic load behavior. With the projected traffic load, max-flow optimized routing is computed and implemented [Kri82], which requires a large computational effort and centralized real-time control of routing tables. Examples of centralized five-minute periodic real-time dynamic routing systems are STAR and DR-5. These methods recompute alternate routing paths every five minutes based on traffic data. WIN is an example of a distributed five-minute periodic real-time dynamic routing system. In WIN, several countries exchange traffic data every five minutes between network management processors, and based on analysis of this data, the network management processors can dynamically select alternate routes to optimize network performance. This method is illustrated in Figure 1.19. Distributed periodic real-time dynamic routing methods may also be based on call completion statistics determined through learning models. One such learning method involves observing the completion rate of calls routed on candidate paths and choosing the paths according to their completion probability, with the path having the highest probability chosen first, on down to the path with the lowest probability.

DCR is a centralized periodic real-time dynamic routing system. In DCR the selection of candidate paths at each switch is recalculated every 10 seconds. The path selection is done by a central routing processor, based on the busy-idle status of all links in the network, which is reported to the central processor every 10 seconds.

As illustrated in Figure 1.19, TSMR also is a centralized periodic routing method with periodic updates based on periodic network status. TSMR routing provides periodic near-real-time routing decisions in the dynamic routing network. The TSMR method involves having an update of the number of idle trunks in each link sent to a network database every five seconds. Routing tables are determined from analysis of the trunk status data using the TSMR dynamic routing method, which provides that the first path choice determined by the erlang flow optimization model is used if a circuit is available. The first-choice path is updated once each hour based on the erlang flow optimization model [ACM81, Ash85]. If the first path is busy, the second path is selected from the list of feasible paths on the basis of having the greatest number of idle circuits at the time; the current second path choice becomes the third, and so on. This path update is performed every five seconds. The TSMR model uses dynamically activated trunk reservation and other controls to automatically modify routing tables during network overloads and failures. TSMR requires the use of network status, traveling class mark, and crankback CCS messages. CCS crankback is used when the second link of a two-link path is blocked at a via switch to allow the originating switch to advance to the next path, and it is also used to implement multiple ingress/egress routing.

In RTNR, the routing computations are distributed among all the ESS processors in the network. This is illustrated in Figures 1.19 and 1.24, the latter of

DYNAMIC NONHIERARCHICAL
ROUTING (DNHR)

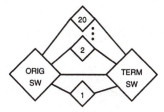

REAL-TIME NETWORK
ROUTING (RTNR)

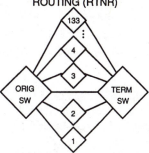

- OFF-LINE SYSTEM PREDETERMINES
  SEQUENTIAL ROUTES BASED ON TRAFFIC

- ROUTING CHANGES WITH HOUR OF DAY

- UP TO 21 PATHS BETWEEN SWITCHES

- CRANKBACK WHEN SECOND
  LINK UNAVAILABLE

- SWITCH SELECTS ROUTE CALL-BY-CALL (DIR. LK.
  FIRST; THEN 2-LINK PATH BASED ON STATUS)

- ROUTING CHANGES IN REAL-TIME

- NETWORK-WIDE PATH SELECTION

- ORIG. SW. QUERIES TERM. SW. FOR
  NETWORK STATUS

Figure 1.24 Comparison of DNHR and RTNR

which contrasts DNHR (preplanned dynamic two-link sequential path routing) with RTNR. RTNR determines the optimum one- or two-link path on a call-by-call basis from among all possible one- and two-link paths, based on real-time status information derived from CCS query-status messages exchanged during call setup. As illustrated in Figures 1.25 and 1.26, RTNR computes required bandwidth allocations by virtual network from ESS-measured traffic flows and uses this capacity allocation to reserve capacity when needed for each virtual network.

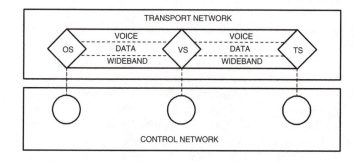

LEGEND

OS - ORIGINATING SWITCH

TS - TERMINATING SWITCH

VS - VIA SWITCH

◇ - TRAFFIC NETWORK
      SWITCH

○ - COMMON CHANNEL
      SIGNALING NETWORK
      SWITCH

STEP 1: IDENTIFY CLASS-OF-SERVICE, VIRTUAL NETWORK,
          AND TERMINATING SWITCH

STEP 2: DETERMINE ROUTING PATTERN FOR VIRTUAL NETWORK

STEP 3: DYNAMICALLY SELECT AVAILABLE PATH CAPACITY

Figure 1.25 Class-of-service RTNR routing method

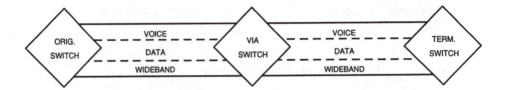

- LINK BANDWIDTH ALLOCATED TO DIFFERENT VIRTUAL NETWORKS (VOICE, DATA, WIDEBAND, ETC.) IN ACCORDANCE WITH BANDWIDTH DEMAND AND PERFORMANCE OBJECTIVES

- NO CONGESTION: ALLOWS FULL SHARING OF LINK BANDWIDTH

- CONGESTION: USES DYNAMIC TRUNK RESERVATION TO ALLOCATED VIRTUAL TRUNK BANDWIDTH TO SERVICES, INHIBIT ALTERNATE ROUTED TRAFFIC, AND GIVE PRIORITY TO KEY SERVICES

Figure 1.26   Class-of-service RTNR-link bandwidth allocation

Any excess traffic above the expected flow is routed to temporarily idle capacity borrowed from capacity reserved for other loads that happen to be below their expected levels. Idle link capacity is communicated to other switches via the CCS query-status messages in the RTNR network, as illustrated in Figure 1.27, and the excess traffic is dynamically allocated to the set of allowed paths that are identified as having temporarily idle capacity. RTNR controls the sharing of available capacity by using dynamic trunk reservation, to protect the capacity required to meet expected loads and to minimize the blocking of traffic for classes-of-service which exceed their expected load and allocated capacity.

Extensions of RTNR class-of-service routing to multiple ingress/egress routing and to real-time internetwork routing are illustrated in Figures 1.28 and 1.29, respectively. In Figure 1.28, a call from the originating switch T_1 destined for the United Kingdom switch UK_1 tries first to access the international links from international switching center T_3 to UK_1. In doing this it is possible that the call could be routing from T_1 to T_3 directly or via T_2. If all circuits from T_3 to UK_1 are busy, the call control can be returned to T_1 with a CCS crankback message, after which the call is routed to T_4 to access the T_4-to-UK_1 circuits. In this manner all ingress/egress connectivity is utilized to a connecting network, maximizing call completion and reliability.

Figure 1.29 illustrates real-time internetwork routing (RINR), which extends class-of-service routing concepts and increased routing flexibility for internetwork routing. RINR implements dynamic class-of-service routing capabilities for real-time dynamic routing between networks. RINR works synergistically with

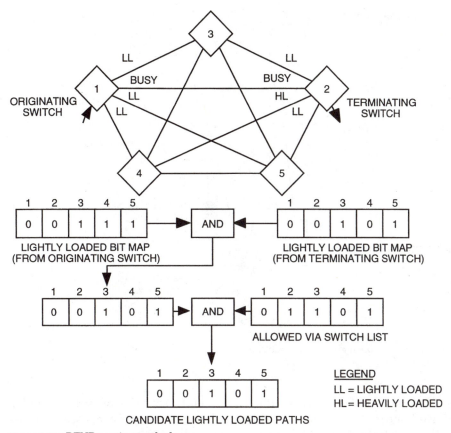

Figure 1.27 RTNR routing method

multiple ingress/egress routing, described above and illustrated in Figure 1.28. RINR uses link status information in combination with call completion history to select paths and reuses dynamic trunk reservation techniques implemented for class-of-service RTNR. RINR extends class-of-service RTNR concepts from the domestic network to routing between the domestic network and all connecting international networks, which previously used bilateral hierarchical routing with minimal flexibility. RINR provides dynamic class-of-service routing features and route selection based on real-time load status, completion performance, and traffic levels. Calls are set up where there is the most available capacity in the network. Traffic status for each class-of-service is continually monitored by switches and used to dynamically allocate link bandwidth. RINR provides (1) partitioning of trunk capacity among virtual networks, each allocated a specific set of services, based on the traffic patterns for various times of day; (2) dynamic trunk reservation to ensure that each service gets its allocated capacity while making unneeded capacity available to other services; (3) key service protection; (4) independent

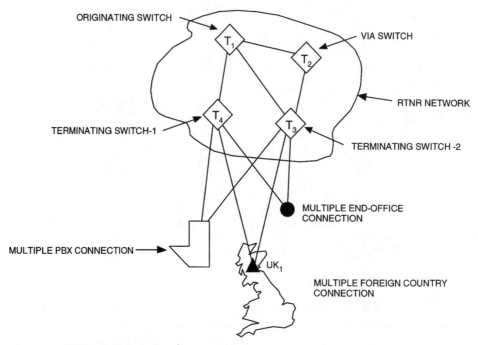

Figure 1.28 RTNR multiple ingress/egress routing

control of incoming and outgoing traffic; (5) automatic selection of overflow routes transiting other countries to increase call completion in times of congestion; (6) circuit capability selection for individual services, such as fiber versus satellite; and (7) automatic provisioning of route lists and other routing information. RINR is now fully deployed to all countries to which AT&T provides service. Further details of RINR are given in Chapter 12, along with results from a call-by-call simulation model that is used to measure the performance of the network with RINR in comparison with the performance with bilateral hierarchical routing. Model predictions of superior RINR performance are confirmed by actual measured performance.

1.3.2.7 Mixed dynamic routing. In a mixed dynamic routing (MXDR) network, many different methods of dynamic routing are used simultaneously. Calls originating at different switches use the particular dynamic routing method implemented at that switch, and different switches could be a mix of STT, TSMR, and RTNR, for example. Studies such as those reported in Section 1.5 show that a mix of dynamic routing methods achieves good throughput performance in comparison with the performance of individual dynamic routing strategies. These results also show that it is unnecessary to standardize a single dynamic routing method to be used uniformly in a given network.

REAL-TIME NETWORK ROUTING (RTNR)

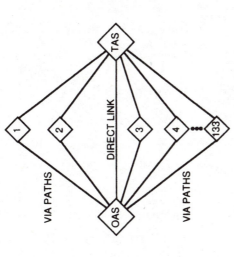

- TRY DIRECT LINK FIRST. IF BUSY, SELECT BEST TWO-LINK PATH BASED ON REAL-TIME ROUTE STATE
- ROUTING CHANGES IN REAL TIME
- NETWORK-WIDE PATH SELECTION
- ORIGINATING SWITCH QUERIES TERMINATING SWITCH FOR NETWORK STATUS
- DIRECT TRAFFIC PROTECTED DURING PERIODS OF CONGESTION

REAL-TIME INTERNETWORK ROUTING (RINR)

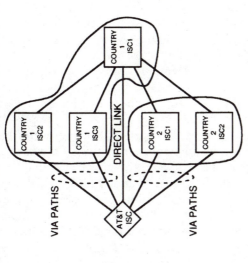

- TRY DIRECT LINK FIRST. IF BUSY, TRY VIA PATHS BASED ON REAL-TIME EGRESS LINK STATUS AND CALL HISTORY
- ROUTING CHANGES IN REAL TIME
- NETWORK-WIDE SELECTION
- ORIGINATING SWITCH USES REAL-TIME EGRESS LINK STATUS AND CALL HISTORY
- DIRECT TRAFFIC PROTECTED DURING PERIODS OF CONGESTION

Figure 1.29 Extending real-time network routing (RTNR) to real-time internetwork routing (RINR)

1.3.3 Dynamic transport routing

As with dynamic traffic routing, the introduction of new technology opens opportunities for dynamic transport routing [ACK89, ACL94, ACM91, AsS90, CED91, KDP95]. Dynamic transport routing can combine with dynamic traffic routing to shift transport bandwidth among switch pairs and services through use of flexible transport switching technology. Dynamic transport routing can provide automatic link provisioning, diverse link routing, and rapid link restoration for improved transport capacity utilization and performance under stress. Models for fixed and dynamic transport networks are discussed in Chapters 14–15. In Chapter 14, we discuss reliable traffic and transport routing networks, and in Chapter 15 we discuss dynamic traffic and transport routing networks.

Figure 1.30 illustrates the difference between the physical transport network and the logical transport network. Trunks are individual logical connections between network switches, which make up the link connections and are routed on the physical transport network. Links can be provisioned in multiples of DS1 units of capacity, which is equivalent to twenty-four 64-kbps channels, called DS0s, which are multiplexed on a DS1 channel with a combined bandwidth of 1.536 mbps. Links can vary in capacity from zero trunks to thousands of trunks, dependent on the level of traffic demand between switches. Figure 1.30 indicates that in the logical transport network, many switch pairs have a "direct" logical link connection where none exists in the physical transport network. A logical link connection is obtained by cross-connecting through transport switching devices. This is distinct from call routing, which switches a call on DS0 trunk channels at

LOGICAL TRANSPORT (TRUNK) NETWORK VIEW

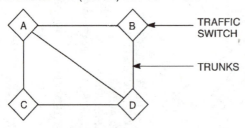

PHYSICAL TRANSPORT NETWORK VIEW

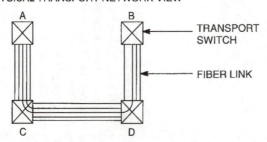

Figure 1.30 Logical and physical transport networks

each ESS switch in the call path. Thus, the trunk network is a logical transport network overlaid on a sparser physical transport network.

Digital cross-connect devices (DCSs) are able to switch transport channels, for example DS1 channels, onto different higher-capacity transport links such as DS3 channels and fiberoptic cables. Transport routes can be rearranged at high speed using digital cross-connect systems, typically within tens of milliseconds switching times. These circuit-oriented digital cross-connect devices can reconfigure logical trunk network capacity on demand for use in preplanned rearrangement of link capacity, such as for peak day traffic, weekly redesign of link capacity, or emergency restoration of capacity under switch or transport failure. Rearrangement of link capacity involves reallocating both transport bandwidth and switch terminations to different links. Digital cross-connect system technology is amenable to centralized network management control providing rearrangeable transport routing and perhaps real-time transport routing.

Figure 1.31 illustrates the concept of dynamic traffic and transport routing from a generalized switching node point of view. At the traffic demand level, or DS0 level in the transmission hierarchy, call requests are switched using dynamic traffic routing by ESS switching logic. At the DS1 and DS3 demand levels in the transmission hierarchy, demands are switched using digital cross-connect systems, which allow dynamic transport routing to route transport demands in accordance with traffic levels or dedicated circuit demand level. Real-time DS1 restoration in combination with DS3 transport level restoration can be performed using digital cross-connect system dynamic transport routing. Similarly,

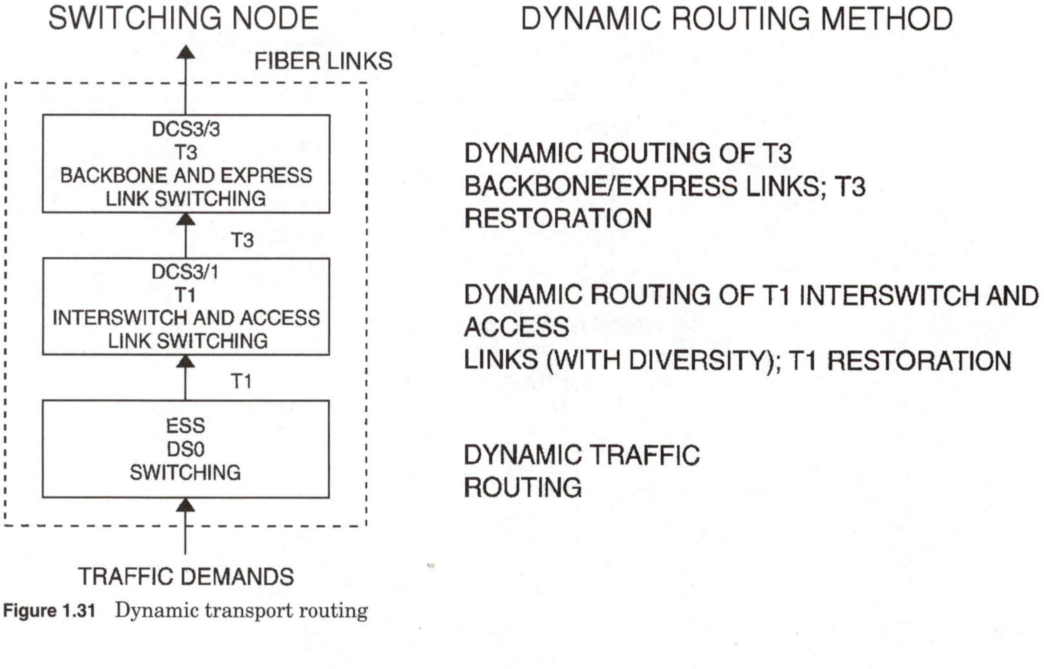

SWITCHING NODE   DYNAMIC ROUTING METHOD

Figure 1.31 Dynamic transport routing

real-time response to traffic blocking can be allowed by digital cross-connect system (DCS) dynamic transport routing to improve network performance.

Other technological capabilities for dynamic transport routing include packetized switching devices, such as the asynchronous transfer mode (ATM) technology now emerging. ATM transport provides greater flexibility than digital cross-connect system channel switching to rearrange transport channels. ATM network transport channels are called virtual channels (VCs), which are analogous to individual DS0 channel connections, and virtual paths (VPs), which are analogous to logical links comprising many virtual channels. In contrast to the fixed DS0/DS1/DS3 transport hierarchy in circuit-switched networks, ATM VCs and VPs have no preassigned bandwidths, and services with widely varying bandwidths and traffic characteristics can be routed over VCs and VPs. ATM packets, called cells, contain in their headers routing instructions known as virtual channel identifiers (VCIs) and virtual path identifiers (VPIs).

Packet-oriented ATM switching allows virtual transport networking—that is, real-time VC/VP logical routing, rearrangement, and restoration—through real-time remapping of VCIs and VPIs in ATM cells. Since these logical VC/VP routing changes can be made without interruption of service connections, traffic and transport dynamic routing might be accomplished with real-time ESS routing design in call processing.

An illustration of dynamic transport routing is given in Figure 1.32, which shows how transport demand is routed according to varying seasonal require-

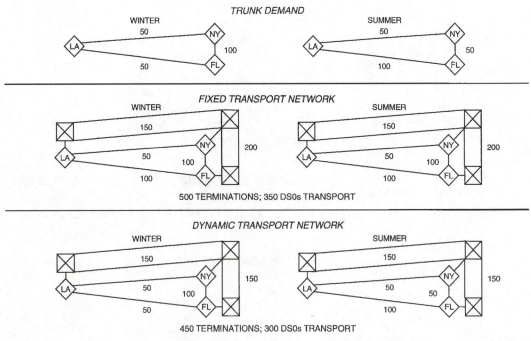

Figure 1.32 Weekly transport network design and rearrangement

ments. As seasonal demands shift, the dynamic transport network is better able to match demands to routed transport capacity, thus gaining efficiencies in transport requirements. Dynamic traffic and transport routing takes advantage of new technology, as did previous stages of routing evolution, and in this case the deployment of transport switching devices allows the traffic and transport networks to be integrated and use dynamic routing. Studies show increases in network utilization efficiency with dynamic traffic and transport routing, especially for networks supporting broadband services, and performance improvements obtained through diverse link routing, rapid transport restoration under failure, and dynamic switch load balancing [ACL94, AsS89]. Further details of dynamic transport routing are discussed in Chapter 15.

We summarize the classifications of routing methods discussed in this book in Table 1.1. As we shall see in Section 1.5, the routing alternatives identified in Table 1.1 are listed in increasing order of network performance and utilization advantage. Mixes of preplanned dynamic routing and real-time dynamic routing also are possible. For example, preplanned dynamic traffic routing tables could be adjusted online in near-real time, as is done in the DNHR network and explained in Chapter 8, to react dynamically to the unpredictable components of traffic load.

1.4   Network Management and Design

As shown in Figure 1.1, network management and design functions include real-time traffic management, capacity management, and network planning. Real-time traffic management ensures that network performance is maximized under all conditions, including load shifts and failures. Capacity management ensures that network capacity is designed and provisioned to meet performance objectives at minimum cost. Network planning ensures that switching and transport capacities are planned and deployed in advance of forecasted traffic growth. Figure 1.1 illustrates real-time traffic management, capacity management, and network planning as three interacting feedback loops around the network, in which the feedback controls function to regulate the service provided by the network through network management controls, capacity adjustments, and routing adjustments. Measurements are taken of the traffic load, which consists of predictable average demand components added to unknown forecast error and load variation components, to drive the network management/design functions. Network management and design implementation experience for dynamic routing networks is discussed in Chapter 17, for the DNHR network implementation in 1984 and RTNR implementation in 1991.

As illustrated in Figure 1.33, real-time traffic management provides monitoring of network performance through collection and display of real-time traffic and performance data, and it allows traffic management controls such as code blocks, call gapping, and reroute controls to be used by network managers when circumstances warrant. Monitoring of network performance is illustrated in Figures 1.34 and 1.35. Figure 1.34 illustrates network managers performing these

TABLE 1.1 Traffic and Transport Routing Methods

Routing classification	Routing table update			Examples	Chapter
	Frequency	Control point			
fixed traffic routing	infrequently (e.g., switch homing change)	network management system		fixed hierarchical	2
				fixed nonhierarchical	4–5
dynamic traffic routing: (preplanned)	weekly (hourly routing patterns)	network management system		dynamic hierarchical	3
				dynamic multilink cyclic path	4
				dynamic multilink sequential path	4
				dynamic two-link cyclic path	4
				dynamic two-link CGH path	4
				dynamic progressive	5
				dynamic two-link sequential path (DNHR)	6–10
(real-time event-dependent)	blocked call	ESS/CCS network		dynamic alternate routing (DAR) state- & time-dependent routing (STR) learning with random routing (LRR) success-to-the-top routing (STT)	12–13
(real-time state-dependent)	minutes	network management system		worldwide international network (WIN) system to test adaptive routing (STAR) dynamic routing—five minutes (DR-5)	12–13
	seconds	network management system		dynamically controlled routing (DCR) trunk status map routing (TSMR)	11, 13 / 11, 13
	call by call	ESS/CCS network		real-time network routing (RTNR)	12–13
dynamic transport routing: (rearrangeable)	minutes	network management system		T1 DCS networking T3 DCS networking (FASTAR)	14–15
(real-time)	seconds call-by-call	ESS/CCS network		VC/VP ATM networking	15

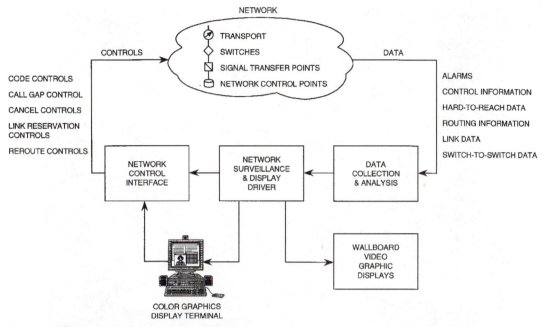

Figure 1.33   Real-time traffic management

Figure 1.34   Network operations center, Bedminster, New Jersey

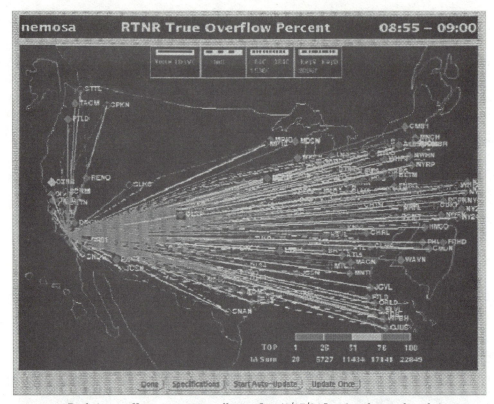

Figure 1.35 Real-time traffic management calls overflow (1/17/94 Los Angeles earthquake)

functions in the network operations center located in Bedminster, New Jersey. Figure 1.35 illustrates real-time congestion performance data for the January 17, 1994, Los Angeles earthquake, displayed on video wall-boards and network managers' terminals.

Heavy traffic overloads or major equipment failures may cause the network to degrade in an objectionable manner. For example, if a network designed to carry a certain traffic load is subjected to severe overload, in the absence of adequate real-time traffic management the realized carried load may fall substantially below the design load. This phenomenon is the result of trunk and switching congestion. As direct routes become fully occupied, calls are forced to follow alternate routes. These generally involve more trunks and switches per call, and hence the network operates at lower efficiency. At higher traffic levels, when alternate routes are blocked the switching and trunk resources used in the process of blocking the call are wasted. A call arriving at a switching system is placed in a queue if all digit receivers are in use. While the call is in a queue, the digit transmitter, or sender, in the preceding switching system is tied up until a digit receiver becomes free. After 20–30 seconds, the sender times out and blocks the call. If a sufficient number of calls receive this treatment, other switching systems become congested and congestion spreads throughout the network.

A number of real-time traffic management controls are used to control network congestion. Directionalization of a link by using trunk reservation can favor calls originating from a disaster area, for example. A call subjected to trunk reservation of k trunks, for example, requires that there be $k + 1$ idle trunks on a link before the call can seize a trunk. Hence, if calls being routed *to* the disaster area are subject to trunk reservation, but calls originating *from* the disaster area are not, then calls from the disaster area are completed more easily than calls to the disaster area. This is a frequently used technique to directionalize a link.

Cancellation of alternate routing removes alternate-routed traffic from a link and thereby reduces the load on the distant switching systems, as well as the average number of links per call. *Cancel-from* and *cancel-to* controls cancel all traffic overflowing from or to the controlled link, respectively. Reroute controls route overflow traffic to a link that is not in the normal routing pattern. This technique is often used in fixed hierarchical networks to allow traffic to complete through use of idle network capacity. Reroute controls are less frequently employed in dynamic routing networks where most or all routes are examined within the call-routing logic.

Code-blocking controls block calls to a particular destination code. This control is particularly useful in the case of focused overloads, especially if the calls are blocked at or near their origination. Code blocking need not be total unless the destination switch is completely disabled through natural disaster or equipment failure. Switches equipped with code-blocking controls can typically control a percentage of the calls to a particular code. The controlled code may be NPA, NXX, NPA-NXX, or NPA-NXX-XXXX, when in the latter case one specific customer is the target of a focused overload.

Figure 1.36 illustrates a call gap control typically used by network managers in a focused call overload, such as sometimes occurs with radio call-in give-away contests. Call gapping allows one call for a controlled code or set of codes to be accepted into the network, by each switch, once every x seconds, and calls arriving

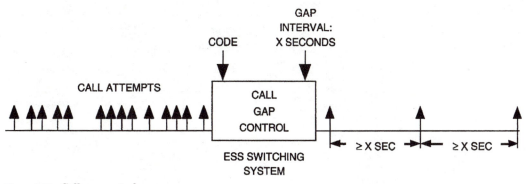

Figure 1.36   Call gap control

after the accepted call are rejected for the next x seconds. In this way call gapping throttles the calls and prevents overloading the network to a particular focal point.

Dynamic overload control (DOC) is an automatic control that senses congestion in a switch—for example, by measuring the queue length of calls waiting for senders—and sends a signal to connecting switches, which cancel some portion of traffic to the congested switch. The control action may last only about 10 seconds. Such automatic controls are preferred over manual controls because they are activated as soon as congestion appears and are removed as soon as congestion disappears. Selective dynamic overload control (SDOC) uses completion statistics by code, and the traffic cancellation is performed on the hard-to-reach codes. This provides a more effective control by allowing traffic that completes to not be cancelled.

As illustrated in Figure 1.37, capacity management provides for projection of demands, including adjustments for business forecasts and projected new service demands, and execution of the network design model to determine the capacity requirements in the forecast horizon. The updated capacity requirements are sent to switching and transport provisioning systems so that capacity expansion is implemented on a scheduled basis to meet the projected demands. As illustrated in Figure 1.37, capacity management provides monitoring of network performance through collection and display of daily traffic and performance data, and if service problems are detected it allows capacity and routing redesign and implementation to alleviate the service problems. Under exceptional circumstances, capacity can be added on a short-term basis to alleviate service problems, but it is normally planned, scheduled, and managed over a period of several months to one year or more.

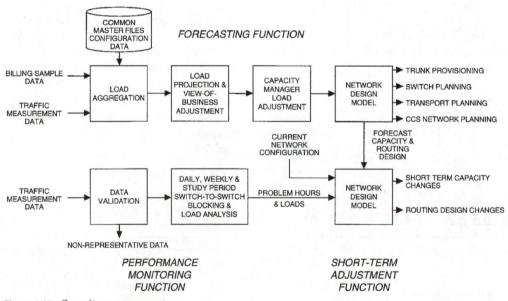

Figure 1.37 Capacity management

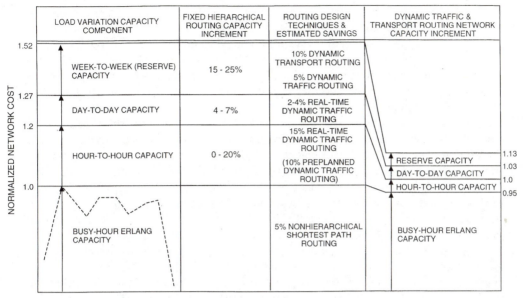

Figure 1.38 Estimated savings with dynamic routing network design for load variation components

Network design encompasses capacity and routing design. In the following sections we discuss network design models developed in the book. Non-alternate-routing networks and fixed hierarchical routing networks, along with associated network design techniques, lead to capacity requirements that can be reduced by improved routing techniques. Upper limits are established on the maximum improvement with improved network routing and design. This is illustrated in Figure 1.38 and discussed in this section. Even though the bounds given in Figure 1.38 are unrealizable limits, experience has shown that practical techniques are available to achieve much of the projected savings. Therefore these limits serve as bounds useful in approximating potential gains. The limits are additive, so that the total range is about 30–50% of the network first cost, of which half or more is realizable. Each area of savings is discussed here, as are the network design models used to realize the projected improvement. These design techniques for realizing savings are also identified in Figure 1.38.

1.4.1 Network design models
for traffic load variations

In this section we discuss load variation models used in network design. In network design, traffic load variations such as instantaneous variations, hour-to-hour variations, day-to-day traffic variations, and week-to-week variations are taken into account in network design to provide sufficient capacity to carry the expected traffic variations so as to meet end-to-end blocking objective levels. Traffic load variations lead in direct measure to the capacity increments illustrated in Figure 1.38 and can be categorized as (1) minute-to-minute instantaneous variations and associated busy-hour erlang load capacity, (2) hour-to-hour variations and associated multihour capacity, (3) day-to-day variations and associated day-to-day capacity, and (4) week-to-week variations and associated reserve capacity.

For each switch pair for a given hour, the load is modeled as a stationary random process characterized by a fixed mean and variance, and therefore a fixed value of peakedness, which equals variance over mean. From hour to hour, the mean loads are modeled as changing deterministically; for example, according to their 20-day average values. From day to day, for a fixed hour, the mean load is modeled as a random variable having a gamma distribution with a mean equal to the 20-day average load. From week to week, the load variation is modeled as a time-varying deterministic process in the network design procedure. The random component of the realized week-to-week load is the forecast error, which is equal to the forecast load minus the realized load. Forecast error is accounted for in short-term capacity management. Table 1.2 summarizes the types of models used to represent the different traffic variations under consideration.

Peakedness methods within the network design procedure account for the mean and variance of the within-the-hour variations of the offered and overflow loads. As one component of the network design procedure, Wilkinson's equivalent random method [Wil56] is used to size links for these two parameters of load, with the details discussed in Chapter 2. This results in the busy-hour erlang capacity, as illustrated in Figure 1.38, which can be reduced by link alternate routing flow optimization, and shortest path selection. These two models are discussed below.

TABLE 1.2   Traffic Models for Load Variations

Traffic variations time constant	Load variation examples	Illustrative traffic model for network design	Capacity impacts
minute to minute	real-time random traffic fluctuations; peaked overflow traffic	stochastic model; variance = peakedness $\times$ mean; Wilkinson equivalent random model	busy-hour erlang load capacity
hour to hour	business day peak; consumer evening peak	deterministic model; 20-day average time-varying mean; multihour dynamic routing design	multihour capacity
day to day	Monday busiest compared to average day	stochastic model; gamma-distributed variance = $.13 \times$ mean$^{\phi}$; $\phi = 1.5/1.7/1.84$ for low/ med./high day-to-day variations; Neal-Wilkinson $\overline{B}$ tables	day-to-day capacity
week to week	winter/summer seasonal variations; forecast errors	stochastic model; variance = $\sigma^2_{\text{forecast-error}}$ maximum flow routing & capacity design	reserve capacity

Multihour dynamic route design accounts for the hour-to-hour variations of the load and, as illustrated in Figure 1.38, accounts for the multihour capacity component in the overall design. As illustrated below, hour-to-hour capacity can vary from zero to 20 percent or more of network capacity. Hour-to-hour capacity can be reduced by multihour preplanned or real-time dynamic routing design models such as the erlang flow optimization, discrete event flow optimization, and transport flow optimization models described below. Hour-to-hour capacity reduction is achieved through maximum flow dynamic routing optimization and results in greater network utilization.

It is known that some daily variations are systematic (for example, Monday is always higher than Tuesday); however, in some day-to-day variation models these systematic changes are ignored and lumped into the stochastic model. For instance, the traffic load between Los Angeles and New Brunswick is very similar from one day to the next, but the exact calling levels differ for any given day. We characterize this load variation in network design by a stochastic model for the daily variation [Wil71], which results in the additional capacity illustrated in Figure 1.38, called day-to-day capacity. Day-to-day capacity is needed to meet the average blocking objective when the load varies according to the stochastic model [HiN76]. Day-to-day capacity is nonzero due to the nonlinearities in link blocking. When the load on a link fluctuates about a mean value, because of day-to-day variation, the mean blocking is higher than the blocking produced by the mean load. Therefore, additional capacity is provided to maintain the blocking probability grade-of-service objective in the presence of day-to-day load variation. In fixed routing design, the day-to-day capacity required is 4–7 percent of the network cost for medium to high day-to-day variations, respectively. This level of day-to-day capacity can be reduced with real-time routing design, as described below, in which greater network utilization and network design savings are achieved.

Reserve capacity, like day-to-day capacity, comes about because load uncertainties—in this case forecast errors—tend to cause capacity buildup in excess of the network design, which exactly matches the forecast loads. Reluctance to disconnect and rearrange link and transport capacity contributes to this reserve capacity buildup. Typical ranges for reserve capacity are from 15 to 25 percent or more of network cost. Once again, preplanned or real-time routing design, through maximum flow optimization models described below, can reduce reserve capacity by achieving higher network utilization.

We now present network design models that address each of these load variations, which include link alternate routing flow optimization with Truitt's model [Tru54], shortest path selection, multihour dynamic route design, day-to-day variation design models, and forecast error/reserve capacity design models.

1.4.2 Link alternate routing flow optimization model

Truitt's link alternate routing flow optimization model [Tru54] is illustrated in Figure 1.39. It is used in some form in essentially all network design models discussed in this book. Figure 1.39 illustrates a three-switch network, sometimes called Truitt's triangle, where a erlangs of traffic are offered from switch A to

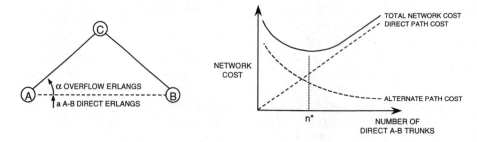

- FIND NUMBER OF HIGH USAGE LINK A-B TRUNKS TO MINIMIZE COST

- TOTAL COST $= C_D \times n + a \dfrac{B(a,n) \times C_A}{\gamma}$

- SET $\dfrac{\partial \text{(Total Cost)}}{\partial n} = 0$, OR

$$CLLT = a\left(B(a,\,n^*\text{-}1) - B(a,\,n^*)\right) = \frac{\gamma}{C_A/C_D} = ECCS$$

WHERE C_D = DIRECT LINK COST
C_A = ALTERNATE PATH COST
a = A-B DIRECT ERLANGS
n = NUMBER OF A-B DIRECT TRUNKS
B = ERLANG B BLOCKING FORMULA
γ = MARGINAL CAPACITY OF ALTERNATE PATH
CLLT = CARRIED LOAD ON LAST TRUNK
ECCS = ECONOMIC CCS

Figure 1.39 Truitt's ECCS method for high-usage link sizing

switch B. The objective is to determine the direct link alternate routing flow and alternate path flow that minimize the network cost of carrying a erlangs on the network and that meet the blocking probability grade-of-service objective. The following quantities are defined:

C_D = Cost of a direct trunk

C_A = Cost of an alternate network trunk

C_R = Cost ratio = C_A/C_D

γ = Marginal capacity, or the erlangs carried by a single alternate network trunk

CLLT = Carried load on the last direct trunk

ECCS = Economic hundred call seconds, or CCS

α = Overflow load from direct link

Here the marginal capacity γ is assumed to be a constant that is independent of the amount of overflow load α. The following equation then describes the total cost of carrying a erlangs of traffic on the direct route and the alternate network:

$$\text{Total cost} = C_\text{D} \times n + \frac{a \times B(a, n) \times C_\text{A}}{\gamma}$$

Here $B(a, n)$ is the Erlang blocking formula for a erlangs offered to n trunks. This formula is derived in Chapter 2. The cost relationship is plotted in Figure 1.39 and shows that as the number of direct trunks is increased, the direct path cost increases linearly as $n \times C_\text{D}$. The alternate path cost decreases as more direct trunks are added, because the overflow load $\alpha = a \times B(a, n)$ decreases, and the cost of carrying the overflow load

$$\frac{a \times B(a, n) \times C_\text{A}}{\gamma}$$

decreases. As illustrated in the figure, an optimum, or minimum, cost condition is achieved with the optimum direct link flow with n^* trunks. If we take the derivative of the above total cost equation with respect to the number of trunks n and set the result to zero, that is,

$$\frac{\Delta(\text{total cost})}{\Delta n} = 0$$

we then derive the minimum cost condition, which is then given by the following expression:

$$a \times [B(a, n^* - 1) - B(a, n^*)] = \text{CLLT} \geq \gamma \times \frac{C_\text{D}}{C_A}$$

or

$$\text{CLLT} \geq \frac{\gamma}{C_\text{R}} = \text{ECCS}$$

Hence the direct trunks are increased until the cost per erlang to carry traffic on the last direct trunk is still lower than the cost per erlang to carry traffic on the alternate network. This is Truitt's economic hundred call seconds, or ECCS, condition and is used in many design models described in the book, be they for fixed hierarchical routing networks or dynamic nonhierarchical routing networks.

1.4.3 Shortest path selection models

Hierarchical routing has limited path choices and low blocking design on final links, which limits flexibility and reduces efficiency. If we choose paths based on cost and relax the hierarchical network structure, a more efficient network results. Additional benefits can be provided in network design by allowing a more flexible routing plan that is not restricted to hierarchical routes but allows the selection of the shortest nonhierarchical paths. Dijkstra's method [Dij59], for example,

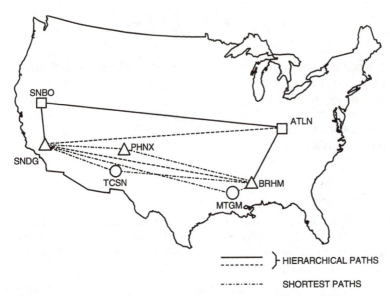

Figure 1.40 Shortest path routing

is often used for shortest path selection. Figure 1.40 illustrates the limitation that hierarchical routing imposes between the SNDG and BRHM switches. The alternate hierarchical paths between these switches go through two regional centers, SNBO and ATLN, providing relatively long paths. Selecting more direct paths—for example, the paths through the TCSN, PHNX, or MTGM switches—provides design benefits by allowing path selection beyond the hierarchical choices to nonhierarchical path choices.

It is estimated from the 28-switch national intercity network model that about 5 percent of the network's first cost can be attributed to designing for a network using fixed hierarchical routing versus nonhierarchical shortest path selection. To show this, we first design a hierarchical network using multihour hierarchical network design techniques, described below and in Chapter 2. Then, to quantify the extra capacity provided, we also design the 28-switch model for the individual hourly loads. These hourly networks are obtained by using each hourly load, and ignoring the other hourly loads, to size a hierarchical network and also a nonhierarchical shortest path design that perfectly matches that hour's load. This procedure results in 17 hierarchical network designs and 17 nonhierarchical shortest path network designs, one for each hour.

Figure 1.41 is a plot of the normalized network cost, including switching and transport costs, required for the multihour hierarchical network design and hourly network designs. On the top line, the multihour hierarchical network design has a network capital cost of one unit to satisfy all 17 hours of load with fixed, hierarchical routing. The 17 hourly networks, shown on the lower curves, represent the normalized capital cost of the circuit miles and trunks required at each hour to satisfy that hour's load, for hourly hierarchical path designs

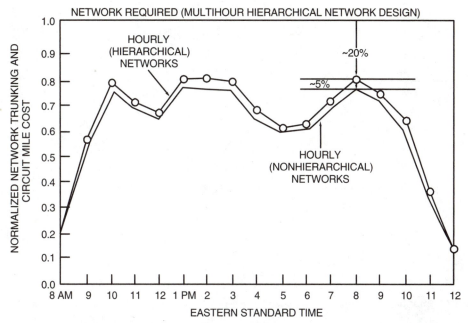

Figure 1.41   Network first cost vs. hour (28-switch network model)

and hourly nonhierarchical shortest path designs. Applied to each hourly load, nonhierarchical shortest path design yields an overall savings of about 5 percent in comparison with the hourly hierarchical networks. This savings potential translates into actual benefits by allowing nonhierarchical shortest path routing in traffic network design.

There are really two components to the 5 percent shortest path selection savings. One component results from eliminating link splintering. Splintering occurs, for example, when more than one switching system is required to satisfy a metropolitan area load, in which multiple links to a distant switch could result, thus dividing the load among links which are less efficient than a single large link. A second component of shortest path selection savings arises from intercity route selection. Routing on the least costly, most direct, or shortest paths is often more efficient than routing over longer hierarchical paths.

1.4.4   Multihour dynamic routing design models

Dynamic route design improves network utilization relative to fixed routing because fixed routing cannot respond as efficiently to traffic load variations that arise from business/residential phone use, time zones, seasonal variations, and other causes. Dynamic routing increases network utilization efficiency by varying routing tables in accordance with traffic patterns. A simple illustration of this principle is shown in Figure 1.42, where there is afternoon peak load demand between switches A and B but a morning peak load demand between switches A

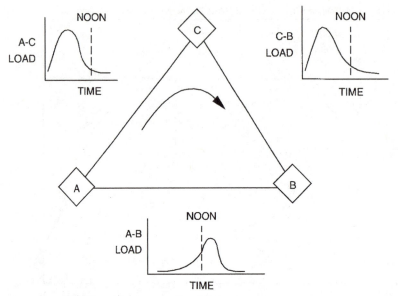

Figure 1.42 Multihour dynamic routing design

and C and switches C and B. Here a simple dynamic route design is to provide capacity only between switches A and C and switches C and B but no capacity between switches A and B. Then the A–C and C–B morning peak loads route directly over this capacity in the morning, and the A–B afternoon peak load uses this same capacity by routing this traffic on the A–C–B path in the afternoon. A fixed routing network design provides capacity for the peak period for each switch pair and thus provides capacity between switches A and B, as well as between switches A and C and switches C and B.

The effect of dynamic route design in a larger network is illustrated again by the 28-switch national intercity network model designs illustrated in Figure 1.41. Here it is shown that about 20 percent of the network's first cost can be attributed to designing for time-varying loads using fixed hierarchical routing. As illustrated in the figure, the 17 hourly networks are obtained by using each hourly load, and ignoring the other hourly loads, to size a hierarchical network that perfectly matches that hour's load. Each hourly network represents the erlang load capacity cost referred to in Table 1.2. The 17 hourly networks show that three network busy periods are visible—morning, afternoon, and evening—and illustrate the noon-hour drop in load and the early-evening drop as the business day ends and residential calling begins in the evening. The hourly network curve separates the capacity provided in the multihour hierarchical network design into two components: Below the curve is the capacity needed in each hour to meet the load; above the curve is the capacity that is available but is not needed in that hour. This additional capacity exceeds 20 percent of the total network capacity through all hours of the day, which represents the multihour capacity cost referred to in Table 1.2 for fixed hierarchical routing design. This gap represents the capacity of the network to meet noncoincident loads with fixed hierarchical routing and suggests a maximum limit on network capacity cost reduction that might be achieved through dynamic traffic routing.

We now discuss the three types of multihour network design models—erlang flow optimization models, transport flow optimization models, and discrete event flow optimization models—and illustrate how they are applied to various fixed and dynamic network designs. For each model we discuss steps that include initialization, routing design, capacity design, and parameter update.

1.4.4.1 Erlang flow optimization models. Erlang flow optimization models are used for fixed and dynamic traffic network design. These models optimize the routing of traffic flows, as measured in erlang loads, and the associated link capacities. Such models typically solve mathematical equations that describe the routing of erlang flows analytically and, for dynamic network design, often solve linear programming flow optimization models. Various types of erlang flow optimization models are distinguished as to how flow is assigned to links, paths, and routes. In hierarchical network design, erlang flow is assigned to high-usage links based on Truitt's ECCS method. Overflow from the high-usage links is routed to intermediate high-usage links or to final links, in which the latter are sized with Neal-Wilkinson methods for a blocking probability grade-of-service objective, normally 0.01 blocking, given the erlang offered load, the peakedness of the traffic, and the day-to-day variation of the traffic. In dynamic network design, erlang flow models are often path based, in which erlang flow is assigned to individual paths, or route based, in which erlang flow is assigned to routes. We now describe HU-final-erlang flow optimization models, route-erlang flow optimization models, and path-erlang flow optimization models.

Fixed hierarchical routing networks. As applied to fixed hierarchical routing networks, HU-final-erlang flow optimization models use switch homing patterns and alternate route patterns, which are designed to accommodate various routing rules, as outlined in Chapter 2. These hierarchical routing tables change slowly. Capacity design for high-usage links uses Truitt's model to determine the number of direct trunks, the Erlang B formula to calculate the overflow erlang load, and Nyquist's formula to determine the peakedness of the overflow traffic. Overflow traffic is accumulated on the overflow intermediate high-usage (HU) links and final links. Intermediate high-usage links again are sized with Truitt's model, and overflow loads determined from Wilkinson's equivalent random theory, Rapp's approximation, and Nyquist's formula. Final links are sized for a blocking probability grade-of-service objective, which is often 0.01 or 1 in 100 calls blocked, using Neal-Wilkinson models that account for peakedness and day-to-day load variation. The Erlang B formula, Wilkinson's equivalent random theory [Wil56], Rapp's approximation [Rap62], Nyquist's formula, and Neal-Wilkinson models [HiN76] are discussed further in Chapter 2. Eisenberg's multihour engineering model [Eis79] is a variation of fixed hierarchical network design that capitalizes on sizing high-usage links to best take advantage of traffic noncoincidence and sharing of capacity across multiple design hours on alternate routes and final links.

Preplanned dynamic routing networks. As applied to preplanned fixed and dynamic nonhierarchical traffic routing networks, erlang flow optimization models do network design based on shortest path selection and linear programming erlang flow optimization. A particular erlang flow optimization model for preplanned dynamic routing networks is illustrated in Figure 1.43. There are two versions of this model: route-erlang flow optimization and path-erlang flow optimization. Shortest least-cost path routing gives calls access to paths in order of cost, such that calls access all direct circuits between switches prior to attempting more expensive overflow paths. Routes are constructed with specific routing selection rules. For example, preplanned dynamic routing erlang flow optimization models construct routes for multilink or two-link path routing by assuming crankback and originating switch control capabilities in the ESS routing. As another example, routes are constructed for switch-to-switch progressive routing by selecting the best outlet switch and saving switch history in ESS routing. The linear programming flow optimization model strives to share link capacity to the greatest extent possible with the variation of loads in the network. This is done by equalizing the loads on links throughout the busy periods on the network, such that each link is used to the maximum extent possible in all time periods. The routing design step finds the shortest paths between switches in the network, combines them into candidate routes, and uses the linear programming flow optimization model to assign erlang flow to the candidate routes.

The capacity design step takes the routing design and solves a fixed-point erlang flow model to determine the capacity of each link in the network. This model determines the flow on each link and sizes the link to meet the design level

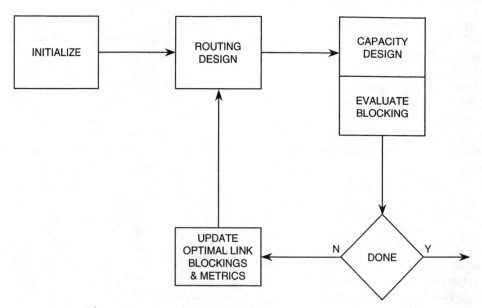

Figure 1.43 Path/route erlang flow optimization model

of blocking used in the routing design step. Once the links have been sized, the cost of the network is evaluated and compared to the last iteration.

If the network cost is still decreasing, the update module (1) computes the slope of the capacity versus load curve on each link, which reflects the incremental link cost, and updates the link "length" using this incremental cost as a weighting factor and (2) recomputes a new estimate of the optimal link blocking using the ECCS method. The new link lengths and blockings are fed to the routing design, which again constructs route choices from the shortest paths, and so on. Minimizing incremental network costs helps convert a nonlinear optimization problem to a linear programming optimization problem. Yaged [Yag71, Yag73] and Knepley [Kne73] take advantage of this approach in their network design models. This favors large efficient links, which carry traffic at higher utilization efficiency than smaller links. Selecting an efficient level of blocking on each link in the network is basic to the route/path erlang flow optimization model. The ECCS approach of Truitt [Tru54] is used in the erlang flow optimization model to optimally divide the load between the direct link and the overflow network, and this determines an optimal level of link blocking.

The erlang flow optimization model is modified for preplanned dynamic two-link sequential path routing networks, as described in Chapters 6–10, in which erlang flow assignments are specially tailored to be assigned directly to paths rather than to routes. This path erlang flow optimization model greatly simplifies and improves the quality of the network design.

Real-time dynamic routing networks. Fixed-point erlang flow optimization models can also be used for real-time dynamic routing designs that attempt to model the detailed link state probabilities and erlang flows on paths selected in the route steps based on real-time link status. Figure 1.44 illustrates the fixed-point flow optimization model. A fixed-point model [Kat67, Kil86, Whi85] is constructed for the particular real-time dynamic routing method. For example, a fixed-point model for RTNR is discussed in Chapter 13 [AsH94]. It consists of the following steps: (1) routing erlang flow to paths and links in accordance with the real-time path selection logic for each routing method; (2) estimating blocking on links according to the solution of link state probability equations with the birth-death model, described in Chapter 2; and (3) evaluating switch-to-switch blocking probability and updating link capacity if needed according to a Kruithof allocation model. The Kruithof model is described in the next section. An example of the erlang flow optimization model applied to real-time dynamic routing networks is given in Chapter 13.

Real-time dynamic transport routing networks. Erlang flow optimization models can be applied to real-time transport routing network design, as discussed in Chapter 15. In this application, shortest path selection is used to determine the primary path for routing erlang traffic flows directly on the transport network. A simple rule is implemented for sizing each link to a link-blocking probability grade-of-service objective that allows an overall maximum path-blocking objective to be met. Once the link-blocking objective is determined in this manner, the link is sized for the total erlang flow with the Erlang B formula.

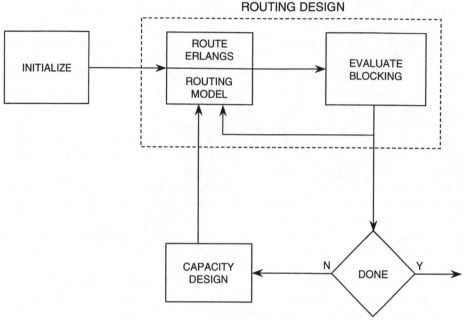

Figure 1.44 Erlang–fixed-point flow optimization

1.4.4.2 Transport flow optimization models. Transport flow optimization models are used for fixed and dynamic traffic and transport network design. These models optimize the routing of transport flows, as measured in units of transport demand such as 64 kbps or 1.5 mbps, and the associated link capacities. For application to traffic network design, transport flow optimization models use mathematical equations to convert traffic demands to transport capacity demands, and the transport flow is then routed and optimized. Figure 1.45 illustrates the transport flow optimization steps. The transport model converts erlang demands to virtual transport demands. This model typically assumes an underlying traffic routing structure. For example, the traffic routing model can be fixed hierarchical routing, as described in Chapter 3, or nonhierarchical dynamic routing, as described in Chapter 13.

Dynamic hierarchical routing networks. In the transport flow optimization model application to dynamic hierarchical network design, as described in Chapter 3, a linear programming transport flow optimization model is used for network design. In this application there is an underlying hierarchical traffic routing. Switch-to-switch erlang traffic demands are converted to high-usage link and final link switch-to-switch transport demands by using the hierarchical network design methods described in Chapter 2, which include the ECCS approach of Truitt [Tru54], Wilkinson's equivalent random method [Wil56], and Neal-Wilkinson [HiN76] models for final link sizing. A linear programming transport flow optimization is solved for network sizing, which routes switch-to-

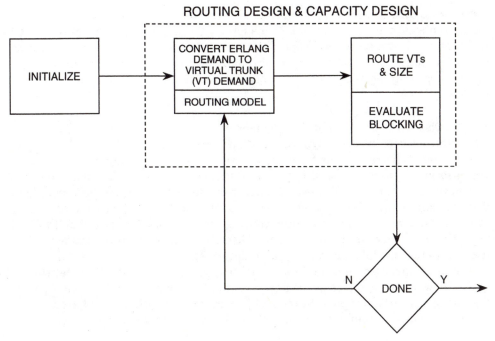

ROUTING DESIGN & CAPACITY DESIGN

Figure 1.45 Transport flow optimization

switch virtual transport load demands by hour on the shortest, least-cost paths and sizes the link to meet the design level of flow.

Real-time dynamic routing networks. In the transport flow optimization model application to real-time dynamic routing network design, as described in Chapter 13, a linear programming transport flow optimization model is used for network design. Switch-to-switch erlang traffic demands are converted to switch-to-switch transport demands by using the ECCS approach of Truitt [Tru54]. Here Truitt's approach is used to optimally divide the load between the direct link and the overflow network, but in this application of the model we obtain an equivalent virtual transport demand, by hour, as opposed to an optimum link-blocking objective. A linear programming transport flow optimization is solved for network sizing, which routes switch-to-switch virtual transport load demands by hour on the shortest, least-cost paths and sizes the link to meet the design level of flow. Once the links have been sized, the performance of the network is evaluated, and if the performance objectives are not met, further modification is made to capacity requirements. Techniques used in the erlang flow optimization model for optimizing erlang flows by hour are reused in the transport flow optimization model for optimizing virtual transport flows, as discussed in Chapter 13.

Reliable transport routing networks. In the transport flow optimization model application to reliable transport network design, as described in Chapter 14, a linear programming transport flow optimization model is used for network design.

Preplanned dynamic transport routing networks. In the transport flow optimization model application to rearrangeable transport routing network design, as described in Chapter 15, the network design determines the minimum-cost primary path and minimum-cost alternate path through use of shortest-path selection. A linear programming transport flow optimization is then solved for the network design.

1.4.4.3 Discrete event flow optimization models.

Discrete event flow optimization models are used for fixed and dynamic traffic network design. These models optimize the routing of discrete event flows, as measured in units of individual calls, and the associated link capacities. Figure 1.46 illustrates steps of the discrete event flow optimization model. The event generator converts erlang traffic demands to discrete call events. The discrete event model provides routing logic according to the particular real-time dynamic routing method and routes the call events according to the dynamic routing logic. Discrete event flow optimization models use simulation models for routing table design to route discrete event demands on the link capacities, and the link capacities are then optimized to meet the required flow. In the discrete event flow optimization models for real-time dynamic routing networks, we generate initial link capacity requirements based on the traffic load matrix input to the model. Based on design experience with the model,

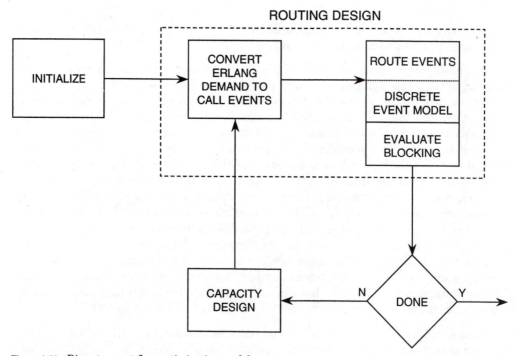

Figure 1.46   Discrete event flow optimization model

an initial total link capacity demand for each switch is estimated based on a maximum design occupancy in the switch busy hour of 0.93. Then the occupancy of the total network link capacity in the network busy hour is adjusted to fall within the range of 0.84 to 0.89, based on design experience with the model. Blocking performance is evaluated as an output of the discrete event model, and any needed link capacity adjustments are determined. Trunks are allocated to individual links in accordance with the Kruithof allocation method [Kru37], which distributes link capacity in proportion to the overall demand between switches.

Kruithof's technique is used to estimate the switch-to-switch requirements p_{ij} from the originating switch i to the terminating switch j under the condition that the total switch link capacity requirements may be established by adding the entries in the matrix $\mathbf{p} = [p_{ij}]$. Assume that a matrix $\mathbf{q} = [q_{ij}]$, representing the switch-to-switch link capacity requirements for a previous iteration, is known. Also, the total link capacity requirements b_i at each switch i and the total link capacity requirements d_j at each switch j are estimated as follows:

$$b_i = \frac{a_i}{\gamma}$$

$$d_j = \frac{a_j}{\gamma}$$

where a_i erlangs is the total traffic at switch i, a_j erlangs is the total traffic at switch j, and γ is the average erlang-carrying capacity per trunk, or switch design occupancy, as given previously.

The terms p_{ij} can be obtained as follows:

$$\text{fac}_i = \frac{b_i}{\sum_j q_{ij}}$$

$$\text{fac}_j = \frac{d_j}{\sum_i q_{ij}}$$

$$E_{ij} = \frac{\text{fac}_i + \text{fac}_j}{2}$$

$$p_{ij} = q_{ij} E_{ij}$$

After the above equations are solved iteratively, the converged steady state values of p_{ij} are obtained.

The discrete event flow optimization model generates traffic call events according to a Poisson arrival distribution with a settable average holding time for exponentially distributed holding times. However, more general arrival streams can easily be used, such as peaked traffic arrivals and nonexponentially distributed holding times, because such models can readily be implemented in the discrete routing table simulation model. Traffic call events are generated in accordance with the traffic load matrix input to the model. These traffic call events are routed on the real-time dynamic routing path choice according to the real-time dynamic routing method, as modeled by a set of routing simulation modules that implement the real-time dynamic routing logic for each routing method. The routing design finds the real-time paths between switches in the network for each call event and flows the event onto the network capacity. Each real-time dynamic routing method attempts to share link capacity to the greatest extent possible in accordance with the distribution of loads in the network, with the objective of maximizing the utilization of network resources throughout the busy periods of the network.

The output from the routing design is the fraction of traffic completed in each time period. From this traffic completion performance, the capacity design determines the new link capacity requirements of each switch and each link to meet the design level of blocking. From the estimate of blocked traffic at each switch in each time period, an occupancy calculation determines additional switch link capacity requirements for an updated link capacity estimate. Such a link capacity determination is made based on the amount of blocked traffic. The total blocked traffic Δa erlangs is estimated at each of the switches, and an estimated link capacity increase ΔT for each switch is calculated by the relationship

$$\Delta T = \frac{\Delta a}{\gamma}$$

where again γ is the average erlang-carrying capacity per trunk. Thus, the ΔT for each switch is distributed to each link according to the Kruithof estimation method described above. The Kruithof allocation method [Kru37] distributes link capacity in proportion to the overall demand between switches and in accordance with link cost, so that overall network cost is minimized. Sizing individual links in this way ensures an efficient level of blocking on each link in the network to optimally divide the load between the direct link and the overflow network. Once the links have been resized, the network is re-evaluated to see if the blocking probability grade-of-service objective is met, and if not, another iteration of the model is performed.

1.4.5 Day-to-day load variation design models

In network design we use the forecast traffic loads, which are actually mean loads about which there occurs a day-to-day variation, characterized by a gamma distribution with one of three levels of variance [Wil58]. Even if the forecast mean loads are correct, the actual realized loads exhibit a random fluctuation from day to day. Studies have established that this source of uncertainty requires the network to be augmented in order to maintain the blocking probability grade-of-service objectives [HiN76].

Network capacity required by the load uncertainties can be reduced by dynamically controlling the traffic routing design to meet the realized traffic loads.

This routing design can be either preplanned dynamic routing or real-time dynamic routing design. A given realization of the traffic loads can be expected to yield some switch-to-switch loads that are higher than average and others that are lower. While part of the network is overloaded, another part might be underloaded. If the routing design can be adjusted to use the idle capacity of the underloaded portion of the network, any required capacity augmentation might be reduced or eliminated.

Uncertain variations in the instantaneous network loads imply that capacity is never perfectly matched to the traffic load. Loads in the network shift from hour to hour and from day to day, and some amount of reserve or idle capacity is always present. Hence, there is an opportunity to seek out this idle capacity in real time. Real-time routing design attempts to find and use idle network capacity to satisfy current loads, and models predict that network blocking probability is substantially reduced with such real-time routing design, as illustrated in Section 1.5. Real-time routing design, therefore, can reduce capacity augmentation required by day-to-day traffic load uncertainties. It deals with day-to-day traffic load variations, unforecasted traffic demand until any needed capacity augmentation is made, and traffic management under overload and failure. Hence, the routing decisions necessary in real time involve conditions as they become known in real time: day-to-day load variations, unforecasted demand, network failures, and network overloads. Real-time traffic routing design can improve network service significantly even with simple procedures, while holding the transport routing policy fixed. As shown in Chapter 13 and illustrated in Figure 1.38, the improvement can also be equated to an equivalent network cost savings in the range of 2 to 4 percent or improved network performance with a higher overall completion rate. In the latter case real-time routing design improves network performance under network failures, especially when reserve capacity is available for redirecting traffic flows to available idle capacity.

Accommodating day-to-day variations in the network design procedure can involve the use of an equivalent load technique that models each switch pair in the network as an equivalent link engineered to the blocking probability objective. On the basis of Neal-Wilkinson engineering [HiN76, Wil58], the number of circuits N that are required in the equivalent link to meet the required blocking probability objective for the forecasted load R with its specified peakedness Z and specified level of day-to-day variation ϕ is

$$N = \overline{\mathrm{NB}}(R, \phi, Z, \mathrm{GOS})$$

where GOS is the required switch-to-switch blocking probability grade-of-service objective, and $\overline{\mathrm{NB}}$ is a function mapping R, ϕ, Z, and GOS into the link requirement N.

Holding fixed the specified peakedness Z and the calculated circuits N, we calculate what larger equivalent load R_e requires N circuits to meet the blocking probability objective if the forecasted load had had no day-to-day variation as

$$N = \overline{\mathrm{NB}}(R_e, 0, Z, \mathrm{GOS})$$

where $\phi = 0$ signifies no day-to-day variation. The equivalent load R_e then produces the same equivalent number of circuits N when engineered for the same peakedness level but in the absence of day-to-day variation.

1.4.6   Forecast uncertainty/reserve capacity design models

Network designs are made based on measured traffic loads and estimated traffic loads that are subject to error. In network design we use the forecast traffic loads because the network capacity must be in place before the loads occur. Errors in the forecast traffic reflect uncertainty about the actual loads that will occur, and as such the design needs to provide sufficient capacity to meet the expected load on the network in light of these expected errors. Studies have established that this source of uncertainty requires the network to be augmented in order to maintain the blocking probability grade-of-service objectives [FHH79].

The network design and routing management process accommodate the random forecast errors in the procedures. When some realized switch-to-switch blockings are found to be larger than their objective values, additional circuits and/or routing changes are provided to restore the network blocking to the objective level. Circuits often are not disconnected in the network design process even when load forecast errors are such that this would be possible without service degradation. As a result, the process leaves the network with a certain amount of reserve or idle capacity even when the forecast error is unbiased.

Network design, then, is based on the forecast traffic loads and the link capacity already in place. Consideration of the in-service link capacity entails a transport routing policy that could consider (1) fixed transport routing, in which transport is not rearranged; (2) rearrangeable transport routing, which allows periodic transport rearrangement including some trunk disconnects; and (3) real-time transport routing, in which transport capacity is adjusted in real time according to transport demands. In rearrangeable transport routing, the trunk disconnect policy may leave capacity in place even though it is not called for by the network design. In-place capacity that is in excess of the capacity required to exactly meet the design loads with the objective performance is called *reserve capacity*. There are economic and service implications of the capacity management strategy. Insufficient capacity means that occasionally link capacity must be connected on short notice if the network load requires it. This is short-term capacity management. There is a trade-off between reserve capacity and short-term capacity management, which we explore in models described in this section. Reference [FHH79] analyzes a model that shows the level of reserve capacity to be in the range of 6–25 percent, when forecast error, measurement error, and other effects are present. In fixed transport routing networks, if links are found to be overloaded when actual loads are larger than forecasted values, additional link capacity is provided to restore the blocking probability grade-of-service to the objective value, and, as a result, the process leaves the network with reserve capacity even when the forecast error is unbiased. Operational studies in fixed transport routing networks have measured up to 20 percent and more for network reserve capacity. Methods such as the

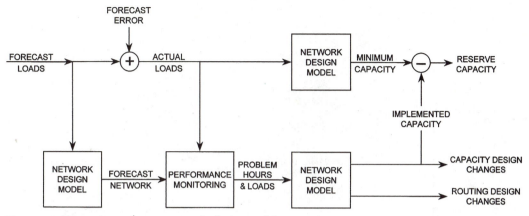

Figure 1.47 Forecast error/reserve capacity design model

Kalman filter [PaW82], which provides more accurate traffic forecasts and rearrangeable transport routing trunk disconnect policies, can help reduce this level of reserve capacity. On occasion, the planned design underprovides link capacity at some point in the network, again because of forecast errors, and short-term capacity management is required to correct these forecast errors and restore service [Sze80].

The model illustrated in Figure 1.47 is used to study network design of a network on the basis of forecast loads, in which the network design accounts for both the current network and the forecast loads in planning network capacity. Network design can make short-term capacity additions, if necessary, if network performance for the realized traffic loads becomes unacceptable and cannot be entirely corrected by routing adjustments alone. Network design tries to minimize reserve capacity while maintaining the design performance objectives and an acceptable level of short-term capacity additions. Network design uses the traffic forecast, which is subject to error, and the existing network. The model assumes that the network design is always implemented, and, if necessary, short-term capacity additions are made to restore network performance when design objectives are not met.

Figure 1.48 illustrates the level of reserve capacity as a function of the transport routing strategy—fixed transport routing, rearrangeable transport routing, or real-time transport routing—with the traffic routing strategy—fixed traffic routing, preplanned dynamic traffic routing, or real-time dynamic traffic routing—as an additional parameter. In the fixed traffic and transport routing network model, network design compares the existing network with a network designed for the forecast traffic loads, which is made without reference to the existing link sizes. When the network design calls for additional trunks on a link, the augments are implemented, and when the network design calls for fewer trunks on a link, a disconnect policy is invoked to decide whether trunks should be disconnected. This disconnect policy reflects a degree of reluctance to disconnect link capacity, so as to ensure that disconnected link capacity is not needed a short time later if traffic loads grow. An erlang flow optimization model

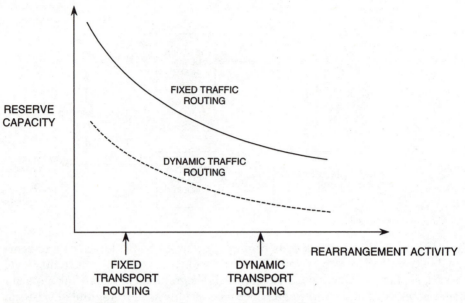

Figure 1.48   Trade-off of reserve capacity vs. rearrangement activity

allows the network design to reflect initial link capacities that are treated as lower bounds on the design link capacities and designs a minimum-cost capacity augmentation to the existing network to meet the forecast traffic loads. In this way, the erlang flow optimization model takes into account the reluctance to remove link capacity by using capacity in place to the extent possible by making both traffic and transport routing design changes. This procedure limits the link capacity augments to those required to meet the forecast traffic loads and, thus, achieves a lower reserve capacity than the fixed routing network. A generalization of this network design procedure is to set minimum link sizes and to allow link capacity disconnects. This is done by deriving, for each link, a lower threshold and an upper threshold for its size and using these thresholds to make initial adjustments to the link size, as described below, prior to the network design.

The size thresholds for a link are based on the forecast traffic loads of the corresponding switch-to-switch traffic and are determined by choosing a maximum value r_{max} for the ratio

$$r_{max} \triangleq \frac{\text{link size}}{\text{switch-to-switch traffic load}}$$

The upper (disconnect) threshold is based on the forecast peak load of the switch-to-switch traffic over the next two years, with some allowance for forecast error, and it corresponds to the ratio r_{max}. Results of using this model for network design are reported in the next sections and, in detail, in Chapter 9.

We find that dynamic traffic routing and fixed transport routing compared with fixed traffic routing and fixed transport routing provide a potential 5 percent reduction in reserve capacity while retaining a low level of short-term capacity management. We investigate the benefits of dynamic traffic routing together with rearrangeable transport routing, in which we find in Chapter 15 an additional 10 percent reduction in reserve capacity, and the benefits of dynamic traffic routing together with real-time transport routing, in which we find in Chapter 15 further utilization and performance benefits if transport switching costs are sufficiently low.

With dynamic traffic routing and dynamic transport routing design models, as illustrated in Figure 1.48, reserve capacity can be reduced in comparison with fixed transport routing, because with dynamic transport network design the link sizes can be matched to the network load. With dynamic transport routing, the link capacity disconnect policy in effect becomes one in which link capacity is always disconnected when not needed for the current traffic loads. Models given in reference [FHH79] predict reserve capacity reductions of 10 percent or more under this policy, and the results presented in Chapter 15, as summarized in Table 1.3, substantiate this conclusion. The table summarizes the network design models discussed in this book.

TABLE 1.3   Network Design Models for Routing Methods

Routing method	Examples	Network design model			Chapter
		Optimization method	Routing design	Capacity design	
fixed traffic	fixed hierarchical	HU-final-erlang flow	homing & alternate routes	Truitt/ECCS; Neal-Wilkinson	2
	fixed nonhierarchical	route-erlang flow	route-linear program	fixed-point model; Truitt/ECCS	4–5
dynamic traffic (preplanned)	dynamic hierarchical	transport flow	homing & alternate routes	virtual transport linear program	3
	dynamic progressive	route-erlang flow	route-linear program	fixed-point model; Truitt/ECCS	4
	dynamic multilink path	route-erlang flow	route-linear program	fixed-point model; Truitt/ECCS	5
	dynamic two-link sequential path (DNHR)	path-erlang flow	path-linear program	fixed-point model; Truitt/ECCS	6–10

(Continued)

TABLE 1.3   (Continued)

Routing method	Examples	Network design model			Chapter
		Optimization method	Routing design	Capacity design	
(real-time event dependent)	learning with random routing (LRR); success-to-the-top (STT)	discrete event flow	routing simulation model	Δ trunks = (blocked erlangs)/γ; Kruithoff model	13
(real-time state-dependent)	trunk status map routing (TSMR)	path-erlang flow	path-linear program; equivalent load model	fixed-point model; Truitt/ECCS	13
	real-time network routing (RTNR)	transport flow	Truitt model	virtual transport linear program	13
		fixed-point erlang flow	fixed-point model	Δ trunks = (blocked erlangs)/γ; Kruithoff model	13
	TSMR; RTNR; worldwide international network (WIN)	discrete event flow	routing simulation model	Δ trunks = (blocked erlangs)/γ; Kruithoff model	13
dynamic transport (rearrangeable)	*T*1/*T*3 DCS networking	transport flow	1-link model; minimum-cost alternate-path model	virtual transport linear program	14–15
(real-time)	VC/VP ATM networking	path-erlang flow	shortest-path selection	Erlang B sizing to link blocking objective	15

1.5   Modeling Results

1.5.1   Network design comparisons

We illustrate network design examples of various routing alternatives. In general we find that all of the alternatives have network design advantages that make them more highly utilized and better performing than fixed hierarchical networks. With the use of the design models described in the previous section, comparisons are made between the described routing techniques. Details of the models and results are given in Chapters 2–15. Results for large network models are in general agreement with results for small models.

In Table 1.4 we illustrate network designs for a 42-switch metropolitan area network model [Fie83], which include the nonalternate routing network design, fixed hierarchical network design, preplanned dynamic two-link sequential path design, success-to-the-top routing, learning with random routing, trunk status map routing with 2-, 5-, 10-, and 300-second updates, real-time network routing, and mixed dynamic routing. These results illustrate the cost savings potentially achievable with designs for metropolitan area dynamic routing in comparison with fixed hierarchical design.

TABLE 1.4 Network Design Comparisons

(42-Switch Metropolitan Area Model)

Routing method	Network design model	Normalized network cost	Savings (%)	Average trunk occupancy (network busy hour)	Average trunk occupancy (trunk group busy hour)
nonalternate (NAR)	final-erlang flow optimization	1.436	−43.6	0.328	0.294
fixed hierarchical (HIER)	HU-final-erlang flow optimization	1.000	—	0.470	0.563
dynamic two-link sequential path (DNHR)	path-erlang flow optimization	0.876	12.4	0.536	0.676
success-to-the-top (STT)	discrete event flow optimization	0.775	22.5	0.608	0.746
learning with random routing (LRR)	discrete event flow optimization	0.972	2.8	0.609	0.746
trunk status map (TSMR 2-second update)	discrete event flow optimization	0.715	28.5	0.658	0.760
trunk status map (TSMR 5-second update)	discrete event flow optimization	0.734	26.6	0.641	0.727
trunk status map (TSMR 10-second update)	discrete event flow optimization	0.755	24.5	0.624	0.688
trunk status map (TSMR 300-second update)	discrete event flow optimization	1.095	−9.5	0.430	0.395
real-time network (RTNR)	discrete event flow optimization	0.715	28.5	0.658	0.770
mixed dynamic (MXDR)	discrete event flow optimization	0.740	26.0	0.636	0.742

Table 1.5 summarizes network design examples of fixed hierarchical routing and various preplanned dynamic routing methods for national intercity network models. These results illustrate the large cost savings potentially achievable with preplanned dynamic routing designs for national intercity dynamic routing in comparison with fixed hierarchical design.

Tables 1.6–1.8 summarize network design examples for nonalternate routing network design, fixed hierarchical network design, preplanned dynamic two-link sequential path design, success-to-the-top routing, learning with random routing, trunk status map routing with 2-, 5-, 10-, and 300-second updates, real-time network routing, and mixed dynamic routing, for 10-switch, 28-switch, and 135-switch national intercity network models, respectively.

In Table 1.9 we illustrate the cost impacts for a 13-switch global international network design, which include nonalternate routing network design, fixed hierarchical network design, preplanned dynamic two-link sequential path design, success-to-the-top routing, learning with random routing, trunk status map routing with 2-, 5-, 10-, and 300-second updates, real-time network routing, and mixed dynamic routing.

In Table 1.10 we summarize the network utilization improvements with the network design models for uncertain traffic loads, as discussed in the previous section. We present results for dynamic traffic routing methods, as well as dynamic transport routing methods. Note that in comparison with the above results for the deterministic average hourly loads, we can see the additional benefit of reduction in reserve capacity among the various designs for traffic

TABLE 1.5   Network Design Comparison of Preplanned Dynamic Routing Methods
(National Intercity Models)

Routing method	Network design model	Normalized network trunking & circuit mile cost	Savings (%)
fixed hierarchical	HU-final-erlang flow optimization	1.000	—
dynamic hierarchical	transport flow optimization	0.900	10.0
fixed two-link sequential path	route-erlang flow optimization	0.890	≤ 11.0*
dynamic progressive	route-erlang flow optimization	0.852	14.3
dynamic multilink cyclic path	route-erlang flow optimization	0.844	15.6
dynamic two-link cyclic path	route-erlang flow optimization	0.855	14.5
dynamic two-link CGH path	path-erlang flow optimization	0.833	16.7
dynamic two-link sequential path (DNHR)	path-erlang flow optimization	0.838	16.2

*Savings are not robust to network model

and transport routing methods. Reserve capacity adds to total network cost, and reduction of reserve capacity through more effective design increases efficiencies.

We now make some general observations and comparisons from the results in Tables 1.4 to 1.10. These results illustrate the cost savings potentially achievable with dynamic network designs in comparison with non-alternate-routing and fixed hierarchical design, across all network configurations. Across a wide range of routing methods and network applications, designs for dynamic routing networks

TABLE 1.6 Network Design Comparisons

(10-Switch National Intercity Model)

Routing method	Network design model	Normalized network cost	Savings (%)	Average trunk occupancy (network busy hour)	Average trunk occupancy (trunk group busy hour)
nonalternate (NAR)	final-erlang flow optimization	1.211	−21.1	0.565	0.636
fixed hierarchical (HIER)	HU-final-erlang flow optimization	1.000	—	0.651	0.827
dynamic two-link sequential path (DNHR)	path-erlang flow optimization	0.841	15.9	0.788	0.894
success-to-the-top (STT)	discrete event flow optimization	0.800	20.0	0.833	0.919
learning with random routing (LRR)	discrete event flow optimization	0.815	18.5	0.822	0.903
trunk status map (TSMR 2-second update)	discrete event flow optimization	0.803	19.7	0.833	0.911
trunk status map (TSMR 5-second update)	discrete event flow optimization	0.806	19.4	0.830	0.908
trunk status map (TSMR 10-second update)	discrete event flow optimization	0.806	19.4	0.825	0.903
trunk status map (TSMR 300-second update)	discrete event flow optimization	0.860	14.0	0.790	0.858
real-time network (RTNR)	discrete event flow optimization	0.799	20.1	0.833	0.910
mixed dynamic (MXDR)	discrete event flow optimization	0.804	19.6	0.830	0.906

TABLE 1.7   Network Design Comparisons

(28-Switch National Intercity Model)

Routing method	Network design model	Normalized network cost	Savings (%)	Average trunk occupancy (network busy hour)	Average trunk occupancy (trunk group busy hour)
nonalternate (NAR)	final-erlang flow optimization	1.236	−23.6	0.551	0.621
fixed hierarchical (HIER)	HU-final-erlang flow optimization	1.000	—	0.644	0.854
dynamic two-link sequential path (DNHR)	path-erlang flow optimization	0.839	16.1	0.786	0.896
success-to-the-top (STT)	discrete event flow optimization	0.799	20.1	0.834	0.935
learning with random routing (LRR)	discrete event flow optimization	0.859	14.1	0.789	0.860
trunk status map (TSMR 2-second update)	discrete event flow optimization	0.796	20.4	0.837	0.922
trunk status map (TSMR 5-second update)	discrete event flow optimization	0.797	20.3	0.836	0.920
trunk status map (TSMR 10-second update)	discrete event flow optimization	0.804	19.6	0.830	0.909
trunk status map (TSMR 300-second update)	discrete event flow optimization	0.861	13.9	0.783	0.854
real-time network (RTNR)	discrete event flow optimization	0.794	20.6	0.838	0.926
mixed dynamic (MXDR)	discrete event flow optimization	0.800	20.0	0.835	0.924

typically achieve from 10–25 percent and more in utilization efficiency improvement and capital cost savings over the non-alternate-routing and hierarchical network designs. These results apply for metropolitan area models, national intercity models, and global international models. The latter applications achieve the greatest utilization improvements because of the global noncoincidence of traffic patterns, but even the metropolitan area applications achieve substantial benefits. For the various preplanned dynamic traffic routing methods—progressive,

TABLE 1.8   **Network Design Comparisons**

(135-Switch National Intercity Model)

Routing method	Network design model	Normalized network cost	Savings (%)	Average trunk occupancy (network busy hour)	Average trunk occupancy (trunk group busy hour)
nonalternate (NAR)	final-erlang flow optimization	1.260	−26.0	0.540	0.609
fixed hierarchical (HIER)	HU-final-erlang flow optimization	1.000	—	0.643	0.863
dynamic two-link sequential path (DNHR)	path-erlang flow optimization	0.834	16.6	0.785	0.894
success-to-the-top (STT)	discrete event flow optimization	0.796	20.4	0.834	0.928
learning with random routing (LRR)	discrete event flow optimization	0.896	10.4	0.765	0.810
trunk status map (TSMR 2-second update)	discrete event flow optimization	0.784	21.6	0.851	0.911
trunk status map (TSMR 5-second update)	discrete event flow optimization	0.788	21.2	0.849	0.904
trunk status map (TSMR 10-second update)	discrete event flow optimization	0.793	20.7	0.844	0.896
trunk status map (TSMR 300-second update)	discrete event flow optimization	0.925	7.5	0.738	0.778
real-time network (RTNR)	discrete event flow optimization	0.782	21.8	0.852	0.912
mixed dynamic (MXDR)	discrete event flow optimization	0.788	21.2	0.847	0.914

two-link path, or multilink path—the results in Table 1.5 show that there are relatively small differences in potential network utilization efficiency and capital cost savings between the various preplanned dynamic routing methods. The results suggest that using CCS crankback and originating switch control saves about one percent in network cost, as seen by comparing progressive routing, which has no crankback capability, with multilink path routing. The reason for this small difference is that most traffic in the various preplanned dynamic

TABLE 1.9 Network Design Comparisons

(13-Switch Global International Model)

Routing method	Network design model	Normalized network cost	Savings (%)	Average trunk occupancy (network busy hour)	Average trunk occupancy (trunk group busy hour)
fixed hierarchical (HIER)	HU-final-erlang flow optimization	1.000	—	0.605	0.691
dynamic two-link sequential path (DNHR)	path-erlang flow optimization	0.678	32.2	0.850	0.959
success-to-the-top (STT)	discrete event flow optimization	0.625	37.5	0.925	0.922
learning with random routing (LRR)	discrete event flow optimization	0.640	36.0	0.906	0.869
trunk status map (TSMR 2-second update)	discrete event flow optimization	0.626	37.4	0.926	0.917
trunk status map (TSMR 5-second update)	discrete event flow optimization	0.627	37.3	0.926	0.914
trunk status map (TSMR 10-second update)	discrete event flow optimization	0.630	37.0	0.923	0.909
trunk status map (TSMR 300-second update)	discrete event flow optimization	0.663	33.7	0.903	0.837
real-time network (RTNR)	discrete event flow optimization	0.626	37.4	0.926	0.917
mixed dynamic (MXDR)	discrete event flow optimization	0.626	37.4	0.927	0.919

routing networks is routed on the same links, because for many switch pairs these routing methods carry a significant amount of traffic on the direct path and on the same two-link, first-alternate path. The dynamic hierarchical routing and fixed nonhierarchical routing methods both show large potential savings, indicating that both the dynamic and the nonhierarchical aspects of the routing contribute significantly to these improvements in utilization efficiency and capital cost savings. However, these methods appear to have less potential network utilization improvement in comparison with the dynamic path routing methods. Furthermore, the savings for fixed nonhierarchical routing do not

TABLE 1.10   Network Design Comparison for Load Forecast Errors

(National Intercity Models)

Routing method	Traffic network design model	Transport network design model	Normalized reserve capacity cost	Savings (%)
fixed traffic, fixed transport	HU-final-erlang flow optimization	fixed transport	0.250	—
dynamic traffic, fixed transport	discrete event flow optimization	fixed transport	0.200	5.0
dynamic traffic, dynamic transport	discrete event flow optimization	dynamic transport flow optimization	0.100	15.0

appear to be robust across the different network models studied, as discussed in Chapters 4, 5, and 6.

Real-time dynamic network design, as illustrated in Tables 1.4 and 1.6–1.10, adds substantial efficiencies above preplanned dynamic traffic routing network design. These results illustrate the cost savings potentially achievable with dynamic network designs for national intercity dynamic routing in comparison with non-alternate-routing and fixed hierarchical design. These design results show a large range of design efficiencies. Learning with random routing (LRR) and trunk status map routing with 300-second updates are the least efficient. Because of the random selection of paths with LRR and possible loss of calls at a via switch, the individual switch-pair-blocking performance is difficult to control. This leads to less efficient network design, unless one is willing to relax the maximum switch-to-switch blocking objective. The Japanese implementation of event-dependent routing, STR, limits the choice of paths to a small subset of total choices, which helps limit the negative effects of random path selection. RTNR achieves the most efficient design, with TSMR with 2-second updates a close second, while MXDR reflects a weighted average of the individual performance of each dynamic routing strategy. The discrete event flow optimization model achieves much better design efficiency than other models for real-time dynamic routing networks. It is the basis for refinement of the real-time dynamic routing detailed implementation, in which alternative real-time dynamic routing approaches are compared. The successful design of such complex networks illustrates the effectiveness of the discrete event flow optimization design approach. Network design methods applied to preplanned dynamic traffic routing, real-time dynamic traffic routing, rearrangeable transport routing, and real-time transport routing, as illustrated in Table 1.10, yield substantial utilization improvements through reduction in reserve capacity that arises from design for load uncertainties.

1.5.2   Performance comparisons

We compare traffic routing methods that include nonalternate routing, fixed hierarchical routing, preplanned dynamic two-link sequential path routing, success-to-the-top routing, learning with random routing, trunk status map routing with 2-, 5-, 10-, and 300-second updates, real-time network routing, and mixed dynamic routing. These methods capture the dynamic routing methods in use throughout the world: DNHR, DAR and STR (LRR), DCR (TSMR), and RTNR. We characterize network performance improvements that these approaches can achieve at the time of network overloads and failures. We also investigate a mixed dynamic routing method (MXDR) in which STT, TSMR, and RTNR are implemented together within the same network. In MXDR, each switch uses one of these three methods.

In Tables 1.11–1.15, we compare dynamic routing methods with fixed hierarchical routing for various network overload and failure conditions. For the global international model, we consider two specific cable failures, those of TAT9 and TPC5. In the figures, we give the average network percent blocking over the simulation hours, the maximum switch-pair percent blocking, and the 90th percentile switch-pair percent blocking. For the nonalternate routing and fixed hierarchical routing examples, two designs representing low day-to-day variation and high day-to-day variation are considered. The DNHR design is used for all the remaining dynamic routing examples.

We now make some general observations and comparisons from these results. We conclude that all dynamic routing methods generally outperform the non-alternate-routing and fixed hierarchical (HIER) routing methods by a large margin under all scenarios. RTNR generally outperforms the other routing methods in large networks. In small networks, STT, DNHR, and TSMR do better in comparison with RTNR. A mix of individual dynamic routing methods (MXDR) achieves synergy and near-maximum throughput performance. These results illustrate the performance gains potentially achievable with dynamic routing methods in comparison with non-alternate-routing and fixed hierarchical routing across all network configurations. Real-time dynamic routing, as illustrated in Tables 1.11–1.15, adds substantial efficiencies above preplanned dynamic traffic routing network performance. These results show a large range of performance efficiencies. Learning with random routing (LRR) and trunk status map routing with 300-second updates are the least efficient. Because of the random selection of paths with LRR and possible loss of calls at a via switch, the blocking performance is not as good as other methods. RTNR achieves the most efficient design, with TSMR with 2-second updates a close second, while MXDR reflects a weighted average of the individual performance of each dynamic routing strategy.

Dynamic routing and originating call control are achieved by permitting information exchange between switches through standardized CCS messages [AKA94, AsH94]. These messages include via switch indicator messages to control route choices at via switches, crankback messages to return call control to originating switches from via switches or terminating switches, query messages to request link status from a switch, and status messages to contain link status information. We conclude also that standardized information exchange is necessary so that switching equipment from different vendors interacts to implement dynamic routing methods in a coordinated fashion, a topic that is discussed in Chapter 16.

1.6   Conclusion

Conclusions on dynamic routing methods and network design, as described in this chapter, are as follows:

- Design for dynamic routing achieves from 10–25 percent and more utilization efficiency improvement and capital cost savings over the non-alternate-routing (NAR) and fixed hierarchical (HIER) network design cost for metropolitan area models, national intercity models, and global international models.

- All dynamic routing methods generally outperform the non-alternate-routing and fixed hierarchical routing methods by a large margin under nearly all scenarios.

- Both the dynamic and the nonhierarchical aspects of the routing approaches contribute significantly to these improvements in utilization efficiency and capital cost savings.

- Routing methods that incorporate crankback and allow path choices to have at most two links achieve nearly the same savings as methods that allow maximum flexibility; that is, any paths with any number of links. In addition to permitting the economic and performance benefits of two-link routing, crankback also allows the originating switch to retain control of a call.

- Among the preplanned dynamic routing alternatives, the simple dynamic two-link sequential path routing (DNHR) technique achieves network design savings that are close to those of more complicated methods.

- Call-by-call real-time state-dependent routing (RTNR) achieves the greatest network efficiency compared with other routing methods, especially in large networks.

- A mix of individual dynamic routing methods (MXDR) achieves synergy and near-maximum throughput performance.

- Design for dynamic traffic routing and dynamic transport routing reduces the effect of traffic load uncertainties leading to additional efficiencies with 5–10 percent or more reduction in reserve capacity.

- Erlang flow optimization models provide efficient designs for fixed and pre-planned dynamic routing networks, transport flow optimization models provide improvement in designs for real-time dynamic networks, and discrete event flow optimization models are found to yield the most efficient network designs for real-time dynamic networks.

In addition to increases in network utilization efficiency and capital cost savings, a preferred routing method must consider implementation factors for differentiation. A routing method such as CGH has additional development and

**TABLE 1.11 Performance Comparisons
(42-Switch Metropolitan Area Model)**

Routing method	30% general overload			Transport failure		
	Average network blocking (%)	Maximum switch–switch blocking (%)	90% switch–switch blocking (%)	Average network blocking (%)	Maximum switch–switch blocking (%)	90% switch–switch blocking (%)
nonalternate (NAR, low day-to-day variation design)	0.87	2.50	1.50	6.39	100.00	100.00
nonalternate (NAR, high day-to-day variation design)	0.54	1.88	1.02	6.20	100.00	100.00
fixed hierarchical (HIER, low day-to-day variation design)	2.93	27.78	8.96	4.98	60.90	27.02
fixed hierarchical (HIER, high day-to-day variation design)	1.96	29.03	5.94	6.29	87.75	39.19
dynamic two-link sequential path (DNHR)	1.61	12.57	3.29	2.50	100.00	4.53
success-to-the-top (STT)	0.95	18.10	2.51	1.77	100.00	2.38

(Continued)

TABLE 1.11 (Continued)

Routing method	30% general overload			Transport failure		
	Average network blocking (%)	Maximum switch–switch blocking (%)	90% switch–switch blocking (%)	Average network blocking (%)	Maximum switch–switch blocking (%)	90% switch–switch blocking (%)
learning with random routing (LRR)	1.27	18.32	3.16	2.10	100.00	3.39
trunk status map (TSMR 2-second update)	0.70	8.83	1.05	1.61	100.00	1.61
trunk status map (TSMR 5-second update)	0.76	7.85	1.42	1.71	100.00	2.31
trunk status map (TSMR 10-second update)	0.94	9.42	1.98	1.86	100.00	3.54
trunk status map (TSMR 300-second update)	3.63	39.29	11.81	4.25	100.00	17.27
real-time network (RTNR)	0.68	8.46	0.64	1.61	100.00	2.38
mixed dynamic (MXDR)	0.81	9.93	1.62	1.71	100.00	2.20

TABLE 1.12 Performance Comparisons
(10-Switch National Intercity Model)

Routing method	10% general overload			Transport failure			Peak day overload		
	Average network blocking (%)	Maximum switch–switch blocking (%)	90% switch–switch blocking (%)	Average network blocking (%)	Maximum switch–switch blocking (%)	90% switch–switch blocking (%)	Average network blocking (%)	Maximum switch–switch blocking (%)	90% switch–switch blocking (%)
nonalternate (NAR, low day-to-day variation design)	0.01	0.17	0.11	1.67	41.61	18.83	5.46	66.72	32.80
nonalternate (NAR, high day-to-day variation design)	0.00	0.04	0.00	1.53	43.88	19.43	1.06	22.31	3.98
fixed hierarchical (HIER, low day-to-day variation design)	0.05	0.78	0.28	1.88	37.11	23.75	3.99	39.02	20.14
fixed hierarchical (HIER, high day-to-day variation design)	0.01	0.26	0.09	2.16	47.58	26.01	3.21	41.70	17.25
dynamic two-link sequential path (DNHR)	0.60	1.77	1.33	1.54	19.97	8.65	4.39	24.02	15.59
success-to-the-top (STT)	0.56	3.98	1.66	1.68	22.76	9.34	3.85	28.97	11.15

(Continued)

TABLE 1.12 (Continued)

Routing method	10% general overload			Transport failure			Peak day overload		
	Average network blocking (%)	Maximum switch–switch blocking (%)	90% switch–switch blocking (%)	Average network blocking (%)	Maximum switch–switch blocking (%)	90% switch–switch blocking (%)	Average network blocking (%)	Maximum switch–switch blocking (%)	90% switch–switch blocking (%)
learning with random routing (LRR)	0.65	3.87	2.11	1.85	25.50	10.47	4.02	31.36	12.06
trunk status map (TSMR 2-second update)	0.46	3.76	1.41	1.52	18.70	9.61	3.87	27.97	12.41
trunk status map (TSMR 5-second update)	0.51	3.61	1.33	1.55	19.71	8.21	3.78	27.80	11.80
trunk status map (TSMR 10-second update)	0.55	3.83	2.17	1.63	21.62	9.44	3.90	27.75	11.73
trunk status map (TSMR 300-second update)	1.12	12.09	3.37	2.42	31.04	12.76	4.57	29.02	13.40
real-time network (RTNR)	0.50	3.91	1.46	1.56	20.31	9.21	3.70	27.66	12.48
mixed dynamic (MXDR)	5.75	4.11	1.80	1.55	19.97	8.65	3.92	29.41	11.45

TABLE 1.13 Performance Comparisons
(28-Switch National Intercity Model)

Routing method	10% general overload			Transport failure			Peak day overload		
	Average network blocking (%)	Maximum switch–switch blocking (%)	90% switch–switch blocking (%)	Average network blocking (%)	Maximum switch–switch blocking (%)	90% switch–switch blocking (%)	Average network blocking (%)	Maximum switch–switch blocking (%)	90% switch–switch blocking (%)
nonalternate (NAR, low day-to-day variation design)	0.01	0.54	0.00	2.52	59.32	25.99	1.51	23.96	4.87
nonalternate (NAR, high day-to-day variation design)	0.00	0.16	0.00	2.30	54.89	2.42	0.75	22.52	3.15
fixed hierarchical (HIER, low day-to-day variation design)	0.25	4.51	0.78	2.29	80.35	13.74	5.50	56.11	22.10
fixed hierarchical (HIER, high day-to-day variation design)	0.09	2.56	0.31	1.95	83.89	10.46	4.11	59.09	15.09
dynamic two-link sequential path (DNHR)	0.71	1.97	1.15	1.19	25.73	8.51	5.01	23.36	15.35
success-to-the-top (STT)	0.30	2.67	0.82	0.97	20.21	6.43	4.40	33.51	14.29

(Continued)

TABLE 1.13 (Continued)

Routing method	10% general overload			Transport failure			Peak day overload		
	Average network blocking (%)	Maximum switch–switch blocking (%)	90% switch–switch blocking (%)	Average network blocking (%)	Maximum switch–switch blocking (%)	90% switch–switch blocking (%)	Average network blocking (%)	Maximum switch–switch blocking (%)	90% switch–switch blocking (%)
learning with random routing (LRR)	0.56	4.85	1.88	1.70	28.55	11.08	5.03	34.73	17.01
trunk status map (TSMR 2-second update)	0.25	2.34	0.76	0.85	15.72	5.13	4.32	31.30	14.26
trunk status map (TSMR 5-second update)	0.28	2.05	0.79	0.92	18.23	5.66	4.39	31.50	13.95
trunk status map (TSMR 10-second update)	0.31	2.64	0.82	1.05	21.32	6.73	4.48	31.34	14.31
trunk status map (TSMR 300-second update)	1.23	18.45	3.29	2.63	42.24	16.68	5.89	42.53	17.55
real-time network (RTNR)	0.22	2.59	0.69	0.83	16.59	5.06	4.27	32.83	14.17
mixed dynamic (MXDR)	0.27	2.18	0.81	0.94	19.73	5.67	4.34	33.47	14.62

TABLE 1.14 Performance Comparisons
(135-Switch National Intercity Model)

Routing method	10% general overload			Transport failure			Peak day overload		
	Average network blocking (%)	Maximum switch–switch blocking (%)	90% switch–switch blocking (%)	Average network blocking (%)	Maximum switch–switch blocking (%)	90% switch–switch blocking (%)	Average network blocking (%)	Maximum switch–switch blocking (%)	90% switch–switch blocking (%)
nonalternate (NAR, low day-to-day variation design)	0.01	0.86	0.0	2.80	62.09	21.30	1.40	59.78	3.31
nonalternate (NAR, high day-to-day variation design)	0.00	0.72	0.00	2.51	58.27	19.10	0.85	59.29	1.77
fixed hierarchical (HIER, low day-to-day variation design)	0.48	33.33	1.26	3.19	85.12	20.76	4.70	66.06	14.53
fixed hierarchical (HIER, high day-to-day variation design)	0.22	20.00	0.57	2.94	89.67	17.11	3.76	72.38	11.33
dynamic two-link sequential path (DNHR)	0.75	50.00	0.99	1.65	75.00	9.48	4.44	44.95	11.37
success-to-the-top (STT)	0.38	50.00	1.27	1.10	100.00	5.89	2.47	37.20	7.51

(Continued)

TABLE 1.14 (Continued)

Routing method	10% general overload			Transport failure			Peak day overload		
	Average network blocking (%)	Maximum switch-switch blocking (%)	90% switch-switch blocking (%)	Average network blocking (%)	Maximum switch-switch blocking (%)	90% switch-switch blocking (%)	Average network blocking (%)	Maximum switch-switch blocking (%)	90% switch-switch blocking (%)
learning with random routing (LRR)	0.73	33.33	2.45	1.99	100.00	10.17	3.68	38.18	11.30
trunk status map (TSMR 2-second update)	0.22	51.25	4.75	0.99	63.64	5.71	2.95	40.47	9.28
trunk status map (TSMR 5-second update)	0.23	16.67	0.78	1.08	100.00	6.57	2.46	30.60	7.95
trunk status map (TSMR 10-second update)	0.28	50.00	0.98	1.24	90.91	7.60	2.65	36.92	8.48
trunk status map (TSMR 300-second update)	1.65	51.25	4.25	3.59	100.00	22.41	5.63	69.27	17.76
real-time network (RTNR)	0.20	16.67	0.73	0.89	72.73	4.99	2.33	35.45	7.06
mixed dynamic (MXDR)	0.30	16.67	1.18	1.03	72.73	6.20	2.51	45.88	8.23

TABLE 1.15 Performance Comparisons
(13-Switch Global International Model)

Routing method	10% general overload			TAT9 transport failure			TPC5 transport failure		
	Average network blocking (%)	Maximum switch–switch blocking (%)	90% switch–switch blocking (%)	Average network blocking (%)	Maximum switch–switch blocking (%)	90% switch–switch blocking (%)	Average network blocking (%)	Maximum switch–switch blocking (%)	90% switch–switch blocking (%)
fixed hierarchical (HIER, low day-to-day variation design)	0.01	0.64	0.17	1.58	6.74	6.16	0.78	10.25	6.54
fixed hierarchical (HIER, high day-to-day variation design)	0.00	0.01	0.00	0.44	3.09	1.95	0.18	3.57	1.82
dynamic two-link sequential path (DNHR)	0.83	2.23	1.36	2.41	11.18	7.34	1.07	12.62	7.56
success-to-the-top (STT)	0.38	2.74	0.93	2.03	10.80	8.16	0.70	8.75	5.72
learning with random routing (LRR)	0.38	2.36	1.04	2.05	11.09	8.31	0.70	10.03	6.37

(Continued)

TABLE 1.15 (Continued)

Routing method	10% general overload			TAT9 transport failure			TPC5 transport failure		
	Average network blocking (%)	Maximum switch–switch blocking (%)	90% switch–switch blocking (%)	Average network blocking (%)	Maximum switch–switch blocking (%)	90% switch–switch blocking (%)	Average network blocking (%)	Maximum switch–switch blocking (%)	90% switch–switch blocking (%)
trunk status map (TSMR 2-second update)	0.37	2.60	0.94	2.03	11.28	8.15	0.67	8.59	5.85
trunk status map (TSMR 5-second update)	0.38	2.84	1.03	2.01	11.20	7.87	0.68	8.59	5.85
trunk status map (TSMR 10-second update)	0.42	2.48	1.05	2.04	10.81	7.92	0.72	15.81	6.10
trunk status map (TSMR 300-second update)	0.87	24.63	4.06	2.36	25.81	8.57	1.37	46.35	12.86
real-time network (RTNR)	0.36	1.99	0.85	2.02	11.19	8.08	0.67	7.72	5.20
mixed dynamic (MXDR)	0.38	2.25	1.09	2.02	11.18	7.34	6.73	8.76	5.28

administrative costs because the switching system needs to store traffic allocation proportions and markers to indicate where the cyclic blocks begin and end, along with the ordered list of paths, and requires real time to generate and process the appropriate random numbers. Preplanned dynamic progressive routing requires a history of visited switches to be sent with each call to prevent looping. Since no central controller or single switch has complete control of a particular call, it would also be quite difficult to measure switch-to-switch blocking. We contrast this to the use of originating switch control in preplanned or real-time two-link path routing, which makes it easier to measure switch-to-switch blocking. The switch-to-switch blocking measurement is necessary for network design in order to adjust routing tables and augment links to satisfy unforeseen loads; the switch-to-switch blocking measurement indicates when corrective action is necessary. With a real-time dynamic routing method such as RTNR or TSMR, no routing tables need to be managed, and thus network management and design are simplified.

Fixed Hierarchical Routing Networks

2.1   Introduction

Present-day hierarchical routing rules are time fixed. In a fixed traffic routing network, such as fixed hierarchical routing networks deployed throughout the world today, there is minimal flexibility to share bandwidth among switch pairs. Fixed hierarchical traffic routing is used in almost all current networks, including most metropolitan area networks, national intercity networks, and global international networks. In 1930 a hierarchy of switching centers was established by the General Toll Switching Plan [Osb30], which limited the number of switches in a call to six in any connection including the metropolitan area switches. As illustrated in Figure 1.16, routing through the network was handled manually. During the late 1940s, a new plan [Pil52] established ten regions and a five-level hierarchy of switching systems, as illustrated in Figure 1.17. By 1951, this five-level hierarchy of switching centers had evolved, with direct distance dialing and automatic alternate routing, enabled by the No. 4A crossbar system, automatic message accounting, and efficient network design techniques for alternate-routing networks.

Figure 1.17 illustrates hierarchical traffic routing in which the hierarchical plan defines regions and a hierarchy of switching systems centering on the regional switches. The switching hierarchy used in the Bell System network prior to divestiture in 1984 had five levels, as illustrated in Figure 1.17, in which successively higher-level switches (also called classes) concentrated traffic from increasingly larger geographical areas. At the highest level of switching were 10 regional switching centers, or Class 1 switches, shown as squares in Figure 1.17. At the next level of switching were the sectional switching centers, then primary switching centers, then toll switching centers, and, at the lowest level, end-offices, or Class 5 end-offices, which are part of the metropolitan area switching network. The approximate numbers of these switching centers at the time of divestiture in 1984 are given in Figure 1.17.

Two types of links connect these switches. *High-usage links* (dashed lines) connect any two switches that have sufficient traffic between them to make a direct route economical. *Final links* are the links between each switch and its immediate superior in the hierarchy, together with the final links interconnecting all the regional centers (solid lines). A switch connected by a final link to a higher-class switch is said to "home" on that switch. High-usage links are sized to handle only a portion of the traffic directed to them, and the switching systems redirect traffic by automatic alternate routing to a different link when all circuits of a high-usage link are busy. At each stage, the alternate-routing plan shifts overflow calls from the more direct route toward the final route. Final links are designed to handle their own direct traffic plus overflow traffic from high-usage links with low blocking, such as an average loss of one call in a hundred. The hierarchical routing pattern precludes switching or looping calls back on themselves or using an excessive number of links on a call.

When the hierarchical network was developed, computers were in their infancy and design rules had to be kept simple to allow manual network design. It was also necessary that routing decisions of the early electromechanical switches be quite limited. Hierarchical traffic routing served for more than 50 years prior to the first conversion to dynamic routing, and it is still in widespread use throughout the world today. In fact, it is the only standard routing method that permits interworking among switching products from multiple vendors. Hierarchical traffic routing is, however, less efficient and beneficial than dynamic traffic routing methods, since the ability to share available bandwidth among switch pairs is minimal.

We present network design models that address traffic load variations, which include link alternate routing flow optimization with Truitt's model [Tru54], equivalent random methods for peaked traffic, and day-to-day variation design models. In Section 2.2 we discuss hierarchical network architecture functions including switching, trunk terminations and hunting, numbering plan and digit translation, signaling, and billing [ATT77]. In Section 2.3 we discuss first- and alternate-route design rules and hierarchical route selection. In Section 2.4 we discuss network management and design principles including the traffic model, the Erlang B formula, the Wilkinson equivalent random method, Truitt's ECCS method for high-usage link sizing, cluster busy-hour sizing, and Neal-Wilkinson final link sizing.

2.2   Network Architecture

Metropolitan area networks are generally two-level networks serving calls originating and terminating within a given metropolitan area. National intercity networks often have more than two levels; for example, the North American hierarchical network had five levels at the time of the Bell System divestiture in 1984. Global international networks generally have bilateral hierarchical configurations.

The basic objective of network design is to determine the way in which network switching systems are to be interconnected. The specific objectives are to determine how calls are to be routed and how much network capacity is required in a future period to meet a given service objective in an economic manner, when given a set of traffic loads. Service objectives for final links, which have no alternate routes, are normally stated in terms of the busy-season average network busy-hour blocking probability. It is normally required that the network be designed to meet these blocking objectives, while at the same time be able to efficiently load all links.

Overall network design objectives, however, are more broad than this. There is also a responsibility to provide satisfactory service during those periods when changes in load levels or load distribution place a strain on the network. Within the appropriate economic constraints, a network should provide maximum flexibility in meeting the continuous variability of offered loads. This flexibility is ordinarily provided by a network that has several ways of getting from point to point. Any network has the requirement that it can be designed, administered, and managed. In meeting this design requirement, there are objectives in addition to the probability of blocking, speed of connection, and length of traffic delay. To meet the overall quality-of-service objective, consideration must be given to

- Transmission objectives concerned with the true reproduction of the original information at the distant end.

- Service protection objectives concerned with safeguards against interruptions to service and with the rapid restoration of service.

- Maintenance objectives concerned with reliable performance and ease of repair.

Capacity management and network design include the accumulation, interpretation, and projection of traffic load data to be carried on the message network. They require the scheduling, procurement, and delivery of traffic load measurement data; determining network requirements for the construction program;

the coordination of all matters concerned with traffic routing, including the requirements for number translation; and preparation of timely reports of network performance in terms of service and loading.

2.2.1 Switching

In telecommunications networks, the principal role of switching is to obtain economy in the provision of transmission facilities. This is done by concentrating small traffic items into traffic loads that are large enough to be routed efficiently. This is accomplished by successive stages of concentration, which have evolved over time. When Alexander Graham Bell had only two telephones, the routing problem was simple, as illustrated in Figure 2.1. When a third telephone was added, two lines had to terminate at each telephone and a line selection and signaling device became necessary, as illustrated in Figure 2.2. As more telephones were added, the number of lines required to directly connect all of them began to be unmanageable, as illustrated in Figure 2.3. It then became economical to serve lines from a central switching system, as determined by a wire-centering procedure—first with a manual switching system, or switchboard (1878), as illustrated in Figure 2.4, and then with a mechanical and later an electronic switching system. Network design determines the optimum economic balance between the provision of transport and switching for a given set of traffic demands and service objectives. In order to meet this balance, additional stages of concentration are built up as follows: For a grouping of switching systems as illustrated in Figure 2.5, a wire-center study determines a central location in which another intermediate switching system can best be placed to most efficiently interconnect the other switching systems. The intermediate switching system is known as a *tandem*. This centering process saves transport cost by investing switching cost.

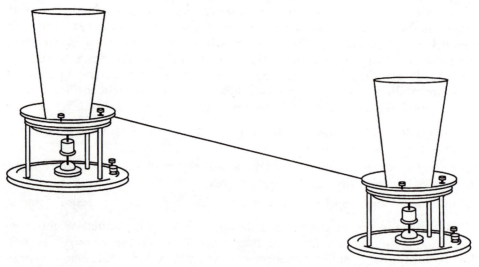

Figure 2.1 Two-telephone network

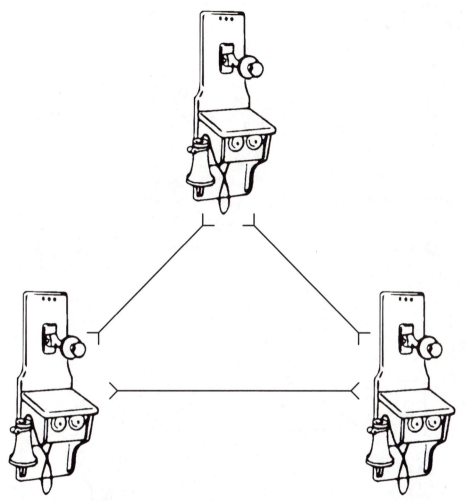

Figure 2.2 Three-telephone network

The cost of a path through a tandem is often higher than the cost of a direct path between the two switching systems. Even so, when the community of interest between two switching systems is low, traffic efficiency favors the tandem route. When the community of interest grows, however, a level is reached at which it is economical to provide both the direct and tandem routes, as illustrated in Figure 2.6. Tandems, in turn, have communities of interest for the groupings of switching systems that they serve, and the process can start over again, as illustrated in Figure 2.7. This second level of concentration may be adequate for the economic justification of a direct link between tandems, such as those serving adjacent cities. As distances increase, however, or where calling rates are low, a third level of concentration may be required in order to accumulate sufficient loads to economically justify a direct link. This process continues with

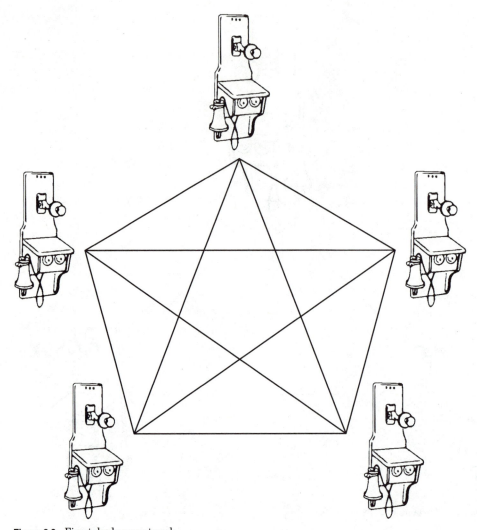

Figure 2.3   Five-telephone network

further increases in distance or lower calling rates requiring a fourth level of concentration; for example, serving an entire state or numbering-plan area. For national intercity calling, several states may be grouped together to form a large geographic area served by a fifth and highest level of concentration—the regional center. All regional switching systems are directly interconnected in the hierarchical network.

Transport investment can be saved by adding switching systems, and, conversely, switching investment is saved by spending on transport, as in the establishment of direct links. Thus, the network design and implementation process is a trade-off between transport and switching, which changes with traffic volume, traffic distribution, and technology. The interconnection of switching systems is

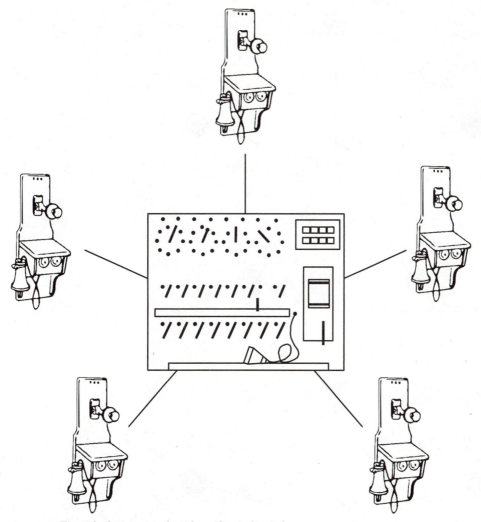

Figure 2.4   Five-telephone network with mechanical switch

accomplished by links between pairs of switching systems whenever there is sufficient traffic to make such connections economically feasible. Smaller items of traffic are switched via connections of two or more links in tandem. Transmission standards impose certain limitations on the number of such links in tandem. In the 1940s, steps were taken to develop a system for handling message traffic on a high-speed automatic switching basis. The network employs a definite pattern for routing traffic between switching systems over interconnecting links, provided in such a manner as to ensure efficient and economical use. The network provides for a low incidence of blockage for traffic during the average busy hour of the various network busy seasons. This is achieved by the pooling of link capacities in handling many traffic items and through the use of automatic alternate

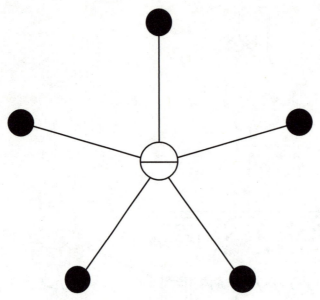

Figure 2.5   Tandem star network

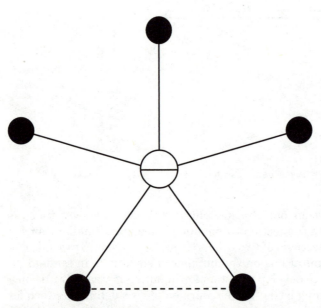

Figure 2.6   Tandem network with high-usage link connection

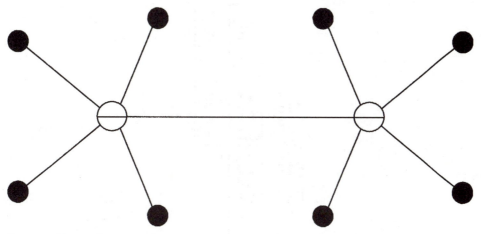

Figure 2.7 Two-tandem network

routing and appropriate techniques for allocating loads between first and alternate routes.

There are various classes of switching systems in a hierarchical network, each identified by a class number. The class number always indicates the highest switching function performed, but higher-class switching systems can, and generally do, perform lower switching functions. Links that interconnect the switching systems are basically of two types from a traffic-routing standpoint: high-usage and final. Final links are the route of last resort in the switching pattern. These links may be final choices in a sequence of links that overflow to each other, or they may be only-route links. A series of final links connected in tandem constitutes a last-choice route chain and thereby defines the network hierarchy. A final link is sized for a low blocking probability in its average busy-season busy hour. Calls failing to find an idle trunk in a final link are blocked and are then either abandoned or retried by the customer. High-usage links may be provided between any two switching systems, regardless of class or location, whenever the volume of routed traffic between those switching systems makes direct handling economical. A high-usage link is designed to overflow to an alternate route a predetermined part of the busy-hour traffic offered it.

2.2.2 Trunk terminations and hunting

The physical embodiment of a trunk requires the interconnection of many hardware components. Figure 2.8 illustrates the hardware components including the trunk equipment, circuit equipment, switching equipment, and transport. Trunks consist of cable, wire, or radio transmission facilities (usually expanded by the use of carrier equipment) and additional items of central office equipment such as channel banks, 4-wire terminating sets, pads, repeaters, signaling units, and impedance compensators. Different trunk designs, utilizing varying combinations of equipment, are necessary for particular systems, depending on the length of

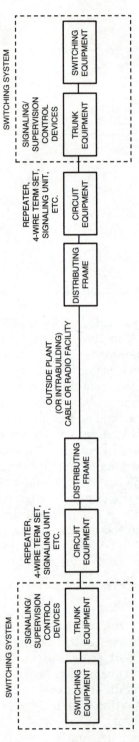

Figure 2.8 Example of trunk arrangement, switching system, and transport equipment components

the trunk, the nature of the transmission facility, the switching system type, and the types of traffic carried. The trunk design is chosen to establish the proper transmission and signaling capability. A switching system interface, known as a trunk termination, is required to connect the trunk described above to the switch fabric. The trunk termination, wired as part of the switching system, is provided on a per-trunk basis. Various types are available, depending on the switching system type, directionality of the trunk, or signaling employed.

With digital ESS switching systems, trunk terminations are used to interface digital transmission facilities (such as T carrier) and ESS switching systems. The use of such digital trunk terminations eliminates the need for the channel-bank terminating equipment, which is a modulator/demodulator device used to convert digitally formatted data to voice frequency analog signals and vice versa. Such digital trunk terminations are significantly less expensive than channel banks and eliminate much of the need for the switching system to decode signals. Calls on analog-type switching systems are processed on a trunk-per-terminal basis; that is, each trunk is assigned to a specific transmission path such as a cable pair or carrier channel. ESS digital switches operate on the digital principle by which each trunk is allotted a time slot on a common facility, and many trunks may be combined onto one wire bus. As the turn for each trunk (time slot) comes to be processed, it is recognized for a change (or no change) of status, and the common control equipment reacts accordingly for further processing. Time slots are a small fraction of a second apart; thus, many trunks may use a common facility within the system simultaneously.

Switching systems, once directed to the desired link, must search for an idle trunk and determine its condition (busy or idle) before connecting to the trunk. The search is referred to as trunk hunting. Knowledge of which trunk is selected first and what selection order follows is necessary in order to minimize glare on two-way trunks and optimize load balance in the terminating switching system. *Glare* is the simultaneous seizure by switching systems at both ends of a two-way trunk. Trunk glare becomes a minimal problem with ESS when routing to another ESS switching system, which is normally arranged for a special hunting configuration on two-way trunks to avoid glare. With two-way trunks, on each call offered to the link, hunting begins on trunk number 1 of trunk block 1 of the link and proceeds to the last trunk of the last block. The reverse order is used at the opposite end of the link. When both systems select the same trunk simultaneously, one of them automatically selects another path for the call.

2.2.3 Numbering plan and digit translation

Another essential architectural element of the network is an addressing system in which each piece of terminal equipment, such as a telephone set, is assigned a unique number. With such a numbering system, customers may use this number to reach the desired destination through the network. This is called destination code routing. The routing codes for dialing within the North American Numbering Plan (NANP) consist of two basic parts: (1) a 3-digit numbering plan area (NPA) code, and (2) a 7-digit number made up of a 3-digit central office code plus a

4-digit station number. Together these ten digits comprise the network "address" or "destination code" for each piece of terminal equipment. Calls between any two main stations in the same NPA can be completed using only the 3-digit central office code plus the 4-digit station number. When the switching system at the originating location receives the seven dialed digits, translation of the 3-digit central office code directs the switching system to select the proper outgoing routing table, which consists of high-usage and final link choices to route the call to or toward the destination. The seven digits, or that part of the seven digits required by the distant switching system, are sent forward. If the distant switching system is a tandem, it translates the code received, identifies the proper outgoing routing table, selects an outgoing link, and signals forward either the full seven digits to another tandem or four or more digits as required to the called end-office. The procedure of translation, selection, and signaling is followed from tandem to tandem until the call reaches its destination, making use of high-usage or final links as necessary within the routing table established for each call. Calls between locations in different NPAs are handled similarly, using the full 10-digit destination code. Both the originating end-office and tandem switching systems make use of the 3-digit NPA code to direct each call to a particular routing table and link to the called NPA. Once a call reaches the called NPA, only the last seven digits are needed to advance the call.

The connections made by a switching system to lines or trunks in the network are initiated by an address code. As discussed above, for the NANP this code is usually made up of a 10-digit number supplied by the calling customer and consists of a 3-digit NPA, a 3-digit end-office code, and a 4-digit customer station number. Dialing to Mexico is arranged for by a 2-digit prefix, a 1-digit routing code, and a 7-digit customer number, thus retaining the 10-digit code. The International Dialing Plan requires 12 digits comprising the country code and national number. Other codes—for example, 3-, 4-, 5-, or 6-digit operator codes— are translated and routed by the switch serving the operator location. At the terminating switch that serves the called station, after the end-office prefix has been identified only the 4-digit station number is usually needed to connect to the called number. Other switches along the route, however, including the originating switch, must interpret an NPA code or end-office code in order to select a route to the terminating switch. When a switch selects a trunk to the end-office where the call terminates, no routing digits (NPA or office code) need be transmitted to that terminating end-office. Only the called station number, and occasionally a directing digit, is required. If, however, the link selected is to a tandem switch, address digits including the NPA or end-office code must be forwarded to that tandem system for its use in selecting a route to, or toward, the terminating switch.

In common-control systems, the interpretation of digits for the purpose of selecting the proper outgoing link is called translation. Either the first three digits (i.e., NPA or end-office code), or the first six digits (i.e., NPA plus the end-office code), are interpreted by a switching system; thus the terms *3-digit translation* and *6-digit translation*. Pretranslation is the process of determining from the first, second, and third digits (the A, B, and C digits) how many total digits the receiving equipment may expect to receive on a particular call. This process takes place prior to translation by common-control equipment for routing purposes; thus the name *pretranslation*. It is required when the total number of digits within the range of customer dialing varies. Pretranslation provides to the switch the desired information on whether or not it may expect more digits before processing the call further. In end-offices not using interchangeable NPA/end-office codes, checking the B digit for a 1 or 0, for example, can tell the switch the first three digits are, or are not, an area code; if a 1 or 0 is found, there are seven additional digits after the first three.

A call may be routed from the first three digits received in the following situations: (1) All traffic to a foreign NPA or other 3-digit code is to be first-routed on the same link. In this case, three digits are sufficient to route the call, and the switching system does not require additional digits before selecting the link. (2) The call is destined for an end-office in the home NPA, and the area code has either been omitted (as on calls originated within the area) or the only routing required is based on the three digits of the end-office code. (3) A 3-digit international country code is used. Where there are two or more available routes to, or toward, a foreign NPA, it is necessary for the switch to examine the first six digits (area code and end-office code) to determine which route should be used. This is called 6-digit translation. When codes received at a switch along the route are not usable at the next switch, equipment can be directed to convert the digits received into the proper digits to be sent forward. This code conversion is accomplished by a combination of (1) deleting digits that were dialed but are no longer required and (2) prefixing new digits. An example of code conversion on a call is as follows. The number NNX-XXXX, where X is any digit, is used at the originating point to route the call through the network to a tandem, which, when it receives the call for NNX-XXXX, translates the first three digits, selects the proper link, and pulses the X-XXXX digits if five digits are required at the terminating end-office or the XXXX digits if only four digits are required. If four digits are required to complete the call via a direct route, and five or more digits if completed through a tandem route, then the switch makes the translation, selects the proper link, and sends the required digits forward. Routing digits sent forward to a given switching system depend upon the requirements of the distant connecting point. For example, extra digits dialed by an operator or prefixed and sent forward by a preceding switch may be required to switch calls through a direct control switch. On a call requiring NPA + NXX digits, if the switch has a link specifically designed to carry only NPA traffic, the first three digits serve for translation and may be omitted from sending forward.

International Direct Distance Dialing (IDDD) originating in the North American Network was introduced in 1970 in New York City, with direct dialing to the United Kingdom. The customer must dial a special access code for IDDD. Station-to-station calls are initiated by dialing 011 + country code + national number. Thus, a call to Japan uses 011 + 81 + 8 (or 9) digits.

2.2.4 Signaling

Signaling is the general term applied to the transmission of information used to establish, maintain, or release a connection through the network or used to indicate the status of a call. Interoffice signals are classified as on-hook, off-hook, or a combination of the two. These terms come from the hook on which the receivers of earlier telephone sets were hung when not in use and are used to designate the two signaling conditions of a trunk. Usually, if a trunk is not in use, it is in an on-hook condition at both ends. Seizure of the trunk at the calling end initiates an off-hook signal transmitted toward the other end. Both off-hook and on-hook signals, when not used to convey numerical information, are often referred to as supervision signals. Supervision is the information transmitted between switching systems regarding the status of a call or the readiness of an item of equipment to process a call attempt.

Pulsing, on the other hand, is the transmission of address signals that represent the digits required to route a call to its destination. Pulsing is usually initiated by the calling customer, operator, or billing equipment and provides address code information in the form of digits dialed on a rotary dial or keyed from a TOUCH-TONE® telephone. The serving end-office or tandem then converts the digits received to the required type of pulsing. The duration, relative time of occurrence, and frequency of repetition are factors in establishing the intelligence conveyed by a signal. On single-link calls, pulsing occurs only once, when the originating switching system pulses the called station number over the trunk to the terminating switching system. On multilink tandem-switched calls, the pulsed digits may be stored and analyzed by the tandem as they are received. When the required digits have been received, they are then regenerated, prefixed, or deleted as required and sent to the terminating end-office. Thus, pulsing can occur repeatedly in the establishment of switched connections.

Switching systems have been developed with different types of pulsing capability. Dial pulsing, multifrequency pulsing (MF), and common-channel signaling (CCS) are examples of the progression of signaling technology. Definitions of these pulsing types are as follows:

- *Dial pulsing* A system of pulsing consisting of short signals sent from an originating source; the number of pulses transmitted corresponds directly to the value of the digits 1 to zero on the dial. The digit 1 equals one pulse, the digit 2 equals two pulses, and so on, to digit zero, which equals ten pulses. The signals may be interruptions of the direct current path or interruptions of tone using single frequency.

- *Single frequency* Dial pulsing using the presence or absence of a single frequency to represent break or make intervals.

- *Multifrequency pulsing (MF)* A system of pulsing embodying a simultaneous combination of two out of six frequencies to represent each of the ten digits zero to 9 and the various start and end signals.

- *Key pulsing (KP)*   An arrangement providing for the transmission of tones or direct current pulses corresponding to each of switchboard keys depressed.

- *Dual-tone multiple frequency (DTMF), or* Touch-Tone   A method of telephone station equipment pulsing, or possibly trunk pulsing, using voice frequency tones. DTMF is used for communication of dialed digits from telephone station equipment to central office equipment. Each signal is composed of two voice-band frequencies, one from each of two mutually exclusive frequency groups of four frequencies each.

- *Common-channel signaling (CCS)*   A system in which the voice and signaling portions of a call are separated at the originating or tandem switch. The voice portion is transmitted in the usual manner over a conventional channel; the signaling portions are encoded by a data set and routed via dedicated or switched data links.

- *CCITT (International Telegraph and Telephone Consultative Committee)*   A system that uses systems different from those in national service. One example is CCITT No. 5, which is an inband signaling system in which signals are within the used frequency spectrum and which is compatible with the transmission facilities used for 3KHz-spaced channel banks, pulse code modulation channel banks, satellite channels, and Time Assignment Speech Interpolation (TASI) systems. Another example is CCITT No. 6, which is similar in many respects to the CCS system. Detailed specifications of CCITT signaling protocols are found in the CCITT *Green Book,* a publication of the International Telecommunications Union (ITU).

CCS is commonly used between stored-program-controlled switching systems such as ESSs. Different types of pulsing do have implications for network design. For example, pulsing limitations of the various switches may restrict the use of a direct link. Not all switches can be equipped for all types of pulsing. Since the pulsing at the two ends of a link must be compatible, some links that might otherwise be justified cannot be established. A partial solution of this problem has been achieved through the ability of some switches to pulse-convert; that is, to send out a different type of pulse than that received. The result is that switches with pulse-conversion capability may serve as intermediate or gateway switches when a pulsing incompatibility exists. This often occurs in routing calls between networks, or internetwork routing.

2.2.5   Billing

Switching systems are equipped with various types of billing equipment. Centralized automatic message accounting (CAMA) systems record call billing details at a tandem. Switches equipped with local automatic message accounting (LAMA) can record call details at the originating end-office. All CAMA calls to be billed must be routed to the recording tandem, which precludes the use of a direct link out of end-offices for this traffic. LAMA-equipped switches can route calls directly to other systems with no requirement for billing equipment in other parts of the network. More flexible routing arrangements are usually possible when LAMA is employed. CAMA systems with operator number identification require an operator associated with the tandem to record the calling party number. Automatic number identification is a process in which the calling party identification is recorded by a mechanized process at the tandem switch.

2.3   Routing Methods

The aim of the hierarchical routing pattern is to carry as much traffic as is economically feasible over direct links between pairs of switches low in the hierarchy. This is accomplished by application of routing procedures to determine where sufficient load exists to justify high-usage links, and then by application of alternate-routing principles that effectively pool the capacities of high-usage links with those of final links, to the end that all traffic is carried efficiently and at a low overall probability of blocking in the average busy hour of the busy season.

2.3.1   First- and alternate-route design rules

In this section we describe the rules for routing traffic in hierarchical alternate-routing networks of two or more levels [ATT77]. There are eight rules that govern the selection of first routes and alternate routes in the network design process. Some of the rules required for multilevel national intercity networks do not apply to two-level networks serving metropolitan areas. A list of the rules and their applicability is shown here:

	Rule	Applicable to	
		Two-level	More than two-level
1.	two-ladder limit	X	X
2.	intraladder direction	—	X
3.	multiple switching function	X	X
4.	one-level limit	—	X
5.	switch low	—	X
6.	directional routing	X	X
7.	single route	X	X
8.	alternate-route selection	X	X

The rules assume that traffic between two points is composed of two separate items; that is, traffic from A to Z and from Z to A. Traffic from A to Z always routes over a path different from that used for the Z-to-A traffic if one-way links are employed. Where links are two-way, however, the rules may result in the routing of both directions of traffic over the same path or over separate paths.

2.3.1.1 Two-ladder limit rule. This rule is that traffic must route only via the routing ladders of the originating and terminating switches. As illustrated in Figure 2.9, traffic from end-office A to end-office P may route A–T1–T3–P. It is not permitted to route A–T2–P because a third ladder would be involved. As illustrated in Figure 2.10, traffic from B to P may route B–C–R–Q–P. It is not permitted to route B–K–P because a third ladder would be involved. This rule, jointly with the intraladder direction rule that follows, defines the permissible

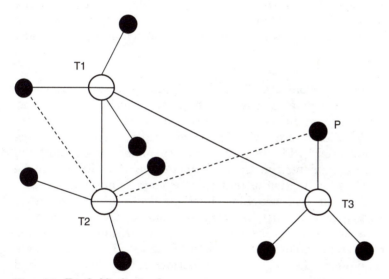

Figure 2.9 Two-ladder limit rule

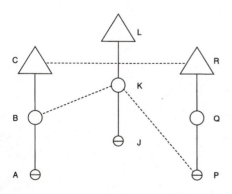

Figure 2.10 Another illustration of the two-ladder limit rule

switching points for any given call and, additionally, the links that can be connected in tandem. These rules limit the number of available routes, making it possible to design and manage the network. In multilevel networks, the rule also functions to help ensure that transmission requirements are met by limiting the number of links in tandem.

2.3.1.2 Intraladder direction rule. This rule is that switched traffic must only route toward the terminating location upward in direction on the originating ladder and downward on the terminating ladder. As illustrated in Figure 2.11, traffic from B to Q may route up the originating ladder to C, across to R, and down the terminating ladder to Q. It is not permitted to route B–A–R–Q. If this were allowed, traffic arriving at A could find the A–R high-usage link busy. It would then be overflowed back to B, where it would again be offered to A. This condition, sometimes called *shuttling* or *looping,* would rapidly tie up all the trunks between A and B, a condition that is not permitted. This rule in conjunction with the two-ladder limit rule defines the permissible switching points for any given call.

2.3.1.3 Multiple switching function rule. This rule is that a switch performing multiple switching functions must be assumed to have a routing ladder internal to the switch extending from its lowest function to its highest function. This rule is applicable whenever a switch performs multiple switching functions. If switch B, illustrated in Figure 2.12, serves customer lines in addition to serving as a Class-4 tandem, the Class-5 function of the switch can be said to home on the Class-4 function of that same system. Similarly, switches C, D, and E all perform multiple switching functions. In each of these cases, an "intraswitch" routing ladder must be assumed from its lowest to highest function. For example, switch E performs the Class-4, -3, -2, and -1 switching functions for end-office H. Switch D performs the Class-4, -3, and -2 switching functions for end-office G, and switch C performs the Class-4 and -3 switching functions for end-office F. Because switching functions are determined by the final links in a routing ladder, the presence or absence of an intraladder

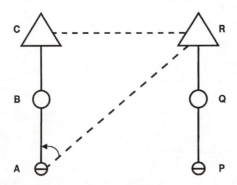

Figure 2.11 Intraladder direction rule

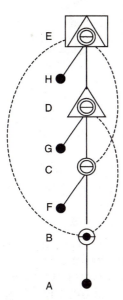

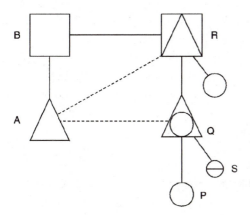

Figure 2.12 Multiple switching function rule

high-usage link such as B–E, B–D, or C–E does not alter this switching function rule. This rule is necessary to prevent the development of more than one route from one switch to another in the application of a routing discipline based on switching function. It also serves to keep switching at lower levels in the network. If the switching function were dependent upon the external routing ladder, the Class-4 switch S shown in Figure 2.13 would home on the function 3 of Q and, in turn, the function 2 of R. If a link A–R were developed, to handle that R function 2 traffic, A-to-S traffic would be included and the route from A to S would be A–R–Q–S. Similarly, if a link A–Q were developed for the Q function 2 traffic, that link would include A-to-P traffic and the route from A to P would be A–Q–P. This results in two different routes from A to Q, depending upon the switching function used at Q, an undesirable arrangement.

Figure 2.13 Another illustration of the multiple switching function rule

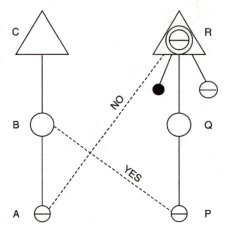

Figure 2.14 One-level limit rule

2.3.1.4 One-level limit rule. This rule is that when evaluating potential candidate links, only those first-route traffic items for which the switching functions performed at each end of the link differ by no more than one level should be considered. The A–R link in Figure 2.14 should not be considered if it is dependent upon the Class-2 function load at R. The initial justification of skip-level links such as A–R must be based on the load resulting from the application of the one-level limit rule, in this case the load between the Class-4 function at A and the Class-4 and -3 functions at R. The routing of the Class-2 function load cannot be determined until consideration is given to other links between A, B, or C and the R Class-2 area. Note that this rule applies to the load that can be considered when evaluating potential candidate skip-level links. The subsequent rules in this section cover the traffic that may properly route over these links, once justified. This rule is required to facilitate the establishment of links at lower levels in the hierarchy and to prevent distant higher-class switches from becoming overused for traffic from remote lower-class switches.

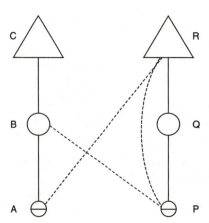

Figure 2.15 Switch low rule

2.3.1.5 Switch low rule. This rule is that switched traffic must route via tandems involving the lowest functional level of switching, considering both routing ladders. As illustrated in Figure 2.15, traffic from A to P should route via B switching function 3 and not via R switching function 2, thereby keeping switching at the lowest functional level. This rule is required to ensure that lower-functional-level switching is used for traffic that can switch at that lower level to make available higher-functional-level switching for traffic that must switch at that higher level. The presence of the R–P link does not change the fact that R performs the function-2 switch for P.

2.3.1.6 Directional routing rule. This rule is that if there is a choice of routes involving switching at the same functional level in each of two routing ladders, the route using that functional level in the terminating ladder should be chosen. In the trunking configuration shown in Figure 2.16, traffic from A2 to P1 has the choice of a route via A or P, each involving a Class-4 function switch. Under the directional routing rule, the preferred route is via the Class-4 function in the terminating ladder, or, in this case, the route is via P. In a similar manner, the preferred route from P1 to A2 is via A. This rule relates to the directional first routing of traffic from one switch to another. It is important to note that the rule deals with the functional level of switching as illustrated in Figure 2.17. In this trunking configuration, link B–R must be offered all the traffic from B to R, Q, and P. The C–Q link must be offered all the traffic from Q to C, D, B, and A. However, the D–R link is only offered the traffic between the Class-4 functions at D and R. The traffic from D to Q and P switches at C, with C performing a Class-3 switching function for this traffic; to switch at R involves a Class-2 switching function.

2.3.1.7 Single-route rule. This rule is that routes must be chosen so that there is only one first-choice route from one switch to another, regardless of the switching functions performed by those switches. In two-level networks, this requirement is met through application of the directional routing rule discussed above. In networks with more than two levels, this rule is necessary to ensure that extra and unnecessary switching is not planned. As illustrated

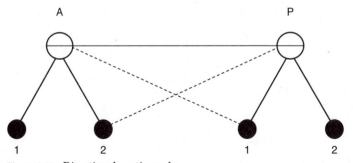

Figure 2.16 Directional routing rule

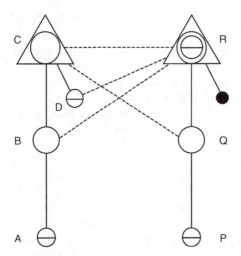

Figure 2.17 Another illustration of the directional routing rule

in Figure 2.18, it is not permitted to first route from switch B to the Class-2 function at switch S via switch C if there is a direct B–S link justified by the Class-4 or -3 function load. Routing via C requires an extra switch and link. The B–S link is justified by the load between the Class-4 function at B and the Class-4 and -3 function at S. In accordance with the one-level limit rule, it cannot be justified by inclusion of load requiring a Class-2 function switch at S. The C–S link may be justified by the load between the Class-4 and -3 functions at C and the Class-4, -3, and -2 functions at S. If there is not enough load to plan links from either B or C to P, Q, or R, it is apparent that all traffic from B to P, Q, R, and S requires a switch at Class-2 switch S. Because there is a direct link between B and S, it provides the one and only first route from B to S, including all traffic that switches at S. It is important to note that the decision to add Class-2 function traffic to

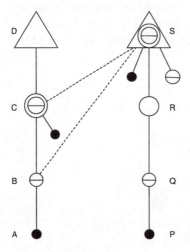

Figure 2.18 Single-route rule

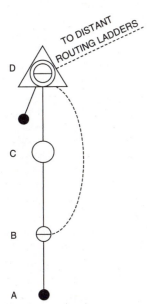

Figure 2.19 Another illustration
of the single-route rule

the B–S link must not be made until it is determined that link C–R or C–Q is not planned. This rule also applies to intraladder links, as illustrated in Figure 2.19. Intraladder link B–D may be justified by Class-4 function traffic between B and D. There may not be enough load to justify a direct link between B or C and other routing ladders, but such links are justified from Class-2 switch D. In such cases, traffic from B to these other ladders should route over the direct intraladder link B–D until such time as the lower-level direct link can be justified.

2.3.1.8 Alternate-route selection rule. This rule is that the alternate route at each end of a high-usage link must be the route the switch-to-switch traffic load between the switches would follow if the high-usage link did not exist. Alternate routing is an extension of first-choice routing. The considerations applicable to one are also applicable to the other. Just as first-choice routing is done on a directional basis, it is also necessary to determine the alternate routes separately for each end of a two-way high-usage link, as illustrated in Figure 2.20. The alternate routes for the A–P link are the routes the A-to-P and P-to-A traffic would follow if the A–P link did not exist. If that were the case, in accordance with the routing rules already discussed, A-to-P traffic would route via Q, and P-to-A would route via B. This situation, then, involves directional alternate routing as discussed under the directional routing rule. Therefore, if the A–P link did exist, the alternate routes for it would be via Q and B, a case of directional alternate routing. The alternate routes for the A–Q link are the routes the A-to-Q and Q-to-A traffic would follow if the A–Q link did not exist. If that were the case, in accordance with the routing rules already discussed, A-to-Q traffic would route via B, and Q-to-A traffic would route via B. Therefore, if the A–Q link did

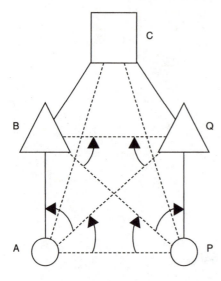

Figure 2.20 Alternate-route selection rule

exist, the alternate routes would be via B, a case of triangular alternate routing. The rules used to determine first-choice routes are also used to determine link-alternate routes. These alternate routes may be directional or triangular, just as first routes may be directional or triangular.

2.3.2 Hierarchical route selection

This section describes procedures to determine the route a call should follow when links and homing arrangements are known. The routing of calls in a hierarchical network involves an originating ladder, a terminating ladder, and links interconnecting the two ladders. In metropolitan networks, a two-level ladder is normally employed. Figures 2.21 and 2.22 illustrate double-tandem and single-tandem arrangements of a two-level network. A five-level ladder, for example, as used in the North American network prior to Bell System divestiture

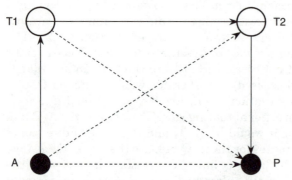

Figure 2.21 Double-tandem two-level metropolitan area network

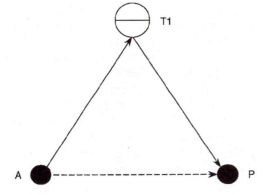

Figure 2.22 Single-tandem two-level metropolitan area network

in 1984, is illustrated in Figure 2.23. In metropolitan networks of two levels, the originating ladder is the final link from end-office to tandem, and the terminating ladder is the final link from tandem to end-office. Links A–P, A–T2, T1–P and T1–T2 in Figure 2.21 are examples of interladder links. In the five-level network illustrated in Figure 2.23, for the routing of a call from A to P, the originating ladder is A–B–C–D–E, the terminating ladder is T–S–R–Q–P, and links D–S, D–T, E–S, and E–T are examples of interladder links. The identification of the proper interladder link for the routing of a given call identifies the originating ladder "exit" point and the terminating ladder "entry" point. Once these exit and entry points are identified and the intraladder links are known, a first-choice path from originating to terminating location can be determined.

Various levels of traffic concentration are used to achieve an appropriate balance between transport and switching. The primary requirement is that

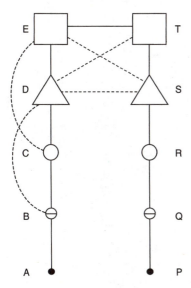

Figure 2.23 Five-level national intercity network

every customer be connectable to every other customer. In a hierarchy having a maximum of five levels, customer lines are terminated on the switching-function-5 level. Switching functions 5, 4, 3, 2, and 1 are provided for concentrating traffic into efficient traffic items for routing. Under this arrangement, there are a maximum of 25 interladder links from the originating ladder to the terminating ladder. The routing procedures provide a specific sequence of first-route selection from among the 25 choices. Figure 2.24 shows all of these links numbered sequentially, indicating the preference arrangement for routing a call from A to P. Note that this illustrates the link choice only for the interladder links. Intraladder routing is described in a later section. For both interladder and intraladder routing, the switching function performed by a given switch is a controlling factor.

Figure 2.24 depicts the maximum interladder link arrangement for a hierarchy having a maximum of five levels, but it is also usable in specifying the link selection sequence within a hierarchy of fewer than five levels, with final links between the highest equal-class switches. For example, in a two-level

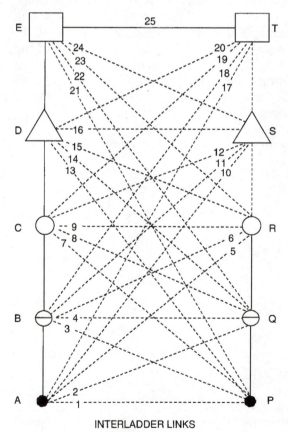

INTERLADDER LINKS

Figure 2.24 Hierarchical route-selection order

metropolitan area network, link B–Q in Figure 2.24 may be a final link, and links to switches C, D, E, R, S, and T are not involved. For a three-level hierarchy, link C–R may be a final link, and links to switches D, E, S, and T are not involved. For a four-level hierarchy, link D–S may be a final link, and links to switches E and T are not involved. In a similar manner, the trunking diagram in Figure 2.24 may also be used to specify the link-selection sequence within a hierarchy of five or fewer levels, where the originating and terminating ladders are connected by final links to a common switch. This can be illustrated by an intrasectional configuration, as shown in Figure 2.25. To summarize, the trunking diagram in Figure 2.24 can be used to determine which interladder link should be selected as the first-choice route for a call from A to P. This can be done by overlaying an actual trunking arrangement on the maximum arrangement per Figure 2.24 and selecting the lowest-numbered route that is available. The generally preferred sequence for the interladder link is

1. A call involving no tandem switch—route A–P.
2. A call involving a function-4 switch—route A–Q, B–P, or B–Q, in that order.
3. A call involving a function-3 switch—route A–R, B–R, C–P, C–Q, or C–R, in that order.
4. A call involving a function-2 switch—route A–S, B–S, C–S, D–P, D–Q, D–R, or D–S, in that order.
5. A call involving a function-1 switch—route A–T, B–T, C–T, D–T, E–P, E–Q, E–R, E–S, or E–T, in that order.

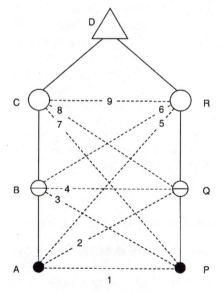

Figure 2.25 Interladder route selection

This procedure provides only the first-choice interladder link from A to P. Calls from P to A often route differently. To determine the P-to-A route requires reversing the diagram, making P–Q–R–S–T the originating ladder and E–D–C–B–A the terminating ladder. In Figure 2.24 the preferred route from P to A is P–A, P–B, Q–A, Q–B, P–C, Q–C, R–A, R–B, R–C, P–D, Q–D, R–D, S–A, S–B, S–C, S–D, P–E, Q–E, R–E, S–E, T–A, T–B, T–C, T–D, T–E, in that sequence. The alternate route for any high-usage link is the route the switch-to-switch traffic load between the switches would follow if the high-usage link did not exist. In Figure 2.24, this is the next-highest-numbered link terminated at one of the common switches. For example, link 16 D–S alternate routes from D over link 20 D–T, the next-highest-numbered link with a switch D common to both the direct and the alternate routes.

Just as there is a preferred first-choice route interconnecting originating and terminating ladders as described above, there is also a preferred routing within each of these ladders. In a five-level hierarchy there can be a maximum of 10 intraladder links, excluding parallel protective high-usage and divided links, and eight possible routes. In a four-level hierarchy, there can be six intraladder links and four possible routes. For a three-level hierarchy or cases with exit or entry at the third level, there can be three intraladder links and two possible routes. For a two-level hierarchy or cases with exit or entry at the second level, there is only the one final route. The routing procedures provide a specific sequence for selecting the preferred first-choice route from among the available routes. Figure 2.26 shows all the possible routes, numbered sequentially by order of preference. Intraladder routing is dependent upon switching function rather than class. In the five-level hierarchy depicted in Figure 2.26, for example, Class-3 switching system C could home directly on Class-1 switching system E. In such an arrangement, E would perform the Class-2 function as well as the Class-1 function for C. Link C–E in that case would be a final link. Because the procedures described here relate to switching functions, they are applicable in this situation and in others involving multiple switching functions. The preferred first-choice route from the originating end-office to the terminating end-office can be determined by overlaying an actual trunking configuration on the maximum configuration, per Figure 2.26, and selecting the lowest-numbered route that is available. The generally preferred sequence for the intraladder route is

1. End-office, Class-1 switching system—route A–E, A–B–E, A–C–E, A–D–E, A–B–C–E, A–B–D–E, A–C–D–E, or A–B–C–D–E, in that order.

2. End-office, Class-2 switching system—route A–D, A–B–D, A–C–D, or A–B–C–D, in that order.

3. End-office, Class-3 switching system—route A–C or A–B–C, in that order.

The alternate routes for intraladder high-usage links are determined by the next-highest-numbered route terminated at both of the common switching systems. For example, referring to Figure 2.26, link 1 A–E alternate routes via B over route 2 A–B–E.

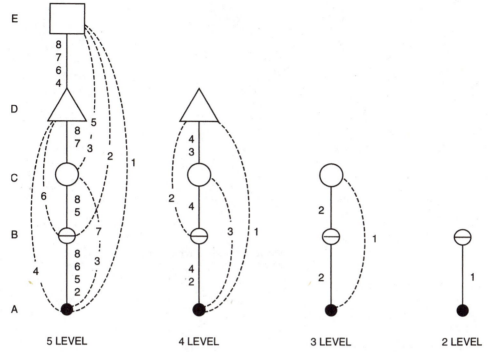

Figure 2.26 Intraladder route selection

By following these procedures, the preferred first-choice path for any end-office-to-end-office call can be ascertained if the trunk network and homing arrangements are known, as follows:

1. Referring to Figure 2.24, identify the lowest-numbered interladder link in existence, thereby identifying originating ladder exit point and terminating ladder entry point.

2. Referring to Figure 2.26, identify the lowest-numbered route from originating end-office to the exit point.

3. Referring to Figure 2.26, identify the lowest-numbered route from terminating ladder entry point to the terminating end-office.

This process identifies the recommended first-choice path from any end-office to any other end-office. In many cases, a given switch performs several different switching functions. If Class-4 switch B in Figure 2.27 is homed directly on E, then E performs switching functions 3, 2, and 1 for B. This is illustrated in Figure 2.28. In the application of the path selection portion of the routing procedures (Figures 2.24 and 2.26), interladder links should be assumed for C–S, D–S, and E–S, and intraladder links should be assumed for A–C,

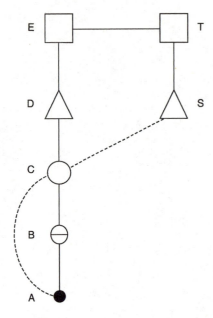

Figure 2.27 Hierarchical route selection

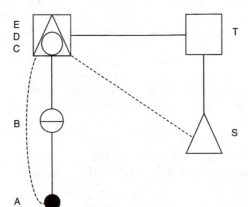

Figure 2.28 Hierarchical route-selection example

A–D, and A–E. In a similar manner, a link terminated at any switch serving multiple switching functions is available for each of those switching functions.

2.4   Network Management and Design

In this section we discuss principles of traffic engineering and then discuss hierarchical network design. We discuss the traffic model, the Erlang B formula, the equivalent random method for peaked traffic, hierarchical network design steps, sizing high-usage links with Truitt's ECCS technique, cluster busy-hour sizing techniques, and sizing final links with Neal-Wilkinson techniques. General references on traffic theory include [BTL77, Coo72, Kle75, and Min74]. General references on traffic engineering include [BTL77, Gir90, Min74, and Pio89].

2.4.1 Traffic model

An offered load a can be defined as the average number of simultaneous calls in progress if none of the offered calls is blocked. If calls arrive at a constant average rate λ and require service for a constant average holding time τ, then the offered load a is given by

$$a = \lambda \times \tau$$

Because λ is given in calls per unit of time and τ is given in time units per call, a is a dimensionless quantity, which we designate as *erlangs* of traffic load. Telephone traffic is not predictable in detail, but fluctuates regardless of the time scale in which it is viewed. Calls arriving with intensity λ are often modeled as occurring in a stationary Poisson process, which is precisely defined mathematically. A Poisson process produces traffic with unit peakedness, which is defined as the ratio of the variance of the calls in progress to the mean. Poisson variability occurs even when the mean load a is constant. Mean load can change as a function of time, even within the hours that are the standard measurement intervals. This is known as within-the-hour variation. Traffic load can vary from day to day in the same hourly interval, which is known as day-to-day variation, and usually there is a cyclic pattern over the days of the week. There is usually week-to-week and seasonal variation, as well. In addition, traffic loads tend to grow from year to year at rates from zero to more than 20 percent.

Patterns of variability differ from place to place and from time to time. A balance is sought between cost and network availability, which is a function of the frequency of occurrence of the load for which the network is designed. For example, designing for average hourly loads would be inadequate because these loads are often exceeded by an unpredictable amount. Conversely, service objectives for performance in the 10-year peak hour are overly conservative and impractical, because it is impossible to predict quantities on the tail of the load distribution, which is more volatile than the mean and variance. Large traffic variability costs additional network investment, because equipment provided to handle peaks is underutilized most of the time.

2.4.2 The Erlang B formula

We discuss a mathematical model of a traffic system and its application to network design. Consider a link with n trunks and call arrivals that occur as a Poisson process. If an arriving call finds at least one trunk idle, it seizes the trunk and service begins. With Poisson arrivals the intervals between adjacent arrivals, the interarrival times, are independent random variables with identical exponential distributions. The probability that two events occur in an interval of Δt goes to zero as $o(\Delta t)$, which means that

$$\lim_{\Delta t \to 0} \frac{o(\Delta t)}{\Delta t} = 0$$

This means that batches of simultaneous events do not occur. The process also has no memory, in that the number of events occurring in disjoint intervals are independent random variables. A Poisson process can be time varying, but in this analysis the arrival process is stationary. The expected number of events per unit of time, or rate, is denoted by λ, and the number of events occurring in an interval of length τ has a Poisson distribution; that is, k events occur with probability

$$[(\lambda\tau)^k/k!]e^{-\lambda\tau}$$

The probability that an arrival occurs in an interval of length Δt is $\lambda\Delta t + o(\Delta t)$, independent of what happens outside the interval. The probability density function of the interarrival times is $\lambda e^{-\lambda t}$.

The service times or holding times of calls are independent and exponentially distributed with density function $\mu e^{\mu t}$ and distribution function $1 - e^{-\mu t}$, where the parameter μ is the hang-up rate. The exponential distribution has no memory, because given that a call has lasted for a time τ, the conditional distribution of its remaining time is still $1 - e^{-\mu t}$. A call in progress terminates in an interval of length Δt with probability $\mu\Delta t + o(\Delta t)$, and if j calls are in progress, one of them ends in the interval of length Δt with probability $j\mu\Delta t + o(\Delta t)$. The probability of two hang-ups in such an interval vanishes as $o(\Delta t)$. The model also assumes that blocked calls are cleared.

The system based on these model assumptions of Poisson arrivals, exponential holding times, and blocked calls cleared is at every instant of time in one of $n + 1$ states. The ith state is that exactly i trunks are busy, where i ranges from 0 to n. For a particular set of arrival times and service times, the history of the system can be represented by a graph showing that state as a function of time. In this case, the state can only increase or decrease one unit at a time for an arrival or departure, respectively, and the corresponding stochastic process is known as a *birth–death process*. In the steady state, we are interested in the fraction of time spent by the process in each state, or the stationary probability of each state. In statistical equilibrium, the average rate-of-flow into any state equals the flow out, or on average the system makes as many transitions per unit of time into a state as out of a state.

Therefore, the net flow across a boundary between states k and $k - 1$ equals zero. The system can reach the upper states only from below, where a call arrival moves it from state $k - 1$ into state k. Let the probability of being in state $k - 1$ be $p_{(k-1)}$. Given that the system is in state $k - 1$, the probability of an upward transition during the next interval of length Δt is $\lambda\Delta t + o(\Delta t)$; in other words, an upward transition from state $k - 1$ occurs at rate λ per unit of time. Thus the unconditional rate of upward transitions from $k - 1$ to k is $\lambda p_{(k-1)}$. Likewise, downward transitions from k to $k-1$ occur at rate $k\mu p_k$, owing to hang-ups. The net flow in and out of state $k-1$ is zero if

$$p_k = \frac{\lambda}{k\mu}p_{(k-1)}, \quad k = 1, 2, \ldots, n$$

The load or traffic intensity a is defined as λ/μ. Here a, measured in erlangs, is the mean number of call arrivals per average holding time, or the mean number of calls in progress if there is no blocking. Then $p_k = (a/k)p_{(k-1)}$, which implies that

$$p_k = \frac{a^k}{k!}p_0$$

Because the sum of the state probabilities must be one,

$$\sum_{k=0}^{n} p_k = p_0 \sum_{k=0}^{n} \frac{a^k}{k!} = 1$$

and

$$p_0 = \left[\sum_{k=0}^{n} \frac{a^k}{k!}\right]^{-1}$$

Hence, solving these equations for the fraction of time or the probability that the system is in state n, with all trunks busy, is given by

$$p_n = \frac{a^n/n!}{\sum_{k=0}^{n}(a^k/k!)} = B(n, a)$$

$B(n, a)$ is the *Erlang B formula*, first derived by Danish engineer and mathematician A. K. Erlang. The behavior of an interoffice link that serves only first-routed traffic is remarkably well described by the Erlang B formula. However, the formula does not account for overflow traffic, which can be non-Poisson and peaked, that is, with its variance exceeding the mean. Erlang B also does not take into account day-to-day traffic variations, in which, due to nonlinearity of blocking as a function of traffic load, the average blocking for variable load exceeds the blocking for the average load. The effects of peakedness and day-to-day variations are discussed next.

2.4.3   Equivalent random method for peaked traffic

We consider an example three-switch network shown in Figure 2.29, where the offered and overflow traffic flows can be represented schematically as in

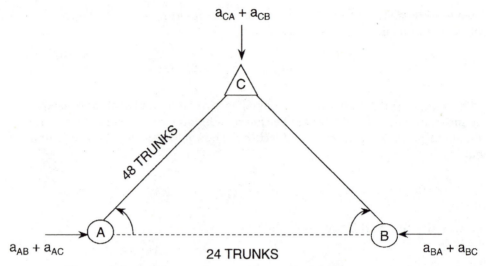

Figure 2.29 Three-switch network example

Figure 2.30. This is further generalized on the left side of Figure 2.31, where a_1 and a_2 denote independent Poisson first-route traffic loads. The equivalent random method is an approximation developed by R. I. Wilkinson [Wil56] to analyze this system in terms of the mean and variance of each of the overflow traffic items. Solution of the birth–death equations for the system in Figures 2.29 to 2.31 is impractical. Also, approximating the overflow traffic α_1 and α_2 as Poisson and solving for the overflow α_3 from link c [$\alpha_3 = (\alpha_1 + \alpha_2)B(c, \alpha_1 + \alpha_2)$] significantly underestimates the quantity α_3. This is because overflow traffic is not Poisson but tends to be more bunched or bursty than Poisson arrivals with the same mean traffic load. That is, the overflow traffic has a variance v_3 greater than the mean

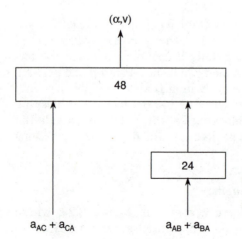

Figure 2.30 Three-switch network example, schematic traffic flow diagram

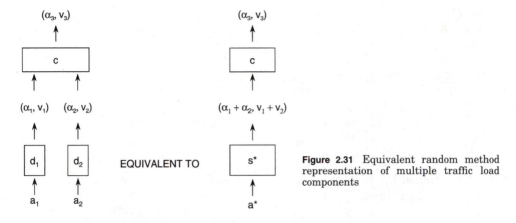

Figure 2.31 Equivalent random method representation of multiple traffic load components

α_3, as compared with Poisson traffic, in which the variance-to-mean ratio is 1. The variance-to-mean ratio for overflow traffic, $z = v/\alpha$, is called the *peakedness* of the overflow traffic. Because the means and variances of the component traffic items overflowing to link c are additive, the mean of the total traffic offered to link c is $\alpha_1 + \alpha_2$, the variance of the total traffic offered to link c is $v_1 + v_2$, and the peakedness of the total traffic offered to link c is therefore

$$\frac{(v_1 + v_2)}{(\alpha_1 + \alpha_2)} > 1$$

The equivalent random method approximates the distribution of $\alpha_1 + \alpha_2$ and $v_1 + v_2$ with a single equivalent link of s^* trunks with Poisson input of intensity a^*. This equivalence is illustrated on the right side of Figure 2.31, so that α_3 can be determined from the Erlang B formula with $\alpha_3 = a^* B(c + s^*, a^*)$. The equivalent random method is very accurate in most overflow systems of interest.

The mean α and variance v of the overflow from s trunks given an offered Poisson traffic of a erlangs are given by the Erlang B formula for the mean

$$\alpha = aB(s, a)$$

and the *Nyquist formula* for the variance [Wil56]

$$v = \alpha \left[1 - \alpha + \frac{a}{s + 1 + \alpha - a} \right]$$

or

$$s = \frac{a}{1 - \dfrac{1}{\alpha + z}} - \alpha - 1 \tag{2.1}$$

Solving Equation 2.1 together with the Erlang B formula for s and a, given the overflow mean α, variance v, and peakedness z, is cumbersome. An approximation

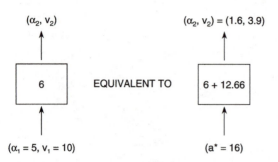

(α_2, v_2) $(\alpha_2, v_2) = (1.6, 3.9)$

$\uparrow$ $\uparrow$

| 6 | EQUIVALENT TO | 6 + 12.66 |

$\uparrow$ $\uparrow$

$(\alpha_1 = 5, v_1 = 10)$ $(a^* = 16)$

Figure 2.32 Example of equivalent random method

for a^* developed by Y. Rapp [Rap62] simplifies computation of the equivalent random load and trunks:

$$a^* = v + 3z(z - 1) \tag{2.2}$$

Equations 2.1 and 2.2 are solved for equivalent random s^* and a^*, given the overflow mean α, variance v, and peakedness z. An example calculation is given in Figure 2.32, where for $\alpha = 5$, $v = 10$; from Equation 2.2, $a^* = 16$; and from Equation 2.1, $s^* = 12.66$. The equivalent system then yields $\alpha_2 = 1.6$ and $v_2 = 3.9$.

2.4.4 Hierarchical network design steps

The overall network design process for hierarchical networks involves six steps:

1. *Selecting a network configuration* Several network configurations are used in hierarchical networks. These include the two-level originating sector, terminating sector, and combined sector tandem networks used in metropolitan area networks, multilevel hierarchical networks for national intercity traffic, and bilateral hierarchical networks for global international traffic. Some of these configurations are shown in Figure 2.33.

2. *Establishing a homing arrangement* Hierarchical network configurations assume a hierarchy of switching systems; therefore, a homing arrangement must be established to indicate which switching systems are subordinate to each higher-class switching system. Each switching system performs a concentration function for the switching systems that subtend it. Switching systems in each homing ladder are connected by final links, which form the backbone of the network. The choice of network configuration and homing arrangement is normally made as a result of planning studies.

3. *Determining traffic volumes* Traffic load data derived from link measurements can be used directly as the base from which to develop link sizes. However, where new links are to be established, traffic is to be rerouted for switching system load-balance purposes, or because of rehoming or new switching systems, link loads must be developed from switch-to-switch data such as message-billing data.

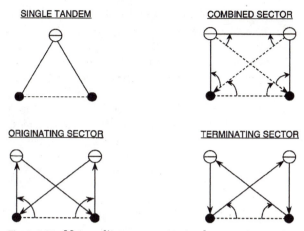

Figure 2.33   Metropolitan area sector tandem arrangements

4. *Identifying candidate high-usage links*   Given a network configuration, homing arrangements, and switch-to-switch traffic loads, the routing discipline discussed here can be used to identify candidate high-usage links, to assign the appropriate traffic to those links, and to determine alternate routes for all high-usage links. High-usage links identified as candidates by this process, and which subsequently are justified by the sizing process, supplement the backbone final links that are specified when the homing arrangement is determined. The identification of candidate high-usage links is accomplished by evaluating switch-to-switch traffic loads to see if there is sufficient traffic volume to meet a threshold level such that direct-link capacity should be considered. The threshold level for direct-link capacity is a function of last-trunk loading requirements and the overflow loads to be added to the first-route load.

5. *Selecting first and alternate routes*   Where the switch-to-switch traffic loads do not meet the threshold value, they are assigned to a first-choice switched route, where they are accumulated with other first-route loads. The accumulated loads are then examined to see if they meet the candidate link threshold. Trunk quantities and overflow volumes are computed in this load summation process, which requires iteration because of link and network interdependencies.

6. *Sizing links*   After all candidate links are identified and first and alternate routes selected, sizing takes into account the first-route loads and the overflow loads, including the peakedness of the overflow traffic. Overflow load means and variances for each high-usage link are computed by the equivalent random method and are accumulated on each intermediate high-usage link and each final link in the alternate path.

The network design starts with the homing arrangements of all switching systems that have been determined by prior study for the forecast periods for which link requirements are being estimated. High-usage links are planned by routing the switch-to-switch traffic in accordance with the hierarchical routing discipline and determining where sufficient load exists for a direct link. For proper sizing of the network links, it is necessary to properly account for overflow traffic. This involves load allocation between direct and alternate routes. The choice of an alternate route for a high-usage link is similar to the choice of a switched route for a switch-to-switch traffic load. Because the same rules apply to the selection of both a first route and an alternate route, the alternate route for a link is the route the switch-to-switch traffic load between two switches would take if the high-usage link did not exist.

2.4.5 Sizing high-usage links with Truitt's ECCS technique

For high-usage link sizing, we determine the portion of a given load offered to a high-usage link that should be carried by that link and the portion that should be overflowed to an alternate route. The aim of this process is to provide a number of trunks in both the high-usage and alternate routes, such that the traffic between a given pair of switching systems is handled at the least cost consistent with service objectives. With the average offered load between switches and the relative costs of the high-usage link and the alternate route, the most economical network arrangement is determined by calculating the amount of load which the last (least efficient) trunk in the high-usage link should carry. Once the objective last-trunk economic hundred call seconds (ECCS) has been calculated by Truitt's method [Tru54], standard capacity tables or formulas are used to determine the number of high-usage trunks required. When the least efficient (or "last") trunk in the high-usage link is carrying the required load, the cost per unit of traffic carried is optimal. In theory, any deviation from this load, whether up or down, results in a more costly arrangement. Studies have indicated, however, that total network costs are not particularly sensitive to the last-trunk load value. Deviations of up to 40 percent do not cause a significant change in network costs. These deviations in the last-trunk load can, however, produce a significant change in the placement of link capacity, with a resulting impact on network administration and on service during periods of overload.

Figure 1.39 illustrates a high-usage link from end-office A to end-office B with an alternate (final) route via a tandem. In general, the direct or high-usage link is shorter and less expensive than the alternate route. However, because each leg of the alternate route is used by other calls, a number of traffic items can be combined for improved efficiency on that route. The design problem is to minimize the cost of carrying the offered load such that there is an optimum balance of direct-link cost and alternate-route cost. Figure 1.39 shows the relationships involved. The graph shows, as a function of the capacity of the high-usage link, the cost of the direct link, the cost of the alternate route, and total cost for serving the given offered load. The high-usage link cost increases in direct proportion to its capacity. If there is no high-usage link capacity, all of the offered traffic must be carried on the alternate route, so the incremental alternate-route cost is high.

As link capacity is added to the high-usage link, less offered traffic is overflowed to the alternate route, so the incremental alternate-route cost decreases. This cost decreases very rapidly as the first trunks are added to the high-usage link, because each of these trunks is very efficient, thereby relieving the alternate route of a substantial amount of load. As more high-usage link capacity is added, each successive high-usage trunk carries less traffic. This principle can be illustrated with a switching system offering a call to a link with ten trunks, tested in order. Trunk 1 is selected first, reselected when idle, and thus kept busy most of the time; trunk 2 is slightly less busy; trunk 3 is used less than trunk 2; and so on to the tenth trunk, which is used only when all prior trunks are busy. As more direct high-usage trunks are added, the alternate-route network is required to carry less overflow traffic. The overflow traffic is carried efficiently on the alternate-route network, but at a higher per-erlang cost than the direct high-usage link, up to a point. Eventually it becomes undesirable to add any more high-usage link capacity. The point at which this threshold occurs is where the total cost, which is the sum of the two curves, is minimized.

A method commonly used to determine the optimum point N is the ECCS design [Tru54]. This method determines the maximum number of high-usage trunks for which the cost per erlang carried on the "last" trunk of the high-usage link is less than or equal to the cost per erlang on an additional alternate-route path. This relationship can be expressed by an equation first derived by Truitt, which is the basis of ECCS engineering. Truitt's link alternate routing flow optimization model, illustrated in Figure 1.39, is used in some form in essentially all network design models discussed in this book. The figure illustrates a three-switch network, sometimes called "Truitt's triangle," where a erlangs of traffic are offered from switch A to switch B. The objective is to determine the direct link alternate routing flow that minimizes the network cost of carrying a erlangs on the network and which meets the blocking probability grade-of-service objective. The derivation given in Chapter 1 is repeated here for convenience, wherein the following quantities are defined:

C_D = Cost of a direct trunk
C_A = Cost of an alternate network trunk
C_R = Cost ratio = C_A/C_D
γ = Marginal capacity, or the erlangs carried by a single alternate network path
CLLT = Carried load on the last direct trunk
ECCS = Economic hundred call seconds
α = Overflow load from direct link

Here, marginal capacity γ is assumed to be constant independent of the amount of overflow load α. The following equation then describes the total cost of carrying a erlangs of traffic on direct link and alternate network:

$$\text{Total cost} = C_D \times n + \frac{a \times \text{B}(a, n) \times C_A}{\gamma}$$

Here, $B(a, n)$ is the Erlang blocking formula for a erlangs offered to n trunks. This formula is derived earlier in this chapter. The cost relationship is plotted in Figure 1.39 and shows that as the number of direct trunks is increased, the direct-link cost increases linearly as $n \times C_D$. The alternate-path cost decreases as more direct trunks are added, because the overflow load $\alpha = a \times B(a, n)$ decreases and the cost of carrying the overflow load

$$\frac{a \times B(a, n) \times C_A}{\gamma}$$

decreases. As illustrated in Figure 1.39, an optimum or minimum cost condition is achieved with the optimum direct link flow with n^* trunks. If we take the derivative of the above total cost equation with respect to the number of trunks n and set the result to zero, that is,

$$\frac{\Delta(\text{total cost})}{\Delta n} = 0$$

we then derive the minimum cost condition, which is then given by the following expression:

$$a \times [B(a, n^* - 1) - B(a, n^*)] = \text{CLLT} \geq \gamma \times \frac{C_D}{C_A}$$

or

$$\text{CLLT} \geq \frac{\gamma}{C_R} = \text{ECCS}$$

Hence, the direct link capacity is increased until the cost per erlang to carry traffic on the last direct trunk is still lower than the cost per erlang to carry traffic on the alternate network. This is Truitt's economic hundred call seconds, or ECCS, condition and is used in many design models described in the book, be they for fixed hierarchical routing networks or dynamic nonhierarchical routing networks.

The equation is solved for the ECCS by the load to be carried by the last or least efficient trunk in the high-usage link. Given the ECCS and the offered load, Truitt's ECCS formula is used to determine the number of trunks required and the estimated amount of overflow. This is the largest number of trunks for which the load carried on the last trunk is not less than the ECCS. Because the equation is solved for the ECCS, the other elements of the equation must be known. The cost ratio is the cost of a path on the alternate route to the cost of a trunk on the direct link, and it is always greater than one. The γ shown in the equation is the incremental capacity of the alternate route, which is the traffic-carrying capacity that is added to the alternate route by the addition of one alternate path. This value is usually assumed to be a constant, such as 0.78 erlang, thereby permitting calculation of the ECCS as a function of a single variable, the cost ratio. Thus it can be seen that with low cost ratios, the ECCS is high and fewer high-usage trunks are provided. Conversely, a low ECCS results from a high cost ratio and a greater number of high-usage trunks are provided. Simply put, the more expensive the alternate route relative to the high-usage link, the less traffic that is overflowed to it. The total cost curve in Figure 1.39 has a rather broad minimum. As a result, errors in ECCS that might result from inaccuracies in the cost ratio or incremental capacity γ do not have a significant impact on network cost.

2.4.6 Cluster busy-hour sizing techniques

In practice, network loads are often noncoincident, and procedures such as cluster busy-hour engineering and multihour engineering [Eis79] can determine the optimum ECCS in these cases. The number of trunks required in a given link is dependent upon the load offered to it in a busy-season and busy-hour characteristic of the "network clusters" of which it is a part. The design approach used is to subdivide the overall network into network clusters, each capable of being designed independently. A network cluster consists of a final link and all the high-usage links having at least one switch in common for which the final link is in the last-choice route chain. Both ends of the final link must be considered. Every high-usage link, then, is a part of two network clusters. Network clusters are sized in a consistent busy hour in which the total sum of high-usage-link offered loads and final-link offered loads is maximum. Normally the *cluster busy hour* for each end of a high-usage link is considered, and the high-usage link is sized for the maximum offered load in the two busy hours.

From a blocking standpoint, the best service is attained when calls are handled end to end over high-usage links. In a given network cluster, the average blocking level is to a large extent a function of the calls handled over high-usage links, as illustrated in Figure 2.34 and the table that follows:

Link	Calls offered			Calls carried	Calls overflowed	Calls blocked	% blocking
	1st route	Overflow received	Total				
A–B	300	—	300	270	30	—	—
A–C	300	—	300	270	30	—	—
A–D	300	—	300	270	30	—	—
A–T2	410	90	500	450	50	—	—
A–T1	950	50	1,000	990	10	10	1
Total cluster	2,260			2,250		10	0.4

If all 2,260 calls are offered to the final link with 1 percent blocking, more than twice as many calls would be blocked. The table depicts the blocking condition for the A–T1 cluster only. From an end-to-end service standpoint, there is also a probability of blocking within each network cluster encountered.

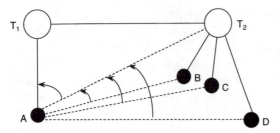

Figure 2.34 Cluster busy-hour sizing

New links are economically justified when the cost of carrying the load on the direct link is lower than the cost of switching the load via a tandem. One important consideration is the fact that establishing any new link involves a *splintering* of loads between at least two links. Some link capacity is then available to fewer customers, and there is less network flexibility to handle changes in load characteristics. Also, the total link capacity requirements are larger than with the combined link, causing some loss in network efficiency. Figure 2.35 illustrates this load-splintering consideration. Assume a link exists between G and H to handle all interarea traffic. As this traffic volume grows in time, a link E–H can be justified, saving switching at G and replacing E–G–H transport and switching capacity with E–H capacity. When this is done, however, it causes a reduction in the size of the G–H and E–G links. Customers served by switches A, D, and G consequently have a smaller link capacity serving their calls to the H sectional center area, and customers served by switches B and E have less link capacity to G. In other words, network splintering has taken place.

Although the creation of new links generally is desirable in order to conserve switching and transport capacity, other factors need to be considered. If the busy hours of the load on the proposed new link and the tandem switching

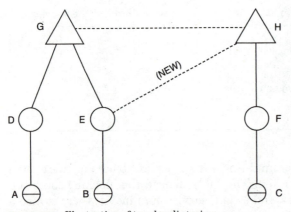

Figure 2.35 Illustration of trunk splintering

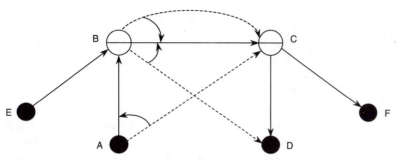

Figure 2.36 Illustration of parallel protective high-usage link

system from which the load is removed are different, no switching capacity may be required specifically to carry the load, even if the high-usage link is not established. In effect, the load gets a free ride on switching capacity provided for other loads in other hours. Where this occurs, the switching-cost portion of the costs used to develop the appropriate objective ECCS is effectively zero. As a result, the high-usage link may not actually be economically justified. Where cost ratios are used to develop objective ECCS values and busy hours are assumed coincident, this situation may not be apparent. If route costs are developed by basing transport costs on airline miles, a link may appear justified. If the actual transport route, however, is via the location through which the traffic was previously handled, the high-usage and alternate-route costs may be nearly identical, indicating that smaller savings result from establishing the high-usage link.

A parallel protective high-usage link is a high-usage link provided parallel to a final link for the purpose of protecting service on first-routed traffic. The first-routed traffic is offered first to the parallel protective high-usage link, which in turn overflows to the final link. Such links are desirable because during an overload in an alternate-route network, the overflow to a final link from subtending high-usage links can increase greatly with modest increases in offered load. The cumulative effect of such a condition is a tendency for the overflow load to monopolize the capacity of the final link to the detriment of service on any traffic first routed there. Offering the first-routed traffic to a parallel protective high-usage link instead minimizes the impact of any network overloads and tends to limit the degree of service degradation. The use of parallel protective high-usage links is illustrated in Figure 2.36, and sizing of parallel protective high-usage links is similar to that of other high-usage links.

2.4.7 Neal-Wilkinson model for sizing final links

The number of high-usage trunks provided in a link depends not only on the ECCS and offered load but also on the variability of the offered load. This variability, for example, can be either peakedness within the hour or day-to-day variation. Such variability can be the result of traffic patterns, as in the case of day-to-day variations, or it can be system induced, as is usually the case with peakedness.

In either event, the effect of such variability is a reduction of the link capacity. Where such variability is present, equivalent random-engineering techniques are used to size high-usage links, and Neal-Wilkinson capacity tables are used to size final links. Traffic volumes reach peaks during certain hours. Link capacity is usually provided to carry the average time-consistent busy-hour loads in the busy season of the year. Where only one link is available, capacity must be provided for the link busy-hour load. If two routes such as a direct and an alternate route are available, the busy hours on each of the two routes are often different. Where this is the case, capacity need only be provided on the direct route to carry that portion of busy-hour offered load that cannot be carried on idle capacity in the alternate route. The alternate route may be sized for a different busy hour and thus is not fully loaded in the busy hour of the direct route.

The average blocking on a final link, $\overline{B}_n$, is given by the average of the hourly blockings $B_i, i = 1, \ldots, n$, for time-consistent busy hours i, typically over a 20-day period:

$$\overline{B}_n = \frac{1}{n} \sum_{i=1}^{n} B_i$$

where if OF_i and OV_i are the number of call attempts and blocked attempts for busy hour i,

$$\overline{B}_n = \frac{1}{n} \sum_{i=1}^{n} \frac{OV_i}{OF_i}$$

We seek to design the final link such that $\overline{B} \leq 0.01$, where

$$\overline{B} = E[\overline{B}_n]$$

Normally, the observed variation from day to day in the number of arrivals OF_i is larger than for the variation induced by a random load λ over multiple hourly periods. To account for the increased variation, the loads $\{\alpha_i, i = 1, \ldots, n\}$ are modeled as a gamma distribution:

$$\Gamma(\alpha \mid \overline{\alpha}, v_d)$$

with mean $\overline{\alpha}$ and variance v_d. In addition, part of the observed day-to-day load variation is due to the finite measurement interval in addition to the variation in the source load [HiN76]. The resulting day-to-day source load variance v_d is

$$v_d = \max\left[0, 0.13\overline{\alpha}^\phi - \frac{2\alpha Z}{t/h}\right]$$

where Z is the traffic peakedness, t is the measurement interval, h is the mean holding time, and ϕ is the parameter describing the level of day-to-day variation;

TABLE 2.1 Comparison of Typical Blocking Probabilities

Blocking	Offered load (erlangs)	Peakedness of offered load	Trunks required	Percent occupancy	Percent idle
B.01L	11.3	1.0	20	56	44
B.02L	11.3	1.0	18	63	37

$\phi = 1.5$ for low day-to-day variation, $\phi = 1.7$ for medium day-to-day variation, and $\phi = 1.84$ for high day-to-day variation. The term $2\overline{a}Z/(t/h)$ is the correction term for the finite measurement interval t. Neal-Wilkinson tables, which apply the above formulas, are used for sizing a final link for a given traffic load mean, variance, and day-to-day variation and an objective mean blocking probability.

Network service objectives are expressed in terms of the percent of calls blocked in the busy-season busy hour. With a given load, the degree of blocking is a function of the amount of link idle time available. When the idle time is low, there is little capacity available to handle new calls and the blocking rate is high. When the idle time is high, there is capacity available to handle new calls and the blocking rate is low. As an example, with a random offered load of 11.3 erlangs, the different link capacities and idle times required to achieve B.01 and B.02 blocking, according to the Neal-Wilkinson theory, are shown in Table 2.1.

It should be noted that with B.01 service there is more idle time than with B.02. Final links sized on a B.01 basis, therefore, do not react as severely to overloads as when higher blocking probability objectives are used. This applies to all levels of offered load. A B.01 objective for sizing a final link does not necessarily mean 1 percent switch-to-switch blocking. If the traffic only routes on the final link, for example, that probability of blocking can be expected in the busy-season busy hour. At other periods of the year or during other hours of the day, probability of blocking is lower. If the B.01 objective is used on an alternate-route final link, the blocking in the network is considerably lower, even in the busy-season average busy hour. For example, if 50 percent of the calls are first offered to high-usage links within the network, only 1 percent of those overflowing to the final link as an alternate link would be subject to link blocking. In such a case, the average blocking in the total network is closer to B.005 than to B.01. Hence, blocking within alternate-route networks is always substantially lower than the blocking objective for final links.

Network capacity is provided such that the probability of blocking, including switching equipment blocking in the connection paths, contributes to a satisfactory overall blocking probability grade-of-service. Based on the expected number of links per connection path and the relative capacity of high-usage and final links, the use of B.01 as the blocking probability objective for final links produces overall blocking probability for traffic that is usually in the B.01 range under average conditions.

2.4.8   Network design model for hierarchical networks

In this section we summarize the methods used in this book for hierarchical network designs, such as those reported in Chapter 1.

Inputs to the model include the homing of each switch and the traffic matrix. The first step in the model is to compute the hierarchical alternate routes for each high-usage link. Each switch level is computed, from level 1, for the highest-level switches, to level 2, for end-offices in a two-level metropolitan area network, or to level 5, for end-offices in a five-level national intercity network. Each candidate link in the network is given a sizing order, which is equal to $10 -$ level (switch i) $-$ level (switch j). Each candidate high-usage link is examined and the routing rules described in Section 2.3 are applied to determine the hierarchical alternate routes. For each high-usage link, the directional alternate routes are determined according to working up the far end hierarchical ladder, with the final alternate route to the final link connecting to the home switch of the high-usage link in question. Each link cost is determined according to a fixed termination cost for each switch end and a transport cost, which is the per-mile transport cost multiplied by the distance in miles.

The engineering hour of each high-usage link is established by determining the cluster busy hour for each end of the high-usage link, as well as the busy hour on each interregional link. The sizing hour for each high usage link $i-j$ is then the hour of maximum high-usage-link offered load in the switch-i cluster busy hour, the switch-j cluster busy hour, or the interregional $i-j$ link busy hour if switch i and switch j are in different regions. High-usage links are sized in the order determined as above for the sizing hour using the ECCS technique given in Section 2.4.5. The equivalent random method, as described in Section 2.4.3, is used to determine the equivalent random offered load and equivalent random link size, which are used in applying the ECCS high-usage link-sizing technique. Overflow loads and variances are computed according to the Erlang B formula and Nyquist formula given in Sections 2.4.2 and 2.4.3, respectively. The overflow means and variances are accumulated on the alternate-route links.

Final links are sized using the Neal-Wilkinson technique, described in Section 2.4.7, in the final link busy hour. A given level of day-to-day variation is used, which is normally medium or high day-to-day variation. In computing the overflow to final links, there is a need to iterate because the directional alternate routing and interactions of links cause the overflow loads to be interdependent. Therefore, a network flow iteration is used in which final link sizes are assumed to have infinite capacity so that there is no overflow from the final links, and the overflow loads from the high-usage links are iteratively determined. That is, given the current estimate of link offered loads, the overflow loads are computed according to the equivalent random method, Erlang B formula, and Nyquist formula and the overflow means and variances accumulated on the alternate-route links for an updated estimate of the link offered load. The overflow loads are then computed again in the next iteration, until the process converges. This is the classical fixed-point calculation [Kat67]. At this point the maximum final link loads are determined and the final links sized using the Neal-Wilkinson methods.

2.5 Conclusion

We present hierarchical routing rules that are time fixed but can be complex, although there is minimal flexibility to share bandwidth among switch pairs. Fixed hierarchical traffic routing is descriptive of almost all current networks, including most metropolitan area networks, national intercity networks, and global international networks. The hierarchical routing plan establishes regions and a hierarchy of switching systems centering on the regional switches (the switching hierarchy used in the Bell System network prior to divestiture in 1984 had five levels), in which successively higher-level switches concentrate traffic from increasingly larger geographical areas. Two types of links connect these switches. High-usage links connect any two switches that have sufficient traffic between them to make a direct route economical, and final links are the links between each switch and its immediate superior in the hierarchy, together with the final links interconnecting all the regional centers. High-usage links are sized to handle only a portion of the traffic directed to them, and the switching systems redirect traffic by automatic alternate routing to a different link when all circuits of a high-usage link are busy. At each stage, the alternate routing plan shifts overflow calls from the more direct route toward the final route. Final links are designed to handle their own direct traffic plus overflow traffic from high-usage links with low blocking.

We present network design models that address traffic load variations and include link alternate routing flow optimization with Truitt's model, equivalent random methods for peaked traffic, and day-to-day variation design models.

Preplanned Dynamic Hierarchical Routing Networks

3.1   Introduction

With the introduction of electronic switching (ESS) and common-channel signaling, new concepts in routing are possible. With hierarchical network design models, as discussed in Chapter 2, we determine link sizes to carry peak load demands within a given blocking objective. However, due to the effects of traffic noncoincidence, a significant amount of reserve or idle capacity exists during the off-peak hours. Day–evening, business–residence, time zone, and seasonal load variations are the prime examples of traffic noncoincidence. By allowing a greater selection of routes, nonhierarchical or optimal path selection techniques can take advantage of the idle capacity to provide a more efficient network and thereby reduce overall network cost. As described in Chapter 1, a dynamic traffic routing method is one in which the optimal path selections are preplanned and time-dependent or state-dependent, and in both cases routing varies in accordance with the traffic load.

An early application of linear programming optimization techniques to the design of a communication network subjected to time-varying demands was made by Gomory and Hu [GoH64]. In this chapter, we investigate a three-stage triad selection method that includes possible savings for all links in a hierarchical network design, regardless of a switch's hierarchical class. It is shown that this method has large potential network cost savings in networks where traffic noncoincidence exists. We investigate this three-stage linear programming design model by determining optimum link sizes for 17 sets of sublink patterns for the 28-switch network model described in Chapter 1. The first step is to determine 17 hourly trunk requirements for all switch pairs in the network. The second step determines a minimum-cost network based on n peak hours ($n = 3$ in this study) allowing triad sublinks. The third step expands the n-hour solution to a 17-hour solution using a maximum-flow technique, producing a 12 percent reduction in network cost.

The 28-switch national intercity network model then is used to study this linear programming design model for dynamic hierarchical routing networks. A linear programming model cannot be used to directly model trunk sizing starting from traffic demands because the traffic engineering constraints are not linear. Therefore, single-hour trunk sizing is used to determine sets of hourly transport demand requirements. These hourly networks are sized according to hierarchical routing and standard design criteria, as described in Chapter 2. Two linear programming models are presented to determine the minimum network cost needed to satisfy all the hourly transport demands while permitting one- or two-link sublinks. The sublink patterns change for each hour, resulting in a dynamic hierarchical routing design. Overall results produce a 12 percent reduction in network cost compared with the hierarchical network design.

3.2   Dynamic Hierarchical Routing Method

Figure 3.1 shows a four-switch network with five possible sublinks identified. For this example, solid lines do not necessarily mean final links; they simply indicate existence of a direct link. Broken lines indicate sublinks. A–B is a high-usage link, A–C a final link. Traffic attempting to route across the A–B link would normally overflow to A–C and be blocked if all trunks are busy on A–C. With sublinks, traffic overflowing A–B can attempt to complete across any one of the three out-of-chain triad sublinks before trying the A–C final link. Likewise, A–C overflow can try two triad sublinks before being blocked. In effect, the high-usage link is composed of the direct link plus its three triad sublinks, and the final link of its direct link plus two triad sublinks. A triad sublink routing example is also illustrated in Figure 1.20.

3.3   Network Management and Design

As has been mentioned, the network design model is a three-step method of triad selection. The 28-switch national intercity network model, shown in Figure 1.12, is used to evaluate multiple-hour triad selection. This model represents a three-level hierarchical network covering four regions and four time zones, and we consider a 17-hour view of the average business-day load pattern.

For a linear programming solution to be possible, the problem must be formulated with linear constraints. A transport flow optimization model is used for

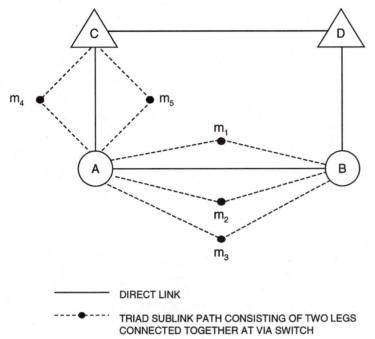

Figure 3.1 Example of dynamic hierarchical routing with triad sublink paths

multihour dynamic routing design, as described in Chapter 1, and accomplishes the linearization of the problem. Distributing offered traffic loads over a set of switches to determine link size requirements is a nonlinear optimization problem, but the problem can be linearized by converting the traffic load demands to transport requirements, utilizing hierarchical network design methodology described in Chapter 2, which comprises step 1 of the triad selection method. Using this methodology on a single-hour basis for each offered traffic load pattern in 17 hours, we obtain a set of trunk requirements per hour—that is, 17 hierarchical network designs. This phase incorporates hierarchical routing and standard blocking constraints for each hour. Figure 3.2 shows the total hourly requirements in normalized network cost for the model network, where the time shown is hour beginning, Eastern Standard Time. From here on, the term *demand* refers to the hourly trunk requirements.

The concept of trunk requirements is fundamental to understanding the design models in the following sections. At this point, the model network has been given an underlying hierarchical structure, but with time-varying trunk requirements. Blocking criteria, transport costs, switching costs, and other necessary inputs are considered, and standard hierarchical routing is employed. The linear programming models now size the links and find a way to satisfy the time-varying demands by forming temporary, longer alternate routes: the triad sublinks.

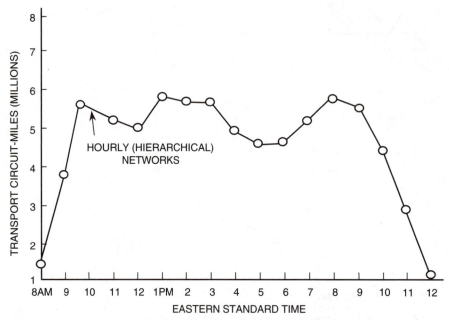

Figure 3.2 Transport circuit-miles for 28-switch network

Steps 2 and 3 of the design procedure involve two different linear programming models: minimum cost and maximum flow. The minimum-cost (MINCOST) model discussed in Section 3.3.1 finds the network that can satisfy the time-varying load demand with minimum network cost. The maximum-flow (MAXFLOW) model discussed in Section 3.3.2 maximizes the amount of demand satisfied by a fixed network configuration; that is, it minimizes the amount of additional trunk capacity required to satisfy the demand. Both models are set up as linear programming models, and the solutions are obtained from a linear programming solution package that solves linear programs by the revised simplex method, with or without an initial feasible basis solution. In this case an initial feasible basis solution of all direct routing is used to accelerate problem solution.

3.3.1 MINCOST linear programming model

Given a set of switches, a routing method, and switch-to-switch demands in trunks per hour, the minimum-cost linear programming model formulation is straightforward. The cost of the total network, a function of link distances and switching costs, must be minimized subject to two sets of constraints. First, for each switch-to-switch transport demand requirement for each hour, the demand must be completely satisfied on up to 27 paths—the direct link and 26 alternate triad sublinks. The model can be set up to provide multiple-link sublinks but in this example is restricted to simple two-link, or triad, sublinks. Second, all transport demands routed via one particular link in an hour, when summed, cannot be greater than the optimum size of that link. If the sum of the routed demand flow in any hour is less than the optimum link size, the "leftover" trunks are labeled spare capacity.

The following variables are used to describe the MINCOST linear programming model:

x_{ij} = the optimum size of the link between switches i and j

c_{ij} = $f_1 \times$ (distance between switches i and j) + f_2, where f_1 is a constant transport cost per mile and f_2 represents the cost of terminal multiplex and switch termination equipment

a_{ij}^h = the trunks required between switches i and j in hour h

p_{ikj}^h = the traffic of i–j demand that is routed through switch k in hour h (p_{ij}^h is the direct-link traffic)

t_{ij}^h = spare capacity in link i–j in hour h (link slack)

Given these variable descriptions, the MINCOST linear programming model is formulated as follows:

$$\text{Minimize} \sum_{i,j} c_{ij} x_{ij}$$

subject to

(1) $\quad p_{ij}^h + \sum_{k \neq i,j} p_{ikj}^h = a_{ij}^h \qquad$ for all switch pairs i–j in hour h

(2) $\quad p_{ij}^h + \sum_{m<i} p_{mij}^h + \sum_{m>i} p_{ijm}^h + t_{ij}^h = x_{ij} \qquad$ for all links i–j in hour h

for $n \leq 17$ hours, where all variables take on nonnegative values. The resulting solution is a set of link sizes and n sets of hourly routing patterns.

A fully connected 28-switch network would have 378 links that yield over 10,000 viable paths. Designing the network for all 17 hours of demand data yields nearly 13,000 constraint equations—that is, a $13,000 \times 13,000$ solution matrix (basis). Because the network model has nonzero demand for only 240 switch pairs, the number of type-(1) constraints is automatically reduced from $378 \times n$ to $240 \times n$, which still yields over 10,000 constraints for 17 hours.

Two steps are taken to increase computation efficiency. First, triads are allowed only if the total distance of the triad sublink is not greater than five times the distance of the direct link. Second, the network is initially designed only for three peak hours. Referring to Figure 3.2, those three hours are 10 A.M., 1 P.M., and 8 P.M., and the MINCOST model solution is denoted as the 10–1–8 network.

Regardless of the specific model used, solving linear programming models is a two-step process. First, an initial feasible basis solution is determined; then that solution is optimized. Solution time can be reduced if the model is given a good initial basis, and for the MINCOST linear programming model we construct an initial basis where all 240 switch-to-switch demands are carried on 240 direct links with no sublinks. As pointed out later, the MAXFLOW linear programming model can be used to produce an improved initial basis.

3.3.2 MAXFLOW linear programming model

The MINCOST linear programming model is used to design the 28-switch network for the three peak hours. It remains to extend this solution to satisfy the demands in the remaining 14 hours by introducing the MAXFLOW linear programming model, step 3 of the design approach. This model assumes that the link sizes are fixed and, by choosing new paths for the traffic demands, tries to maximize the demand satisfied in some particular hour h on existing link capacity and thereby determines minimum required augment capacity to satisfy the side-hour demands. The MAXFLOW linear programming model is very similar to MINCOST, with one new variable introduced: s_{ij}, which represents the demand on path i–j that cannot be satisfied in hour h (path slack). The MAXFLOW linear programming model follows:

$$\text{Minimize} \sum_{i,j} s_{ij} c_{ij}$$

subject to

$$(1) \quad p_{ij} + \sum_{k \neq i,j} p_{ikj} + s_{ij} = a_{ij} \qquad \text{for all switch pairs } i\text{–}j \text{ in hour } h$$

$$(2) \quad p_{ij} + \sum_{m < i} p_{mij} + \sum_{m > i} p_{ijm} + t_{ij} = x_{ij} \qquad \text{for all links } i\text{–}j \text{ in hour } h$$

where all variables take on nonnegative values.

The x_{ij}'s are the fixed link sizes determined by the MINCOST solution for the 10–1–8 network. Constraint (1) states the s_{ij} definition: Path slack is that portion of path demand a_{ij} that cannot be carried on the network, which also represents necessary augment capacity to satisfy the demand. Constraint (2) states that the sum of the demands carried on link i–j cannot be greater than the existing link capacity.

The MAXFLOW linear programming model can be used to find a better starting basis for the MINCOST problem. Recall that the initially chosen solution routes all the demands in each hour exclusively on direct links. The MAXFLOW linear programming model is run twice to produce a nonoptimal three-hour network with some sublinks. By running the 1 P.M. demand against the 10 A.M. network (which exactly meets the 10 A.M. demand with direct trunks), a two-hour network results. Running the 2 P.M. demand against the 10–1 network produces a three-hour solution, which is then offered to MINCOST as an initial basis. Because this new basis already contains some sublinks, the MINCOST solution reaches optimality more quickly. The 17-hour solution is then extrapolated using the MAXFLOW linear programming model as before.

There are also inefficiencies due to linearizing the problem early in the design process. These arise because of sublinks and also because of possible inefficient routing of switch-to-switch traffic demands versus routing of transport demands. The route-erlang flow optimization model presented in Chapter 4 eliminates these inefficiencies.

3.4 Modeling Results

The MINCOST linear programming model for designing a dynamic hierarchical routing network produces a set of link sizes and three sets of routing tables. Overall network cost savings are on the order of 12 percent, but some links exhibited particularly dramatic results. The Los Angeles–Freehold link is an excellent example. Figure 3.3 shows the hourly trunk requirements for this link. For the three hours evaluated, the transport demands are 17 trunks (10 A.M. set), 25 trunks (1 P.M.), and 9 trunks (2 P.M.). The maximum requirement for this link over all 17 hours is 25 trunks—the demand at 1 P.M. and 2 P.M. By coincidence, the hierarchical network design solution also results in exactly 25 trunks, which

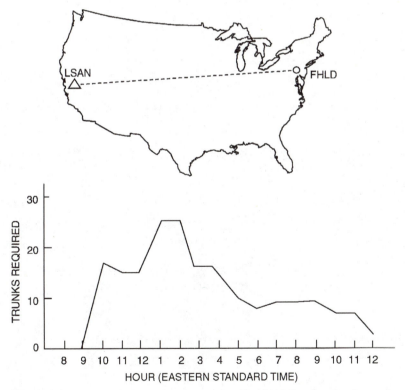

Figure 3.3 Hourly network trunk requirements (Freehold–Los Angeles demand)

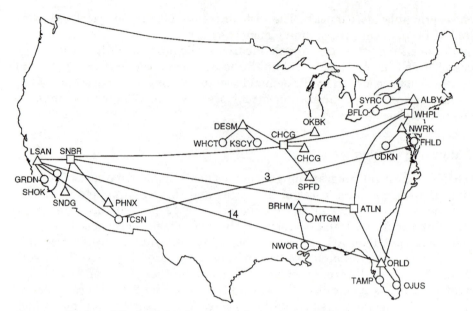

Figure 3.4 Routing of Freehold–Los Angeles trunks at 10 A.M. (trunks required = 17)

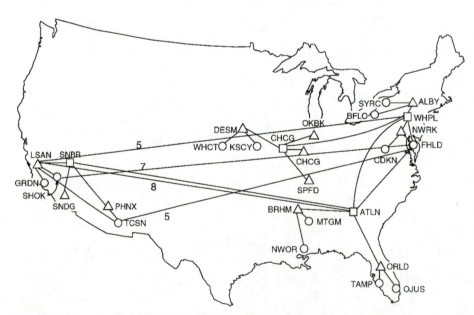

Figure 3.5 Routing of Freehold–Los Angeles trunks at 1 P.M. (trunks required = 25)

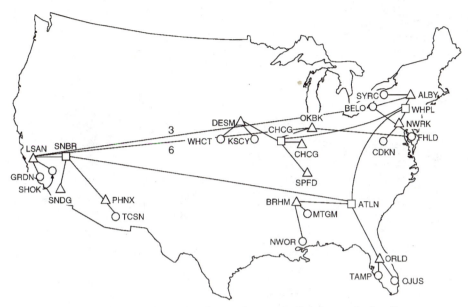

Figure 3.6 Routing of Freehold–Los Angeles trunks at 8 P.M. (trunks required = 9)

illustrates how fixed hierarchical network design techniques provide nearly the maximum requirement in order to meet peak demand with a fixed routing method, whereas the MINCOST linear programming model solution indicates 0 trunks to be installed between Los Angeles and Freehold.

Figures 3.4–3.6 show the individual hourly solutions; Figure 3.7 shows the composite three-hour solution. The numbers in these figures are rounded to integers; the actual results are summarized in Table 3.1.

Figure 3.4 shows the particular solution for 10 A.M. The 17-trunk requirement is divided between two paths, one through Tucson (2.8 trunks) and one through Orlando (14.2 trunks). Figure 3.5 shows the 1 P.M. solution. The demand of 25 trunks is split among four paths: Tucson (5.2 trunks), Atlanta (8.3 trunks), Sherman Oaks (7.0 trunks), and White Plains (4.5 trunks). Figure 3.6 shows the results for the third hour, 8 P.M. Again, only two paths are required to route all the demand: Oakbrook (6.4 trunks) and Buffalo (2.6 trunks). Figure 3.7 shows a composite three-hour result for the Los Angeles–Freehold link. During the three hours, seven different paths are chosen, but the direct route is always omitted. This appears to be a great savings, removing a 3,000-mile link, but the capacity of other links must be increased to carry the Los Angeles–Freehold demand. Because the larger links have their capacity more effectively shared, the overall solution shows a significant reduction in network cost.

The MAXFLOW linear programming model is then applied to the next five highest demands in descending order: 2 P.M., 3 P.M., 9 P.M., 7 P.M., and 11 A.M.

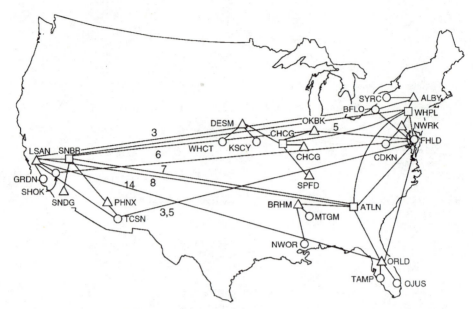

Figure 3.7 Composite 3-hour routing of Freehold–Los Angeles trunks at 10 A.M. (17 trunks), 1 P.M. (25 trunks), and 8 P.M. (9 trunks)

TABLE 3.1   Los Angeles–Freehold 3-Hour Routing Patterns

10 A.M.	
via Tucson	2.8 trunks
via Orlando	14.2 trunks
	17.0 trunks = demand

1 P.M.	
via Tucson	5.2 trunks
via Atlanta	8.3 trunks
via Sherman Oaks	7.0 trunks
via White Plains	4.5 trunks
	25.0 trunks = demand

8 P.M.	
via Oakbrook	6.4 trunks
via Buffalo	2.6 trunks
	9.0 trunks = demand

TABLE 3.2   2 P.M. MAXFLOW Results

Link	Additional trunks
Oakbrook–Buffalo	8.2
Chicago 59–Syracuse	8.2
Cedar Knolls–Newark	8.9
Chicago 11–Freehold	3.6
Chicago 59–Freehold	8.0
Chicago 11–Cedar Knolls	2.1
Orlando–Tampa	12.2
Chicago 11–Springfield	79.5
Chicago 11–Chicago 59	28.2

When evaluating the three-hour MINCOST linear programming model solution against the 2 P.M. demand, it is found that some demand could not be satisfied. For those $s_{ij} > 0$, s_{ij} trunks are added to x_{ij}. Table 3.2 shows the specific values obtained for the 2 P.M. run. The table compares the 10–1–8 network with the 2 P.M. demand and shows that additional capacity is needed on nine links. For instance, the Chicago 11–Springfield final link is increased by 79.5 trunks. At 3 P.M. no additional trunks are needed. For the 10–1–8–2–3 network design and the 9 P.M. demand, an increase in capacity is required on only one link, Phoenix–Tucson (8.0 trunks). For the 10–1–8–2–3–9 network design and the two remaining hours, it is found that this network can, indeed, satisfy the demand in all the outside hours.

For this combined MINCOST-MAXFLOW linear programming network design, Figure 3.8 shows the overall trunk-sizing results. The top line represents the total network transport cost for the hierarchical network design for the 28-switch network model. The curve represents the transport network cost for the hierarchical network design for each hourly demand pattern. The maximum hourly demand occurs at 8 P.M. Because the optimum network must be able to carry this peak demand, the 8 P.M. network can be considered a lower limit on the minimum-sized network possible. As discussed in Chapter 1, the difference between this limit and the multihour hierarchical design shows that the upper limit on achievable savings is about 20 percent. This bound is not achievable because the 8 P.M. network cannot, in fact, satisfy the demand at the other hours. The line labeled Dynamic Hierarchical Routing Network represents the MINCOST-MAXFLOW linear programming model solution, which has a cost approximately 12 percent lower than the hierarchical network design.

Multihour engineering [Eis79] techniques for hierarchical network design yield savings due to noncoincidence of about 1 to 2 percent in the 28-switch model compared with the cluster busy-hour hierarchical design techniques discussed in Chapter 2.

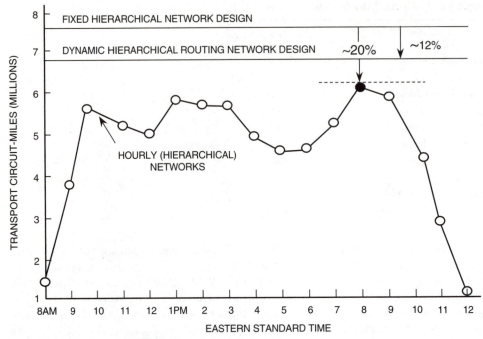

Figure 3.8 Transport circuit-miles for 28-switch network

3.5 Conclusions

A 28-switch intercity network model is used to study a linear programming model to design dynamic hierarchical routing networks. A linear programming model cannot be used to directly model trunk sizing starting from traffic demands because the traffic engineering constraints are not linear. Therefore, single-hour trunk sizing is used to determine sets of hourly transport demand requirements. These hourly networks are sized according to hierarchical routing and standard design criteria. Two linear programming models are presented to determine the minimum network cost needed to satisfy all the hourly transport demands while permitting one- or two-link sublinks. The sublink patterns are changed for each hour, resulting in a dynamic hierarchical routing design. Overall results produce a 12 percent reduction in network cost compared with the hierarchical network design.

Preplanned Dynamic Multilink Path Routing Networks

4.1   Introduction

This chapter considers a route-erlang flow optimization (EFO) model to design near-minimum cost networks that allow preplanned dynamic multilink path routing. We focus on the application of this model to preplanned dynamic multilink path routing, and in Chapter 5 we consider the application of the route-EFO model to preplanned dynamic progressive routing networks. This model attempts to incorporate nearly all concepts known to yield efficient network design into a single design procedure. These concepts include

- The economic-hundred-call-second (ECCS) method for efficient blocking levels

- Preplanned, time-sensitive, dynamic routing to take advantage of traffic non-coincidence

- Favoring the establishment of larger, more efficient links

- Routing over least-cost paths
- Minimizing incremental network cost

The route-EFO model designs near-minimum-cost, preplanned dynamic path routing networks meeting a required switch-to-switch blocking probability grade-of-service in one or more design hours. Input parameters for the model are the required switch-to-switch blocking probability grade-of-service objective and the switch-to-switch offered load in each hour of interest.

We show that substantial savings are possible with the route-EFO model. In the single-hour stationary load case, network design savings are on the order of 5 to 7 percent. These savings are due to the improved routing made possible by allowing a nonhierarchical routing plan instead of constraining the design to hierarchical routing. Network design savings for multihour nonstationary loads are even larger due to economies made possible by traffic noncoincidence. The total savings obtained for intercity network models are on the order of 15 percent. Network design results indicate only a very small cost penalty for constraining path selection in the route-EFO model to paths with a maximum of two links. As there are many more possible two-link paths in a larger network such as a 28-switch network than in a smaller network such as a 10-switch network, one would expect that the penalty for the two-link constraint would decrease for larger networks. Indeed, this is true, as almost no cost penalty is found for employing two-link routing in a 28-switch network as opposed to allowing multilink unconstrained paths. Hence, for networks larger than 28 switches, it could be expected that two-link path selection will provide networks comparable in cost to those networks designed using the much more complicated multilink path routing. Because the routing possible with multilink paths is about the most flexible routing possible, we find that little is gained by employing anything more complicated than two-link dynamic routing. Implementation of two-link routing is quite simple, as discussed in this chapter.

4.2   Preplanned Dynamic Multilink Path Routing Methods

Terminology used in the route-EFO model is illustrated in Figure 1.2. The term *link* refers to a connection between two switches; shown in Figure 1.2 is a link between switches A and B. The term *path* denotes one or more links arranged in tandem to route a call between the endpoints of a path. A two-link path between switches A and B composed of links A–C and C–B is shown in Figure 1.2. Because a blocking probability grade-of-service objective must be met and the blocking seen by a call traversing one individual path may not be acceptable, it is necessary to combine paths in parallel to form routes. Hence, the term *route* refers to a collection of paths with a routing discipline for routing calls between paths so that the overall blocking probability grade-of-service objective is met. For example, Figure 1.2 shows a route comprising three paths (A–B, A–D–B, and A–E–B). If a call is blocked on path A–B, it tries path A–D–B. If a call is blocked on

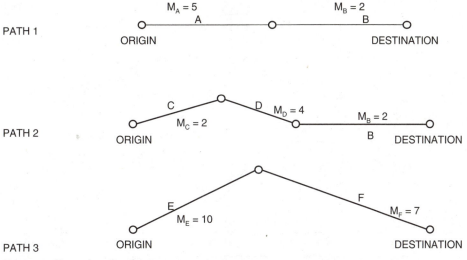

Figure 4.1 Examples of paths

path A–D–B, either on link A–D or link D–B, it returns to the originating switch A and tries the third path, A–E–B. Calls that are blocked on the third path are lost. Routing discussed in this chapter is unconstrained multilink and two-link path-to-path routing. In Chapter 5, the route-EFO model is used to study routing methods following hierarchical rules and progressive routing rules. Indeed, the generality of the route-EFO model is such that almost any routing method can be incorporated into it.

For each switch pair, unconstrained path-to-path routing selects paths; for example, Figure 4.1 shows the first three paths from a particular origin to a destination switch pair that could be selected. Note that a path is one or more links arranged to go from the origin to the destination. A sequence of these paths then forms a route, which is defined as a collection of paths arranged so that the percent of traffic finding all paths busy in the route is less than or equal to the required blocking probability grade-of-service objective for the design traffic loads. The resulting routing table from this combination of paths is termed unconstrained, or "multilink," path-to-path routing because overflow from one path is offered directly to another path and there are no constraints on the selection of paths. As an example, Figure 4.2 shows the route resulting from the addition of the three paths in Figure 4.1. CCS crankback and originating switch control is assumed with this routing method because, when a call is blocked on any link, it is cranked back to the originating switch (or the nearest common switch between the path it is on and the next path) and tries the next path. If a call finds all paths busy, it is blocked. Notations shown in Figures 4.1 and 4.2 are further explained in Section 4.3.

Multilink path routing allows any path, with or without common links with other paths, to be included in the route. A constraint that greatly simplifies the

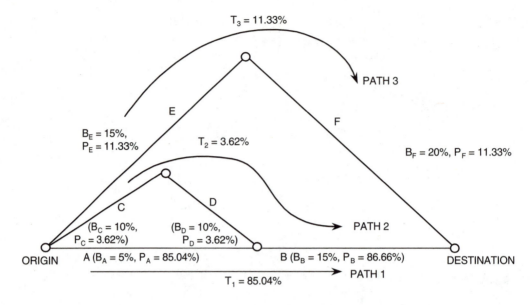

B_I = BLOCKING OF I^{TH} LINK

T_I = PROPORTION OF ROUTE CARRIED LOAD CARRIED BY PATH I

P_I = PROPORTION OF ROUTE CARRIED LOAD CARRIED BY LINK I

(B_I IS GIVEN, P_I AND T_I CALCULATED AS IN TEXT)

Figure 4.2 Route example

routing method is to allow only one or two links in any path. This eliminates common links and associated routing complexities.

Generally, more than one route is used by each switch pair for different design hours in order to take advantage of traffic noncoincidence. One way to accomplish this is with ordered routing that uses as its first route $1, 2, \ldots, M_1$, where the notation (x, y, z) means all traffic is offered first to path x, which overflows to path y, which overflows to path z. Here, M_1 denotes the number of the last path added to the first route so that the switch-to-switch blocking probability grade-of-service objective is met. The second route could then be $2, 3, \ldots, M_2$, in which case the third route would be $3, 4, \ldots, M_3$, and so on. Another approach is cyclic routing, which has the same first route as ordered routing; that is, the route $1, 2, \ldots, M_1$. The second route of cyclic routing is a cyclic permutation of the first route: $2, 3, \ldots, M_1, 1$. The third route is likewise $3, 4, \ldots, M_1, 1, 2$, and so on. The route blockings of cyclic routes are identical; what is varying from route to route is the proportions of flow on the various links so that the linear programming (LP) step in the route-EFO model can allocate traffic to equalize link loads.

In path-to-path routing, the objective is for a call to be routed from the origin to the destination along the first free path. If the call encounters a busy link

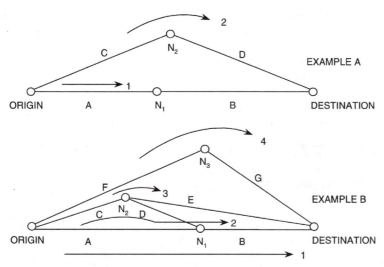

Figure 4.3 Path-to-path routing examples

in the path, it may be cranked back to the originating switch. The originating switch may then release all unneeded links on the incomplete path and try the next path or else block the call if all paths in the route have been tried. A CCS message conveys the crankback information to previous switches. For example, a call blocked on link B (on path 1) of the route shown in Figure 4.3A is cranked back from switch N_1 to the origination switch and will then try path 2, consisting of links C and D. A slightly different situation is shown in Figure 4.3B, where a call blocked on link D in path 2 really need not be cranked back to the origin but can be sent directly over link E to the destination, because switch N_2 is also on path 3. If link E is busy, the call is returned to the origin to try path 4. Another situation is also shown in Figure 4.3B, where a call blocked on link B of path 1 need not be offered to path 2 because link B is also on path 2; instead, such a call is routed directly to path 3. Multiple-link crankback in which the call is returned to the originating switch or an intermediate switch over several links is also a possibility. Crankback, then, is the process of returning control of a call to a previous switch in a path and returning unneeded trunks to the idle state. If call control is returned to the originating switch, it is said that originating switch control is used, because the originating switch will make the decisions regarding the progress of the call.

What complicates unconstrained path-to-path routing is that paths of more than two links are allowed and may cause problems with common links, as shown in Figure 4.4B. Note that the route shown in Figure 4.4B has calls traveling in both directions over the link from switch N_1 to switch N_2. Which link is to be used from switch N_2 or switch N_1 depends on which path the call is on. When

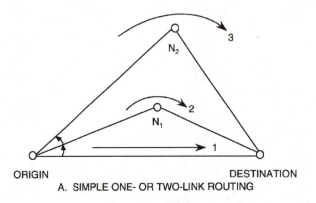

A. SIMPLE ONE- OR TWO-LINK ROUTING

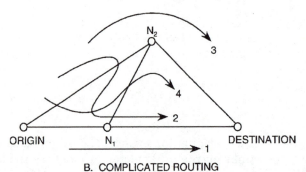

B. COMPLICATED ROUTING

Figure 4.4   Simple and complicated routes

allowance is made for permutations of the paths to form various routes, it can be seen that the information required to route a call at an intermediate switch of an unconstrained path-to-path route can be considerable, unless the call is always cranked back to the source and tries another path.

From this discussion it appears that path-to-path call routing can be rather complicated due to links common to two or more paths. That is not always the case for the simple route shown in Figure 4.4A, where the first path is the direct link and each alternate path contains two links. With such a route there can be no common links and an intermediate or via switch need only crank a blocked call back to the originating switch. The originating switch can have complete control and may advance any blocked calls to the next path. Central to this routing method is the assumption that most traffic is carried by routes of one or two links. Network designs for a single hour show that about 98 percent of the traffic in a multilink path design is carried on one- or two-link paths, while only about 2 percent of the traffic is carried on paths of greater length. Since so little traffic is carried by paths prohibited by this constraint, network cost and efficiency are nearly unchanged from unconstrained path-to-path routing.

4.3   Network Management and Design

4.3.1   Network design concepts

Several concepts have led to well-designed networks in the past, which are included in the route-EFO design model. A brief description of each concept is given here, along with references to applications that employ each concept.

4.3.1.1   Favor large trunk groups.   Previous studies show routing techniques are most efficient when they allow calls access to all circuits between the two switches, thereby effectively combining the traffic-handling capability of several splinter links into one larger, more efficient link. It is known from Erlang's model that there is a larger inherent efficiency of large links in terms of average link occupancy at a given level of link blocking. This basic approach is also used in the route-EFO model as now explained. Figure 4.5 shows a qualitative relation between the number of trunks, N, required to carry a particular offered load, a, at constant blocking. From the shape of the curve comes the fact that at constant blocking the number of additional trunks required to carry an increment of offered load decreases as the link size increases. In other words, the slope of the curve is decreasing as a increases. Hence, it is advantageous to combine several traffic items into one large traffic item to be routed over a

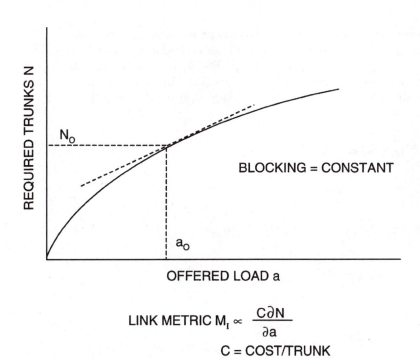

Figure 4.5   Efficiency of large links

large link because one large link is inherently more efficient than several smaller links. In the route-EFO model, establishment of larger links is favored through the use of a link-incremental cost metric proportional to the slope $\partial N / \partial a$ of the Figure 4.5 curve. Attached to the Ith link is a link metric number M_I, which indicates the attractiveness of this link to carry additional traffic. As discussed in Section 4.3.3, the actual link metric expression used for path selection contains an additional term intended to account for the cost of carrying overflow traffic, but the link metric is still dominated by the $\partial N / \partial a$ term.

4.3.1.2 Route on least-cost paths.

Figure 4.6 gives a simple example of routing calls along the least-cost path to their destination. In city 1, switch A homes on switch B, which in turn homes on switch C. City 2 has the single switch D. A call arriving at B in city 1 tries the B–D link first, as is normal. However, if B–D is busy, the call could be routed to switch A in order to try link A–D instead of being routed to C and following the normal hierarchical routing table from C, which is perhaps more expensive. It is assumed that if link A–D is busy, the call could be returned to B and then sent to C, which is the process called crankback. Note that the originating switch B can be said to have control of the call because if the call is cranked back to B, B will decide to route the call to C, because link B–D has already been tried. Hence, the use of crankback and originating switch control can ensure that paths to the destination are tried in the most economical order.

4.3.1.3 Use time-sensitive routing to take advantage of traffic noncoincidence.

We now emphasize procedures that take advantage of traffic noncoincidence. The multihour engineering method of Eisenberg [Eis79] for hierarchical networks is an attempt to equalize the traffic load on final links by considering the time variation of traffic offered to a high-usage link in sizing that link. Here

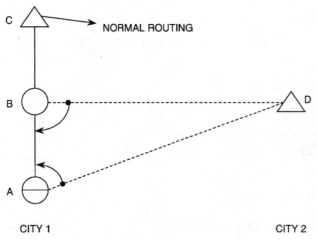

Figure 4.6 Route on least-cost paths

the routing remains invariant with time. In Chapter 3 we considered dynamic hierarchical routing to exploit traffic noncoincidence by connecting two unused links in tandem and assigning the resulting capacity to a heavily loaded link in a triad arrangement. This design method determines triad arrangements among high-usage and final links, in which the number of required trunks for each hour is still determined by hierarchical design methods but the routing of those trunks is changed by hour to take advantage of traffic noncoincidence. An example of the effect of traffic noncoincidence is shown in Figure 4.7. Here the three switches A, B, and C are fully interconnected. The traffic load on links A–C and B–C peak in the morning, whereas the load on A–B peaks in the evening when the other loads are minimal. Assuming the A–C and B–C links are sized for their morning busy hour, one way to save link capacity in this network is to install only enough link capacity on link A–B to meet the small A–B morning load, and meet the evening A–B traffic requirement by routing A–B calls through C because there is spare capacity in the A–C and C–B links in the evening. Hence, changing the routing for A–B calls from trying the A–B link alone in the morning to trying the A–B link and then attempting A–C–B can result in a decrease in size of the required A–B link. In effect, this routing change tends to equalize the link loads during the various load periods and to reduce spare capacity by keeping all trunks busy at all times. Dynamic hierarchical routing, described in Chapter 3, takes advantage of noncoincidence in much the same manner as the example: Sublinks from A–C and C–B are connected together at C in the evening, and the resulting trunks are added to the A–B link. Eisenberg's multihour engineering approach also tries to equalize link loads, but in this case the routing remains fixed while the sizes of high-usage links are chosen to result in an approximately equal final load in all hours.

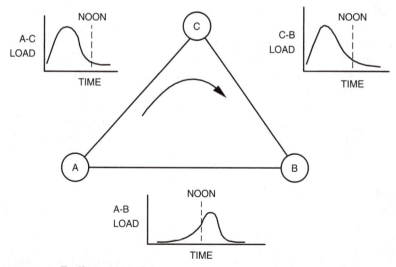

Figure 4.7   Traffic noncoincidence

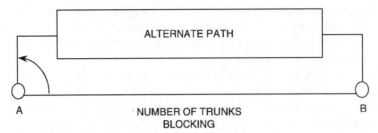

Figure 4.8 Efficient blocking levels

4.3.1.4 Use efficient blocking levels. Figure 4.8 illustrates a common question arising when designing an alternate routing network; given a direct link between A and B, and the alternate path that overflow calls follow, what is the optimum value of blocking (or, equivalently, the link capacity) to handle the offered load at a minimum cost? This question was first answered by Truitt [Tru54], who derived the concept of an economic hundred call seconds (ECCS) based on the direct-path-to-alternate-path cost ratio and the marginal capacity of the alternate path. Truitt's ECCS method, which is derived in Chapter 2, is commonly used in hierarchical network design for metropolitan, intercity, and international networks. The concept that is important in this case is the existence of an optimal blocking level for links, which can be calculated based on network parameters.

4.3.1.5 Minimize incremental network cost. Network cost and performance are related by very nonlinear relationships. Hence, the network design problem is inherently a nonlinear programming problem. To avoid the complexities associated with nonlinearities, the network cost function can be linearized around the present operating point and the linearized cost function (incremental cost) minimized to yield a minimum-cost network. This approach of minimizing the incremental cost network function has been successfully used by others and has led to near-minimum-cost networks. In particular, Yaged [Yag71,Yag73] uses this technique to find a near-minimum-cost network for the problem of routing communications channels (trunks) over links displaying a concave transport-cost-versus-channel-capacity relationship. Yaged shows this technique satisfies the Kuhn-Tucker conditions for local optimality. Much the same approach is used by Knepley [Kne73], who applies the minimal incremental cost concept to the design of the AUTOVON network.

**4.3.2 Overview of the route-erlang
flow optimization model**

In this section the methodology of the route-EFO model is explained. The route-EFO model is an iterative method, and Figure 4.9 shows the iteration loop of the route-EFO model. Basic input parameters include link cost, switch-to-switch offered loads, and required switch-to-switch blocking probability grade-

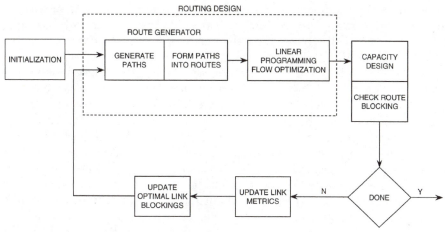

Figure 4.9   Route-erlang flow optimization model

of-service objective. The iteration starts with an initialization step that sets all link blockings equal to an input parameter and sets the metric for each link (M_I^* for the I th link) equal to the cost of a trunk on that link. The next stage in the iteration method is the routing design, whose function is to find efficient ways to route traffic through the network being designed. The routing design step has two parts: a route generator and a linear programming (LP) model.

Inputs to the route generator are the optimal link blockings, B_{opt}^I, the link metrics, M_I^*, and also the required switch-to-switch blocking probability grade-of-service objective. Using these data, the route generator constructs minimum-cost traffic routes through the network that meet the required blocking probability grade-of-service in each hour. The route generator also calculates the proportion of any traffic load carried on each route that flows on each link in the route. In the multihour case, the route generator constructs several different routes for each switch pair for the purpose of providing the next stage of the iteration (the LP model) with a choice of where to assign traffic in each hour. It should be noted that the route generator does not assign traffic to the generated routes or consider traffic noncoincidence; that function is left to the LP model.

The inputs to the LP model following the route generator are route information including links in each route and proportion of total route carried load on each link, the link metrics M_I, and the hourly switch-to-switch offered loads. The difference between the metric used by the LP model and the metric used for path selection is explained in Section 4.3.3. The LP model will then choose the hourly assignment of route offered loads to minimize incremental cost. The output from the LP model is the amount of traffic to be carried by each route in each hour and the associated hourly link flows.

The next step in the route-EFO model is to size the network and determine the required number of trunks in each link. Inputs to the capacity design step are the route traffic assignments and link flows from the LP model, in particular the maximum link flow (a_I for the Ith link). If a_I is zero, the Ith link has been assigned no flow and is eliminated from the network design at this time because the cost of the eliminated link has risen to the point that it is too expensive relative to the other links. Because the route generator and LP model have used B_{opt}^I as the blocking of the Ith link, it is natural for the capacity design process to pick N_I such that the blocking on N_I trunks equals B_{opt}^I when the offered load is a_I. In other words, the network is sized to achieve the previously used blocking level for the maximum hourly load. In the multihour case, setting N_I to match the required blocking in one hour may cause a difference between achieved blocking and the blocking assumed in the other hours; hence, it is necessary in the multihour case to iterate within the capacity design step until the off-hour blocking and link offered loads agree. The output of the capacity design step is the number of trunks in the network, and, because the blockings are all equal to or less than the assumed blocking in the routing design step, the network should satisfy the blocking probability grade-of-service requirement.

If the route-EFO model has not converged to a minimum-cost network, the next step is to iterate to improve the network design. We calculate new metrics to reflect the new size and loads of each link in the network design. Links that have become larger should have a smaller metric because they should be more efficient, and links that have become smaller should have a larger metric. The last step is to calculate an updated set of optimal link blockings that account for the cost of alternate routing versus the cost of carrying traffic on the direct link. The ECCS approach to Truitt [Tru54], as derived in Chapter 2, is used to minimize network cost given the current network parameters. An average ECCS is calculated over all routes using each link in the busiest hour of that link. Once a new set of blockings and metrics is obtained, another iteration begins.

4.3.3 Detailed description of the route-erlang flow optimization model

4.3.3.1 Initialization. The initialization phase is quite simple: The network is assumed fully connected with all link blockings equal to an input parameter (usually 1.5 percent) and with each link metric set equal to the cost of a trunk on that link. The routing design and LP model steps are then executed, but the capacity design step is skipped. The reason for initially skipping over the capacity design phase is that there is little reason to spend time converging on a feasible solution at this point because the initial blockings are probably far from optimum. Instead, new blockings and metrics are calculated from the flows obtained from the LP model and the initial routing. These flows are only approximate because the capacity design step has not produced a network of consistent blocking and flows in all hours. However, the flows are sufficiently accurate to be used to compute much better blocking values and much better link metrics. The resulting blockings and metrics are then passed to the routing design step and the iteration proceeds normally.

4.3.3.2 Routing design. The routing design step is composed of two operations, a route generator and an LP model, which are discussed separately.

Route generator. The route generator discussed here is an unconstrained multilink path-to-path route generator. A special case of this type of route generator is one constrained to one- or two-link paths, which provides an easier implementation. Other route generators follow hierarchical rules, as discussed later, or progressive routing rules, as discussed in Chapter 5.

For each switch pair, the unconstrained path-to-path route generator begins by selecting the shortest path, second shortest path, third shortest path, ..., kth shortest path from the origin to the destination in the current network based on the metric of each link. For example, Figure 4.1 shows the first three paths from an originating switch to a destination switch that could be selected. Note that a path is one or more links arranged to go from the origin to the destination and each path is selected and rank-ordered based on the incremental cost (M_I^* for link I) of the links it contains. A standard k-shortest path model is used here, in which the path selection considers noncoincidence by having the link metric reflect traffic noncoincidence as discussed later in this section.

The next task of the route generator is to combine these paths into a route. As discussed earlier, a route is a collection of paths and a routing discipline arranged so that the percent of traffic finding all paths busy in the route is less than or equal to the required blocking probability grade-of-service. One way of arranging the k-shortest paths is to offer all the traffic to the shortest path and, if the blocking on the shortest path is too high to satisfy the desired blocking probability grade-of-service, take the overflow traffic from the shortest path and offer it to the second-shortest path. Then, if the blocking of the first two paths in parallel is still too high, take the overflow traffic load from the second-shortest path and offer it to the third-shortest path, and so on until the blocking probability grade-of-service objective is satisfied.

The resulting routing pattern from this combination of paths is termed unconstrained or multilink path-to-path routing because overflow from one path is offered directly to another path and there are no constraints on the selection of paths. As an example, Figure 4.2 shows the route resulting from the addition of the three paths in Figure 4.1. CCS crankback and originating switch control is assumed with this routing method because, when a call is blocked on any link, it is cranked back to the originating switch, or the nearest common switch between the path it is on and the next path, and tries the next path. If a call finds all paths busy, it is blocked. As the switch-to-switch blocking of a route utilizing such a path-to-path routing method is the same in both directions, it is only necessary to construct one set of routes for each switch pair. A constraint that simplifies the path routing method is to allow only one or two links in any path, which eliminates common links and their associated routing complexities. Section 4.4 gives network cost comparisons between two-link path designs and multilink path designs.

As noted previously, it is necessary to generate more than one route for each switch pair in the multihour cases in order to give the LP model a choice of routes to take advantage of traffic noncoincidence. Multiple-candidate routes can be generated with an ordered route design or a cyclic route design, both of which are explained in Section 4.2. The cyclic route design has computational advantages because its cyclic structure requires considerably fewer calculations to find the proportions for all routes than does a collection of paths as found in the ordered route design.

Calculation of route blocking and link proportions. Once a route has been found, such as the one in Figure 4.2, the proportion of route carried load that is carried by each link is calculated. Figure 4.2, although not at all a typical route (a typical route usually involves two-link paths and no common links), does illustrate that some links may be common to more than one path and hence route-blocking calculations and route carried flow calculations can become complex. To calculate route blocking and link carried load proportions for multilink path-to-path routing, we first calculate the probability that a free circuit exists in at least one path when links may be common to one or more paths. The resulting graph is not necessarily series-parallel [Lee55]. We first give the methodology followed by the example shown in Figure 4.2.

To define notation, let

$$
\begin{aligned}
S_i &= \{\text{Event there is a free circuit in path } i\} \\
S_i \cup S_j &= \{\text{Event there is a free circuit in path } i \ or \ \text{path } j\} \\
S_i S_j &= \{\text{Event there is a free circuit in path } i \ and \ \text{path } j\} \\
p_a &= \text{Blocking probability of link } a \\
q_a &= 1 - p_a = \text{Connectivity of link } a
\end{aligned}
$$

Assuming that all blocking probabilities are constant and independent of offered load and of each other, and that all traffic loads experience the same blocking probability on each link, we wish to calculate $P(S_1 \cup S_2 \ldots \cup S_N)$, where S_i, $1 \le i \le N$, are the N paths in the route. This can be accomplished by first expressing the union probability in terms of intersection probability [Fel57]:

$$
P(S_1 \cup S_2 \ldots \cup S_N) = \sum_i P(S_i)
$$

$$
- \sum_{i \neq j} P(S_i S_j) + \sum_{i \neq j \neq k} P(S_i S_j S_k) - \cdots \pm P(S_1 S_2 \ldots S_N)
$$

$$
1 \le i, j, k \le N \tag{4.1}
$$

The intersections shown are thus taken over all possible combinations of the set intersections taken $1, 2, 3, \ldots, N$ at a time.

We now illustrate with the route shown in Figure 4.2, which is constructed from the paths shown in Figure 4.1. The route blocking for the three-path route is then

$$F_3 = P(S_1 \cup S_2 \cup S_3)$$
$$= P(S_1) + P(S_2) + P(S_3) - P(S_1 S_2) - P(S_1 S_3) - P(S_2 S_3) + P(S_1 S_2 S_3)$$

$$(4.2)$$

where F_j denotes the blocking of the first j paths taken together. To compute the intersection probabilities of these three paths, note that path 1 has links (A, B), path 2 has links (C, D, B), and path 3 has links (E, F). Using the link-blocking probabilities shown in Figure 4.2, we find the probability of a free circuit in a single path to be

$$P(S_1) = q_A q_B = (.95)(.85) = .8075$$
$$P(S_2) = q_C q_D q_B = (.9)(.9)(.85) = .6885$$
$$P(S_3) = q_E q_F = (.85)(.8) = .68$$

To compute the probability of a free circuit in path 1 and path 2, or $P(S_1 S_2)$, note that due to the independence assumption the result still involves a product of the connectivities of all the links in both paths, but once the connectivity of a link is included in this product that link is considered idle for all other paths. Hence, the required probability is the product of the connectivities of all the links on both paths, with the stipulation that no connectivity term is repeated. Then,

$$P(S_1 S_2) = q_A q_B q_C q_D = (.95)(.85)(.9)(.9) = .6541$$

and likewise,

$$P(S_1 S_3) = q_A q_B q_E q_F = (.95)(.85)(.85)(.8) = .5491$$
$$P(S_2 S_3) = q_C q_D q_B q_E q_F = (.9)(.9)(.85)(.85)(.8) = .4682$$
$$P(S_1 S_2 S_3) = q_A q_B q_C q_D q_E q_F = (.95)(.85)(.9)(.9)(.85)(.8) = .4448$$

In general,

$$P(S_1 S_2 \ldots S_M) = \prod_{(a \in S_i, 1 \le i \le M)}^{*} q_a$$

where $\prod^*$ signifies that no q_a is repeated. Then the blocking for the three-path route shown in Figure 4.2 is [from (4.2)]

$$F_3 = P(S_1 \cup S_2 \cup S_3)$$
$$= .8075 + .6885 + .68 - (.6541 + .5491 + .4682) + .4448$$
$$= .9495$$

The route blocking probability for this three-path route is then $1 - .9495 = .0505$.

To compute the proportion of total route carried load carried on each link, the load proportion carried by each path is first computed. Then the total link carried load proportion is obtained as the sum of the path proportions for those paths containing the link in question. As an example, consider again the route shown in Figure 4.2. The load carried by the third path alone is the difference between the load carried by all three paths and the load carried by the first two paths. The proportion of offered traffic load carried by the first two paths is (from Equation 4.1)

$$F_2 = P(S_1 \cup S_2) = P(S_1) + P(S_2) - P(S_1 S_2)$$
$$= .8075 + .6885 - .6541 = .8419$$

The proportion of total route carried load carried by path 3 is then

$$T_3 = \frac{F_3 - F_2}{F_3} = \frac{.9495 - .8419}{.9495} = .1133$$

where T_j denotes the proportion of total route carried load carried by path j. Likewise, the proportion of offered load carried by path 1 above is

$$F_1 = P(S_1) = .8075$$

The proportion of total route carried load carried by path 2 is then

$$T_2 = \frac{F_2 - F_1}{F_3} = \frac{.8419 - .8075}{.9495} = .0362$$

Likewise,

$$T_1 = \frac{F_1}{F_3} = \frac{.8075}{.9495} = .8504$$

Let P_A denote the proportion of total route carried load carried by link A, and so on. Then the required proportions are

$$P_A = T_1 = .8504$$

$$P_B = T_1 + T_2 = .8866$$

$$P_C = P_D = T_2 = .0362$$

$$P_E = P_F = T_3 = .1133$$

Linear programming model. The second step in the routing design step is the LP model, which assigns the offered traffic to the previously generated routes in such a manner as to minimize the incremental network cost. The link-incremental cost factors used by the LP model are explained in Section 4.3.3.6. Here we use the following notation:

L = number of links
K = number of switch pairs
H = number of design hours
J_k^h = number of routes for switch pair k in hour h

v_k^h = offered load for switch pair k in hour h

P_{jk}^{ih} = proportion of carried load on route j for switch pair k on link i in hour h

M_i = incremental link cost metric in terms of dollar cost per erlang of carried traffic

b_{jk}^h = blocking on route j for switch pair k in hour h

B_i^h = blocking on link i in hour h

B_{opt}^i = optimal link blocking objective on link i

R_{jk}^h = carried load on route j for switch pair k in hour h

a_i = maximum carried load capacity on link i

Then the LP model step selects the R_{jk}^h and the resulting a_i so as to

$$\text{Minimize} \sum_{i=1}^{L} M_i a_i$$

subject to

$$\sum_{j=1}^{J_k^h} \sum_{k=1}^{K} P_{jk}^{ih} R_{jk}^h \leq a_i \qquad \begin{aligned} i &= 1, 2, \ldots, L \\ h &= 1, 2, \ldots, H \end{aligned}$$

$$\sum_{j=1}^{J_k^h} \frac{R_{jk}^h}{1 - b_{jk}^h} = v_k^h \qquad \begin{aligned} h &= 1, 2, \ldots, H \\ k &= 1, 2, \ldots, K \end{aligned}$$

$$R_{jk}^h \geq 0 \qquad a_i \geq 0$$

Inputs to the LP model are P_{jk}^{ih} and b_{jk}^h from the route generator, M_i from the previous metric calculation, the link blockings B_i^h, and the traffic demands v_k^h. Outputs from the LP model are R_{jk}^h, the assignment of offered load to the routes, and a_i, the associated link capacity, which is the maximum carried traffic load. Standard LP solver packages are used in the route-EFO model to find the LP model solution. All switch-to-switch traffic is first assigned to the least expensive route to form an initial feasible solution for the LP model.

4.3.3.3 Capacity design. After the LP model assigns traffic to routes, the network is sized to achieve a link blocking no higher than the assumed blocking used as input to the routing design module. In this way, the blocking probability grade-of-service objective is satisfied, or at least the route blocking is no worse than that calculated by the routing design step. If the route blocking is still not satisfactory, it can be corrected by the check route blocking step.

For the multihour case, the blocking values used by the routing design step are assumed to be equal in all hours. Hence, for a multihour network design it is sufficient to size to the hour of maximum offered load. This approximation is inaccurate, however, because the blocking in nonpeak hours may now differ substantially from the peak blocking. Hence, the overflow from the sized link may be considerably reduced, thus affecting the sizing of other links. To arrive at a consistent set of hourly blockings and offered loads, the iteration method shown in Figure 4.10 is used. Essentially, the iteration operates by using the present estimates of the offered loads to size each link in its peak hour of flow to the optimal link blocking objective, B_{opt}^i, and calculate actual blockings in each side hour. After all links have been sized, new proportions of carried load flow are calculated using the actual blocking in all hours. The routing given by the LP model is held fixed in the iteration, and the link flows are then recalculated and the process repeated. The iteration is continued until the sum total absolute change of blocking over all links in all hours is less than a prescribed limit. That is,

$$\sum_h \sum_i |B_i^{h'} - B_i^h| < \varepsilon$$

where $B_i^{h'}$ and B_i^h represent the blocking on the ith link in hour h for the previous and present iteration, respectively. For network design results discussed in Section 4.4, ε is set at 0.1. Sizing is accomplished either by ignoring peakedness and using the Erlang B blocking relation or by including the peakedness methods described in Chapter 6. Fractional trunks are allowed so as to achieve the required blocking exactly. These procedures simplify the iteration loop and speed convergence. Design for peakedness and integer trunks is discussed in Chapter 6.

If a link is unused in an iteration and hence is assigned zero flow, it is eliminated from the current network design and from subsequent consideration. The rationale behind this is that the process of routing according to incremental cost tends to favor larger, more efficient links over smaller, less efficient links. Hence, traffic tends to be more and more concentrated on larger links. When a link has zero flow, a means has been found to route traffic around the link at lower cost; hence, there is no need to keep the link because its cost will probably not

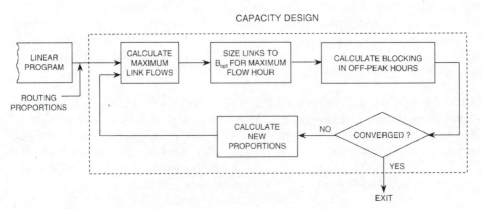

Figure 4.10 Multihour capacity design iteration

decrease and is now too high. This approach allows the fully connected network at the start of the design model to become more efficient through the elimination of inefficient links.

4.3.3.4 Check route blocking.

The philosophy behind the check route blocking step is straightforward: If the route blocking of the network exceeds a threshold, say B_{max}, the blocking on the first path is decreased until the route blocking is equal to B_{max}. The additional traffic that must be carried to reduce the route blocking to B_{max} is then carried on the path that has the minimum incremental cost, and the network cost increase required to correct the route blocking is minimized.

Let B_{old} be the route blocking of the route requiring correction, and let G_i be the blocking of the i th path. Assuming there are N paths in the route and the paths are constrained to a maximum of two links,

$$\prod_{i=1}^{N} G_i = B_{old} \tag{4.3}$$

or

$$G_1 \prod_{i=2}^{N} G_i = B_{old} \tag{4.4}$$

Letting G_1^* be the new first path blocking after blocking correction, it is desired that

$$G_1^* \prod_{i=2}^{N} G_i = B_{max} \tag{4.5}$$

or from (4.4) and (4.5),

$$G_1^* = \frac{B_{max}}{B_{old}} G_1 \tag{4.6}$$

Hence, the required blocking on the first path is thus found, and if the first path has a single link, its blocking is directly obtained. If the first path has two links, a choice must be made as to which link should have its blocking corrected. In this case, the link with the lowest incremental cost is increased in size, the incremental costs are recalculated, and the process repeated until the path blocking equals G_1^*. Once a link blocking is decreased to correct a route blocking, the blocking of that link is not increased again.

The capacity design step is allowed to converge to a solution before the route blockings are checked for possible corrective action. This is done because in the multihour case many route blockings are improved because of low blockings on some links during nonpeak hours. Once a network solution is obtained, the route blockings needing correction are rank-ordered and the highest route blocking is corrected first. Routes are processed in this order because the worst route blocking must be corrected anyway, and if its first path has any common links with the other routes needing correction, it is possible that these other routes may be given an acceptable blocking at the same time. After the new link blockings are obtained, the new number of required trunks is calculated based on the link blockings. The link blockings in the other hours are recalculated, and then new route blockings are calculated based on the new link blockings. Routes are once again checked for blocking violations, and the entire process is repeated until a network solution is found that does not violate the maximum route-blocking constraint. In the multihour case the blocking correction is part of the capacity design iteration shown in Figure 4.10 after initial convergence is obtained, in which blocking violations are corrected as the loop reconverges.

4.3.3.5 Convergence test. After the capacity design step a test is made as to whether this network is optimal or whether the design process should continue for another iteration. The iterative process is stopped when the new network design cost is no lower than the network design obtained one iteration earlier, which is then used as the final network design.

4.3.3.6 Update of link metrics. Once a newly sized network is obtained, new link-incremental costs or metrics are calculated to reflect link economies in the present network design. There are two terms in the link metric used for path selection, the first reflecting the incremental cost of carrying traffic directly on the link in question, and the second approximating the cost of carrying overflow traffic from the link in question on its various overflow paths. The metric used in the LP model only reflects the cost of carrying traffic directly on the link in question. The LP model uses proportions relating the route carried traffic to the resulting link flows on all links in the route to account for the flow and cost of overflow traffic.

Figure 4.11 shows the model used for the metric calculation. Here two routes use the link in question (link I). Route 1 has an offered load a_1 to link I, and route 2 has an offered load a_2 to link I. Both routes overflow traffic from link I to alternate paths: Route 1 overflows α_1 erlangs to path 1, and route 2 overflows α_2 erlangs to path 2. Each alternate path has its own incremental cost; these alternate-path costs are M_1 for route 1 and M_2 for route 2 in the example considered. The method for computing these alternate-path costs is identical to that used to compute the ECCS alternate-path costs; the individual link metrics used are from the previous iteration. There are N_I trunks in link I and the offered load is a_I.

The first term in the expression for the path selection link metric is $C_I \, \partial N_I / \partial a_I$, or the cost per trunk multiplied by the rate of change of trunks required to keep the blocking constant with a changing offered load. Hence, the first term is the incremental cost to carry an increment of offered load at constant blocking on link I itself. In particular, the partial derivative is calculated as

$$\frac{\partial N_I}{\partial a_I} = -\frac{\partial B_I / \partial a_I}{\partial B_I / \partial N_I} \tag{4.7}$$

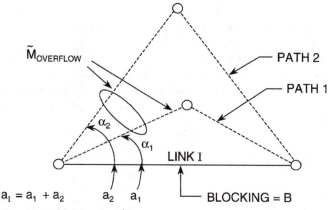

Figure 4.11 Metric calculation

about the current operating point (B_I, a_I). The numerator of (4.7) can be obtained from [Jag74]:

$$\partial B_I / \partial a_I = (N_I / a_I - 1 + B_I) B_I \tag{4.8}$$

As the denominator may involve expansion about a noninteger number of trunks, it can be approximated as

$$-\partial B_I / \partial N_I = [B(N_I - 0.1, a_I) - B(N_I, a_I)]/0.1 \tag{4.9}$$

where $B(N, a)$ is the Erlang B loss formula for N trunks offered a erlangs. In other words, the required derivative is obtained as the difference in blocking caused by the removal of one-tenth of a trunk.

The second term in the metric expression is intended to reflect the cost of carrying traffic blocked by the link in question on alternate paths. Letting $\widetilde{M}_1$ denote the cost of alternate path I, this cost is as follows:

$$\text{Total alternate cost} = \alpha_1 \widetilde{M}_1 + \alpha_2 \widetilde{M}_2$$

$$= (a_1 \widetilde{M}_1 + a_2 \widetilde{M}_2) B_I$$

To express this as a cost per erlang of offered load, define

$$\widetilde{M}_{\text{overflow}} = \frac{a_1 \widetilde{M}_1 + a_2 \widetilde{M}_2}{a_I} \tag{4.10}$$

as the average alternate path cost per erlang. The total path selection link metric M_I^*, which is used in the k-shortest paths selection in the route generator, then becomes

$$M_I^* = C_I \frac{\partial N_I}{\partial a_I} + B_I \widetilde{M}_{\text{overflow}} \tag{4.11}$$

It is noted that the average for $\widetilde{M}_{\text{overflow}}$ is taken only over those routes that overflow to an actual alternate path. If link I is a link on the last path on a route, the route would not be counted in the calculation of $\widetilde{M}_{\text{overflow}}$.

Because the LP model accounts for the cost of carrying overflow traffic by directly calculating the flow of overflow traffic on all links in each route, the expression for the metric used by the LP model only contains the first term of (4.11), which is divided by $(1 - B_{\text{opt}}^I)$ to yield the incremental cost per carried erlang in the hour of maximum flow, that is,

$$M_I = C_I \frac{\partial N_I}{\partial a_I} \bigg/ (1 - B_{\text{opt}}^I) \qquad (4.12)$$

where M_I is the LP model metric for link I.

4.3.3.7 Update optimal link blockings.

At this point in the iterative loop, a network design is available, and it is desirable to lower network cost by selection of more efficient blocking levels. The parameters used for these calculations, such as incremental link cost, are all obtained from the current network design. That is, the current network operating point in terms of trunks, loads, and routing is used to calculate a new, more efficient operating point.

The ECCS approach of Truitt [Tru54] is modified to consider the fact that with path-to-path routing several different routes may use the same link and that each overflow traffic load may use a different path. Here we consider only the first alternate path for each overflow traffic load.

Figure 4.12 shows the model used to calculate ECCS values in the route-EFO model. Two routes are shown: Each route sees the same blocking probability on the link in question (link I) and overflows to a different path. The first route overflows to a path comprising links A and B, and the second route overflows to

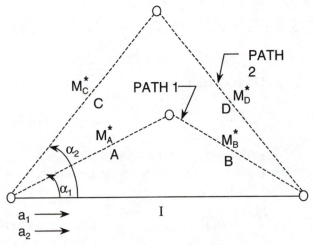

Figure 4.12 Calculation of optimal link blocking

a path consisting of links C and D. The terms used in the model are as follows:

a_i = offered load to link I for route i

α_i = overflow load from link I for route i

$\widetilde{M}_1 = M_A^* + M_B^*$ = the cost per erlang of the alternate path for route 1

$\widetilde{M}_2 = M_C^* + M_D^*$ = the cost per erlang of the alternate path for route 2

M_A^* = the link metric or the incremental cost of link A

C_I = cost of a trunk in link I

B_{opt}^I = the optimal link blocking objective on link I (to be calculated)

N_I = the desired number of trunks on link I (to be calculated)

The objective is to calculate N_I (and hence B_{opt}^I), given the a_i and $\widetilde{M}_i$, that will minimize the total cost of carrying the a_i offered load over the combination of the direct path and alternate paths. To do this the network cost is first written as

$$\text{Cost} = C_I N_I + \alpha_1 \widetilde{M}_1 + \alpha_2 \widetilde{M}_2$$

$$= C_I N_I + a_1 B_I \widetilde{M}_1 + a_2 B_I \widetilde{M}_2 \qquad (4.13)$$

where use has been made of the relation $\alpha_i = B_I a_i$. A partial derivative is taken of (4.13) with respect to N_I, and the resulting expression is set equal to zero to obtain a minimum. This yields

$$0 = C_I + (a_1 \widetilde{M}_1 + a_2 \widetilde{M}_2) \frac{\partial B_I}{\partial N_I}$$

Hence, N_I should be selected so that

$$\frac{-\partial B_I}{\partial N_I} = \frac{C_I}{a_1 \widetilde{M}_1 + a_2 \widetilde{M}_2} \qquad (4.14)$$

for minimum cost. When multiplied by the total offered load, the quantity on the right-hand side of (4.14) corresponds to the economic CCS (ECCS) for link I. The quantity on the left when multiplied by the total offered load corresponds to the load on the last trunk for link I. Once N_I is found, B_{opt}^I is calculated from the Erlang B formula.

For the calculation of the alternate-path cost $\widetilde{M}_i$, Figure 4.13 shows a situation in which the direct path consists of three links and the alternate path has a link

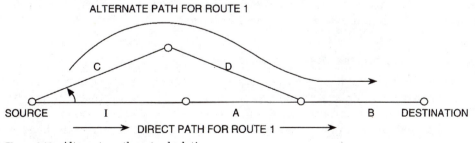

Figure 4.13 Alternate-path cost calculation

in common with the direct path. Because traffic blocked on link I in this case need not use capacity on link A and still must use link B, the alternate-path cost is $M_C^* + M_D^* + M_B^* - (M_A^* + M_B^*) = M_C^* + M_D^* - M_A^*$, or the difference in total metric between the two paths (leaving M_I^* out of the direct-path cost). Also note that the M_i used here contains an allowance for the cost of carrying alternate-routed traffic from link i, as explained in the previous section. This may reflect the fact that the links in the alternate path may themselves overflow to other paths.

The calculation of ECCS is done for the hour in which the offered load to the link in question is maximum. As an approximation, the value of blocking thus obtained is used as the link blocking in all hours for the routing design and LP model steps in the next iteration, regardless of load variations on the link in the current network. Calculation of the load on the last trunk is done at first for integer values of trunks. Once the least integer is found that satisfies equation (4.14), a linear interpolation is used to find a fractional value of trunks that is very near to that required by (4.14). Such fractional values of trunks are allowed to aid in limiting the blocking change from one iteration to another. Through experience with the route-EFO model, the amount that an optimal link blocking could change in one iteration is limited to 300 percent of the previous design blocking for the first two iterations and 20 percent of the previous design blocking thereafter. This is done to speed convergence by limiting the amount of change during any one iteration.

4.4   Modeling Results

Figure 1.12 shows the 28-switch national intercity network model that has been used to study network design. A 10-switch subset of this 28-switch network, also shown in Figure 1.12, is used for some comparisons. As can be seen from the figure, the 10-switch study model has three regions, whereas the 28-switch network has four regions. Three hours of switch-to-switch loads for both networks are used, which include 10 A.M., 1 P.M., and 8 P.M. (hour beginning, Eastern Standard Time). These hours correspond to the most expensive single-hour hierarchical design periods in the 28-switch network. Standard switching and transport costs are used, and the required switch-to-switch blocking probability grade-of-service objective is set to .005, which is a reasonable value for purposes of comparison with hierarchical network designs. The hierarchical network designs use techniques described in Chapter 2, in which the homing pattern used in the two model networks is the actual pattern as shown in Figure 1.12.

4.4.1   Multilink routing results—10-switch model

The route-EFO model designs a network for a single hour by assigning all the traffic for a particular switch pair to the least expensive route for that pair. There is no need to use the LP model in this case because there are no savings to be obtained from traffic noncoincidence. In fact, there is really no need to

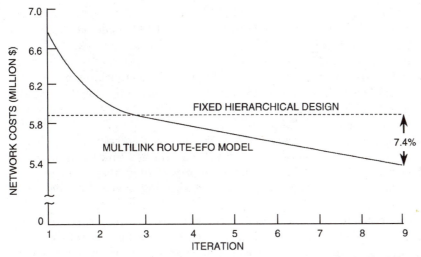

Figure 4.14 10-switch route-erlang flow optimization model single-hour results (10 A.M.)

generate more than one route for each switch pair, because the route formed by path 1 overflowing to path 2 overflowing to path 3, and so forth, is nearly always the least expensive. Figure 4.14 shows the network design cost as a function of iteration for a single traffic load (10 A.M.) using the multilink route-EFO model. The hierarchical network design cost for the same load is also shown. As can be seen, the multilink route-EFO model design cost decreased monotonically, and after nine iterations saved about 7.4 percent of the hierarchical network design cost.

Table 4.1 shows single-hour network design data for all three hours considered (10 A.M., 1 P.M., and 8 P.M.), together with the percent savings achieved by the multilink route-EFO model in comparison with the hierarchical network design. The weighted average of the network switch-to-switch blocking probability is also shown for the multilink route-EFO model designs. For the hierarchical network designs, an average blocking probability of about 0.009 is obtained in each hour.

The savings data shown in Table 4.1 indicate that the multilink route-EFO model savings are not very dependent on the hour considered and are all in the

TABLE 4.1 10-Switch Multilink Network Results (Single-Hour Solutions)

Hour	Hierarchical network design cost	Multilink path routing			Savings
		Design cost	Average blocking		
10 A.M.	$5,949,500	$5,511,100	0.005		7.4%
1 P.M.	5,839,300	5,378,340	0.005		7.9%
8 P.M.	5,516,600	5,103,380	0.005		7.5%

range of the 7.4 percent savings figure for the 10 A.M. loads. The average network blocking probability attained by the multilink route-EFO model is somewhat better than the hierarchical network designs. Because the largest switch-to-switch loads in the 10-switch network are applied to final links that are sized to a blocking of 0.01, this accounts for the rather high average blocking of 0.009 for the hierarchical network designs. To compare further blocking probability data, a distribution of blockings as seen by individual switch pairs is given in Figure 4.15 for the multilink route-EFO model and hierarchical network designs for the 10 A.M. network. Blocking in excess of 0.01 in the hierarchical network design corresponds to calls that traverse a final link and another link in tandem. As shown, the blocking probability distributions are very similar, although one small switch-pair traffic load in the route-EFO model design experienced a blocking of about 0.036.

The reason that the route-EFO model design can save about 7 percent over a hierarchical network design is that it has a much better choice of routing. One aspect of this improved routing is that the model can choose more direct paths, as demonstrated by the fact that the model utilized paths from Los Angeles to Orlando that passed through Birmingham and Phoenix, along with the paths through San Bernardino and Atlanta, which are available to the hierarchical network design. Another aspect of the more efficient routing employed by the route-EFO model is that there are no final links to be sized for 1 percent blocking, hence the average trunk occupancy can be higher. For example, the Atlanta-to-White Plains link is sized to 16 percent blocking by the route-EFO model, and paths through the subtending sectional centers in the hierarchical network

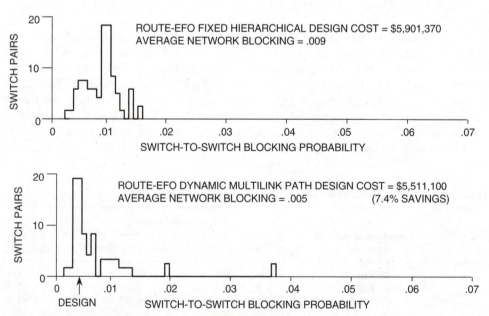

Figure 4.15 Switch-to-switch blocking probability comparison

design are used to carry overflow Atlanta–White Plains traffic so that the overall switch-to-switch blocking is 0.004. In fact, the average blocking on links that are interregional final links in the hierarchical network design is about 21 percent in the route-EFO model design. This results in much higher occupancy of these expensive interregional links.

Multihour results are obtained by designing networks with the hierarchical network design model and with the multilink route-EFO model for the three hours of traffic loads at 10 A.M., 1 P.M., and 8 P.M. Two strategies are employed in the route-EFO model for combining paths into routes: a cyclic routing design and an ordered routing design. Recall the cyclic routing design forms the first route as $1, 2, \ldots, M_1$; that is, path 1 overflowed to path 2, which overflowed to path 3, and so on, until enough paths M_1 are employed in the route to satisfy the blocking probability grade-of-service objective. The other routes are constructed as cyclic permutations of the paths in this route. The ordered routing design, on the other hand, constructs the same first route as the cyclic routing design, but the second route is composed of paths $2, 3, \ldots, M_2$, the third of $3, 4, \ldots, M_3$, and so on. The ordered routing design thus always presents traffic to paths in order of increasing cost; no path is overflowed to a less expensive path. Considering these two routing strategies allows a comparison to be made of different routing techniques. In each case, five routes are constructed for each switch pair.

Figure 4.16 shows the network cost as a function of iteration for both route-EFO model routing designs, along with the three-hour hierarchical network design. As

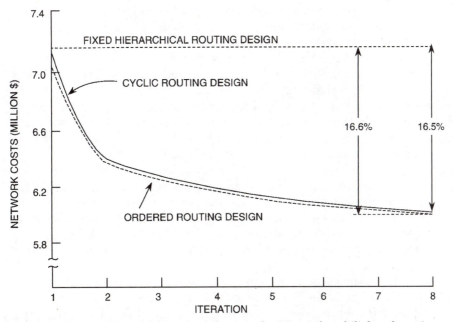

Figure 4.16 Route-EFO model design multihour results (10-switch multilink path routing network)

TABLE 4.2   10-Switch Multilink Route-EFO Model Multihour Results

Design	Cost	Savings	Average blocking	Hour
fixed hierarchical	$7,160,000	—	—	—
ordered multilink path	5,970,600	16.6%	0.003	10 A.M.
			0.001	1 P.M.
			0.006	8 P.M.
cyclic multilink path	5,980,100	16.5%	0.002	10 A.M.
			0.001	1 P.M.
			0.002	8 P.M.

shown in the figure, the two route-EFO model network design methods yield networks that monotonically decrease in cost as the number of iterations increases. The two routing designs yield remarkably similar networks both in structure and in cost, indicating that the design process is not particularly sensitive to the method for combining paths into routes. After eight iterations, both design procedures yield very efficient low-cost networks with good blocking probability distributions. The ordered routing design saves about 16.6 percent of total network cost compared with the hierarchical network design, and the cyclic routing design saves about 16.5 percent of the total hierarchical network design cost.

Table 4.2 shows the average network blocking for the multilink route-EFO model designs together with cost data for the multilink route-EFO model designs and the hierarchical network design that appear quite acceptable.

Figure 4.17 shows the switch-pair blocking probability distribution for the cyclic routing design network, where the maximum switch-pair blocking is less than .015, a figure comparable to traversing one final link and another link in tandem within the hierarchical design. The excellent blocking probability distribution is partially a characteristic of the sizing process as it iterates to find the required number of trunks to meet a maximum blocking objective. The routes are constructed using the maximum blocking in all hours; however, when the network is actually sized, some blockings in nonpeak hours are much less than the maximum, and this tends to improve the blocking distribution.

An example of the choices made by the LP model among the available routes between Birmingham, Alabama, and Albany, New York, is shown in Table 4.3, where the percentage of traffic offered to each route is given. A cyclic routing design is used for this example, and although only the first route is used at 10 A.M., two or three routes are used during the other two hours. Approximately two-thirds of the switch pairs utilize hourly routing changes to take advantage of traffic noncoincidence.

Figure 4.18 shows the cyclic routing network design. It is interesting to note that there is no link between San Bernardino and White Plains, which is an interregional final link for the hierarchical design and therefore must be provided. All switch-to-switch traffic between these two switches is carried via other paths,

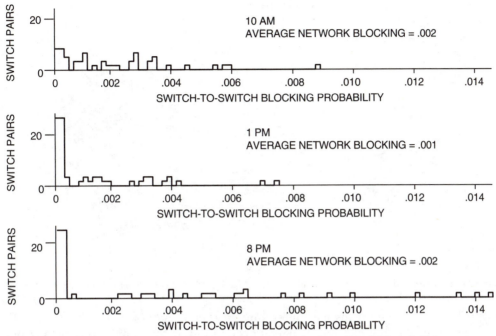

Figure 4.17 Switch-to-switch blocking probability comparison by hour (10-switch multilink path routing design)

TABLE 4.3   Multihour Routing Example, Birmingham–Albany

Hour	Percent traffic	Route Number	Paths
10 A.M.	100	1	$1, 2, \ldots, 9$
	0	2	$2, 3, \ldots, 1$
	0	3	$3, 4, \ldots, 2$
	0	4	$4, 5, \ldots, 3$
	0	5	$5, 6, \ldots, 4$
1 P.M.	71	1	
	29	2	
	0	3	
	0	4	
	0	5	
8 P.M.	35	1	
	0	2	
	43	3	
	21	4	
	0	5	

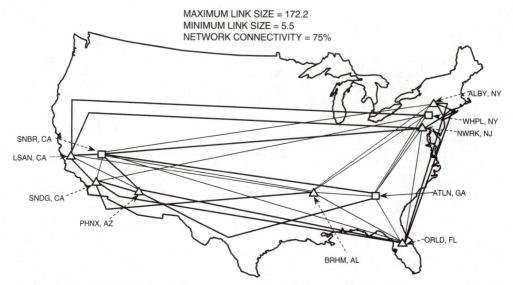

MAXIMUM LINK SIZE = 172.2
MINIMUM LINK SIZE = 5.5
NETWORK CONNECTIVITY = 75%

Figure 4.18   Route-EFO model multilink path routing design (10-switch multipath network)

which is equivalent to using subtending sectional centers in the hierarchical design. This network is about 75 percent connected, with a maximum link size of 172.2 and a minimum link size of 5.5.

4.4.2   Multilink routing results—28-switch model

The same methodology employed in the 10-switch network design is applied to the 28-switch network design. Table 4.4 gives a comparison of the network cost with the hierarchical network design and that using the multilink route-EFO model for each of the three design hours. The average network blocking probability is also shown for the route-EFO model; in each hour the average blocking probability is about .004 for the network designs selected for cost comparisons. As shown in the table, the savings for each single-hour network are around 4.5 percent, a value about 3 percentage points less than the savings in the 10-switch network. One explanation of this difference is that the 28-switch hierarchical network design has a higher proportion of its traffic carried on high-usage links, as opposed to

TABLE 4.4   28-Switch Multilink Route-EFO Model Single-Hour Results

Hour	Hierarchical network cost	Multilink path routing		Savings
		Design cost	Average blocking	
10 A.M.	$37,910,000	$36,196,200	0.004	4.5%
1 P.M.	38,595,100	36,854,300	0.004	4.5%
8 P.M.	38,345,400	36,516,400	0.004	4.8%

TABLE 4.5   28-Switch Multilink Route-EFO Model Multihour Results

Design	Cost	Savings	Average blocking	Hour
fixed hierarchical	$48,430,100	—	—	—
cyclic multilink path	42,745,800	11.7%	0.002	10 A.M.
			0.002	1 P.M.
			0.003	8 P.M.

final links, than does the 10-switch network. Hence, the savings from increasing link occupancy due to allowing alternate routing from final links is decreased.

Table 4.5 shows the results obtained from the multilink route-EFO model as compared with the hierarchical network design. A savings of 11.7 percent is obtained, and the average network blocking probability for each of the three hours is about .002. The potential multihour, multilink route-EFO model savings are probably greater than these three hour results indicate. For example, in the dynamic hierarchical network design presented in Chapter 2, we find very little increase in the cost of the 28-switch network design in going from a three-hour design to a 17-hour design, whereas the hierarchical network design cost increases by about 3 percent. This is reasonable because the dynamic hierarchical routing network design meets off-peak traffic loads by changing routing rather than by adding capacity. The same result can be expected from the route-EFO model design, and because the 17-hour 28-switch hierarchical network design costs $49.9 million, the potential route-EFO model multihour savings are close to 15 percent.

4.4.3   Two-link routing results—10-switch model

In this section, results are presented for 10-switch networks designed by the route-EFO model, with the added constraint that any path may have a maximum of two links. Such routes always appear as shown in Figure 4.1, with the added possibility that the first path may contain two links.

Table 4.6 compares 10-switch results for the two-link routing design with the multilink route-EFO model design and the hierarchical network design. The cost penalty of the two-link designs with respect to the multilink designs is also given. The network without blocking probability grade-of-service correction (no B_{max}) costs about 0.8 percent more than the multilink route-EFO model design and gave about the same average blocking probability. This two-link network design still saves about 6.6 percent over the hierarchical network design. The two-link design has three switch pairs with blockings in excess of 0.02, whereas the multilink route-EFO model design has only one. Constraining the two-link design to a maximum route blocking of .02 does not affect the two-link network cost, but it does improve the average network blocking. It appears that when $B_{max} = .02$, capacity is taken away from large switch-pair traffic loads with low

TABLE 4.6   10-Switch Two-Link Route-EFO Model Single-Hour Results (10 A.M.)

Design	Cost (savings)	Penalty*	Average blocking
fixed hierarchical	$5,949,500	—	—
multilink path	5,511,100 (7.4%)	—	0.005
two-link path (no B_{max})	5,555,700 (6.6%)	0.8%	0.005
two-link path ($B_{max} = 0.02$)	5,555,900 (6.6%)	0.8%	0.005
two-link path ($B_{max} = 0.01$)	5,612,100 (5.7%)	1.8%	0.004

*With respect to multilink route-EFO model total cost

blockings and given to smaller traffic loads with higher blockings. A maximum route blocking of .01 requires a further increase of about 1 percent in network cost, but this level of maximum blocking provides a service level better than that given by the hierarchical design.

Table 4.7 compares 10-switch two-link designs for various values of B_{max} with a multilink route-EFO model design and the hierarchical network design. A two-link design that does not correct route blocking (no B_{max}) costs about 1.2 percent more than the unconstrained route-EFO model design and gives only slightly poorer average network blocking. Ensuring that the route blocking of the two-link design does not exceed .02 requires a total cost penalty of 1.4 percent. Using a maximum route blocking of .01 further increases the network cost penalty to 1.9 percent. Two-link design savings are still about 15 percent.

TABLE 4.7   10-Switch Two-Link Route-EFO Model Multihour Results

Design	Cost (savings)	Penalty*	Average blocking	Hour
fixed hierarchical	$7,160,000	—	—	—
multilink path	5,980,100 (16.5%)	—	0.002	10 A.M.
			0.001	1 P.M.
			0.002	8 P.M.
two-link path (no B_{max})	6,054,600 (15.4%)	1.2%	0.003	10 A.M.
			0.002	1 P.M.
			0.003	8 P.M.
two-link path ($B_{max} = 0.02$)	6,064,300 (15.3%)	1.4%	0.003	10 A.M.
			0.002	1 P.M.
			0.002	8 P.M.
two-link path ($B_{max} = 0.01$)	6,095,700 (14.9%)	1.9%	0.002	10 A.M.
			0.001	1 P.M.
			0.001	8 P.M.

*With respect to multilink route-EFO model total cost

4.4.4 Two-link routing results—28-switch model

Table 4.8 shows results for the 10 A.M. 28-switch network using a two-link routing design, where the 28-switch single-hour multilink design is included for comparison. Results are shown for a no-blocking-correction (no B_{max}) two-link design and a two-link design using the blocking correction method in which a blocking maximum of .02 is used. As can be seen, the savings obtained by the two-link route-EFO model are about 4.7 percent, and the cost penalty due to using a two-link routing design with blocking correction is about 0.2 percent. These values almost exactly equal the savings obtained by the multilink 28-switch design. The two-link design without blocking correction achieves a slightly higher savings figure with a slightly better average blocking. These data suggest that the penalty for using paths constrained to a maximum of two links is negligible in the 28-switch network, whereas the 10-switch network shows a slight penalty. This result is not surprising, because there are many more possible two-link paths in the 28-switch network than in the 10-switch network.

Table 4.9 shows 28-switch results for a two-link route-EFO model multihour design; the 28-switch multilink route-EFO model solution and the hierarchical network design are shown for comparison. Here, the two-link solution is allowed a wider choice of routes than the multilink solution, which increases the design efficiency. A 28-switch two-link design that utilizes the more constrained route

TABLE 4.8   28-Switch Two-Link Route-EFO Model Single-Hour Results (10 A.M.)

Design	Cost (savings)	Penalty*	Average blocking
fixed hierarchical	$37,910,000	—	—
multilink path	36,196,200 (4.5%)	—	0.004
two-link path (no B_{max})	36,129,300 (4.7%)	−0.2%	0.003
two-link path ($B_{max} = 0.02$)	36,291,300 (4.3%)	0.2%	0.003

*With respect to multilink route-EFO model total cost

TABLE 4.9   28-Switch Two-Link Route-EFO Model Multihour Results

Design	Cost	Savings	Average blocking	Hour
fixed hierarchical	$48,430,100	—	—	—
multilink path	42,745,800	11.7%	0.002	10 A.M.
			0.002	1 P.M.
			0.002	8 P.M.
two-link path ($B_{max} = 0.02$)	42,503,100	12.2%	0.004	10 A.M.
			0.001	1 P.M.
			0.001	8 P.M.

parameters similar to those used in the multilink path design saves about 11.3 percent, a savings value almost identical to the multilink design savings. Hence, as in the single-hour 28-switch network designs, there appears to be little penalty for constraining paths to a maximum of two links.

4.5 Conclusion

A model is developed that designs near-minimum-cost nonhierarchical networks using dynamic routing. This model is termed the route-EFO model, and it combines into one procedure such network optimization concepts as (1) favoring larger, more efficient links, (2) utilizing efficient link-blocking levels determined by the ECCS method, (3) routing traffic along the least-cost paths, (4) using preplanned time-dependent dynamic routing to take advantage of traffic noncoincidence, and (5) minimizing incremental network cost. Because crankback and originating office control are assumed and traffic may be split among several routes in different proportions depending on the time of day, this routing method utilizes new routing capabilities made possible by ESS and CCS.

In a 28-switch national intercity network model, results comparing route-EFO model designs with hierarchical network designs indicate a potential capital cost savings of about 5 percent in network first cost for the single-hour (stationary) load case and 12 percent for a three-hour (nonstationary) load case. Total savings of up to 15 percent are expected when designing for all design hours, as found with the network optimization methods involving triad sublink routing described in Chapter 3. Constraining the route-EFO model to use paths with a maximum of two links results in a small cost penalty compared with allowing paths containing any number of links. Implementation of such a two-link network is greatly simplified. The average switch-to-switch blocking probability of the route-EFO model designs is about 0.2 percent, and is approximately equal to the blocking of hierarchical designs. A method to guarantee a maximum switch-to-switch blocking probability to all traffic loads is included and increases network cost only slightly.

Chapter

5

Preplanned Dynamic Progressive Routing Networks

5.1 Introduction

This chapter presents preplanned dynamic progressive routing methods and the design of networks employing such routing tables with the route-erlang flow optimization (route-EFO) model introduced in Chapter 4. In progressive routing, a call progresses through the network one switch at a time until it either reaches its destination or arrives at an intermediate switch from which it has no outlets. It is assumed that a call does not predetermine the busy or idle status of all links in its desired path from the originating switch. Nor does a call, if blocked at an intermediate switch along its desired path to the destination, have the ability to utilize crankback. That is, in contrast with preplanned dynamic path routing, a call blocked at a switch in progressive routing cannot retrace any part of the path it has already traveled. Alternate routing must then be used to flow the call through the intermediate switch, or else it is blocked from the network.

Fixed hierarchical routing is a progressive routing method, call connections being established one switch at a time. However, there are two significant differences between hierarchical routing and the preplanned dynamic progressive routing methods discussed in this chapter, which we refer to simply as progressive routing. First, progressive routing is nonhierarchical. Routing is not limited by a homing pattern that, in a hierarchical network, dictates a subset of switches to route to, along with a specified order in which this must be done. Second, progressive routing is dynamic. The routing can change with time in an attempt to take advantage of the noncoincidence of loads in the network, whereas hierarchical routing is normally fixed over time.

Of interest is how progressive routing compares with preplanned dynamic path routing, discussed in Chapter 4. In Section 5.2 we discuss route control, which is necessary both to prohibit shuttling and looping of calls and to promote efficient link utilization. Implementation considerations for flowing calls over multiple routes in the progressive routing method are discussed, and a comparison is made between progressive routing and preplanned dynamic path routing. In regard to implementation considerations, a simplified version of preplanned dynamic path routing, in which paths of only one or two links are allowed, offers benefits in its ease of implementation, overall network performance, and network management and design. The modifications to the route-EFO model necessary to design networks with progressive routing are discussed in Section 5.3. Differences in route construction and appropriate blocking probability calculations are the significant changes. Section 5.3 discusses the issues involved with implementing a progressive routing method. Section 5.4 presents results for a 10-switch national intercity network model. Progressive routing networks are compared with hierarchical networks designed for the same loads, and progressive routing is seen to yield sizable savings, on the order of 15 percent. Approximately 11 percent savings is shown to be achievable solely by employing fixed progressive routing. Next, progressive routing is compared with preplanned dynamic path routing with crankback to determine the value of crankback in a nonhierarchical routing environment. Crankback appears to provide a relatively modest economic benefit, on the order of 1 to 2 percent.

5.2 Preplanned Dynamic Progressive Routing Methods

Earlier work on progressive routing is discussed, and then preplanned dynamic path routing is briefly reviewed. Segal [Seg64] considers the design of nonhierarchical networks, both with and without crankback. Inputs to Segal's model include switch-to-switch offered traffic loads and a switch-to-switch blocking probability objective. Note that when designing nonhierarchical networks, one is forced to consider a switch-to-switch blocking probability grade-of-service objective. With hierarchical routing, a maximum blocking, normally 0.01, is set on all final links in an attempt to ensure satisfactory switch-to-switch blocking probability. However, no final links exist in a nonhierarchical network. An additional input to the design is a set of routing tables for all switch pairs. This routing is not optimized in Segal's procedure but rather remains fixed. The main assumption of Segal's design lies in setting the blockings equal on all links used by the same switch pair. This blocking probability is determined so as to meet the switch-to-switch blocking probability grade-of-service objective. If a link is used by more than one switch pair, the blocking is set equal to the weighted average (weighted by carried load) of its blockings in all routes in which it is used. The design now reduces to the iterative solution of a set of simultaneous equations to derive a consistent set of link blockings and link flows. The required number of trunks is then directly calculated from the Erlang B blocking formula.

Katz [Kat72] studies design of nonhierarchical progressive routing networks, one of which is the U.S. Department of Defense's Autovon network [She75]. The design process consists of a sequence of network synthesis programs. Among the inputs are switch locations, switch-to-switch offered loads, and a switch-to-switch blocking probability grade-of-service objective. First, a network connectivity program is run, which establishes which links should be allowed. Next, a set of programs is executed to determine the routing. Finally, given the network connectivity and routing, a capacity design program is run. In this approach the above three steps (connectivity, routing, and capacity design) are considered separately, although they are actually interdependent. Each step can at best yield an optimization of the problem restricted by the previous step(s). Additionally, there is no explicit consideration of network cost. The assumption is that equalization of switch-to-switch blocking probability produces a near-minimum-cost solution.

Knepley [Kne73] introduces new concepts in nonhierarchical network design. His procedure consists of an iterative loop that aims at minimizing the incremental network cost. Additionally, routing is not a fixed input but is optimized in the design. In Knepley's design the blocking probability is set to be equal on all links.

All of the above nonhierarchical network design procedures leave the routing fixed over time. The progressive routing studied in this chapter allows routing to change dynamically in preplanned routing tables, as now described. In a progressive routing environment, a call progresses through the network one switch at a time, as illustrated in Figures 5.1A and 5.1B. In this example, paths 1, 2, and 3 exist between switches A and Z, the order denoting increasing incremental cost; that is, path 1 is the least-cost path to switch Z, path 2 is the second-least in cost, and so on. If a call progresses first to switch B, then gets blocked on the B–Z link, the call is not allowed to crank back to switch A in order to route on path 2, the next path choice. Instead, it will attempt to arrive at Z on the next-best path, given that it must route through switch B. In this case, the call will attempt to route to switch D. Observe that the routing is actually switch-to-switch routing, in the sense that each switch must know an ordered set of switches to route to for a given destination. This is somewhat different from path routing with crankback. As illustrated in Figure 5.1B, a call blocked on link B–Z is able to crank back to switch A to attempt the next path choice, A–C–Z. The routing in this case is path routing, each switch pair having an ordered set of paths to route on, with path-to-path overflow.

Because progressive routing is switch-to-switch, each switch, for a given destination, must have an ordered list of switches to route to, which is called a *routing*

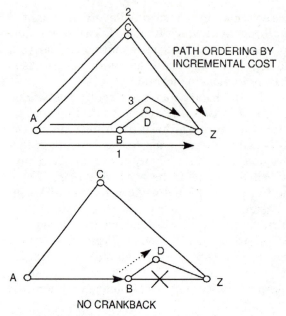

PATH ORDERING BY
INCREMENTAL COST

NO CRANKBACK

Figure 5.1A Routing without crankback

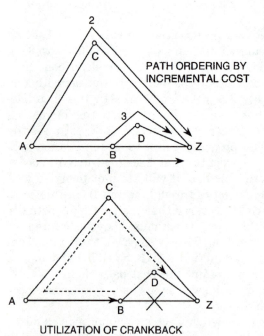

PATH ORDERING BY
INCREMENTAL COST

UTILIZATION OF CRANKBACK

Figure 5.1B Routing with crankback

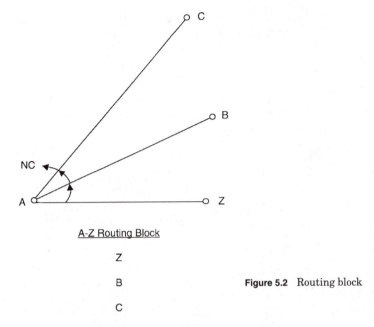

A-Z Routing Block

Z

B   **Figure 5.2**   Routing block

C

block. This is illustrated in Figure 5.2. A call arriving or originating at switch A and destined for switch Z first attempts to route to switch Z. If the A–Z link is busy, switch B is tried, then switch C. If all three links are busy, the call is blocked from the network and receives a no-circuit (NC) announcement. Thus, for any progressive routing method, a routing block must be constructed for each pair of switches. An example is given in Figure 5.3. To route from switch A to switch Z, the A–Z routing block is first consulted. This block indicates to a call at switch A destined for switch Z to first attempt to go to switch Z directly. If the A–Z link is busy, the call next attempts to go to switch B. If link A–B is busy, the call next attempts to go to switch C. If link A–C is also busy, the call is blocked from the network. When a call arrives at an intermediate switch, that switch's routing block to the destination is used to determine the next switch in its path. For example, if a call finds the A–Z link busy and therefore arrives at switch B (assuming link A–B is idle), the B–Z routing block now directs the call to route directly to switch Z. If this is not possible (link B–Z is busy), then switch C is tried. If link B–C is also busy, the call is blocked from the network.

In order to achieve dynamic progressive routing, as designed in the route-EFO model, the routing design step must construct more than one route for each switch pair. In each hour, each route is allocated a certain fraction of the total switch-to-switch offered load. These allocations can change by hour in an attempt to take advantage of the noncoincidence of traffic loads. Routing blocks, together with these time-dependent allocations, constitute a routing table, as discussed in Chapter 1. A set of routes is generated for each pair of switches. Route 1 is constructed by following the usual ordering of the routing blocks, as in the

ROUTING BLOCKS:

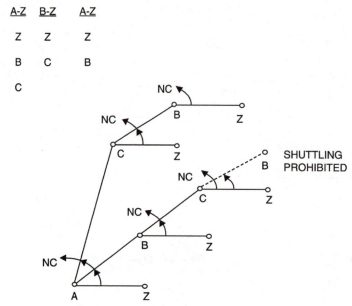

Figure 5.3   Construction of a route

example above. Additional routes are constructed through a cyclic rotation of the originating switch's routing block, as illustrated in Figure 5.4. Route 2 is constructed by following the originating switch's routing block, starting from the second switch, proceeding all the way down to the bottom, and then routing to the first switch. Route i is constructed by following the originating switch's routing block starting from the i th switch, proceeding down to the bottom, then routing from switch 1 through switch i -1. Once the call leaves the originating switch, the routing blocks are followed in their usual order.

An exception to the above generation of routes, which follows directly from the routing blocks or a cyclic permutation of the originating switch's routing block, must be noted. Sometimes we choose to exercise route control in order to (1) prohibit shuttling and looping of calls and (2) promote efficient link utilization. Shuttling is when a call travels back and forth between a pair of switches. Looping, or "ring around the rosie," is when a call travels three or more links only to arrive back at the same switch, as illustrated in Figure 5.5. Both can result in indefinitely tying up trunks in the network without successful call completion. This does not occur in hierarchical routing because the hierarchical routing structure inherently prohibits it. One way to prohibit this in progressive routing is, through the capabilities of ESS and CCS, to send the complete switch history of a call as it is routed from switch to switch and to prohibit the call from visiting the same switch more than once. Note that in Figure 5.3 this is so. For the case when a call travels from switch A to switch B to switch C, the choice of

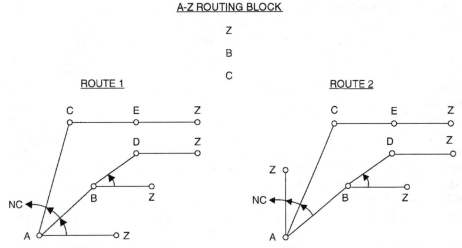

Figure 5.4 Construction of multiple routes

switch B in the C–Z routing block is not allowed because this would result in a B–C–B shuttle. In this case, if the call cannot complete directly to switch Z, it is blocked from the network.

The second reason to exercise route control is to promote efficient link utilization. Basically, this is an attempt to prohibit excessive alternate routing, which can result in calls routing on paths of too many links. The more this happens, the

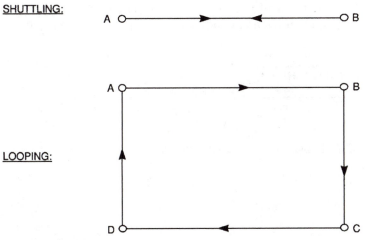

Figure 5.5 Shuttling and looping

more these calls "steal" trunks from calls that can complete on one or possibly a small number of links. This has a cascading effect and can result in inefficient link utilization, with fewer call completions than is otherwise possible.

The way we prevent this is by choosing which paths we want to allow in the network and which paths we want not to allow in order to inhibit the above problem from occurring. With the use of thresholds α and β, as discussed later in this section, we eliminate circuitous paths that are unnecessary to meet the desired blocking objective. In addition, the restriction of paths to promote efficient link utilization turns out to be computationally quite useful. In a 10-switch fully interconnected network, there are 109,601 different paths between any two switches. If we do not restrict paths, the problem arises as to how to realistically model a progressive routing network, because it would not be computationally feasible to include all 109,601 different paths between all switch pairs.

The general nature of all previously utilized route-control plans lies in exercising control using one or two techniques—route-control digits [Kat72, Kne73, She75] or nonlooping routing blocks. The concept of route-control digits is illustrated in Figure 5.6. A call arrives at switch X with route-control digit d. In order to route the call to the next switch, switch X consults its routing block to the destination. The digit d controls switch X's choice of switches from the routing block. For instance, in Figure 5.6, digit d prohibits routing to switches n_3 and n_4. Additionally, switch X must output a control digit d' for the next switch. In this example, if outlet switch n_1 is selected, digit d'_1 is sent with the call. If n_2 is selected, d'_2 is sent along. Under a route-control digit method, the idea is to construct a consistent set of routing blocks and route-control digits in such a way as to meet the above route-control objectives. However, there is only so much information that can be transmitted by a single route-control digit. Because one digit cannot contain the complete history of a call, upon receiving this digit it cannot be known specifically which switches in a given routing block will result in shuttling, looping, or inefficient paths through the network. Thus, the construction of consistent routing blocks and route-control digits, if only to prohibit shuttling and looping, introduces additional routing restrictions. Due to

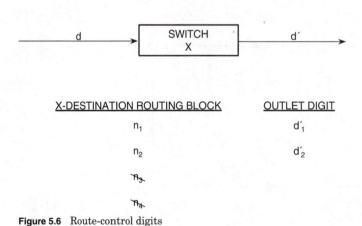

Figure 5.6 Route-control digits

both the difficulty and the additional routing restrictions of this approach, Weber [Web64] proposes a variation of a route-control digit method in which a limited amount of looping is allowed. Under his "free routing" plan, a route-control digit is used to count the number of links that have already been traversed. This digit is initialized to the maximum number of links to be allowed and decremented by one as each successive link in the path is selected. If this digit reaches zero, the corresponding call is blocked from the network. Weber estimates the degradation in switch-to-switch blocking due to looping, which turns out to be small.

Another method of route control is the construction of routing blocks that are inherently nonlooping. This eliminates the need to send along a route-control digit. One example of this is hierarchical routing, where, as discussed in Chapter 2, the rule of routing up the destination switch's homing chain and down the originating switch's homing chain, which are built into the routing tables at each switch, prohibits looping. Another example of nonlooping routing blocks is to route first on the direct link to the destination, and then choose a via switch K to alternate route I–J traffic if this ensures a nonlooping routing table. Therefore, an additional looping detection and correction method must be used. This route-control method appears to have the same general problem as a route-control digit method, in that it introduces additional unwanted routing restrictions. The problem lies in the construction of routing blocks general enough to prohibit all possible looping. For example, switch Y must be eliminated from the X–Z routing block if it is at all possible for a call at switch X destined for switch Z to have previously visited switch Y.

There are consequences to overexercising route control—that is, introducing routing restrictions beyond the minimum that are necessary to meet the desired route-control objectives. First, because this results in limiting the freedom of the routing plan, not all of the least-cost paths through the network can be accessed, and this increases the network cost. Second, it is quite possible for some switch pairs to be unable to achieve satisfactory blocking performance because they may now have only a limited number of paths connecting them. In order to specifically prohibit shuttling and looping, in progressive routing we must restrict a call from visiting the same switch more than once. To do this, we could send with the call via CCS its complete history—that is, a list of switches that the call has already visited. A switch is then prohibited from routing that call to any switch already on this list. For a network of 250 switches, 8 bits can uniquely identify each switch. Therefore, the CCS call setup message needs to transmit a maximum of nine 8-bit words with a call to represent a complete history of that call, where nine is the maximum number of links in a five-level hierarchical network. However, very few calls actually need to have this many switches sent along, because the majority of calls traverse only one or two links.

In order to promote efficient link utilization, we want to eliminate the cascading effects of excessive alternate routing of calls on paths of overly many links. As one means to achieve this, we can define thresholds α and β such that an A–Z path is classified as good if the A–Z peak-hour load offered to the path is greater than α erlangs *or* if the fraction of the total A–Z peak-hour load offered to the path is

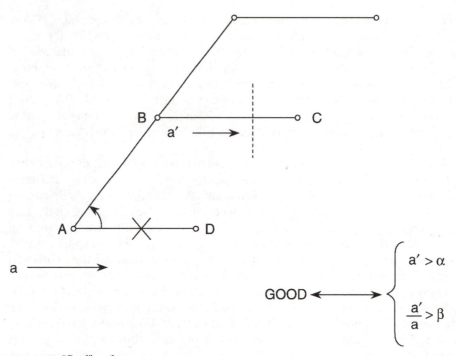

Figure 5.7 "Good" paths

greater than β, as illustrated in Figure 5.7. This α, β method illustrates one way to eliminate inefficient paths. The path starting at switch A through switch B is continued through switch C if a', the A–Z peak-hour load offered to link B–C, is greater than α erlangs or a'/a (a = total A–Z peak-hour offered load) is greater than β. If neither of these conditions is met, any call traveling on this branch of the route (i.e., a path starting A–B–C) is blocked from the network at switch B. What is required to implement this method is that a and a' be known as the call progresses switch by switch through the network. Because a is equal to the A–Z peak-hour offered load, each switch needs a list of its peak-hour offered load to every other switch. Obtaining a', the A–Z peak-hour load offered to the current path, is more complicated because a' is equal to the fraction of the A–Z peak-hour offered load that is offered to the current path. Thus, a' equals the product of the A–Z peak-hour offered load and the connectivity of the current path to the next switch. Path connectivity is the probability of a call traveling on that path, which can be computed switch by switch as a function of the link blockings. Assuming each switch knows the designed blocking level of all links leaving that switch, the connectivity could be calculated switch by switch as the running product of the blocking probability of each link that the call found busy and the blocking probability of each link the call has found idle and therefore seized. Thus, the A–Z peak-hour offered load must be multiplied by the appropriate link blockings and connectivities as the call progresses through the network. Referring back to Figure 5.7, we multiply a by the blocking probability of link A–D, (1 − blocking probability of link A–B), and $(1 - \text{blocking probability of link B–C})$ to obtain a' at switch C. Observe that we can calculate this product at switch B and therefore would not need to send the call to switch C if the path starting A–B–C failed the test for being good.

As to the choice of specific values for α and β, networks are preferred to have a blocking probability grade-of-service objective comparable to a hierarchical network. For a network designed using $\alpha = 0.01$, $\beta = 0.001$, this restriction results in blocking slightly more than 4 erlangs from the 10-switch network shown in Figure 1.12 out of a total network load of over 17,000 erlangs, or less than 0.25 percent of the total network load. Although this seems like a small amount of traffic, these 4 erlangs are equal to half of the total network blocked load, the other half being either restricted by a 30-path maximum per route or blocked from the network due to a call arriving at an intermediate switch with all of that switch's outlets being busy. If we relax the route control, setting $\alpha = \beta = 0.0001$, and increase the allowed maximum number of paths per route from 30 to 300, instead of completing more traffic in the network by allowing more paths, as one might initially think would happen, the cascading effects of allowing many inefficient paths result in an increase in the amount of blocked traffic from 4 erlangs to 35 erlangs. This reinforces the fact that route control is necessary in addition to prohibiting shuttling and looping; eliminating long, inefficient paths; and therefore promoting efficient network utilization.

With progressive dynamic routing for a given switch pair, traffic is allocated among a set of routes. We construct different routes by a cyclic rotation of the originating switch's routing block to the destination. Therefore, once the originating switch allocates the call to a route with a given probability, in a given hour, no additional information as to which route the call is using need be sent with the call. After the call reaches its first intermediate switch, it travels the same sequence of switches no matter which route it has been assigned to.

Nonhierarchical path-to-path routing must have the ability, if a call is blocked from a desired path, to crank the call back to the originating switch in order to overflow to the next path. Each intermediate switch must know the next switch to route to when it receives a call on a given path in a given route for a given switch pair. These additional capabilities are nontrivial. From an implementation viewpoint, two-link path routing is much simpler than the more general multilink path routing, and it offers several implementation benefits over progressive routing. In two-link path routing, the originating switch need only know a list of via switches to route to. Each via switch requires knowledge only of the destination, which is essential in any routing method, and the originating switch to which it will crank back if necessary. The originating switch can be determined from the CCS setup message or by looking at which link the call arrived on.

In addition to its ease of implementation, two-link path routing has other implementation benefits over progressive routing. Shown in Figure 5.8 are the average and distribution of the average number of links per call by switch pair for progressive and two-link routing for the 10-switch network model. As can be seen,

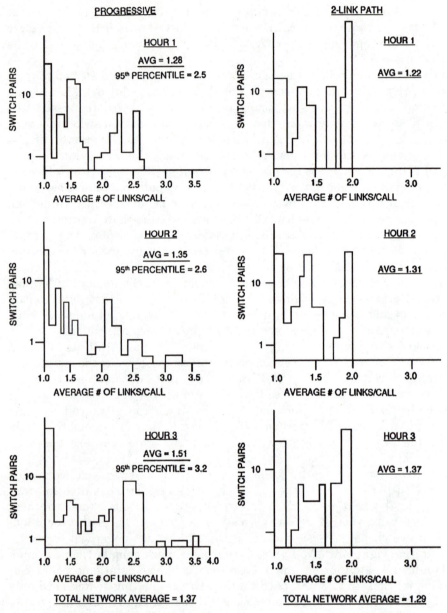

Figure 5.8 Average number of links per call

TABLE 5.1   Distribution of Number of Links per Call by Network Carried Load
(Entries Indicate Erlangs of Carried Load)

Routing	Number of links						
	1	2	3	4	5	6	7
Hour 1							
progressive	487.7	102.1	23.6	5.1	1.4	0.07	—
2-link path	486.6	133.4	—	—	—	—	—
Hour 2							
progressive	437.8	117.9	33.2	5.8	1.4	0.11	—
2-link path	413.0	183.4	—	—	—	—	—
Hour 3							
progressive	317.6	109.7	49.6	10.5	2.4	0.25	—
2-link path	307.3	182.8	—	—	—	—	—

although the averages are quite close, the two-link distributions are significantly better, because a call is restricted to a maximum of two links. For progressive routing, in each hour, between 30 and 40% of the switch pairs see an average number of links per call greater than 2. Table 5.1 shows the distribution of the number of links per call by network carried load. Here, it can be seen that for progressive networks, there are some calls that occupy as many as six links. The better distribution of the number of links per call of two-link routing can have several implementation benefits in the area of overall network performance, such as transmission quality and ESS/CCS load. In the area of real-time traffic management, under overload conditions the two-link distribution has all switch pairs seeing the number of links per call less than or equal to 2, as opposed to progressive routing, where one call could conceivably occupy a large number of links.

Other implementation benefits arise for path routing from the fact that the originating switch has control of a call. This can benefit real-time traffic management, where overloads can be controlled at their source. Also, it is easier to measure switch-to-switch blocking, which is necessary in a nonhierarchical routing method, because the originating switch always knows the status of a given call.

5.3   Network Management and Design

As discussed in Chapter 4, the design of preplanned dynamic routing networks is a complex problem. Major difficulties are the nonlinear relationships of cost, traffic, and performance. One example of this is the relationship of link size and load at a fixed blocking level, as given by the Erlang B loss formula. An additional complication is the complex flow of traffic, where links can overflow back to each other. Here we describe the route-EFO design model to generate preplanned dynamic progressive routing networks. In Chapter 4 we discussed the use of the route-EFO model to design preplanned dynamic path routing networks; here we focus on the necessary extensions to the route-EFO model to design networks with preplanned dynamic progressive routing. Differences in route construction, along with appropriate blocking probability calculations, are the significant changes.

The route-EFO model is used to design the progressive routing networks discussed in this chapter. This model can be used to design near-minimum-cost networks with general routing methods. Figure 4.9 gives an overall view of the route-EFO model. The design procedure is an iterative loop that seeks to converge to a near-minimum-cost network. After the initialization step, the model enters the routing design step, which consists of a route generator and a linear programming model. The inputs to this step include the optimal link blocking objectives, link-incremental costs, and a required switch-to-switch blocking objective. First, the route generator constructs a set of routes for each switch pair. The linear programming model then allocates, for each switch pair in each hour, the offered traffic load to the set of possible route choices. This allocation is done in such a way as to minimize the total incremental network cost. The reason for using more than one route per switch pair is that each route results in a different distribution of traffic to the links in the network. Even for two routes that consist of the same links but different overflow rules, the proportion of route load that flows on a given link in the two routes is different. These link proportions are calculated by the route generator and are used by the linear programming model to determine an optimal routing allocation and the link capacities needed for this allocation. An example linear programming model allocation is the following—for switch pair A–Z, in hour 1, route 1 receives 45 percent of the A–Z offered traffic load, route 2 receives 15 percent, and route 3 receives 40 percent; in hour 2, route 2 receives 100 percent. Note that the allocation can change by hour to take advantage of load noncoincidence.

Next, a capacity design with consistent link blockings and flows is determined from the routing output from the routing design step. An iterative loop is necessary here. For each link, the peak-hour capacity (in erlangs) determined by the linear programming model and the optimal link-blocking objective are used to calculate an initial link size. Blockings in the nonpeak hours are computed and then checked against their previous values, which, for the first iteration, are equal to the optimal link blocking level. If these blockings match on all links, convergence to a network with a consistent set of link blockings and flows has been reached; if not, link flows are recalculated based on the new blocking values and each link is resized for its peak-hour flow and its input optimal link blocking objective. This loop continues until the link blockings match in all hours. Note that only one iteration is needed for a single-hour design problem.

The network design generated is then tested to determine if it is a near-minimum-cost network. If network improvement is called for, we compute new link-incremental cost metrics that represent the cost of offering an additional erlang to a link while remaining at the same optimal link-blocking level. Also, a new optimal blocking level is computed for each link, through use of the ECCS method, as discussed in Chapter 4. Now the route-EFO model branches back to the routing design step, completing its loop, in order to generate new routes based on the new optimal link blockings and link-incremental cost metrics.

To design preplanned dynamic progressive routing networks with the route-EFO model, all significant changes are to the route generator within the routing design step, which generates paths and forms paths into routes, as illustrated in Figure 4.9. The major outputs of the route generator are (1) a set of routes for each switch pair and (2) the proportion of route load that flows on each link in the route, which is needed by the linear programming model to determine how traffic flows through the network for a given routing allocation of the switch-to-switch traffic loads.

5.3.1 Progressive route construction

In progressive routing, a call progresses through the network one switch at a time, as illustrated in Figures 5.1A and B, discussed earlier. Because progressive routing is switch-to-switch, each switch for a given destination must have an ordered list of switches to route to, called a routing block. This is illustrated in Figure 5.2. A call arriving or originating at switch A destined for switch Z first attempts to route to switch Z. If the A–Z link is busy, switch B is tried, then switch C. If all three links are busy, the call is blocked from the network.

Thus, for any progressive routing method, a routing block must be constructed for each pair of switches. We do this in such a way that calls first attempt to route on the least-cost paths to the destination. The first step of route construction is, for a fixed destination switch Z, to find the shortest path from all switches to switch Z. This is done through an application of Dijkstra's shortest-path method [Dij59]. The distance function is that of incremental cost, in keeping with the objective of minimizing incremental network cost. Next, for a fixed originating switch A, all switches n connected to A are found. These are the switches that appear in the A–Z routing block. To determine the order of entry, the shortest path from A to Z through n is constructed for each switch n. This can be done directly from the shortest-path tree. In ordering these paths by ascending distance, the ordering of all switches n in the A–Z routing block is obtained. Thus, switch I is entered before switch J if there is a path to switch Z through I that is shorter than all paths through switch J.

The procedure is illustrated in Figure 5.9. In step 1, a tree is constructed that gives the shortest path from all switches to a given destination switch Z. Step 2 finds, for a fixed switch A, all switches n connected to A. In this example, switch A is connected to switches E, C, and Z. Next, the shortest path from switch A to switch Z through each of the switches E, C, and Z is constructed. By referring to the tree of step 1, these paths are seen to be A–E–B–Z, A–C–Z, and A–Z, respectively. These paths are now ordered by path-incremental cost, which, for a given path, is equal to the sum of the incremental cost of all links in the path. In this example, the ordering is A–Z, A–C–Z, and then A–E–B–Z. The

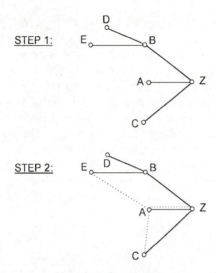

STEP 3: ORDER SWITCHES n BY THEIR DISTANCE A THROUGH n TO Z

    1   d (A-Z)

    2   d (A-C-Z)

    3   d (A-E-B-Z)

STEP 4: CONSTRUCT A-Z ROUTING BLOCK

A-Z

Z

C

E

Figure 5.9 Construction of routing blocks

last step is simply to enter switches E, C, and Z in the A–Z routing block in the order corresponding to this path ordering—first switch Z, then switch C, then switch E.

After routing blocks are constructed for all switch pairs, the generation of routes is a matter of tracing those paths through the network, which follows directly from the routing blocks. The stack algorithm is used to construct the progressive routes. For a given source and destination, the stack algorithm utilizes a sequence of stacks to generate an ordered set of paths and, in addition, to calculate for each path the connectivity that equals the probability of a call traveling on it.

The steps are now outlined, and the step numbers refer to the example given in Figure 5.10. The blocks in Figure 5.10 refer to blocks of the flow chart in Figure 5.11.

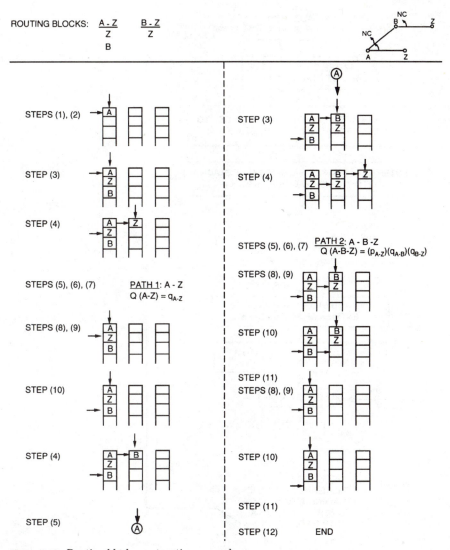

Figure 5.10 Routing block construction example

The stack algorithm begins with entering the source at the top of stack 1 (step 1) and entering the source–destination routing block, in order, below it (step 3). A cyclic rotation of the source–destination routing block is used when constructing routes other than route 1. A pointer in stack 1 is set to the first switch of this routing block—call this switch $n_{2,1}$, which in this case is switch Z. The notation $n_{\text{row}\,i,\text{stack}\,j}$ indicates the entry in row number i and stack number j in Figure 5.10, where rows i are numbered starting from the top to the bottom of stack number j and stacks j are numbered from the left to the right. Switch $n_{2,1}$ is now entered at the top of stack 2 (step 4), which then becomes entry $n_{1,2}$. If $n_{1,2}$ is, in fact, the destination, then path 1 is complete. This happens to be true in the example

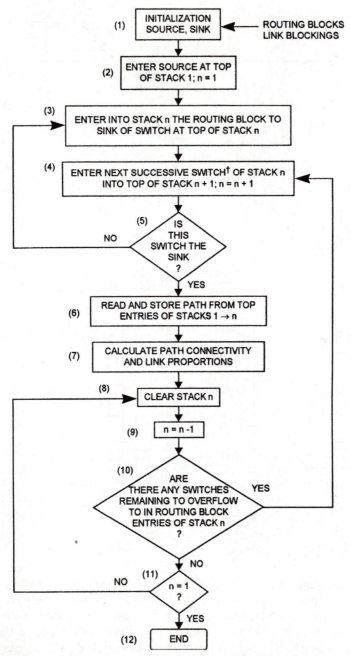

Figure 5.11 Routing block construction steps

of Figure 5.10, where path 1 is read off the tops of the stacks as path A–Z. If $n_{1,2}$ is not the destination, the $n_{1,2}$–destination routing block is entered in stack 2 below $n_{1,2}$. This is the case in step 3 on the right side of Figure 5.10, where Z is now entered in position $n_{2,2}$. A pointer in stack 2 is set to the first switch of this routing block, which is switch $n_{2,2}$. Switch $n_{2,2}$ is now entered at the top of stack 3, becoming entry $n_{1,3}$ (step 4 on the right). If $n_{1,3}$ is the destination, then the current path, $n_{1,1}$–$n_{1,2}$–$n_{1,3}$ which is path A–B–Z, is complete and appears along the tops of the stacks. In this case the path A–B–Z is stored in a separate array, along with its path connectivity (steps 5, 6, and 7). Below we give the path connectivity calculation. Stack 3 is now cleared (steps 8 and 9) and the pointer of stack 2 incremented by 1. If another switch exists below $n_{2,2}$ in stack 2 (this is not the case in the example of Figure 5.10), then call this switch $n_{3,2}$. It is entered at the top of stack 3, and the above path-generation procedure continues until once again a complete path to the destination appears at the tops of the stacks. The resulting path is then stored in a separate array along with its connectivity.

Path connectivity is equal to the product of the connectivity of each link in the path and the blocking of each link that must be busy for a call to travel on that path. In the example of Figure 5.10, the connectivity of path 2, A–B–Z, is equal to $(p_{A-Z})(q_{A-B})(q_{B-Z})$ because link A–Z must first be busy for switch A to attempt to route to switch Z on path 2, and then the links in the path, A–B and B–Z, must be idle. In general, all of the links in the path can be read off the tops of the stacks. The links that must be busy can also be known from the stacks. All of the links that must be busy for a call to travel to switch n at the top of stack i are found by looking at the switches above switch n in stack $i - 1$. Thus, the stack representation of a path is useful in calculating the connectivity of a path.

Actually, path connectivity is not calculated at the termination of generating a path but rather at each step of the stack algorithm. The connectivity through the current switch is always known and is constantly updated with each entry or removal of a switch in the path being generated. This turns out to be quite valuable, because as each new switch is generated for the current path, we want to check whether the path carries a significant amount of that route's load, as defined by thresholds α and β. The connectivity through the switch must be known in order to do this. The step-by-step updating of the path connectivity, along with the threshold tests, has been omitted in Figures 5.10 and 5.11 for clarity.

Continuing with the path-generation procedure, if, after a path is constructed, the pointer in a stack i is incremented to the point that no switches exist in that stack (as in step 10 at the bottom on the right), then all switches of the routing block of stack i have been used for path generation. Stack i is then cleared, and the pointer in stack $i - 1$ incremented. The path generation continues until all good paths are constructed.

Hierarchical network designs are also generated by the route-EFO model in which the stack algorithm described above is modified to convert the hierarchical homing pattern into the corresponding routing blocks. The most significant change is in the construction of hierarchical routes. The stack algorithm has as one of its inputs the routing blocks for all switch pairs. A method of converting the input hierarchical homing pattern into its corresponding routing blocks is the only change needed to construct hierarchical routes from the above stack algorithm. Note that because crankback is also not utilized in hierarchical design, the path connectivity calculations reflect this constraint. A flow chart of the method used to convert a hierarchical homing pattern into its corresponding routing blocks is shown in Figure 5.12. Other modifications for hierarchical route generation are that only one hierarchical route for each switch pair is used at all times, as opposed to the dynamic routing design produced by the route-EFO model. Final links, which are determined from the input homing pattern, are also introduced and are sized for a blocking level of 0.01. Final links are not allowed the flexibility of an economic optimization of blocking level. The ability to construct hierarchical routes enables the route-EFO model to be used in several possible applications. One, which is presented here, is generating a fixed hierarchical network. Another possible design is a dynamic hierarchical network, in which the homing pattern changes with the time of day. The route-EFO model can be applied to solve this design problem.

An example progressive route output from the route generator is given in Figure 5.3. To construct a route from switch A to switch Z, first the A–Z routing block is consulted. Thus, a call at switch A destined for switch Z first attempts to go to switch Z directly. If the A–Z link is busy, the call next attempts link A–B, and if link A–B is busy, it next tries link A–C. If link A–C is also busy, the call is blocked from the network. When a call arrives at an intermediate switch, the routing block to the destination determines the next switch in its path. For example, if a call finds the A–Z link busy and arrives at switch B over link A–B, the B–Z routing block indicates first to route directly to switch Z. If this is not possible because link B–Z is busy, link B–C is tried, and if link B–C is also busy, the call is blocked from the network.

In order to design dynamic progressive routing in the route-EFO model, the route generator must construct more than one route for each switch pair. In each hour, each route is allocated a certain fraction of the total switch offered load. These allocations can change by hour to take advantage of the noncoincidence of loads. In the designs discussed in this chapter, a set of five routes is generated for each pair of switches. Route 1 is constructed by following the usual ordering of the routing blocks, as in the example above. Additional routes are constructed through a cyclic rotation of the originating switch routing block only, as illustrated in Figure 5.4. Route 2 is constructed by following the originating switch's routing block starting from the second switch, proceeding all the way down to the bottom, and then routing to the first switch. Route i is constructed by following the originating switch's routing block starting from the i th switch, proceeding down to the bottom, then routing from switch 1 through switch $i - 1$. Once the call leaves the originating switch, the routing blocks are followed in their usual order.

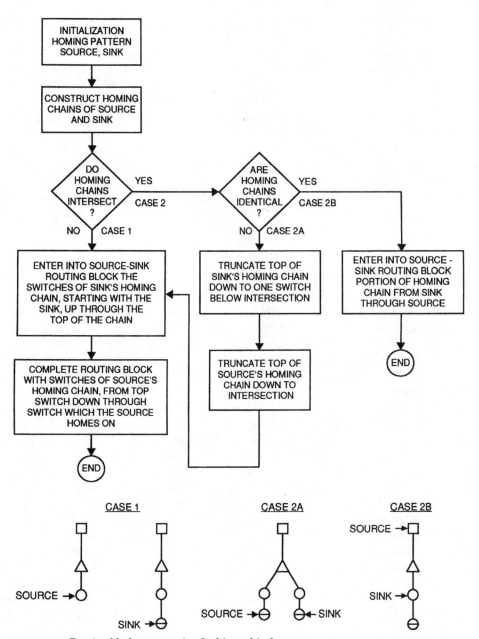

Figure 5.12 Routing block construction for hierarchical routes

As discussed in Section 5.2, with the use of thresholds α and β we choose to exercise route control in order to (1) prohibit shuttling and looping of calls and (2) promote efficient link utilization. We impose a further restriction of allowing a maximum of 30 paths per route. A 10-switch network design with the 30-path restriction is compared with a design that eliminates this restriction to see what effect it has on the accuracy of the design. The difference in carried load of the networks with and without the 30-path restriction is less than 1/10 of an erlang, out of a total network offered load of over 1,700 erlangs. Hence, the 30-path restriction had little effect on the accuracy of the design. Curiously enough, the network without the 30-path restriction actually carries less traffic. This is, in fact, degradation resulting from allowing many paths consisting of a large number of links, which "steal" trunks from calls that can potentially complete on paths with a small number of links. This is the same phenomenon that generates the need for the α, β route-control method.

5.3.2 Blocking calculations

The route generator also calculates route-blocking probabilities and the proportions of route load that flow on each link in the route, which are needed to flow traffic through the network. Among the inputs to the route generator are blocking probabilities for all links in the network. Inherent in the blocking calculations is the fact that crankback is not allowed. Calls are set up one switch at a time, not being able to test all links in the desired path ahead of time. If a call finds a link busy, it must attempt to continue routing from the switch it is currently at. In a network in which crankback is allowed, a route is equivalent to a set of paths in parallel, resulting in different blocking probabilities.

For computing blocking probabilities in the progressive routing environment, it is best to think of the concept of connectivity rather than blocking. We define the connectivity of a path P, $Q(P)$, as the probability of a call traveling on that path. Observe that a call traveling on path P of route R is equivalent to the set of links in route R being in a particular busy or idle state. This is shown in Figure 5.13. In order for a call to travel on path P = A–D–E–F–Z, link A–B must be busy, links A–D and D–E must be idle, link E–Z must be busy, and links E–F and F–Z must be idle. Because the link blockings are assumed to be independent of each other, $Q(P)$ is then the product of these probabilities. In general, $Q(P)$ is equal to the product of the link blocking of each link that must be busy in order for a call to travel on path P. Path connectivities are calculated by the stack algorithm described above while routes are being constructed.

In order to calculate the proportion of route offered load that is carried on a given route, this is equal to the sum of the path connectivities of all paths (in the route) in which the link appears. To determine the blocking of a route, this is $1.0 -$ route connectivity. Route connectivity is the fraction of the offered load of route R that is carried on route R, which is the sum of the path connectivities over all paths of route R.

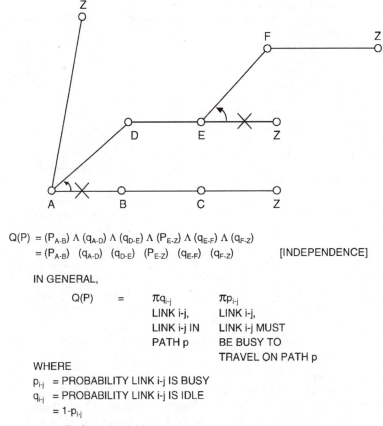

$$Q(P) = (P_{A-B}) \wedge (q_{A-D}) \wedge (q_{D-E}) \wedge (P_{E-Z}) \wedge (q_{E-F}) \wedge (q_{F-Z})$$
$$= (P_{A-B})\ (q_{A-D})\ (q_{D-E})\ (P_{E-Z})\ (q_{E-F})\ (q_{F-Z}) \qquad \text{[INDEPENDENCE]}$$

IN GENERAL,

$$Q(P) \quad = \qquad \pi q_{i-j} \qquad\qquad \pi p_{i-j}$$

	LINK i-j,	LINK i-j,
	LINK i-j IN	LINK i-j MUST
	PATH p	BE BUSY TO
		TRAVEL ON PATH p

WHERE

p_{i-j} = PROBABILITY LINK i-j IS BUSY

q_{i-j} = PROBABILITY LINK i-j IS IDLE

$\quad = 1 - p_{i-j}$

Figure 5.13 Path connectivity

5.4 Modeling Results

Progressive routing networks are designed for the 10-switch national inter-city network model, illustrated in Figure 1.12 and also studied in Chapter 4. Networks are designed for a single hour of traffic load and for multiple hours. For the single-hour design, three peak hours—10 A.M., 1 P.M., and 8 P.M.—are chosen from the available 17 hours of switch-to-switch traffic load data. The multiple-hour design uses the same three peak hours. We compare the progressive network designs with the route-EFO model with hierarchical network designs used in Chapter 4 for the same data. One objective here is to determine the cost savings of progressive routing compared with those of hierarchical routing and path routing and, as a byproduct, to determine the economic value of path routing and crankback in a nonhierarchical routing environment.

To determine the cost savings of progressive routing over hierarchical routing, networks are generated for both routing methods and compared. The hierarchical designs use the design methodology given in Chapter 2; however, hierarchical network designs are also generated by a modified route-EFO model in which

TABLE 5.2   Fixed Hierarchical vs. Progressive Routing Single-Hour Design

Routing	10 A.M.		1 P.M.		8 P.M.	
	Cost	Average blocking	Cost	Average blocking	Cost	Average blocking
fixed hierarchical	$5,949,500	0.00865*	$5,839,300	0.00786*	$5,516,600	0.00809
progressive	5,567,800 (6.4% savings)	0.00396	5,463,700 (6.4% savings)	0.00408	5,121,000 (7.2% savings)	0.00525

*Average blocking from route-EFO model design

the stack algorithm described above is modified to convert the hierarchical homing pattern into the corresponding routing blocks, as illustrated in Figure 5.12. This provides additional verification of the route-EFO model, because we find the link sizes for the differently generated hierarchical designs to be quite close, as are the network costs. Single-hour network costs are within 1 percent of each other—multihour designs within 2 percent. A second reason for generating hierarchical designs with the route-EFO model is to obtain a comparative basis for switch-to-switch blocking distribution available from the model.

The network costs of the progressive and hierarchical networks are shown in Tables 5.2 and 5.3. In the three single-hour cases, the progressive routing design saves between 6 and 7 percent over the hierarchical routing design. The average network switch-to-switch blocking is also lower in each hour for the progressive network designs. For the 10 A.M. loads, the progressive routing design yields a 0.00396 switch-to-switch blocking, as compared with 0.00865 for the hierarchical routing design. The distribution of blocking for each switch pair is shown in Figure 5.14. As the figure suggests, the relatively large number of final links for the 10-switch network, each of which corresponds to a switch pair receiving 0.01 blocking, tends to raise the average network blocking. For a larger network with a lower ratio of final links to total links, a better blocking distribution is expected for the hierarchical networks.

TABLE 5.3   Fixed Hierarchical vs. Progressive Routing Multihour Design

Routing	Cost	Average blocking	Hour
fixed hierarchical	$7,160,000	0.00658*	10 A.M.
		0.00054*	1 P.M.
		0.00231*	8 P.M.
progressive	6,043,100 (15.6% savings)	0.00320	10 A.M.
		0.00220	1 P.M.
		0.00251	8 P.M.

*Average blocking from route-EFO model design

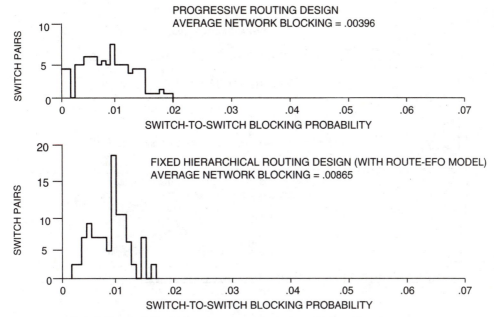

Figure 5.14 Switch-to-switch blocking comparison (single-hour design, 10 A.M.)

For the multihour network designs, progressive routing saves 15.6 percent over hierarchical routing. With respect to switch-to-switch blocking, although progressive routing does not provide better service in all hours, it does provide a more uniform blocking distribution over the three hours. This is due to the route-EFO model's ability to change routing by hour to take advantage, in a given hour, of spare capacity that is provided for another hour. The multihour blocking distributions for the progressive and hierarchical routing designs are shown in Figures 5.15 and 5.16. All switch pairs receive less than 0.02 blocking in all hours for both designs.

There are two reasons for this 15.6 percent savings in the multihour dynamic progressive routing network design. One reason is the restrictive nature of hierarchical routing, where the selection of outlet switches and their order are dictated by a homing pattern. In the progressive network design, routing is constructed solely as a function of cost. A second reason for the savings is the dynamic nature of progressive routing, where routing is allowed to change with time in a preplanned way. In a given hour, traffic is allocated over multiple routes for each switch pair. This allocation is optimized for each hour to take advantage of the noncoincidence of loads. The hierarchical design has fixed routing over all hours—the same route is used at all times.

In order to distinguish the savings resulting from each of the two reasons, the progressive routing design is modified so that, for a given switch pair, only the one least-cost progressive route is used in all hours. The comparison of fixed

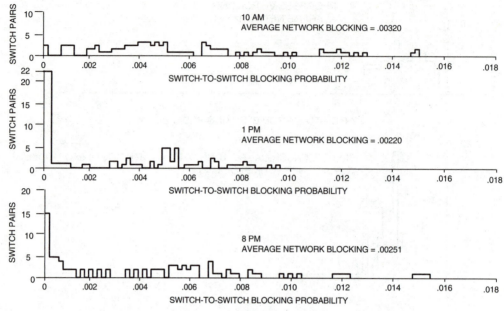

Figure 5.15 Progressive routing switch-to-switch blocking probability (multihour design)

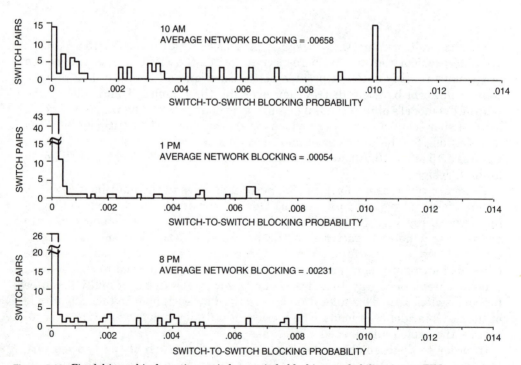

Figure 5.16 Fixed hierarchical routing switch-to-switch blocking probability (route-EFO model multihour design)

TABLE 5.4 **Fixed Progressive Routing vs Fixed Hierarchical Routing Design**

Routing	Cost	Average blocking	Hour
fixed hierarchical	$7,160,000	0.00658*	10 A.M.
		0.00054*	1 P.M.
		0.00231*	8 P.M.
fixed progressive	6,351,300	0.00109	10 A.M.
	(11.3% savings)	0.00034	1 P.M.
		0.00067	8 P.M.

*GOS for route-EFO design hierarchy

progressive with fixed hierarchical routing appears in Table 5.4. Fixed progressive routing still yields savings of 11.3 percent. However, although similar savings for fixed path routing are achieved in the 10-switch network, the savings decrease considerably for fixed routing design in the 28-switch national intercity network model. Hence, the percent savings of fixed progressive routing over hierarchical routing is not robust to the network model. Thus, for the 10-switch network, the incremental value of dynamic routing based on the dynamic versus fixed progressive network is $15.6 - 11.3 = 4.3$ percent.

In order to determine the economic value of crankback in a nonhierarchical routing environment, we compare progressive routing design with the path routing designs discussed in Chapter 4. The results are shown in Tables 5.5 and 5.6. For three single-hour networks, crankback is worth between 1 and 2 percent. As shown in Figure 5.17, when a multilink path routing network is chosen with a more comparable blocking probability grade-of-service objective, the savings decrease. In the multihour case, multilink path routing saves slightly over 1 percent. It also provides a somewhat better blocking performance in all hours, with the distributions appearing quite similar, as shown in Figures 5.15 and 5.18. Thus, it appears that crankback provides a modest economic benefit in a preplanned dynamic routing environment. A reason for this modest benefit is that, in comparing the two types of networks, most traffic is routed on the same

TABLE 5.5 **Progressive vs. Multilink Path Routing Single-Hour Design**

	10 A.M.		1 P.M.		8 P.M.	
Routing	Cost	Average blocking	Cost	Average blocking	Cost	Average blocking
progressive	$5,567,800	0.00396	$5,463,700	0.00408	$5,121,000	0.00525
multilink path	5,458,100 (2.0% savings)	0.007	5,367,600 (1.8% savings)	0.006	5,053,400 (1.3% savings)	0.007

TABLE 5.6   Progressive vs. Multilink Path Routing Multihour Design

Routing	Cost	Average blocking	Hour
progressive	$6,043,100	0.00320	10 A.M.
		0.00220	1 P.M.
		0.00251	8 P.M.
multilink path	5,980,100	0.002	10 A.M.
	(1.04% savings)	0.001	1 P.M.
		0.002	8 P.M.

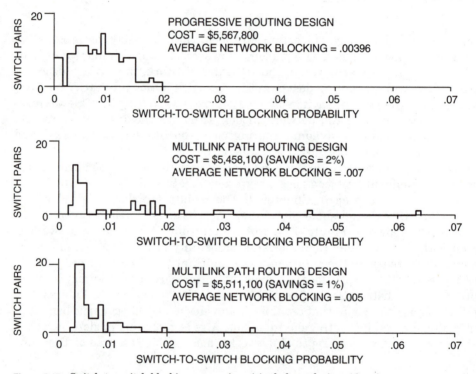

Figure 5.17   Switch-to-switch blocking comparison (single-hour design, 10 A.M.)

links. This is because, for many switch pairs, both routings carry a significant amount of the traffic on the direct path and the same two-link first alternate path.

An example of the similar traffic flow in the 10-switch network is illustrated in Figure 5.19. For traffic originating at switch A destined for switch Z, multilink path routing first routes on path A–Z and then overflows to path A–B–Z. Similarly, progressive routing first routes traffic to switch Z and then overflows to switch B. At switch B, the traffic first routes to switch Z. From these routing plans, together with link-blocking probabilities, the amount of traffic that flows on these paths can be calculated. Under either method, more than 95 percent of the A–Z traffic

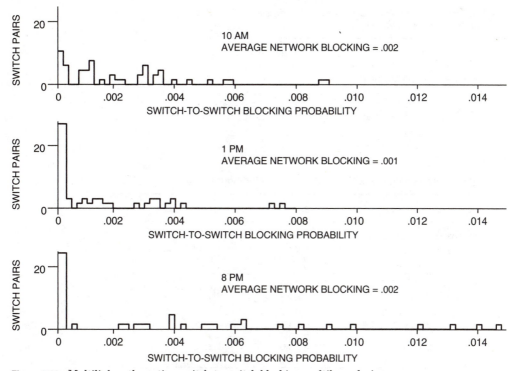

Figure 5.18 Multilink path routing switch-to-switch blocking multihour design

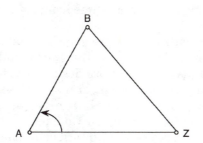

FRACTIONS OF TRAFFIC ROUTING ON A-Z AND A-B-Z

	10 AM		1 PM		8 PM	
	PROG.	MULTILINK PATH	PROG.	MULTILINK PATH	PROG.	MULTILINK PATH
A-Z	.938	.969	.933	.930	.948	.907
A-B-Z	.049	.027	.055	.045	.047	.056
BOTH PATHS	.987	.996	.988	.975	.995	.963

Figure 5.19 Routing comparison example

**TABLE 5.7   Progressive vs. 2-Link Path Routing
Single-Hour Design (10 A.M.)**

Routing	Cost	Average blocking
progressive	$5,567,800	0.00396
2-link path	5,555,700	0.00470 (unconstrained)
	5,555,900	0.00537 ($B_{max} = 0.02$)

TABLE 5.8   Progressive vs. 2-Link Path Routing Multihour Design

Routing	Cost	Average blocking	Hour
progressive	$6,043,100	0.00320	10 A.M.
		0.00220	1 P.M.
		0.00251	8 P.M.
2-link path	6,054,600	0.00316	10 A.M. (unconstrained)
		0.00237	1 P.M.
		0.00257	8 P.M.
	6,064,300	0.00256	10 A.M. ($B_{max} = 0.02$)
		0.00171	1 P.M.
		0.00200	8 P.M.

flows on either path A–Z or A–B–Z in all three hours, with more than 90 percent traveling on the direct link A–Z.

The results of two-link path routing are shown in Tables 5.7 and 5.8, which show results for networks in which 0.02 is the maximum blocking B_{max} allowed for any switch pair. As can be seen, network design costs for progressive routing design and two-link path routing design are very close, within 1 percent of each other. Overall, it appears that two-link path routing offers advantages in implementation, as discussed further in Chapter 6. In general, preplanned dynamic routing, either progressive or path oriented, has potential to result in significant network efficiency and performance improvements.

5.5   Conclusions

This chapter presents methods for preplanned dynamic progressive routing. In a progressive routing method, a call progresses through the network one switch at a time until it either reaches its destination or arrives at an intermediate switch from which it has no outlets. A call does not predetermine the busy or idle status of all links in its first-choice path from the originating switch. Nor does a call, if blocked at an intermediate switch along its desired path to the destination, have the ability to utilize crankback; that is, it cannot retrace any part of the path it has already traveled. Alternate routing must then be used to flow the call through the intermediate switch, or else it is blocked from the network.

Progressive routing networks have been designed by the route-erlang flow optimization model, which combines various savings concepts into a single design method. This model allows for preplanned, time-varying dynamic routing design that is feasible in the ESS/CCS environment.

For a 10-switch national intercity network model, progressive routing yields sizable savings, on the order of 15 percent, over hierarchical routing. Approximately 11 percent savings is shown to be achievable solely by employing fixed progressive routing. In addition, when compared with preplanned dynamic path routing (which uses crankback), progressive routing is economically comparable. Hence, preplanned dynamic progressive routing has potential to result in significant savings. As discussed further in Chapter 6, a simplified version of preplanned dynamic path routing, in which paths of only one or two links are allowed, offers benefits in its ease of implementation, in overall network performance, and in network management and design.

Preplanned Dynamic Two-Link Path Routing Networks

6.1 Introduction

The route-erlang flow optimization (route-EFO) model, described in Chapters 4 and 5, is a systematic procedure for designing near-minimum-cost preplanned dynamic routing networks. Chapters 4 and 5 show that the route-EFO model can design preplanned dynamic routing networks with a wide variety of routing methods to take advantage of traffic noncoincidence, taking into consideration network efficiency and efficient blocking levels to iteratively reduce network cost. This formulation of the erlang flow optimization model decides *a priori* on possible routes a call might take before assigning traffic.

That is, this route-EFO model first selects candidate paths that a call might take and then combines these paths into candidate routes. The next step uses a linear programming model to assign traffic to the candidate routes in a minimum-cost manner, wherein the choice of routes must be limited because of the large number of possible routes. For example, the number of sequential routes that can be formed from 10 paths is 10!, or over 3 million routes. Hence, the restricted choice of routes in the route-EFO model can result in suboptimality, because a better route not contained in those generated *a priori* may exist.

This chapter describes a somewhat different implementation of the erlang flow optimization model, called the path-EFO model, which is particularly well suited for preplanned dynamic two-link path routing design. For these designs, the path-EFO model achieves lower final network cost and greatly reduces the storage and run time compared with the route-EFO model. In the path-EFO model the linear programming model is given the permissible paths on which to assign flows, and then the path flows selected by the linear programming model are used to construct the routes. Thus, the linear programming model in essence selects the routing from all possible choices and has much greater flexibility and much lower dimensionality.

In Section 6.2 we discuss various preplanned dynamic two-link path routing methods, which are all amenable to the path-EFO model technique that forces feasibility of the flows chosen by the revised linear programming model. These techniques range in complexity from one that allows very general flow realizations (CGH routing) to one that consists of merely forming one possible sequential route for each traffic load. In the latter case the order of paths within the route is determined by ordering from the largest to the smallest offered loads to the paths assigned by the linear programming model; this method is called sequential routing and is particularly simple. Section 6.3 reviews the route-EFO model and highlights the differences from the path-EFO model formulation. Section 6.4 presents results using the different routing methods. It is found that the path-EFO model results in significant savings in storage and run time and saves an additional 1.5 percent in network cost compared with the route-EFO model for preplanned dynamic two-link path routing designs. Additional design efficiency occurs because the linear programming model can obtain a more optimal solution in the given time and is allowed greater freedom in flow assignment. A maximum cost penalty of 0.5 percent is found for using the simple sequential routing technique as compared with techniques requiring more complex preplanned dynamic two-link path routing.

6.2   Preplanned Dynamic Two-Link Path Routing Methods

As an outcome of the path-EFO model design, preplanned dynamic two-link path routing is constructed from the linear programming model flows. A flow realization model takes the linear programming model flows as inputs and designs the routing to approximate those flows as output. Three methods for preplanned dynamic two-link path routing to realize these flows are discussed here. These methods differ in their computational complexity and their flexibility in approximating the desired flows. Each method treats the desired flows in each design hour independently, hence the routing changes from hour to hour. Numerical results comparing the ability of these techniques to realize the desired flows and thus reduce network cost are given in Section 6.4.

6.2.1 CGH routing method

The following routing method is due to F.R.K. Chung, R. L. Graham, and F. K. Hwang [CGH81] and is termed *CGH routing* after its originators. Mathematical analysis supporting the routing method is given in [CGH81]; here we give a brief description of the method. For simplicity, subscripts dealing with demand pairs and design hours have been suppressed. Let

$$f_i = \text{desired flow on path } i$$
$$B_i = \text{blocking of path } i$$
$$Q_i = 1 - B_i = \text{connectivity of path } i$$
$$\sigma_i = f_i/Q_i = \text{desired offered load to path } i$$
$$\delta_i = B_i \sigma_i = \text{desired overflow load from path } i$$
$$J = \text{total number of paths}$$

CGH routing is composed of "cyclic blocks." As an example, suppose there are seven paths with desired flows f_i. One possible cyclic block realization of the seven f_i is

$$(1)\,(2\,3\,4)\,(5\,6)\,(7)$$

The notation means that all the offered load to this route is first offered to path 1. The overflow from path 1 is then offered to a cyclic block composed of paths 2, 3, and 4. The term *cyclic block* means that a proportion β_i^k of the total load offered to the kth block is offered to cyclic permutation i, where cyclic permutation i is selected so that the ordering within the block is preserved but a different path appears first. In the cyclic block under consideration, a proportion β_1^2 of the input traffic is offered to the paths in the order 2, 3, 4 (path 2 overflows to path 3, which overflows to path 4). Likewise, a proportion β_2^2 of the input traffic is offered to the paths in the order 3, 4, 2, and a proportion β_3^2 of the input traffic is offered to paths in the order 4, 2, 3. Offering traffic in this manner is accomplished by generating a random number as a call is offered to the cyclic block, with probability β_i^2 that the call is offered first to the ith path in the block and then searches all other paths in a cyclic order. Note that all calls see the same blocking probability within the cyclic block because all paths are searched. In the same manner, calls that overflow the cyclic block 2, 3, 4 are offered to cyclic block 5, 6, with a proportion β_1^3 of the incident traffic being offered to paths in the order 5, 6 and a proportion β_2^3 being offered to paths in the order 6, 5. All the overflow from the cyclic block 5, 6 is offered to path 7, which, like the first path, constitutes a cyclic block with only one path. Calls that find path 7 busy are blocked.

The realization method must define the contents of each cyclic block and calculate the proportions β_i^j associated with the jth cyclic block. The steps to accomplish this are as follows:

Step 1: Calculate σ_i and δ_i. Sort and relabel the σ_i, if necessary, so that $\sigma_1 \geq \sigma_2 \geq \sigma_3 \ldots \geq \sigma_J$.

Step 2: The first path in the cyclic block to be formed is the as yet unused path i with the largest σ_i.

Step 3: Insert the as yet unused path i with largest remaining σ_i after a path j with $\sigma_i > \delta_j$ if such a path j exists. Repeat this step until no such j exists.

Step 4: The current kth cyclic block ends with the last path inserted. If there is only one path in the block, set its coefficient to 1.0 and go to step 5. If there is more than one path in the block, let $\alpha_l^k = \sigma_{m(l)} - \delta_j$, where $m(l)$ refers to the path in the lth position in the kth block and j refers to the path preceding it in the cyclic ordering of the block. Note that the method guarantees that all α_l^k are positive. Then calculate $\beta_l^k = \alpha_l^k/(\sum_{l=1}^{L} \alpha_l^k)$, which is the cyclic coefficient associated with the lth path in the kth block, assuming there are l paths in the block.

Step 5: If there are remaining $x_i > 0$, return to step 2.

Step 6: Add single-path cyclic blocks at the end of the route, if necessary, until the blocking probability grade-of-service objective is satisfied.

6.2.2   Skip-one-path routing method

An example of *skip-one-path routing* is shown at the top of Figure 6.1. A switch implementation of skip-one-path routing is achieved by generating a random number before a call is offered to the next path in the list, such that with a predetermined probability the call skips over that path without being offered to it and proceeds to the next path in the routing sequence. The other information shown in Figure 6.1 is explained later in the chapter.

6.2.3   Sequential routing method

A very simple method to realize desired path flows is termed *sequential routing*. Traffic is routed sequentially from path to path with no probabilistic methods being used to get the realized flows closer to the desired flows. Sequential routing has the highest error in flow realization of all the three routing methods studied, which is expected because sequential routing is the least flexible of the three routing methods. Although sequential routing has the highest error in flow realization, the effect of this flow inaccuracy on network cost is minimal, as discussed in Section 6.4. Among the three preplanned dynamic two-link path routing alternatives, the sequential routing technique achieves network design savings within 0.5 percent of those of much more complicated methods such as CGH routing.

A routing method such as CGH routing has additional costs in switching system development because CGH routing has to store traffic allocation propor-

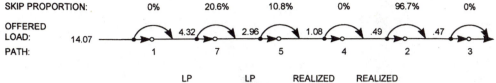

SKIP PROPORTION: 0%   20.6%   10.8%   0%   96.7%   0%

OFFERED
LOAD: 14.07   4.32   2.96   1.08   .49   .47
PATH: 1   7   5   4   2   3

PATH NUMBER	PATH BLOCKING	LP CARRIED LOAD	LP OFFERED LOAD	REALIZED OFFERED LOAD	REALIZED CARRIED LOAD	| ERROR |
1	.307	9.75	14.07	14.07	9.75	0
7	.603	1.36	3.43	3.43	1.36	0
5	.287	1.88	2.64	2.64	1.88	0
4	.453	1.06	1.94	1.08	0.59	0.47
2	.239	0.012	0.016	0.016	0.012	0
3	.309	0	0	0.47	0.32	0.32
6	.339	0	0	0	0	0
8	.488	0	0	0	0	0

TOTAL: 0.79

EXCEPT FOR PATH NUMBER AND PATH BLOCKINGS, ALL ENTRIES ARE IN ERLANGS

Figure 6.1 Skip-one-path routing example

tions and markers to indicate where the cyclic blocks begin and end, along with the ordered list of paths. Sequential routing, on the other hand, needs only the ordered list of paths. Also, applying traffic allocation techniques such as those needed by CGH routing requires switch real time to generate and process the appropriate random numbers. Sequential routing needs no such traffic allocation and hence has a real-time advantage. Simplicity is the greatest advantage of sequential routing. Such simplicity can benefit network traffic managers, whose job includes finding and correcting network problems, which is easier if they do not have to consider traffic splitting among several paths. The network management and design data collection and display systems that must process and display network data in near-real time for easy understanding are also simplified.

In two-link routing such as sequential routing, it is sometimes advantageous that the direct link not be the first path tried but that a two-link path might in some situations be tried prior to a direct link. An example is shown in Figure 6.2 for a four-switch network, in which the load patterns for each of two hours are shown between switches A–C, A–B, B–C, B–D, and C–D. In this example, in order to avoid putting in direct trunks between switches A and C, the A–C load in hour 1 needs to be routed through other network capacity. It is possible to route the A–C load in hour 1 through path A–B–C if the B–C load in hour 1 can somehow be routed elsewhere. This is possible in this example if the B–C load in hour 1 is routed on path B–D–C, in which there is idle capacity on both links B–D and C–D in hour 1. Therefore, it is advantageous in this example to route the B–C

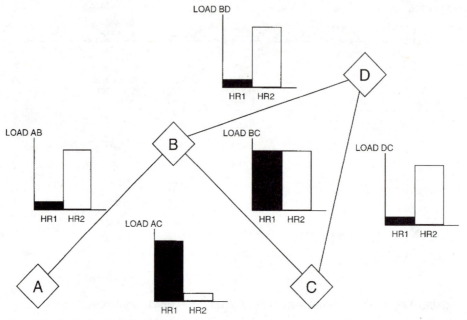

Figure 6.2   Example of two-link-path-first routing

traffic first to the two-link path B–D–C, so as to carry most or all of the B–C traffic on that path. By doing this, the A–C traffic in hour 1 can be routed on the A–B–C path, which now has idle capacity on both links A–B and B–C in hour 1, so that direct trunks can be omitted from the A–C switch pair. Note that in hour 2, the B–C traffic routes first on the direct link B–C. These kinds of traffic patterns are unusual but can occur. Normally the direct link is the first choice in the designs made by the path-erlang flow optimization model, but on occasion in situations such as that just described a two-link path could be selected prior to the direct path.

6.3   Network Management and Design

Figure 4.9 shows a block diagram of the route-EFO model, which is discussed in detail in Chapters 4 and 5. Figure 6.3 shows the path-EFO model, which is identical to the route-EFO model except for the routing design step, in which the path-EFO model forms routes after the linear programming model step is complete. Hence, the only change in the path-EFO model is that the route formation and linear programming model steps have been interchanged from the route-EFO model. Once routes have been formed, the routing is passed to the capacity design step and the path-EFO model continues as before. Only the routing design stage in the path-EFO model is described here because the other stages are virtually identical to those used in the route-EFO model, as described in Chapter 5.

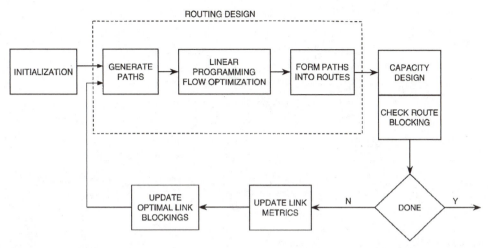

Figure 6.3 Path-erlang flow optimization model

The first step in routing design is to generate the required number of one- and two-link paths, which considers traffic noncoincidence in path selection. These paths are then passed to the linear programming model, which is somewhat different in structure than that of the route-EFO model. This difference arises because the amount of flow that can be carried on a particular path depends on the blocking on that path and on the flow assigned to all other paths composing the particular route. For instance, if the blocking probability on a path is .2 and the offered load is 100 erlangs, it would be impossible to carry more than 80 erlangs on this path. Hence, some method is needed to determine upper limits on path flow so that the resulting flows assigned by the linear programming model are feasible. Such questions of feasibility did not arise in the route-EFO model because the proportions of route-carried load calculated to flow on its links reflected the blocking probability of those links and the flow on all other paths in the route.

6.3.1 Flow feasibility model

An iterative method of utilizing upper bounds to force flow feasibility is shown in Figure 6.4. Here, flow feasibility constraints have been incorporated into the routing design stage. Immediately after the generation of paths, initial upper bounds on path flows are set for use by the first linear programming model iteration. At this point, nothing is known about the amount of flow that is optimal on any path. Hence, it is desirable to constrain the linear programming model as little as possible. For this reason, the initial upper bound on flow on any path is set according to the following formula, in which reference to the design hour is suppressed for clarity:

$$\text{UPBD}_j = A(1 - B_j)$$

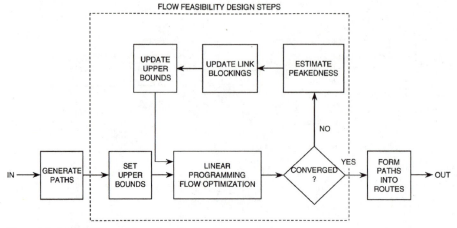

Figure 6.4 Path-EFO model routing design steps

where UPBD_j = upper bound on path j
   A = offered load for this switch pair
   B_j = blocking probability on path j

Hence, the initial upper bound on flow is set assuming that the entire offered load is offered to any path independently of the load offered to any other path. The resulting flows can thus be infeasible, because there might not be enough offered load to simultaneously achieve the desired flow on all paths for the same switch pair. For instance, suppose

$$B_1 = 0.2, \qquad B_2 = 0.1, \qquad B_3 = 0.2$$

$$A = 10 \text{ erlangs}$$

Then

$$\text{UPBD}_1 = 8 \text{ erlangs}, \qquad \text{UPBD}_2 = 9 \text{ erlangs}, \qquad \text{UPBD}_3 = 8 \text{ erlangs}$$

It is assumed that the required blocking probability grade-of-service objective is .005 so that the flow on all three paths should total 9.95 erlangs; this required flow is feasible because an overall blocking of $B_1 B_2 B_3 = 0.004$ is possible should all paths be used. Now suppose that the linear programming model chooses the optimal flows for this switch pair as

$$f_1 = 8 \text{ erlangs}, \qquad f_2 = 1.95 \text{ erlangs}, \qquad f_3 = 0$$

where f_1 is the carried flow on path i. The only way to realize the desired flow of 8 erlangs on path 1 is to offer path 1 the entire 10 erlangs. This means that 2 erlangs overflow from path 1. These 2 erlangs can then be offered to path 2 but can result in a maximum flow of 1.8 erlangs due to the blocking on path 2, and hence the desired flows are infeasible. A method to force these flows toward a feasible solution is discussed in the next section.

We focus now on the linear programming model used in the path-EFO model to optimize path flows, in which there are now upper-bound constraints on path flows:

$$\text{Minimize} \sum_{i=1}^{L} M_i a_i$$

subject to

$$\sum_{k=1}^{K} \sum_{j=1}^{J_k^h} P_{jk}^{ih} f_{jk}^{h} \le a_i \qquad \begin{aligned} i &= 1, 2, \ldots, L \\ h &= 1, 2, \ldots, H \end{aligned}$$

$$\sum_{j=1}^{J_k^h} f_{jk}^{h} = g_k^{h} \qquad \begin{aligned} h &= 1, 2, \ldots, H \\ k &= 1, 2, \ldots, K \end{aligned}$$

$$f_{jk}^{h} \le \text{UPBD}_{jk}^{h} \qquad \begin{aligned} h &= 1, 2, \ldots, H \\ k &= 1, 2, \ldots, K \\ j &= 1, 2, \ldots, J_k^h \end{aligned}$$

$$f_{jk}^{h} \ge 0, \qquad a_i \ge 0$$

where L = number of links
K = number of switch pairs
H = number of design hours
J_k^h = number of paths for switch pair k in hour h
g_k^h = total load required to be carried for switch pair k in hour h
P_{jk}^{ih} = 1 if path j for switch pair k uses link i in hour h
P_{jk}^{ih} = 0 otherwise
M_i = incremental link cost metric in terms of dollar cost per erlang of carried traffic
f_{jk}^h = carried load on path j for switch pair k in hour h
a_i = maximum carried load capacity on link i

As in the route-EFO model, a heuristic solution is used to solve the linear programming model. The total carried load for switch pair k in hour h, g_k^h, is related to the total offered load for switch pair k in hour h, v_k^h, as follows. Let B_{jk}^h be the blocking of path j for switch pair k in hour h. Then the minimum blocking that can be achieved is

$$E_k^h = \prod_{j=1}^{J_k^h} B_{jk}^h$$

Let GOS = the desired blocking probability grade-of-service objective and G_k^h = the maximum (E_k^h, GOS). Then $g_k^h = v_k^h(1 - G_k^h)$. Thus, the required carried flow is determined by the blocking objective, unless the minimum blocking is greater than the objective. If the blocking objective cannot be met, all paths are required to be at their maximum flow to minimize the blocking. A route-blocking correction model similar to that used in the route-EFO model described in Chapter 4 is used in the capacity design step to correct routes with unacceptable blocking.

If the flow feasibility model is not converged, the next step shown in Figure 6.4 after the linear programming model is to update the link blockings in all hours based on the current link flow. This is done by calculating the link size so that the maximum allowed blocking in any hour is not exceeded and then calculating the blocking in all hours. After the blockings are updated, the upper bounds are recalculated based on the current desired flows, as determined by the linear programming model, and on the resultant blockings in order to force a more feasible solution.

The method employed to recalculate the upper bounds is now illustrated by an example. The data in Figure 6.5 show how the routing method described above, called skip-one-path routing, is used to set upper bounds that force more feasible flows while still allowing the linear programming model some flexibility in choosing new flow patterns. Basically, this routing method works by keeping track of the offered load available and utilizing this load to realize the desired flows in a sequential manner. The data in Figure 6.5 are obtained from the output of the linear programming model, of which the complete results are discussed

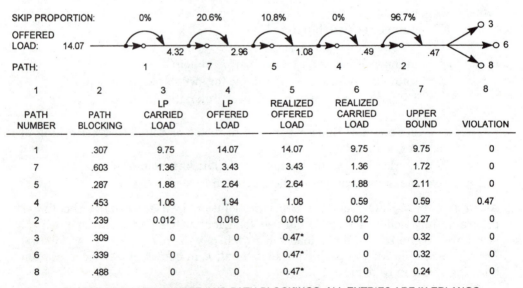

PATH NUMBER	PATH BLOCKING	LP CARRIED LOAD	LP OFFERED LOAD	REALIZED OFFERED LOAD	REALIZED CARRIED LOAD	UPPER BOUND	VIOLATION
1	2	3	4	5	6	7	8
1	.307	9.75	14.07	14.07	9.75	9.75	0
7	.603	1.36	3.43	3.43	1.36	1.72	0
5	.287	1.88	2.64	2.64	1.88	2.11	0
4	.453	1.06	1.94	1.08	0.59	0.59	0.47
2	.239	0.012	0.016	0.016	0.012	0.27	0
3	.309	0	0	0.47*	0	0.32	0
6	.339	0	0	0.47*	0	0.32	0
8	.488	0	0	0.47*	0	0.24	0

EXCEPT FOR PATH NUMBER AND PATH BLOCKINGS, ALL ENTRIES ARE IN ERLANGS

* OFFERED LOAD ASSIGNED FOR UPPER BOUND CALCULATION AS DESCRIBED IN TEXT

Figure 6.5   Upper-bound determination using skip-one-path routing

in Section 6.4. These results are the flows selected by the linear programming model using the initial upper bounds to route the 14.07 erlangs of load offered by a particular demand pair in a particular hour.

The first step in the model is to calculate the load that must be offered to each path to realize the flow selected by the linear programming model. This offered load, given in the fourth column of Figure 6.5, has been calculated from the carried load on each path selected by the linear programming model (column 3) divided by the path connectivity, which is 1 minus the blocking on the path (column 2). The next step is to sort on the linear programming model path offered loads in order of largest to smallest, because it is most important to accurately realize the larger path flows. This sort has been done for the data in Figure 6.5; note that path number 7 follows path number 1 in terms of offered load.

Once the path ordering is determined, the flow feasibility model proceeds as follows. Because the largest offered load assigned by the linear programming model is equal to the total offered load of 14.07 erlangs for the demand pair and hour of interest, all of the load must be offered to path 1, as shown in the diagram in Figure 6.5. Hence, none of the offered load is "skipped over" path 1. Applying the load in this way realizes the desired flow on the first path. Note that path 1, which carries the greatest flow, is at its new upper limit, which is a common situation. With the given blocking of path 1, the overflow from path 1 is 4.32 erlangs.

Thus, the offered and carried loads desired on path 1 can be achieved, as shown in columns 5 and 6 in Figure 6.5. Because the total traffic load is available for path 1 and the blocking is assumed constant for this example, the upper bound on path 1 flow remains constant. The last column in Figure 6.5 gives the "violation," or amount by which the desired flow exceeds the new upper bound. In this case the violation is zero.

Now consider path 7, which is next in order of offered load. The desired offered load to this path is 3.43 erlangs, which is less than the overflow from path 1. The difference between these two loads, which is $4.32 - 3.43 = 0.89$ (20.6 percent of 4.32), is skipped over path 7, and 3.43 erlangs is offered to path 7. That is, with probability .206, the call skips path 7 and is offered to the next path. A call that does not skip over path 7 is offered to path 7.

Thus, the desired flow on path 7 is realized. The upper bound on path 7 is calculated assuming that the entire offered load not yet used, which is $3.43 + 0.89 = 4.32$ erlangs, is offered to path 7. Note that this allows for more flow to be assigned to path 7 if desirable on the next iteration of the linear programming model. The skip-one-path routing method gives an actual offered load to path 7 of 3.43 erlangs with 2.07 erlangs overflow, because $(0.603)(3.43) = 2.07$. The total available load for any other path is now $2.07 + 0.89 = 2.96$ erlangs.

Now consider path 5, which needs 2.64 erlangs of offered load. The amount of traffic to be skipped is $2.96 - 2.64 = 0.32$, or 10.8 percent of the 2.96 erlangs available. The upper bound on path 5 flow is based on 2.96 erlangs, which is the total available load at present that could be offered to path 5.

A different situation arises, however, when attempting to realize the desired offered load to path 4 of 1.94 erlangs. The total of the overflow from path 5 (0.76) and the 0.32 that was skipped over path 5 is 1.08 erlangs, which is the maximum load that can be offered to path 4. Hence, the linear programming model has assigned more flow than can be realized. The maximum possible flow is calculated as 0.59; likewise the upper bound is 0.59. Thus, there is a bound violation of $1.06 - 0.59 = 0.47$ erlangs.

The process continues as shown in Figure 6.5 until the last path with a nonzero linear programming model flow has been dealt with. At this point, all unused load—0.47 erlangs in this example—is assumed to be available as offered load for all paths with a zero flow assigned by the linear programming model. The upper bounds are then calculated in the same way that the initial upper bounds are set. The idea here is to allow maximum choice on the least desirable paths for the next iteration of the linear programming model.

Once the upper bounds are calculated, the linear programming model is again executed to reoptimize the path flows. The sum total of bound violations is available as a measure of flow feasibility. It is not necessary to start the linear programming model at the beginning because the current routing patterns, with upper bounds updated to reflect the new flows, can be used as a starting basis.

6.3.2   Flow realization techniques

Once the flow assignment model has converged, it is necessary to realize the final linear programming model flows, as shown in Figure 6.4. The flow realization model has the desired linear programming model flows as its inputs and has routing selected to approximate those flows as an output. The three flow-realization routing methods discussed in Section 6.2 are illustrated here: the CGH routing method, the skip-one-path routing method, and the sequential routing method. These methods differ in their computational complexity and their flexibility in approximating the desired flows, as discussed in Section 6.2.

Figure 6.6 shows an example of the CGH routing method and the resulting routing. The data for this example are identical to those used for the example of Figure 6.5. The path order in Figure 6.6 has already been sorted on offered load σ_i. As described in Section 6.2.1, path 1 becomes the first path in the first cyclic block because σ_1 is the largest; it is also the only path in the first cyclic block because no other σ_i is larger than δ_1.

Path 7 begins the next cyclic block because σ_7 is the largest remaining offered load. Path 5 follows path 7 in the second cyclic block because σ_5 (2.64) is greater than δ_7 (2.07). Likewise, path 4 follows path 5. The second block ends with path 4 because no other unused path has an offered load greater than δ_4.

The coefficients of the second cyclic block are computed as follows:

$$\alpha_1^2 = \sigma_7 - \delta_4 = 3.43 - 0.88 = 2.55$$

$$\alpha_2^2 = \sigma_5 - \delta_7 = 2.64 - 2.06 = 0.58$$

$$\alpha_3^2 = \sigma_4 - \delta_5 = 1.94 - 0.76 = \underline{1.18}$$

$$4.31$$

PATH NUMBER	PATH BLOCKING	LP CARRIED LOAD	LP OFFERED LOAD	LP OVERFLOW LOAD	REALIZED CARRIED LOAD	|ERROR|
	(B_i)	(F_i)	(σ_i)	(δ_i)		
1	.307	9.75	14.07	4.32	9.75	0
7	.603	1.36	3.43	2.06	1.26	0.10
5	.287	1.88	2.64	0.76	1.74	0.14
4	.453	1.06	1.94	0.88	0.98	0.08
2	.239	0.012	0.016	0.0037	0.26	0.25
3	.309	0	0	0	0.06	0.06
6	.339	0	0	0	0	0
8	.488	0	0	0	0	0

TOTAL: 0.63

ROUTE:

(1) (7,5,4) (2) (3)

COEFFICIENTS:

(100%) (59.2%, 13.4%, 27.4%) (100%) (100%)

EXCEPT FOR PATH NUMBER AND PATH BLOCKING, ALL ENTRIES ARE IN ERLANGS

Figure 6.6 CGH routing example

Then

$$\beta_1^2 = \frac{2.55}{4.31} = 59.2\%$$

$$\beta_2^2 = \frac{0.58}{4.31} = 13.4\%$$

$$\beta_3^2 = \frac{1.18}{4.31} = 27.4\%$$

The coefficients β_i^k for the four blocks are shown below the route shown in Figure 6.6 in order of starting path; thus, 59.1 percent of the traffic offered to the second block starts with path 7.

Path 2 forms a one-member cyclic block, because it is the only path left with a positive σ. Note that path 3 is included to decrease the overflow load from 0.08 to 0.02 erlang and hence the blocking from 0.0057 to 0.0014, thus meeting the blocking probability grade-of-service objective of 0.005. As illustrated in the figure, the total path flow error in matching the CGH routing flows with the linear programming model flows is 0.63 erlangs.

The skip-one-path routing method can be used to realize path flows, as well as calculate upper bounds as discussed in the previous section. An example

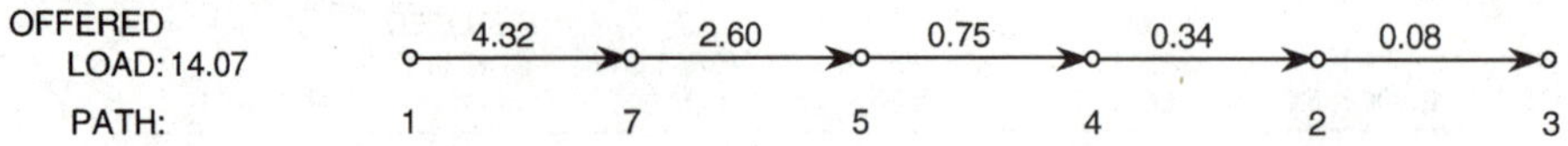

OFFERED
LOAD: 14.07

| PATH: | 1 | 7 | 5 | 4 | 2 | 3 |

Flow: 4.32 → 2.60 → 0.75 → 0.34 → 0.08

PATH NUMBER	PATH BLOCKING	LP CARRIED LOAD	LP OFFERED LOAD	REALIZED OFFERED LOAD	REALIZED CARRIED LOAD	|ERROR|
1	.307	9.75	14.07	14.07	9.75	0
7	.603	1.36	3.43	4.32	1.72	0.35
5	.287	1.88	2.63	2.60	1.86	0.02
4	.453	1.06	1.94	0.75	0.41	0.65
2	.239	0.012	0.016	0.34	0.26	0.25
3	.309	0	0	0.08	0.06	0.05
6	.339	0	0	0	0	0
8	.488	0	0	0	0	0

TOTAL: 1.32

EXCEPT FOR PATH NUMBER AND PATH BLOCKINGS, ALL ENTRIES ARE IN ERLANGS

Figure 6.7 Sequential routing example

of skip-one-path routing is shown in Figure 6.1. With a predetermined probability, a call skips over a path without being offered to it and proceeds to the next path in the routing sequence. Once again, the first step is to sort the paths by offered load. The method used to calculate the amount of offered traffic to skip the next path is discussed above in connection with Figure 6.5. Note that path 3 has been added to meet a blocking probability grade-of-service objective of .005. Also, Figure 6.1 shows that the path flow error in matching skip-one-path flows with linear programming model flows is 0.79, which is greater than the 0.63 path flow error given by the CGH routing method. This may be expected, because the CGH routing method has more flexibility in meeting path flows than does the skip-one-path routing method.

Figure 6.7 shows a sequential routing example. The given blockings and desired flows are identical to those used in the other routing examples. Note that in this particular example, sequential routing has the highest error in flows of all the three routings studied. In general, this is expected because sequential routing is the least flexible of the three realization methods discussed here. The reason that sequential routing works well is that most flow in the path-EFO model design is carried on the first one or two paths, and because these paths are loaded to their maximum flow, errors in meeting flow on later paths are not significant. The effect of this flow inaccuracy on network design cost efficiency is considered in Section 6.4.

6.3.3 Heuristic linear programming solution method

A heuristic linear programming solution method is used to solve the path-EFO model linear programming model formulated above. We discuss first the three basic ideas underlying the heuristic linear programming method and provide a brief overview. We then illustrate the performance of the heuristic method as compared with the optimal linear programming solution.

6.3.3.1 Rerouting of traffic.

The first concept concerns the rerouting of traffic. A reroute is a reassignment of flow of a particular switch pair from one path to another in a single design hour. Given an initial assignment of path flows for each switch pair, the heuristic linear programming method progresses to its final solution by a sequence of reroutes. Thus, each iteration of the heuristic method affects the flow on a few links and in only one design hour.

6.3.3.2 Marginal costs.

Next, we discuss a concept that allows us to evaluate the potential cost savings of any reroute. The marginal link cost is an estimate for the rate of change of the total network cost function relative to the change in flow on a link during a particular design hour. The heuristic linear programming method uses an UPCOST and a DOWNCOST, indicating the predicted cost change if we increase or decrease the link flow during a particular hour. We maintain marginal costs for every link during every design hour.

The rules for determining the marginal link costs of a link are simple. For a particular link we examine the flow during each design hour. If the peak flow on the link occurs in only one hour, then increasing or decreasing the flow in that hour increases or decreases the capacity of the link. We then set the UPCOST and DOWNCOST in the peak hour equal to the metric of the link. If the peak flow occurs in more than one hour, then increasing the flow in one of the peak hours increases the link capacity, while decreasing the flow in one of the peak hours leaves the capacity unchanged. We then set the UPCOST in all peak hours equal to the link metric and set the DOWNCOST in all peak hours equal to zero. In all design hours where the link flow is below the peak flow, we set the UPCOST and DOWNCOST equal to zero, because increasing or decreasing the flow does not affect the link capacity.

Once the marginal link costs are computed, we determine the marginal cost of diverting flow from one path to another in the following way. We first sum the UPCOSTs of the path that will gain flow and then sum the DOWNCOSTs of the path that will lose flow. Subtracting the latter sum from the former sum yields the marginal cost of the reroute. If this cost is negative, then the reroute is profitable. Once we decide to perform a particular reroute whose marginal cost indicates that it is profitable, we then determine the amount of flow to divert. The rule for finding this quantity is to continue rerouting until the marginal cost of the reroute changes.

Figures 6.8 and 6.9 give an example of how the marginal link costs are determined and how they are modified when traffic rerouting occurs. Figure 6.8 shows two paths between switches A and D in the first of two design hours. Each

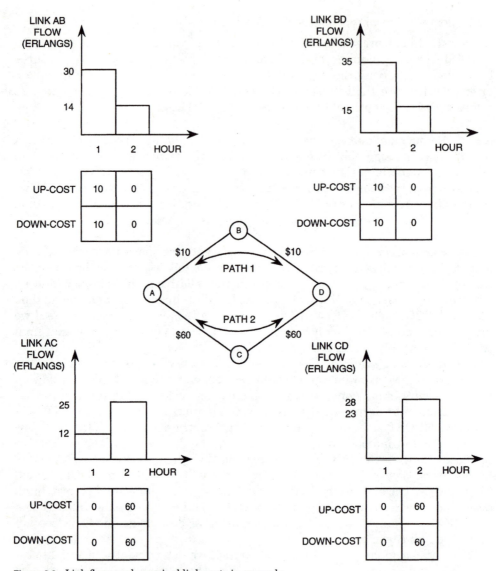

Figure 6.8   Link flows and marginal link costs in example

path has two links. The metrics for links A–B, B–D, A–C, and C–D are 10, 10, 60, and 60, respectively. Figure 6.8 also shows the initial flow and marginal costs for each of the links in each of two design hours. For example, link A–B carries 30 erlangs in hour 1 and only 14 erlangs in hour 2. Because it has a unique peak in hour 1, the UPCOST and DOWNCOST of link A–B in hour 1 are set equal to 10, the metric of link A–B. In hour 2, they are set equal to zero.

We can now use the marginal link costs to determine the marginal cost for rerouting traffic from path 1 to path 2. Summing the UPCOSTs of links A–C and

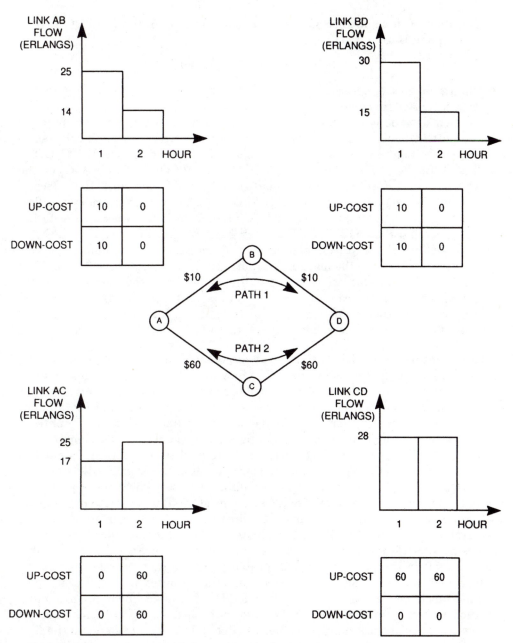

Figure 6.9 Link flows and marginal link costs in example

C–D in hour 1 and subtracting from this sum the sum of the DOWNCOSTs of links A–B and B–D in hour 1 yields

$$(0 + 0) - (10 + 10) = -20$$

the marginal cost of diverting from path 1 to path 2 in hour 1. Therefore, the reroute is a profitable one.

We next determine the amount of flow to divert. Now the marginal profit of the reroute will hold until one of the marginal link costs in the above calculation changes. We then reroute as much flow as possible until either the DOWNCOST of link A–B, the DOWNCOST of link B–D, the UPCOST of link A–C, or the UPCOST of link C–D changes in hour 1. Figure 6.9 describes the effect of rerouting 5 erlangs of flow in hour 1. Link C–D has gained 5 erlangs of flow and now has two peak hours. The UPCOST in each hour is now equal to the link metric, and the DOWNCOST in each hour is zero. Because the UPCOST of link C–D has changed from zero to 60, 5 erlangs is the total amount of flow that we reroute. If after the marginal reroute cost is reevaluated we find that the reroute is still profitable, we continue to divert flow until the marginal reroute cost changes again. We continue in this manner until either the reroute is no longer profitable, there is no more flow assigned to path 1, or the flow on path 2 reaches its upper bound. We then search for another profitable reroute.

6.3.3.3 Candidate list. The last concept for the heuristic linear programming method concerns the method for deciding how many candidate reroutes to evaluate before actually performing a particular reroute. In selecting a reroute pair, there is a trade-off between the quality of the reroute found and the amount of time spent searching for it. Although we would like to find very profitable reroutes, the heuristic linear programming method should also be computationally efficient. The heuristic linear programming method uses a candidate list to find the next reroute to perform. This concept works in the following way. The first M switch pairs are searched for profitable reroutes. The K most profitable reroutes are put into a candidate list, and the most profitable reroute in the list is selected and performed. Once this particular reroute is no longer profitable, the remaining members of the list are evaluated and the most profitable reroute is selected and performed. This process continues until there are no more profitable reroutes left in the list. The next M switch pairs are then searched for profitable reroutes, a new list is generated, and the reroutes in the list are performed until they are no longer profitable. The heuristic linear programming method continues in this manner. Whenever the last switch pair in the set of all switch pairs is encountered, the next switch pair to be considered is the first switch pair in the set. The heuristic then "wraps around" the set of all switch pairs. The heuristic linear programming method finally terminates when there are no profitable reroutes among all switch pairs.

The three concepts described in this section—the rerouting of traffic, the marginal costs, and the candidate list—are used together in the heuristic linear programming method. After an initial feasible solution is selected, the marginal link costs are determined. A group of switch pairs is searched and a list of the most profitable reroutes is formed. Each reroute is performed until it is no longer profitable. When there are no more profitable reroutes in the list, a new group of switch pairs is searched and a new list is formed. The heuristic solution method continues in this manner until there are no profitable reroutes. The next section describes typical computational results for the solution method.

6.3.3.4 Computational results for heuristic linear programming solution method. We compare the heuristic linear programming solution method with an optimal solution determined by the revised simplex method implemented in MPSX/370 using a linear programming problem that is generated by the path-EFO model. The problem is derived from a 30-switch national intercity network model, in which the average number of paths per switch pair is 9.4. The corresponding linear programming model had 1,402 rows, 7,630 columns, and a density of 0.17 percent. We use as a reference point an optimal solution obtained by MPSX/370 after 5,340 CPU seconds. The heuristic solution method progresses very rapidly until it is within 2.37 percent of the optimal MPSX/370 solution, which is obtained after five CPU seconds. In contrast, MPSX/370 requires 750 CPU seconds to produce a solution of similar quality. We see that the heuristic linear programming solution method can produce a near-optimal solution much more quickly than MPSX/370. Also, the path-EFO model contains many approximations, so that an optimal solution to the linear program is not necessary. In fact, there is little penalty in the network cost if the heuristic linear programming method is used. To design a 200-switch network with six design hours, the path-EFO model requires about 20 million bytes of memory, much of which is needed to solve the linear programming model. Tests with a 190-switch national intercity network model for six design hours indicate that the path-EFO model requires less than one hour of CPU time to design a large network. Half of this time is spent by the heuristic linear programming method. With continued advances in computer hardware, the solution time will continue to decrease.

6.3.4 Peakedness methods

For an accurate network design, the peakedness of overflow traffic must be considered in the model. In the path-EFO model, peakedness is considered in the routing design step, as illustrated in Figure 6.3, and also in the capacity design step, as illustrated in Figure 6.10. Therefore, peakedness is considered in the path-EFO model, in which there are three blocking values computed for each link: the time congestion (time), the average blocking (PB), and the blocking as seen by the overflow calls that have been blocked on other links (BR). An approximate value for the time congestion is calculated directly from the Erlang B blocking formula and the equivalent random method, given the link offered load and peakedness. The alternate-routed traffic blockings BR are calculated using the quantity of first-routed traffic to a link, which gives the time congestion and the average blocking probability as follows:

Let F = amount of first-routed traffic offered to the link
A = total traffic offered to the link

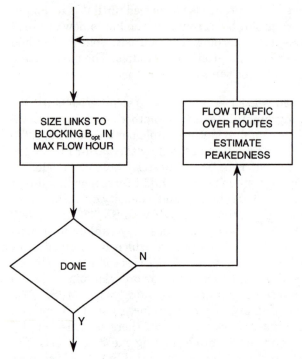

Figure 6.10 Capacity design steps

time = time congestion of the link
PB = average blocking on the link

Then the amount of first-routed traffic blocked is

$$F * \text{time}$$

and the amount of alternate-routed traffic blocked is

$$R = A * \text{PB} - F * \text{time}$$

and hence the alternate-routed traffic blocking must be

$$\text{BR} = \frac{R}{A - F}$$

The variance of the offered load to each link is taken to be the sum of the variances of all traffic loads offered to the link. The variance of each individual traffic load and the peakedness of a traffic load overflowing a path to another path are computed in the estimate peakedness step of the flow feasibility design illustrated in Figure 6.4 and the capacity design illustrated in Figure 6.10.

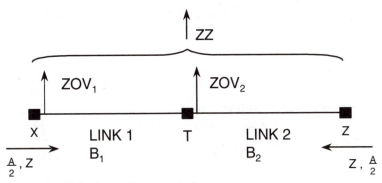

Figure 6.11 Methods to estimate peakedness

6.3.4.1 Estimate peakedness.   This step in the path-EFO model calculates an estimate of offered peakedness on all links based on the current linear programming model solution.

We first calculate the "reduced" path offered load to each path as the desired carried load divided by the path connectivity. The average blocking is used to compute the connectivity because at this point the model does not have any knowledge of which path comes first in the route. The paths are then sorted in decreasing order of reduced offered load. The calculation of offered peakedness then proceeds using the path ordering determined by the sort and the following general idea: Given the offered peakedness Z to a path, one can compute the offered variance V to the links of that path and the overflow peakedness ZZ from these links. The overflow peakedness ZZ, once computed, becomes the offered peakedness to the next path of the route, and the procedure continues for all paths of the route. The offered peakedness of the first path is assumed to be $Z = 1$. The procedure for calculating the offered variance to the links of a path is as follows:

Let Z = the offered peakedness to the path
     A = offered load to the path

Then for a single-link path, the variance offered to that link is

$$V = Z * A$$

For a two-link path, assume that $A/2$ erlangs are offered to each direction of the path, as shown in Figure 6.11. Let B_i = blocking of link i $(i = 1, 2)$. Then the variance offered to link 1 due to the load of $A/2$ erlangs offered to the path in the order (link 1, link 2) is

$$V_1^1 = Z \frac{A}{2}(1 - B_2)$$

Assuming the offered and the overflow variances to a link are independent, the variance offered to link 2 due to the same load is

$$V_2^1 = Z\frac{A}{2} - \text{ZOV}_1\frac{A}{2}B_1$$

where $\text{ZOV}_1 =$ the overflow peakedness from link 1.

Similarly, due to the load of $A/2$ erlangs offered to the path in the order (link 2, link 1), the variance offered to link 2 is

$$V_2^2 = Z\frac{A}{2}(1 - B_1)$$

The variance offered to link 1 is

$$V_1^2 = Z\frac{A}{2} - \text{ZOV}_2\frac{A}{2}B_2$$

Therefore, the total offered variances of link 1 and link 2 are

$$V_1 = V_1^1 + V_1^2 = A*Z\left[1 - \frac{B_2}{2}\left(1 + \frac{\text{ZOV}_2}{Z}\right)\right]$$

$$V_2 = V_2^1 + V_2^2 = A*Z\left[1 - \frac{B_1}{2}\left(1 + \frac{\text{ZOV}_1}{Z}\right)\right]$$

The overflow variance is computed as follows. Again assume that $A/2$ erlangs are offered to each direction of the two-link path where the blockings of the links are B_1 and B_2, respectively. For the load of $A/2$ erlangs offered to the path in the order (link 1, link 2), then, the overflow variance of link 1 is

$$\text{OV}_1^1 = \text{ZOV}_1\frac{A}{2}B_1$$

where $\text{ZOV}_1 =$ the average overflow peakedness from link 1. The overflow variance of link 2 due to the same load is

$$\text{OV}_2^1 = \text{ZOV}_2\frac{A}{2}(1 - B_1)B_2$$

Therefore, the overflow variance of the path due to this same load is

$$\text{OV}^1 = \text{OV}_1^1 + \text{OV}_2^1 = \text{ZOV}_1\frac{A}{2}B_1 + \text{ZOV}_2(1 - B_1)B_2\frac{A}{2}$$

For the load of $A/2$ erlangs offered to the path from the other direction, the path overflow variance is then

$$\mathrm{OV}^2 = \mathrm{ZOV}_2 \frac{A}{2} B_2 + \mathrm{ZOV}_1 (1 - B_2) B_1 \frac{A}{2}$$

The total overflow variance of the path is

$$\mathrm{OV} = \mathrm{OV}^1 + \mathrm{OV}^2 = A \left[\mathrm{ZOV}_1 B_1 \left(1 - \frac{B_2}{2} \right) + \mathrm{ZOV}_2 B_2 \left(1 - \frac{B_1}{2} \right) \right]$$

Because the mean overflow load is $\mathrm{OA} = A(B_1 + B_2 - B_1 B_2)$, the overflow peakedness of the path is

$$ZZ = \frac{\mathrm{OV}}{\mathrm{OA}} \frac{\mathrm{ZOV}_1 B_1 (1 - B_2/2) + \mathrm{ZOV}_2 B_2 (1 - B_1/2)}{B_1 + B_2 - B_1 B_2}$$

For a single-link path, the overflow peakedness of the path is the overflow peakedness of the link. The offered variances of each link in each hour are accumulated, and the offered peakedness is determined by dividing the accumulated variance by the average offered load of that link in that hour. The mean carried flow is accumulated for each link in each hour.

6.3.4.2 Capacity design. Once the routing is finalized by the sequential routing step, the capacity design step illustrated in Figure 6.10 uses an iterative procedure to get a consistent set of blockings and flows. As discussed in Chapter 5, the size links procedure updates the link blockings based on the current link flow, the current optimal link blocking, and the current peakedness. The flow traffic procedure takes the new blockings to calculate the new link flow parameters. The process converges when the blocking error is less than a threshold value. Here we describe the peakedness methods incorporated in the capacity design step.

The size links procedure starts by finding the required number of trunks TK in the design hour in order to satisfy the optimal link blocking objective B_{opt}. The hour for sizing each link is the hour with the maximum equivalent random offered load. Once the number of trunks TK is found, the blocking in all hours is determined. The blocking of a link as seen by the overflow traffic is

$$\mathrm{BR} = \frac{A * \mathrm{PB} - \mathrm{time} * F}{A - F}$$

where A = average offered load to the link
  PB = average blocking of the link
   F = first routed offered load to the link

Then the average blocking is

$$PB = \frac{A^*}{A} B(A^*, TK + S)$$

where A^* = equivalent random load
S = equivalent random link size
A = offered load
TK = link size
B = Erlang B function

The time congestion is time $= (B\mathrm{up} + B\mathrm{low})/2.0$, with

$$B\mathrm{up} = \left(1.0 + \frac{S}{TK}\right) * PB * \frac{A}{A^*}$$

$$B\mathrm{low} = \left[1.0 + \frac{1}{TK}(S - A^* + A)\right] PB \frac{A}{A^*}$$

The average overflow peakedness is computed using Nyquist's formula:

$$ZOV = 1.0 - A * PB + \frac{A^*}{S + TK + 1 + A * PB - A^*}$$

The flow traffic procedure calculates new link flows, and in case of peakedness, it also calculates the offered peakedness factors using the current blockings. The order of the paths is set in the routing design step. The peakedness method is the same as the one used in the estimate peakedness step except that use is made of the time and BR blockings depending on the path order. The offered link variance for each hour is accumulated, and the offered peakedness is determined by dividing this quantity by the total link offered load.

6.3.5   Day-to-day variation methods

We just discussed the methods for increasing trunk requirements to account for peakedness values Z greater than 1 for offered and overflow traffic loads, and now we discuss methods for increasing trunk requirements to account for a specified nonzero value ϕ of day-to-day variation. Day-to-day variation models and the parameter ϕ were discussed in Chapter 2. We can take day-to-day variation into account by adjusting the switch-to-switch traffic loads input to the path-EFO model by applying an equivalent load method separately to each switch-to-switch load, without reference to other switch-to-switch loads. In making the calculation, we first hypothesize that the traffic load with day-to-day variation offered to a route is equivalent to the same traffic load with no day-to-day variation being offered to an "equivalent" link meeting the same blocking objective. We can then calculate, on the basis of Neal-Wilkinson methods [HiN76] discussed in Chapter 2, the number of trunks N that are required in the equivalent link to meet the required blocking probability grade-of-service objective for the traffic load A, with its specified peakedness Z and specified day-to-day variation ϕ.

The equivalent link requires the following number of trunks N to satisfy the traffic load A:

$$N = \overline{\mathrm{NB}}(A, \phi, Z, \mathrm{GOS})$$

where GOS is the required link-blocking probability grade-of-service objective and $\overline{\mathrm{NB}}$ is a function mapping A, ϕ, Z, and GOS into the trunk requirement N.

Holding fixed the specified peakedness Z and the calculated number N of trunks, we next calculate what larger load A' requires N trunks to meet the blocking objective if the traffic load had had *no* day-to-day variation instead of the actual level ϕ. A' must then satisfy the following equation:

$$N = \overline{\mathrm{NB}}(A', 0, Z, \mathrm{GOS})$$

where we have set $\phi = 0$, signifying no day-to-day variation. If we then adjust the original offered traffic load A by a factor A'/A, the adjusted load, which equals A', then produces the same number of trunks N when sized for the same peakedness level but in the absence of day-to-day variation.

In applying the path-EFO model to design the dynamic routing network, we replace the triple (A, Z, ϕ) characterizing the switch-to-switch traffic load by the triple $(A', Z, 0)$. The adjusted traffic load is equivalent to the original traffic load in terms of the number of trunks that it requires to meet the blocking probability objective.

6.3.6 Modular engineering methods

In a digital or an analog network, links are economically sized to modular quantities such as the T1 transmission rate (1.536 mbps) in a digital network [Els79]. In the path-EFO model, links are sized to modular quantities in a postprocessor to the steps described in the previous sections. In the modular sizing technique, methods very similar to those used in the capacity design step described above and in Chapter 5 are used, in which the optimal link blocking objective is adjusted to suit modular link quantities, such that the optimal link-blocking objective is raised by as much as 40 percent in comparison with the nonmodular objective link blocking.

6.4 Modeling Results

6.4.1 Comparison of solutions of path linear programming and route linear programming

To compare the size and convergence of the path linear programming model with the route linear programming model, data for a 28-switch route-EFO model design are used to construct a path-EFO model design. The 28-switch model is shown

TABLE 6.1   Path Choices in Path-EFO Model and Route-EFO Model Designs

Switch-to-switch transport distance (miles)	Path-EFO model design (number of paths)	Route-EFO model design (number of routes)
0–750	5 (if blocking meets objective)	5
>750	all available paths	maximum possible (up to 27)

in Figure 1.12. The paths and optimal link blocking objectives available for the path-EFO model are identical to those used in the route-EFO model. The number of paths allowed for the path-EFO model is selected to match the dimensionality of the route linear programming model. The number of paths for a particular switch pair varies with the distance between the endpoints of that switch pair in the same way that the number of routes varies. Table 6.1 compares the number of path choices in the path-EFO model design with the number of route choices used in the route-EFO model design. The reason for limiting the choice of paths and routes in this manner is to give the longer, more expensive paths and routes more choices, while limiting the resulting dimensionality. Note that because the starting network for the route-EFO model design is not fully connected, not all paths are available.

Table 6.2 compares the path and route linear programming models. The data are obtained using the MPSX-370 linear programming package for 30 minutes. As shown, the path linear programming problem is less dense than the route linear programming problem. The reason for this density decrease is that each path has a maximum of two links, whereas each route can be composed of many paths and hence contain many links. Note that the path linear programming model is able to decrease the initial cost by 14.6 percent in 14,587 pivots, and the route linear programming model decreases cost by 9.4 percent in 3,429 pivots. The difference in efficiency arises because a path linear programming pivot needs to reevaluate link flows on many fewer links than does the route linear programming pivot. Hence, it appears that the linear programming model in the path-EFO model is easier to solve than the linear programming model in the route-EFO model.

TABLE 6.2   Comparison of Route Formulation Linear Program with Path Formulation Linear Program

Design model	Rows	Columns	Density	Size (million bytes)	Objective function decrease*†	Pivots*
route-EFO	1,399	9,429	0.95%	2.4	9.4%	3,429
path-EFO	1,399	9,433	0.17%	1.04	14.6%	14,587

*Obtained with 30 minutes execution of MPSX-370
†Linear program objective function (incremental network cost) change

Another advantage of the path-EFO model is that there are no flow proportions to store. Because all flow on a path is carried by each link on the path, the corresponding coefficients for the path-EFO model are all 1s.

6.4.2 Comparison for a single flow realization

In order to compare the three flow realization techniques, the four iterations of the linear programming model flow assignment are executed. The output of the second and fourth iterations are saved as representative flow assignments against which to test the flow realization techniques. Once a realization technique has determined the appropriate routing, the network is sized to the given objective blocking levels holding the routing constant. This tests the ability of the flow realization methods to meet the desired flows and the resulting network cost. Hence, only the routing design and capacity design steps of the route-EFO model (Figure 4.9) are executed. The MPSX-370 linear programming package is used to generate these results, and the total linear programming model solution time is held fixed for both the path-EFO and the route-EFO model solutions.

Table 6.3 shows the results for each of the four iterations of the flow feasibility model. Included are the objective function value at which the linear programming model started, normalized to the initial starting value; the normalized objective function value where the linear programming model stops after 15 minutes of computer time; the resulting total decrease in objective function value from the initial linear programming model starting basis; and the total flow violation. Of particular interest is that the decrease in the total flow violation, which is a measure of the feasibility of the linear programming flows, indicates that the model does force the flows toward feasibility and converges to a fixed point. Note that the linear programming model starting value decreases with each iteration, indicating that the recomputed starting basis is making good use of information from the last linear programming model solution. The total linear programming model design cost savings stabilizes at about 12.8 percent after the first iteration.

Table 6.4 compares network cost data for the three flow realization techniques. The desired path flow is obtained from the first two iterations of the flow feasibility model. The quantities being compared include the realized incremental network cost; this quantity is given with respect to the network incremental cost achieved

TABLE 6.3   Flow Feasibility Algorithm Results

Iteration	LP start (normalized)	LP* stop (normalized)	Decrease in initial LP cost	Total flow violation (erlangs)
1	1.00	0.874	12.6%	924
2	0.920	0.872	12.8%	125
3	0.904	0.872	12.8%	80
4	0.899	0.872	12.8%	68

*15-minute execution of MPSX-370

TABLE 6.4 Network Cost Comparisons of Flow Realization Techniques (2 Iterations of Flow Feasibility Algorithm)

Routing method	Realized incremental network cost*	Total path flow error	Final network cost	Network cost compared with CGH routing
CGH	+5.6%	7.22%	$41,634,000	—
skip-one-path	+6.4%	7.50%	$41,670,100	+0.09%
sequential	+7.2%	8.50%	$41,694,400	+0.14%

*Incremental network cost achieved by each routing technique with respect to the incremental cost achieved by the linear program

by the linear programming model and shows how much the linear programming model solution cost is altered by the realization process. The total path flow error is the sum of the absolute value of the difference between the desired flow and the realized path flow, divided by the sum of the realized path flows. The final network cost is the cost of the network after being sized by the capacity design step. The last column compares the network costs of the skip-one-path and sequential routing techniques with that of the CGH method. The network obtained by the CGH method is selected as the basis for comparison because the CGH realization is the most accurate of the routing methods and hence produces the least costly network.

From Table 6.4 it is apparent that CGH routing is the most accurate in realizing the desired path flows, followed in order by skip-one-path routing and sequential routing. It should be noted that although there is a difference of about 1 to 2 percent between the three methods in realized network incremental cost and total path flow error, the difference in the final network cost is very small. Indeed, the difference in final network cost between the best (CGH) and the worst (sequential) realization techniques is on the order of one-tenth of 1 percent.

Table 6.5 shows the same data as Table 6.4, using the network obtained after four iterations of the flow feasibility model. The realized networks now have incremental cost and path flows very close to those of the desired linear

TABLE 6.5 Network Cost Comparisons of Flow Realization Techniques (4 Iterations of Flow Feasibility Algorithm)

Routing method	Realized incremental network cost*	Total path flow error	Final network cost	Network cost compared with CGH routing
CGH	1.2%	1.68%	$41,518,600	—
skip-one-path	1.1%	1.09%	$41,539,900	+0.05%
sequential	1.9%	3.44%	$41,565,500	+0.1%

*Incremental network cost achieved by each routing technique with respect to the incremental cost achieved by the linear program

programming model solution. This indicates that the flow feasibility model is performing well. Once again, the difference between CGH routing and sequential routing in final network cost is very slight, about one-tenth of 1 percent.

The route-EFO model gives a final network cost of $42,409,600. Hence, the path-EFO model technique for this case of a single iteration saves about 2 percent with respect to the route-EFO model. As the corresponding hierarchical network cost for three design hours is $48,430,100, the preplanned dynamic routing savings has increased from 12.4 percent to 14.3 percent, a difference of 1.9 percent. This improvement can be attributed to the more efficient and accurate assignment and realization of flows in the path-EFO design model.

6.4.3   Comparisons for multiple iterations

Here the complete path-EFO model shown in Figure 6.3 is used to compare the CGH routing method and the sequential routing method to quantify the difference between the most accurate and the least accurate routing methods. Experience with CGH routing indicates that the advantage of CGH routing over sequential routing in realizing path flows occurs on the first few paths. That is, the majority of sequential routing path flow realization error is due to flow errors on the first one or two paths; the flow errors on remaining paths tend to be small. In view of this, for some cases of interest a variation of CGH and sequential routing is tried in which the first three paths could be formed into one CGH block; sequential routing is then employed among the remaining paths. Such a routing method, called single-block CGH routing, is intended to allow more flexibility in meeting flows on the first few paths, while using the simplicity of sequential routing for later paths.

A 30-switch national intercity network model, shown in Figure 6.12, is used for these results, in addition to the 28-switch network model. Tables 6.6 and 6.7 show results from the path flow realization techniques for both network models. Six iterations of the path-EFO model, each including three iterations of the flow feasibility model, are used to generate these results. The networks are each designed for the three peak hours of load. Note that the CGH routing method is again more accurate than sequential routing, but the difference in final network cost of 0.3 percent is very small and probably not significant. Note also that the single-block CGH routing method yields a 28-switch network slightly less expensive than the full CGH routing method. In the 30-switch network model, however, the single-block CGH network design is slightly more expensive than the full CGH network design, and the difference appears insignificant. Hence, the maximum difference between sequential routing and any other routing method is about 0.6 percent.

The route-EFO model 30-switch network cost is $15,349,000; hence, the path-EFO model saves an additional 1.6 percent of the cost of the route-EFO model network for the 30-switch network model. The cost of the corresponding hierarchical network for three design hours is $131,740,900. Hence, the use of the path-EFO model increases the dynamic routing savings from 12.4 percent to 13.8 percent, a difference of 1.4 percent.

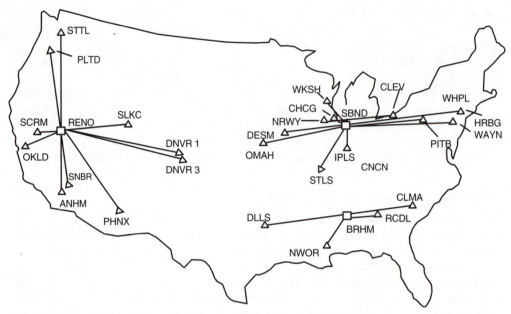

Figure 6.12 30-switch national intercity network model

TABLE 6.6 28-Switch Network (Three-Hour Designs)

Routing	Total path flow error	Final network cost	Difference with respect to CGH
CGH	1.42%	$41,712,700	—
single-block CGH	1.58%	$41,568,300	−0.3%
sequential	1.83%	$41,822,700	+0.3%

TABLE 6.7 30-Switch Network (Three-Hour Designs)

Routing	Total path flow error	Final network cost	Difference with respect to CGH
CGH	1.07%	$113,510,200	—
single-block CGH	1.37%	$113,631,300	+0.1%
sequential	1.62%	$113,725,500	+0.2%

TABLE 6.8   28-Switch Network, Eight Design Hours

Routing	Total path flow error	Network cost	Difference with respect to CGH
CGH	1.38%	$42,352,300	—
single-block CGH	1.91%	$42,434,700	+0.2%
sequential	3.73%	$42,566,000	+0.5%

TABLE 6.9   28-Switch Network, Three Design Hours (Peakedness Considered)

Routing	Total path flow error	Network cost	Difference with respect to single-block CGH
single-block CGH	2.27%	$43,781,900	—
sequential	2.56%	$43,993,400	+0.5%

Tables 6.8 and 6.9 show results for the 28-switch network designed for eight hours of data and three hours of data (with peakedness considered). These are the eight peak load hours, and a network designed for these eight hours suffices for all 17 hours. Once again, sequential routing achieved a network cost within 0.5 percent of that of CGH routing, even though the path flow error is greater than that for CGH routing. Hence, the inflexibility of sequential routing does not appear to be a major factor in designs for a large number of hours.

Table 6.10 shows an example of sequential routing for the 30-switch network. The route shown is from Anaheim to Cleveland; routing in the other direction is the same. The routing for each design hour is shown in the path-EFO-model-selected sequence. Note that the direct path is not first in order for hour 1; this occurs because IPLS, the first choice, had some excess capacity in hour 1. Note also that the path choices and ordering, together with the number of paths in the route, change from design hour to design hour.

TABLE 6.10   30-Switch Sequential Routing Example
(Anaheim-to-Cleveland Routing)

Order	Hour 1 via	Hour 2 via	Hour 3 via
1	IPLS	CLEV	CLEV
2	CLEV	DNVR	IPLS
3	SNBR	IPLS	SBND
4	PHNX		DESM
5	OMAH		NRWY
6	SCRM		DLLS
7	BRHM		

6.5 Conclusion

The route-EFO model, which designs preplanned dynamic routing networks for general classes of preplanned dynamic multilink path routing networks, is reformulated to allow the allocation of traffic directly to two-link paths instead of predetermined routes. This path-EFO model is specifically oriented to design preplanned dynamic two-link path routing networks. It greatly reduces storage and computer run time required by the model and improves the design efficiency by an additional 1.5 percent in the dynamic routing savings.

As part of the path-EFO model, a flow feasibility model forces the assigned near-optimum path flows to be realizable. Several methods to realize these flows are compared. One of the flow realization methods considered, called the CGH algorithm, is very flexible and sophisticated in its ability to meet network flows. Another method, called the sequential routing method, is less flexible but allows greatly simplified routing. Sequential routing offers traffic to an ordered list of paths, with the overflow from one path being offered to the next path. The list of paths changes by time of day to take advantage of traffic noncoincidence.

Using both 28-switch and 30-switch national intercity network models, results indicate at most a 0.5 percent cost penalty in using sequential routing instead of CGH routing. Because of its simplicity, efficiency, and performance, sequential routing is the method that was selected for initial implementation of dynamic routing in the AT&T national intercity network, starting in 1984.

Chapters 7 to 10 further develop essential elements of dynamic routing networks in relation to dynamic routing design for a four-switch network example (Chapter 7), real-time routing design (Chapter 8), dynamic routing design under forecast uncertainty (Chapter 9), and dynamic routing design for multiservice integrated networks (Chapter 10). Although we use preplanned dynamic two-link path routing for illustration in these chapters, the principles developed are generally applicable to all types of dynamic routing networks, which will be further elaborated as these topics are developed.

7

Four-Switch Example of Dynamic Routing Network Design

7.1 Introduction

This chapter illustrates the steps in the path-erlang flow optimization (path-EFO) model for designing preplanned dynamic two-link path routing networks. Because the model is illustrated in detail for a small network, the workings of the model and principles applied are made very clear. These principles, which are illustrated include shortest path selection, multihour dynamic routing design, multihour capacity optimization, and link flow optimization. As discussed in Chapter 1, these are the basic principles of network design which are in fact generally applicable to the design of all dynamic routing networks. The outline of the chapter is as follows: First, we give a high-level review of the path-EFO model and the preplanned dynamic two-link path routing tables it designs. Next, we review the major functional areas of the path-EFO model. Finally, we explore in some depth the concepts employed in each of the major functional areas of the path-EFO model. Throughout the chapter we use a four-switch network example to illustrate the methods employed in the path-EFO model.

Figure 1.2 illustrates the terminology employed in the path-EFO model and used throughout this chapter. A link is the connection between two switches; the link between switches A and B is shown in Figure 1.2. In preplanned dynamic two-link path routing, a path consists of one or possibly two links in tandem connecting the originating switch A with the terminating switch B. A route is an ordered set of paths connecting originating switch A with terminating switch B. In the example shown in Figure 1.2, there are three paths in the route. The first path consists of the direct link from A to B. The second path consists of the two-link path from A through C to B. And finally the third path in the route consists of the two-link path from A through switch D to B.

7.2   Review of Preplanned Dynamic Two-Link Path Routing Methods

Figure 1.22 illustrates preplanned dynamic two-link path routing, which in this chapter is sequential routing. The figure presents an example of routing from the originating switch at San Diego (SNDG) to the terminating switch at White Plains (WHPL), with possible two-link paths between SNDG and WHPL through Albany (ALBY), through Phoenix (PHNX), and finally through Newark (NWRK). For the example shown in Figure 1.22, the route from San Diego to White Plains consists of the ordered set of paths 1, 2, 3, and 4. The first path is the direct one-link path from San Diego to White Plains, the second path is the two-link path from San Diego through Albany to White Plains. Path 3 is the path from San Diego to Phoenix to White Plains, and path 4 is the path from San Diego to Newark to White Plains.

Preplanned dynamic two-link path routing operates as follows: Suppose that a call arrives at San Diego destined for White Plains. The originating switch at San Diego searches the trunks in the direct link connecting San Diego and White Plains. If there is an available trunk, it is seized and the call is set up. Suppose, however, that there are no trunks available in the first path, which is the direct path. The originating switch then moves to the second path and searches the trunks in the link connecting San Diego and Albany. If there is an available trunk, it is seized and a call setup message is sent to the via switch at Albany through the common-channel signaling (CCS) network notifying Albany that the call is to be set up to White Plains, and only on the direct link from Albany to White Plains. Suppose that a trunk is available in the Albany–to–White Plains link. The Albany switch seizes the trunk, sets up the tandem connection through the Albany switch, and sends a call setup message again through the CCS network to White Plains notifying White Plains that a call is arriving from San Diego via Albany. If there were no trunks available in the Albany–to–White Plains link, the Albany switch would then initiate a crankback message on the CCS network back to San Diego notifying San Diego that there were no trunks available. San

Diego would advance the call setup process to the third path in the route, the path through Phoenix. And again, the call setup through Phoenix would work just as it had through Albany. The call is advanced finally to the last path in the route if there had been no free trunks in the first, second, or third paths. If a call is blocked on the fourth path—that is, there are no trunks available in either the San Diego-to-Newark or the Newark-to–White Plains link—the call is blocked from the network. Figure 1.22 also illustrates the nonhierarchical aspect of the dynamic routing method, in that there is no hierarchical relationship among the switches.

We now illustrate the preplanned dynamic nature of preplanned dynamic two-link path routing. The routing is dynamic in that it is possible to change the routing between each pair of switches in the network by the time of day. As illustrated in Figure 1.22, four separate routing sequences are available from San Diego to White Plains. Routing sequence number 1, the same routing sequence described above, is employed in the morning hours. Routing sequence number 2, consisting now of only two paths—that is, the direct path from San Diego to White Plains with overflow to the two-link path through Phoenix—is employed in the afternoon. Routing sequence number 3, which has the first path as the direct path, the second path as the path through Newark, and the last path as the path through Albany, is employed in the evenings. And finally, routing sequence number 4 is employed on the weekends, when there are different traffic patterns in the network.

We should note two additional things from Figure 1.22. First, the number of paths in a route may vary as a function of the time of day. The number of paths in a route is determined on the basis of a desired switch-to-switch blocking probability grade-of-service objective. Second, the first path in a route need not always be the direct path. For instance, routing sequence number 4 employs as its first-choice path the path through Phoenix to White Plains with overflow to Newark. Finally, still referring to Figure 1.22, one additional term that we use throughout this chapter is the term *load set period*. Load set period refers to a number of contiguous hours in the day, consisting of one or possibly several adjacent hours, in which traffic patterns in the network are sufficiently similar to each other that the same routing is employed throughout all those hours that compose the load set period. In this chapter, we use the terms *load set period* and *hour* interchangeably. However, keep in mind that a load set period can consist of several contiguous hours.

7.3   Design Example: Four-Switch Path-Erlang Flow Optimization Model

In Figure 7.1 we present a high-level overview of the path-EFO model. There are three fundamental steps within the path-EFO model that are described in depth in this chapter. First, the routing design step is used to determine the paths and routing between each pair of switches in the network for all load set periods in the day. In this step, we determine the assignment of traffic to each path. Second, the

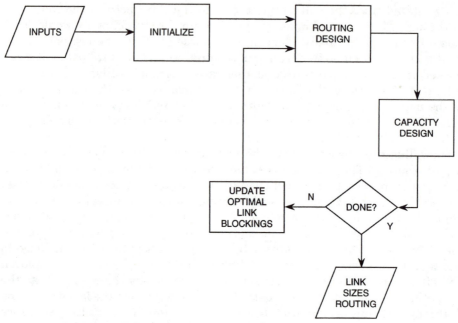

Figure 7.1 Path-EFO model: Top-level view

capacity design step is used to size each link in the network. Finally, the update optimal link blockings step is used to determine the optimal link blockings used in the routing design step and capacity design step.

Note from Figure 7.1 that the path-EFO model is an iterative method in which we proceed in a serial fashion from the selection of routing to the sizing of the network and finally to the updating of optimal link blockings to be used in both the routing design step and in the capacity design step in which we size the network. On the first iteration of the path-EFO model, the network must be initialized to an initial starting point from which the design will proceed. As discussed later in the chapter, the decision to terminate the iterative loop is made on the basis of network cost from iteration to iteration of the fundamental loop in the path-EFO model. When the path-EFO model has converged, the outputs are the number of trunks in each link and the routing to be employed in the network throughout all load set periods in the day.

Figure 7.2 presents in more detail the major functional areas of the path-EFO model. In particular, the step labeled Routing Design in Figure 7.1 has been expanded to show five separate steps that actually make up the routing design step. The routing design step consists of the steps labeled Generate Candidate Paths, Determine Path Flow Constraints, Assign Flow to Paths, Update Blockings, and Define Sequential Routes. The step that had been labeled Capacity Design in Figure 7.1 has actually been expanded into two steps here: Capacity Design and Check Route Blocking in the network. The heart of the path-EFO model is the routing design step (the five steps shown in Figure 7.2).

ROUTING DESIGN

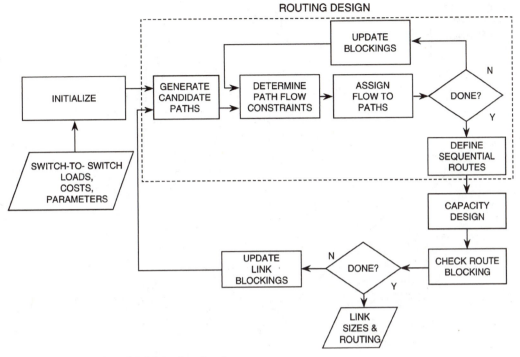

Figure 7.2 Path-EFO model: Major functional areas

It is the five areas that make up the routing design step that we explore in the most detail in this chapter.

Figure 7.3 illustrates a four-switch network example that we use to illustrate the methods employed in the path-EFO model in this chapter. It shows the four switches, along with the switch-to-switch loads between each pair of switches, with loads in two separate hours. Once again, keep in mind that although we use the term *hours* here, we are actually referring to load set periods that may consist of several hours. The loads for this example network are given in erlangs. For instance, the offered load between switch 1 and switch 2 is 10 erlangs in hour 1 and 15 erlangs in hour 2, and the offered load between switches 2 and 4 is 10 erlangs in both hours 1 and 2. For illustrative purposes, we assume that the cost of a trunk in each of the six links in the network is identical and is set to 19 units of cost per trunk and that the link metric is equivalent to the cost per trunk.

7.3.1   Initialize

Figure 7.4 lists the type of data that are required as input to the path-EFO model. The first inputs required are switch-to-switch traffic loads between each pair of switches in the network for each load set period or time period throughout the day. Second, we input the objective switch-to-switch blocking probability

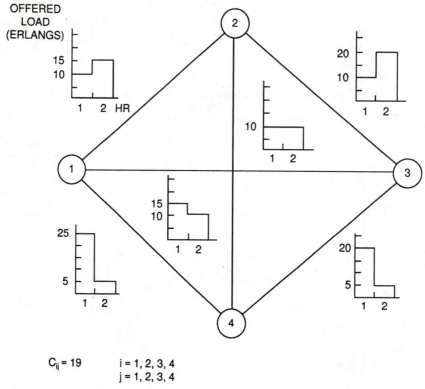

$$C_{ij} = 19 \qquad i = 1, 2, 3, 4$$
$$j = 1, 2, 3, 4$$

Figure 7.3   Offered loads for example network

grade-of-service, which is the blocking probability to which we design the network. In preplanned dynamic two-link path routing, that objective is typically .005 blocking probability between each pair of switches. The third set of inputs to the path-EFO model is the trunk costs for each link in the network. Fourth, we provide distances between each originating and terminating switch in the network. Distances are used typically in the implementation of routing constraints in the network. Certain constraints are applied on two-link paths in the network

> ■ Switch-to-switch loads for each load set period (LSP)
>
> ■ Switch-to-switch blocking probability objective
>
> ■ Link costs
>
> ■ Distances
>
> ■ Routing constraints:
>   Alternate path-distances
>   Tandem prohibitions
>
> ■ Allowed number of paths per route

Figure 7.4   Path-EFO model input

to ensure a satisfactory transmission grade-of-service objective in the network. For example, one routing constraint requires that the total distance on a two-link path not exceed twice the distance of the direct one-link path. The final input is the allowed number of paths per route. This number is typically set to 5–15 paths per route, although in practice we find that most of the traffic is carried on the first two or three paths in any route for each switch pair in the network.

Figure 7.5, Figure 7.6, and Table 7.1 describe the initialization step of the path-EFO model. Initialization takes the form of defining an initial operating point of the network, stated in terms of estimates of optimal blockings for each link in the network and the estimated number of trunks for each link. Given our initial estimate of the optimal blocking for a link, we size the link for the direct switch-to-switch load offered to the link and eliminate links with trunks less than some minimum initial link size, typically 6 to 12 trunks. The philosophy here is that a link should exist only if the direct switch-to-switch load offered to it is large enough to justify the link. Further, by initially eliminating the small links, we find that the execution time of the path-EFO model is reduced.

• DEFINE AN INITIAL OPERATING POINT FOR THE NETWORK

    – ESTIMATE $\hat{B}_{opt}$ FOR EACH LINK

    – SIZE EACH LINK TO $\hat{B}_{opt}$

    – ELIMINATE LINKS WITH $\hat{T}$ < MINIMUM INITIAL LINK SIZE

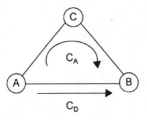

• FOR EACH LINK, FIND THE LEAST COST ALTERNATE PATH
• SATISFYING ROUTING CONSTRAINTS AND APPLY TRUITT'S
• ECCS METHOD

$$CLLT \le \frac{\gamma}{C_A / C_D}$$

USE MAXIMUM A-B SWITCH-TO-SWITCH LOAD

Figure 7.5 Path-EFO model initialization

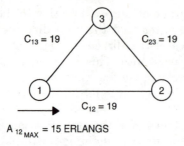

$$A_{12_{MAX}} = 15 \text{ ERLANGS}$$

- ASSUME $\gamma = 28 \text{ CCS} = 0.778 \text{ ERLANGS}$

- MINIMIZE $C_{TOT} = C_{12} N_{12} + \dfrac{(C_{13} + C_{23}) A_{12_{MAX}} B(N_{12}, A_{12_{MAX}})}{\gamma}$

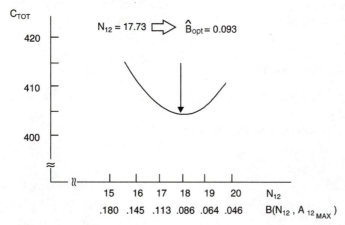

Figure 7.6   Initialization for link 1–2

TABLE 7.1   Initial Estimates of Link Blockings

Link	$\hat{B}_{opt}$	B (hour 1)	B (hour 2)
1–2	0.093	0.008	0.093
1–3	0.093	0.093	0.008
1–4	0.070	0.070	~0
2–3	0.079	~0	0.079
2–4	0.117	0.117	0.117
3–4	0.079	0.079	~0

The method used to determine the initial estimate of the optimal blocking for each link is illustrated in the bottom half of Figure 7.5. This is an application of Truitt's economic CCS method [Tru54], with the load on the last trunk in the link determined by gamma, the marginal capacity, divided by the cost ratio of the alternate and direct paths. Because Truitt's method was originally formulated for hierarchical networks, the analogy here for preplanned dynamic two-link path routing networks is that the aggregation of the overflow paths is modeled as an equivalent final link. Hence, for each link in the network we find the least-cost alternate path satisfying the routing constraints and apply Truitt's ECCS method. In the example shown in Figure 7.5 for determining the optimal blocking on link A–B, we have assumed that the two-link path A–C–B represents the least-cost alternate path.

Figure 7.6 illustrates the determination of the optimal blocking for link 1–2, that is, the link connecting switches 1 and 2 in the four-switch example. The cost of the direct path is the cost of one trunk connecting switches 1 and 2, which is 19 cost units in our example, and the cost of the alternate path, from switch 1 to switch 2 through switch 3, which has a total cost of 38 units. In determining the optimal blocking for link 1–2 we assume that the marginal capacity gamma of the alternate path is 28 CCS or 0.778 erlang carried per trunk. As in Truitt's method, the objective is to minimize the total cost of carrying the traffic between switches 1 and 2, which is the sum of the cost of a trunk between switches 1 and 2 multiplied by the number of trunks connecting them and the cost of carrying the overflow from path 1–2 on the two-link alternate path through switch 3. In the expression for total cost, the function $B(N, A)$ is the erlang blocking function, and A_{12} is the offered load between switches 1 and 2. The bottom of Figure 7.6 shows that the minimum cost for carrying the traffic between switches 1 and 2 occurs when the link connecting 1 and 2 is provisioned with 17.73 trunks. From the erlang blocking function, for a link with 17.73 trunks and 15 offered erlangs, we determine that the optimal blocking on link 1–2 is 0.093.

Table 7.1 summarizes our initial estimates of optimal link blockings and actual realized link blockings in each of the two hours in our four-switch, two-hour example (Figure 7.3). The column labeled $\hat{B}$opt is the estimate of the optimal link blockings. The third and fourth columns, labeled B (hour 1) and B (hour 2) are the realized blockings on those links if (1) the link is sized to its optimal blocking and (2) the link is offered only the direct switch-to-switch traffic across the link. For instance, link 2 has an optimal blocking of 0.093. As we determined from Figure 7.6, that optimal link blocking is realized in hour 2, in which the offered load between switches 1 and 2 is maximum at 15 erlangs. The realized blocking in hour 1, when the offered load between switches 1 and 2 is 10 erlangs, is 0.008. We note that two links in this example, links 1–2 and 1–3, share the same optimal blocking, and, similarly, links 2–3 and 3–4 share the same optimal blocking. This identity of the optimal link blocking is simply the result of the symmetry in our four-switch example network, where all links share a common cost of 19 units. This identity also comes about because the direct switch-to-switch loads between switches 1–2 and 1–3 and between switches 2–3 and 3–4 are equal. Of course, in general this is not true.

This completes the discussion of the initialization step of the path-EFO model. Most importantly, we have determined estimates for the optimal blocking for each link in the network. These estimates of optimal link blocking remain unchanged throughout the first iteration of the path-EFO model. As we proceed through the model, we constantly size each link in the network to that optimal blocking. Only after the first iteration of the path-EFO model is complete do we revise our estimates of optimal blocking for each link in the network based on our then-current view of the operating conditions of the network, that is, the loads carried by each link in the network.

7.3.2 Generate candidate paths

Returning to Figure 7.2, we now want to discuss the step labeled Generate Candidate Paths. The procedure for selection of candidate paths between each pair of switches in the network is illustrated in Figures 7.7 and 7.8, which address the methods used in the first iteration of the path-EFO model.

Different methods are used on second and subsequent iterations of the path-EFO model; they are discussed near the end of the chapter. An n-switch network

---

- An n-switch network has $(n - 1)$ one- and two-link paths for each switch pair.
- Generate k "shortest" paths for each switch pair ($k \leq 15$).

    Storage

    Run time

- k shortest paths: k least-cost one- or two-link paths satisfying routing constraints.

    Cost = cost of 1 trunk in each link in path

    Constraints:

    Distance (alternate path) $\leq 2 \times$ distance (direct path)

    Distance (two-link path without echo control) $\leq 2,400$ miles

---

Figure 7.7 Generate candidate paths (first iteration)

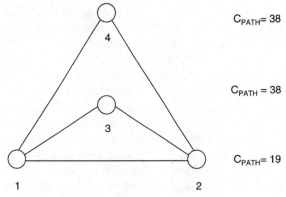

$C_{PATH} = 38$

$C_{PATH} = 38$

$C_{PATH} = 19$

Figure 7.8 Candidate paths for switch pair 1–2

has a maximum of $n - 1$ one- and two-link paths between each pair of switches in the network. In the Generate Candidate Paths step, we wish to choose for each pair of switches in the network the k "shortest" paths as the potential candidate paths for carrying traffic between each pair of switches and eliminate the $n - 1 - k$ other paths between each pair of switches in the network.

Why do we wish to eliminate $n - 1 - k$ paths from consideration in the design? First, we must keep in mind that dynamic network design and the path-EFO model need to consider networks consisting of up to about 200 switches or more, so we are dealing with a very large problem from a design standpoint. In the heuristic linear programming model stage, consideration of $n - 1$ paths for each switch pair in a 200-switch network would require massive amounts of computer storage. Consideration of 199 possible paths between each pair of switches in a 200-switch network would also greatly increase the run time of the path-EFO model. And finally, from an implementation standpoint, it is not feasible to store in a switching system large numbers of paths for each switch pair in the network. Therefore, to reduce computer storage requirements, to improve the run time of the path-EFO model, and to be realistic from an implementation standpoint, we wish to initially select only the k best paths for each pair of switches in the network and eliminate the remaining paths from consideration in the heuristic linear programming model.

In the implementation of the path-EFO model, k actually varies between 5 and 15 paths, depending on the distance between a pair of switches. However, an average of k equal to about 10 is reasonable to think about in terms of our discussion of the path-EFO model. For the first iteration of the model, the metric used to define shortest paths is simply the cost of one trunk in each link. The k shortest paths are selected subject to transmission constraints on alternate-path distance relative to direct-path distances. One constraint is that an alternate path consisting of two links cannot have a total distance greater than twice the distance of the direct path. This constraint is imposed to help ensure satisfactory performance with the network from a transmission grade-of-service objective standpoint. There is also a constraint on the maximum distance of a two-link path without echo control. Such constraints depend on whether the transmission technology is analog or digital, whether echo cancellation is applied on all links or not, and other factors. These constraints can be relaxed—for example, with all digital transmission and echo cancellation deployed on all links.

In Figure 7.8, and referring once again to our four-switch example network, we illustrate the only three paths connecting switches 1 and 2 in the network. The direct path is the shortest in terms of cost, with a cost of 19 units. Each of the two alternate paths through switches 3 and 4 has a cost of 38 units, the sum of the cost of one trunk in each of the two links in each path.

7.3.3 Routing design

We turn now to Figure 7.9, which presents an overview of the routing design step in the path-EFO model. The objective is to assign flow or carried loads to paths for each switch pair in the network in each load set period or time

- ASSIGN FLOW (CARRIED LOAD) TO PATHS FOR EACH SWITCH PAIR IN EACH LSP SO AS TO MINIMIZE NETWORK COST

- CONSTRAINTS

  - CARRY ENOUGH LOAD FOR EACH SWITCH PAIR IN EACH LSP TO MEET REQUIRED SWITCH-TO-SWITCH BLOCKING OBJECTIVE

  - IF 10 ERLANGS OFFERED, MUST CARRY 9.95 ERLANGS

  - ASSIGN NO MORE FLOW TO EACH PATH THAN CAN BE CARRIED

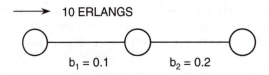

10 ERLANGS

$b_1 = 0.1$ $b_2 = 0.2$

$b_{PATH} = 0.1 + 0.2 - (0.1)(0.2) = 0.28$

FLOW $\leq (1.0 - 0.28) \times 10 = 7.2$ ERLANGS

- GIVEN "OPTIMIZED" PATH FLOWS, FIND A SEQUENTIAL ROUTE FOR EACH SWITCH PAIR IN EACH HOUR THAT REALIZES THE OPTIMIZED FLOWS AS CLOSELY AS POSSIBLE

Figure 7.9   Routing design overview

period in the day so as to minimize network cost. In assigning flow to paths, two constraints must be observed. The first constraint is the blocking probability grade-of-service objective. That is, we wish to carry enough of the offered load between each switch pair in the network to meet our required switch-to-switch blocking objective. As we described earlier, our objective is to block no more than one-half of 1 percent of the offered load between any switch pair in the network. That is, if there are 10 erlangs of offered traffic, we wish to carry at least 9.95 erlangs. The second constraint is a realizability constraint. It basically says that we can assign no more flow to a path than can be carried by the path given our view of the optimal blockings for links in the path. For the two-link path shown in Figure 7.9, we have 10 erlangs of traffic offered with blocking on the first link equal to 0.1 and blocking on the second link equal to 0.2. If we assume independence of blocking between the two links, the blocking for the path is 0.28. Given this blocking of 0.28, we should assign no more than 7.2 of the 10 erlangs to this path. Finally, given our estimates of optimized path flows, we determine a sequential route or ordering of one- and two-link paths for

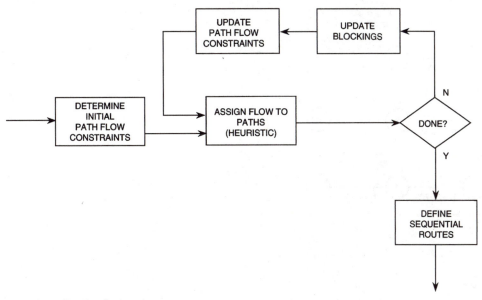

Figure 7.10 Routing design steps

each switch pair in each time period that realizes the optimized flows as closely as possible.

Figure 7.10 provides a closer look at the routing design step in the path-EFO model. First, we note that the step is itself an iterative process. Earlier, we described the path-EFO model as an iterative process sequentially moving from the selection of routing to the network sizing to updating estimated optimal link blockings. In Figure 7.10, we see that the routing design step is also an iterative process within an iterative process. Within the routing design module we see that there are really four steps. First is the determination of path flow constraints, which take the form of initial and updated constraints. Second is the assignment of flow to paths. Third is updating of link blockings in the network, and last is determination of sequential routing tables to realize the optimal path flows as closely as possible. The stopping criterion for the routing design iterative loop is simply three passes through this iterative loop. This number of iterations was determined through experimentation.

7.3.3.1 Determine initial path flow constraints and initial flow assignment. Figures 7.11 and 7.12, and Table 7.2 illustrate the methods used to determine path flow constraints and to determine initial flows assigned to each path for each switch pair for each hour in the network. Figure 7.9 illustrates the determination of path flow constraints by use of our four-switch network example. In particular, we focus on the paths for switch pair 1–2 in hours 1 and 2. Looking at the top half of Figure 7.11, we have illustrated the three paths between switches 1 and 2 in the network and labeled each link in each path with its estimated blocking in

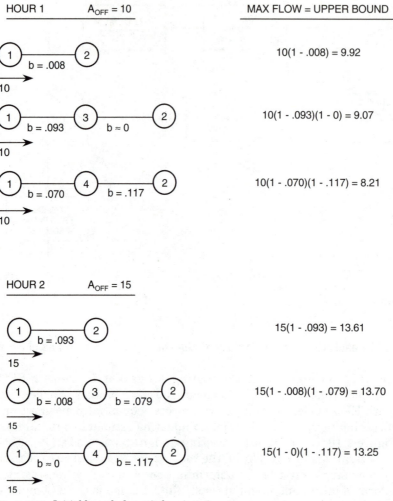

HOUR 1 $A_{OFF} = 10$ MAX FLOW = UPPER BOUND

10(1 - .008) = 9.92

10(1 - .093)(1 - 0) = 9.07

10(1 - .070)(1 - .117) = 8.21

HOUR 2 $A_{OFF} = 15$

15(1 - .093) = 13.61

15(1 - .008)(1 - .079) = 13.70

15(1 - 0)(1 - .117) = 13.25

Figure 7.11 Initial bounds for switch pair 1–2

hour 1. These estimated blockings are those that we discussed earlier in Table 7.1. Recall that we are in the initialization phase of the routing design step and have not yet assigned flow to any path. Therefore, we have no estimate of the ordering of the paths. To assign initial bounds on path flow, we assume that any one of the three paths can just as well be the first path in the route. That is, we assume that the full offered load in the first hour, 10 erlangs, for traffic load 1–2 could be offered to the one-link path or either of the two-link paths first. The right-hand column of Table 7.1 illustrates the upper bounds on path flow. For instance, for the one-link path directly connecting switches 1 and 2, the blocking on link 1–2 is 0.008 in hour 1, and, therefore, the maximum flow that can be assigned to this path is the product of the offered load of 10 erlangs and the complement of the

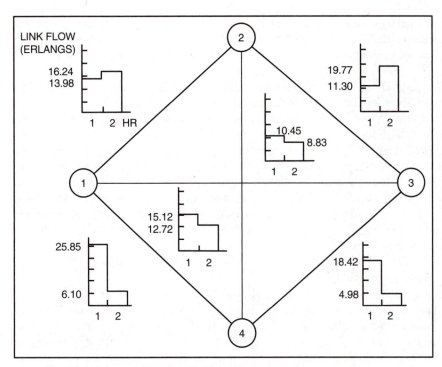

Figure 7.12 Link flows resulting from initial path flows

TABLE 7.2 Initial Flows for Switch Pair 1–2

Path	Via	Offered load	Path connectivity	Bound	Flow possible	Flow assigned	Cumulative flow
Hour 1, $A = 10$							
1	2	10	$1 - 0.008$	9.92	9.92	9.92	9.92
2	3	0.08	$(1 - 0.093)(1 - 0)$	9.07	0.07	0.03	9.95
3	4	*		8.21	*	0	9.95

$$\text{Average Blocking} = \frac{10 - 9.95}{10} = 0.005$$

Path	Via	Offered load	Path connectivity	Bound	Flow possible	Flow assigned	Cumulative flow
Hour 2, $A = 15$							
1	2	15	$1 - 0.093$	13.61	13.61	13.61	13.61
2	3	$15 - 13.61 = 1.39$	$(1 - 0.008)(1 - 0.079)$	13.70	1.27	1.32	14.93
3	4	*		13.25	*	0	14.93

$$\text{Average Blocking} = \frac{15 - 14.93}{15} = 0.005$$

*Preceding paths carry enough flow to meet blocking probability objective.

blocking, resulting in a maximum flow of 9.92 erlangs. Similarly, the maximum flows are determined for the other two paths. In the bottom half of Figure 7.11, we repeat the process for the second hour of our four-switch problem, again for switch pair 1–2.

Table 7.2 illustrates the methods that we used to initially assign flows to each path in the network. The example focuses on switch pair 1–2. The upper half of Table 7.2 addresses assignment of initial path flows for the first hour in our two-hour problem. The lower half addresses the assignment of flows for the second hour. Let us first discuss the assignment of flow to paths in hour 1 for switch pair 1–2. Here, we have chosen to order the paths simply by ascending cost: The direct one-link path, being the least-cost path, is the first path; the second and third paths are two-link paths. For the purpose of the initial assignment of path flows, we assume that this ordering of paths is used in the route to connect switches 1 and 2. The first path—the direct path—between switches 1 and 2 is offered the full 10 erlangs of traffic. Given our initial estimates of blocking, illustrated previously in Table 7.1, this first path carries 9.92 of the 10 erlangs. We assign all 9.92 erlangs to the path.

Assuming that 9.92 erlangs are assigned to the first path, 0.08 of the 10 erlangs overflow and are offered to the second path, the path through switch 3. Given our initial estimates of blocking on the links in this path, it is possible for the path to carry 0.07 of the 0.08 erlangs. However, to meet the blocking probability grade-of-service objective, which is a switch-to-switch blocking of 0.005, this second path can only carry 0.03 erlangs, and that is all that is assigned. The process is repeated for the second hour of our two-hour problem, and of course this is repeated for all switch pairs in the network. The next-to-last column in Table 7.2, labeled *Flow assigned,* represents our initial operating point of the network. It is these flows that we wish to adjust so as to minimize the cost of the network.

Figure 7.12 depicts the flows on each of the six links in our four-switch network example. These are the link flows resulting from our initial assignment of flow to each of the paths in the network in each of the two hours for each of the six switch pairs in the network.

7.3.4 Assignment of flow to paths

The problem of assigning flow to paths can be stated as a linear program on the assumption that capacity required on a link is a linear function of flow carried on the link, as shown in Figure 7.13. M_i is the incremental link cost per erlang of carried traffic on link i, and a_i is the maximum over all hours of the load carried on link i. There are four types of constraints in the linear program. The first constraints shown in Figure 7.13 are the equations for determining flow on link i, and L is the number of links in the network. The variables here are as follows; a_i is the maximum carried load on link i. Once again, the maximum is taken over all hours h ranging from 1 to H, the total number of hours in the problem. Link flow is the sum of the flow from all paths that use link i in hour h. The variable P_{jk}^{ih} takes the value of 1 if

Minimize network "cost":

$$\sum_{i=1}^{L} M_i a_i$$

subject to

$$\sum_{k=1}^{K} \sum_{j=1}^{J_k^h} P_{jk}^{ih} f_{jk}^h \le a_i \qquad i = 1, \ldots, L$$
$$h = 1, \ldots, H$$

$$\sum_{j=1}^{J_k^h} f_{jk}^h = g_k^h \qquad k = 1, \ldots, K$$
$$h = 1, \ldots, H$$

Carry enough flow for each demand in each hour to meet the blocking probability objective (= GOS).

$$0 \le f_{jk}^h \le \text{UPBD}_{jk}^h \qquad h = 1, \ldots, H$$
$$k = 1, \ldots, K$$
$$j = 1, \ldots, J_k^k$$

Flow on each path cannot exceed the realizable flow (upper bound).

$$a_i \ge 0 \qquad i = 1, \ldots, L$$

Link flows are nonnegative.

Figure 7.13 Assign flow to paths—linear programming flow optimization

link i appears in path j for switch pair k in hour h. J_k^h is the number of paths for switch pair k in hour h, and K is the number of switch pairs in the network.

The second set of constraints shown in Figure 7.13 consists of constraints related to blocking probability grade-of-service objective. These constraints simply say that we must carry enough flow g_k^h for each switch pair k in each hour h such that the blocking probability grade-of-service is satisfied. Once again, the blocking probability grade-of-service objective is typically .005. The third set of constraints shown in the figure simply states that the flow f_{jk}^h on a path (1) must be positive and (2) must not exceed the upper bound UPBD_{jk}^h or capacity of path j for demand pair k in hour h. The process for determining the upper bounds is illustrated in Figure 7.9. Finally, the last set of constraints are nonnegativity constraints for link flows.

Use of a linear program is not really feasible for the size of problems that we consider with dynamic routing. For instance, for a problem with 140 switches and ten load set periods throughout the day, there are in excess of 1 million variables in the problem. This problem is too large for any commercially available linear programming package, and, even if the size of the problem could be accommodated by a linear program, the program would be unacceptable with respect to computation time. Therefore, it was necessary to develop a heuristic linear programming model to solve the assignment of flow to paths.

In Figures 7.14 to 7.23 we describe the methods employed by the path-EFO heuristic linear programming model for assigning flows to paths. The example used here is different from the original four-switch example; it was chosen to better illustrate the principles used in the heuristic. The problem under consideration is defined in Figure 7.14. Here, once again, we are dealing with a four-switch, two-hour (two–load set period) example network. We focus on the assignment of traffic to path 1 or path 2 for switch pair A–D in the first of the two hours of the problem. There are two paths defined for switch pair A–D. Path 1 is a two-link path through switch B; the unit cost of capacity in each of the two links in path 1 is $10. Path 2 is a two-link path through switch C—that is, A–C–D—with a unit cost of capacity of $60 per unit in each of the two links. The question to be answered is how much of the flow is to be assigned to each path for hour 1.

TWO PATHS FOR AD PAIR IN DESIGN HOUR 1

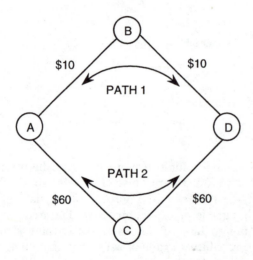

PATH FLOW INFORMATION

	INITIAL FLOW (ERLANGS)	UPPER BOUND (ERLANGS)
PATH 1	20	20
PATH 2	0	15

Figure 7.14 Heuristic example

The bottom half of Figure 7.14 states the initial conditions and upper bounds for the problem. The initial flow assigned to path 1 is 20 erlangs and to path 2 is zero erlangs. This flow assignment is determined through the methods previously illustrated in Table 7.2. The upper bound on path flows is 20 erlangs for path 1 and 15 erlangs for path 2. These upper bounds are determined through methods previously illustrated in Figure 7.11.

Figure 7.15 states the remainder of the initial conditions for the problem. It illustrates the link flows on each of the four links for the four-switch network

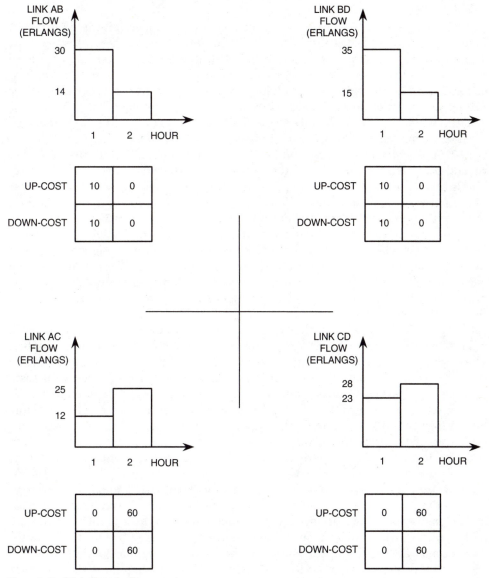

Figure 7.15 Link flows before reroute

in each of the two hours. These link flows are determined by the sum of the path flows for all switch pairs in this four-switch problem. For instance, on link A–B the total flow in hour 1 is 30 erlangs and in hour 2 is 14 erlangs. These flows include the initially assigned flows for switch pair A–D. In our particular example, in hour 1, 30 total erlangs carried on link B include 20 erlangs for switch pair A–D as defined in Figure 7.14 and 10 erlangs for all other switch pairs in the network. In addition, an important concept is illustrated on Figure 7.15: the concept of up-cost and down-cost. Simply stated, up-cost is defined as the unit cost for increasing flow on a link; down-cost is the unit savings for decreasing flow on a link.

We illustrate this by considering link A–B in the upper left-hand corner of Figure 7.15. Link A–B has a unit cost of $10 per unit of capacity—that is, per unit of carried load. Increasing the flow on link A–B in hour 1 requires us to increase the capacity of the link and to incur a cost of $10 for each additional unit of flow added to it in hour 1. Similarly, in hour 1, we can reduce the cost of link A–B by $10 for each unit of flow removed. Note that in hour 2, though, both the up-cost and the down-cost for link A–B are 0. This is true because link A–B must be sized to accommodate the larger of the flows in hours 1 and 2—that is, the flow in hour 1. Adding load to link A–B in hour 2 results in no additional cost, provided we add no more than 16 additional erlangs of flow; therefore, the up-cost is zero. Similarly, removing flow from link A–B in hour 2 does nothing to reduce the cost of the network. That is because link A–B has been sized for its maximum flow, the flow that occurs in hour 1. Therefore the down-cost in hour 2 is zero.

To summarize the initial conditions for this problem, the initial path flows in hour 1 for switch pair A–D are illustrated at the bottom of Figure 7.14, and the accompanying link flows are illustrated in Figure 7.15. We now address the question of determining a better assignment of flow to paths 1 and 2 for pair A–D in hour 1, where *better* is defined in terms of the flow assignment that reduces the cost to the network.

In Figure 7.16, we ask whether it is possible to reduce the cost of the network in hour 1 by diverting some of the flow of switch pair A–D from path 1 to path 2. The path 1 down-cost is the sum of the down-costs for links A–B and B–D, the two links that form path 1; the sum is 20. This value represents network costs that are saved for each unit of flow (that is, for each erlang) for switch pair A–D that is removed from path 1. The path 2 up-cost is zero; that is, it is possible to add flow to links A–C and C–D in hour 1 without increasing their necessary capacities. The difference between the up-cost on the path gaining flow, path 2, and the down-cost on the path losing flow, path 1, is −20, implying that it is possible to divert traffic from path 1 to path 2 and reduce the cost of the network.

Given that it is possible to divert some traffic from path 1 to path 2, the next question that arises is, How much traffic should be diverted from path 1 to path 2? The answer is the minimum of three values. The first value is the total flow on path 1; that is, we cannot remove more than the total offered traffic of 20 erlangs. The second value is the upper bound on the path gaining flow, path 2, which is 15—path 2 cannot carry more than 15 erlangs. The third value is the flow that

- For switch pair A–D in hour 1, is a reroute profitable?

    Path 1 down-cost = $10 + 10 = 20$.

    Path 2 up-cost = $0 + 0 = 0$.

    Reroute cost = $0 - 20 = -20$. (Reroute is profitable.)

- How much flow should be diverted from path 1 to path 2?

    Total flow on path 1 = 20.

    Upper bound of path 2 = 15.

    Flow that changes reroute cost = 5.

- Divert the minimum: 5 erlangs.

Path	Flow	Bound
1	15	20
2	5	15

Figure 7.16   Path-EFO model heuristic flow optimization

changes the reroute cost. That flow is 5 erlangs. Why does the flow of 5 erlangs added to path 2 change the reroute cost? For a moment, refer back to Figure 7.15. Path 2 is composed of links A–C and C–D. Examine link C–D in hour 1. Before any rerouting of flow on switch pair A–D is made, the flow on link C–D in hour 1 is 23 erlangs and in hour 2, 28 erlangs. If we add 5 erlangs of flow to link C–D in hour 1, the flow on link C–D is then 28 erlangs in both hours 1 and 2. Once this flow has been equalized in both hours, the addition of any flow beyond 28 erlangs on link C–D in hour 1 requires an accompanying increase in capacity on link C–D to accommodate the extra flow beyond 28 erlangs. Therefore, the up-cost, the cost of adding flow to link C–D, would change from zero to $60 per unit of flow. This change of up-cost from zero to 60 changes the reroute cost; therefore, we choose to divert only 5 erlangs of flow from path 1 to path 2. Once again, an addition of 5 erlangs of flow to path 2 is the largest flow that can be diverted without change to the reroute cost. After we divert 5 erlangs of flow for switch pair A–D in hour 1, as shown in the bottom of Figure 7.16, there are 15 erlangs of flow now assigned to path 1 and 5 erlangs of flow assigned to path 2.

Figure 7.17 displays the link flows on each of the four links in each of the two hours after we have diverted 5 erlangs of flow from path 1 to path 2 for switch pair A–D in hour 1. Note that we have reduced the maximum flow on link A–B from 30 to 25 erlangs, thereby reducing the capacity that must be provided on link A–B. A similar reduction has been achieved on link B–D. Examining link A–C, the flow in hour 1 has increased from 12 erlangs to 17 erlangs. However, it has not been necessary to increase the capacity of link A–C, because its maximum

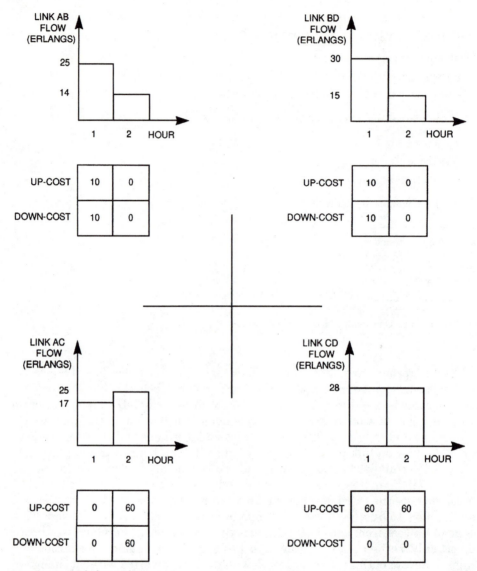

Figure 7.17 Link flows after reroute

flow occurs in the second hour. As we discussed previously, we note that with this assignment of flow to the two paths, the flow on link C–D is equal in both hours to 28 erlangs. Notice that in the bottom right-hand corner of Figure 7.17 the values of the up-costs and down-costs for link C–D have changed. It is no longer possible to add additional flow to link C–D in either hour 1 or hour 2 without incurring a need for additional capacity, and therefore additional cost, on link C–D.

In Figure 7.18, we ask the question, Is it possible to divert more flow from path 1 to path 2? We diverted only as much traffic as could be allowed without

---

- Update link up/down costs.
- Could we divert more flow?

  Path 1 down-cost = $10 + 10 = 20$.
  Path 2 up-cost = $0 + 0 = 0$.
  Reroute cost = $+60 - 20 = +40$ (reroute is not profitable).

- Examine other switch pairs and other hours.
- Continue until no further profitable reroutes exist, then stop.

Figure 7.18   Path-EFO model heuristic flow optimization

changing the reroute costs. We first update the link up-costs and down-costs. (The link up-costs and down-costs shown in Figure 7.17 are updated—compare them with the up-costs and down-costs shown in Figure 7.15.) Is it possible to divert more flow from path 1 to path 2 for switch pair A–D in hour 1, such that the cost in the network is reduced? The down-cost—cost saved by diverting an erlang of flow from path 1—is unchanged, at $20 per erlang of flow diverted. However, the up-cost—the cost associated with adding capacity to accommodate more flow on path 2, the path gaining flow—is now $60 per erlang diverted. The reroute cost—the difference between the up-cost and down-cost—is positive, indicating that a reroute is not profitable. Therefore, we stop the process for switch pair A–D in hour 1 and leave our flow assignments at 15 erlangs of flow on path 1 and 5 erlangs of flow on path 2.

The process now continues by examining all other demand pairs in the network for all load set periods, or design hours, in the network. The process of examining each demand pair in each hour terminates when we find that there are no more profitable reroutes in the network. One of the disadvantages of the heuristic implementation of the flow assignment path-EFO model, as compared with the implementation as a linear program, is that the heuristic takes a rather localized view, and at any one time it is examining cost savings associated with diverting flow for only a single switch pair at a time. With a linear programming path-EFO model, the linear program in essence has the ability to consider performing flow diversion for several switch pairs in the network simultaneously. This single-pair view of the heuristic linear programming model, contrasted with the multiple-pair view possible with the linear program, can cause the heuristic linear programming model to terminate prematurely, with the network flows assigned by the heuristic representing a local minimum rather than a global minimum. A perturbation technique called zero-profit reroutes, or ZPR, can be implemented as part of the heuristic linear programming model to overcome this difficulty.

Starting with Figure 7.19, we illustrate the ZPR methodology. The figure shows a slight extension to the four-switch example we are considering by supposing that in fact we did not have a four-switch problem but rather a five-switch problem, with switch E added. Here, we have shown only a portion of the five-switch problem with switches C, D, and E, with the interconnecting links having

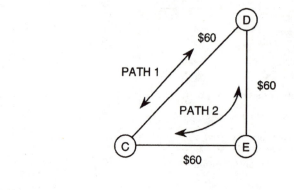

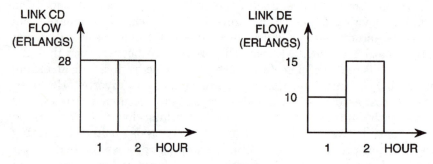

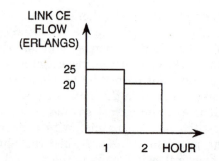

IN HOUR 2

      DOWN-COST PATH 2 = 0 + 60 = 60

      UP-COST PATH 1 = 60

      REROUTE COST = +60 - 60 = 0

    DIVERT 5 ERLANGS FROM PATH 2 TO PATH 1

Figure 7.19 Zero-profit reroutes (ZPR)

identical unit costs of $60 per erlang. We consider two paths, path 1 and path 2, for traffic load C–D in hour 2. In the middle of Figure 7.19, we illustrate the loads on links C–D, D–E, and C–E in each of the two hours of the problem. Here, we assume that for switch pair C–D in hour 2, the current path assignment of flow is 10 erlangs of flow on path 1, the direct path C–D, and 10 erlangs of flow on path 2, the C–D–E path.

Let us examine the question of changing the flow assignment for switch pair C–D in hour 2 by diverting flow from path 2 to path 1. In hour 2, the down-cost on path 2—the path from which we propose to remove flow—is a total of $60, which consists of a zero down-cost on link C–E in hour 2 (that is, no cost is saved by removing flow) and $60 per unit removed on link D–E. Note that link D–E's capacity is determined by the flow in hour 2, and therefore reduction of flow in hour 2 reduces the cost of the link. The up-cost on link C–D in hour 2 is 60 units ($60 per unit of flow). Note that the flows on link C–D are the same as illustrated in Figure 7.17. The cost of rerouting traffic from path 2 to path 1 for switch pair C–D in hour 2 is the difference between the up-cost and the down-cost of the two paths involved. In this case, the difference is 0. That is, it is possible to reroute or divert flow from path 2 to path 1 without a change in network cost. Although there is no cost saved by diverting flow, there is no cost added, so we choose to divert flow from path 2 to path 1, in the hope that diversion of flow for switch pair C–D in hour 2 makes it possible to adjust flows on other paths for other switch pairs, possibly in different hours, which will actually lead to a reduction in network cost.

Figure 7.20 focuses once again on the four links in the original problem, that is, links A–B, B–D, A–C, and C–D. The only change in Figure 7.21 relative to Figure 7.18 is the change of flow on link C–D in hour 2 from 28 to 33 erlangs. That change came about because of the diversion of flow for traffic load C–D that we just discussed. Note that the up-cost and down-cost for link C–D have now changed. In hour 1, they are both zero because the capacity of link C–D is determined by the flow in hour 2 and not the flow in hour 1. Now, once again returning to our original switch pair—the A–D switch pair in hour 1—we ask, Is it possible to divert more flow from path 1 to path 2 such that we reduce the cost of the network? Recall that path 1 consists of links A–B and B–D, and path 2 consists of links A–C and C–D. In Figure 7.21, we calculate the down-cost—the money saved by removing flow from switch pair A–D's first path in hour 1. That down-cost is $20 per erlang, and the cost for adding flow to path 2 in hour 1, the up-cost, is 0 dollars per erlang. The reroute cost—the difference between the up-cost and down-cost—is negative, implying that we can reduce the cost of the network by further diverting flow from path 1 to path 2.

To summarize, the zero-profit reroute technique is a perturbation technique that allows the heuristic to move closer to the global optimum solution and prevents the heuristic from terminating prematurely on a local optimum. The principles illustrated in Figures 7.14 to 7.21 are indeed the methods implemented in practice in the heuristic linear programming model for flow assignment to paths in the path-EFO model. When one considers the use of heuristic path-EFO models

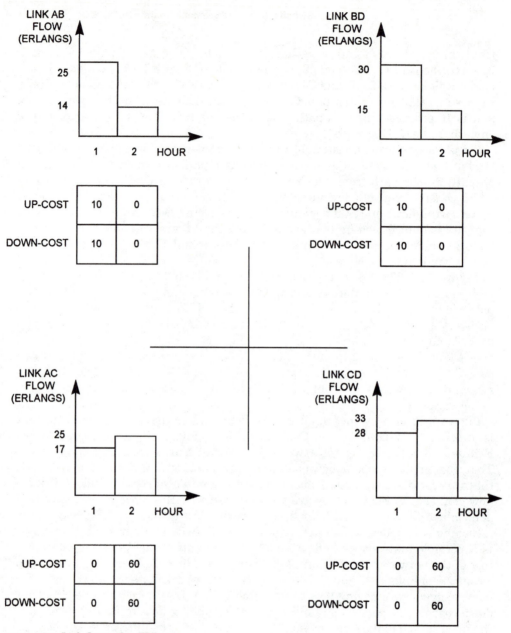

Figure 7.20 Link flows after ZPR

- Reexamine switch pair A–D in hour 1.

  Path 1 down-cost = 10 + 10 = 20.
  Path 2 up-cost = 0 + 0 = 0.
  Reroute cost = +0 − 20 = −20 (reroute is profitable).

- Perform reroute.

Figure 7.21 Path-EFO model heuristic flow optimization

in place of exact path-EFO models, one must address the critically important question of the accuracy of the solution, that is, the cost of solutions determined by the heuristic path-EFO model compared with the cost of exact solutions derived through a linear program.

Figure 7.22 compares the quality of the solution obtained with the heuristic linear programming model implemented in the path-EFO model relative to the quality of the solution with a linear program implementation of the flow assignment problem. We have shown in Figure 7.22 a plot of the value of the objective function—the network cost—for the heuristic implementation and for the linear programming implementation, as a function of CPU time. The problem considered here is a 30-switching switch, 3–load set period problem, the largest problem that could be considered with the MPSX-370 linear programming package (MPSX-370 implements the simplex method). The upper curve—the curve for the linear program—shows that the initial starting solution for the linear program has a cost of about 12 percent above the optimal cost and that the linear program converged to an optimal solution after 5,340 seconds of CPU time, or roughly 89 minutes. The lower curve plotted in Figure 7.22 shows that the heuristic implementation, starting from the same initial feasible solution with a cost of about 12 percent above optimal, reached its best solution after 5 seconds of CPU time. Its best solution is 2.4 percent above the optimal solution. In contrast, the linear program requires 750 seconds of CPU time to reach a solution with the same cost as the heuristic's best solution. Therefore, we judge the heuristic to be approximately 150 times faster than the simplex method.

This run-time performance of the heuristic is crucial to the implementation of preplanned dynamic two-link path routing. Were it not possible to solve the flow assignment problem in two orders of magnitude less time than is required with a linear program, it would not be feasible to design a dynamic routing network for large-scale networks consisting of hundreds of switches.

Table 7.3 and Figure 7.23 return us to our initial four-switch network. In particular, Table 7.3 illustrates the assignment of flow for switch pair 1–2, in both hours 1 and 2. In the top half of the table we illustrate the flow assignment of the three paths for switch pair 1–2 in hour 1. The column labeled *Initial flow* shows the flows described in Table 7.2. The last column, labeled *Heuristic solution flow*, lists the flows assigned by the heuristic. We note that for hour 1 the heuristic found the initial feasible flow to be a satisfactory solution for switch pair 1–2. Looking at the bottom half of Table 7.3, we note that the heuristic diverts 1.62 erlangs of flow from path 1 to path 3 while leaving the flow on path 2 at its initial feasible solution.

We can see that the assignment of flow given us by the heuristic for switch pair 1–2 in hour 2 is an eminently reasonable assignment. If, for a moment, we refer back to Figure 7.12, we recall that path 1 for switch pair 1–2 is the direct path between connecting switches 1 and 2, and path 3 is the two-link path from switch 1 to switch 4 to switch 2. Note that for the first path, link 1–2, the capacity of link 1–2 and therefore its cost was determined by the flow in hour 2, and both links

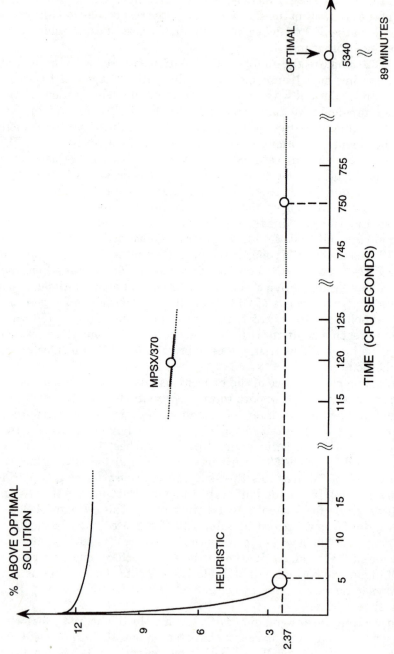

Figure 7.22 Heuristic speed and accuracy: 30-switch, three-hour test

TABLE 7.3   Heuristic Solution for Switch Pair 1–2

Hour	Path	Via	Bound	Initial flow	Heuristic solution flow
1	1	2	9.92	9.92	9.92
	2	3	9.06	0.03	0.03
	3	4	8.21	0	0
2	1	2	13.61	13.61	11.99
	2	3	13.70	1.32	1.32
	3	4	13.25	0	1.62

In hour 2, 1.62 erlangs are diverted from path 1 to path 3.

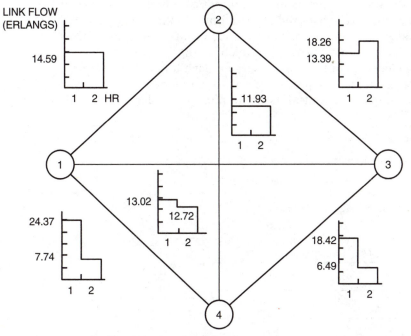

Figure 7.23   Link flows resulting from heuristic (first pass)

1–4 and 4–2 had spare capacity in hour 2. Therefore, it was possible to divert flow from the first path and reduce its capacity without adding that flow to path 3 and without increasing the capacity required on path 3.

Indeed, turning to Figure 7.14, we see the link flows in our four-switch, six-link example resulting from the first application of the heuristic linear programming model. In comparing Figures 7.14 and 7.12, we note that the heuristic was able to reduce the maximum flow, and therefore the capacity required, on four links, that is, on links 1–2, 1–4, 1–3, and 2–4. The heuristic increased the maximum flow on link 2–4 and found no change in the maximum flow on link 3–4.

Summing the peak link flows that existed from our initial solution, shown in Figure 7.12, and comparing that sum with the sum of the peak flows—the heuristic solution—shown in Figure 7.23, we see that the sum of the peak loads was reduced by slightly more than 5 erlangs.

7.3.4.1 Update link blockings. We now examine the remaining steps of the routing design step in the path-EFO model. Let us refer back to Figure 7.10. So far, we have discussed the determination of the initial path flow constraints, the initial assignment of flows to paths, and the heuristic for optimizing path flows. At this point in our discussion, the heuristic has terminated after its first iteration. We now want to move on to the update link blockings step.

In Figure 7.24, the path flows assigned by the heuristic change the link blockings. We now want to determine what the actual blockings on each link of the network should be in each hour. Recall that throughout the path-EFO model, we are holding our estimate of the optimal link blockings constant. We therefore now wish to size each link in the network such that in the hour of the maximum required trunks for a link, the optimal link blocking is realized. We then calculate the resulting blockings in the other hours for each link in the network.

Figure 7.25 illustrates the process of updating link blockings for link 1–2 in our four-switch example. Our initial estimate of optimal blocking for this link was

---

- Path flows assigned by heuristic have changed link blockings.

- For each link, find the hour of maximum required trunks and size the link to $\widehat{B}_{\mathrm{opt}}$ in that hour.

- Calculate resulting blockings in the other hours.

Figure 7.24   Routing design: update link blockings

Link 1–2: $\widehat{B}_{\mathrm{opt}} = 0.093$, initial trunks = 17.73

Hour 1	Hour 2

Initial estimate

$B = 0.008$	$B = 0.093$

From heuristic

flow = 14.59   flow = 14.59

$$A_{\mathrm{off}} = \frac{14.59}{1 - 0.008} \doteq 14.71 \qquad A_{\mathrm{off}} = \frac{14.59}{1 - 0.093} = 16.08$$

Find number of trunks to realize $B = 0.093$ with $A_{\mathrm{off}} = 16.08$:

- Result: Trunks = 18.80.

- New blockings: $B(\text{hour 1}) = B(14.71,\ 18.8\ \text{TKS}) = 0.062$,
  $B(\text{hour 2}) = B(16.08,\ 18.8\ \text{TKS}) = 0.093$.

Figure 7.25   Update link blockings

0.093, and we initially sized this link to 17.73 trunks. This sizing led to initial blocking estimates of 0.008 in the first hour and 0.093 in the second hour. Recall also that these initial blockings were calculated by assuming that only the direct traffic between switches 1 and 2 was offered to link 1–2. The heuristic linear programming model has assigned traffic to link 1–2 from a number of different switch pairs in the network. In fact, it has assigned the flows shown in Figure 7.25: 14.59 erlangs in hour 1 and the same amount of traffic in hour 2. Our estimate of the offered load to realize those flows is 14.71 erlangs in the first hour and 16.08 erlangs in the second hour. Neglecting the effects of peakedness, the number of required trunks for link 1–2 is determined by the maximum offered load, that is, the load in hour 2 of 16.08 erlangs. Once again holding constant our estimate of the optimal link blocking of link 1–2 to 0.093, we find that link 1–2 now requires 18.8 trunks. We therefore recalculate the blockings on link 1–2 in both hours by use of the erlang blocking function. This results in the optimal link blocking being the realized blocking in hour 2, and the realized blocking in hour 1 is 0.062.

7.3.4.2 Update path flow constraints. With our new estimate of link blockings and our first assignment of flow to paths by the heuristic, we are now in the position to revise our bounds on path flows. We can start to estimate a good ordering of the paths within each route so that the realized flows on a path closely approximate the optimal flows determined by the heuristic. On the first pass through the heuristic solution, we had no knowledge of path order, so the upper bounds were set high enough so that any path could be first.

Based on the results of the heuristic solution, we start to get an estimate of a good path ordering, one that will allow us to take advantage of noncoincidence in the network. The process for updating flow constraints and updating and estimating path ordering is then to (1) estimate the path order, (2) set the flow constraints that are related to sequential routing, and (3) reassign flows. This process is illustrated in Figure 7.26. In the top half of the figure, we focus once again on the flows for traffic load 1–2 in the second hour. First we estimate the proper ordering of the three paths. We are given as inputs the flows on the three paths determined by the heuristic and our new estimates of the link blockings. Based on that information we calculate the load that should be offered to the path to realize the flows assigned by the heuristic, and we order the paths in terms of descending offered load. Note that we now estimate that the first path for switch pair 1–2 in hour 2 should be the direct one-link path. The second path should be the two-link path through switch 4, and the third path should be the two-link path through switch 3.

The bottom half of Figure 7.26 illustrates the process of revising the upper bounds on path flow. These bounds are now set so as to cause the implementable flows to match the heuristic flows. For the first path, we calculate the bound on path flow as the path connectivity $(1 - 0.093)$ times the offered load of 15 erlangs, which results in a bound of 13.61 erlangs. For the second path, which is the path through switch 4, we estimate the offered load here to be the difference between 15 erlangs, the load offered to the first path, and 11.99 erlangs, the flow assigned

Example for switch pair 1–2 in hour 2:

1. Find path order.

Path	Via	Heuristic flow	Path connectivity	Path offered load	Order
1	2	11.99	$1 - 0.093$	13.22	1
2	3	1.32	$(1 - 0.056)(1 - 0.079)$	1.52	3
3	4	1.62	$(1 - 0)(1 - 0.117)$	1.83	2

2. Set new bounds.

Path	Via	Offered load	Path connectivity	Upper bound*	Heuristic flow	Cumulative flow
1	2	15	$1 - 0.093$	13.61	11.99	11.99
2	4	$15 - 11.99 = 3.01$	$(1 - 0)(1 - 0.117)$	2.66	1.62	13.61
3	3	$15 - 13.61 = 1.39$	$(1 - 0.056)(1 - 0.079)$	1.21	1.32	14.93

*For second and subsequent paths, this is the bound if the flows assigned by the heuristic were actually realized on the preceding paths.

Figure 7.26 Update path flow constraints

to the first path. The difference is 3.01 erlangs, and we calculate its flow bound using our current estimates of link blockings. Similarly, for the third path, we estimate the offered load to the third path to be the difference between the total offered load for the traffic load and the flows assigned by the heuristic to the first two paths; that is, we estimate that 1.39 erlangs should be offered to the third path.

Finally, in Figure 7.27, we determine a new initial feasible solution for the second pass through the heuristic using the following process. First, we use the path ordering determined through the process in the top of the figure. We set

Reassign flows:

Path	Via	Upper bound	Assigned flow	Cumulative flow
1	2	13.61	13.61	13.61
2	4	2.66	1.32	14.93
3	3	1.21	0	14.93

$$\text{Average blocking} = \frac{15 - 14.93}{15} = 0.005$$

Figure 7.27 Update path flow constraints

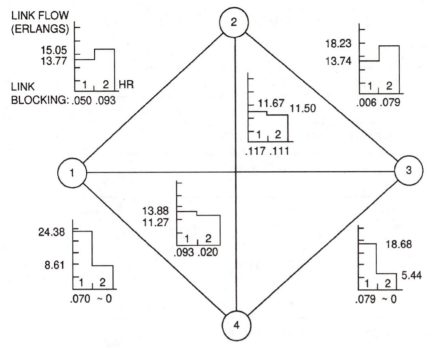

Figure 7.28   Link flows and blockings resulting from heuristic (third pass)

the flow on the first path to the upper bound on flow, and we assign enough flow to the second path so that the blocking probability grade-of-service objective, .005, is realized. The iterative loop consisting of the heuristic for assigning flows to paths, updating link blocking, and updating path flow constraints is executed a total of three times, after which the loop terminates.

The operating point of our four-switch network after three passes through this loop is illustrated in Figure 7.28 with the flows assigned to each of the six links in each of the two hours, as well as the estimated blockings on the links. Note once again that we have held the optimal blockings for each link constant and simply calculated blockings in the side hours. Comparing the link flows shown in Figure 7.28 for this current solution with the link flows in the initial solution shown on Figure 7.12, we see that we have reduced the maximum link flows on four of the links—links 1–2, 1–4, 1–3, and 2–3—and increased the maximum link flows on links 2–4 and 3–4. Summing up the peak link flows in the current network illustrated in Figure 7.28, we find that the sum of the peak link flows, which is approximately the capacity of the network, has been reduced by roughly 4 erlangs compared with our initial solution.

7.3.4.3 Define sequential routes.   Referring to Figure 7.10, we see that now we have discussed all but the define sequential routes step within the routing design step of the path-EFO model, and the methods used in this step are illustrated in Figures 7.29 and 7.30. Figure 7.29 illustrates the process for defining path

Switch pair 1–2, hour 1, offered load = 10 erlangs.

1. Order paths.

Path	Via	Path connectivity	Heuristic flow	Offered load	Order
1	2	$1 - 0.050$	9.45	9.95	1
2	3	$(1 - 0.093)(1 - 0.006)$	0.50	0.55	2
3	4	$(1 - 0.070)(1 - 0.117)$	0	0	3

2. Flow traffic.

Path	Via	Offered load	Path connectivity	Resulting flow	Cumulative flow	Heuristic flow	|Error|
1	2	10	$1 - 0.050$	9.50	9.50	9.45	0.05
2	3	$10 - 9.50 = 0.50$	$(1 - 0.093)(1 - 0.006)$	0.45	9.95	0.50	0.05
3	4	*		0	9.95	0	0
							0.10

*Preceding paths carry enough flow to meet blocking probability objective.

$$\text{Average blocking} = \frac{10 - 9.95}{10} = 0.005$$

$$\text{Error} = \frac{0.10}{9.95} \times 100\% = 1\%$$

Figure 7.29   Realize flows—form a sequential route (hour 1)

ordering in the route for switch pair 1–2 in hour 1. The first step is to determine the order of the path. Given information includes the flow assigned by the heuristic of 9.45 erlangs to path 1, the direct one-link path, and 0.5 erlangs to path 2, a two-link path that is switched through switch 3. Furthermore, at this point, we have current estimates of link blockings; those estimates were shown in Figure 7.28. Given the carried loads assigned by the heuristic and our current estimates of link blockings, we calculate the offered load to each path that is required to realize the heuristic assigned flows. We order the paths in order of descending offered loads. We note that for switch pair 1–2 in hour 1, the rank ordering based on offered load has not changed the order of the paths.

We now compare the realizable flow on each path with the heuristic's optimal flows. This process is illustrated in the bottom half of Figure 7.29. First, the entire offered load of 10 erlangs is offered to the first path, the direct one-link path. Given our current estimates of blocking on that path, the resulting flow is 9.50 erlangs, and this flow differs from the heuristic's optimal flow by a mere

Switch pair 1–2, hour 2, offered load = 15 erlangs.

1. Order paths.

Path	Via	Path connectivity	Heuristic flow	Offered load	Order
1	2	$1 - 0.093$	12.51	13.79	1
2	3	$(1 - 0.020)(1 - 0.079)$	0	0	3
3	4	$(1 - 0)(1 - 0.111)$	2.42	2.72	2

2. Flow traffic.

Path	Via	Offered load	Path connectivity	Resulting flow	Cumulative flow	Heuristic flow	|Error|
1	2	15	$1 - 0.093$	13.61	13.61	12.51	1.10
2	4	$15 - 13.61 = 1.39$	$(1 - 0)(1 - 0.111)$	1.24	14.85	2.42	1.18
3	3	$15 - 14.85 = 0.15$	$(1 - 0.020)(1 - 0.079)$	0.13	14.98	0	0.13
							2.41

$$\text{Average blocking} = \frac{15 - 14.98}{15} = 0.0013$$

$$\text{Error} = \frac{2.41}{14.93} \times 100\% = 16.1\%$$

Figure 7.30   Realize flows—form a sequential route (hour 2)

0.05 erlangs. With 9.50 erlangs carried on the first path, the traffic offered to path 2, the two-link path via switch 3, is 0.5 erlang, and the second path carries 0.45 erlang. We note two things: (1) the two paths in this route carry enough traffic to satisfy our 0.005 blocking constraint and (2) the realized flows on the first two paths very closely match the optimal flows assigned by the heuristic.

Figure 7.30 illustrates the process of path ordering in the route for the same switch pair 1–2 in hour 2. The methods are exactly the same as those used for the first hour. There are two things to note in Figure 7.30. First, the realized average blocking probability for switch pair 1–2 in hour 2 is actually better than our objective. The estimated blocking for switch pair 1–2 is 0.0013. Second, we note that, unfortunately, we are not able to realize flows that match the optimal flows determined by the heuristic quite as closely as we were in the first hour.

Referring back to Figure 7.1, we can now complete a discussion of the routing design step of the path-EFO model. We now move on to the capacity design step.

7.3.5 Capacity design

At the completion of the routing design step we have two forms of information. First, we have the ordering of the paths determined in the routing design step, and second, we have retained our initial estimates of optimal link blockings. We now wish to size the network so that the optimal blocking on each link is the realized blocking in the hour of maximum required trunks. It is illustrated in Figure 7.31 that this process is also an iterative process in which we alternately size each link to calculate blockings and flow traffic over each of the paths in each route. We continue this process until our estimates of blockings on all links in all hours become sufficiently stabilized in terms of differences in blockings calculated from one iteration to the next. The convergence criterion is stated formally at the bottom of Figure 7.31; the small-value epsilon is typically set to 0.01, based on experience with the model. As our estimates of blockings on links improve—that is, change—the resulting flows on each of the links in the network also change and start to differ from the optimal flows determined by the heuristic. This is an implicit result of the nonlinear nature of the actual network design problem, in contrast with the linear assumptions that were made in the heuristic.

- GIVEN ROUTING AND $\hat{B}_{opt}$ FOR EACH LINK, SIZE EACH LINK TO $\hat{B}_{opt}$ IN THE HOUR OF MAXIMUM REQUIRED TRUNKS

- CALCULATE RESULTING BLOCKINGS IN OTHER HOURS

- ITERATIVE PROCEDURE:

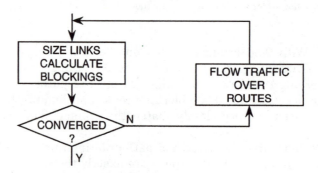

- CONVERGENCE CRITERION

$$\frac{\displaystyle\sum_{i=1}^{L}\ \sum_{h=1}^{H}\ \left|B_{OLD}(i,h) - B_{NEW}(i,h)\right|}{\displaystyle\sum_{i=1}^{L}\ \sum_{h=1}^{H}\ B_{OLD}(i,h)} < \varepsilon$$

Figure 7.31 Path-EFO model: capacity design

Figures 7.32 to 7.35 illustrate changes in actual link flows and link blockings as we iterate among the capacity design steps of sizing the network, recalculating the link blockings, and flowing traffic in the network. The methods illustrated in these figures are the same methods that we have discussed in connection with previous figures. Two points should be made in connection with these figures, however. Figure 7.33 shows the link blockings and link sizes at convergence of the capacity design step. First, if we sum the peak link flows on the six links of the network shown in Figure 7.33, we find that they total 103.35 erlangs. Contrast this with the heuristic's optimal flows shown in Figure 7.32, where the sum of the peak flows was 101.89 erlangs. However, we should keep in mind that we still have reduced the sum of the peak link flows by more than 2.5 erlangs compared with our initial feasible solution. That is, we have reduced the cost of the network by roughly 2.5 percent. The second point to be noted, in connection with Figure 7.35, is that the realized blocking probability for switch pair 1–2 in both of the two hours is actually much better than our objective blocking of 0.005. In the first hour, the realized blocking probability is .003, and in the second hour it is .0007.

These results are typical of the results found in virtually all networks studied for preplanned dynamic two-link path routing; that is, many switch pairs in the network actually experience blocking considerably better than the objective.

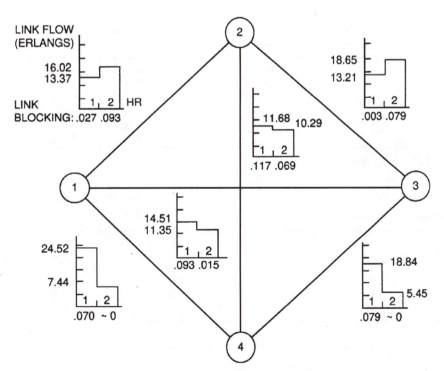

Figure 7.32 Link flows and blockings resulting from sequential routes

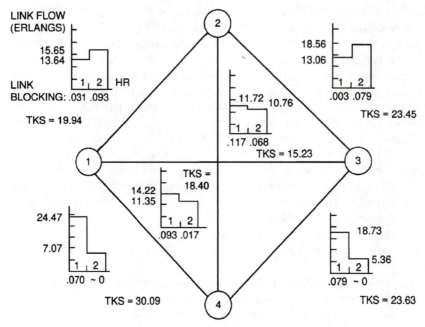

Figure 7.33 Link flows and blockings resulting from revised path flows

Switch pair 1–2, hour 1, offered load = 10 erlangs.

Path	Via	Offered load	Path connectivity	Flow	Cumulative flow	Previous flow
1	2	10	1 − 0.027	9.73	9.73	9.50
2	3	10 − 9.73 = 0.27	(1 − 0.093)(1 − 0.015)	0.24	9.97	0.45
3	4	*		0	9.97	0

*Preceding paths carry enough flow to meet blocking probability objective.

$$\text{Average blocking} = \frac{10 - 9.97}{10} = 0.003$$

Hour 2, offered load = 15 erlangs.

Path	Via	Offered load	Path connectivity	Flow	Cumulative flow	Previous flow
1	2	15	1 − 0.093	13.61	13.61	13.61
2	4	15 − 13.61 = 1.39	(1 − 0)(1 − 0.069)	1.30	14.91	1.24
3	3	15 − 14.91 = 0.09	(1 − 0.015)(1 − 0.079)	0.08	14.99	0.13

$$\text{Average blocking} = \frac{15 - 14.99}{15} = 0.0007$$

Figure 7.34 Revised path flows (based on revised blockings)

Switch pair 1–2, hour 1, offered load = 10 erlangs.

Path	Via	Offered load	Path connectivity	Flow	Cumulative flow	Previous flows	
						Initial	Second
1	2	10	$1 - 0.031$	9.69	9.69	9.50	9.73
2	3	$10 - 9.69 = 0.31$	$(1 - 0.093)(1 - 0.003)$	0.28	9.97	0.45	0.24
3	4	*		0	9.97	0	0

*Preceding paths carry enough flow to meet blocking probability objective.

$$\text{Average blocking} = \frac{10 - 9.97}{10} = 0.003$$

Hour 2, offered load = 15 erlangs.

Path	Via	Offered load	Path connectivity	Flow	Cumulative flow	Previous flows	
						Initial	Second
1	2	15	$1 - 0.093$	13.61	13.61	13.61	13.61
2	4	$15 - 13.61 = 1.39$	$(1 - 0)(1 - 0.068)$	1.30	14.91	1.24	1.30
3	3	$15 - 14.91 = 0.09$	$(1 - 0.017)(1 - 0.079)$	0.08	14.99	0.13	0.08

$$\text{Average blocking} = \frac{15 - 14.99}{15} = 0.0007$$

Figure 7.35 Path flows at convergence on iteration number 1

We now turn to the last major functional area in the path-EFO model—the determination of optimal link blockings.

7.3.6 Update optimal link blockings

The method employed in the path-EFO model to update our estimate of optimal link blockings is illustrated starting in Figure 7.36. In the path-EFO model, we have generalized Truitt's economic CCS method to a multihour environment. Figure 7.36 reviews Truitt's method for a three-switch example—that is, the Truitt triangle. Here we consider the problem of selecting the number of trunks between switches E and F so that the total cost of carrying E-to-F offered load A erlangs is minimized. Traffic between switches E and F has a direct path E–F and an alternate path E–G–F. Truitt's method is mathematically stated in Figure 7.36, where $C(T_{EF})$ is the cost of carrying E-to-F traffic, which is composed of the sum of the products of the trunks on the direct path E–F times the unit cost of trunks on the E–F path plus the sum of the trunks on the alternate path E–G–F and their costs. Gamma once again is the marginal capacity—that is, the carried load per incremental trunk on the alternate path. A is the offered load and B

- TRUITT'S ECCS METHOD IS GENERALIZED TO A MULTI-HOUR ENVIRONMENT

- TRUITT'S METHOD

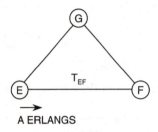

CHOOSE T_{EF} SUCH THAT $C(T_{EF})$ IS MINIMIZED

$$C(T_{EF}) = C_{EF} \, T_{EF} + \left[\frac{C_{EG}}{\gamma_{EG}} + \frac{C_{GF}}{\gamma_{GF}} \right] \times A \times B(A, T_{EF})$$

γ = MARGINAL CAPACITY

$B(.,.)$ = ERLANG BLOCKING FORMULA

ASSUMES ALTERNATE LINKS ARE BUSY IN SAME HOUR AS DIRECT LINK

Figure 7.36   Path-EFO model update optimal link blockings

is the erlang B blocking formula. A fundamental assumption in Truitt's method is that the links in the alternate path are busy; that is, sized for the same hour as the link in the direct path. However, in the preplanned dynamic two-link path routing environment, we are clearly considering the multihour design problem, and if the busy hours of the links in the alternate path (links E–G and G–F) are not time-coincident with the busy hour of link E–F, then spare capacity exists on the alternate path. This spare capacity is free, and its existence should be recognized when one tries to minimize the total cost of carrying traffic from E to F.

Figure 7.37 expresses the spare capacity on link E–G as the difference between the number of trunks on the link and the ratio of the link's background load to its marginal capacity. The background load on link E–G is the load carried on it in the hour for which we are trying to size link E–F. In the bottom half of Figure 7.37, we state mathematically our generalization of Truitt's method, where we have now introduced the concept of spare capacity that is free on the alternate links. As in Truitt's method, we choose the number of trunks in link E–F so that the total cost of carrying E-to-F traffic has been minimized. In choosing the number of trunks on link E–F, we implicitly determine the optimal blocking for the link. As noted in the bottom of Figure 7.38, we have chosen in the path-EFO model to place a limit on the difference between our updated estimate of the optimal blocking on a link relative to our previous estimate of the optimal blocking on

- IF BUSY HOURS NOT TIME COINCIDENT, THEN SPARE CAPACITY EXISTS ON THE ALTERNATE PATH THAT IS "FREE"

e.g.

$$T_{SPARE_{EG}} \approx T_{EG} - \frac{BG_{FG}}{\gamma_{EG}}$$

$$BG_{EG} = \text{BACKGROUND LOAD ON LINK EG}$$

- CHOOSE T_{EF} TO MINIMIZE

$$C(T_{EF}) = C_{EF} T_{EF}$$

$$+ C_{EG} \times MAX \left[0, \left[\frac{A \times B(A, T_{EF})}{\gamma_{EG}} - \left(T_{EG} - \frac{BG_{EG}}{\gamma_{EG}} \right) \right] \right]$$

$$+ C_{GF} \times MAX \left[0, \left[\underbrace{\frac{A \times B(A, T_{EF})}{\gamma_{GF}}}_{\substack{\text{NUMBER OF} \\ \text{TRUNKS} \\ \text{TO CARRY} \\ \text{OVERFLOW}}} - \underbrace{\left(T_{GF} - \frac{BG_{GF}}{\gamma_{GF}} \right)}_{\substack{\text{NUMBER} \\ \text{OF} \\ \text{SPARE} \\ \text{TRUNKS}}} \right] \right]$$

- CHOOSING T_{EF} DETERMINES $\hat{B}_{OPT_{EF}}$
- LIMIT: $0.8 \times \hat{B}_{opt_{old}} \le \hat{B}_{opt_{new}} \le 1.2 \times \hat{B}_{opt_{old}}$

Figure 7.37 Update optimal link blockings, part 2

a link. This limitation was determined based on experience with the model and aids in the convergence of the path-EFO model as a whole.

Figure 7.38 provides a numerical example of our extension of Truitt's method. The example illustrated here comes from the four-switch problem we have been considering throughout the chapter. Here, we wish to update our estimate of the optimal blocking on link 1–2; we calculate the optimal link blocking in the hour of maximum offered load between switches 1 and 2—that is, in the second hour in which 15 erlangs are offered. The alternate path considered for optimal link blocking determination is the first alternate path in the route between switches 1 and 2 in hour 2; that is, path 1–4–2. In Figure 7.38, we have illustrated the link flows and trunks on links 1–4 and 4–2 in both hour 1 and hour 2. These link flows and trunks are exactly those link flows and trunks shown in Figure 7.33.

In working through the example in Figure 7.38, there are several things to note. First, the background load on link 1–4 is defined to be the flow on link 1–4 in hour 2, that is, 7.07 erlangs. We make an analogous assumption for link 4–2. Second, we have made the assumption that the marginal capacity of both links 1–4 and 4–2 is one erlang carried per trunk. Last, from the beginning we have assumed a unit cost of $19 per trunk on each link in the network. We note in Figure 7.38 that our initial estimate of the optimal blocking on link 1–2 was 0.093. Our generalized Truitt method implies that the new estimate of optimal blocking for this link should be 0.7. This rather high optimal link blocking for

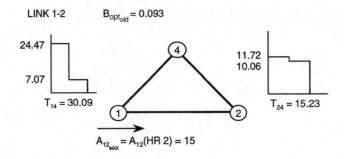

CHOOSE T_{12} TO MINIMIZE

$$C_{12} T_{12} + C_{14} \text{ MAX} \left\{ 0., \left[\frac{A_{12} \times B(A_{12}, T_{12})}{\gamma_{14}} - \left(T_{14} - \frac{BG_{14}}{\gamma_{14}} \right) \right] \right\}$$

$$+ C_{24} \text{ MAX} \left\{ 0., \left[\frac{A_{12} \times B(A_{12}, T_{12})}{\gamma_{24}} - \left(T_{24} - \frac{BG_{24}}{\gamma_{24}} \right) \right] \right\}$$

OR

$$19 T_{12} + 19 \times \text{MAX} \left\{ 0, [15 \times B(15, T_{12}) - (30.09 - 7.07)] \right\}$$

$$+ 19 \times \text{MAX} \left\{ 0, [15 \times B(15, T_{12}) - (15.23 - 10.06)] \right\}$$

FINDING $T_{12} \Rightarrow \hat{B}_{opt_{new}}$

$\hat{B}_{opt_{new}} \approx 0.70 > 1.2 \times \hat{B}_{opt_{old}} = 0.111$

THEREFORE $\hat{B}_{opt_{new}} = 0.111$

Figure 7.38   Update optimal link blockings—example

link 1–2 is in part influenced by the considerable spare capacity available on the alternate path in the second hour. Also, note that our limitation on changes in optimal link blocking results in limiting the new optimal link blocking to no more than a 20 percent increase in the old optimal link blocking; that is, our new estimate of optimal link blocking, 0.111, is 20 percent greater than our old estimate of optimal link blocking of 0.093.

7.3.7   Generate candidate paths

Referring back to Figure 7.2, our overview flow diagram of the path-EFO model, there remain two areas to be discussed. The first is the method by which we generate candidate paths in second and subsequent iterations, and this is the subject of Figure 7.39. Recall that on the first iteration of the path-EFO model, we selected k candidate paths for each switch pair in the network strictly on the basis of the cost of a trunk in each link in the network. As a result of the first iteration of the path-EFO model, we typically find that of the k candidate

- SUPPOSE j < k PATHS IN A ROUTE HAVE FLOW ASSIGNED

- RETAIN j PATHS AND CONSIDER REPLACING (k-j) UNUSED PATHS WITH "BETTER" PATHS

- FIND ALL TWO-LINK PATHS IN THE NETWORK THAT ARE NOT CURRENTLY USED AND THAT SATISFY THE ROUTING CONSTRAINTS. RANK ORDER THESE PATHS ON THE BASIS OF "SLACK COST" AND SELECT THE (k-j) BEST PATHS

- FOR SWITCH PAIR EG IN HOUR 1:

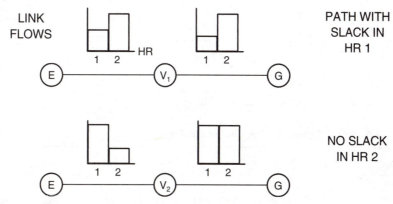

Figure 7.39 Selection of candidate paths (second and subsequent iterations)

paths for any switch pair, only j paths, almost always considerably less than k, have flow assigned to them. Those $k - j$ paths without flow were deemed by the path-EFO model to be unattractive to carry flow for some particular traffic load. Our strategy now is to retain the j paths that have flow assigned to them and replace the $k - j$ unused paths with paths that are in some sense better.

Our metric for determining a better path is actually very similar to the metric used in Truitt's ECCS method, generalized to a multihour environment. We look for paths in which there is considerable spare capacity available. For instance, referring to the bottom of Figure 7.39, we have two candidate paths for carrying flow from switch E to switch G. Let us consider hour 1 of our two-hour problem. The first path shown, E–V_1–G, is seen to have spare capacity on both links in hour 1, whereas the second path, E–V_2–G, has no spare capacity in hour 1. On the basis of slack as a metric, we clearly prefer path E–V_1–G to path E–V_2–G.

7.3.8 Check route blocking

The last step in the path-EFO model to be discussed is the check route blocking step illustrated in Figure 7.2 and, in more detail, in Figure 7.40 and Tables 7.4 and 7.5. In the process of checking the blocking probability grade-of-service

• EXAMINE ROUTING AND FLOWS FOR EACH SWITCH PAIR IN EACH HOUR

• EXAMPLE: SWITCH PAIR 1-2 IN HOUR 1 ON ITERATION #6

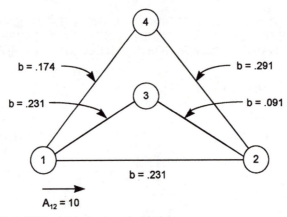

Figure 7.40   Path-EFO model: check route blocking

TABLE 7.4   Path-EFO Model: Check Route Blocking

Path	Via	Offered load	Path connectivity	Flow	Cumulative flow
1	2	10	$(1 - 0.231)$	7.69	7.69
2	3	$10 - 7.69 = 2.31$	$(1 - 0.231)(1 - 0.091)$	1.62	9.31
3	4	$10 - 9.31 = 0.69$	$(1 - 0.174)(1 - 0.291)$	0.40	9.71

Average blocking $= (10 - 9.71)/10 = 0.029 > 0.005$—not satisfactory.

TABLE 7.5   Check Route Blocking

Correct by reducing blocking on first path; add trunks to link 1–2 to reduce blocking to 0.040.

Path	Via	Offered load	Connectivity	Flow	Cumulative flow
1	2	10	$(1 - 0.040)$	9.6	9.6
2	3	$10 - 9.6 = 0.4$	$(1 - 0.231)(1 - 0.091)$	0.28	9.88
3	4	$10 - 9.88 = 0.12$	$(1 - 0.174)(1 - 0.291)$	0.07	9.95

Average blocking $= (10 - 9.95)/10 = 0.005$—OK.

objective, we wish to ensure that each switch pair in the network in each hour is provided with satisfactory blocking probability grade-of-service.

Figure 7.40 illustrates for our four-switch network example the route for switch pair 1–2 in the first hour of our two-hour problem, after six iterations of the path-EFO model have been completed. Recall that the offered load between switches 1 and 2 is 10 erlangs, and we have listed in the figure the blockings on each link in each path in hour 1. Now let us check the blocking probability grade-of-service provided to this traffic load. Refer to Table 7.4. Ten erlangs of load are offered to the first path; this is multiplied by the connectivity of the first path, the complement of the blocking on the one-link path, which yields 7.69 erlangs of flow (2.31 erlangs overflow and are offered to the second path). The second path carries 1.62 erlangs, and paths 1 and 2 together carry a total of 9.31 erlangs of the 10 offered erlangs (0.69 erlang overflows to the last path, the path via switch 4). Given the current blockings on links 1–4 and 4–2, 0.40 of that 0.69 offered erlang is carried, and in total the three paths in the route carry 9.71 of the 10 offered erlangs. Relative to the 10 offered erlangs, we have blocked 2.9 percent of the traffic. We find this blocking to be unsatisfactory and therefore take corrective action. The corrective action illustrated in Table 7.5 takes the form of reducing the blocking on the first path in the route for switch pair 1–2 such that the overall blocking exactly meets our 0.005 blocking requirement. Through straightforward calculations, we find it necessary to reduce the blocking on this one-link path from 0.231 to 0.040, and Table 7.5 illustrates that if we reduce the blocking on link 1–2, the switch pair 1–2 experiences exactly 0.005 blocking in hour 1, which we deem to be satisfactory.

7.3.9 Convergence and final solution of the path-EFO model

As we have discussed throughout the chapter, the path-EFO model is an iterative model. We mentioned earlier that our stopping criterion is to terminate the path-EFO model when we find no further improvement in network cost from iteration to iteration. Figure 7.41 illustrates the improvement in network cost as a function

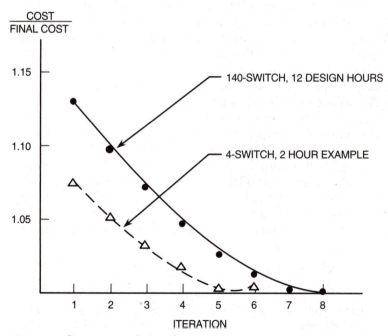

Figure 7.41 Convergence of network cost

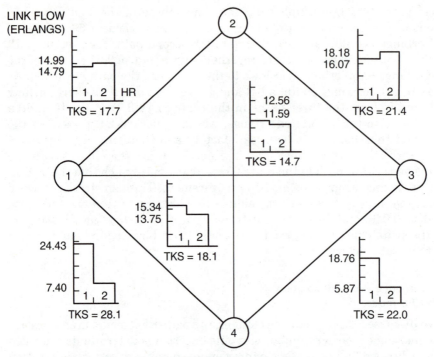

Figure 7.42   Final network (iteration 5) link flows and trunks

of the number of iterations of the path-EFO model for two examples. The lower example, shown by the dashed line, is the four-switch, two-hour example that we have been considering throughout the chapter. At the end of the first iteration of the path-EFO model, the cost of our four-switch two-hour example network is roughly 8 percent above the best cost observed after five iterations. In practice, we would choose to implement the network found after five iterations of the path-EFO model. The second example shown in Figure 7.41 is for a realistic size problem, that is, for 140 switches for 12 design hours. The best cost for this network is found after eight iterations of the path-EFO model, which is roughly 14 percent less than the cost found after the first iteration.

Figure 7.42 and Table 7.6 illustrate the final solution for the four-switch, two-hour example that we have been considering. Figure 7.42 shows the number of trunks required on each of the six links. Table 7.6 details the routing for each traffic load in the network in each hour. As we discussed earlier, many of the switch pairs experience a blocking much better than the one-half percent objective blocking.

7.4   Conclusion

This concludes the four-switch example of dynamic routing design and capacity optimization. We have shown how the principles of shortest-path selection,

TABLE 7.6   Final Network Routing (Iteration 5)

Switch pair	Hour	Offered load	Path	Via
1–2	1	10	1	2
			2	3
			3	4
	2	15	1	2
			2	4
			3	3
1–3	1	15	1	3
			2	2
			3	4
	2	10	1	3
			2	4
1–4	1	25	1	4
			2	2
			3	3
	2	5	1	4
2–3	1	10	1	3
			2	1
			3	4
	2	20	1	3
			2	1
			3	4
2–4	1	10	1	4
			2	1
			3	3
	2	10	1	4
			2	3
			3	1
3–4	1	20	1	4
			2	2
			3	1
	2	5	1	4
			2	1

multihour dynamic routing design, multihour capacity optimization, and link flow optimization are applied within a unified design model for dynamic routing networks. As discussed in Chapter 1, these are basic design principles that can be applied to the design of dynamic routing networks of all types.

Real-Time Dynamic Routing Design

8.1 Introduction

As discussed in Chapters 1 and 3 to 7, dynamic traffic routing follows the network load dynamically and often uses nonhierarchical routing tables to minimize network cost. The term *dynamic* describes routing techniques that are either preplanned time-varying, as opposed to fixed hierarchical routing rules, or real-time state-dependent, which is the limiting case of dynamic traffic routing. The preplanned dynamic routing methods use routing design in advance to follow a prescribed set of predicted traffic patterns, but they can also use real-time routing adjustments to improve network performance under unexpected traffic patterns, overloads, and failures.

In this chapter, we study dynamic routing design under random load variations, in which real-time routing methods are combined with preplanned dynamic routing. We use preplanned dynamic two-link path routing as a basis for study, but the principles discussed apply in general to all dynamic routing networks. In particular, this chapter discusses analytical and simulation models for real-time dynamic routing techniques that include trunk reservation, call gapping, and real-time path selection. Such models and techniques are applied and extended in later chapters to real-time state-dependent and event-dependent routing networks (Chapters 11–13) and dynamic transport routing networks (Chapters 14 and 15).

It is known that some daily load variations are systematic (e.g., Monday is always higher than Tuesday), but in some day-to-day variation models these systematic changes are ignored and lumped into the stochastic model. For instance, the load between Los Angeles and New Brunswick will look very similar from one day to the next, but the exact calling levels differ for any given day. We characterize this load variation by a stochastic model that provides within capacity management design a "day-to-day capacity," which is needed to meet a specified average blocking objective when the load varies according to the stochastic model for day-to-day variations [HiN 76, Wil58, Wil71]. Each forecasted traffic load is actually a mean load about which there occurs a day-to-day variation, characterized by a gamma distribution with one of three levels of variance. Forecast loads are subject to error, but even if the forecast mean loads are correct, the actual realized loads exhibit a random fluctuation from day to day. Hence, there are two sources of load uncertainty: forecast error and day-to-day variation. Earlier studies have established that each of these sources of uncertainty requires the network to be augmented in order to maintain the blocking probability grade-of-service. Day-to-day capacity is nonzero due to the nonlinearities in link blocking, and in fixed routing design for the 28-switch model, the day-to-day capacity required is 4 to 7 percent of the network cost, for medium to high day-to-day variations. This level of day-to-day capacity can be reduced with real-time routing table updates, in which greater network utilization and capacity management design savings are achieved. Reserve capacity is the difference between actual in-place capacity and the ideal capacity design for the current network loads. As shown in Chapter 9, reserve capacity arises because of forecast error and trunk administration policies, and can reach 15–25 percent or more of total network capacity. Dynamic routing table update, based either on preplanned dynamic routing or real-time dynamic routing table updates, can lead to reduction in reserve capacity. In addition to possible economic advantages of real-time routing table updates, significant advantages are improved network switch-to-switch blocking performance and enhanced network survivability under failure and overload, as illustrated in this chapter.

Uncertain variations in the instantaneous network loads imply that capacity is never perfectly matched to demand. Loads in the network shift from hour to hour and from day to day, and some amount of reserve or idle capacity is almost always present. Hence, there is an opportunity to seek out this idle capacity in real time.

We discuss real-time routing table update models that find and use idle network capacity to satisfy current loads. These models predict that network blocking probability is substantially reduced with such real-time routing table updates. Another question is, to what extent can the capacity augmentation required by the uncertainties be reduced by real-time control of the routing tables to meet the realized loads? A given realization of the traffic loads can be expected to yield some loads that are higher than average and others that are lower, and hence part of the network is overloaded while another part might be underloaded. If the real-time routing tables can be adjusted to use the idle capacity of the underloaded portion of the network, the required capacity management augmentation might be reduced. Real-time routing table updates deal with day-to-day traffic load variations, unforecasted traffic demand until any needed capacity augmentation is made, and real-time traffic management under overload and failure. Hence, the routing decisions necessary in real time involve conditions as they become known in real time: day-to-day load variations, unforecasted demand, network failures, and network overloads. Real-time routing table updates can improve network performance significantly, even with simple procedures—the improvement can also be equated to an equivalent trunk cost savings in the range of 2 to 3 percent in day-to-day capacity, as discussed in Chapter 13, and a larger possible reduction of 5 percent or more in network reserve capacity, as discussed in Chapter 9. Real-time routing can also lead to improved network performance with a higher overall completion rate. In the latter case, real-time routing table updates improve network performance under network failures, especially when reserve capacity is available for redirecting traffic flows to available idle capacity.

The basic dynamic two-link path routing method considered in this chapter is illustrated in Figure 1.22. The example is for national intercity dynamic traffic routing but applies equally well to metropolitan area dynamic routing or global international dynamic routing. The dynamic routing method capitalizes on two factors: (1) selection of minimum-cost paths between the originating and terminating switches, and (2) designing optimal preplanned time-varying routing tables to achieve minimum-cost design by capitalizing on noncoincident network busy periods. The preplanned dynamic, or time-varying, nature of the routing method is achieved by introducing several route choices, where the routes consist of different orderings of the available paths (in this case, four paths). Each path consists of one or at most two links in tandem. The originating switch (SNDG) in Figure 1.22 retains control over a dynamically routed call until it is either completed to its destination or blocked from the network. Destination in the national intercity network means the terminating intercity tandem switch in the dynamic routing portion of the network; destination in the metropolitan area network means the terminating end-office in the dynamic routing portion of the network; and destination in the global international network means the terminating international switching center in the dynamic routing portion of the network. A call overflowing the second link of a two-link connection (e.g., the ALBY–WHPL link of the SNDG–ALBY–WHPL path) is returned to the originating switch (SNDG) for possible further alternate routing. Control is returned by using the CCS crankback signal sent from the via switch to the originating switch.

Each of the four routing sequences illustrated in Figure 1.22 uses a different order of the four paths. Each routing sequence results in a different allocation of link flows, but all satisfy the switch-to-switch blocking objective. Allocating traffic to the optimum route choice during each time period or load set period leads to design benefits due to the noncoincidence of loads. This route selection changes with time as shown in the columns on the right in the figure—thus it is preplanned dynamic. The example shown indicates that in the morning the routing method is to offer the SNDG–WHPL traffic to routing sequence number 1 (starting with the direct link to WHPL overflowing to the two-link connection through ALBY) and in the afternoon to routing sequence number 2 (overflowing to the two-link connection through PHNX). In the evening, routing sequence number 3 is used.

As discussed in Chapter 6, design of the dynamic network depends primarily on performing off-line calculations for routing tables and link sizing. The off-line calculations select the optimal routing tables from a very large number of possible alternatives in order to minimize the network cost. Real-time dynamic routing entails an extensive search for the optimal routing design to be performed in real time. This extensive search is in fact being made in the preplanned dynamic traffic routing method, but most of the searching is performed off line, in advance, using an appropriate network design system. The effectiveness of the off-line design depends on how accurately we can forecast the expected traffic load on the network. Routing table designs are updated in the ongoing routing design process, as discussed in Chapters 1 and 9.

Hence, the preplanned dynamic routing design accounts for predicted, systematic load variations in the preplanned routing tables and link sizes. Real-time routing decisions account for conditions that become known in real time, such as day-to-day load variations, network failures, and network overloads. These components of load variation are not systematic or easily predictable. Reasonably accurate traffic load patterns can be predicted months in advance. The unforecasted demands can then be identified and routing design corrected over a period of a few weeks as the loads develop, but daily load variations must ultimately be identified in real time. As discussed earlier, the network design sizes the network to accommodate all expected load patterns, including day-to-day load variations. Sizing the network for day-to-day variations guarantees that some capacity must stand idle at least some of the time. If routing tables were *totally* preplanned, no advantage could be taken of temporarily idle network capacity to complete calls that might otherwise be blocked. For this reason, methods of real-time routing updates combined with preplanned routing patterns are investigated in this chapter.

With a more efficient dynamic routing network design, response to network failures and overloads is affected. However, real-time routing responds automatically to network failures and overloads and counterbalances the effect of reduced capacity. This application of real-time routing to improve network survivability is also a subject of this chapter. Survivability refers to the ability of the network to continue to operate satisfactorily in the face of a network overload or a switching or transport failure. As discussed in Chapter 14, survivability improvement is an area of concern, especially with the increasing concentration of network resources in large-capacity switching and transport network elements. Real-time routing improves network performance in the event of overloads or failures, especially when some amount of reserve capacity is available for redirecting traffic flows from their usual patterns. The presence of reserve capacity enhances real-time routing capabilities, even during hours of peak network load, because there are more opportunities to find and use available capacity. Manual rerouting through real-time traffic management control is a common practice under overload and failure conditions. Real-time routing can aid in the selection of reroutes to access available capacity, but network managers must still consider the large number of expansive control possibilities if they have not already been tested by real-time routing.

Several procedures for real-time routing are discussed in Section 8.2, which include a sequential method identical to two-link path routing design and two cyclic routing methods in which calls are rotated among the real-time paths. It is found that the sequential method provides equal or better performance than the others and thus is preferred because of its ease of implementation. This method appends to each sequence of two-link "engineered" paths designed for the expected network load additional two-link "real-time" paths to be used only after the normal sequence is exhausted, and only when idle capacity is available. The method for identifying idle network capacity is dynamic trunk reservation, which permits access to trunks on a particular link only after a specified number of trunks—the "reservation" level—are available. This procedure guarantees that capacity is truly idle and accessing it will produce minimal interference with normal traffic.

Results are derived via analytic models and simulation models discussed in Section 8.3, and modeling results are given in Section 8.4. Three criteria are used to evaluate real-time routing: (1) call completion probability, (2) interference of real-time routing with normal traffic performance, and (3) switching and signaling effort required for real-time routing. Results are obtained for network overloads and failures, and the results show that real-time routing improves average network blocking performance somewhat in each case, with some increased interference with normal traffic performance for some switch pairs. Call gapping is investigated as a means of protecting normal traffic blocking performance. This feature limits the maximum rate of calls being routed to the real-time paths. It is found that trunk reservation provides sufficient protection and that the addition of call gapping is unnecessary. The total switching system work in terms of crankbacks per originating attempt can be reduced with a path turn-off method. However, the relatively small additional effort to route real-time calls without turn-off makes it questionable whether the cost of implementing this feature is desirable.

8.2   Real-Time Dynamic Routing Methods

In this section we discuss several possible methods for implementing real-time routing that employ call-by-call routing decisions.

8.2.1 Automatic out-of-chain routing

The 4ESS switch provides an automatic expansive control for fixed hierarchical routing called automatic out-of-chain routing [Mum76]. With automatic out-of-chain routing, overflow calls from a final link, which are otherwise blocked, are sent to real-time routing paths for possible completion, as illustrated in Figure 8.1. Up to seven real-time routing paths can be identified for each final link, and the 4ESS switch spreads the overflow traffic uniformly over these routes as capacity is available. This is accomplished by using a cyclic routing method where each call tries the path following the previously attempted path. All real-time routing traffic is accompanied by a CCS traveling class mark so that it receives special treatment at the via switch to prevent shuttling. CCS signaling procedures for dynamic routing are discussed in Chapter 17. If a call fails to find a free trunk leaving the via switch, the call is blocked at the originating switch and the particular real-time routing path is "turned off" from further attempts for a period of about 30 seconds.

8.2.2 Learning methods

Another decentralized routing method involves the use of learning automata [KuN79, McN78, NaT78, NaT80, NWM77]. A learning automaton performs actions constantly modified by feedback from its environment, as shown in Figure 8.2. The updating procedures used to modify the actions of the different types of automata determine their learning characteristics. Studies of these methods have shown that the best choice seems to be the linear reward-inaction (L_{R-I}) automaton. In this automaton, if a particular action gets a

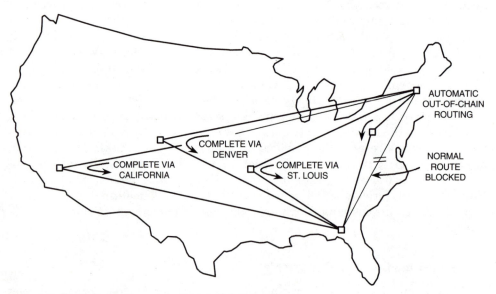

Figure 8.1 Example reroute: automatic out-of-chain routing

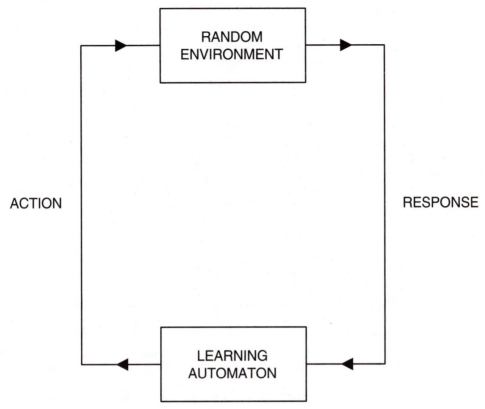

Figure 8.2 Automaton–environment interaction

positive response from the environment—that is, a call is completed—the probability of choosing this action for subsequent calls is increased. However, if a negative response is received, the action probabilities are not modified. A detailed mathematical model for the $L_{R\text{-}I}$ automaton can be found in [NaT78]. Models for the sample mean (M) automaton and the linear reward-penalty ($L_{R\text{-}P}$) automaton have also been developed by Narendra and others. Simulation studies have been made using simplified networks and show that these automata perform better than a fixed hierarchical routing method. These learning automata models permit paths of more than two links, and progressive routing versus path routing is used. Progressive routing transfers call control from one switch to the next as it proceeds along its path, and no crankback or return of control to previous switches is used.

Other learning techniques for network routing include state- and time-dependent routing [KaI95, YMI91], deployed in the NTT Japan network in 1992, and dynamic alternative routing [Gib86, Mee86, StS87], deployed in the British Telecom network in 1996.

8.2.3 Centralized real-time routing

Trunk status map routing (TSMR), which is a centralized approach to real-time state-dependent two-link path routing, is discussed in Chapter 11. The path selection is done by a centralized trunk status map network management processor, based on the busy/idle status of all trunks in the network. With TSMR the selection of candidate paths at each switch is recalculated every n seconds, where n varies from 1 second to 30 seconds. Another centralized routing system is dynamically controlled routing (DCR), which is implemented by Stentor Canada in the Canadian intercity network and by Bell Canada in the Ontario metropolitan area network [BNR86, Cam81, Car88, CGG80, HSS87, LaR91, RBC83, RBC95, SzB79, WaD88]. With DCR the selection of candidate paths at each switch is recalculated every ten seconds by a central network management processor. Five-minute dynamic routing (DR-5) [CKP91, KrO88], developed by Bellcore, is another centralized dynamic routing method, based on five-minute traffic data updates and routing updates. Another approach to centralized real-time routing uses all traffic loads for the entire network together with optimization methods, which satisfies the Kuhn-Tucker conditions for optimality, to solve for the optimal routing design [Kri82]. Therefore, a large amount of centralized computation must be performed, which makes this method difficult for large networks.

A centralized real-time routing enhancement for DNHR is provided by the real-time traffic management system called NEMOS (Network Management Operations System), which supports the network operations center in Bedminster, New Jersey. NEMOS gathers data from the network every five minutes and can augment the preplanned DNHR routes in the network to deal with network events such as failures, overloads, and other problems that cannot be fully foreseen or planned for in advance. NEMOS automatically puts in reroutes to solve blocking problems, in which the NEMOS reroute model searches everywhere in the network for additional available capacity and adds up to 7 additional paths to the 14 existing engineered and real-time paths used by DNHR at any given point in time (engineered and real-time paths are discussed in this chapter) to provide up to 21 total paths. In this way NEMOS, on a five-minute basis, can look at every possible alternate route through the network, check them for available capacity, and, if finding that available capacity, can add those paths to the routing tables within the five-minute interval. In Chapter 18 we give an example of NEMOS operation for Easter 1988, when DNHR was in operation in the AT&T network, and the overall performance shows that the automatic NEMOS reroutes were able to add significant additional call completions to improve performance.

8.2.4 Distributed real-time routing

Real-time network routing (RTNR) [ACF91, ACF92, AsC93, Ash95, AsH93, MGH91], which is a distributed approach to real-time state-dependent two-link path routing, is discussed in Chapter 12. RTNR is now implemented in the AT&T intercity network. RTNR uses real-time link status exchange between switches on a call-by-call basis to select the optimal path through the network.

8.2.5 Combined preplanned and real-time dynamic routing methods

In this section we discuss various possible implementations of real-time routing combined with preplanned two-link path routing. We first discuss various aspects of the real-time routing methods, which include (1) selection of via switches, (2) protection for normal traffic, and (3) real-time path selection methods. In Section 8.4 we present results from models of real-time routing and draw conclusions regarding the optimum selection of parameters.

8.2.5.1 Selection of via switches. One consideration involves the number, selection, and updating of real-time routing paths for each switch pair. This can be done, for example, as a natural extension of the path-erlang flow optimization model discussed in Chapters 6 and 7, which recognizes noncoincidence factors and can identify links that are expected to have slack capacity at a particular time. The real-time paths can be selected from a list of candidate paths produced by the design model. This list contains a large number of potential paths to be used in forming the normal route. The real-time paths are chosen from those paths not already used as part of the normal route. It should also be noted that the larger the allowed number of real-time routing paths, the lower the real-time routing blocking for an individual switch pair. However, there are administrative costs, storage limitations, and real-time penalties restraining the number of allowed paths. The network is designed to give a desired blocking performance on the engineered paths, whereas no specific blocking objective is set for the real-time paths.

8.2.5.2 Call gapping and trunk reservation protection of normal traffic. Protection must be provided for the normal traffic on the real-time routing paths, because it is undesirable to complete real-time routing calls at the expense of normal calls, which might not then receive their objective blocking. Call gapping and trunk reservation are investigated as ways to ensure proper service to the normal traffic.

Call gapping is analogous to a gate that allows a call through and then closes for a specified time interval. During this interval calls are not allowed through the gapper and are blocked from using a real-time path or a set of real-time paths. This limits access to these paths to a certain maximum rate determined by the length of the gap. There are two types of gappers considered, a master gapper and a link gapper. The master gapper limits the arrival rate to the entire set of real-time routing paths at once, whereas the link gapper does it on a path-by-path basis. Figure 8.3 illustrates the difference between the two types of gappers. The link gapper is similar to the automatic out-of-chain routing turn-off feature implemented in the 4ESS switch. Once a call is blocked on the second link of a particular real-time path, the path is turned off for the desired gap interval. The master gapper turns off all real-time routing paths for the gap interval after a call attempts the real-time routing paths and fails to complete.

Trunk reservation gives preference to the normal traffic by allowing it to seize any trunk in a link, while only allowing the real-time routing traffic to seize a trunk if there are at least $r + 1$ free trunks, where r is the reservation level.

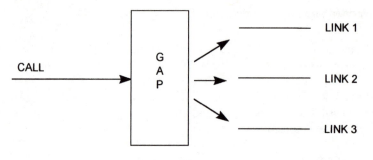

MASTER GAPPER

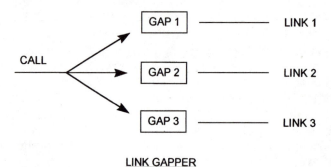

LINK GAPPER

Figure 8.3 Gappers

Call gapping and trunk reservation are studied separately and in tandem in order to see the effects of each on protecting normal traffic. The following equations describe the behavior of trunk reservation. They were derived originally by P. J. Burke [Bur61] from solution of the birth–death equations for the trunk reservation model. As discussed in Chapter 2, the birth–death process equations are solved for the stationary probability of each state. In statistical equilibrium, the average rate of flow into any state equals the flow out, or on average the system makes as many transitions per unit of time into a state as out of a state. Therefore, the net flow across a boundary between states k and $k - 1$ equals zero. Both the preferred traffic of intensity λ_1 and the nonpreferred traffic of intensity λ_2 can access states up to $s - r - 1$. For these states, the net flow in and out of state $k - 1$ is zero if

$$p_k = \frac{\lambda_1 + \lambda_2}{k\mu} p_{(k-1)} \qquad k = 1, 2, \ldots, s - r - 1$$

Only the preferred traffic can access states of $s - r$ up to state s. For these states, the net flow in and out of state $k - 1$ is zero if

$$p_k = \frac{\lambda_1}{k\mu} p_{(k-1)} \qquad k = s - r, s - r + 1, \ldots, s$$

The preferred load or traffic intensity is defined as λ_1/μ. The nonpreferred load or traffic intensity α is defined as λ_2/μ.

These equations are then solved for the preferred and nonpreferred traffic blocking probabilities:

$$B_{\text{preferred}} = \frac{K(\alpha + a)^{s-r} \alpha^r}{s!}$$

$$B_{\text{nonpreferred}} = K(\alpha + a)^{s-r} \sum_{k=s-r}^{s} \frac{\alpha^{k-(s-r)}}{k!}$$

where α = preferred traffic (erlangs)
$\quad a$ = nonpreferred traffic (erlangs)
$\quad s$ = number of trunks in link
$\quad r$ = number of reserved trunks in link
$\quad K$ = normalizing constant

It can be seen from the above two equations that the blocking level for the nonpreferred traffic is higher than that for the preferred traffic. Figure 8.4 illustrates the blocking of preferred and nonpreferred traffic on a typical link with 10 percent traffic overload. It is seen that the preferred traffic blocking is near zero, whereas the nonpreferred traffic blocking is much higher, and this situation is maintained across a wide variation in the percentage of the preferred traffic load. Hence, trunk reservation protection is robust to traffic variations and provides significant dynamic protection of particular streams of traffic.

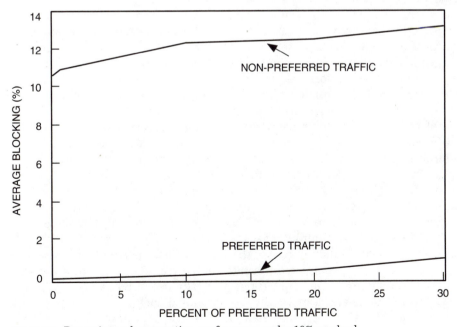

Figure 8.4 Dynamic trunk reservation, performance under 10% overload

Trunk reservation is a crucial technique used in nonhierarchical networks to prevent "instability," which can severely reduce throughput in periods of congestion, perhaps by as much as 50 percent of the call-carrying capacity of a network. The phenomenon of instability has an interesting mathematical solution to network flow equations, which has been presented in several studies [NaM73, Kru82, Aki84]. It is shown in these studies that nonhierarchical networks exhibit two stable states, or bistability, under congestion and that networks can transition between these stable states in a network congestion condition that has been demonstrated in simulation studies. A simple explanation of how this bistable phenomenon arises is that under congestion, a network is often not able to complete a call on the direct link. If alternate routing is allowed, such as on two-link paths, then the call might be completed on a two-link path selected from among a large number of two-link path choices, only one of which needs an idle trunk on both links to be used to route a call. Because this two-link call now occupies resources that could perhaps otherwise be used to complete two one-link calls, this is a less efficient use of network resources under congestion. In the event that a large fraction of all calls cannot complete on the direct link but instead occupy two-link paths, the total network throughput capacity is reduced by one-half because most calls take twice the resources needed. This is one stable state; that is, most or all calls use two links. The other stable state is that most or all calls use one link, which is the desired condition in a mesh network with a reasonably high logical link connectivity.

Trunk reservation is used to prevent this unstable behavior by having the preferred traffic on a link be the direct traffic and the nonpreferred traffic, subjected to trunk reservation restrictions as described above, be the two-link alternate-routed traffic. In this way the alternate-routed traffic is inhibited from selecting two-link paths when sufficient idle trunk capacity is not available on both links of a two-link call, which is the likely condition under network and link congestion. Mathematically, the studies of bistable network behavior have shown that trunk reservation used in this manner to favor direct one-link calls eliminates the bistability problem in nonhierarchical networks and allows such networks to maintain efficient utilization under congestion by favoring one-link completed calls. For this reason, dynamic trunk reservation is universally applied in nonhierarchical, and often in hierarchical [Mum76], networks. There are differences in how and when trunk reservation is applied, however, such as whether the trunk reservation for direct-routed calls is in place at all times or whether it is dynamically triggered to be used only under network or link congestion. This is a complex network throughput trade-off issue, because trunk reservation can lead to some loss in throughput under normal, low-congestion conditions. This loss in throughput arises because if trunks are reserved for one-link calls, but these calls do not arrive, then the capacity is needlessly reserved when it might be used to complete alternate-routed traffic that might otherwise be blocked. However, under network congestion, the use of trunk reservation is critical to preventing network instability, as explained above.

In the studies described in this chapter, and also in Chapters 11 and 12, which discuss real-time dynamic routing methods, dynamically triggered trunk reservation techniques, where trunk reservation is triggered only under network congestion, are shown to be effective in striking a balance between protecting network resources under congestion and ensuring that resources are available for sharing when conditions permit. In Chapter 12, the phenomenon of network instability is illustrated through simulation studies, and the effectiveness of trunk reservation in eliminating the instability is demonstrated. Trunk reservation is also shown in Chapter 12 to be an effective technique to share bandwidth capacity among services integrated on a direct link, where the reservation in this case is invoked to prefer direct link capacity for one particular service as opposed to another service when network and link congestion are encountered.

8.2.5.3 Real-time path selection. Three options are considered for real-time path selection: two forms of cyclic routing and one type of sequential routing. Both cyclic methods try all the real-time paths in a specified order. The difference between them is the way in which the starting point is chosen. In one case, if a call is completed along a path, the next call will start on the same path. This is a success-to-the-top (STT) method. For example, if a call completes along path 1 in Figure 8.5, the next call begins its search with path 1. This is done to take advantage of links that exhibit spare capacity. The other cyclic method is employed in automatic out-of-chain routing. In this case, if a call is completed, the next call begins by trying the path following the one on which the last completion is made. Once again referring to Figure 8.5, if a call completes along path 1, the next call begins its search with path 2. The advantage of this method is that it will more evenly distribute the calls among the real-time routing paths. The sequential method consists of a pure overflow system and always starts at the same place

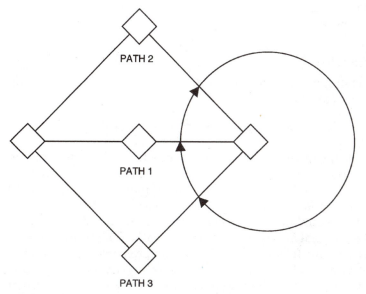

Figure 8.5 Success-to-the-top (STT) and cyclic routing examples

and tries the paths in the same order. This is the same way in which normal calls are processed in the dynamic two-link sequential path routing method. The main virtue of this form of real-time routing is its ease of implementation.

8.3   Network Management and Design

We consider here mathematical models and simulation models of real-time routing.

8.3.1   Mathematical models

Several assumptions are made in developing the mathematical models to investigate real-time routing:

- All traffic is Poisson; however, we show that peakedness does not significantly affect the results.

- The network has the designed number of trunks with no reserve capacity; however, we show that reserve capacity has a small effect on the results.

- The holding time for calls has a negative exponential distribution with a mean of five minutes.

- All real-time paths have two links, each with a comparable number of trunks. The size of each link ranges from 10 to 170 trunks.

- The mathematical analysis is done using three real-time routing paths for a particular switch pair.

The models consider (1) protection of normal traffic, (2) real-time routing performance, (3) real-time load on the switch, and (4) traffic peakedness. Each of these factors is now discussed.

8.3.2   Protection of normal traffic

If a call attempts a two-link path with blocking B_1 on the first link and blocking B_2 on the second link, its path blocking is

$$B_1 + B_2 - B_1B_2$$

assuming B_1 and B_2 are independent. To simplify the notation, only the mathematical expressions for the blocking on the first link of each path are given here. The blocking on the second link is calculated using Equations (8.1) and (8.2) below, and the path blocking is computed using the above expression.

We need to determine the combined effects of trunk reservation and call gapping on blocking for both the normal and real-time routing traffic. The traffic streams include the real-time routing traffic, which is the nonpreferred traffic and is subject to trunk reservation, and the normal traffic, which is the preferred traffic and is free to use all the trunks in a link. The blocking for the preferred normal traffic and nonpreferred real-time routing traffic is calculated as follows. For this analysis, let

> r_i = number of reserved trunks on link i
> k_i = number of busy trunks at which reservation becomes effective in link i $(= s_i - r_i)$
> P_n = probability of n trunks busy
> α_i = preferred normal traffic on link i, in erlangs
> a_i = nonpreferred real-time routing traffic passed through the gapper on link i, in erlangs
> s_i = number of trunks in link i

To illustrate the above definitions, a flow diagram of a real-time routing call is shown in Figure 8.6.

In the rest of this section we will drop the index i to simplify the notation. Thus, for a particular link, the steady-state equations are

$$P_0 = \left[\sum_{j=0}^{k} \frac{(\alpha + a)^j}{j!} + (\alpha + a)^j \sum_{j=k+1}^{s} \frac{a^{j-k}}{j!} \right]^{-1}$$

$$P_{k+j} = a^j (\alpha + a)^k \frac{P_0}{(k+j)!} \qquad j = 0, \ldots, s - k$$

$$P_n = (\alpha + a)^n \frac{P_0}{n!} \qquad n = 0, 1, \ldots, k \quad (0! = 1)$$

$$P_0 + P_1 + \cdots + P_s = 1$$

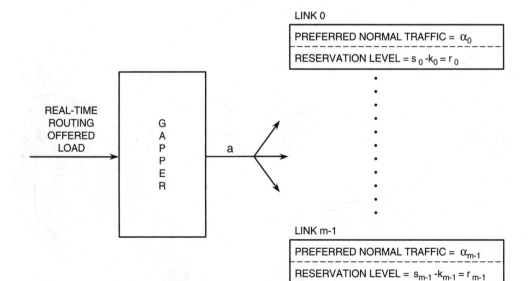

Figure 8.6 Real-time routing model

Solving the preceding equations for the P_n allows us to compute the blocking of the normal traffic and real-time routing traffic. Below we denote the probability of the real-time routing traffic being blocked by $B_{ik}(s, a + \alpha)$ and the probability for the normal traffic by P_s.

$$B_{ik}(s, \alpha + a) = P_k + P_{k+1} + \cdots + P_s \tag{8.1}$$

$$P_s = \alpha^{s-k}(\alpha + a)^k \frac{P_0}{s!} \tag{8.2}$$

As shown in Figure 8.6, the first source of blocking seen by a real-time routing call is the call gapper. There are two types of call gappers considered: a master gapper and a link gapper. The master gapper limits the arrival rate to the entire set of real-time routing paths at once, whereas the link gapper does it on an individual path basis. (Figure 8.3 illustrates the differences between the gappers.)

Both gappers are modeled as single-server queues with constant service time. If the real-time routing traffic load has an average arrival rate of λ calls per second, its blocking is

$$B(1, \bar{a}) = \frac{\bar{a}}{1 + \bar{a}} \qquad \bar{a} = \lambda g \qquad g = \text{gap interval in seconds} \tag{8.3}$$

We assume that the probability of a call being blocked by the call gapper and by a link with trunk reservation are independent. Using Equations (8.1) and (8.3), we see that the probability of blocking for the real-time routing traffic on link i, β_i, can be written as

$$\beta_i = B(1, \bar{a}) + B_{ik_i}(s_i, a_i + \alpha_i) - B(1, \bar{a})B_{ik_i}(s_i, a_i + \alpha_i)$$

$$= \frac{\bar{a}}{1 + \bar{a}} + B_{ik_i}(s_i, a_i + \alpha_i)\left(1 - \frac{\bar{a}}{1 + \bar{a}}\right)$$

$$= \frac{\bar{a} + B_{ik_i}(s_i, a_i + \alpha_i)}{1 + \bar{a}}$$

The above expression assumes that all the real-time routing traffic tries to route on link i and is affected by the call gapper.

At this point the different call gappers and real-time routing methods begin to look different and need to be treated separately. We first treat the call gappers and then, in the next section, the routing methods. Let

m = number of paths in the real-time route
A_i = probability of a call attempting to use link i
S_i = probability of a call starting with link i

For the master gapper case, β_i is still given by

$$\beta_i = \frac{\bar{a} + B_{ik_i}(s_i, a_i + \alpha_i)}{1 + \bar{a}} \tag{8.4}$$

This is because the master gapper acts on all calls attempting any real-time routing path, as shown in Figure 8.3. For the link gapper case, not all calls route to any one link gapper. Figure 8.3 shows that the probability of a call being blocked by a link gapper depends on the probability of the call attempting to use that particular link. It is assumed that all gap intervals have the same length. Therefore, for the link gapper β_i can be written as

$$\beta_i = \frac{A_i \bar{a} + B_{ik_i}(s_i, a_i + \alpha_i)}{1 + A_i \bar{a}} \tag{8.5}$$

8.3.3 Real-time routing performance

In the above analysis we use the real-time routing traffic attempting each link a_i in computing the probability of a call being blocked by trunk reservation and the different gappers. To compute a_i we need to find A_i and S_i, the probabilities of attempting and starting with a particular link. These quantities are also necessary to develop the real-time routing models.

The steady-state equations for A_i are the same for both cyclic routing methods using a master gapper:

$$
\begin{aligned}
A_i = {} & \sum_{j=0}^{i-1} S_j \prod_{n=j}^{i-1} B_{nk_n}(s_n, a_n + \alpha_n) + S_i \\
& + \sum_{j=i+1}^{m-1} S_j \prod_{n=j}^{i-1+m} B_{n|m\ k_{n|m}}(s_{n|m}, a_{n|m} + \alpha_{n|m})
\end{aligned} \tag{8.6}
$$

$$i = 0, 1, \ldots, m - 1$$

where $n \mid m$ means n modulo m.

The first and last terms on the right-hand side of equation (8.6) represent the probability of a call starting at path $j = i$ and being blocked until it tries path i. The middle term represents the probability of starting at path i. The difference between the two cyclic methods is in their probability of starting at a particular link. In both cases, we assume that if all the trunks are busy, the next call starts with link 0. If a call starts on the link following the one on which the last call is completed and a master gapper is used, the steady-state equations are

$$
\begin{aligned}
S_i = {} & \left[1 - B_{i-1\ k_{i-1}}(s_{i-1}, a_{i-1} + \alpha_{i-1})\right] \\
& \times \left[S_{i-1} + \sum_{n=0}^{m-2} \prod_{j=i+n|m}^{i-2} B_{jk_j}(s_j, a_j + \alpha_j)S_{i+n|m}\right] \\
& + \delta_{i0} \prod_{j=0}^{m-1} B_{jk_j} \qquad i = 0, 1, \ldots, m - 1
\end{aligned} \tag{8.7}
$$

$$S_0 + S_1 + \cdots + S_{m-1} = 1$$

$$\delta_{i0} = \begin{cases} 1, & i = 0 \\ 0, & \text{otherwise} \end{cases}$$

If a call starts on the link that had the last completion, and a master gapper is used, the steady-state equations are

$$S_i = [1 - B_{ik_i}(s_i, a_i + \alpha_i)] \times \left[S_i + \sum_{n=1}^{m-1} \prod_{j=i+n|m}^{i-1} B_{jk_j}(s_j, a_j + \alpha_j) S_{i+n|m} \right]$$

$$+ \, \delta_{i0} \prod_{j=0}^{m-1} B_{jk_j}(s_j, a_j + \alpha_j) \qquad i = 0, 1, \ldots, m - 1$$

$$S_0 + S_1 + \cdots + S_{m-1} = 1$$

$$\delta_{i0} = \begin{cases} 1, & i = 0 \\ 0, & \text{otherwise} \end{cases} \tag{8.8}$$

If a link gapper is used, then the steady-state equations for A_i are given in equation (8.9). β_i is given in equation (8.5), and the terms on the right-hand side have the same meaning as in the case of the master gapper. The starting probabilities for the link gapper are derived from the equations above by substituting β_i for $B_{ik_i}(s_i, a_i + \alpha_i)$.

$$A_i = \sum_{j=i+1|m}^{i-1|m} S_j \prod_{k=j|m}^{i-1|m} \beta_k + S_i \tag{8.9}$$

The sequential routing method is the easiest to model. The probability of starting with the first path is 1: $S_0 = 1$. The probability of attempting the other paths is the product of the probabilities of being blocked on the preceding paths: $\prod_{j=0}^{i-1} \beta_j$. The starting probability is independent of the type of gapper used. However, the probability of attempting the other links depends on the type of gapper. If a master gapper is used, β_j is given by equation (8.4), but if a link gapper is used, β_j is given by equation (8.5).

A summary of all of the above models is given in Table 8.1.

8.3.4   Real-time load on the switch

We model the real-time load on a switch as proportional to the number of crankbacks used on a call before the call is either completed or blocked. The expected number of paths attempted, E (SW), is given by

$$E(\text{SW}) = \sum_{j=0}^{m-1} \text{PS}(\,j\,)j$$

TABLE 8.1   Summary of Steady-State Equations

Gapper type	Cyclic 1 routing	Cyclic 2 routing	Sequential routing
master	A_i: Eq. (8.6) S_i: Eq. (8.7)	A_i: Eq. (8.6) S_i: Eq. (8.8)	$A_i = \prod_{j=0}^{i-1} \beta_i$, using β_j Eq. (8.4); $S_0 = 1, S_i = 0$ if $i \neq 0$
link	A_i: Eq. (8.9) S_i: Eq. (8.7), using β_j Eq. (8.5)	A_i: Eq. (8.9) S_i: Eq. (8.8), using β_j Eq. (8.5)	$A_i = \prod_{j=0}^{i-1} \beta_j$, using β_j Eq. (8.5); $S_0 = 1, S_i = 0$ if $i \neq 0$

where $PS(j)$ = the probability of attempting j paths per call and

$$PS(1) = \sum_{j=0}^{m-1} S_i(1 - \beta_i)$$

$$PS(2) = \sum_{i=0}^{m-1} S_i\beta_i(1 - \beta_{i+1|m})$$

$$\vdots$$

$$PS(j+1) = \sum_{i=0}^{m-1} S_i \prod_{k=i}^{i+j-1} \beta_{k|m}(1 - \beta_{i+j|m})$$

$$PS(m) = \sum_{i=0}^{m-1} S_i \prod_{k=i}^{i+m-2} \beta_{k|m}$$

A call generates a crankback if it succeeds in seizing a trunk on the first link of a real-time path but gets blocked at the via switch. This is a straightforward calculation using the previous results. Let

β = probability of being blocked on the first link
B = probability of being blocked on the second link
CB = probability of a crankback message

Then

$$CB = (1 - \beta)B$$

8.3.5   Peakedness models

In the preceding analysis we assumed that all traffic was Poisson. We now investigate the effect of peaked traffic in our analysis. We need only calculate blocking for the peaked traffic subject to trunk reservation and also the peakedness of the overflow traffic from the real-time routing paths. The model for peaked traffic uses the interrupted Poisson process, in which the peaked traffic is the output of a random gate whose input is Poisson. The gate is called random because the length of time the gate is open or closed is exponentially distributed. It is this exponential assumption that makes the analysis tractable. The analysis is similar to that for Poisson traffic, but now the state of the system depends on the state of the gates, as well as the number of busy trunks.

We make the approximation that the peakedness of the overflow traffic from the real-time paths is the same as the peakedness of the original overflow traffic to the real-time paths, which can be justified as follows. First, we use Wilkinson's equivalent random method [Wil56] to find the equivalent number of trunks that yields the observed blocking for the given peakedness and load of the real-time routing traffic, assuming that the traffic has full access to all trunks in the equivalent link. The observed blocking of the original real-time routing traffic actually includes the effects of trunk reservation. Second, we determine the peakedness of the overflow traffic from the equivalent link. We assume that the resulting peakedness is the actual peakedness of the overflow traffic. The blockings observed on the real-time routing path are usually 0.5 or higher. We find that when the blocking on real-time paths is high, the approximation that the overflow peakedness is equal to the offered peakedness is quite good, as shown in Section 8.4.

8.3.6   Simulation models

A call-by-call simulation model is implemented to study real-time routing performance, which assumes transparent switches—that is, no queuing or blocking is modeled for the switching systems in the network. Poisson arrivals are used to model originating calls, together with an exponential holding time distribution having a mean of five minutes. A call-by-call simulation model has a simulated clock that controls the time at which events are executed. Individual call arrivals are put on an event list, and at the simulated time of their arrival they are routed on the simulated links in the model using the exact routing logic of the routing method being studied. For a typical routing logic, a call is first routed on the direct simulated link, and if the direct link is busy, the call is routed on the next path in the routing table, which for two-link sequential path routing is normally an engineered path. If either the first or second simulated link of the path is busy, the call is then routed on the next path in the routing table. In the simulation, if the second link in a path is busy, this results in a simulated crankback message, and such events are counted in the model. If a call finds an idle path, the simulated trunks are made busy, the call holding time is determined according to the exponential holding time distribution (other distributions can easily be modeled in simulations), and the end time of the call is noted in the event list for a later disconnect event at the simulated time the call is disconnected. At the simulated disconnect time, the simulated trunks occupied by the call are made idle once again, and the calls-in-progress counter for the affected switch pair is decremented. For a real-time path in a preplanned dynamic routing sequential list, trunk reservation is applied in selecting trunks on each link. When master gapper, link gapper, or path turn-off logic is included in the routing method, the detailed timing and blocking mechanisms are exactly modeled in the simulation. Hence the simulation model can very accurately model complex network routing logic and thereby give accurate estimates of network performance for such logic. It can also be used to validate analytical models such as the ones presented in this section. However, call-by-call simulations can go far beyond the ability of analytical models to represent complex network behavior.

This model complements simulation models such as the one described in [HKO83], which includes switch dynamics and is used to study switch-congestion effects in a dynamic routing network. Day-to-day variations are modeled with a gamma distribution with a variance equal to $0.13a^\phi$, with $\phi = 1.5$ to model low daily variations. The parameter a represents the switch-to-switch offered load in erlangs. Reserve capacity is modeled using a uniform distribution on the link size centered about the average level of reserve capacity. A 7 percent average reserve capacity is used in most cases and varies uniformly between 3 and 11 percent. It is taken as a typical level of reserve capacity for the dynamic routing network. As discussed in Section 8.1 and also in Chapter 9, the presence of reserve capacity is to be expected and enhances the performance of real-time routing.

8.4   Modeling Results

In this section we present results from the mathematical models and then the simulation model.

8.4.1   Results from mathematical models

The network configurations investigated for the various mathematical models are given in Table 8.2. Because the conclusions are the same over the entire range of link sizes, most of the results presented corresponded to three real-time routing paths with 100, 90, and 80 trunks on both links of each path (model number 8 in Table 8.2). This model is illustrated in Figure 8.7. The engineered paths between switches A and B are treated as a single combined link of 80 trunks. The overflow from this combined link uses the real-time paths with via switches C, D, and E. We assume that links A–C and B–C each have 100 trunks with normal offered traffic of 84.0 erlangs, links A–D and B–D have 90 trunks with a normal traffic of 74.7 erlangs, and links A–E and B–E have 80 trunks with normal traffic of 68.9 erlangs. Link A–B also has 80 trunks with a normal load of 60 erlangs between switches A and B; thus, for this configuration the blocking probability for A–B traffic meets the blocking probability objective of 0.005. Later, we consider a 20 percent increase of the A–B load that results in 72 erlangs of offered load and produces an overflow of about 3 erlangs.

Three criteria are used to evaluate the merits of various configurations of real-time routing: (1) protection of normal traffic, (2) real-time routing performance, and (3) switch real-time load.

TABLE 8.2   Real-Time Routing Models

Model number	Number of trunks									
	Path 1		Path 2		Path 3		Path 4		Path 5	
	Leg 1	Leg 2	Leg 1	Leg 2	Leg 1	Leg 2	Leg 1	Leg 2	Leg 1	Leg 2
1	11	18	8	9						
2	12	11	13	18						
3	26	15	29	10						
4	26	18	10	9						
5	16	11	12	11	10	9				
6	29	27	26	15	10	9				
7	56	56	50	50	40	40				
8	100	100	90	90	80	80				
9	161	155	150	150	170	145				
10	170	145	150	150	29	27				
11	170	145	150	150	29	27	26	15	10	9
12	161	155	150	150	170	145	26	15	10	9
13	161	155	150	150	16	11	10	9	8	9

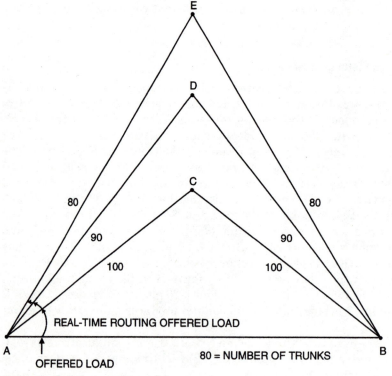

Figure 8.7   Real-time routing model

The probability of a real-time routed call being blocked is the product of the blockings for each path in its real-time route. Thus, the completion probability for a real-time routed call is 1 minus this product. Interference with normal traffic on a particular link is defined as follows:

B = blocking of normal traffic when no real-time routing traffic uses the link

B' = blocking of normal traffic when real-time routing traffic also tries to use the link

I = interference $\triangleq B' - B$

The real-time switching load required to complete a real-time call is reflected by the number of added switch attempts. In particular, each crankback required per real-time call is approximately equal to one added switch attempt. Hence, the number of real-time routing crankbacks per originating call is used in the analysis to measure the extra real-time load of the switch. For the simulation results presented in Section 8.4, we compare total network (normal engineered path plus real-time path) crankbacks, as well as total network switch attempts, to describe the work of the switch.

8.4.1.1 Protection of normal traffic.

Protection for the normal traffic is provided by trunk reservation and call gapping. As defined above, interference is a measure of the increase in blocking seen by the normal traffic caused by the real-time-routed calls. A comparison is made between the link gapper and the master gapper using a cyclic routing pattern. It is found that both gappers behave in almost the same way, giving much the same blocking for a given traffic level. Figure 8.8 shows a sample comparison between the gappers. This analysis assumes a gap interval of 10 seconds and uses the model shown in Figure 8.7. We observe that the blocking curves remain close together for both types of gappers over a wide range of real-time routing loads. For this reason, the results presented in the remainder of this section only reflect the behavior of the master gapper.

We next analyze the trade-off between trunk reservation and call gapping. A certain tolerance level for interference with the background traffic is set and kept constant. Given this tolerance level, various gap intervals are tried and the reservation level is varied to achieve the selected interference level. If there is too much interference with the normal traffic, the reservation level is increased. The increase in reservation level further limits the access of real-time routing calls to the links. If the reservation level is sufficiently high, the interference is then smaller than the tolerance level. It is found that the required reservation level decreases as the gap interval increases.

Given a tolerance level for the interference with normal traffic, one must decide whether trunk reservation and call gapping enhance each other's performance. An important decision to be made concerning the gapper is the length of the gap interval. First, it is shown that large intervals are undesirable because they lead to higher blocking for real-time routed calls. Smaller intervals are then considered. However, the improvement in real-time call blocking gained by using

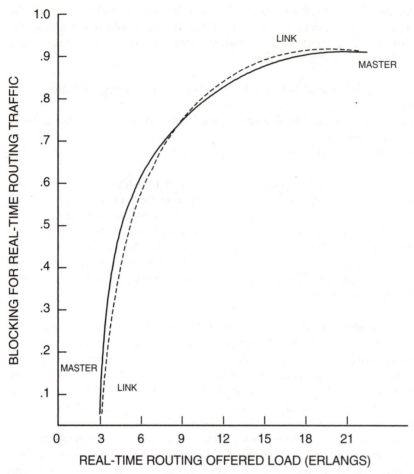

Figure 8.8 Master gapper vs. link gapper (10-second gap)

a gapper is small and only realizable in a limited range of loads. Furthermore, the optimum gap setting varies as a function of real-time traffic load. This is illustrated in Figure 8.9. For a 20 percent overload of 3 erlangs delivered to the real-time paths, the blocking increases as the gap interval increases. However, when the real-time routing load is increased to 6 erlangs, there is a slight decrease in blocking when the gap interval is increased from 0 to 90 seconds.

These two observations are explained by the interaction between the length of the gap and the required reservation level. A longer gap interval allows less real-time traffic to be offered to the real-time paths. Therefore, a lower reservation level is required. When 3 erlangs are offered to the real-time paths, almost no reservation is necessary and no gap is required. Thus, increasing the gap interval increases the blocking because there is no corresponding decrease in the required reservation level. On the other hand, when 6 erlangs are offered to the real-time

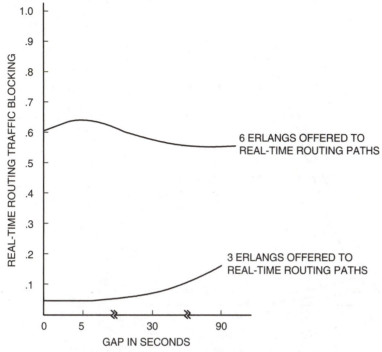

Figure 8.9 Real-time routing traffic blocking vs. gap length

paths, there is a substantial amount of reservation necessary when no gap is used. In this case, the combination of increasing the gap and decreasing the reservation level produces a net decrease in blocking. However, a sufficiently large gap requires no trunk reservation, and any further increase in the gap interval beyond this point only further increases the blocking. Thus, for ease of implementation and manageability, we conclude that it is best not to have call gapping. In the remaining analysis, it is assumed that no call gapping is used for real-time routing.

We now investigate the proper reservation level without the presence of call gapping. In choosing the reservation levels, interference is allowed to vary so that we can compare different levels of trunk reservation. In order to investigate the trade-off between real-time routed calls completed and interference with the normal traffic, varying percentages of trunks on the links are reserved. For example, if there are 100 trunks on a link and a 10 percent reservation level is imposed, then the reservation setting is 10. Typical results are shown in Figure 8.10. In this case, the sequential routing method is used. The cyclic routing cases give analogous results. The reservation level varies between 5 and 25 percent. The graph clearly illustrates the increase in blocking and decrease in interference as the amount of reservation is increased. In the extreme case of 25 percent reservation there is almost total blocking of real-time traffic and no

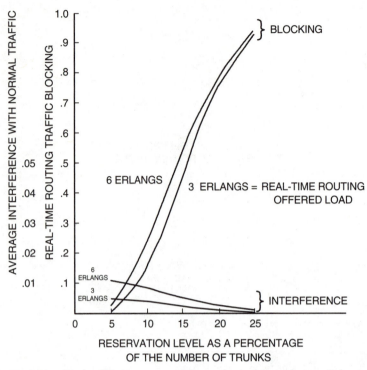

Figure 8.10   Trade-off between reservation and interference (sequential routing)

interference with normal traffic. It is concluded after considering a wide range of link sizes that a 5 to 10 percent reservation level provides a good balance between real-time call blocking and interference with normal traffic.

8.4.1.2 Real-time routing performance.   We compare the two cyclic routing methods under various conditions. The method that starts on the path following the one on which the last completion was made performs slightly better than the cyclic method that starts from the same real-time path on which the last completion was made. Hence, the former version of cyclic routing is used in the comparisons made with the sequential routing method. The sequential and cyclic methods are found to give about the same blocking when imposing either (1) a fixed tolerance for interference or (2) a range of fixed percentage levels for trunk reservation. Figure 8.11 shows the blocking found using each type of routing with a fixed tolerance for interference. It is seen that each method has a very small advantage in certain ranges of the load. The results are similar for the case of a fixed percentage of reserved trunks, with the sequential method doing slightly better.

A second consideration is the number of times a call is cranked back before being either completed or blocked. This represents the amount of added real-time load carried by the switch. Table 8.3 shows the results of one such calculation. It

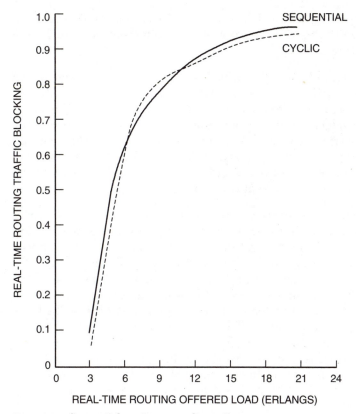

Figure 8.11   Sequential routing vs. cyclic routing

TABLE 8.3   Work of Switch

Real-time routing offered load (erlangs)	Expected number of real-time routing crankbacks per originating call	
	Cyclic	Sequential
3	0.013	0.015
6	0.031	0.037
9	0.053	0.062

is found that under the same circumstances, the expected number of crankbacks using the sequential method is approximately the same as the number obtained with the cyclic method. The table shows that the difference between the two methods is at most 0.01 crankbacks per call, depending on the real-time traffic load. When the real-time traffic load is increased, the expected number of crankbacks also increases because the probability of being blocked on each path goes up.

In the preceding paragraphs, it is shown that the sequential method has about the same blocking characteristics as the cyclic methods. It also has about the same real-time load impact on the switch. However, the ease of implementation is an advantage of sequential routing, and the sequential method is therefore the preferred real-time routing method.

8.4.1.3 Switch real-time load. As we saw in the last section, the impact on switch real-time load varies primarily with the number of real-time paths and, hence, the amount of potential searching for a free path. Table 8.4 illustrates the trade-off involved, in which models 8 and 12 from Table 8.2 are used to compare the crankbacks required per originating call. As can be seen, the crankbacks required increase with the number of real-time paths and the traffic load, but at the same time the real-time call blocking decreases. An acceptable number of real-time paths must therefore consider switch real-time limitations.

8.4.1.4 Peaked traffic and reserve capacity. In the above analysis it is assumed that all traffic is Poisson. However, the real-time routing traffic is peaked because it results from the overflow from several engineered paths. From an analysis of the 28-switch intercity network model, it is found that traffic peakedness for engineered path overflow that is between 1.0 and 1.5 is typical; hence, this range of peakedness is assumed in the analysis. We also assume that the peakedness of the overflow traffic from the real-time paths remains the same as the peakedness of the offered real-time routing traffic. This is a good approximation because of the high blocking experienced by the real-time routing traffic. An intuitive argument to justify this approximation is based on the fact that if all the traffic is blocked, the overflow traffic would have exactly the same characteristics as the original traffic. For example, Table 8.5 considers the case of 6 erlangs of

TABLE 8.4   Blocking and Crankbacks for Different-Length Real-Time Routes

Real-time routing offered load (erlangs)	Real-time routing crankbacks per originating call		Real-time routing blocking	
	5 real-time paths	3 real-time paths	5 real-time paths	3 real-time paths
3	0.018	0.015	0.015	0.02
6	0.042	0.037	0.027	0.045

TABLE 8.5   Approximation of Peakedness

Peakedness of offered real-time routing traffic	Real-time routing traffic (erlangs)	Blocking of real-time routing traffic	Equivalent number of trunks	Peakedness of overflow
1.5	6	0.62	3	1.85
	6	0.51	4	1.96
	6	0.41	5	2.07
2.5	6	0.66	3	2.80
	6	0.57	4	2.89
	6	0.48	5	2.96

real-time routing traffic being offered to the real-time paths. It is clear from the table that as the blocking increases, the approximation gives the expected result: If all the traffic is blocked, the overflow traffic has the same peakedness as the offered traffic. In particular, if the blocking is at least 0.5, the offered and overflow traffic peakedness differ by no more than 0.6 in both cases. Hence, we assume that offered and overflow peakedness are equal for real-time routing traffic.

Figure 8.12 shows the effect of varying the peakedness of the real-time traffic on the trade-off between real-time call blocking and interference with normal traffic. The peakedness of the normal traffic is 1.0 in this case. Both the blocking and interference curves are stable over a wide range of peakedness values for the real-time routing traffic. The real-time call blocking does not significantly increase as the peakedness is varied because it is already quite high. A normal traffic peakedness of 1.5 yields the same-shaped curves. Thus, our results remain valid for peaked traffic.

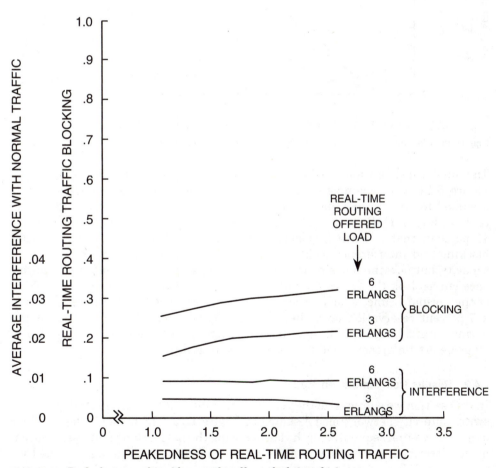

Figure 8.12 Peakedness results with normal traffic peakedness of 1.0

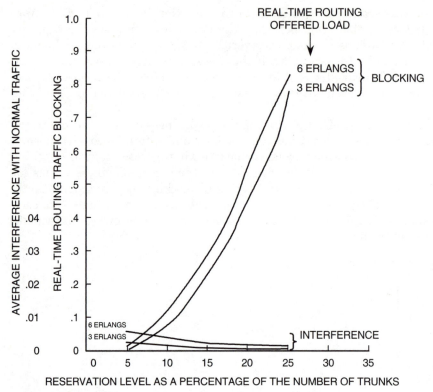

Figure 8.13 Blocking and interference with an average 7% reserve capacity

The final consideration is the addition of reserve capacity to the network. Figure 8.13 is an example of the results. The real-time traffic peakedness is assumed to be 1.5, and the average reserve capacity is assumed to be 7 percent, with each realization chosen from the uniform distribution between 3 and 11 percent; that is, each link size is increased in this range. The curves for blocking and interference look very similar to those obtained without reserve capacity, but now the blocking curves are lower. Note that although the interference curves look the same, they are now referenced to the new, lower blocking of the normal traffic, which arises from the reserve capacity. In fact, with only a 5 percent reservation level there is a negligible increase in blocking for the normal traffic above the design objective. Hence, with reserve capacity, a 5 to 10 percent trunk reservation level is quite conservative.

8.4.2 Results from simulation models

In this section we summarize the results of a call-by-call simulation using the 28-switch intercity network model shown in Figure 1.12. The model is designed for dynamic two-link sequential path routing using the path-erlang flow optimization model described in Chapters 6 and 7. The simulation results are obtained for the 10 A.M. E.S.T. morning busy-hour load and routing design. Various trade-off studies are performed comparing the behavior of real-time routing under various conditions, which include (1) the protection of normal traffic, in which we investigate the trunk reservation technique, call-gapping method, and path turn-off method; (2) the real-time routing performance; (3) switch real-time load; and (4) network overload and failure performance. Each of these items is discussed in the following sections.

8.4.2.1 Protection of normal traffic.

Various approaches to trunk reservation are tried. All involve two basic parameters: the minimum number of reserved trunks in each link (R_{min}) and the fraction of reserved trunks in each link (P). Therefore the maximum of R_{min} or P times the number of trunks in the link sets the reservation level for the link.

Table 8.6 illustrates the performance of real-time routing for various values of R_{min} and P. The table lists the network performance criteria used in the simulation to compare various real-time routing design parameters. These criteria are as follows:

- Average network blocking, which indicates the effectiveness of completing real-time calls.

- Maximum switch-pair blocking, which indicates the interference of real-time calls with normal traffic.

- Crankbacks per originating call and switch attempts per originating call, which indicate the switch real-time load to complete real-time routed calls. Total normal and real-time call crankbacks are counted. Switch attempts include originating, terminating, crankback, and tandem-completing.

The table shows that very good performance is achieved for $R_{min} = 2$ and $P = 0.05$. However, only moderate loss of performance is found using other

TABLE 8.6 Performance of Various Reservation Methods (Low Day-to-Day Variations, 7% Reserve Capacity)

Reservation					
Minimum number of trunks (R_{min})	Fraction of trunks (P)	Average network blocking	Maximum switch-pair blocking	Crankbacks per originating call	Attempts per originating call
No real-time routing		0.00249	0.017	0.0207	2.17
0	0	0.00047	0.010	0.0237	2.18
0	0.05	0.00047	0.008	0.0235	2.18
0	0.10	0.00068	0.007	0.0242	2.18
0	0.15	0.00070	0.005	0.0245	2.18
1	0	0.00042	0.011	0.0243	2.18
1	0.05	0.00072	0.012	0.0248	2.18
1	0.10	0.00058	0.009	0.0230	2.17
1	0.15	0.00073	0.011	0.0261	2.18
2	0	0.00068	0.012	0.0252	2.18
2	0.05	0.00040	0.005	0.0238	2.18
2	0.10	0.00084	0.009	0.0253	2.18

values of reservation. Higher reservation values are important for acceptable performance when the network is under stress, as we shall see in later results. Hence, the desirability of a 5 to 10 percent reservation level is confirmed by the simulations. Notice also that real-time routing has little impact on total switch attempts.

Finally, with the improved blocking performance from real-time routing in which blocked calls are reduced by about two-thirds, consideration can be given to raising the switch-to-switch blocking objective or reducing the capacity provided for day-to-day variation in view of real-time routing capabilities. Raising the blocking objective results in network capacity savings while retaining objective network performance. Figure 8.14 illustrates the improvement in blocking performance with real-time routing as a function of reserve capacity. We can see that the effectiveness of real-time routing increases as reserve capacity increases, which is to be expected.

The analysis model results discussed above show that the master-gapping method is about equivalent in performance to the link-gapping method. Simulations are performed of master call gapping to see if network performance can be improved when gapping is combined with trunk reservation. These simulation re-

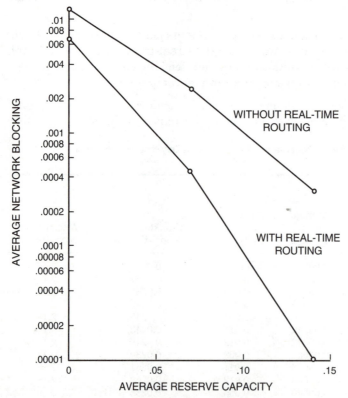

Figure 8.14 Average network blocking vs. average reserve capacity (low day-to-day variation; $R_{min} = 0$, $P = 0.05$)

TABLE 8.7 Performance of Gapping Methods (10% Overload, Gap = 1 Arrival Time, 7% Reserve Capacity)

Gapping used?	Reservation		Average network blocking	Maximum switch-pair blocking	Crankbacks per originating call	Attempts per originating call
	Minimum number of trunks (R_{min})	Fraction of trunks (P)				
No	0	0	0.00748	0.086	0.0747	2.26
Yes	0	0	0.00819	0.061	0.0560	2.24
No	0	0.05	0.00622	0.044	0.0575	2.24
Yes	0	0.05	0.00667	0.040	0.0474	2.22
No	1	0.10	0.00691	0.041	0.0542	2.22
Yes	1	0.10	0.00677	0.033	0.0451	2.21

sults, given in Table 8.7, are determined from a general network overload of 10 percent, in which all switch-pair loads are increased by 10 percent, because it is believed that the advantages of call gapping are greater under overload conditions. A gap interval of one call arrival time is used for these simulations. The addition of call gapping to the trunk reservation method appears to make small improvements in performance when an optimum combination of gapping and reservation is found, but it also degrades network performance in some respects. Hence, the analysis and simulation results both show that call gapping seems unnecessary and difficult to optimize when trunk reservation is used.

8.4.2.2 Real-time routing performance. As discussed previously, three real-time routing methods are investigated:

1. *Sequential:* Calls overflow from the normal engineered paths directly to a fixed sequence of real-time paths.

2. *Cyclic:* Calls overflow from the normal paths to the real-time paths, starting at the real-time path following the last real-time path tried for the previous real-time call.

3. *Cyclic (starting with the last path tried):* Calls overflow from the normal paths to the real-time path, starting at the last real-time path tried on the previous real-time call.

The results are compared in Table 8.8, which shows that the performance among the various routing methods is comparable. The second cyclic method tends to produce fewer crankbacks, although this is not particularly significant when comparing total switch attempts. In view of its comparable performance and its ease of implementation, sequential routing is judged to be the best choice in agreement with the mathematical analysis.

8.4.2.3 Switch real-time load. Various methods are simulated for turning a path off after a call fails to complete on the second link of a real-time path, thus

TABLE 8.8 Comparison of Real-Time Routing Methods (Low Day-to-Day Variations, 7% Reserve Capacity)

Routing method	Reservation		Average network blocking	Maximum switch-pair blocking	Crankbacks per originating call	Attempts per originating call
	Minimum number of trunks (R_{min})	Fraction of trunks (P)				
sequential	0	0.05	0.00047	0.008	0.0237	2.17
cyclic	0	0.05	0.00049	0.009	0.0228	2.18
cyclic (last path tried)	0	0.05	0.00040	0.011	0.0208	2.17
sequential	1	0.10	0.00058	0.009	0.0230	2.17
cyclic	1	0.10	0.00038	0.004	0.0226	2.17
cyclic (last path tried)	1	0.10	0.00064	0.010	0.0213	2.17

resulting in a crankback. The turn-off is applied for a fixed period of time following a second-link blockage, with the hope of reducing crankback messages that occur if the path is tried again before the required number of trunks become free. Various turn-off intervals and network overload levels are tried, and the results are summarized in Table 8.9. It can be seen that the work of the switch can be reduced by a turn-off method with some degradation in blocking performance. It is doubtful, however, that the moderate reduction in switching resources is worth the additional complexity and administrative effort of using path turn-off.

The primary factor affecting the switch real-time load impact is the number of real-time paths per switch pair. Table 8.10 shows the results when this number is varied over a range of values. The results show that more real-time paths

TABLE 8.9 Comparison of Path Turn-Off Methods (Low Day-to-Day Variations, 7% Reserve Capacity, $R_{min} = 1$, $P = 0.10$)

Turn-off interval (seconds)	Overload percent	Average network blocking	Maximum switch-pair blocking	Crankbacks per originating call	Attempts per originating call
0	0	0.00058	0.009	0.0230	2.17
15	0	0.00054	0.008	0.0234	2.18
30	0	0.00070	0.008	0.0222	2.17
60	0	0.00071	0.009	0.0231	2.18
300/N *	0	0.00064	0.007	0.0229	2.17
0	10	0.00691	0.041	0.0542	2.22
300/N	10	0.00673	0.035	0.0463	2.22
0	20	0.0525	0.203	0.180	2.35
300/N	20	0.0531	0.216	0.151	2.32
0	30	0.112	0.312	0.286	2.41
300/N	30	0.112	0.327	0.243	2.36

*N = number of trunks on second link

TABLE 8.10 Effect of Number of Real-Time Paths per Switch Pair (Low Day-to-Day Variations, 7% Reserve Capacity, $R_{min} = 1$, $P = 0.10$)

Maximum real-time paths per switch pair	Average network blocking	Maximum switch-pair blocking	Crankbacks per originating call	Attempts per originating call
20	0.00058	0.009	0.0230	2.17
15	0.00058	0.009	0.0230	2.17
10	0.00092	0.017	0.0244	2.18
5	0.00148	0.010	0.0244	2.17
0	0.00249	0.017	0.0207	2.17

improve blocking performance at the expense of increased crankback messages. The effect on overall network attempts, however, is minimal.

8.4.2.4 Network overload and failure performance. Real-time routing should not degrade network performance under overload and failure conditions. The results presented in this section reflect network congestion under general overload, focused overload, link failure, and switch failure conditions.

During typical Monday busy hours, traffic can average 5 to 8 percent above the normal weekly average. Figure 8.15 illustrates that the average Monday loads

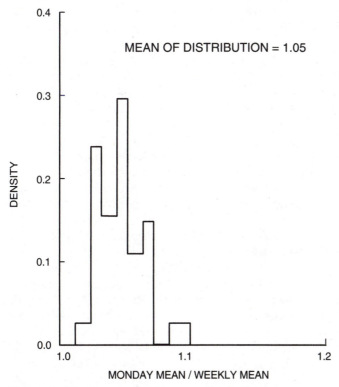

Figure 8.15 Monday mean/weekly mean distributions

TABLE 8.11 Performance under General Overload (7% Reserve Capacity, $R_{min} = 1$, $P = 0.10$)

Real-time routing used?	Overload percent	Average network blocking	Maximum switch-pair blocking	Crankbacks per originating call	Attempts per originating call
No	0	0.00249	0.017	0.0207	2.17
Yes	0	0.00058	0.009	0.0230	2.17
No	10	0.00878	0.037	0.0342	2.19
Yes	10	0.0069	0.041	0.0542	2.22
No	20	0.0547	0.181	0.0880	2.23
Yes	20	0.0525	0.203	0.180	2.35
No	30	0.116	0.324	0.144	2.24
Yes	30	0.112	0.312	0.286	2.41

are in this range. Under unusual conditions, overloads can range much higher. For this reason, real-time routing is investigated under such general overload conditions.

Simulation results are given in Table 8.11, in which a 10 percent overload indicates that all switch-to-switch loads are increased by 10 percent above their design values. The results show that real-time routing improves average blocking performance even up to a 30 percent overload, but other performance characteristics degrade. These simulations are made using a minimum trunk reservation (R_{min}) of one trunk and minimum fraction (P) of 10 percent.

Additional results are obtained to determine the network blocking performance when reservation is applied to all alternate-routed traffic [Aki83, Aki84, Kru82, NaM73], including normal traffic and real-time traffic. Traffic on the first path choice of the engineered paths has no reservation restrictions applied to it. The results, which are given in Table 8.12, show that average blocking performance can be improved under larger general overloads by using trunk reservation on engineered paths. However, this improvement is at the expense of higher blocking levels for some switch pairs under design traffic conditions—particularly for longer-distance switch pairs. Full-time use of trunk reservation should therefore be reflected in network design to achieve correct switch-pair blockings. Design for full-time use of trunk reservation increases network cost and also increases network design complexity.

These problems can be avoided and the advantages of engineered path trunk reservation attained if trunk reservation on engineered paths is triggered automatically. Three versions of automatic trunk reservation (ATR) are simulated in the 28-switch network. Switch-to-switch blocking is used to trigger engineered trunk reservation at a 10 percent switch-pair-blocking threshold. In ATR #1, once the threshold is exceeded two actions are taken. First, the triggered switch pair is subjected to engineered trunk reservation as described above. Second, traffic attempting to alternate-route over the direct link of the triggered switch pair is also subjected to engineered trunk reservation. Action one protects other switch pairs from interference from the triggered switch pair; action two protects the triggered switch pair from interference from other switch pairs. In ATR #2, only

TABLE 8.12 Performance under General Overload with Trunk Reservation (No Reserve Capacity, $R_{min} = 1$, $P = 0.10$)

Engineered path reservation used?	Overload percent	Average network blocking	Maximum switch-pair blocking	Crankbacks per originating call	Attempts per originating call
No	0	0.00217	0.015	0.0325	2.19
Yes	0	0.00681	0.059	0.0411	2.18
Yes (ATR #1)	0	0.00217	0.015	0.0325	2.19
Yes (ATR #2)	0	0.00217	0.015	0.0325	2.19
Yes (ATR #3)	0	0.00217	0.015	0.0325	2.19
No	10	0.0380	0.149	0.146	2.33
Yes	10	0.0275	0.171	0.187	2.33
Yes (ATR #1)	10	0.0263	0.156	0.164	2.33
Yes (ATR #2)	10	0.0293	0.113	0.176	2.35
Yes (ATR #3)	10	0.0309	0.143	0.172	2.34
No	20	0.103	0.301	0.277	2.41
Yes	20	0.0749	0.303	0.303	2.39
Yes (ATR #1)	20	0.0693	0.373	0.266	2.36
Yes (ATR #2)	20	0.0690	0.279	0.275	2.38
Yes (ATR #3)	20	0.0717	0.325	0.186	2.28
No	30	0.164	0.410	0.354	2.42
Yes	30	0.124	0.421	0.375	2.40
Yes (ATR #1)	30	0.120	0.439	0.351	2.38
Yes (ATR #2)	30	0.120	0.347	0.358	2.39
Yes (ATR #3)	30	0.120	0.420	0.181	2.20

action two is triggered. In ATR #3, action two is triggered and real-time routing is canceled for triggered switch pairs.

We see from Table 8.12 that all versions of ATR are effective in reducing the network blocking under overload and that ATR #2 performs somewhat better than ATR #1. ATR #3 significantly reduces crankback attempts over ATR #1 and ATR #2. These methods impose trunk reservation only when the network requires it, and ATR can be left on continuously because it does not affect traffic under normal traffic conditions.

The performance of real-time routing is also investigated under a three-to-one focused overload condition on the Newark switch in Figure 1.12. All loads to the Newark switch are increased by a factor of 3. The results, which are given in Table 8.13, indicate that real-time routing can improve network blocking and individual switch-pair blockings under focused overload conditions. In this particular case, the NWRK–CDKN switch pair sustains the maximum blocking, which results from the large additional overflow to this link, including attempted real-time calls routing on this link. We see that ATR #2 improves blocking performance but also increases crankback messages. ATR #3 is effective in limiting the use of crankback and is judged to be the best overall control tested.

A link failure of the LSAN–NWRK link (23 trunks) is simulated, with the results shown in Table 8.14. The results show that the average network blocking improves using real-time routing with a slight increase in switch real-time impact.

TABLE 8.13 Performance under Focused Overload (3 × Focus on Newark; 7% Reserve Capacity, R_{min} = 1, P = 0.10)

Real-time routing used?	Average network blocking	Maximum network blocking	Crankbacks per originating call	Attempts per originating call
No	0.0857	0.615	0.118	2.22
Yes	0.0789	0.605	0.216	2.34
Yes (ATR #2)	0.0746	0.517	0.299	2.38
Yes (ATR #3)	0.0763	0.526	0.136	2.21

TABLE 8.14 Performance under Link Failure (LSAN–NWRK Failure; 7% Reserve Capacity)

Reservation		Average network blocking	Maximum switch-pair blocking	Crankbacks per originating call	Attempts per originating call
Minimum number of trunks	Fraction of trunks				
No real-time routing		0.00264	0.023	0.0258	2.17
1	0.10	0.00062	0.009	0.0291	2.18

A switch failure of the NWRK switch is simulated, with the results shown in Table 8.15. The high-blocking switch pair in this case is ALBY–WHPL. NWRK is normally a via point for this switch pair. Additional real-time calls using the ALBY–WHPL link also contribute to the higher blocking. However, it is significant that the overall network blocking improves when real-time routing is applied. All calls destined for NWRK overflow all engineered and real-time paths, causing a large number of real-time path attempts and crankback messages. Normally, real-time traffic management controls would cancel such attempts to alleviate this situation. Another interesting phenomenon is that most switch pairs not normally using NWRK as a via point achieved better than normal service. This happened because the links normally carrying NWRK traffic are relatively free, because NWRK traffic cannot complete. Hence, other traffic can make use of these relatively lightly loaded links to achieve better than normal performance.

TABLE 8.15 Performance under Switch Failure (NWRK Switch Failed; 7% Reserve Capacity)

Reservation		Average network blocking*	Maximum switch-pair blocking*	Crankbacks per originating call	Attempts per originating call
Minimum number of trunks	Fraction of trunks				
No real-time routing		0.00319	0.082	0.183	2.19
1	0.10	0.00099	0.089	0.187	2.20

*Excluding traffic to the NWRK switch

8.5 Conclusions

In this chapter, we study dynamic routing design under random load variations in which real-time routing methods are combined with preplanned dynamic routing. We use preplanned dynamic two-link path routing as a basis for study, but the principles discussed apply in general to all dynamic routing networks. In particular, this chapter discusses analytical and simulation models for real-time dynamic routing techniques that include trunk reservation, call gapping, and real-time path selection.

The analysis and simulations of real-time routing for dynamic routing networks lead to the following general conclusions:

- Reservation levels of about 5 to 10 percent of the trunks in each link appear adequate.

- Call gapping does not enhance performance of real-time routing.

- A sequential real-time routing method achieves about the same level of performance as more complicated cyclic methods and is preferred because of the simplicity of implementation.

- Switch real-time load is not significantly increased by real-time routing, especially when automatic real-time traffic management controls are used.

- Path turn-off can reduce the switch real-time impact but does not seem worth the additional complexity.

- Real-time routing under various overload and failure conditions significantly improves network performance.

The dynamic routing design models and techniques analyzed in the chapter are applied and extended in later chapters to real-time state-dependent and event-dependent routing networks (Chapters 11–13), and dynamic transport routing networks (Chapters 14 and 15). When designing networks for real-time dynamic routing, an important consideration is meeting the blocking performance objective under day-to-day traffic load variations, and as described in Chapter 13, dynamic routing table updates based on real-time dynamic routing design can lead to a 2–3 percent reduction in day-to-day capacity. A larger possible reduction of 5 percent or more in network reserve capacity is achieved with routing table updates under forecast uncertainty, as discussed in Chapter 9. Further reduction of up to 10 percent or more in network reserve capacity is gained with real-time transport routing table updates under week-to-week load variations, as discussed in Chapter 14.

Dynamic Routing Design under Forecast Uncertainty

9.1 Introduction

As discussed in Chapter 1, Figure 1.1 illustrates a model for network routing and network management and design. The central box represents the network, which can have various configurations, and the traffic and transport routing tables within the network. Network configurations include metropolitan area networks, national intercity networks, and global international networks, which support both hierarchical and nonhierarchical structures and mixes of the two. Routing tables describe the route choices from an originating switch to a terminating switch for a connection request for a particular service. Hierarchical and non-hierarchical traffic routing tables are possible, as are fixed and dynamic routing tables. Routing tables are used for a multiplicity of traffic and transport services on the telecommunications network.

Figure 1.1 illustrates network management and design as interacting feedback loops around the network. The input driving the network ("system") is a noisy traffic load ("signal") consisting of predictable average demand components added to unknown forecast error and load variation components. The feedback controls function to regulate the service provided by the network through capacity and routing adjustments. Network management functions include traffic performance management and capacity management. Traffic performance management ensures that performance objectives are met under all conditions including load shifts and failures, and capacity management ensures that network designs meet performance objectives at minimum cost. Traffic performance management includes real-time and daily performance monitoring, real-time traffic management, and routing design. Real-time traffic management provides monitoring of network performance through collection and display of real-time traffic and performance data and allows traffic management controls such as code blocks, call gapping, and reroute controls to be inserted when circumstances warrant. Routing design takes account of the capacity provided by capacity management and on a weekly or possibly real-time basis adjusts traffic and transport routing tables as necessary to correct service problems. The updated routing tables are sent to the switching systems via an automated routing update system or are derived within the switch itself. Capacity management includes network design and operates over a multiyear forecast interval to drive network capacity expansion. Under exceptional circumstances, capacity can be added on a short-term basis to alleviate service problems, but it is normally planned, scheduled, and managed over a period of one year or more.

The fixed hierarchical routing structure and method of managing the network lead to capacity requirements that can be reduced by improved routing techniques. As discussed in Chapter 1, upper limits can be established on the maximum improvement in network efficiency with capacity management and improved network design, as well as with traffic performance management and improved preplanned and real-time routing table design and update. This is illustrated in Figure 1.38 and discussed in this section. Even though the bounds given in Figure 1.38 are unrealizable limits, experience has shown that practical techniques are available to achieve much of the projected design savings. Therefore, these limits serve as bounds that are useful in approximating potential gains in network cost reduction. The limits are additive, so that the total range is from about 20 to 50 percent of the network first cost, of which half or more is realizable.

In Chapter 2, we discussed the network design principles for busy-hour erlang capacity design in hierarchical networks, including Truitt's ECCS technique, equivalent random methods, and final-link sizing methods. In Chapters 4 to 7, we extended these design principles to nonhierarchical networks, in which the value of shortest-path routing is also illustrated. Hierarchical routing has limited path choices and low-blocking design on final links, which limits flexibility and reduces efficiency. If we choose paths based on cost and relax the rigid hierarchical network structure, a more efficient network results. The upper limit on improvement in network design efficiency due to nonhierarchical shortest-path routing, as illustrated in Figure 1.38, is a reduction of about 5 percent in total network cost.

As discussed in Chapters 4 to 7, the route-erlang flow optimization (route-EFO) model and path-EFO model combine the above design principles and others into a single, unified approach for achieving network design savings. The steps of the path-EFO model for preplanned dynamic routing network design, which incorporate these design principles, are illustrated in Figure 6.3 and show it to be an iterative technique consisting of a routing design step, a capacity design step, and link parameter update steps. These route-EFO and path-EFO models were applied to preplanned dynamic routing networks and, in particular, the value of multihour network design was illustrated. Multihour design provides an opportunity to improve network design efficiency by incorporating dynamic traffic routing techniques into the design. Because of its fixed nature, hierarchical routing cannot respond efficiently to traffic load variations that arise from business or residential phone use, time zones, seasonal variations, and other causes. Dynamic traffic routing increases network utilization efficiency by varying routing patterns in accordance with traffic patterns, as reflected in the design. Multihour network design savings are discussed in Chapter 1 and Chapters 4 to 7 and are in the range of 10 percent for metropolitan area networks, 15 percent for national intercity networks, and 25 percent for global international networks.

The effects of day-to-day variation, unlike those of forecast error, are taken into account in the initial design of the network. They arise from the nonlinear relation between link load and blocking. When the load on a link fluctuates about a mean value, because of day-to-day variation, the mean blocking is higher than the blocking produced by the mean load. Therefore, additional capacity is provided to maintain the blocking probability grade-of-service in the presence of day-to-day load variation. The multihour traffic load is characterized by a forecast mean load about which there occurs a day-to-day variation, characterized by a gamma distribution with one of three levels of variance [Wil58]. Even if the forecast mean loads are correct, the actual realized loads exhibit a random fluctuation from day to day. Earlier studies have established that this source of uncertainty requires the network to be augmented in order to maintain the blocking probability grade-of-service objective [HiN76], which leads to the 4–7 percent day-to-day capacity shown in Figure 1.38. Normally, it is not practical to update preplanned dynamic routing tables to anticipate daily traffic variations. This is because the required interval between measuring the changing load patterns and updating the routing tables to reflect the changed loads is usually much shorter than the period over which the preplanned routing patterns are held fixed, which is typically at least one hour. Real-time routing methods combined with preplanned routing are discussed in Chapter 8, which include real-time paths searched using real-time routing methods and real-time routing path changes provided by the real-time traffic management system (NEMOS). These real-time routing methods applied to preplanned dynamic routing networks then provide a basis for improved network design for day-to-day variations, which could possibly result in some reduction in the day-to-day capacity illustrated in Figure 1.38. However, the real-time dynamic routing methods discussed in Chapter 1 and discussed further in Chapters 11 to 13 lend themselves to improved design efficiency for day-to-day load variations. For example, the path-EFO model and discrete event flow optimization model described in Chapters 1 and 13 can be applied to real-time dynamic network design and incorporate techniques for achieving day-to-day capacity design savings in such networks. These savings are shown to be about

2–4 percent of the network design cost, or approximately half of the day-to-day capacity required without such design techniques.

Dynamic network design for week-to-week load variations, which include load forecast uncertainties, is the subject of this chapter and is discussed in the following sections. Although the basis for analysis is preplanned dynamic two-link path routing, these methods and results apply to dynamic routing networks in general and in particular to real-time dynamic routing networks. Reserve capacity is the difference between actual in-place capacity and the ideal capacity design for the current network loads. As shown in this chapter, reserve capacity, which arises because of forecast error and trunk administration policies, can reach 25 percent or more of total network capacity, and through dynamic routing table updates with either preplanned or real-time dynamic routing, reduction of 5 percent or more in reserve capacity can be achieved. Further reduction of up to 10 percent or more in network reserve capacity is gained with real-time transport routing table updates under week-to-week load variations, as discussed in Chapter 14.

Fixed routing performance management cannot easily incorporate preplanned routing changes that respond to unpredicted network load shifts, whereas in a dynamic routing network such a mode of operation is possible. Fixed routing performance management leads to an additional network "reserve" capacity, which as illustrated in Figure 1.38 is estimated to range from 15–25 percent depending on (1) the accuracy and stability of the traffic load forecast, (2) the trunk-disconnect policy, and (3) the short-term capacity design method. Prior analyses [FHH79] have shown levels of reserve capacity up to 25 percent when forecast bias, measurement error, and other effects are introduced into the model. Operational studies in fixed routing networks have measured up to 20 percent and more for network reserve capacity. Innovations such as the Kalman filter [PaW82] and more aggressive disconnect policies can help reduce this level of reserve capacity.

Insufficient capacity means that occasionally trunks must be connected on short notice if the network load requires it. This process is called short-term capacity design, and it is an aspect of capacity management. Routing changes in preplanned or real-time routing table design provide a more efficient use of the network by allowing traffic flow on underutilized links to be an alternative to providing additional capacity based on current routing tables. This yields capital cost savings because of the increased utilization of the network and reduced expenses because of reduced manual rearrangement activity. The upper limits on savings in reserve capacity with routing design procedures are now discussed.

9.2 Dynamic Routing Design under Forecast Uncertainty

Forecast errors are always present in network designs; errors arise either in overestimating or underestimating the future load on the network. To correct for forecast errors, we use a routing design procedure that behaves like a fast-acting inner feedback loop, which as illustrated in Figure 1.1 covers a period on the order of one week for preplanned dynamic routing or is done in real time for real-time dynamic routing. Routing design models estimate the current loads and determine where routing changes should be made. In short-term capacity design for fixed hierarchical routing networks, reports are made on links that are either underloaded or overloaded. Overloaded links can lead to service problems and are usually quickly augmented. In contrast, immediate action is not often taken to reduce capacity on underloaded links. This comes from a reluctance to disconnect trunks that may be needed in the relatively near future, such as within one year, and also from the fact that the link may not be in its busy season. The reluctance to disconnect trunks, together with the forecast errors, causes an overall tendency to provide extra reserve capacity in the network.

If the future loads on the network are completely known, then it would be sufficient to design the minimum-cost network to meet these loads—for example, by applying the path-EFO design model described in Chapters 6 and 7. In actuality, various categories of load uncertainty are present that necessitate the design models for each category, which are summarized in the previous section. Network demands are continually growing and shifting, which means that we must forecast, design, and plan the required capacity far enough in advance, approximately one to two years, to meet the load. Of course, these forecasts are subject to error, and the recognition of this error influences our network design method in various ways. The goal is to provide sufficient capacity to meet the expected load on the network. In implementing the capacity management function, the capacity manager plans the network based on the forecast loads and the trunks already in place. Consideration of the in-service trunks results in a disconnect policy that may leave capacity in place even though it is not called for by the design.

There are, however, economic and service implications of the capacity design process. Insufficient capacity means that occasionally trunks must be connected on short notice if the network load requires it. This process is called short-term capacity design. There is a trade-off between reserve capacity and short-term capacity design, which we explore in this section. The routing design models for preplanned dynamic routing described in Chapter 6 are enhanced to provide a potential 5 percent reduction in reserve capacity while retaining a low level of short-term capacity design.

Hence, capacity management is driven by the load forecast, which is subject to error, and the existing network. Trunk disconnects are planned in capacity design when the forecast predicts declining or shifting loads and we are reasonably sure the trunks will not be needed in the next one to two years. This procedure reflects some reluctance to disconnect trunks and results in a certain amount of reserve capacity being left in the network. On occasion, the planned design underprovides trunks at some point in the network—again, because of forecast errors—and short-term capacity design is required to correct these forecast errors and restore service [Sze80]. In fixed routing networks, when links are found to be overloaded when actual loads are larger than forecasted values, additional trunks are provided to restore the blocking probability grade-of-service to the objective value, and, as a result, the process leaves the network with reserve or idle capacity even when the forecast error is unbiased.

The question is, to what extent can the network capacity required by the uncertainties be reduced by dynamically controlling the preplanned routing patterns to meet the realized loads? A given realization of the loads can be expected to yield some switch-to-switch traffic loads that are higher than average and others that are lower. While part of the network is overloaded, another part might be underloaded. If the preplanned or real-time routing pattern can be adjusted to use the idle capacity of the underloaded portion of the network, the required capacity augmentation might be reduced. The model illustrated in Figure 1.48 is used to study capacity management of a network on the basis of forecast loads, in which capacity management accounts for both the current network and the forecast loads in planning network capacity, and short-term capacity design makes routing and link size adjustments if network performance under the realized loads becomes unacceptable because of errors in the forecast. As discussed earlier, the routing design models try to minimize reserve capacity while maintaining an acceptable level of short-term capacity design.

The model in Figure 1.48 assumes that capacity design is an annual process that predicts the required network capacity to meet the future demand. It considers both the traffic forecast, which is subject to error, and the existing network. In dealing with forecast errors, capacity design attempts to provide sufficient network capacity to meet these traffic loads with a minimum of short-term capacity augmentation. The model assumes that the capacity design is implemented immediately and, if necessary, routing design and short-term capacity design are invoked to restore network service when shortages are detected.

In a fixed routing network, capacity design compares the existing network with a network designed for the forecast loads, which is made without reference to the existing link sizes. When the capacity design calls for additional trunks on a link, the augments are usually implemented, and when the forecast calls for fewer trunks, a disconnect policy is invoked to decide whether trunks should be disconnected (this policy reflects a degree of reluctance to disconnect trunks). In a preplanned or real-time dynamic routing network, it is possible to design a minimum-cost augmentation to the existing network capacity to meet the traffic loads. For preplanned dynamic routing, modifications to the path-EFO model allow a capacity design and routing design with initial link capacities as lower bounds on the designed link capacities. In effect, the model takes into account the reluctance to remove trunks. This procedure limits the trunk augmentations to those required to meet the forecast loads and, thus, achieves a lower reserve capacity. A generalization of this capacity management design procedure is to set minimum link sizes and to allow trunk disconnects. This is done by deriving, for each link, a lower threshold and an upper threshold for its size and using these thresholds to make initial adjustments to the link size (as described below) prior to the capacity management design.

The size thresholds for a link are based on the forecast loads of the corresponding direct traffic and are determined by choosing a minimum value $r_{\min}$ and a maximum value $r_{\max}$ for the ratio $r \triangleq$ (link size)/(direct traffic load). The lower (reserve) threshold for the link size is based on the forecast peak load of the direct traffic in the next year and corresponds to $r_{\min}$. The upper (disconnect) threshold

TABLE 9.1   Limits on $r = T/L = $ (Link Size)/(Direct Traffic Load)

Load L (erlangs)	$r_{\min}(L)$	$r_{\max}(L)$
0–5	0.3	4.5
5–10	0.4	4.5
10–25	0.5	3.0
25–50	0.7	3.0
50–100	0.95	2.5
>100	1.05	1.5

is based on the forecast peak load of the direct traffic over the next two years, with some allowance for forecast error, and it corresponds to $r_{\max}$.

The limits $r_{\min}$ and $r_{\max}$ are chosen by examining the range of values of the ratio in a typical preplanned dynamic traffic network design. In general, the ratio has a smaller spread of values for large direct traffic loads than for small direct traffic loads. With large traffic loads, the corresponding link can be quite efficient carrying just the direct load; hence, its size to a large extent depends just on the direct traffic. For small loads, however, the link size is less dependent on the direct load and is influenced more by the alternate-routing traffic carried on that link; hence, the ratio is expected to have a wider range of values.

Table 9.1 shows the limits for the ratio (link size)/(direct traffic load) as a function of the direct load.

The lower threshold $T_{\min}$ and upper threshold $T_{\max}$ for a link are determined in terms of the forecast traffic loads for the corresponding direct load:

Let L_i = peak forecast load for the direct traffic load in year i, $i = 1, 2$.

  β_i = forecast uncertainty factor for year i, $i = 1, 2$, introduced to allow for probable error in the load forecast. The results presented here are obtained with $\beta_i = 1.15, \beta_2 = 1.3$, corresponding to a 0.15 coefficient of variation in the forecast.

Then,

$$T_{\min} \overset{\Delta}{=} r_{\min}(\beta_1 L_1) * \beta_1 L_1$$

$$T_{\max} \overset{\Delta}{=} \max[r_{\max}(\beta_1 L_1) * \beta_1 L_1, r_{\max}(\beta_2 L_2) * \beta_2 L_2]$$

where $r_{\min}$ and $r_{\max}$ are the appropriate limits established for ratio (T/L), such as those in Table 9.1.

With these lower and upper thresholds, we then define an initial size for each link, which depends on its current size as follows:

- If the current size of a link is between its lower and upper thresholds, its initial size equals its current size.

- If the current size of the link is below its lower threshold, its initial size equals the lower threshold.

- If the current size of the link is above its upper threshold, its initial size equals the upper threshold.

We use the initial network defined in this manner as the starting network for capacity design, and comparing the result with the current network, we determine the actual augmentations and disconnects that must be made to implement the capacity design. Under normal growth conditions, the current trunks, rather than the upper and lower trunk limits, are most often used in the initial network, and the primary effect is to route traffic on the actual trunks in place and, thus, minimize rearrangements.

As illustrated in Figure 6.3, the path-EFO model is an iterative procedure with four basic steps: selection of cost-effective traffic paths, optimization of path flows, sizing the links to correspond to the optimum flows, and updating of incremental link cost metrics and optimal link blockings for the next iteration. The procedures for capacity design with preplanned routing design involve modifications to the routing design and capacity design steps to allow the existing link capacities to be used as lower bounds on the designed link capacities.

To describe the routing design model, let

$$
\left.
\begin{array}{rl}
L =&\ \text{number of links} \\
H =&\ \text{number of design hours} \\
B_{\text{opt}}^i =&\ \text{optimal link (maximum) blocking on link } i \\
B_i^h =&\ \text{blocking of link } i \text{ in hour } h, h = 1, \ldots, H \\
y_i^h =&\ \text{carried load on link } i \text{ in hour } h, h = 1, \ldots, H \\
a_i =&\ \text{capacity of unaugmented link } i, \text{ in carried load,} \\
&\ \text{at blocking } b_i^{\max} \\
\Delta a_i =&\ \text{capacity augmentation, in carried load, on link } i \\
M_i =&\ \text{incremental link cost metric in terms of dollar cost} \\
&\ \text{per incremental erlang of carried traffic on link } i
\end{array}
\right\} \quad i = 1, \ldots, L
$$

The objective is to allocate the traffic flow of each hour among its admissible paths so as to minimize the cost of the required link capacity augmentations. On each link, for a given number of existing trunks and the optimum link-blocking objective determined from economic considerations, there is a maximum load that can be carried on that link without augmentation; this is the unaugmented initial capacity of that link. The preplanned routing design problem is then a modified linear program in which the decision variables are the flow assignments and the augmentations Δa_k above the existing link capacities a_k, instead of total link capacities as in the path-EFO design model described in Chapters 6 and 7, and the cost to be minimized is the incremental cost of augmentation

$$
\sum_{k=1}^{L} M_K \Delta a_k
$$

This formulation ensures that efficient use is made of existing link capacities, by means of preplanned routing design that minimizes capacity augmentation.

In the capacity design step we are given the routing design and traffic loads, and we find the needed augmentation to links that exceed their maximum permitted optimal link blockings. This is accomplished by the following iterative procedure:

1. Begin with assumed link blockings $\hat{B}_i^h \leq B_{opt}^i, i = 1, \ldots, L$.

2. Calculate the corresponding carried link loads y_i^h under the known routing and assumed link blockings.

3. If for all h, $y_i^h \leq a_i$, the capacity of the unaugmented link at its maximum optimal link blocking, then the link needs no augmentation; if $y_i^h > a_i$, the required augmentation Δa_i is determined by sizing the link for load y_i^h at blocking B_{opt}^i.

4. From the link loads y_i^h and link sizes computed in steps 1 and 2, we recalculate all the link blockings B_i^h; if $|B_i^h - \hat{B}_i^h|$ is not sufficiently close to zero for all i in all hours h, redefine $\hat{B}_i^h = B_i^h, i = 1, \ldots, L, h = 1, \ldots, H$, and return to step 2.

If, after the capacity design including the preplanned routing design is implemented, the realized loads exceed forecast values and cause unacceptable blocking, quick corrective action in the form of short-term capacity design is needed. In fixed routing networks, short-term capacity design is limited to link augmentations. In preplanned dynamic traffic routing networks, however, the preplanned routing patterns can be modified in the short-term capacity design to reduce network augmentation.

For real-time dynamic routing, an analogous model is formulated. For example, the discrete event flow optimization (DEFO) model described in Chapters 1 and 13 can be formulated to incorporate existing link capacities and to do a minimum-cost link augmentation to meet blocking probability grade-of-service objectives. Note that the routing table updates are made in the network automatically, given a real-time dynamic routing method, but any required link augmentations must be determined in the short-term capacity design. For example, as illustrated in Figures 1.45 and 13.8, given the link sizes, current traffic loads, and real-time routing method, the DEFO model evaluates the blocking probability. Given the switch-to-switch blocking and traffic loads, the DEFO model estimates the capacity augmentation required in the capacity design step for each switch pair to meet its blocking probability grade-of-service objective. As discussed in Chapter 13, the augmentation capacity is allocated to links according to a Kruithoff model to achieve a minimum-incremental-cost design.

Short-term capacity design therefore consists of three steps:

1. Detecting the need for short-term capacity design; that is, determining whether or not all traffic loads are receiving adequate service.

2. If short-term capacity design is needed, determining the best combination of preplanned routing changes (in the case of preplanned dynamic routing) and link augmentations to restore the desired grade-of-service at minimum cost.

3. Implementing the preplanned routing changes and link augments.

These steps are discussed in more detail below.

Switch-to-switch blocking measurements determine the level of service being provided and, thus, detect the existence of service problems. Because of measurement errors and day-to-day traffic variations, such blocking measurements will have an inherent statistical variability, which must be allowed for by establishing acceptable bounds for the measured blockings. In short-term capacity design, the routing tables are designed for the realized traffic loads, with minimum link augmentations required to limit link blockings to their maximum permitted values. In the case of preplanned dynamic routing, an automatic routing update system then implements the routing table updates. In the case of real-time dynamic routing, the routing table updates are made automatically in the network according to the real-time dynamic routing method.

9.3 Modeling Results

9.3.1 Results for dynamic routing design under forecast uncertainty

The model in Figure 1.48 is used to simulate the capacity management and preplanned routing table design process for the case of preplanned dynamic routing on the 28-switch national intercity network model (illustrated in Figure 1.12). The network model is taken through 10 simulated years, each iteration consisting of a capacity design for the forecast traffic loads followed by a short-term capacity design for the realized traffic loads. The forecast traffic loads grow at a 5 percent annual rate, and the realized traffic loads in each year are assumed to be normally distributed about the forecast loads, with a 15 percent coefficient of variation.

The following four methods are compared:

1. Capacity design with no trunk disconnect and no preplanned dynamic routing design changes.

2. Capacity design with no trunk disconnects and preplanned dynamic routing design changes.

3. Capacity design with limited trunk disconnects and preplanned dynamic routing design changes.

4. Capacity design with full trunk disconnects and traffic and transport routing design changes.

In methods 1 and 2, no disconnects are allowed in capacity design, and they simulate complete reluctance to disconnect trunks, whereas in method 3, some trunk disconnects are allowed according to the model above. In method 4, all trunk disconnects are implemented, reflecting flexible transport network routing in which link capacity rearrangements are automatically and quickly implemented. Method 4 corresponds to dynamic traffic and transport routing design, which is further discussed in Chapter 15. In methods 1–3, no disconnects are implemented in short-term capacity design, whereas in method 4 all disconnects are implemented.

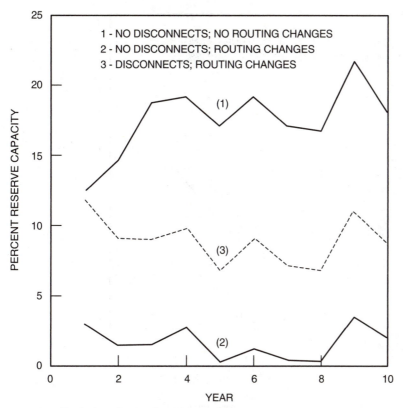

Figure 9.1   Evolution of network reserve capacity

Figure 9.1 shows the evolution of network reserve capacity for methods 1–3, measured by the percentage difference in cost between the realized network and an ideal network designed for the realized loads. Figure 9.2 shows the cumulative short-term capacity design trunk augments for the three methods as percentages of the number of trunks in the starting network. Figure 9.3 shows the level of short-term capacity design in each year as measured by the trunk augments in short-term capacity design as a percentage of trunks in the realized network.

We note from Figures 9.1 and 9.2 that method 2 achieves a lower reserve capacity than method 1 but requires more short-term capacity design augments in response to forecast error. Method 3 falls between the other two, both in reserve capacity and in short-term capacity design. Compared with method 1, it achieves a striking reduction in reserve capacity for a modest increase in short-term capacity design. Figure 9.3 shows that, in all three methods, short-term capacity design rearrangements in each year are in the range of 1 to 4 percent of the trunks in the network, which is quite favorable in comparison

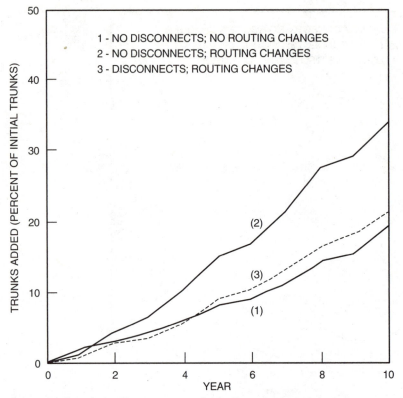

Figure 9.2 Cumulative short-term capacity design rearrangements

with the short-term capacity design level in fixed routing networks, which is typically about 10 percent of trunks in the network.

Figure 9.4 shows the cumulative *total* rearrangements (consisting of augments and disconnects in the forecast capacity design and augments in short-term capacity design) for the three methods as percentages of the number of trunks in the starting network. Method 2 produces the fewest rearrangements and method 1 the most. It is impractical, however, to have a large number of short-term capacity design rearrangements, which are called for on short notice, to correct existing service problems.

Figures 9.1 and 9.2 show the trade-off between reserve capacity and short-term capacity design rearrangements. By multiplying the factors $r_{\min}$ in Table 9.1 by a factor $\alpha \geq 0$, we can parameterize the resulting levels of reserve capacity and short-term capacity design. The value $\alpha = 1$ corresponds to curve 1 in Figures 9.1 to 9.4, $\alpha < 1$ results in lower reserve capacity and increased short-term capacity design, and $\alpha > 1$ leads to higher reserve capacity and reduced short-term capacity design. Figure 9.5 is a curve of average reserve capacity versus the average level of short-term capacity design in method 3, with α as the parameter.

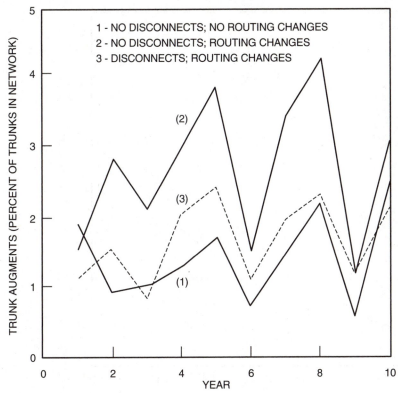

Figure 9.3 Trunk augments in short-term capacity design as a percentage of trunks in network

For comparison, the two points corresponding to methods 1 and 2 are also plotted. Point 2 is almost the same as the limiting case $\alpha = 0$, and point 1 lies above the curve, showing that a more favorable trade-off can be obtained with method 3 than with 1. This curve is a quantitative expression of the trade-off between reserve capacity and short-term capacity design. We conclude that routing table changes in both capacity management (with trunk disconnects allowed) and short-term capacity design are an efficient method of controlling reserve capacity and the level of short-term capacity design in the network.

In Table 9.2, we summarize the network utilization improvements with the network design models for uncertain traffic loads, as discussed in this section, and also include results on dynamic transport routing, discussed in Chapter 15. Method 4, with full disconnects, corresponds to the case when the transport network can be dynamically rearranged to match actual link capacity demands. With dynamic transport routing, a lower reserve capacity can be achieved.

We present results for Method 4 with combined dynamic traffic and transport routing and compare them with Methods 1 and 3, which have fixed transport

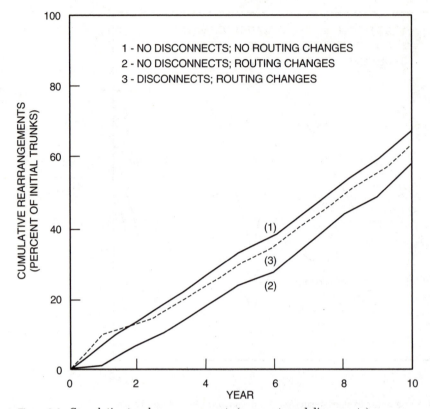

Figure 9.4 Cumulative trunk rearrangements (augments and disconnects)

routing. In the model, we have considered week-to-week variations and the effect on reserve capacity in national intercity network models for the three routing and capacity design methods. We can see the additional reduction in reserve capacity when dynamic transport routing is used. In particular, an additional 10 percent reduction in reserve capacity is achieved with Method 4 compared with Method 3 because of the ability of dynamic transport routing to quickly rearrange link capacity to match the current traffic demand. In addition, Method 3 once again achieves a 5 percent reserve capacity decrease with respect to Method 1 because of the incremental network efficiencies achieved by dynamic traffic routing.

9.3.2 Field example of preplanned dynamic routing design

Next, we look at performance examples with preplanned routing design. Figure 9.6 gives an example of service between two cities in the DNHR network—Atlanta, Georgia, to Miami, Florida—during October 1987. We can see that on this

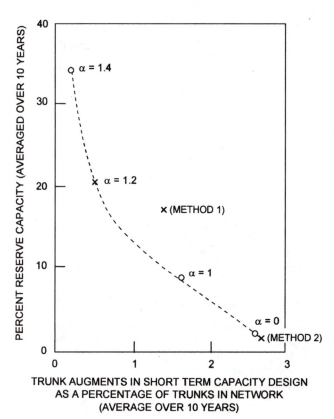

Figure 9.5 Trade-off between reserve capacity and short-term capacity augments

TABLE 9.2 Network Design Comparison for Load Forecast Errors (National Intercity Models)

Routing method	Normalized reserve capacity cost	Savings (%)
fixed traffic routing and fixed transport routing (Method 1)	0.250	—
dynamic traffic routing and fixed transport routing (Method 3)	0.200	5.0
dynamic traffic routing and dynamic transport routing (Method 4)	0.100	15.0

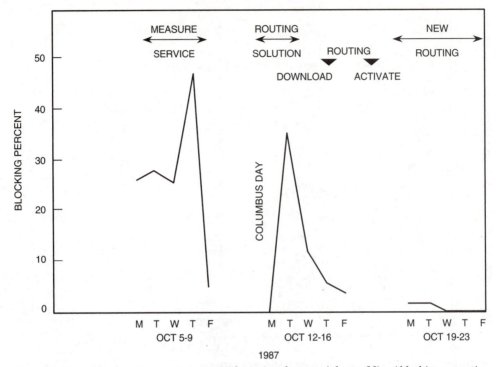

Figure 9.6 Example of service improvement with routing changes: Atlanta–Miami blocking correction

particular route there is high blocking between these cities. The data collected during two weeks is transmitted to the routing design system to process. The routing design system analyzes and solves the blocking problem, not by adding trunks but by reoptimizing and then changing the routing table in the network. After the new routing tables are downloaded to the network, the blocking problem is essentially eliminated. This is one of the operational advantages of routing design, in contrast to the fixed hierarchical network, where we move trunks around to follow the traffic load. In the dynamic network, we move the traffic around to where there is capacity in the network. This is simpler, less expensive, and faster.

9.4 Conclusion

Dynamic routing design brings a benefit to network performance in terms of improved quality and reliability at reduced cost. Network design for dynamic routing is made using the forecasted network loads. Load uncertainties arising from errors in the forecast and from daily variations in network load give rise to reserve or idle network capacity not immediately needed by current network demands. The reserve capacity is reduced by the use of flexible dynamic routing design methods, which allow routing flexibility to help control the network flow under load uncertainties. We illustrate techniques for changing network routing tables in routing design to counteract the effects of forecast errors. Included in the benefits are a reduction both in reserve capacity, estimated to be about 5 percent of network first cost, and in trunk rearrangements. These results apply to both preplanned and real-time dynamic routing. When dynamic transport routing design is combined with dynamic traffic routing design, an additional reduction of up to 10 percent in reserve capacity is estimated.

Dynamic Routing Design for Multiservice Integrated Networks

10.1   Introduction

In this chapter we discuss dynamic traffic routing for multiservice integrated networks. Integrated services digital networks (ISDNs) combine a multiplicity of voice services and switched digital data services on integrated transport networks. Switches interconnected by a flexible transmission network provide connections for voice, data, and wideband services. These connections are distinguished by estimated resource requirements, traffic characteristics, and design performance objectives. It has been suggested that perhaps there should be no "N" in ISDN; that is, there is no network aspect within the ISDN concept. This concept, as defined by the ITU interface standards, says nothing about how the transport network between the ISDN interface points should be configured—only that it must transport the services defined by the interface standard. The reality is, however, that ISDN transport networks can be configured as multiservice integrated networks. For example, in the case of the AT&T network, integration of voice services and switched digital data services has been in place since the initial ISDN implementation.

This chapter describes dynamic routing design principles employed in multiservice integrated networks, which include bandwidth allocation strategies, dynamic routing call setup, network management procedures, and integrated network design models. These principles are illustrated for preplanned dynamic routing networks and for real-time dynamic routing networks and apply in general to all dynamic routing networks. In Chapters 11 to 13 we elaborate on the application of these principles to multiservice integrated networks with real-time dynamic routing. In this chapter we present a dynamic routing method that includes real-time control of routing patterns. Bandwidth allocation procedures manage network bandwidth according to a virtual trunk concept in which limits are placed on the number of connections for each service category. The routing plan provides common-channel signaling (CCS) call setup of the logical connections and network resource management based on near-real-time traffic data. We describe how these integrated network designs can be managed in actual network operation. Routing functions include call setup, real-time routing, network bandwidth allocation, link bandwidth allocation, and congestion control. Network management and design functions include data collection, forecasting, capacity design, and routing design.

The integrated network design models determine the routing and bandwidth capacity required to satisfy all service demands simultaneously on an integrated network. Each service demand meets its blocking and other performance objectives. In this chapter we describe network design models for integrated voice, data, and wideband services networks, which are extensions of the route-EFO and path-EFO models discussed in Chapters 4–7. Our approach is to characterize all service demands in common units, that is, by the mean, variance, and day-to-day variation of both the number of circuits and bandwidth per circuit demanded for each virtual network that carries multiple classes-of-service. These integrated design methods incorporate a network bandwidth allocation procedure to allocate bandwidth to the various virtual networks so that they can meet their performance objectives. These methods apply to both multiservice integrated circuit-switched networks and to multiservice integrated packet-switched networks. The network design methods are illustrated for multiservice voice and switched digital data services networks and wideband packet transport networks.

10.2 Multiservice Integrated Dynamic Routing Methods

Dynamic routing brings three principal changes to the network plan, which include the network configuration, routing method, and network management and design. A typical national dynamic traffic routing network has only one class of switching system, and end-offices home on dynamic routing switches by way of exchange access networks. Dynamic routing rules are used between pairs of dynamic routing switches, and fixed hierarchical routing rules are used between all other pairs of switches. These hierarchical switches home directly on the dynamic routing switches.

The dynamic routing method illustrated in Figure 1.22 uses preplanned dynamic two-link sequential path routing (DNHR) and capitalizes on two factors: selection of minimum-cost paths and design of optimal time-varying routing patterns, as designed by the route-erlang flow optimization (route-EFO) and path-EFO models described in Chapters 4–7, to achieve minimum-cost trunking by capitalizing on noncoincident network busy periods. We achieve the preplanned dynamic, or time-varying, nature of the routing method by varying the route choices with time. The routes, which consist of different sequences of paths, are designed to satisfy a switch-to-switch blocking requirement. Each path consists of one or, at most, two links in tandem. Paths used for routes in different time periods need not be the same. In Figure 1.22, the originating switch at San Diego (SNDG) retains control over a dynamically routed call until it is either completed to its destination at White Plains (WHPL) or blocked. The control of a call overflowing the second leg of a two-link connection (for instance, the ALBY–WHPL link of the SNDG–ALBY–WHPL path in routing sequence #1) is returned to the originating switch (SNDG) for possible further alternate routing. Control is returned when the via switch (ALBY in the example) sends a CCS crankback signal to the originating switch. Fifteen time periods are used to divide up the hours of an average business day and weekend into contiguous routing intervals called load set periods. The real-time, traffic-sensitive component of the dynamic routing method uses real-time routing paths, as described in Chapter 8, for possible completion of calls that overflow the engineered paths, as illustrated in Figure 1.22. The engineered paths are designed to provide the objective blocking performance. The real-time routing paths, which are also determined by the path-EFO model embedded in the central network capacity management design system, can be used only if the number of idle trunks in a link is greater than a specified number of trunks—the reservation level—before the connection is made. This prevents calls that normally use a link from being swamped by real-time-routed calls. Real-time routing for multiservice integrated networks could have each switch send status update messages to a virtual trunk status map (VTSM) every few seconds of the number of idle VTs in each virtual trunk sublink. In return, the VTSM periodically sends ordered routing table updates to be used by the originating switches to perform call setups until the next update is received. These routing tables are determined by the VTSM in real time using, for example, the trunk status map routing (TSMR) method described in Chapter 11. Separate routing tables are generated for each virtual network, and congestion control measures can be implemented by the VTSM real-time routing procedure, as described in Chapter 11.

Figure 10.1 illustrates the multiservice integrated network routing method. We have illustrated three virtual networks that are distinguished by their traffic characteristics, bandwidth requirements, and design performance objectives. These virtual networks include (1) 64-kbps voice circuit mode connections (2) 64-kbps data circuit mode connections, and (3) 384-kbps data circuit mode connections.

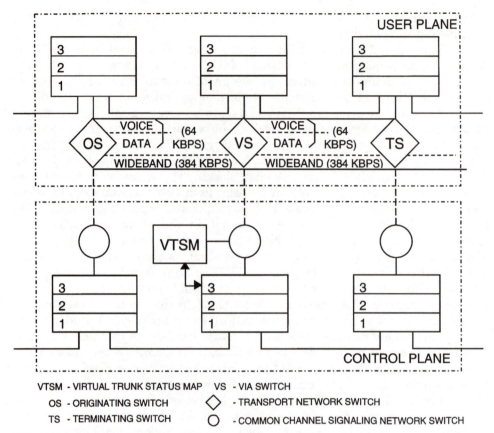

Figure 10.1   Multiservice integrated network routing

The concept is applied to both circuit mode connections and virtual circuit packet mode connections. We now discuss bandwidth allocation and routing procedures.

10.2.1   Bandwidth allocation and routing method

In the bandwidth allocation method, service demands are converted to elements of bandwidth, or virtual trunks (VTs). There are two levels of bandwidth allocation:

1. *Network bandwidth allocation:* When virtual connections are requested, a routing procedure is used to determine on which network path there exists sufficient bandwidth for the service; if no such path exists the connection is blocked.

2. *Link bandwidth allocation:* A maximum number of virtual trunk connections, and hence a maximum allowed bandwidth, is allowed for each virtual network on each link. A link bandwidth allocation procedure is also used to ensure that a minimum allotted bandwidth is provided for each virtual network.

In the network bandwidth allocation method, virtual connections are converted to VTs by the equivalence

$$VT_{ik}^h = \rho_i \times \frac{LBW_{ik}^h}{r_i} \qquad (10.1)$$

where LBW_{ik}^h = bandwidth allocated to virtual network i on link k in hour h

VT_{ik}^h = number of virtual trunks or maximum number of virtual connections for virtual network i on link k in hour h

r_i = average bandwidth per virtual connection (VC) for virtual network i

ρ_i = maximum allowed utilization of bandwidth allocated to virtual network i

The quantity LBW_{ik}^h is determined in network design as described below. The quantity r_i is known for each virtual network for purposes of network design. The quantity ρ_i is determined from load–service relationships, such as those for packet data or for packetized voice, and is used to ensure that packet-delay, bit-dropping, and packet-dropping objectives are met.

With the size VT_{ik}^h of each virtual trunk sublink determined, the integrated network dynamic routing method is used to set up switched virtual connections over these virtual trunk sublinks. This call setup procedure for calls on virtual connections in the integrated network is very similar to call setup on trunk connections in single-service voice networks and uses one- and two-link dynamic path routing as described above. A real-time routing procedure is used to adjust the routing patterns in real time according to network traffic conditions.

Routing functions include call setup, real-time routing, network bandwidth allocation, and link bandwidth allocation. These functions are now discussed.

10.2.2 Call setup

A CCS call setup procedure is used to establish virtual connections. Call setups for virtual trunk connections build on dynamic routing logic. Each virtual network has a dynamic routing table that is used to select an idle VT path as follows:

1. The originating switch translates the customer-specified information, including dialed digits and other class-of-service data, using a local or centralized database to determine the terminating switch, virtual network, and routing table data.

2. The originating switch selects the originating-switch-to-terminating-switch routing table from the originating switch memory based on the terminating switch and virtual network.

3. Each switch defines a virtual trunk sublink for each virtual network: Each virtual trunk sublink has a maximum number of VTs that has been initially determined in the network design but is adjusted in the link bandwidth allocation procedure described below.

4. VTs associated with these virtual trunk sublinks are treated in the same way as trunks are treated in the originating switch; that is, a resource counter in the originating switch keeps a count of idle VTs or total calls in progress as they are set up or disconnected for each virtual network on each link.

5. In CCS call setup, the originating switch uses two-link dynamic path selection for VT paths. For example, the originating switch first selects an idle VT on the first link of the first path choice by decrementing the idle-VT counter for that virtual network. Then, the via switch attempts to select an idle VT on the second link of the first path choice. If the second link VTs are all busy, the call is cranked back and the originating switch tries the next path choice in a similar manner, and so on. This procedure provides originating switch control of every call in setting up VT paths.

6. The real-time routing procedure, as discussed in Chapter 8, is used and switch-to-switch traffic data are collected for each virtual network. As an alternative, a centralized real-time routing procedure can be used that employs a VT status map (VTSM), as illustrated in Figure 10.1 and discussed in Chapter 11, to adjust routing patterns in real time according to network traffic conditions.

10.2.3   Network bandwidth allocation and real-time routing

The function performed continuously in network bandwidth allocation is the assignment of virtual trunks to calls with the use of the dynamic routing procedure described above. The real-time routing concept uses the real-time routing procedure described above and in Chapter 8, or it could have each switch send status update messages to a VT status map every few seconds that indicate the number of idle VTs in each virtual trunk sublink. In return, the VT status map periodically sends ordered routing sequences to be used by the originating switches to perform call setups until the next update is received in a few seconds. These routing sequences are determined by the VT status map in real time, using, for example, the trunk status map dynamic routing method described in Chapter 11. Separate routing sequences are generated for each virtual network. Congestion control measures can be implemented by the VT status map, for example, using a real-time routing procedure, also described in Chapter 11.

10.2.4   Link bandwidth allocation

The above network bandwidth allocation method allocates the network bandwidth to all services in a systematic, equitable fashion. We also use a link bandwidth management method to ensure that the link bandwidth used by each virtual network on each link does not exceed the capacity allocated to it unless there is unused capacity from other virtual networks. The link bandwidth allocation is initially determined in network design but is adjusted in the network control procedure described below.

The link bandwidth allocation procedure is used to control bandwidth utilization on links. This procedure groups VTs of a like nature—that is, for each virtual network—onto separate virtual trunk sublinks. As discussed above, there is a virtual trunk sublink defined for each virtual network on each link. Each virtual trunk sublink has a minimum allowed bandwidth limit. The minimum bandwidth limit is controlled through adjustment of bandwidth control parameters used by the originating switch, which in effect reserve a minimum number of VTs for the virtual network should they be required given the traffic level. This procedure allows for both dedicated and shared bandwidth for each virtual network, which is necessary for efficient use of link bandwidth. The minimum number of VTs is determined from the design link bandwidth LBW_{ik}^{h}, which is periodically updated through use of a method that is now explained.

The allowed bandwidth for each virtual network on each link is adjusted periodically. An average spare capacity on the link is computed based on the difference between the total link bandwidth available, TLBW_{k}, and the average link bandwidth in use, $\overline{\mathrm{TLBW}}_{k}^{h}$, in hour h:

$$\mathrm{SPAREBW}_{k}^{h} = \mathrm{TLBW}_{k} - \overline{\mathrm{TLBW}}_{k}^{h}$$

The spare capacity is then allocated to the virtual networks in proportion to their average bandwidth usage $\overline{\mathrm{LBW}}_{ik}^{h}$ on each link k in each hour h:

$$\mathrm{SPAREBW}_{ik}^{h} = \mathrm{SPAREBW}_{k}^{h} \frac{\overline{\mathrm{LBW}}_{ik}^{h}}{\sum_{j=1}^{N} \overline{\mathrm{LBW}}_{jk}^{h}}$$

where we have assumed there are N virtual networks. This additional bandwidth is added to the design link bandwidth LBW_{ik}^{h}, and the new minimum VTs and bandwidth control parameters associated with the virtual network are used to update the minimum allowed bandwidth on the virtual trunk sublink. A longer-term analysis (for example, a one-week analysis) is performed to determine if spare capacity exists for all hours over an average business day. If it does, the design link bandwidth LBW_{ik}^{h} for each virtual network is increased by an amount given by

$$\mathrm{SPAREBW}_{ik} = \min_{h}(\mathrm{SPAREBW}_{k}^{h}) \frac{\sum_{h}(\overline{\mathrm{LBW}}_{ik}^{h})}{\sum_{h}(\sum_{j=1}^{N} \overline{\mathrm{LBW}}_{jk}^{h})}$$

These new design link bandwidth values are used to change the minimum VTs and bandwidth control parameters.

10.3 Network Management and Design

The third principal change brought about by dynamic routing is in network management and design. Several network management and design systems provide centralized functions such as data collection, real-time traffic management, routing design, traffic forecasting, capacity design, switch planning, and transport network planning. These network and management design impacts for dynamic routing networks are discussed in Chapter 17. Embedded within the network design systems is a network design model such as the path-EFO model for preplanned dynamic routing network design, which simultaneously determines the capacity and routing for the entire dynamic routing network. Alternatively, for real-time dynamic routing network design, methods discussed in Chapter 13 can be employed. These include use of the path-EFO model described in this section but modified to employ a day-to-day traffic load variation model appropriate for application to real-time dynamic routing networks.

Network management functions are performed by these centralized network control systems. The objective of the data collection function is to collect load and performance data by virtual network. Load data include calls per second, bits per second, and simultaneous virtual connections for each virtual network; performance data include VT and virtual connection blocking by virtual network. The functions of capacity management are to (1) monitor the service being provided in the integrated network and (2) plan and schedule all capacity and routing changes that may be necessary to maintain objective network performance. Network service is monitored by direct measurement of network performance parameters, as provided by the data measurements specified above. When measured performance is below design objectives, the network design model is run to generate capacity and routing table changes to restore objective service. The routing table changes generated by the design are automatically implemented into the switches' routing databases through a routing assignment function. The capacity management function also provides a multiyear forecast of capacity requirements for the integrated network, which specifies link capacities within the integrated network. In order to forecast capacity requirements for the integrated network, switch-to-switch traffic data are gathered from the originating switches for each virtual network.

Inputs to the integrated network design consist of the originating-switch-to-terminating-switch traffic, blocking, and service quality objectives for each virtual network. The design methods determine the routing patterns and minimum-cost network for all virtual networks given these traffic inputs and the blocking and service quality constraints.

The path-EFO design procedure simultaneously determines the routing and link capacities between all switch pairs in the network. The objective is to minimize the global network cost while meeting the switch-to-switch blocking objective and service quality levels between all switch pairs in all load set periods. The steps of the design model are illustrated in Figure 7.1. The design model is an iterative procedure consisting of an initialization step, a routing design step, a capacity design step, and an update optimal link blockings step. Because there is a nonlinear relationship among the routing, capacity, and link-blocking variables in the network optimization process, we solve for these variables iteratively. In this way the solution of the very large nonlinear optimization problem is broken down into the solution of a large linear program (for the optimal routing), the solution of a large number of small nonlinear programs (for the optimal link blockings), and the solution of a large set of simultaneous nonlinear equations (for the optimal link bandwidth capacities).

Input parameters to the design model include link cost, switch-to-switch offered load, and maximum switch-to-switch blocking level. The initialization step estimates the optimal link blockings for each link in the network. Within the routing and capacity design steps, the current estimates of the optimal link blockings are held fixed. The routing design step determines the optimal routes for each design load set period by executing the three steps shown within the routing design step in Figure 7.1. The optimal routing is then provided to the capacity design step, which determines the number of VTs and modular bandwidth required on each link to meet the same optimal link-blocking objectives assumed in the routing design step. Once the links have been sized, the cost of the network is evaluated and compared with that of the last iteration. If the network cost is still decreasing, the update optimal link blockings step computes new estimates of the optimal link blockings. The new optimal link blocking objectives are fed to the routing design step, which again selects optimal routing, and so on.

Each of the virtual networks has the following loads as input to the design process:

Virtual connection traffic load (erlangs) By average business day and weekend hour h for each switch pair; mean, variance, and level of day-to-day variation are used.

Bandwidth per virtual connection (kbps) By single stationary average value for each virtual network; bandwidth allocated per VC = r_i, as defined in equation (10.1). The variance and day-to-day variation of the bandwidth per virtual connection are used in the selection of the bandwidth utilization parameters ρ_i.

The maximum allowed utilization ρ_i of the bandwidth allocated to virtual network i is selected for each virtual network such that all performance criteria are met (ρ_i is computed in the iterative design loop, described below). Our notation for the average virtual connection traffic loads is VC_{ij}^h, where $h = 1, H; j = 1, K;$ and $i = 1, N$ designate loads in hour h for switch pair j associated with virtual network i. $\mathrm{NNBW}_{ij}^h = \mathrm{VC}_{ij}^h \times r_i$ is used to denote the total bandwidth in hour h for switch pair j associated with traffic for virtual network i. The notation VT_{ik}^h is used to denote the virtual trunk requirement for virtual network i on link k in hour h. Virtual trunk requirements are determined in the design procedure.

Steps in the path-EFO model design procedure shown in Figure 7.1 are the same as described in Chapters 6 and 7, except for the following additions.

10.3.1 Routing design

This step is unchanged except that the spare VC capacity determined in the modularize link bandwidth function in the capacity design step, described below, is used to set the existing link capacity variables in the heuristic optimization method used in the linear programming step. This spare virtual connection capacity is also considered in the shortest-path selection procedure incorporated in the routing design step. These spare VC capacity variables in general are a function of the design hour h.

10.3.2   Capacity design

This step is modified to size the virtual trunk capacity of each link to the optimal link-blocking objective in each design hour (side hours as well as busy hour). This step therefore creates a variable VT_{ik}^h requirement for virtual network i on link k in hour h.

10.3.3   Modularize link bandwidth

This function incorporated within the capacity design step rounds the virtual link sizes on each link k so as to obtain a modular link bandwidth. The procedure first determines the total, nonmodular link bandwidth requirement, TLBW_k, that satisfies the "background" load for other virtual networks on the link plus the requirements for virtual network i currently being sized for. The rounding procedure subtracts, on each successive path-EFO model iteration, 0.6 DS1, 0.4 DS1, 0.2 DS1, and 0 DS1, where DS1 equals the modular bandwidth size of 1,544 kbps, from the total required link bandwidth (nonmodular), with the background load as a lower bound, and rounds the result up to the nearest DS1 multiple to obtain the modular TLBW_k. That is,

$$\mathrm{TLBW}_k = \left(\frac{\max[\mathrm{BGL}_{ik}^h, \max(\mathrm{BGL}_{ik}^h + \mathrm{LBW}_{ik}^h) - f \times 1{,}544]}{1{,}544}\right) \times 1{,}544$$

where f = 0.6 (first iteration), 0.4 (second iteration), 0.2 (third iteration), 0 (fourth and subsequent iterations)

BGL_{ik}^h = background load on link k in hour h from other virtual networks considered up to this point (an expression is given below for BGL_{ik}^h)

LBW_{ik}^h = bandwidth required for virtual network i on link k in hour h $(\mathrm{LBW}_{ik}^h = \mathrm{VT}_{ik}^h \times r_i/\rho_i)$

r_i = bandwidth per virtual connection for virtual network i

ρ_i = maximum allowed utilization of bandwidth allocated to virtual network i

The link bandwidth will thus monotonically increase to an optimal modular value. Spare bandwidth from the rounding procedure is then used to compute spare virtual connection flow capacity on each link k in each design hour h, as follows:

$$\mathrm{SPAREVC}_{ik}^h = (\mathrm{TLBW}_k - \mathrm{BGL}_{ik}^h) \times \rho_i/r_i$$

The spare virtual connection capacity is used to set the existing link capacity variables in the routing design step. The initial routing of virtual connection flows in the heuristic optimization method is the same routing used in the modularize link bandwidth step.

10.3.4 Update optimal link blockings and optimal bandwidth utilizations

The update optimal link blockings step is unchanged except that the SPAREVC capacity, as computed above, is ignored in this computation. That is, the nonmodular VT_{ik}^h values are used for purposes of determining the optimal link blockings. The update optimal bandwidth utilization step finds the ρ_i parameters such that the constraints on end-to-end delay, bit dropping, and packet dropping are satisfied and the network cost is minimized. Various methods can be adapted for this step, where the variance and day-to-day variation of the virtual connection bandwidth requirements are considered in determining the ρ_i parameters.

We now describe how the above steps are used to size the multiservice integrated network for all virtual networks:

1. Order the virtual connection demands by decreasing total network bandwidth requirement in the network busy hour, that is, from largest to smallest of

$$\max_h \sum_j \text{NNBW}_{ij}^h = \max_h \sum_j \text{VC}_{ij}^h \times r_i / \rho_i.$$

Call these virtual networks $i_1, i_2, \ldots, i_N$.

2. For each virtual network i_n, $n = 1, N$, run the path-EFO model illustrated in Figure 7.1 to determine the virtual connection flow on each link in hour h, $VC_{i_n k}^h$. Use the modular bandwidth engineering procedure described above to modularize the total link bandwidth with

$$\text{BGL}_{i_n k}^h = \sum_{m=1}^{n-1} \text{VT}_{i_m k}^h \times r_{i_m} / \rho_{i_m}$$

Note that the summation is defined to be zero when $n = 1$. $VT_{i_n k}^h$ is sized to the optimal link blocking objective in all hours, as described above. The parameters ρ_i are updated in the update link parameters step of the path-EFO model, based on economic and service criteria. The sizing procedure for the VT requirements considers mean, variance, and day-to-day variation of each virtual connection demand for each virtual network, as in the path-EFO model for single (voice) service network design.

3. Compute spare virtual connection flows on each link k for the next virtual network i_{n+1} to be designed:

$$\text{SPAREVC}_{i_{n+1}k}^h = (\text{TLBW}_k - \text{BGL}_{i_n k}^h - \text{LBW}_{i_n k}^h) \times \rho_{i_{n+1}} / r_{i_{n+1}}$$

Use these spare virtual connection flows to initialize the existing link capacity variables in the routing design step of the path-EFO model, and return to step 2 for virtual network i_{n+1}.

These steps yield (1) the number of VTs required for each virtual network, by hour, for each link; (2) the design link bandwidth requirement for each virtual trunk sublink associated with each virtual network, by hour, for each link; and (3) the routing, by hour, for each virtual network.

Once again, for real-time dynamic routing network design, methods discussed in Chapter 13 can be employed. These include use of the path-EFO model described in this section but modified to employ a day-to-day traffic load variation model appropriate for application to real-time dynamic routing networks. Other models are also presented in Chapter 13 that build on the methods described in this section for multiservice integrated network design.

10.4   Modeling Results

We discuss design results that illustrate applications to multiservice ISDN networks and multiservice wideband packet networks. These examples illustrate how integrated network design principles lead to cost-effective networks and services. We consider here both circuit and packet networks. We have designed dynamic routing circuit-switched networks for data services of varying bandwidth requirements and circuit-switched networks that carry voice, data, and wideband services. In Chapter 15 we design packet-switched networks based on wideband packet technology that carry both voice and wideband data up to 1.5 megabits per second. The results we discuss here demonstrate the advantages that are found in integrated network design. Some of the results are found in terms of network costs, but it is clear that one could design equal-cost integrated networks that carry more traffic and are more flexible than separate networks; hence, the integrated networks are more efficient. Integration of network services allows the network to more effectively carry unexpected load that has not been forecast.

We illustrate dynamic routing design for a multiservice integrated ISDN switched digital services network. This network is designed to handle end-to-end digital needs, which include 64-kbps, 384-kbps, and 1,536-kbps switched digital data services. The integrated network design is compared with the 64-kbps network design, in combination with a minimum-cost separate network designed to carry the 384-kbps traffic. The 384 traffic (erlang) loads are approximately 10 percent of the 64 traffic (erlang) loads. The results in Table 10.1 show that the

TABLE 10.1   Integrated Network Design Results

Network	Cost (millions)
64-kbps network	$7.28
384-kbps network	0.63
integrated network	7.28

integrated design requires no additional capacity compared with the separate 64-kbps design, therefore saving all of the cost to build a separate, minimally connected 384-kbps network.

Network simulations show that the integrated network meets the switch-to-switch blocking probability objective for all services and that the integrated networks are robust to load variations. We also study in Chapter 15 integrated network designs that carry voice and high-bandwidth data in addition to low-bandwidth data. These network designs confirm that the advantages of integrated design are not limited to the lower-bit-rate services but are also advantageous for broadband networks.

Packet transport is a candidate mode for future integrated network design. In Chapter 15, we study both wideband networks—for example, 1.5-megabits-per-second transport rate—and broadband networks—for example, services of 45 megabits per second on fiberoptic lines. Furthermore, broadband-ISDN is evolving in the direction of packet transport with asynchronous transport mode, or ATM. Designs of integrated broadband packet networks to carry voice and low-bit-rate data, as discussed in Chapter 15, again show that integrated design provides advantages over separate networks.

10.5 Conclusions

We have presented techniques for dynamic routing and network design within multiservice integrated networks and have illustrated preplanned dynamic routing, real-time dynamic routing, call setup, and bandwidth allocation methods applicable to such networks. As such, these methods provide the advantages of dynamic routing in reduced network investment, improved customer service, increased network flexibility, and centralized and automated network management and design support. These dynamic routing design principles for multiservice integrated network bandwidth allocation, dynamic routing call setup, network management, and integrated network design apply in general to all dynamic routing networks. In Chapters 11–13 we discuss the application of these principles to multiservice integrated networks with real-time dynamic routing. These methods also provide a framework for integrated network design in future broadband integrated services networks, as discussed in Chapter 15.

Centralized Real-Time
Dynamic Routing Networks

11.1 Introduction

As discussed in Chapters 3–10, preplanned dynamic routing incorporates time-varying routing and associated network design procedures to minimize network cost. Trunking requirements and routing patterns that provide a near-optimum network design to meet the network loads are determined in the network design procedures incorporating, for example, the path-erlang flow optimization (path-EFO) model. The network design model sizes the network to accommodate all expected load patterns including day-to-day load variations [Wil58, Wil71, HiN76]. Sizing the network for day-to-day variations guarantees that some capacity stands idle at least some of the time. And if planned routing patterns are totally preplanned, no advantage can be taken of temporarily idle network capacity to complete calls that might otherwise be blocked. Real-time dynamic routing seeks out and utilizes this idle network capacity by making real-time routing table updates. As discussed in Chapter 1, real-time routing implies that routing table updates are made at a frequency of one call holding time or faster. In the limiting case, routing table updates could be made on a call-by-call basis. The real-time routing method implemented in the DNHR network, as discussed in Chapter 8, allows the preplanned sequence of one- and two-link paths that make up the engineered route to be augmented by a sequence of additional two-link paths called the real-time paths. The real-time paths are searched sequentially to determine whether the number of idle circuits on each link exceeds a reservation threshold. If a real-time path is found that has idle capacity exceeding the reservation threshold on both links, then the call is completed on that path.

In Chapter 1 we identified several real-time routing methods, which include real-time event-dependent and real-time state-dependent routing. Within real-time state-dependent routing, we have centralized methods that provide routing updates on the order of seconds, or perhaps minutes, and distributed methods that provide routing updates as frequently as call by call. Examples of real-time event-dependent routing are dynamic alternate routing (DAR), implemented within the British Telecom UK network, and state- and time-dependent routing (STR), implemented within the NTT Japan network. An example of centralized real-time state-dependent routing is dynamically controlled routing (DCR), implemented within the Stentor Canada network, Bell Canada network, MCI USA network, and Sprint USA network. Examples of distributed real-time state-dependent routing are worldwide international network (WIN) routing, implemented within a consortium of international service providers; real-time network routing (RTNR), implemented within the AT&T USA network; and real-time internetwork routing (RINR), implemented within the AT&T international network.

In this chapter we study centralized real-time state-dependent routing methods, and in Chapter 12 we study distributed real-time state-dependent routing methods and event-dependent routing methods. In Chapter 13 we study network management and design for all these real-time dynamic routing methods. In the case of the preplanned dynamic routing methods discussed in Chapters 3–10, recall that we studied the DHNR method (otherwise known as preplanned dynamic two-link sequential path routing) in the greatest depth and used DNHR as a representative example of the preplanned dynamic routing methods. Similarly, in this chapter and in Chapter 12, we study in detail representative examples of each of the three real-time routing methods. The methods and principles applied to the design and evaluation of these representative examples then also apply to the many possible variations of each method. In applying these same methods and principles, the reader will be able to explore in depth any of the possible variations of these routing methods.

Hence, in this chapter we do an in-depth study of a representative example of a centralized real-time state-dependent routing method. This method is based on a centralized trunk status map concept [Ash85] in which the busy/idle status of each trunk in the network is known in real time. The method routes traffic to the least congested part of the network by assigning calls to the least loaded path in a set of path choices. We call the real-time state-dependent procedure least-loaded routing, or trunk status map routing (TSMR). We study the performance of

TSMR under realistic network load patterns, which deviate from forecasted values and include expected daily load variations, as well as errors in the forecast loads. Such load conditions are conditions under which the real-time TSMR routing patterns and associated link sizes provided by the network design models must maintain objective network performance.

Several methods for implementing TSMR are investigated, but the following procedure established itself as the most attractive. The first path determined by the path-EFO model is used if a circuit is available. If the first path is busy, the second path is selected from the list of other engineered and real-time paths determined by the path-EFO model on the basis of having the greatest number of idle circuits. A variation on this least-loaded-routing procedure is also tested, in which the second path is selected from a larger list consisting of all candidate paths considered by the path-EFO model in making its final selection of engineered and real-time paths.

The results show a significant improvement in network blocking performance with this TSMR method in comparison with other real-time routing procedures. In cases where short-term capacity design action is required in the network employing TSMR, the results show that the path-EFO model is able to recompute routing patterns and determine needed trunk additions to restore the objective blocking performance. Hence, the path-EFO model is able to design the TSMR network when least-loaded routing is employed.

11.2 Trunk Status Map Routing Methods

In Chapter 8 we showed that the real-time routing method that involves augmentation to a sequential routing pattern provides substantial improvement in the network performance. Figure 8.13 illustrates the improvement in blocking performance with real-time routing as a function of reserve capacity. We see that the effectiveness of real-time routing increases as reserve capacity increases. Reserve capacity is the amount of actual network capacity in excess of the capacity required by the current demand. It arises because of various administrative procedures used in provisioning the network, as discussed in Chapter 9, and its presence substantially improves the performance of real-time routing, even during hours of peak network loads. This real-time routing method provides superior performance to that of hierarchical routing, even under high-day loads. For example, the average network blockings for the morning, afternoon, and evening busy hours on a high day are shown in Table 11.1. The morning busy hour, which is also the network busy hour, displays the largest difference between the

TABLE 11.1 Average Network Blocking

Routing method	Morning	Afternoon	Evening
sequential plus real-time	0.0116	0.0073	0.0013
hierarchical	0.0164	0.0110	0.0043

two networks. This analysis also shows that (1) the real-time routing network has a larger fraction of its switch pairs receiving better than objective blocking than does the hierarchical network and (2) switch pairs in the real-time routing network usually receive lower maximum blocking than they do in the hierarchical network.

The real-time routing method is a decentralized method, as opposed to centralized techniques considered earlier. One method of centralized real-time routing modeled in [Kri82] satisfies the Kuhn-Tucker conditions for optimality, which are based on analogous results for data networks derived in [Gal77]. Because all network loads are needed for this computation, a large amount of centralized processing must be performed. This model also shows that additional design efficiencies made possible by centralized real-time routing can achieve an additional 1 to 2 percent of network savings. In DCR the selection of candidate paths at each switch is recalculated every 10 seconds. The path selection is done by a central routing processor, based on the busy/idle status of all trunks in the network. The results show that DCR provides more uniform and better service characteristics than fixed hierarchical routing, especially during periods of network overload.

Hence, the centralized approach to real-time routing seems to be worthy of additional investigation. The trunk status map routing method involves setting up a network database that tracks the busy/idle status of each trunk and also uses this status information to select routes for alternate-routed calls. In the least-loaded-routing concept investigated in this chapter, the first path, as determined by the path-EFO model, is used if there is a circuit available on that path; if there is no circuit available on the first path, then the second path is selected according to the least-loaded criterion of the path having the greatest number of idle circuits.

11.2.1   Trunk status map routing architecture

A view of the dynamic routing network with the trunk status map is shown in Figure 11.1. Each switch in the dynamic routing network periodically sends a special CCS message once every T seconds to be switched through the CCS network to the trunk status map. These CCS messages indicate the number of idle trunks in each link connected to the switch and are sent only for links whose status has changed. The trunk status map updates the status of the links identified by the CCS message and is stored in the trunk status memory.

The switch maintains a TSMR routing table for each destination, which is made up of two functional parts. The first part consists of a single time-varying path, called the first-choice path, which is updated by the trunk status map once each load set period and is made to correspond to the path-EFO-model-selected first-choice path. A skip flag is associated with the first-choice path, which allows a two-link first-choice path to be skipped if it is busy. The second part of the routing table consists of the remaining paths, called least-loaded routing paths. These paths are updated every T seconds according to a least-loaded criterion applied to the current network status. For each route sequence that needs to be

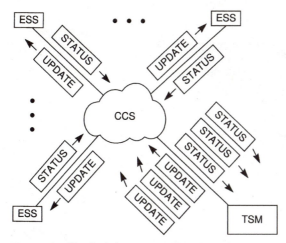

Figure 11.1 Trunk status map routing concept

updated, the trunk status map determines the paths and skip flags in the current route sequence that need to be changed and transmits these changes to the switch.

When a call to be routed by TSMR arrives at the originating switch, the switch translates the dialed digits to obtain the terminating switch. As illustrated in Figure 11.2, the TSMR procedure requires that the first-choice path for the load set period, as determined by the path-EFO model, be used if there is an idle circuit available on that path. The first-choice path is not necessarily the direct path. If there is no circuit available on the first-choice path, then the second-choice path is selected, according to the least-loaded criterion, as the path having the greatest number of idle circuits, which is path A–D–B in Figure 11.2. Hence, there is a preplanned time-varying component of the routing table, associated with first-choice paths, and a real-time state-dependent dynamic component of the routing table, associated with least-loaded routing paths.

The call processing required to implement the TSMR method is shared between the switch and the trunk status map in the following manner: The trunk status map controls the first-choice path in the switch routing table so that it corresponds to the first-choice path required by the path-EFO model. There are two cases: The first-choice path is either the direct path (case 1 in Figure 11.2) or a two-link path (case 2 in Figure 11.2). If the first path assigned by the path-EFO model for a particular load set period is the direct path, then the first-choice (direct) path is transmitted to the switch at the beginning of the load set period and is not marked "skip" by the trunk status map for the entire load set period. If there is a trunk available on that direct path, the switch completes the call in the normal manner. If the first-choice path assigned by the path-EFO model for a particular load set period is a two-link path, the trunk status map sends a message to the switch at the beginning of the load set period that causes the first-choice path to be set to the path-EFO-model-required first-choice (two-link) path for that load set period. The trunk status map also periodically determines whether this first-choice

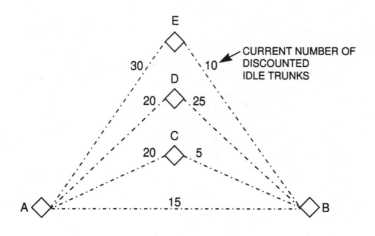

CASE 1: FIRST-CHOICE PATH IS THE DIRECT PATH
CASE 2: FIRST-CHOICE PATH IS A TWO-LINK PATH

Figure 11.2 Example calculation of TSMR

(two-link) path is busy. If it is busy, the trunk status map causes this path to be marked "skip" in the switch for a T-second interval by sending an appropriate message to the switch. Use of this skip procedure avoids possible crankback messages if the switch attempts call setups over the busy path.

The above description pertains to the preplanned time-varying component of the TSMR method, associated with the manipulation of the first-choice path. Recall that the fully dynamic part of the TSMR method requires that the second-choice path (first least-loaded routing path) be set to the least loaded path (path A–D–B in Figure 11.2). The trunk status map determines the least loaded path for each switch pair from a list that averages about 20 candidate paths. This list represents the union of all engineered and real-time paths selected by the path-EFO model over all load set periods. In an 84-switch network design, this list ranges from one path to about 40 paths, with an average of about 20 paths. In comparison with manipulating all the path-EFO model paths for each load set period, such a merged list of paths simplifies routing administration, reduces trunk status map storage requirements, and improves network performance. The merged list of candidate paths is stored in the trunk status map and is distinct from the actual path sequence stored in the switch, which has a smaller maximum number of paths. If this least loaded path differs from the second-choice path currently in the switch, the trunk status map sends an appropriate message to the switch that changes the contents of the least-loaded routing paths to reflect the new least loaded path and changes the order of the other paths in the route sequence (see Figure 11.2). If a second link of a two-link path in the route sequence becomes busy over the T-second update interval and a call attempt is made on that path, the switch encounters a crankback message and alternate-routes the call to subsequent paths in the path sequence.

With this implementation of TSMR, the switch does not require any time-varying routing capability of its own. All such dynamic routing capabilities are controlled by the trunk status map, which changes the time-varying first-choice path every load set period and also controls the least-loaded routing paths in a dynamic manner. The switch is, however, the only place where individual trunks are selected and assigned to a particular call. The switch sets up calls over path choices using the two-link routing procedure with crankback. Crankbacks, however, are reduced by the use of TSMR.

11.2.2 Trunk status map processing logic

On a regular schedule over the T-second routing update interval, the trunk status map receives the link status messages, maintains the status of all links, and overwrites the current link status in its trunk status memory with the new information. Recall that the switch has a sequence of path choices to each terminating switch, and the trunk status map would control the contents of the routing sequences in the switch. A flowchart of the TSMR logic is given in Figure 11.3A.

The trunk status map first retrieves the number of idle trunks on each link, for each path choice, from the trunk status memory (step 2 in Figure 11.3A). In the simulations we find that discounting the number of idle circuits by a small fraction of the link size protects large links from being selected as via-path candidates by a disproportionately large number of switch pairs when these large links have temporarily idle trunks. Large links tend to have short periods with a relatively large number of idle trunks, and overselection of those links for via traffic during these short periods can be detrimental to the direct traffic routed on the large links. The simulation studies indicate that discounting the number of idle trunks by $N/36$ trunks, where N is the link size, works reasonably well in protecting direct traffic on the large links. (Links with fewer than 36 trunks are not affected.)

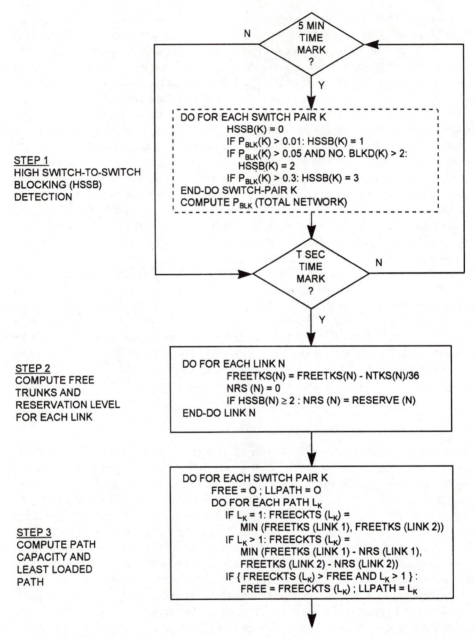

STEP 1
HIGH SWITCH-TO-SWITCH
BLOCKING (HSSB)
DETECTION

STEP 2
COMPUTE FREE
TRUNKS AND
RESERVATION LEVEL
FOR EACH LINK

STEP 3
COMPUTE PATH
CAPACITY AND
LEAST LOADED
PATH

Figure 11.3A   TSMR logic

Using the discounted idle trunk values, the trunk status map then determines (step 3 in Figure 11.3A) the number of idle circuits on each path choice and selects the path having the greatest number of idle circuits (i.e., the least loaded path). This discounting procedure is applied only in the trunk status map path selection logic and is not applied in the path selection logic of the switch.

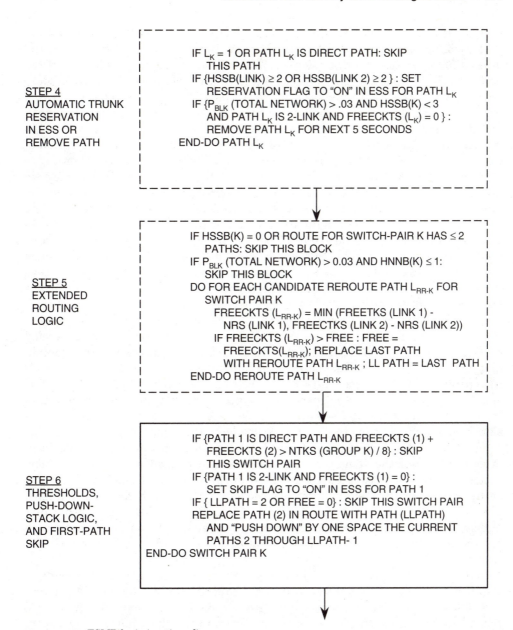

Figure 11.3B   TSMR logic (continued)

For each switch pair, the trunk status map applies a thresholding method to determine whether a routing update is needed (step 6 in Figure 11.3B). This thresholding method is applied only for the typical case in which the direct path is the first choice and a two-link path is the second choice. If the total number of idle circuits on these two choices is greater than a threshold number that is sufficient to permit completion of all calls likely to arrive over the T-second update interval, then there is no need to recompute the least loaded path because the calls can be routed with a low probability that any crankbacks will be generated. A threshold that works reasonably well is $N/8$, where N is the number of trunks on the direct link to the terminating switch. This logic reduces the number of switch pairs requiring updated least-loaded routing paths over each T-second update interval, but at the same time it concentrates the routing updates on those switch pairs for which the update is most beneficial. If the first path is a two-link path and if the path has zero free circuits, then the trunk status map sets the skip flag to "on" in the switch. The effect of this action is to avoid the generation of crankback messages on the first-choice path over the next T-second interval.

If the new least loaded path is the same as the old path or if there are no idle circuits on any candidate path, then a routing update is not sent to the switch because there would be no advantage in doing so (step 6 in Figure 11.3B). Simulations of this logic in combination with the idle circuit thresholds described above predict that only about 10 to 15 percent of switch pairs need routing table updates returned to the switch over a T-second update interval in the busy hour. This trunk status map logic would therefore tend to minimize the use of CCS and switch resources. When the least loaded path does change, then a message is sent to the switch, changing the contents of the routing sequence in the switch. As an example, consider case 2 in Figure 11.2. The current route to switch B has a maximum of four path choices, of which three are least-loaded routing path choices—A–C–B, A–D–B, and A–B—in addition to the two-link, first-choice path A–E–B. The trunk status map determines that the second-least-loaded routing path choice, A–D–B, has become the least loaded path. At this point the trunk status map sends least-loaded routing path choices 1 and 2 to the switch in the order A–D–B and A–C–D. Path choice A–B is not sent to the switch because it remains the third-least-loaded routing path choice. We call this a "push-down" logic, because the new least loaded path replaces the old least loaded path, and the remaining path choices are pushed down one slot. The switch replaces its current least-loaded routing path choices 1 and 2 with the new paths and leaves the current third-least-loaded routing path choice unchanged. The switch also translates the via switch identities into its own internal link numbers. On average, the trunk status map transmits about four entries in the route sequence list with each routing update, according to the simulation results.

To illustrate the magnitude of the trunk status map processing load, consider a TSMR network consisting of 200 switches, 1 million trunks, 20,000 switch pairs, and 20,000 links. For this analysis we assume T equals 5 seconds. About 600,000 erlangs of traffic are offered to such a network in the busy hour. According to simulation results, approximately half of the 20,000 links change state over each five-second status update interval in the busy hour. Therefore, about 10,000 link status messages are sent to the trunk status map every five seconds, if we assume that status messages are sent from only one end of a link. After applying the idle capacity thresholds described earlier and also applying the logic to see whether the least loaded path has changed, the number of switch pairs needing routing table updates is about 2,000–3,000 (requiring 4,000–6,000 switch updates) in each five-second interval.

Each switch pair has an average of approximately 20 candidate paths. Hence, the routing database memory has approximately 400K bytes of active storage, on the assumption that we store via switches at one byte each. Additional trunk status map storage is required to store the following items: (1) the first-choice paths for each switch pair by load set period, (2) the maximum number of paths for each switch pair, (3) the number of trunks in each link, (4) the high switch-to-switch blocking status of each switch pair, and (5) the reservation level of each link.

11.2.3 TSMR congestion control

TSMR routing table updates implement automatic congestion control logic similar to real-time traffic management controls implemented in the 4ESS switch [Mum76] to include selective dynamic overload control (SDOC), selective trunk reservation (STR), and automatic reroutes. The TSMR congestion control logic for trunk reservation, called automatic trunk reservation, and automatic reroutes, called extended routing logic, are shown in the dotted blocks in Figures 11.3A and 11.3B. The automatic trunk reservation logic begins with step 1 in Figure 11.3A, in which high switch-to-switch-blocking (abbreviated HSSB in Figures 11.3A and 11.3B) switch pairs are identified. There are three levels of severity defined for high switch-to-switch-blocking switch pairs, which correspond to 1 percent, 5 percent, and 30 percent switch-pair blocking levels. Also, the total network blocking is computed in this step. In step 2, if the high switch-to-switch blocking level of a switch pair is 2 or 3, trunk reservation is imposed on the direct link associated with the switch pair. In step 3, trunk reservation is imposed by having the trunk status map subtract the number of reserved trunks from the free trunks on each link if the switch pair associated with the link has a high switch-to-switch blocking level of 2 or 3. The reservation level used is approximately 3 percent of the number of trunks in the link.

In step 4, reservation is "turned on" in the switch for paths that include one or more links whose associated switch pairs have a high switch-to-switch blocking level of 2 or 3. Also, if the total network blocking exceeds 5 percent, paths with zero free trunks are canceled for the next T seconds, except for pairs that have a high switch-to-switch blocking level equal to 3.

The extended routing logic, step 5, is performed for all switch pairs with a high switch-to-switch blocking level greater than zero. Also, if the total network blocking exceeds 3 percent, the high switch-to-switch blocking level for the switch pairs must be 2 or greater. The extended routing logic searches for a least loaded path among an expanded list of candidate reroute paths. If such a path is found it replaces the last path of the route in question. This least-loaded reroute path is then moved up to the second path in the route in the push-down stack logic in step 6. The candidate reroute paths are additional paths identified by having the path generator in the path-EFO model output all acceptable paths; the candidate reroute paths are those paths that are not already candidates.

Dynamic overload control is implemented in the trunk status map by having the switches send congested switch signals to the trunk status map. The

TABLE 11.2   Trunk Status Map Dynamic Overload Control

Congestion signal	Trunk status map response
CSO	Clears a previously transmitted congested switch message.
CS1	Removes the congested switch as a via switch for all switches.
CS2	Removes the congested switch as a via switch for all switches. Cancels a percentage of the direct traffic to the congested switch from all switches that have switch-pair blocking above a threshold (i.e., high switch-to-switch blocking traffic).
CS3	Cancels all traffic to the congested switch.
CS4	Removes the failed link from all routing sequences sent to all switches.

trunk status map responds by implementing routing restrictions, as shown in Table 11.2. Note that a provision is made to respond to a CS4 message, which indicates a total failure of a link, as indicated to the switch by a carrier group alarm generated by the transmission system.

It may be necessary to provide a duplicate trunk status map to ensure high reliability. However, the particular implementation discussed above provides a basic routing capability throughout the network, even with a total failure of the central trunk status map or any of the links to it. This is because the switch can use the most recent routing sequence obtained from the trunk status map and could therefore route calls in a decentralized mode. Also, in the event that the switch becomes overloaded, as signified perhaps by the delay to process a particular call, the switch can revert to decentralized routing again and avoid the real-time load needed to process trunk status map messages, while using the most recent routing sequence obtained from the trunk status map. Similarly, if the trunk status map senses it is becoming overloaded, it can signal the connecting switches to curtail routing requests for a short period of time. This control mechanism is similar to dynamic overload control (DOC) used by real-time traffic management.

11.2.4   Real-time TSMR method

In order to implement a real-time TSMR method, illustrated in Figure 11.4, which effectively achieves an update interval T of zero seconds, the call processing required could be shared between the switch and the trunk status map in the following manner: If the first path assigned by the path-EFO model is a one-link (direct) path, and if there is a trunk available on that path, the switch completes the call in the normal manner without use of the trunk status map. If either of these conditions is not met, the central trunk status map is called upon to assist with the routing choices by performing the following steps (refer to Figure 11.4):

1. Receives the terminating switch identification from the originating switch.

2. Retrieves the appropriate routing pattern from its memory based on the switch pair and the current time period.

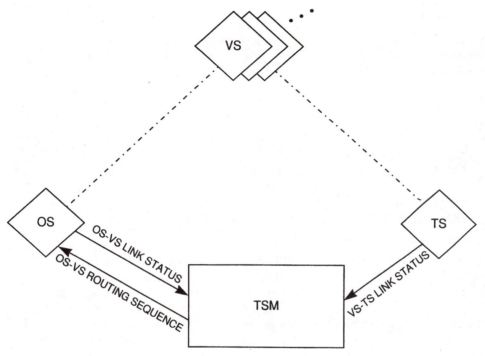

Figure 11.4 Real-time trunk status map routing concept

3. Requests from the originating and terminating switches the status of each link in the routing pattern.

4. Determines the least loaded path in the routing pattern through use of the trunk status obtained from the originating and terminating switches; it also determines the second and higher-numbered least loaded paths in the route.

5. Returns the sequence of least loaded paths to the originating switch (if the first path-EFO-model-selected path is a two-link path, that path is placed first if it has one or more idle circuits).

The originating switch then replaces its routing contents with the new sequence corresponding to the least loaded paths and sets up the call using normal dynamic routing processing. Crankback is eliminated under normal conditions because the sequence of paths starts with paths having the greatest number of idle circuits.

Simplifications of this real-time TSMR approach are possible. For example, path sequences might be determined from status inquiries, not on a call-by-call basis but less often, say once every other call or every third call for each switch pair. A maximum update interval of, say, 30 seconds could also be used. Such an approach might produce results that are close to the limiting performance demonstrated by the call-by-call dynamic routing method but allow reduced processing by the trunk status map.

11.3   Modeling Results

First, a call-by-call simulation model is used to compare the performance of the following three basic routing strategies and to verify that they meet design performance objectives:

1. *Engineered routing (ENR)*—Sequential routing (DNHR) only over the engineered paths, with no real-time paths.

2. *DNHR*—Preplanned sequential routing first over the engineered paths and then over the real-time paths, as described in Chapter 1 and Chapters 6–10.

3. *TSMR*—Choose the least loaded path from the set of all engineered paths and real-time paths (except when the first-choice path determined by the path-EFO model is available, in which case the first-choice path is used). The real-time TSMR method illustrated in Figure 11.4 is used unless otherwise noted.

In the simulation, the route choices are changed for each time period according to the path-EFO model routing design. Figure 11.5 shows the 25-switch DNHR network implemented in 1986 that is used for the simulation studies. The 25-switch model is designed by the path-EFO model for 16 hours throughout the day (from 8 A.M. through 11 P.M.). Figure 11.6 shows the total network load for the 25-switch model. The behavior of the network is investigated over a typical two-week period consisting of 10 average business days. For this purpose, the network is designed for day-to-day load variations by using a path-EFO model enhancement described in Chapter 13. Low daily load variations are applied to each switch-pair load in the network simulation. In addition, a typical variation

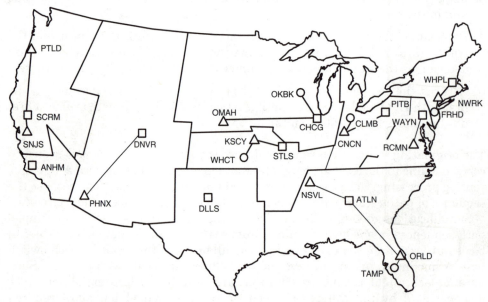

Figure 11.5   25-switch model

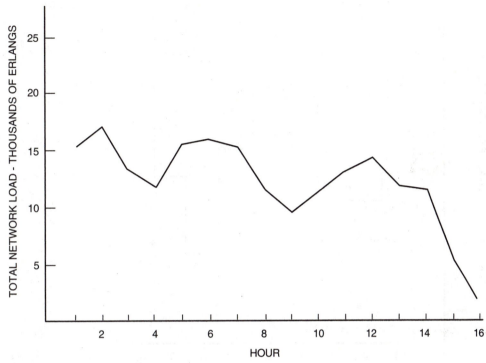

Figure 11.6   Total network load vs. hour (25-switch model)

of the total network load is superimposed on the low daily variations according to expected load patterns over an average business week. The daily total network load compared to the 10-day average total network load is shown in Figure 11.7.

The first set of results for the 25-switch network is obtained through simulation of the network designed for the average loads without day-to-day variations. All three routing methods—ENR, DNHR, and TSMR—are simulated. The results are displayed in Figures 11.8 and 11.9.

Figure 11.8 shows the 10-day average network blocking of the three routing methods as a function of the hour of the day. As expected, TSMR provides the best service and ENR the worst.

For each hour, Figure 11.9 shows the 10-day average switch-pair blocking for the switch pair in the network with the highest average blocking during that hour. ENR, DNHR, and TSMR usually provide better than objective performance, but clearly some switch-pair blockings exceed the 1 percent objective, especially in the late evening hour. This perhaps is indicative of significant changes in calling volumes during the deep discount period, which begins at 11 P.M. Almost all switch-pair blockings are within the threshold blocking level when sampling error is considered. Because of sampling error, the measured blocking occasionally exceeds the blocking probability objectives, even when the link is perfectly designed [Nea80]. Figure 11.10 shows the theoretical 90th percentile of the 10-day average switch-pair blocking as a function of load when sampling error is

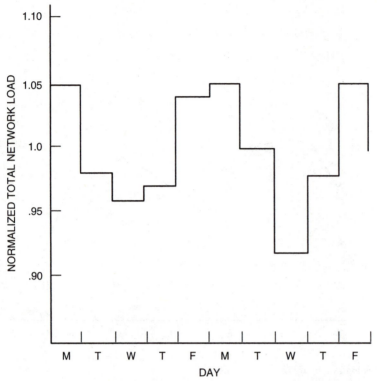

Figure 11.7 Daily total network load by day for 10-day simulations (normalized to 10-day average)

considered. The 90th percentile is calculated for a link size for 1 percent blocking on the assumption of no day-to-day variations.

Here we evaluate the binomial distribution for the 90th percentile confidence interval. Suppose that for a traffic load of A erlangs in which calls arrive over the designated time period of stationary traffic behavior, there are on average m blocked calls out of n attempts. This means that there is an average observed blocking probability of

$$p1 = m/n$$

where, for example, $p1 = .01$ for a 1 percent average blocking probability. Now, we want to find the value of the 90th percentile blocking probability p such that

$$E(n,m,p) = \sum_{r=m}^{n} C_r^n p^r q^{n-r} \geq .90$$

where

$$C_r^n = \frac{n!}{(n-r)!r!}$$

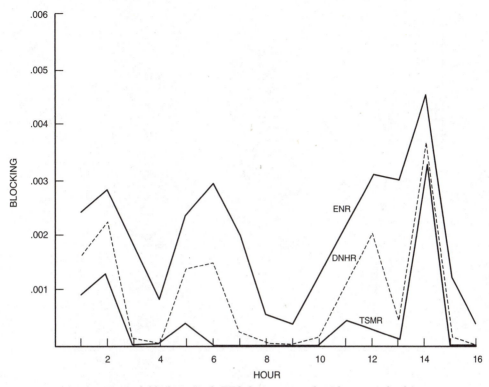

Figure 11.8 Average network blocking (path-EFO design network with average loads)

is the binomial coefficient, and $q = 1 - p$. Then the value p represents the 90th percentile blocking probability confidence interval. That is, there is a 90 percent chance that the observed blocking will be less than or equal to the value p. Methods given in [Wei63] are used to numerically evaluate the above expressions.

Also shown in Figure 11.10 is the 90th percentile of measured switch-pair blockings as determined from the simulation for each of the three routing methods. We find that the 90th percentile of the 10-day average switch-to-switch blockings is well within the 90th percentile blocking probability threshold for all three routing methods. With respect to this criterion, therefore, the network is performing somewhat better than what we would expect for a perfectly designed network. The indication is that the path-EFO model provides designs that meet the blocking objectives for all three routing methods.

DNHR and TSMR provide somewhat better performance than ENR for the design network. The reason that TSMR performs better than DNHR is that it takes greater advantage of real-time network status to distribute calls more intelligently in the network. This result is consistent with the conclusions of earlier studies [Bul75] that investigated analogous routing methods in a switching network. There is a good qualitative rationale for why least-loaded routing achieves lower blocking levels. A routing policy that routes calls through the least congested portion of the network gives a smoothness to the distribution

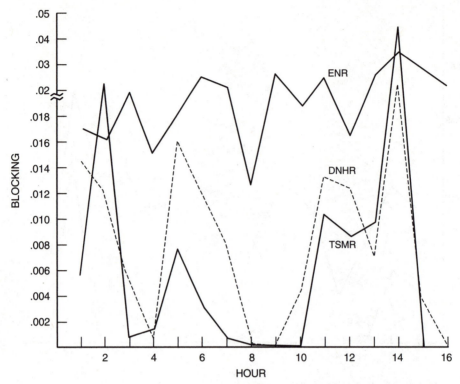

Figure 11.9 Maximum switch-pair blocking (path-EFO design network with average loads)

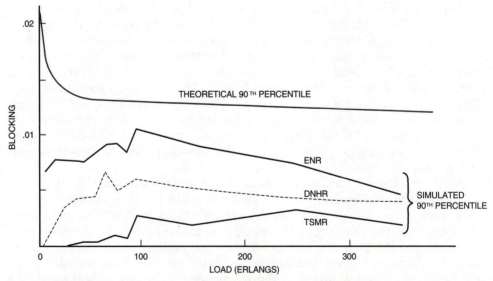

Figure 11.10 90th percentile switch-pair blocking vs. load (path-EFO design network with average loads)

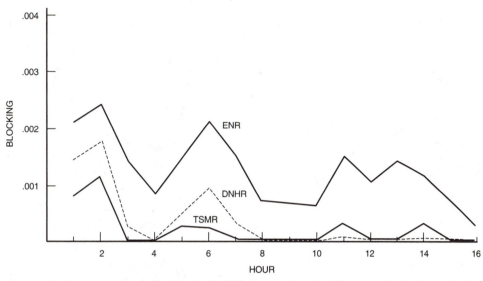

Figure 11.11   Average network blocking (path-EFO design network with day-to-day load variation)

of calls throughout the network. This means that there are fewer regions of localized blocking in the network. The fixed-order search policy of ENR and DNHR is more likely to give rise to localized blocking conditions and a peaked distribution of calls than the TSMR policy. The TSMR policy tends to ensure that some idle capacity is retained on a maximum number of switch-to-switch routes. This reduces the worst-case blocking level as observed in the results.

Figure 11.11 shows the 10-day average network blocking for the three routing methods as a function of hour for a network design including additional capacity for day-to-day load variation. The design method used is described in Chapter 13. The simulation load includes variation in the network average load (Figure 11.7) and low day-to-day variation on all switch-pair loads. Once again, TSMR provides the best service and ENR the worst; however, all three routing methods provide better service than for the case of the network designed and subjected to the loads without day-to-day variation, as discussed above.

The maximum average switch-pair blockings as a function of hour are shown in Figure 11.12. Here again, objective blocking is sometimes exceeded for ENR but not for DNHR or TSMR, and when sampling error is considered, all three routing techniques provide better blocking performance than we would expect from a network that is perfectly designed. Figure 11.13 shows the simulated 90th percentile of the 10-day average switch-to-switch blockings as a function of traffic load. The theoretical 90th percentile is also shown and has been calculated for a link sized for 1 percent blocking on the assumption of low day-to-day variation over a 10-day period, when sampling error is considered. Here, methods given in [HiN76] are used to consider the effect of day-to-day variation on the blocking probability distribution. We again find that the 90th percentile of the 10-day

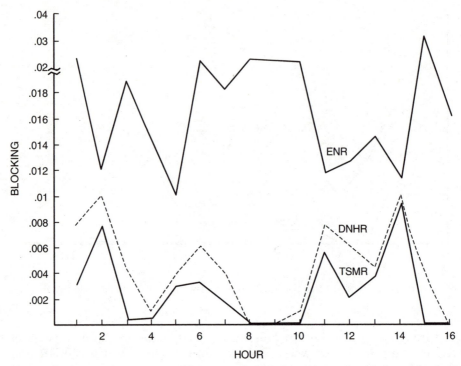

Figure 11.12 Maximum switch-pair blocking (path-EFO design network with day-to-day load variation)

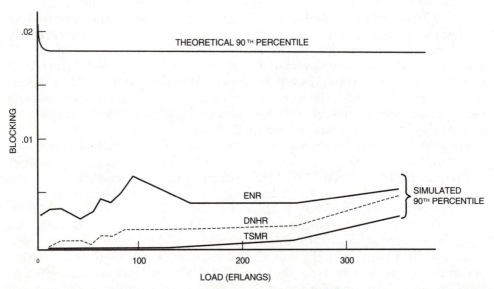

Figure 11.13 90th percentile switch-pair blocking vs. load (path-EFO design network with day-to-day load variation)

average switch-to-switch blocking is well below the theoretical 90th percentile threshold. Again the network is performing somewhat better than we would expect for a designed network.

11.3.1 Alternative approaches to TSMR

Having established that TSMR meets design objectives and provides performance benefits in comparison with ENR and DNHR, we are now ready to investigate various techniques to determine the most efficient TSMR implementation method. These methods include the following:

1. Route each call on the least loaded path (the path having the greatest number of idle circuits) among all candidate paths.

2. Route each call first on the direct path, if it exists and is available, or else select the least loaded path.

3. Route each call on the first path assigned by the path-EFO model, if available, or else select the least loaded path.

4. Compute the path-EFO model routing sequences that maximize carried traffic for a short-term estimate of the network loads and then apply method 3, using the new routing sequences.

Methods 1 and 2 do not provide adequate performance for the following reasons. Method 1 favors the path having the largest available capacity, not necessarily the first path, and usually a two-link path. This tendency to favor two-link paths results in a significant redistribution of flows and relatively poor network performance. Method 2 performs considerably better than method 1 but falls short of the performance of method 3, which we describe shortly. Method 2 has a problem similar to that of method 1: Because it always selects the direct path as the first choice, and because the path-EFO model does not always design the direct path to be the first choice, method 2 makes the path order too different from that designed by the path-EFO model. Hence, performance of methods 1 and 2 is degraded because the actual realization of network flows deviates too far from the network flow patterns designed by the path-EFO model.

Method 3, however, performs quite well. It reflects sufficiently accurately the design of the path-EFO model by assigning first-path traffic to the design first path. Flow on the first-choice path accounts for about 80 to 90 percent of the total network traffic flow, and hence the routing of this flow must correspond well to the placement of trunks for the network to behave properly. Flows on the second and higher-numbered paths are better assigned by a least-loaded selection method than by a preplanned sequential method. This statement is supported by the simulation results shown in Figures 11.14 and 11.15.

Figure 11.14 shows the 10-day average hourly blocking for TSMR method 3 and for DNHR. Here, the TSMR method uses periodic updates, as illustrated in Figure 11.1, with the status and routing update interval T equal to 5 seconds. Figure 11.15 shows the 99th percentile hourly switch-pair blocking for TSMR

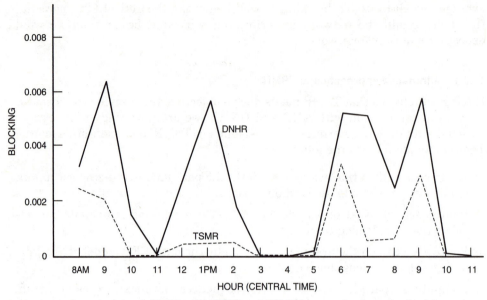

Figure 11.14 Average network blocking for DNHR and TSMR (average business-day loads)

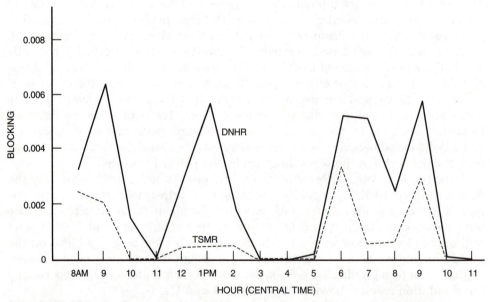

Figure 11.15 99th percentile switch-pair blocking for DNHR and TSMR (average business-day loads)

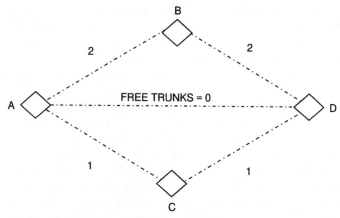

Figure 11.16   Illustration of DNHR vs. TSMR

and DNHR. These results are obtained with a network designed for DNHR and clearly demonstrate the benefits of TSMR in comparison with DNHR.

A simple intuitive explanation of why TSMR method 3 might complete more calls than DNHR is illustrated by the four-switch example shown in Figure 11.16. The current number of idle circuits in each link is shown. If the DNHR routing sequence for A–D calls is A–D, A–C–D, A–B–D, then the next A–D call arrival will block links A–C and C–D, and either an A–C or a C–D call arrival will then be blocked. However, the TSMR routing sequence for A–D calls in this network state is A–D, A–B–D, A–C–D. Therefore, the next A–D call arrival under the TSMR method will not block any additional links. As illustrated by this simple example, TSMR tends to leave capacity on links throughout the network distributed as uniformly as possible. Because of this property of TSMR, calls arriving in various parts of the network will have a greater chance of being completed than under the hypothesized DNHR routing sequence. Bulfer [Bul75] has established the optimality of the least-loaded routing method for a class of two-stage concentrators.

TSMR method 4 has greater adaptivity to load shifts than method 3, but is more complex to implement and, as such, must have better performance to be justified. The details of method 4 are now discussed. To calculate optimum routing sequences with the path-EFO model, as required by method 4, we assume that these routing patterns are computed in advance of their actual use and that the average load in the future load set period is known precisely. This second assumption, of course, is an idealization of reality and represents an upper limit on the possible performance of method 4. We used two methods based on the path-EFO model to determine the routing sequences that maximize network flow in the existing network. For each method we find the minimum incremental network capacity required to carry the future load-set-period load given the existing trunks as available network capacity. That is, as described in Chapter 9, we minimize

$$\sum_{i=1}^{L} \Delta a_i$$

where Δa_i are the augmentations above the existing link capacities a_i and L is the number of links in the network. In this formulation of the path-EFO model, we set all incremental link costs to 1 in order to transform the objective function from incremental network cost to incremental network capacity.

In the first method (which we will call method 4A), the maximum switch-pair blocking is held at 1 percent and the routing and capacity augmentations Δa_i that minimize the objective function are determined. We implement the routing solution, but the capacity augmentations Δa_i produced by the optimization cannot actually be added to the network. Hence, the traffic that would have been carried on these augmentations will actually be blocked. But because we have minimized these hypothetical capacity augmentations, this routing solution approximates the minimization of total blocked traffic. In the second method (method 4B) the switch-pair blocking is raised until there are zero capacity augmentations Δa_i produced by the optimization. This second solution approximates the minimization of the maximum switch-pair blocking.

A comparison of the average network blocking performance of method 3, method 4A, and method 4B is shown in Figure 11.17. For these results, the update interval T is 5 seconds and the network is designed for TSMR using the path-EFO model described in Chapter 13. As can be seen, there is no apparent improvement gained from the more complex methods over method 3, which suggests that method 3 may achieve nearly the maximum flow performance. TSMR method 3 is therefore selected for further study to determine the best switching control logic and trunk status map logic to implement TSMR.

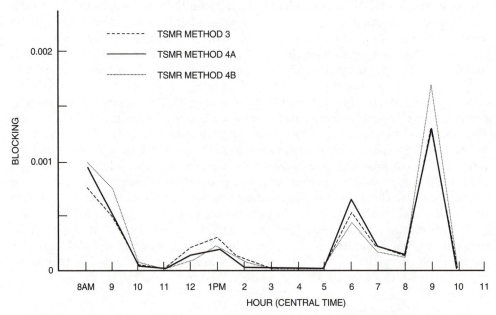

Figure 11.17 Average network blocking comparison for TSMR methods 3, 4A, and 4B

11.3.2 Status and routing update interval *T*

Four 10-day simulations are made with values of the status and routing update interval, T, equal to 2, 5, 10, and 30 seconds. For these four values of T, Figure 11.18 compares the TSMR results on the basis of three performance measures. The first measure is average network blocking, which is the 10-day total number of blocked calls divided by the 10-day total number of originating attempts. The second measure is the average number of crankbacks per originating attempt, which is the 10-day total number of simulated crankbacks divided by the 10-day total number of originating attempts. The third measure is the trunk status map workload, which is the daily total number of least-loaded-path searches averaged over the 10 days. We can see from the results that the status and routing update interval should be as short as possible within the limitations of switch and trunk status map processing if we wish to maximize completions and minimize crankbacks.

There is approximately a 2 percent penalty in switch capacity due to the real time needed to process crankback messages associated with the implementation of DNHR in the 4ESS switch, as discussed in Chapter 16. If we apply this penalty to a forecast of switches, and if we assume that the switch cost penalty for the TSMR method can be allocated in direct proportion to the total number of crankbacks generated, then we obtain the crankback cost as a function of T,

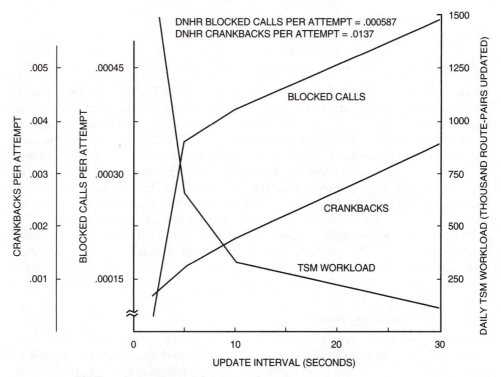

Figure 11.18 TSMR performance vs. update interval

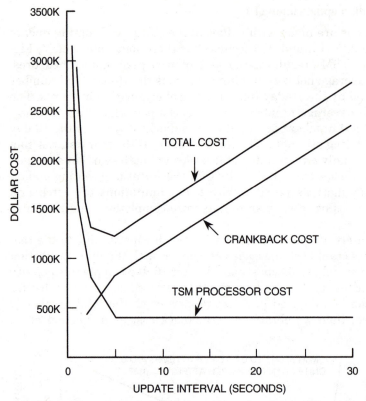

Figure 11.19 Processing and crankback cost vs. update interval T

which is shown in Figure 11.19. Also shown in Figure 11.19 is the total cost of processors needed to support the trunk status map processing level shown in Figure 11.18, under the assumption that two processors are needed to support a five-second update interval for a 180-switch TSMR network. Here, we assume that each processor costs $100,000. As shown in Figure 11.19, no fewer than four processors are required because the trunk status map is assumed to be duplicated for reliability purposes, as are signal transfer points in the CCS network, and at least two processors are required at each trunk status map, also for reliability purposes. The results shown in Figure 11.19 suggest that an update interval T in the range of two to eight seconds will minimize the total cost. The tradeoff analysis pictured in Figure 11.19 assumes that the switch real-time processing load required in addition to the crankback load is not a strong function of the update interval T.

11.3.3   TSMR congestion control

In the face of network overloads, failures, or other causes of network congestion, the trunk status map needs to adjust its routing method to best accommodate such conditions. Here we investigate the use of automatic trunk reservation controls, busy-path removal controls, and extended routing logic controls that are implemented by the trunk status map to help alleviate network congestion. Automatic trunk reservation is used to protect the direct link of overloaded switch pairs from excess overflow from other switch pairs. For traffic subjected to trunk reservation, access to trunks on the direct link is allowed only if the number of idle trunks in the link is greater than a specified number called the reservation level. Busy-path removal controls eliminate paths through congested parts of the network when heavy overload conditions exist. Extended routing logic allows overloaded switch pairs to search out excess capacity available in the network to complete the excess calls. The simulation studies we present next have investigated a range of automatic trunk reservation, busy-path removal, and extended routing logic strategies.

Automatic trunk reservation is triggered automatically for a switch pair if the average switch-pair blocking over a five-minute interval is greater than 5 percent and at least two calls are blocked during the five-minute interval. If automatic trunk reservation is triggered, then trunk reservation is applied to that switch pair for the next five minutes. Once triggered, automatic trunk reservation operates in the following manner. Traffic attempting to alternate-route over the direct link of a triggered switch pair is subjected to trunk reservation. We need a reservation level of approximately 5 percent of the number of trunks in each link. This action protects traffic on the direct link of the triggered switch pair (if it exists) from interference from traffic on other switch pairs. Reservation is applied at both the trunk status map and the switch. Reservation at the trunk status map is implemented by having the trunk status map subtract the number of trunks reserved from the number idle in the process of determining the least loaded path. In no case is trunk reservation applied at the trunk status map or switch for a one-link (direct) path.

Busy-path removal controls are triggered on total network blocking, and, once triggered, busy-path removal removes all busy two-link paths from all route sequences. That is, when the average network blocking over a five-minute period exceeds 3 percent, two-link paths (excluding the first path) that have zero free circuits are removed from the switch route patterns for the next five-second interval. This method results in a uniformly applied restrictive control on the network; all switch pairs are triggered at once and are prevented from routing calls through congested parts of the network. Busy-path removal controls are also used when a switch signals the trunk status map that it is congested or has failed. In the case of switch congestion, the trunk status map removes the switch as a via point in all routing patterns, and in the case of switch failure the trunk status map removes both direct and via routing to the failed switch.

With the extended routing logic, extended searches for idle capacity are triggered for a switch pair whenever the blocking for the switch pair exceeds 1 percent over a five-minute period. When the total network blocking threshold used for busy-path removal is triggered, however, the extended routing logic is used only when the switch-pair blocking over a five-minute interval exceeds 5 percent, with a minimum of two blocked calls. When extended routing logic is triggered, the search for the least loaded path for the triggered switch pair is extended to include a larger set of candidate paths.

11.3.4 Comparisons of TSMR
congestion control alternatives

Here we study various options for automatic trunk reservation, busy-path removal, and extended routing logic under various network overload and failure conditions.

Two versions of automatic trunk reservation are simulated in the 25-switch network. In the first triggering method, automatic trunk reservation is triggered for a switch pair if the average switch-pair blocking over a five-minute interval is greater than 5 percent and at least two calls are blocked during the five-minute interval. If automatic trunk reservation is triggered for a switch pair, then trunk reservation is applied to that pair for the next five minutes, as will be explained shortly. In the second triggering method, automatic trunk reservation is triggered at a switch if the average blocking of calls originating at the switch over a five-minute interval is greater than 5 percent and at least two calls originating at the switch are blocked during the five-minute interval. If automatic trunk reservation is triggered for a switch, it is applied to all originating calls from that switch.

Two methods of implementing automatic trunk reservation are simulated in the 25-switch network. For ATR-1, once a switch pair is triggered, two actions are taken. First, the triggered switch-pair's traffic is subjected to trunk reservation on all paths except the first path. We use a reservation level of approximately 5 percent of the number of trunks in each link. Second, traffic attempting to alternate-route over the direct link of a triggered switch pair is subjected to trunk reservation. Action 1 protects other switch pairs from interference from the triggered switch pair. Action 2 protects the triggered switch pair from interference from other switch pairs when the direct link exists for a triggered switch pair. We constrain the direct link, if it exists, to be a candidate path for a switch pair. For ATR-2, only action 2 is triggered.

Finally, two methods of applying reservation are studied. In the first method, it is applied only at the trunk status map. This is accomplished by having the trunk status map subtract the number of trunks reserved from the number idle. Any path except the first path not having more free trunks than the reservation criterion on each link is deleted from the list of paths sent to the switch from the trunk status map for a given routing update. In the second method, reservation is applied both at the trunk status map, as described above, and at the switch. For each triggered switch pair under the ATR-1 method, the trunk status map marks all two-link paths sent to the switch, except the first path, to have reservations applied by the switch. Similarly, for both ATR-1 and ATR-2, all two-link paths that use the direct link of a triggered switch pair are marked by the trunk status map to have reservation applied by the switch. In no case is trunk reservation applied for a one-link direct path.

We simulate each of the eight combinations of the following parameters: two automatic trunk reservation triggering mechanisms, two automatic trunk reservation methods, and two reservation application methods. We consider each case over the three-hour morning busy period for an average daily load, as well as for

10, 20, and 30 percent general overloads in which each switch-pair traffic load is increased by the overload percentage. We also consider a focused overload on the WHPL switch in which each switch-pair load to WHPL is increased by a factor of 3. The results are given in Tables 11.3 to 11.7, in which the measures used reflect averages over the three hours of the simulation:

- The average network blocking is the total blocked attempts in three hours divided by the total originating attempts.

TABLE 11.3 Performance of Automatic Trunk Reservation Methods under 0% General Overload

Case	Automatic trunk reservation type	Switch reservation	Automatic trunk reservation trigger	Average blocking	99th-percentile blocking	Crankbacks per call
A	no controls			0.00002	0.0	0.00268
B	ATR-1	no	switch–switch blocking	0.00002	0.0	0.00268
C	ATR-2	no	switch–switch blocking	0.00002	0.0	0.00268
D	ATR-1	yes	switch–switch blocking	0.00002	0.0	0.00268
E	ATR-2	yes	switch–switch blocking	0.00002	0.0	0.00268
F	ATR-1	no	switch blocking	0.00002	0.0	0.00268
G	ATR-2	no	switch blocking	0.00002	0.0	0.00268
H	ATR-1	yes	switch blocking	0.00002	0.0	0.00268
I	ATR-2	yes	switch blocking	0.00002	0.0	0.00268
J	ATR-2, busy-path removal	yes	switch–switch blocking	0.00002	0.0	0.00268

TABLE 11.4 Performance of Automatic Trunk Reservation Methods under 10% General Overload

Case	Automatic trunk reservation type	Switch reservation	Automatic trunk reservation trigger	Average blocking	99th-percentile blocking	Crankbacks per call
A	no controls			0.01090	0.05778	0.02253
B	ATR-1	no	switch–switch blocking	0.00794	0.07436	0.00964
C	ATR-2	no	switch–switch blocking	0.00812	0.04839	0.01363
D	ATR-1	yes	switch–switch blocking	0.00794	0.03963	0.01579
E	ATR-2	yes	switch–switch blocking	0.00719	0.03524	0.01541
F	ATR-1	no	switch blocking	0.00878	0.05966	0.01264
G	ATR-2	no	switch blocking	0.00892	0.05242	0.01343
H	ATR-1	yes	switch blocking	0.00862	0.04435	0.01820
I	ATR-2	yes	switch blocking	0.00852	0.04032	0.01897
J	ATR-2, busy-path removal	yes	switch–switch blocking	0.00719	0.03524	0.01541

TABLE 11.5   Performance of Automatic Trunk Reservation Methods under 20% General Overload

Case	Automatic trunk reservation type	Switch reservation	Automatic trunk reservation trigger	Average blocking	99th-percentile blocking	Crankbacks per call
A	no controls			0.06912	0.21390	0.10726
B	ATR-1	no	switch–switch blocking	0.03739	0.20513	0.01806
C	ATR-2	no	switch–switch blocking	0.04129	0.18482	0.02902
D	ATR-1	yes	switch–switch blocking	0.03959	0.18972	0.06548
E	ATR-2	yes	switch–switch blocking	0.04086	0.18647	0.07003
F	ATR-1	no	switch blocking	0.03943	0.19424	0.01885
G	ATR-2	no	switch blocking	0.03845	0.18156	0.01965
H	ATR-1	yes	switch blocking	0.03921	0.18217	0.06533
I	ATR-2	yes	switch blocking	0.04054	0.20543	0.06839
J	ATR-2, busy-path removal	yes	switch–switch blocking	0.03650	0.16891	0.01705

TABLE 11.6   Performance of Automatic Trunk Reservation Methods under 30% General Overload

Case	Automatic trunk reservation type	Switch reservation	Automatic trunk reservation trigger	Average blocking	99th-percentile blocking	Crankbacks per call
A	no controls			0.12675	0.33314	0.19436
B	ATR-1	no	switch–switch blocking	0.07281	0.33054	0.02705
C	ATR-2	no	switch–switch blocking	0.07570	0.28463	0.04715
D	ATR-1	yes	switch–switch blocking	0.07493	0.27703	0.11976
E	ATR-2	yes	switch–switch blocking	0.07726	0.29730	0.12420
F	ATR-1	no	switch blocking	0.07283	0.32497	0.02779
G	ATR-2	no	switch blocking	0.07282	0.34884	0.03007
H	ATR-1	yes	switch blocking	0.07394	0.28338	0.12509
I	ATR-2	yes	switch blocking	0.07385	0.28447	0.12411
J	ATR-2, busy-path removal	yes	switch–switch blocking	0.07268	0.24134	0.04628

- The 99th percentile blocking is the 99th percentile switch-pair blocking where each switch-pair blocking is averaged over the three simulation hours.

- The average crankbacks per originating attempt is the total simulated crankbacks over three hours divided by the total originating attempts.

In each table, we first present case A, which corresponds to the situation with no automatic congestion controls, and then present cases B through I, which

TABLE 11.7   Performance of Automatic Trunk Reservation Methods under 3-to-1 Focused Overload on WHPL

Case	Automatic trunk reservation type	Switch reservation	Automatic trunk reservation trigger	Average blocking	99th-percentile blocking	Crankbacks per call
A	no controls			0.01344	0.20455	0.03924
B	ATR-1	no	switch–switch blocking	0.01569	0.33251	0.00810
C	ATR-2	no	switch–switch blocking	0.01347	0.26910	0.01305
D	ATR-1	yes	switch–switch blocking	0.01479	0.29261	0.04078
E	ATR-2	yes	switch–switch blocking	0.01313	0.25455	0.03482
F	ATR-1	no	switch blocking	0.01468	0.26147	0.01716
G	ATR-2	no	switch blocking	0.01463	0.26453	0.01804
H	ATR-1	yes	switch blocking	0.01481	0.23919	0.02531
I	ATR-2	yes	switch blocking	0.01440	0.24577	0.02490
J	ATR-2, busy-path removal	yes	switch–switch blocking	0.01313	0.25455	0.03482

correspond to the eight combinations of parameters described above. In Table 11.3, we see that all cases give the same results because no automatic congestion controls are triggered for 0 percent overload. Also, case E (switch-pair trigger, ATR-2, reservation applied in switch) performs best for the 10 percent overload case and performs well for all the other cases, although not necessarily best for every case. There is some evidence, especially for the 20 to 30 percent overload cases, that methods that trigger reservation for more switch pairs have somewhat better performance.

From these results we conclude that case E shows promise but needs additional trunk reservation or possible other restrictive controls that take effect at heavier overloads. Several modifications to case E are investigated, all of which include (1) an additional automatic triggering mechanism that triggers only at heavier overloads and (2) an additional trunk reservation method or other restrictive control that is applied when the additional mechanism triggers. With each modification to case E, ATR-2 continues to trigger in the manner described earlier if the switch-pair blocking level over a five-minute period exceeds 5 percent.

One of the candidate modifications to case E uses an additional switch-triggering mechanism to trigger a higher level of trunk reservation for switch pairs originating at the triggered switch. The switch-triggering mechanism triggers when the average blocking of calls originating at the switch over a five-minute period is greater than 10 percent. For the switch pairs triggered by this mechanism, we apply a higher level of trunk reservation on the direct link and use the ATR-2 reservation method already embedded in case E. The reservation level is approximately equal to 10 percent of the trunks in each link.

We find that the additional switch-triggering mechanism with the higher level of trunk reservation tends to be too uneven in its help to the network. Although it does reduce blocking for some switch pairs that receive additional protection, the higher level of reservation tends to be overly restrictive to other switch pairs by denying them access to available network capacity. We conclude that in order for additional restrictive controls to benefit the entire network under heavier overloads, these controls must somehow be applied more uniformly throughout the network.

We investigate another modification to case E that achieves more uniformity of control application. Of the various modifications to case E that are investigated, this approach provides the best overall performance. We use an additional triggering mechanism based on total network blocking to trigger the cancellation of busy two-link paths, which we call busy-path removal. The additional busy-path removal mechanism triggers when the average network blocking over a five-minute period exceeds a threshold that varies between 2 and 10 percent. Once the network blocking threshold is exceeded, all switch pairs whose blocking over a five-minute period is less than 30 percent are subjected to the busy-path-removal method. The switch pairs with blocking larger than 30 percent are subjected to the busy-path-removal method in order to give them some additional advantage in completing calls. The busy-path-removal method implemented by the trunk status map first determines the number of free circuits in each path; trunk reservation is included in this determination through the use of the ATR-2 method embedded in case E. Then, the two-link paths, excluding the first path, that have zero free circuits are removed by the trunk status map for the next five-second interval. This method results in a uniformly applied restrictive control on the network because, with the network blocking triggering mechanism, nearly all switch pairs are triggered at once and in that way mutually help each other. Hence, we restrict most switch pairs from causing interference to other switch pairs by eliminating attempts to route calls through parts of the network that are congested. We found that a network blocking threshold of 3 percent for the network triggering mechanism performs best for this method.

In summary, we add two features to case E based on these results:

1. An additional triggering mechanism based on a 3 percent network blocking threshold that triggers all switch pairs with blocking less than 30 percent over a five-minute period.

2. Removal of paths with zero free trunks (with reservation considered) for all switch pairs triggered by the additional mechanism.

The results for this modified version of case E are given in Tables 11.3 to 11.7, case J.

We also simulate extended routing logic alternatives that can be implemented by the trunk status map. In one alternative, extended routing logic is triggered for a switch pair whenever the blocking for the pair exceeds 1 percent over a five-minute period. When the total network blocking threshold is triggered, however, the extended routing logic is used only when the switch-pair blocking over a five-minute interval exceeds 5 percent, with a minimum of two blocked calls. When extended routing logic is triggered, the search for the least loaded path for the triggered switch pair is extended to include a larger set of candidate paths. We consider this extended routing logic combined with the modified case E for 0, 10, 20, and 30 percent general overloads, and also for the focused overload on WHPL. We also consider an alternative extended routing logic procedure that allows a search for the second-choice path over all candidate paths considered by the path-EFO model. The first-choice path is again the first path assigned by the path-EFO model. To get a first indication of the performance of the modified TSMR method, we considered only the first Monday load pattern and obtained the results shown in Figure 11.20. The indication is that the modified TSMR technique provides improved network performance in comparison with the more restricted TSMR method. However, there is some loss of performance in the late evening hours, perhaps because of a significant shift in network load patterns during the late evening (11 P.M.) discount period. Overall, though, this modified TSMR procedure completes more calls, and this increase in network completions increases network revenues.

We simulate the automatic trunk reservation, busy-path removal, and extended routing logic controls over the three-hour morning busy period for an average daily load, as well as for 10, 20, and 30 percent general overloads in which each switch-pair traffic load is increased by the overload percentage. We also consider a focused overload on the White Plains switch in which each load to White

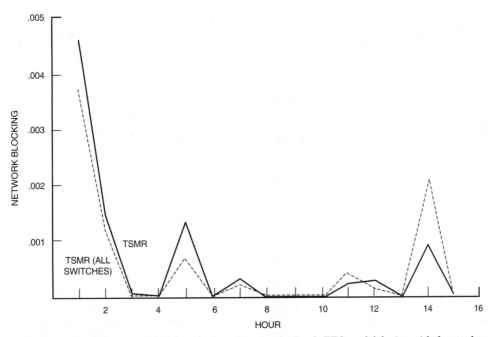

Figure 11.20 Average network blocking vs. hour (day 1 only) [path-EFO model design with day-to-day load variation TSMR and TSMR (all switches)]

Plains is increased by a factor of 3. Finally, we consider two failure situations: a Dallas–Wayne (DLLS–WAYN) link failure and an Anaheim–White Plains (ANHM–WHPL) link failure. The automatic congestion controls for DNHR differ from those of TSMR in three important ways:

1. A switch trigger with a 10 percent blocking threshold is used to trigger trunk reservation in the skip mode (i.e., if a call is blocked on a path subjected to trunk reservation, the call then tries the next path).

2. Both a 20 percent switch-blocking threshold and a 25 percent switch-pair-blocking threshold must be exceeded in order to trigger trunk reservation in the cancel mode (i.e., if a call is blocked on a path subjected to trunk reservation, the call is canceled).

3. Trunk reservation is applied only to two-link paths, not one-link paths, as opposed to only alternate paths (no first paths) for TSMR.

The results are given in Table 11.8. For comparison, we have included results for DNHR. The results indicate that TSMR (with automatic congestion controls) provides improved average blocking performance and comparable or better 99th-percentile blocking performance for all the cases studied. The number of crankback messages is also significantly lower with TSMR.

As another indication of peak load performance, Figure 11.21 shows the hourly blocking performance for TSMR and DNHR under an average Monday load

TABLE 11.8   Performance Comparison of TSMR (with Congestion Controls in Effect) and DNHR

Routing	% overload or failure	Average blocking	99th-percentile blocking	Crankbacks per call
DNHR	0	0.00044	0.00215	0.01131
TSMR	0	0.00002	0.00010	0.00257
DNHR	10	0.01385	0.03396	0.03387
TSMR	10	0.00479	0.01230	0.00931
DNHR	20	0.04933	0.12261	0.07125
TSMR	20	0.03016	0.13793	0.01855
DNHR	30	0.08516	0.22813	0.09920
TSMR	30	0.06524	0.22876	0.05332
DNHR	focus	0.01840	0.23590	0.04474
TSMR	on WHPL	0.00808	0.12265	0.02865
DNHR	DLLS–WAYN	0.00101	0.00806	0.01556
TSMR	failure	0.00004	0.00055	0.00314
DNHR	ANHM–WHPL	0.00187	0.01651	0.03013
TSMR	failure	0.00001	0.00000	0.00167

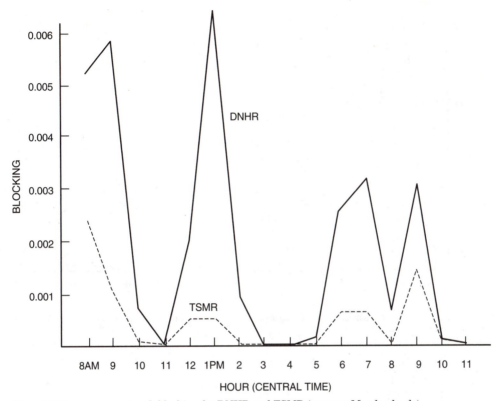

Figure 11.21 Average network blocking for DNHR and TSMR (average Monday loads)

pattern (these loads are normally the highest loads of the week). These results also demonstrate the ability of TSMR congestion controls to increase network flow in comparison with DNHR. Both DNHR and TSMR performance would be improved further by automatic real-time traffic management controls present in the switch and in the traffic management support system.

11.3.5 Correction of service problems

As discussed in Chapter 9, a problem for network capacity management is the ability to correct service problems arising from forecast errors. There is a significant interaction with the routing method because techniques such as real-time routing and least-loaded routing have the ability to complete more calls over a range of load variations, including variations resulting from forecast errors. To simulate a service problem, we selected a single switch pair, Freehold, New Jersey (FRHD), to Chicago, Illinois (CHCG), in hour 2 and inflated the load by a factor of 3 (from about 95 erlangs to about 285 erlangs). The results for the average network blocking and maximum switch-pair blockings are shown in Figures 11.22 and 11.23, where we consider ENR in addition to DNHR and

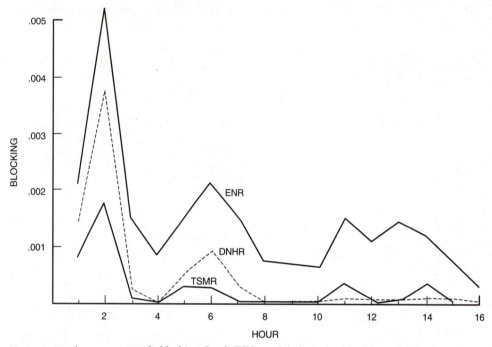

Figure 11.22 Average network blocking [path-EFO model design with day-to-day load variation (forecast error in hour 2)]

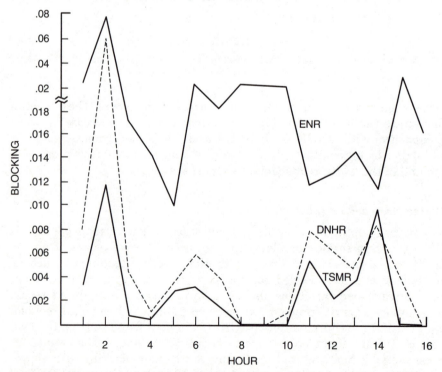

Figure 11.23 Maximum switch-pair blocking [path-EFO model design with day-to-day load variation (forecast error in hour 2)]

TSMR. We observe that hour 2 has notably poorer performance than before, especially the FRHD–CHCG load for the case of ENR. It is also notable that TSMR has effectively found a way to carry almost all the extra load somewhere in the network and therefore comes close to meeting objective performance for all switch pairs. Herein lies the interaction between short-term capacity design and real-time routing: With ENR and DNHR, corrective action is required; with TSMR, however, no action is required. Hence, more flexible routing strategies such as TSMR may reduce the need for short-term capacity redesign.

There still remains the question of whether the path-EFO model can provide corrective action, if needed, for TSMR. To test this, we use the path-EFO model to find the new routing and link sizes required to meet the hour 2 loads. The resulting network involves a number of routing changes and trunk augments. The largest increase in trunks (87 trunks) is provided on the link between CHCG and FRHD, as expected, although there are several other smaller increases on other links. The new routing and link sizes are then "implemented" in the simulation model to test whether or not the service problem is corrected. The results for the average network blocking and maximum switch-pair blocking are shown in Figures 11.24 and 11.25. As can be seen, the service problem in hour 2 is no longer evident. This shows that the path-EFO model with day-to-day variation design is able to correct service problems in the TSMR network.

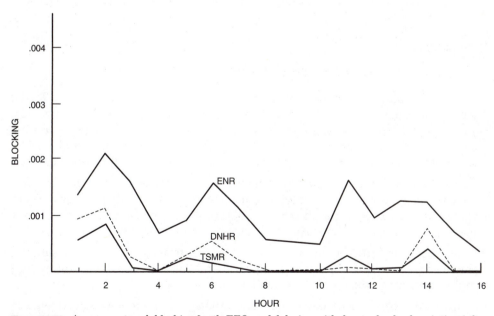

Figure 11.24 Average network blocking [path-EFO model design with day-to-day load variation (after short-term design adjustment in hour 2)]

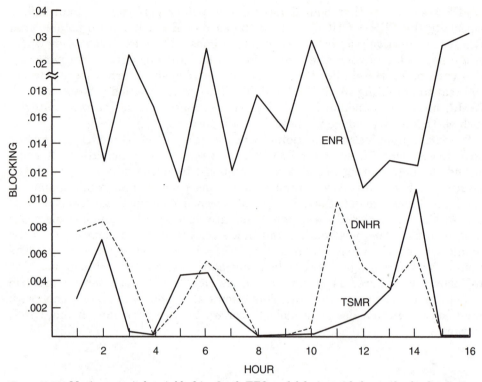

Figure 11.25   Maximum switch-pair blocking [path-EFO model design with day-to-day load variation (after short-term design adjustment in hour 2)]

11.4   Conclusions

Conclusions from the analysis presented in this chapter are as follows:

- Centralized real-time dynamic routing networks, as illustrated by the TSMR method, provide a significant improvement in network service in comparison with either preplanned engineered routing or preplanned engineered routing combined with real-time routing (DNHR) when the network is subjected to design loads, loads with day-to-day variations, and loads with forecast errors.

- Centralized real-time dynamic routing provides uniformly better blocking performance than preplanned dynamic routing and a significantly reduced number of crankback messages; both network blocking and the number of crankback messages decrease as the status and routing update interval T decreases.

- Centralized real-time dynamic routing methods provide service that satisfies the network blocking objectives and provide network flow patterns and performance consistent with path-EFO model routing design.

- Centralized real-time dynamic routing methods can implement effective automatic congestion control strategies using design principles such as automatic trunk reservation, extended routing logic, and busy-path removal.

- An evaluation of centralized real-time dynamic routing processing cost and crankback cost as a function of the update interval T places the optimum value of T in the range of two to eight seconds.

Procedures for design of centralized real-time dynamic routing networks are discussed in Chapter 13. They provide improved design procedures for hour-to-hour and day-to-day traffic variations that satisfy the network blocking probability grade-of-service objective.

The results show that a centralized real-time dynamic routing capability has significant benefits in comparison with preplanned dynamic routing. Distributed real-time routing techniques providing further improvement and advantages are discussed in Chapter 12.

Distributed Real-Time Dynamic Routing for Multiservice Integrated Networks

12.1 Introduction

In this chapter we study distributed real-time state-dependent routing methods and event-dependent routing methods. As discussed in Chapter 1, real-time routing implies that routing table updates are made at a frequency of one call holding time or faster, and we identified several real-time routing methods, which include real-time event-dependent, centralized real-time state-dependent, and distributed real-time state-dependent routing. In Chapter 11 we studied trunk status map routing (TSMR) as a representative method for centralized real-time state-dependent routing, which can provide routing updates on the order of seconds or minutes. In the limiting case of distributed real-time state-dependent routing, routing table updates can be made on a call-by-call basis.

In this chapter we study in detail representative examples of distributed real-time state-dependent routing and distributed real-time event-dependent routing. Implemented examples of real-time event-dependent routing are (1) dynamic alternate routing (DAR), implemented within the British Telecom UK network, and (2) state- and time-dependent routing (STR), implemented within the NTT Japan network. Here we study a learning method called learning with random routing (LRR) and an extension of LRR called success-to-the-top (STT) routing. Implemented examples of distributed real-time state-dependent routing are (1) worldwide international network (WIN) routing, implemented within a consortium of international service providers; (2) real-time network routing (RTNR), implemented within the AT&T USA network; and (3) real-time internetwork routing (RINR), implemented within the AT&T international network. Here we study RTNR and RINR as the representative examples of such networks and emphasize the ability of these networks to provide multiservice network integration, as discussed in Chapter 10 for preplanned dynamic routing networks. Recall that in Chapter 1 we also studied distributed real-time state-dependent routing with five-minute updates, analogous to the routing implementation in WIN. The methods and principles applied to the design and evaluation of these representative examples then also apply to the many possible variations of each method. In applying these same methods and principles, the reader will be able to explore in depth any of the possible variations of these routing methods.

In accordance with the above discussion, in this chapter we do an in-depth study of RTNR and RINR as representative examples of distributed real-time state-dependent routing methods for multiservice integrated networks.

RTNR provides switches a simple way of exchanging link status information, thereby determining the availability and load conditions of the direct and two-link routes to the destination. We analyze the RTNR concept and apply it to integrated ISDN networks that provide a multiplicity of services on an integrated transport network. Switches interconnected by a flexible transport network provide connections for voice, data, and wideband services. These connections are distinguished by estimated resource requirements, traffic characteristics, and design performance objectives. This chapter provides models and results for RTNR design for these multiservice integrated networks.

Motivations for real-time state-dependent routing such as RTNR include

1. Introduction of new services using dynamic routing
2. Network reliability improvement
3. Simplification of network management and design

Dynamic routing is the routing method of choice for new services in the future. However, there is a lengthy list of existing and new services that are constantly being planned for introduction. Examples include 64-, 384-, and 1,536-kbps switched digital data services; international switched transit service; survivable defense communication services; virtual private network services; ISDN User Part (ISUP) preferred service; global software-defined network service; and others. Such needs have led to a plan to introduce a class-of-service routing feature to standardize service classification and dynamic routing treatment to all network services. There is also a benefit to integrate voice and switched digital services network services onto a common shared network. One approach to class-of-service routing is to use preplanned dynamic routing, such as preplanned two-link sequential dynamic routing, to introduce the new services. However, there is a large network management and design administrative overhead in maintaining large, and even increasing, numbers of routing tables for each service for each load set period. Preplanned dynamic routing quickly expands beyond its capacity to support all existing and new services, and therefore another approach requiring less administrative overhead is needed.

There is considerable importance in developing a robust routing method for reliable, self-healing network design, which is discussed in Chapter 14. A self-healing network is one that responds in near-real time to network stress, such as failure or overload, and continues to provide connections to customers with essentially no perceived interruption of service. Real-time state-dependent routing is a step toward the desired robust routing method for self-healing networks.

A third goal is to simplify the network management and design environment, because network management and design costs can be significant. There is therefore motivation for an efficient decentralized routing method that minimizes the need for centralized network management and design support.

The above needs can be met by various real-time routing methods, and in this chapter we use RTNR as an illustrative example. A number of real-time routing methods have been proposed that include RTNR, DNHR with real-time routing, TSMR, dynamically controlled routing (DCR), and DAR. Here we briefly summarize each of these approaches.

For each call entering an RTNR network, the originating switch analyzes the called number of the call and determines the terminating switch for the call. The originating switch will always try to set up the call on the direct path and, if a direct trunk is not available, try to find an available two-link path by first querying the terminating switch through the CCS network as to the busy/idle status of all links connected to the terminating switch. The terminating switch replies with this state information to the originating switch through the CCS network. The originating switch then determines a via switch for the call by comparing its own link busy/idle status information to that from the terminating switch in order to find the most lightly loaded two-link path to route the call over.

DNHR uses a hybrid preplanned time-varying routing and real-time routing control, as discussed in Chapter 8, to respond to network load variations and incorporates one- and two-link path routing between originating and terminating switches [AKK83]. Preplanned time-varying routing allows prespecified routing patterns to change as frequently as every hour to respond to expected traffic patterns. Real-time routing methods are incorporated in the DNHR method through supplementary real-time routing, which searches for idle capacity on a call-by-call basis if needed. The real-time routing method appends to each sequence of two-link paths designed by the path-erlang flow optimization (path-EFO) model, described in Chapters 6 and 7, additional two-link real-time paths to be used only when the engineered paths are unavailable and idle capacity is available on the real-time paths. Dynamic trunk reservation is used to recognize idle network capacity. Access to trunks on a particular link is allowed only after a specified number of trunks—the reservation level—are available. Reservation guarantees that capacity is idle and that accessing it will produce minimal interference with normal traffic. Candidate real-time paths to be considered by the originating switch at the time of call setup are selected by the path-EFO model and periodically downloaded into the switch-routing tables. Real-time paths can also be identified by the real-time traffic management system, as described in Chapter 8, on the basis of real-time network blocking and appended on a five-minute basis to the routing sequence for additional possible call completion. Network blocking performance is improved by real-time routing under a variety of overload and failure conditions, as demonstrated by network simulation studies discussed in Chapter 8.

TSMR, discussed in Chapter 11, is an extension of the DNHR concept to a centralized trunk status map that provides real-time routing decisions in the DNHR network [Ash85]. The trunk status map concept involves having an update of the number of idle trunks in each link sent to a network database every T seconds. These updates are sent by each switch only when the link status has changed. In return, the trunk status map periodically sends the switches ordered routing sequences to be used until the next update in T seconds. These routing table updates are determined by the trunk status map using the TSMR dynamic routing method. The TSMR method provides that the first path choice determined by the path-EFO model is used if a circuit is available. The first-choice path is updated once each load set period. If the first path is busy, the second path is selected from the list of other paths determined by the path-EFO model on the basis of having the greatest number of circuits at the time. Hence, the TSMR approach is a hybrid of preplanned time-variable routing and centralized real-time state-dependent routing. Trunk reservation and automatic congestion control methods are used in the TSMR method to respond to network overloads and failures. TSMR provides uniformly better performance than preplanned dynamic routing under a variety of network stress conditions, including overloads and failures.

DCR is a routing system developed by Bell Northern Research and Northern Telecom Incorporated that uses a central processor to track the busy/idle status of network trunks and determine the best alternate route choices based on status data every 10–15 seconds [BNR86]. The network is fully interconnected, and the direct link is selected first; overflow to the best alternate path choice is made upon an all-trunks-busy condition on the direct link. The alternate path choice is selected at random according to a set of probability values that are proportional to the relative number of idle circuits on each candidate path. The selected best choice is sent by the central processor to each network switch and used by each switch until a fresh update is received. No crankback is used in the DCR system, so a call blocked at the via switch is lost. Upon failure of the central processor, the network reverts to a hierarchical routing system. DCR is available commercially as a feature of Northern Telecom's DMS 100/200 switches and dynamic network controller system and is used in the Stentor Canada national network, the Bell Canada metropolitan area network in Toronto, the Sprint USA national network, and the MCI USA national network.

DAR is a routing system developed by British Telecom that uses a simple decentralized approach to real-time dynamic routing [Gib86]. Once again, the network must be fully interconnected, and upon an all-trunks-busy condition on the direct link, the alternate path last used successfully is used once again for the next path choice. If the alternate path is busy, the call is blocked and a new alternate path is selected at random as the alternate route choice for the next call overflow from the direct link. No crankback is used in the DAR system, so a call blocked at a via switch will be lost. DAR is developed in the Plessey system X switch and implemented in the British Telecom UK national network. STR is a version of DAR deployed in the NTT Japan national network [YMI91].

The above real-time dynamic routing methods have various advantages and disadvantages. RTNR has the advantage of being a simple decentralized method that attains the performance gains demonstrated for TSMR, but it does not have the disadvantage of needing the development of a trunk status map central processor. It also avoids the risks associated with a central processor controlling the routing of the entire network. Hence, RTNR is a promising example of distributed real-time dynamic routing in multiservice integrated networks, which is described in the next section.

12.2 Distributed Real-Time Routing Methods for Multiservice Integrated Networks

12.2.1 Overview

We begin with a brief overview of the RTNR method and then describe its details for multiservice integrated networks. For each call entering the RTNR network, as shown in Figure 1.27, the originating switch analyzes the called number of the call and determines the terminating switch. The originating switch first tries to set up the call on the direct link, if one exists, to the terminating switch. If the direct link is available and has idle trunks, the originating switch sets up the call on a direct trunk to the terminating switch. If a direct trunk is not available, the originating switch tries to find an available two-link path by first querying the terminating switch through the CCS network as to the busy/idle status of all links connected to the terminating switch. The terminating switch maps each link's busy/idle status into one of six states (maximum lightly loaded, medium lightly loaded, minimum lightly loaded, heavily loaded, reserved, or busy) and replies with this state information to the originating switch through the CCS network. The originating switch then determines a via switch to which to connect the call by comparing its own link's busy/idle status information with that from the terminating switch in order to find the most lightly loaded two-link path to route the call over. It then compares the resulting paths to an allowed via switch list, which specifies allowed paths that meet transmission requirements. The major differences between RTNR and DNHR are summarized in Figure 1.24.

Through the use of trunk reservation and congestion control techniques, illustrated in Figures 12.1 and 12.2, this routing system provides good network performance under normal and abnormal operating conditions for all services sharing the integrated network. In the multiservice network, a call on each individual service is assumed to consume an average bandwidth equal to r_i, using a single unit of capacity denoted as one virtual trunk (VT) on each of several virtual networks. For example, each VT would have r_i equal to 64 kbps of bandwidth for voice calls or 64-kbps switched digital service data calls, r_i equal to 384 kbps of bandwidth for 384-kbps switched digital service data calls, and r_i equal to 1,536 kbps of bandwidth for each 1,536-kbps switched digital service data call.

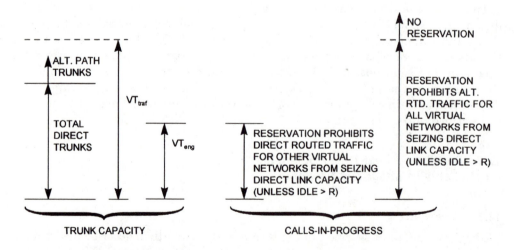

Figure 12.1 RTNR trunk reservation

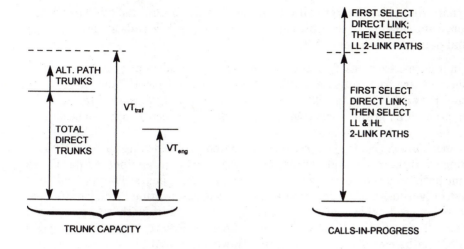

VT$_{traf}$ - TRUNKS REQUIRED TO MEET BLOCKING PROBABILITY OBJECTIVE

VT$_{eng}$ - TRUNKS ALLOCATED ON DIRECT LINK

Figure 12.2   RTNR path selection

Under normal nonblocking network conditions, all services fully share all available capacity. As illustrated in Figure 12.1, when blocking occurs for virtual network i, RTNR trunk reservation acts to prohibit alternate-routed traffic and traffic from other virtual networks from seizing direct link capacity designed for service i. There are three quantities of particular importance to this mechanism:

1. VT$_{traf}$, the number of virtual trunks required to meet the blocking probability grade-of-service objective. VT$_{traf}$ is estimated online by the switch and can exceed the total number of direct trunks, meaning that trunks from two-link alternate paths are needed to meet the blocking probability grade-of-service objective.

2. VT$_{eng}$, which is the number of direct trunks allocated by the network design process to a particular virtual network.

3. The blocking level, which also is estimated online by the switch.

When blocking is detected, three responses are triggered with respect to service i:

1. Trunk reservation is activated on the direct link.

2. Traffic from other virtual networks is prohibited by trunk reservation from seizing direct link capacity as long as the calls in progress for virtual network i are below the engineered level VT$_{eng}$.

3. Alternate-routed traffic from all virtual networks is prohibited by trunk reservation from seizing direct link capacity as long as the calls in progress for virtual network i are below VT_{traf}.

If calls in progress exceed VT_{eng}, the allocated engineered capacity is used up and other services are free to use available direct link capacity. If calls in progress for virtual network i exceed VT_{traf}, reservation is no longer needed to meet the blocking probability grade-of-service objective, and all direct trunks can be shared by all traffic.

These mechanisms replace the role of engineered and real-time paths in DNHR. Triggering of trunk reservation automatically converts an "engineered path" to a "real-time path" and then back again when blocking disappears. That is, real-time paths can only be used when idle trunks exceed the reservation level. Engineered paths can be used whenever there are idle trunks, but with the additional restrictions illustrated in Figure 12.2. As shown in Figure 12.2, lightly loaded (LL) RTNR paths can always be selected. However, heavily loaded (HL) RTNR paths are selected only under the conditions that (1) blocking is detected and (2) the calls in progress are less than VT_{traf}. When calls in progress exceed VT_{traf}, heavily loaded paths should not be required to meet the blocking probability grade-of-service objective, and thus heavily loaded capacity can be used by other traffic in need of additional capacity.

RTNR multiservice integrated network routing is summarized in Figure 1.25 and consists of the following three steps:

1. At the originating switch, the terminating switch and class-of-service information are determined through the digit translation database and other service information available at the originating switch.

2. The terminating switch and class-of-service information are used to access the appropriate routing table, virtual network, and path sequence between the originating switch and terminating switch.

3. The call is set up over the first available path in the path sequence with the required transmission resource selected based on the class-of-service data.

RTNR multiple ingress/egress routing is summarized in Figure 1.28, in which a call is routed either on the direct link or, if not available, via a two-link path through any one of the other switches from an originating switch to an international switching center (in the AT&T national network, for example, there are up to 133 two-link via path choices). A destination country could be served by more than one international switching center, in which case multiple ingress/egress routing is used. As illustrated in Figure 1.28, with multiple ingress/egress routing, a call from the originating switch T_1 destined for the United Kingdom switch UK_1 tries first to access the international links from international switching center T_3 to UK_1. In doing this it is possible that the call could be routing from T_1 to T_3 directly or via T_2. If all circuits from T_3 to UK_1 are busy, the call control can be returned to T_1 with a CCS crankback message, after which the call is routed to T_4 to access the T_4-to-UK_1 circuits. If the international call cannot be completed on circuits connecting terminating switch T_3 to UK_1, the call can return to the originating switch T_1 through use of a CCS crankback message for possible further routing to another international switching center at T_4, which also has circuits to UK_1. In this manner all ingress/egress connectivity is utilized to a connecting network, maximizing call completion and reliability.

Once the call reaches international switching center T_3, this switch determines the routing to the destination country UK_1 and routes the call accordingly. In completing the call to UK_1, international switching center T_3 can use real-time internetwork routing to dynamically select a direct link, a multiple-link path through an alternate switch in the United Kingdom, or perhaps a multiple-link path through an alternate switch in another country, as illustrated in Figure 1.29.

Real-time internetwork routing (RINR) extends class-of-service routing concepts and increased routing flexibility for internetwork routing. It works synergistically with multiple ingress/egress routing, and, as illustrated in Figure 1.29, it uses link status information in combination with call completion history to select paths and reuses dynamic trunk reservation techniques implemented for class-of-service RTNR.

The RTNR and RINR methods have the advantage of being simple distributed real-time dynamic routing methods that attain and even exceed the performance gains demonstrated by centralized real-time routing systems such as TSMR. These benefits are illustrated in Section 12.3.

Event-dependent real-time routing methods represent another class of distributed real-time dynamic routing methods. Such event-dependent routing methods may use learning models such as LRR, STT routing, DAR, and STR.

In a mixed dynamic routing (MXDR) network, many different methods of dynamic routing are used simultaneously. Calls originating at different switches use the particular dynamic routing method implemented at that switch, and different switches could be a mix of, for example, STT, TSMR, and RTNR. Studies show that a mix of dynamic routing methods achieves good throughput performance in comparison with the performance of individual dynamic routing strategies.

Further details of each step of RTNR call setup are discussed in the following sections. Further details of RINR, LRR, and STT are discussed in Sections 12.2.9 and 12.2.10.

12.2.2   Determination of class-of-service

Determination of class-of-service begins with translation at the originating switch, as depicted in Figure 12.3. Digits are translated to determine the network switch number (NSN) of the terminating switch. As illustrated in Figure 12.3, if multiple ingress/egress routing is used, multiple destination network switch numbers are derived for the call.

Other data derived from call information, such as link characteristics, Q.931 message information elements, Information Interchange digits, and network control point routing information, are used to derive the class-of-service for the

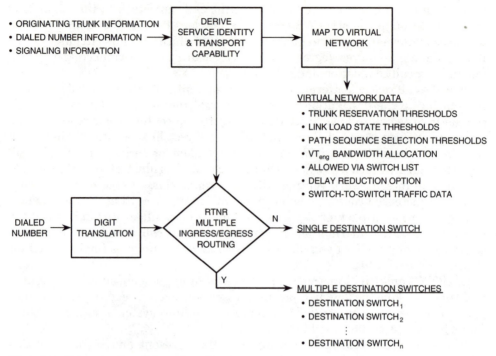

- ORIGINATING TRUNK INFORMATION
- DIALED NUMBER INFORMATION
- SIGNALING INFORMATION

DERIVE SERVICE IDENTITY & TRANSPORT CAPABILITY

MAP TO VIRTUAL NETWORK

VIRTUAL NETWORK DATA

- TRUNK RESERVATION THRESHOLDS
- LINK LOAD STATE THRESHOLDS
- PATH SEQUENCE SELECTION THRESHOLDS
- VT_{eng} BANDWIDTH ALLOCATION
- ALLOWED VIA SWITCH LIST
- DELAY REDUCTION OPTION
- SWITCH-TO-SWITCH TRAFFIC DATA

DIALED NUMBER

DIGIT TRANSLATION

RTNR MULTIPLE INGRESS/EGRESS ROUTING

N → SINGLE DESTINATION SWITCH

Y → MULTIPLE DESTINATION SWITCHES

- DESTINATION SWITCH$_1$
- DESTINATION SWITCH$_2$
  ⋮
- DESTINATION SWITCH$_n$

Figure 12.3 Determination of class-of-service

call. Class-of-service consists of four subparameters: service identity (SI), which describes the actual service associated with the call; transport capability (TC), which describes the specific link and facility type, such as voice or digital data transport, that the call should access; the virtual network (VN), which describes the bandwidth allocation and routing table parameters to be used by the call; and the circuit capability, which describes the circuit hardware capabilities such as fiber, radio, satellite, and digital circuit multiplexing equipment (DCME), that the call should require, prefer, or avoid. The combination of service identity, transport capability, virtual network, and circuit capability constitute the class-of-service, which together with the network switch number is used to access routing table data and the allowed via switch list, which are discussed further in the sections below.

Figure 12.4 illustrates service identity derivation, which uses the type of origin, type of destination, signaling service type, and dialed number service type to derive the service identity. The type of origin can be derived normally from the type of incoming trunk, connecting either to a directly connected (also known as nodal) customer equipment location, a switched access local exchange carrier, or an international carrier location. Similarly, based on the dialed numbering plan, the type of destination is derived and can be a directly connected (nodal) customer location if a private numbering plan is used (for example, within a software-defined network), a switched access customer location if a North

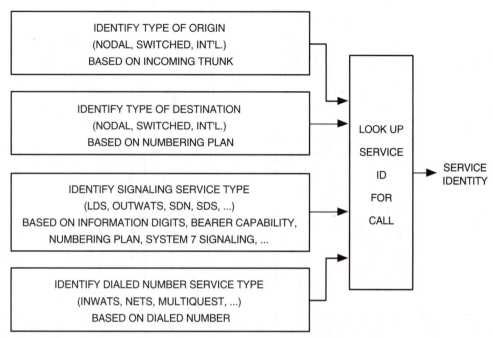

Figure 12.4   Service identity derivation

American Numbering Plan (NANP) number is used to the destination, or an international customer location if the international E.164 numbering plan is used. Signaling service type is derived based on bearer capability within signaling messages, information digits in dialed digit codes, numbering plan, or other signaling information and can indicate long-distance service (LDS), software-defined network (SDN) service, switched digital service (SDS), and other service types. Finally, dialed number service type is derived based on special dialed number codes such as 800 numbers or 900 numbers and can indicate 800 (INWATS) service, 900 (MULTIQUEST) service, and other service types. Type of origin, type of destination, signaling service type, and dialed number service type are then all used to derive the service identity. Transport capability is derived based on signaling information such as bearer capability and numbering plan information and can indicate voice, 64-kbps data, 384-kbps data, 1,536-kbps data, and other transport capabilities.

Figures 12.5–12.8 give examples of the derivation of class-of-service parameters. Figure 12.5 illustrates the derivation of a long-distance service identity, number 7, which is derived from the following information:

- The type of origination is switched access local exchange carrier, because the call originates from a local exchange carrier switch.

- The type of destination is switched access local exchange carrier, based on the the NANP dialed number.

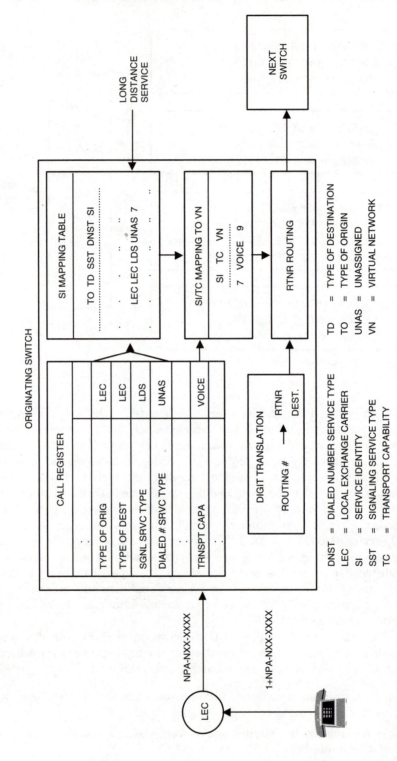

Figure 12.5 Example of class-of-service call processing: long-distance service (LDS)

- The signaling service type is long-distance service, based on the numbering plan (NANP).
- The dialed number service type is not used to distinguish long-distance service identity 7.

As shown, the service identity mapping table uses the above four inputs to derive the service identity. This table is changeable by administrative updates, in which new service information can be defined without ESS software modifications. The transport capability of voice is derived based on the numbering plan (NANP). From the service identity and transport capability the SI/TC-to-virtual network mapping table is used to derive virtual network number 9. Figure 12.6 illustrates the derivation of software-defined network service, which is mapped to virtual network 1. Figure 12.7 illustrates the derivation of 800 service, which is also mapped to virtual network 1. In this illustration, the dialed number service type is used to distinguish service identity 4. Figure 12.8 illustrates the derivation of international long-distance service, which is mapped to virtual network 2.

Figure 12.9 illustrates the virtual network table. Here the service identities and transport capabilities are mapped to individual virtual networks. Note that several virtual networks can share the same transport capability. For example, there can be normal voice service virtual networks such as 1, 2, and 9, and priority voice service virtual networks such as 6 and 8, all sharing the voice transport capability. Routing parameters for priority or key services are discussed further in the sections below.

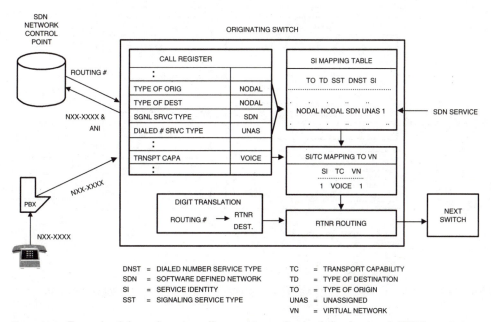

Figure 12.6 Example of class-of-service call processing: software-defined network (SDN) service

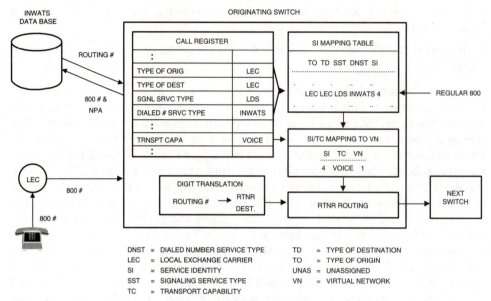

DNST = DIALED NUMBER SERVICE TYPE   TD   = TYPE OF DESTINATION
LEC = LOCAL EXCHANGE CARRIER   TO   = TYPE OF ORIGIN
SI   = SERVICE IDENTITY   UNAS = UNASSIGNED
SST = SIGNALING SERVICE TYPE   VN   = VIRTUAL NETWORK
TC   = TRANSPORT CAPABILITY

Figure 12.7   Example of class-of-service call processing: 800 service

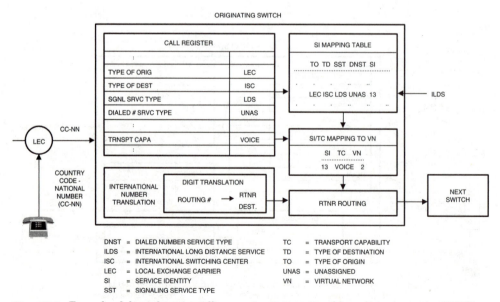

DNST = DIALED NUMBER SERVICE TYPE   TC   = TRANSPORT CAPABILITY
ILDS = INTERNATIONAL LONG DISTANCE SERVICE   TD   = TYPE OF DESTINATION
ISC = INTERNATIONAL SWITCHING CENTER   TO   = TYPE OF ORIGIN
LEC = LOCAL EXCHANGE CARRIER   UNAS = UNASSIGNED
SI   = SERVICE IDENTITY   VN   = VIRTUAL NETWORK
SST = SIGNALING SERVICE TYPE

Figure 12.8   Example of class-of-service call processing: international long-distance service (ILDS)

VN INDEX	VN NAME	SERVICE IDENTITIES
VN-1	VOICE	SDN, M800, MEGACOM, INWATS, WATS, MULTIQUEST, 800 CANADA, SRAS
VN-2	VOICE-INT'L	ILDS OUTBOUND, I800 OUTBOUND, GSDN OUTBOUND, WORLD CONNECT, MEGACOM INTERNATIONAL, TRANSIT
VN-3	64 - DATA	SDS, SDI OUTBOUND AND TRANSIT, GSDN OUTBOUND, SDN
VN-4	384 - DATA	SDS, SDI, SDN
VN-5	1536 - DATA	SDS, SDN
VN-6	VOICE-KEY	ILDS INBOUND, GSDN INBOUND AND TRANSIT, I800 INBOUND AND TRANSIT
VN-7	64-KEY DATA	SDN-KEY, GSDN INBOUND AND TRANSIT, SDI INBOUND
VN-8	VOICE - KEY	MEGACOM 800 GOLD, 800 GOLD, SDN-KEY
VN-9	VOICE	LDS (BUSINESS & RESIDENCE)
VN-10	VOICE	GETS, GETS-INTERNATIONAL

Figure 12.9 Virtual network table

Circuit capability selection allows calls to be routed on specific transmission circuits that have the particular characteristics required by these calls. A call can require, prefer, or avoid a set of transmission characteristics such as fiber transmission, radio transmission, satellite transmission, or compressed voice transmission. Circuit selection requirements for the call can be determined by the service identity of the call or by other information derived from the signaling message or from the routing number. The trunk hunt logic allows the call to skip those trunks that have undesired characteristics and to seek a best match for the requirements of the call.

12.2.3 Determination of link load states

The virtual network routing pattern selected for the service designates which network capacity is allowed to be selected for each call. In using the virtual network routing pattern to select network capacity, the originating switch always first tries to route the call on the direct link. Failing that, the originating switch then routes the call on a two-link path through a via switch. The via switch is selected by exchanging the link load status information between the originating switch and the terminating switch. The originating switch sends a query message to the terminating switch to obtain the bit map(s) of link load states. The specific load state information to be included in the query response is a function of the virtual network routing table, the switch-to-switch blocking from the originating switch to the terminating switch, and the total office congestion from the originating switch to all other switches measured over the last periodic update interval. The terminating switch sends the bit map of link states to the originating switch in its query response message, which is sent over the CCS network. As illustrated in Figure 1.27, the originating switch then "ands" its own bit map(s) of link states with the bit map(s) received from the terminating switch. The originating switch also "ands" the resulting bit map with the allowed via switch list to obtain the allowed paths for the call.

Hence, the determination of the link load states is fundamental to the selection of network capacity, on either the direct link or via paths. We discuss determination of link load states first and then discuss the selection of direct link and via path capacities. In identifying link load states, we use the six-state link model illustrated in Figure 12.10. The six states are maximum lightly loaded (LL1), medium lightly loaded (LL2), minimum lightly loaded (LL3), heavily loaded (HL), reserved (R), and busy (B). These link states are used in RTNR direct link and via path capacity selection, as described below.

Before we proceed further, we discuss the determination of the reservation bandwidth on each link. The logic for RTNR path selection implements a trunk reservation logic that is used to favor direct link traffic in situations of switch-to-switch and link congestion. If switch-to-switch blocking is detected for a virtual network over a periodic update interval, trunk reservation is triggered and the reservation level R^i is set for virtual network i according to the level of switch-to-switch congestion. In this manner traffic attempting to alternate-route over the direct link of a triggered switch pair is subject to trunk reservation, and the direct traffic is favored for that triggered switch pair. At the same time, the lightly loaded and heavily loaded state thresholds are raised accordingly in order to accommodate the reserved bandwidth capacity R^i for virtual network i. Trunk reservation blocking models and performance analysis are presented in

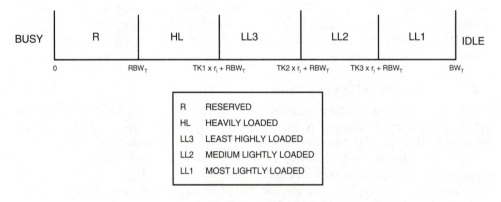

Figure 12.10 RTNR six-state link model

Chapter 8. Example results are presented in Figure 8.4, which illustrates the robustness of dynamic trunk reservation in protecting the preferred traffic across wide variations in traffic conditions.

Hence there is a reservation level R^i calculated for each virtual network on each link. It is based on the blocking level for the virtual network and the level of estimated switch-to-switch traffic. The switch-to-switch blocking level is equal to the overflow count (OV^i) divided by peg count (PC^i) over the last periodic update interval for virtual network i, which is typically three minutes. If the blocking for the virtual network exceeds a threshold value, the reservation level R^i is calculated based on the level of estimated offered traffic load for the virtual network, TL^i, which in turn is used to estimate a minimum virtual trunk requirement for the virtual network to meet its blocking objective, which is denoted as VT^i_{traf}. That is,

$$VT^i_{traf} = 1.1 \times TL^i$$

and the reservation level R^i in virtual trunks (VTs) is computed according to Table 12.1. Similar tables are used for $r_i = 384$ kbps and $r_i = 1,536$ kbps. The method to estimate TL^i is detailed later in this section. In addition, the threshold level for the next level of trunk reservation is computed each periodic update interval, and if the number of blocked calls for a given virtual network and network switch number destination exceeds the threshold, the switch immediately triggers that next level of trunk reservation. For example, if the current trunk reservation level is 0, as given in Table 12.1, and if the number of blocked calls reaches the number needed for trunk reservation level 1, then level 1 is triggered immediately. In this way reservation protection is applied immediately following a load surge or transport failure, which greatly aids in improving the switch and network responses to load variation and failure, as is demonstrated in the simulation results presented in Section 12.3.

There are tables similar to Table 12.1 for the link reservation levels for other transport capabilities, including 64-kbps data, 384-kbps data, and 1,536-kbps data. Note here that the reservation level for each virtual network is at most R^i, as defined above, but is also bounded above by $VT^i_{traf} - CIP^i$. That is,

$$R^i_{traf} = n^i + min[R^i, max(0, VT^i_{traf} - CIP^i)]$$

TABLE 12.1   Link Reservation Level ($r_i = 64$ kbps)

Switch–switch blocking, % threshold	Reservation level	R^i (VTs)	R^i_{min} (VTs)	R^i_{max} (VTs)
[0,1]	0	0	0	0
(1,5]	1	$0.05 \times VT^i_{traf}$	2	10
(5,15]	2	$0.1 \times VT^i_{traf}$	4	20
(15,50]	3	$0.15 \times VT^i_{traf}$	6	30
(50,100]	4	$0.2 \times VT^i_{traf}$	8	40

where R^i_{traf} is the number of VTs reserved for virtual network i, n^i is the number of VTs continuously reserved for virtual network i, and CIP^i is the number of switch-to-switch calls in progress for virtual network i. The total reserved bandwidth capacity RBW_T on the integrated link is given by

$$\text{RBW}_T = \sum_{j=\text{VN}_1}^{\text{VN}_N} R^j_{\text{traf}} \times r_j$$

In the above equation, trunk reservation is only carried from the higher-data-rate transport capabilities to the equal- or lower-data-rate transport capabilities, not from lower- to higher-data-rate transport capabilities. For example, for the voice transport capability, the total number of reserved VTs used to determine the link load state is the sum of the VTs reserved for the voice transport capability, the 64-kbps transport capability, and the 384-kbps transport capability. However, only the 384-kbps transport capability reserved VTs are used to compute the 384-kbps link load state. This allows more opportunity for higher-data-rate traffic to be completed because higher-data-rate calls require contiguous 64-kbps time slots (DS0s) for call completion.

Illustrative values of the VT thresholds to determine the different states in the six-state model are summarized in Table 12.2. Similar tables are used for $r_i = 384$ kbps and $r_i = 1,536$ kbps. Given these load state thresholds and the idle link bandwidth, ILBW,

$$\text{ILBW} = \text{BW}_T - \text{busy DS0s}$$

where BW_T is the total direct link bandwidth, we then define the link load states as illustrated in Table 12.3. Therefore, the reservation level and state boundary thresholds are proportional to the estimated offered traffic load level,

TABLE 12.2 Link Load State Thresholds ($r_i = 64$ kbps)

Threshold	Number of VTs	Minimum	Maximum
TK1	$0.05 \times \text{VT}^i_{\text{traf}}$	2	10
TK2	$0.1 \times \text{VT}^i_{\text{traf}}$	11	20
TK3	$0.2 \times \text{VT}^i_{\text{traf}}$	21	40

TABLE 12.3 Link Load State Definitions

Level of idle link bandwidth	State
$0 \leq \text{ILBW} < r_i$	busy
$r_i \leq \text{ILBW} \leq \text{RBW}_T$	R
$\text{RBW}_T < \text{ILBW} \leq \text{TK1} \times r_i + \text{RBW}_T$	HL
$\text{TK1} \times r_i + \text{RBW}_T < \text{ILBW} \leq \text{TK2} \times r_i + \text{RBW}_T$	LL3
$\text{TK2} \times r_i + \text{RBW}_T < \text{ILBW} \leq \text{TK3} \times r_i + \text{RBW}_T$	LL2
$\text{TK3} \times r_i + \text{RBW}_T < \text{ILBW}$	LL1

which means that the number of trunks reserved and the number of idle trunks required to constitute a "lightly loaded" link rise and fall with the traffic load, as, intuitively, they should. Furthermore, the reservation is applied only up to the VT^i_{traf} level of calls in progress for virtual network i, which we recall is the estimated capacity required for the virtual network to meet its blocking objective. In other words, once the number of calls in progress for virtual network i has reached its estimated level of required virtual trunks VT^i_{traf} to meet its blocking objective, trunk reservation is no longer required to assist the services using the virtual network in meeting the blocking objective. Once reservation is deactivated, the link load state increases and the link capacity is again made available for via calls. Hence, reservation is activated and deactivated as the call level rises and falls for the virtual network when there is switch-to-switch blocking detected for the virtual network, that is, when $R^i > 0$.

12.2.4 Selection of direct link capacity

We now discuss the selection of direct link capacity, as illustrated in the flowchart of Figure 12.11 and depicted schematically in the diagram shown in Figure 12.12.

SELECTION OF DIRECT LINK CAPACITY

ILBW = BW_T - BUSY DS0s
FOR NORMAL VIRTUAL NETWORKS:
- IF CIP < VT_{eng} and ILBW $\geq$ r + RBW_{eng} (KEY VIRTUAL NETWORKS), SELECT DIRECT LINK, OR
FOR KEY VIRTUAL NETWORKS:
- IF $CIP < VT_{eng}$ AND ILBW $\geq$ r, SELECT DIRECT LINK, OR
- IF $CIP \geq VT_{eng}$ AND ILBW $\geq$ r + RBW_{eng} (ALL VIRTUAL NETWORKS), SELECT DIRECT LINK,
ELSE
- GO TO VIA PATH SELECTION

Figure 12.11 Selection of direct link capacity

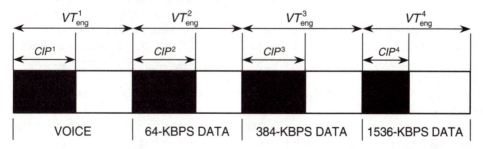

Figure 12.12 Direct link bandwidth allocation

The model illustrated in these figures shows that there is a minimum bandwidth on the direct link defined for each virtual network, denoted VT_{eng}^i, which is a minimum guaranteed VT bandwidth for the virtual network if there is blocking and sufficient traffic to use the VT_{eng}^i bandwidth. If a virtual network is meeting its blocking objective, other virtual networks are free to share the VT_{eng}^i bandwidth allocated to that virtual network. This sharing process on the direct link is implemented by allowing a service on a virtual network to always seize a VT on the direct link if the calls in progress for virtual network i (CIP^i) are below the level VT_{eng}^i. If CIP^i is equal to or greater than VT_{eng}^i, the service can seize a trunk on the direct link, as long as the idle bandwidth on the link is greater than the bandwidth reserved by other virtual networks that are not meeting their blocking objectives. That is, if

$$CIP^i \geq VT_{eng}^i$$

and

$$ILBW \geq r_i + RBW_{eng}$$

where

$$RBW_{eng} = \sum_{j=VN_1}^{VN_N} \{n^j + \min[R^j, \max(0, VT_{eng}^j - CIP^j)]\} \times r_j$$

then we select a VT on the direct link. The second expression means that a call on a virtual network that is above its designed call level can be routed on the direct link if there is capacity in excess of what is reserved for other virtual networks that are below their designed call levels. That is, the reserved bandwidth on the direct link for any virtual network is at most $VT_{eng}^i - CIP^i$, and furthermore this capacity is reserved only if reservation is triggered on the switch pair for that virtual network (that is, R^i is greater than 0). Once the number of calls in progress reaches the VT_{eng}^i level, calls can route on the direct link only when bandwidth is not reserved for direct routing of other virtual networks, which may be below their VT_{eng}^i values and experiencing switch-to-switch blocking. Here again the rule is used that trunk reservation is only carried from higher-data-rate transport capabilities (for example, r_i = 384 kbps or r_i = 1,536 kbps) to equal- or lower-data-rate transport capabilities, not from lower- to higher-data-rate transport capabilities.

A further refinement of direct link capacity selection is the capability to designate key services within each transport capability (identified by designating virtual networks as "key virtual networks"). Key services are given preferential treatment on the direct link as follows. The quantity RBW_{eng} is kept separately for key virtual networks, as well as for all virtual networks. As mentioned above, for key virtual networks when $CIP^i < VT_{eng}^i$, an idle VT on the direct link can always be seized. However, an additional restriction is imposed in selecting direct link capacity for calls on normal virtual networks; that is, if $CIP^i < VT_{eng}^i$, we select a VT on the direct link only if

$$\text{ILBW} \geq r_i + \text{RBW}_{\text{eng}} \text{ (key virtual networks)}$$

This additional restriction allows preferential treatment for key services, especially under network failures in which there is insufficient capacity to complete all calls.

Figure 12.13 illustrates the functioning of the capacity reservation for key and nonkey services. As shown, under normal load conditions, virtual networks share the allocated bandwidth if there is no congestion but protect their allocated bandwidth with trunk reservation protection when there is overload congestion. When there is failure of link capacity, the reservation mechanisms are such that the key services get their allocated bandwidth in preference to nonkey services, as illustrated on the right side of Figure 12.13.

The quantities VT^i_{eng} are chosen in the network design process so that their sum over all services sharing the bandwidth of the direct link is equal to the total bandwidth BW_T on the link. In this manner each virtual network is ensured a minimal level of network throughput performance determined by the network design process, and the computed VT^i_{eng} values are input to the switches after they are determined by the design process.

If direct link bandwidth cannot be accessed in the way described above, the call is routed to possible via paths, as described in the next section.

12.2.5 Selection of via path capacity

We now describe the process of determining a via path to be used by a call through the generation of a query to the terminating switch and through use of a stored via switch, as illustrated in Figure 12.14. As described above, if the direct link is busy or does not exist, the originating switch sends a query to the terminating switch to obtain the bit map of link state information (for example, the LL1 state). The terminating switch sends the bit map to the originating switch over the CCS network. The originating switch then "ands" its own bit map of links in the LL1 state with the bit map received from the terminating switch, as illustrated in Figure 1.27. The originating switch also "ands" the resulting bit map with the allowed via switch list that identifies switches that meet transmission quality parameters to obtain the allowed LL1 paths. The paths that do not meet all transmission quality requirements are designated as grade-2 paths. The originating switch selects the allowed LL1 path that is next after the previously selected path on a rotating network switch number basis.

At this point, we discuss the process of selecting the most desired via path through the network from all the available candidates. The process is outlined in the flowchart of Figure 12.14, in which the processes of path sequence selection and query for a via path are illustrated. There are two path sequences defined, which are illustrated in Table 12.4. The first path sequence consists of the direct path, if it exists, and the most lightly loaded path, medium lightly

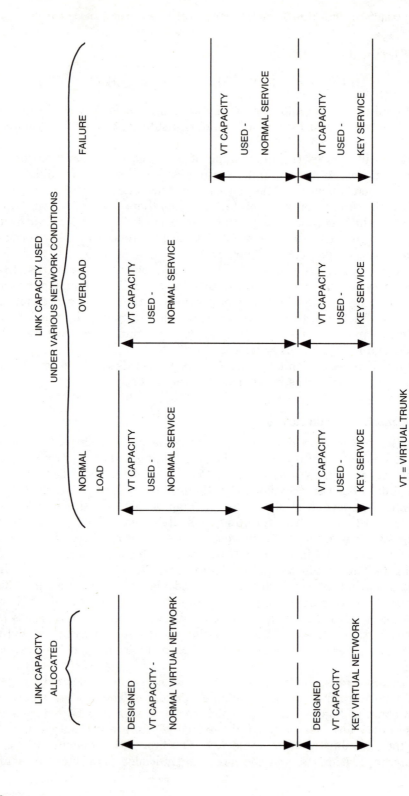

Figure 12.13 Direct link bandwidth allocation and sharing

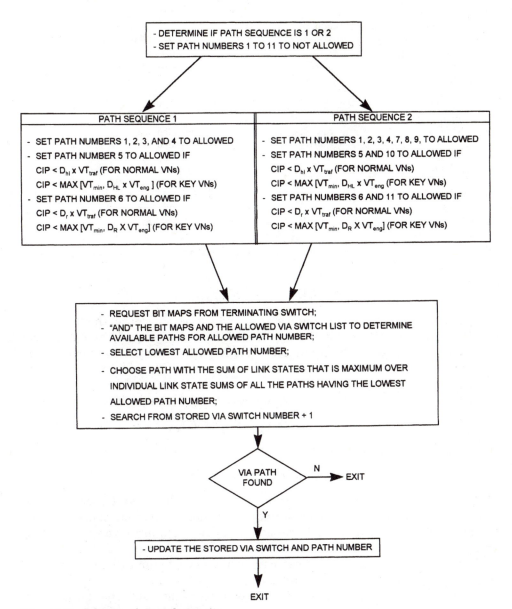

Figure 12.14 Selection of via path capacity

loaded path, least lightly loaded path, heavily loaded path, and reserved path. Two thresholds, D_{hl}^i and D_r^i, are defined to be associated with the path sequence. These thresholds allow a "controlled use" of the associated path, as follows. For normal services, if

$$\text{CIP}^i < D_{hl}^i \times \text{VT}_{\text{traf}}^i$$

TABLE 12.4   Path Sequence Definition

Path sequence number	1	2
direct	1	1
via LL1	2	2
via LL2	3	3
via LL3	4	4
via HL	5 (controlled use)	5 (controlled use)
via R	6 (controlled use)	6 (controlled use)
via LL1 (grade 2)	-	7
via LL2 (grade 2)	-	8
via LL3 (grade 2)	-	9
via HL (grade 2)	-	10 (controlled use)
via R (grade 2)	-	11 (controlled use)

then calls for that switch pair can be routed using heavily loaded via paths. Also, if

$$\text{CIP}^i < D_r^i \times \text{VT}_{\text{traf}}^i$$

then calls for that switch pair can use reserved via paths. Hence, the factors D_{hl}^i and D_r^i are factors controlling the depth of search for a via path on a given call on a given virtual network. Similarly, a second path sequence is defined, which also includes the grade-2 via paths that do not meet all the transmission quality requirements.

For key services, calls can be routed on heavily loaded paths if

$$\text{CIP}^i < \max(\text{VT}_{\text{min}}^i, D_{\text{HL}}^i \times \text{VT}_{\text{eng}}^i)$$

and on reserved paths if

$$\text{CIP}^i < \max(\text{VT}_{\text{min}}^i, D_R^i \times \text{VT}_{\text{eng}}^i)$$

where D_{HL}^i and D_R^i are fixed lower bounds on the depth factors D_{hl}^i and D_r^i, and VT_{min}^i is a fixed lower bound on key service search depth, all of which are provided in the administration process. Here again, key services are given preferential treatment, but only up to a maximum level of key service traffic. This choking mechanism for selecting via path capacity is necessary to limit the total capacity actually allocated to key service traffic.

In general, greater search depth is allowed if blocking is detected for a switch pair, because more alternate route choices serve to reduce the blocking of the switch pair. This greater search depth is inhibited, however, if the total office blocking reaches a high level, indicating that a general overload condition obtains, in which case it is more advantageous to reduce alternate routing. Also, if there are no direct trunks or the switch-to-switch traffic is small, say with $\text{VT}_{\text{traf}}^i$ less than 15 VTs, the search depth is again increased because trunk reservation becomes ineffective or even impossible because of the lack of direct trunks, and greater dependence on alternate routing is needed to meet network blocking objectives.

TABLE 12.5   Path Sequence Selection Thresholds (Voice Transport Capability)

Item			Traffic conditions		Path sequence selection		
	VT_{traf}^i	VT_{eng}^i	Total switch blocking, %	Switch–switch blocking, %	Path sequence	D_{hl}^i	D_r^i
A	>15	>0	[0, 3]	[0, 1] (1, 100]	1 2	0 1.0	0 0
B	≤15	>0	[0, 3]	[0, 1] (1, 100]	2 2	1.0 1.0	0 1.0
C	all	>0	(3, 10]	[0, 50] (50, 100]	1 2	0 1.0	0 0
D	all	>0	(10, 100]	[0, 50] (50, 100]	1 2	0 0	0 0
E	all	=0	[0, 3]	[0, 100]	2	8.0	8.0
F	all	=0	(3, 100]	[0, 100]	2	8.0	0

Hence, the path sequence selection, as well as the depth factors D_{hl}^i and D_r^i, is controlled by various blocking thresholds, the existence of direct trunks, and the level of offered switch-to-switch (direct) traffic VT_{traf}^i, as illustrated in Table 12.5. This path sequence selection table is defined for each virtual network i. The selection of paths is given by the congestion state based on the total switch blocking over a periodic update interval, switch-to-switch blocking over an update interval, and whether or not direct link capacity exists. In this manner the blocking performance of the virtual network can be controlled by the depth of access to via paths in the heavily loaded and reserved states. Simulations have shown that the values of these parameters, illustrated in Table 12.5, provide a good allocation of available network capacity across virtual networks for switch pairs that are experiencing blocking.

A better blocking probability grade-of-service can be achieved if greater search depth is allowed on a selected switch-pair basis in order to reduce switch-to-switch blocking for high-blocking switch pairs. For example, in Table 12.5 the heavily loaded depth thresholds can be set to higher values for items A, B, C, and D, and the reserved depth thresholds can be set to higher values for items A, B, and C for specific switch pairs, which are input from the capacity management system through the routing administration system. This is accomplished by raising the lower bounds on the depth factors, D_{HL}^i and D_R^i, to appropriately higher values. This routing technique is employed to reduce the severity of blocking for problem switch pairs during the limited time period over which additional trunk capacity is being provisioned for them. Simulation results for this capability are presented in Section 12.3.

Figures 1.27 and 12.15 to 12.17 illustrate the RTNR via switch selection process. Once the RTNR path sequence is determined, the originating switch requests the

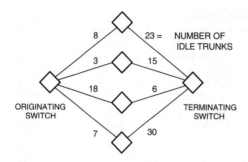

- NO. OF TRUNKS (IN ALL LINKS)   = 100
- BLOCKING < 1%
- THRESHOLDS:
  - MOST LIGHTLY LOADED   = 20
  - MODERATE LIGHTLY LOADED   = 10
  - LEAST LIGHTLY LOADED   =   5

Figure 12.15   RTNR path selection example

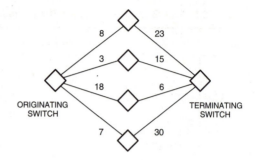

- MOST LIGHTLY LOADED
  THRESHOLD = 20
- INCLUDE LINKS WITH IDLE
  TRUNKS ≥ 20
- NO MATCHES

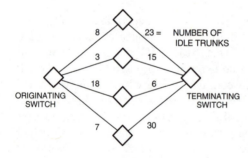

- MODERATE LIGHTLY LOADED
  THRESHOLD = 10
- INCLUDE LINKS WITH
  IDLE TRUNKS ≥ 10
- NO MATCHES

Figure 12.16   RTNR path selection example

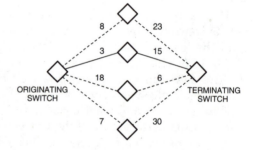

- LEAST LIGHTLY LOADED
  THRESHOLD = 5
- INCLUDE LINKS WITH IDLE
  TRUNKS ≥ 5
- 3 CHOICES (DASHED LINE
  PATHS)

appropriate load status bit maps from the terminating switch through a query message sent to the terminating switch. Upon receiving the bit maps from the terminating switch, the originating switch "ands" the bit maps with the allowed via switch list, resulting in a set of paths. The originating switch then identifies the paths with the lowest allowed path number, and on each of these identified paths it computes the sum of the states of both links in the path. The path with the highest sum, that is, the path with the maximum total idle capacity, is then

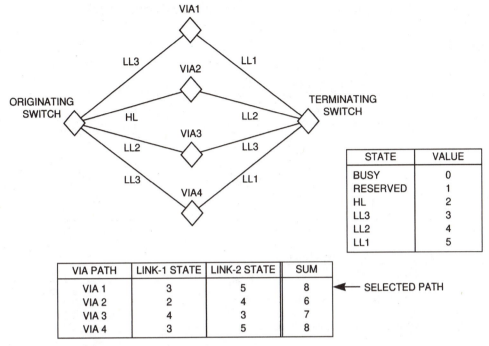

STATE	VALUE
BUSY	0
RESERVED	1
HL	2
LL3	3
LL2	4
LL1	5

VIA PATH	LINK-1 STATE	LINK-2 STATE	SUM
VIA 1	3	5	8
VIA 2	2	4	6
VIA 3	4	3	7
VIA 4	3	5	8

◄—— SELECTED PATH

Figure 12.17   RTNR path selection example

chosen. Figures 12.15 to 12.17 illustrate the operation of bit maps for selection of idle paths. The left side of Figure 12.15 illustrates the number of idle (virtual) trunks on a set of four two-link paths and the thresholds for LL1, LL2, and LL3 link states. As can be seen on the right side of Figure 12.15, none of the four paths has both links with idle trunks exceeding the LL1 threshold of 20 trunks—therefore none of the paths is in the LL1 state. The left side of Figure 12.16 illustrates that none of the four paths has both links with idle trunks exceeding the LL2 threshold of 10 trunks—therefore none of the paths is in the LL2 state. The right side of Figure 12.16 illustrates that three of the four paths have both links with idle trunks exceeding the LL3 threshold of five trunks, and the path choice is between the bottom and top of the three paths, because, as shown in Figure 12.17, the second link for each of those two paths is in the LL1 state, and the second link for the other path is in the LL2 state. Because there are equal paths to choose from, the next path in cyclic rotation from the previous path is selected. That is, if the bottom one of the four paths were the last path selected, the next path above it would be chosen for this call. In this case, the path choice would then be the top path in Figure 12.16. Figure 12.17 illustrates this selection in a numerical representation of the link states and path selection.

Figure 12.18 illustrates path selection in the presence of trunk reservation. If in this example there is blocking and hence trunk reservation between the originating switch and switch V1, the thresholds for the link state determination on the originating-switch-to-V1 link are raised accordingly; in this case the VT_{traf}

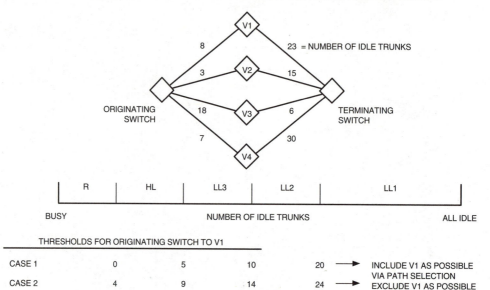

Figure 12.18 RTNR trunk reservation operation

level is such that the threshold is raised by four, which is the number of trunks reserved on the link. Therefore, the path through switch V1 is put into the heavily loaded (HL) state and is excluded from consideration because there is an LL3 path to select instead (through V3 or V4). This action protects the paths through switch V1 from being used by via traffic; when the blocking subsides on the originating-switch-to-V1 link, the reservation is dropped back to zero trunks and the path can then be selected as before.

12.2.6 Call setup delay reduction

The above process of via path selection has some additional post-dialing delay associated with a via call. This is because the originating switch must launch a query message to the terminating switch, wait for the query response, and analyze the bit map(s) before it can route a via call. In order to reduce the volume of calls with this delay, a delay reduction technique is used. As illustrated in Figure 12.19, this delay reduction method uses the concept of a stored-via-switch register that is continually updated and used to select a via switch to complete two-link calls. When a two-link path is required (i.e., the direct path is busy), the originating switch performs the following two steps simultaneously:

1. The originating switch sets up the two-link call using the current stored via switch. If the call is blocked on either link, the originating switch then attempts to set up the call on the updated stored via switch, once step 2 below has been completed.

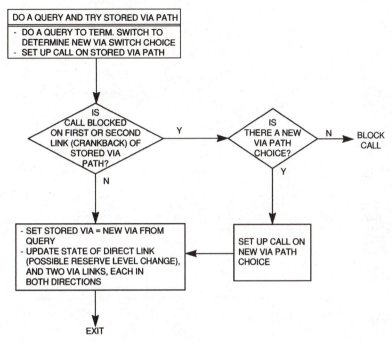

Figure 12.19 Call setup delay reduction

2. The originating switch determines an updated stored via switch each time a two-link path is required—that is, every time step 1 above is executed—by sending a query message to the terminating switch as described above. Once this updated stored via switch is determined, it replaces the current stored via switch in the stored-via-switch register.

Note that a crankback from the via switch to the originating switch is required in step 1 if the second link of the two-link path is busy.

Figure 12.20 illustrates the workings of the delay reduction technique. In this example, for calls from the originating switch to the terminating switch, RTNR attempts to route the call using the direct link. Once the last trunk in the link is used (by the $n - 1$th call), the originating switch sends a query message to the terminating switch; that is, the originating switch executes step 2 above. The originating switch then uses the link status information to derive an updated stored via switch, VN3, for the next call. As the n th call arrives, the originating switch tries to use via switch 3 to connect the call if the direct link is busy. As soon as the originating switch attempts to set up the n th call through via switch 3, the originating switch also initiates a query to the terminating switch for link status bit maps. The comparison of bit map(s) results in an updated stored via switch (via switch 5), possibly for the $n + 1$th call. If the originating switch is successful in connecting the n th call through via switch 3, there is no additional delay for the call.

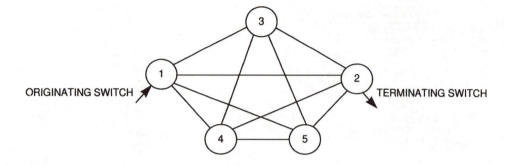

CALL ARRIVAL	VIA FROM LAST CALL	VIA FROM LAST CALL BUSY?	NEW VIA DETERMINED	VIA USED FOR CURRENT CALL	DELAY?
n	3	NO	5	3	NO
n + 1	5	YES	4	4	YES
n + 2	4	NO	3	4	NO
.					
.					

Figure 12.20 Illustration of call setup delay reduction

When the $n + 1$th call arrives, the originating switch, as usual, first tries to use the direct link. If the direct link is busy, the originating switch then routes the call through via switch 5 and also initiates a query message to the terminating switch. It is possible that the path through via switch 5 is busy—if either the link between the originating switch and via switch 5 is busy or the link between via switch 5 and the terminating switch is busy, or both. If the link between the originating switch and via switch 5 is busy, the originating switch awaits the query response from the terminating switch for link status bit maps to derive an updated stored via switch (via switch 4). Via switch 4 is then used for the $n + 1$th call, and there is a delay for this call. If the link between via switch 5 and the terminating switch is busy, via switch 5 sends a crankback to the originating switch. Because the query for an updated stored via switch (via switch 4) is sent as the call is set up through via switch 5, the updated stored via switch (via switch 4) is used to connect the call. In this case, the delay is equivalent to the query delay because the query and the crankback take about the same amount of time.

As discussed in Section 12.3 on RTNR performance, the delay reduction technique is very effective in eliminating delays associated with RTNR queries.

12.2.7 Trunk selection

The actual trunk selection is made on the links associated with the selected path. The trunk is chosen from the list of trunk sublinks associated with the link that are capable of the associated transport capability. Trunk sublinks within

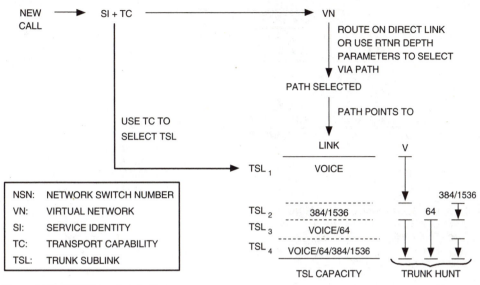

Figure 12.21 RTNR trunk selection

the link are designated by which transport capabilities they are able to carry. The originating switch selects the trunk from the allowed trunk sublinks and launches a CCS initial address message to the via switch containing the network switch number of the terminating switch, class-of-service information, dialed digits associated with the call, and other information.

The selection of a trunk from the trunk sublink is illustrated in Figure 12.21. Here the various trunk sublinks in the list are searched for an idle circuit, and the first available trunk that has the allowed trunk sublink corresponding to the transport capability of the call is selected.

Network control for RTNR uses three types of trunk sublinks: voice only, voice/data, and voice/data/wideband. Trunk hunt is performed first for dedicated trunk sublinks, then shared trunk sublinks, as illustrated in Figure 12.21. Furthermore, some of the voice/data/wideband capacity is reserved at all times for wideband traffic in order to meet network blocking objectives. Also, as illustrated in Section 12.3, network performance analysis shows that this sharing of wideband capacity during peak periods provides additional throughput for voice traffic when wideband traffic is minimal.

12.2.8 Periodic update

The update process is executed each periodic update interval to determine the switch-to-switch blocking level OV^i/PC^i, switch-to-switch reservation level R^i, switch-to-switch estimated traffic load and virtual trunk requirement TL^i and VT^i_{traf}, switch-to-switch path sequence with the D^i_{hl} and D^i_{r} values, and total switch-blocking level TSB^i. The estimate of the offered traffic load in period n is

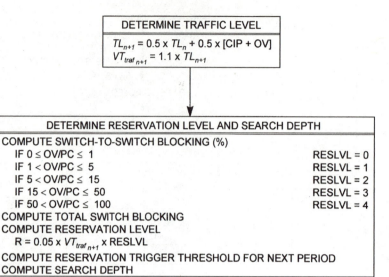

Figure 12.22 Periodic update

given by

$$\mathrm{TL}_n^i = 0.5 \times \mathrm{TL}_{n-1}^i + 0.5 \times (\mathrm{CIP}_n^i + \mathrm{OV}^i)$$

in which OV^i is the switch-to-switch overflow count over the period n for virtual network i and, as before, the virtual trunks required to meet the blocking objective are estimated by

$$\mathrm{VT}_{\mathrm{traf}}^i = 1.1 \times \mathrm{TL}_n^i$$

Hence, TL_n^i is estimated in the switch with an approximation for the blocking term given above. The reservation levels and the path sequence for each switch pair and the D_{hl}^i and D_{r}^i values are determined by applying the $\mathrm{VT}_{\mathrm{traf}}^i$ parameters in Tables 12.1 and 12.5. Figure 12.22 illustrates the periodic update process.

12.2.9 Real-time internetwork routing methods

As illustrated in Figure 1.29, real-time internetwork routing (RINR) extends class-of-service RTNR concepts to distributed real-time dynamic routing, for example, between a national network and all connecting international carrier networks, all connecting local exchange carrier networks, and all connecting customer networks. Like class-of-service RTNR, RINR uses a virtual network concept that enables service integration by allocating bandwidth for services and using real-time reservation controls. Therefore, trunks can be fully shared among virtual networks in the absence of congestion. When a certain virtual network encounters blocking, trunks are reserved to ensure that the virtual network reaches its allocated bandwidth. RINR provides class-of-service routing capabilities including key service protection, directional flow control, circuit selection capability, automatically updated time-variable bandwidth allocation, VT_{eng}, and greatly expanded transit routing capability through the use of overflow paths and control parameters such as RINR load set periods. Circuit selection capability allows specific trunk hardware characteristics, such as fiber transmission, to be preferentially selected. RINR improves performance and reduces the cost of the access/egress network with flexible routing capabilities.

Figure 1.29 illustrates how RINR extends RTNR capabilities to dynamic internetwork routing in the international application, on the links between international switching centers (ISC), gateway switches, and foreign country switches. With RINR, international calls are routed first to a direct engineered path, if it exists and is available, then to a list of alternate paths through alternate switches in the same country (ISC2 and ISC3 in Country 1), and finally to a list of alternate paths through switches in other countries, such as those in Country 2. Therefore, the RINR paths are divided into three types: the direct link or engineered path, alternate paths in the same country, and alternate or transit paths through other countries. Alternate paths use trunks to a via area to get to the destination area.

RINR tries to find an available alternate path based on load state and call completion performance, in which the originating switch uses its link busy/idle status to the via switch, in combination with the call completion performance from the via switch to the destination switch, in order to find the least-loaded, most available path to route the call over. For each path, a load state and a completion state are tracked. The load state indicates whether the trunks from the ISC switch to the destination area are lightly loaded, heavily loaded, reserved, or busy. The completion state indicates whether a path is achieving above-average completion, average completion, or below-average completion. The selection of a via path is based on the load state and completion state. Alternate paths in the same country and in a transit third country are each considered separately. Within a category of via paths, selection is based on the load state and completion state. During times of congestion, the trunks to a destination area may be in a reserved state, in which case the remaining trunks are reserved for traffic to the destination area. During periods of no congestion, capacity not needed by one virtual network is made available to other virtual networks that are experiencing loads above their allocation.

RINR uses four discrete load states for links; lightly loaded, heavily loaded, reserved, and busy. The number of idle trunks in a link is compared with the load state thresholds for the link to determine its load condition. This determination is made every time a trunk in the link is either seized or released. The load state thresholds used for a particular link are based on the current estimates of the same four quantities as maintained by RTNR for switch OS_j for each virtual network i and for each terminating area TA_k that can include one or an aggregation of a number of terminating switches: (1) the current number of calls in progress, CIP_k^i, which is the number of active calls from OS_j to TA_k for a particular virtual network i ; (2) the current switch-to-area blocking level NN_k^i for calls from OS_j to TA_k for a particular virtual network i ; (3) the offered traffic load TL_k^i to each of the other terminating areas in the network, which is based on the number of calls in progress CIP_k^i and the blocking rate to each terminating area, measured over the last several minutes; and (4) VT_{trafk}^i, which is the number of virtual trunks required to meet the blocking probability grade-of-service (GOS^i) objective for the current offered traffic load TL_k^i from OS_j to TA_k for a particular virtual network i.

The load state thresholds for the lightly loaded and heavily loaded states are set to fixed percentages of the VT_{trafk}^i estimate. As such, the load state thresholds rise as the VT_{trafk}^i estimate to that switch increases. Higher load state thresholds reduce the chances that the link is used for alternate path connections for calls to other switches; this enables the link to carry more direct traffic and therefore better handle the call load between the switches connected by the link. The reserved state threshold is based on the reservation level R_k^i calculated on each link, which in turn is based on the switch-to-area blocking level.

As mentioned previously, completion rate is tracked on the various via paths by taking account of the signaling message information relating either the successful completion or noncompletion of a call through the via area. A noncompletion, or failure, is scored for the call if a signaling release message is received from the far end after the call seizes an egress trunk in a link, indicating a network incompletion cause value. If no such signaling release message is received after the call seizes an egress trunk, then the call is scored as a success. There is no completion count for a call that does not seize an egress trunk. Each ISC keeps a call completion history of the success or failure of the last 10 calls using a particular via path, and it drops the oldest record and adds the call completion for the newest call on that path. Based on the number of call completions relative to the total number of calls, a completion state is computed using the completion rate thresholds given below.

The completion state is dynamic in that the call completions are continuously tracked, and if a path suddenly experiences a greater number of call noncompletions, calls are routed over the path whose completion rate represents the highest rate of call completions. Table 12.6 defines the completion states employed in RINR, where

- τ_c is the time in seconds since the last call failure recorded for a given path.

- PATH is specified by a path type, a destination area, a via area, and a transport capability (TC).

TABLE 12.6 Completion State Definitions

Completion state	Condition
NRFH (no recent failure history)	$\tau_c(\text{PATH}) > 300$
HC (high completion)	$\tau_c(\text{PATH}) \leq 300$ AND $C(\text{PATH}) \geq \text{HCthr}(\text{PTYPE,TC,AREA})$
AC (average completion)	$\tau_c(\text{PATH}) \leq 300$ AND $\text{LCthr}(\text{PTYPE,TC,AREA}) \leq C(\text{PATH}) < \text{HCthr}(\text{PTYPE,TC,AREA})$
LC (low completion)	$\tau_c(\text{PATH}) \leq 300$ AND $C(\text{PATH}) < \text{LCthr}(\text{PTYPE,TC,AREA})$

- PTYPE is the path type, which is either a first-choice alternate path through the same country; a second-choice alternate path; a first-choice overflow path through a third, or transit, country; or a second-choice overflow path.

- C(PATH) is the completion rate for the path, which is the number of completed calls out of the last 10 calls recorded.

- HCthr(PTYPE,TC,AREA) and LCthr(PTYPE,TC,AREA) are the high completion rate threshold and low completion rate threshold, respectively, for each path type, TC, and area, which are based on Cavg(PTYPE,TC,AREA) as indicated in Table 12.7.

- Cavg(PTYPE,TC,AREA) is the average number of call completions (rounded to the nearest integer) out of the last 10 calls for all paths with the specified PTYPE, TC, and area.

Table 12.7 lists typical completion rate thresholds. Every three minutes, the Cavg(PTYPE,TC,AREA) is computed and the thresholds HCthr(PTYPE,TC,AREA) and LCthr(PTYPE,TC,AREA) are determined. On a per-call basis, the call completion C(PATH) is compared with the appropriate HCthr(PTYPE,TC,AREA) and LCthr(PTYPE,TC,AREA) to select the route.

From the completion states defined in Table 12.6, calls are normally routed on the first path that experiences no recent failure or a high completion state with a lightly loaded egress link. A path experiencing no recent failure is one that, within a prescribed interval (say 300 seconds), has had no call failures on it. If such a path does not exist, then a path having an average completion state with a lightly loaded egress link is selected, followed by a path having a low completion state with a lightly loaded egress link. If no path with a lightly loaded egress link is available, and if the search depth permits the use of a heavily loaded egress link, the paths with heavily loaded egress links are searched in the order of no recent failure, high completion, average completion, and low completion. If no such paths are available, paths with reserved egress links are searched in the same order, based on the call completion state, if the search depth permits the use of a reserved egress link.

The rules for selecting direct and via paths for a call are governed by the availability of direct trunks and switch-to-area blocking. The path sequence consists of the direct path, if it exists, lightly loaded via paths, heavily loaded via paths, and reserved via paths. In general, greater path selection depth is allowed if blocking is detected to TA_k, because more alternate path choices serve to reduce

TABLE 12.7   Typical Completion Rate Thresholds

Cavg(PTYPE,TC,AREA)	0	1	2	3	4	5	6	7	8	9	10
HCthr(PTYPE,TC,AREA)	5	5	6	6	7	8	8	9	10	10	11
LCthr(PTYPE,TC,AREA)	3	3	3	3	4	4	4	5	5	5	6

the blocking to TA_k. The use of heavily loaded and reserved links for alternate path connections is relaxed for calls between two switches that are not connected by a direct link. In this manner the blocking performance can be controlled by the depth of access to via paths in the heavily loaded and reserved states. Links in the reserved state are normally used only for direct traffic but may be used for via routing when higher-priority calls are placed under heavy network congestion.

RINR implements class-of-service routing, which is similar to RTNR class-of-service routing and includes the following steps for call establishment: (1) The class-of-service (including the service identity, transport capability, virtual network, and circuit capability) and terminating area TA_k are identified; (2) the class-of-service and TA_k information are used to select the corresponding virtual network data, which include path sequence selection, reservation levels, load state thresholds, and traffic measurements; and (3) an appropriate path is selected through execution of path selection logic and the call is established on the selected path. The various virtual networks share bandwidth on the network links. On a weekly basis, RINR allocates a certain number of direct virtual trunks between OS_j and TA_k to virtual network i, which is referred to as VT^i_{engk}.

The ISC switch automatically computes the VT^i_{engk} bandwidth allocations once a week. A different allocation is used for each of 36 two-hour load set periods: 12 weekday, 12 Saturday, and 12 Sunday. The allocation of the bandwidth is based on a rolling average of the traffic load for each of the virtual networks, to each destination area, in each of the 36 load set periods. VT^i_{engk} is based on average traffic levels and is the minimum guaranteed bandwidth for virtual network i, but if virtual network i is meeting its GOS^i blocking objective, other virtual networks are free to share the VT^i_{engk} bandwidth allotted to it.

Switch OS_j uses the quantities VT^i_{trafk}, CIP^i_k, NN^i_k, and VT^i_{engk} to dynamically allocate link bandwidth to virtual networks. Under normal nonblocking network conditions, all virtual networks fully share all available capacity. Because of this, the network has the flexibility to carry a call overload between two switches for one virtual network if the traffic loads for other virtual networks are sufficiently below their design levels. An extreme call overload between two switches for one virtual network may cause calls for other virtual networks to be blocked, in which case trunks are reserved to ensure that each virtual network gets the amount of bandwidth allotted. This reservation during times of overload results in network performance that is analogous to having a number of trunks between the two switches dedicated for each virtual network.

As shown in Figure 12.13, sharing of bandwidth on the direct link is implemented by allowing calls on virtual network i to always seize a virtual trunk on the direct link if the calls in progress CIP^i_k are below the level VT^i_{engk}. However, if CIP^i_k is equal to or greater than VT^i_{engk}, calls on virtual network i can seize a virtual trunk on the direct link only when the idle-link bandwidth (ILBW) on the link is greater than the bandwidth reserved by other virtual networks that are not meeting their blocking objectives.

As in RTNR, key services in RINR are given preferential treatment on the direct engineered path, on which the reserved bandwidth RBW_{eng} is kept separately for key virtual networks, as well as for all virtual networks. For key virtual networks, if $\mathrm{CIP}_k^i < \mathrm{VT}_{engk}^i$, then an idle VT on the direct engineered path can always be seized. An additional restriction, however, is imposed in selecting engineered path capacity for calls for normal services. That is, if $\mathrm{CIP}_k^i < \mathrm{VT}_{engk}^i$, then we select a VT on the engineered path only if

$$\mathrm{ILBW} \geq r^i + \mathrm{RBW}_{eng} \text{ (key virtual networks)}$$

This additional restriction allows preferential treatment for key services, especially under network failures in which there is insufficient capacity to complete all calls, which is illustrated in Figure 12.13.

For key services, calls can be routed on heavily loaded paths if

$$\mathrm{CIP}_k^i < \max(\mathrm{VT}_{min}^i, D_{HL}^i \times \mathrm{VT}_{engk}^i)$$

and on reserved paths if

$$\mathrm{CIP}_k^i < \max(\mathrm{VT}_{min}^i, D_R^i \times \mathrm{VT}_{engk}^i)$$

where D_{HL}^i and D_R^i are fixed lower bounds on the depth factors D_{hl}^i and D_r^i and VT_{min}^i is a fixed lower bound on key service search depth. Here again, key services are given preferential treatment, but only up to a maximum level of key service traffic. This choking mechanism for selecting via path capacity is necessary to limit the total capacity actually allocated to key service traffic.

In general, greater search depth is allowed if blocking is detected from an ISC switch to an area, because more alternate path choices serve to reduce the blocking. If there are no direct engineered path trunks or the switch-to-area traffic is small, say with VT_{trafk}^i less than 15 VTs, then the search depth is again increased because trunk reservation becomes ineffective or even impossible, and greater dependence on alternate routing is needed to meet network blocking objectives. The key service protection mechanism provides an effective network capability for service protection. A constraint is that key service traffic should be a relatively small fraction (preferably less than 20 percent) of total network traffic. As in RTNR class-of-service routing administration, the provisioning of normal services and key services class-of-service routing logic for existing and new services can be flexibly supported via the RINR administrative process, without ESS switch development, once the marketing/service decision is made.

Circuit capability selection allows calls to be routed on specific transmission circuits that have the particular characteristics required by these calls. In general, a call can require, prefer, or avoid a set of transmission characteristics such as fiberoptic or radio transmission, satellite or terrestrial transmission, or compressed or uncompressed transmission. The circuit capability selection requirements for the call can be determined by the service identity of the call or by other information derived from the signaling message or from the routing number. The trunk hunt logic allows the call to skip trunks that have undesired characteristics and to seek a best match for the requirements of the call. For any service identity, a set of circuit capability selection preferences may be specified for the call. Circuit capability selection preferences override the normal order of selection of links, which is automatically derived by the RINR logic or can be provisioned by input parameters. Circuit capability selection preferences allow specific services to use the links in a different order.

For any call, circuit capability selection preferences may be set based on the service identity of the call. If a characteristic is required for a call, then any link that does not have that characteristic is skipped. If a characteristic is preferred, links with that characteristic are used first. Links without the preferred characteristic will be used next, but only if no links with the preferred characteristic are available. A preference can be set for the presence or absence of a characteristic. For example, if the absence of satellite is required, then only trunks with Satellite = No are used. If we prefer the absence of satellite, then trunks with Satellite = No are used first, then trunks with Satellite = Yes. A first preference and a second preference can be specified if needed. The first preference is considered more important than the second preference, so the order of selection is

1. Links with both first and second preference

2. Links with first but not second preference

3. Links with second but not first preference

4. Links with neither first nor second preference

RINR therefore extends class-of-service distributed real-time dynamic routing concepts to internetwork routing, such as the international network, which until now has used bilateral hierarchical routing with minimal flexibility. Performance results for RINR are presented in Section 12.3.

12.2.10   Event-dependent real-time routing methods

As discussed in Section 12.2.1, event-dependent real-time routing methods may use learning models such as LRR and STT routing. As illustrated in Figure 1.19, LRR is a distributed call-by-call routing method with routing updates based on random routing selection. LRR uses a simplified distributed learning method to achieve this flexible dynamic routing. With LRR, the direct link is used first if available, and a fixed alternate path is used until it is blocked, in which case a new alternate path is selected at random as the alternate route choice for the next call overflow from the direct link. No crankback is used at a via switch or egress switch in LRR, so a call blocked at a via switch or egress switch will be lost. Dynamically activated trunk reservation is used under call-blocking conditions. STT routing is an extension of the LRR method, in which crankback is allowed when a via path is blocked at the via switch, and the call advances to a new random path choice. In a limiting case of STT, all possible one- and two-link path choices can be tried by a given call before the call is blocked.

In the DAR and STR learning approach, as in LRR and STT, the alternate path last tried that was successful is tried again until blocked, at which time another alternate path is selected at random and tried on the next call. With DAR, a fixed trunk reservation technique is used. In the STR enhancement of DAR, the set of allowed path choices is changed with time in accordance with changes in traffic load patterns. Event-dependent real-time routing methods perform well but are not as efficient as state-dependent real-time dynamic routing methods, as illustrated in Section 12.3.

12.3 Modeling Results

12.3.1 Simulation model design of routing parameters

A call-by-call simulation model is used to measure the performance of the network under various routing strategies: DNHR, DNHR plus NEMOS centralized real-time path selection (as described in Chapter 8), RTNR three-state, RTNR four-state, RTNR five-state, and RTNR six-state models are simulated. In addition, an extensive number of different network scenarios are simulated to test threshold and parameter settings for the RTNR method. Extensive simulation is also made of the automatic congestion control method for RTNR networks.

A 65-switch and a 103-switch network model are used for the study. The models are designed by the path-EFO model for DNHR and simulated for a variety of traffic conditions to include average business day traffic, high-day traffic, Christmas day traffic, Mother's Day traffic, link failure scenarios, and focused overload scenarios. Five load set periods are used for the average business day and high-day studies: two morning busy hours, the afternoon busy hour, the evening busy hour, and the weekend (Sunday night) busy hour. Peak-day loads are modeled by the evening load set period of each peak day.

12.3.2 Models of load states

Before describing routing table modeling data, we discuss the link state models illustrated in Figure 12.23. An example of a six-state model is given at the top of the figure. The six states are maximum lightly loaded (LL3), medium lightly loaded (LL2), minimum lightly loaded (LL1), heavily loaded (HL), reserved (R), and busy (B). Illustrative values of the thresholds that define these six states are given in the figure. Five-state, four-state, and three-state models are defined that omit the LL3 state, the LL2 state, and the R state, respectively.

Table 12.8 illustrates the performance of RTNR for a high-day network load pattern. The average business day loads for the 65-switch DNHR network deployed in April 1987 were inflated uniformly by 30 percent. This inflation was necessary to illustrate the relative performance of DNHR and RTNR, because no blocking was observed for either routing method under the average business day loads. The table gives the average hourly blocking in load set periods or hours 2, 3, 5, 8, and 15, which correspond to the two early-morning busy hours, the afternoon busy hour, the evening busy hour, and the weekend (Sunday night) busy hour, respectively. Also given at the bottom of the table are the average blocking over these five load set periods, then the average blocking over the five load set periods for the switch pair with the highest average blocking, the switch pair that had the 99th percentile highest average blocking, and the switch pair that had the 90th percentile highest average blocking. The columns of Table 12.8 give results for

- 6-STATE MODEL

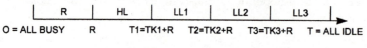

NUMBER OF IDLE TRUNKS

AUTOMATIC TRUNK RESERVATION	
SW. - SW. BLK, %	R(TRUNKS)
≤ 1	0
1 - 5	TK1
5 - 15	2 x TK1
> 15	3 x TK1

LOAD STATUS THRESHOLD			
	# TRUNKS	MIN	MAX
TK1	0.05 x T	2	10
TK2	2 x TK1	11	20
TK3	4 x TK1	21	40

- 5-STATE MODEL: NO LL3 STATE
- 4-STATE MODEL: NO LL3 AND LL2 STATES
- 3-STATE MODEL: NO LL3, LL2, AND RESERVED STATES

LEGEND: R (RESERVED), LL (LIGHTLY LOADED), AND HL (HEAVILY LOADED)

Figure 12.23   RTNR three-state, four-state, five-state, and six-state link models

TABLE 12.8   65-Switch Network Blocking Performance Comparison Under High-Day Network Traffic Load (%)

Hour	DNHR	DNHR + NEMOS	RTNR (3 states)	RTNR (4 states)	RTNR (6 states)
2	1.718	1.546	12.193	0.467	0.225
3	1.741	1.567	22.381	0.579	0.178
5	0.779	0.701	18.903	0.435	0.237
8	0.001	0.001	0.003	0.000	0.000
15	0.362	0.326	0.000	0.311	0.076
average	1.019	0.917	12.047	0.389	0.157
maximum	12.135		51.351	23.032	11.848
99%	5.923		41.238	7.203	2.869
90%	1.391		28.532	1.290	0.066

DNHR, DNHR plus NEMOS, RTNR with three link states, RTNR with four link states, and RTNR with six link states. The performance for DNHR plus NEMOS was estimated based on typical performance for NEMOS based on actual network performance data and simulation models.

The threshold values of the link states are defined in Figure 12.23. Below, we discuss sensitivities of network performance to these link state thresholds. We can see from the results of Table 12.8 that RTNR performance improves as more link states are used. The performance of the RTNR three-state model is far worse than the performance of DNHR or the other RTNR models. The reason for this poor performance is due to the lack of reserved VT capacity to favor direct one-link paths under network congestion conditions; without reservation, nonhierarchical networks can exhibit unstable behavior in which essentially many connections are established on two-link as opposed to one-link paths, which greatly reduces network throughput and increases network blocking [Aki84, Kru82, NaM73]. If we add the reserved link state, as in the four- and six-state models, performance of the RTNR network is greatly improved. In fact, the six-state model exhibits performance that is measurably better than DNHR plus NEMOS in both average blocking and the distribution of blocking.

Results for high-day conditions with the 103-switch simulation model are given in Table 12.9. These results closely parallel the results for the 65-switch model, although we have not included results for an RTNR three-state model but rather the results for an RTNR four-state model. We can see from comparison of the four-, five-, and six-state models that network performance is monotonically improved as more states are added, as would be expected. However, additional simulation studies show that having more than six states does not measurably improve performance. An explanation for the improved performance of RTNR in comparison with DNHR plus NEMOS is that RTNR has networkwide, nearly instantaneous response to congestion, and through link status information it can access the least congested parts of the network. This improvement in performance is consistent with the studies of the TSMR routing techniques, presented in

TABLE 12.9   103-Switch Network Blocking Performance
Comparison Under High-Day Network Traffic Load (%)

Hour	DNHR	DNHR + NEMOS	RTNR (4 states)	RTNR (5 states)	RTNR (6 states)
2	2.138	1.924	0.155	0.034	0.017
3	2.407	2.166	1.028	0.800	0.763
5	2.199	1.979	0.820	0.499	0.453
15	0.061	0.055	0.003	0.000	0.000
average	1.793	1.614	0.543	0.363	0.337
maximum	22.172		19.249	15.356	15.015
99%	8.844		7.717	5.452	4.731
90%	2.463		1.569	0.614	0.517

Chapter 11, that use centralized network state information in the route selection, which was shown to be a decided advantage in routing path selection.

12.3.3 Models of routing parameters

Sensitivities of the results to various parameters were investigated, which included link state thresholds, the amount of reserved VT capacity R, the triggering mechanism for congestion detection, and others. We now discuss the results for these sensitivities.

Table 12.10 illustrates the network performance under different automatic trunk reservation (ATR) thresholding methods, which are candidates for the thresholds presented in Table 12.1. The trunk reservation associated with ATR-1 is that for switch-to-switch blocking of 0.5 percent, 2.5 percent, and 10 percent—there will be 5 percent (in the range of 2 to 10 trunks), 10 percent (in the range of 4 to 20 trunks), and 15 percent (in the range of 6 to 30 trunks) of trunks reserved for protecting the direct traffic, respectively. With ATR-2, the same ranges of reservation are applied for switch-to-switch blocking of 1 percent, 5 percent, and 15 percent. Note that these ranges of trunk reservation are the ones that performed best in the simulation runs. Results for the 65-switch network performance indicated that ATR-2 is somewhat better for peak-day traffic but a little bit worse for the high-day traffic.

In order to determine which automatic trunk reservation method to use, simulation using the 103-switch network was performed. Table 12.11 presents the results for high-day conditions. They reveal that ATR-2 performs slightly

TABLE 12.10   Effect of ATR on 65-Switch RTNR Network Blocking Performance (%)

Hour	ATR-1	ATR-2
	High day	
2	0.441	0.480
3	0.591	0.586
5	0.416	0.433
8	0.000	0.000
15	0.187	0.149
average	0.359	0.364
maximum	24.500	25.800
90%	1.238	1.220
	Mother's Day	
20	17.039	16.949
maximum	97.557	97.496
90%	27.889	27.557

TABLE 12.11 Effect of ATR on 103-Switch RTNR
Network Blocking Performance (High Day, %)

Hour	ATR-1	ATR-2
2	0.164	0.172
3	1.167	1.130
5	0.904	0.899
15	0.000	0.000
average	0.605	0.596
maximum	21.294	22.588
90%	1.940	1.873

better in average blocking but slightly worse in maximum blocking. With this comparison, the ATR-2 method was chosen for further simulation sensitivity study because it gave better throughput for the 103-switch network.

Table 12.12 summarizes the effect of the switch-to-switch blocking update interval for automatic trunk reservation on network performance. In most instances, the 60-second switch-to-switch blocking update interval gives better performance in throughput and blocking distribution than the 30-second update interval. Additional simulations indicated that the 60-second update interval resulted in better performance than the 300-second update interval. The explanation for these results is that the 300-second update interval is not responsive enough,

TABLE 12.12 Effect of ATR Update Time on 65-Switch
RTNR Network Blocking Performance (%)

Hour	30 seconds	60 seconds
	High day	
2	0.488	0.479
3	0.591	0.580
5	0.427	0.436
8	0.000	0.000
15	0.268	0.277
average	0.397	0.386
maximum	23.529	20.464
99%	7.687	7.393
90%	1.520	1.464
	Mother's Day	
20	16.926	16.949
maximum	97.838	97.496
90%	26.627	27.557

TABLE 12.13 Effect of ATR Triggering Mechanism on RTNR Network Blocking Performance (High Day, %)

Hour	Blocked call based	Time based (60 seconds)
2	0.049	0.017
3	1.098	0.763
5	0.902	0.453
15	0.000	0.000
average	0.558	0.337
maximum	18.706	15.015
99%	8.217	4.731
90%	0.730	0.517

whereas the 30-second update interval is not long enough to eliminate statistical variation for this automatic trunk reservation method.

Table 12.13 demonstrates the effect of an automatic trunk reservation triggering mechanism using a 60-second update interval and the calls blocked. The method based on calls blocked provides a weighted estimate of switch-to-switch and total switch blocking, which is updated on a call-by-call basis. That is, switch-to-switch blocking is estimated as $b_i^k = (1 - \alpha)r_i + \alpha b_{i-1}^k$, where b_i^k is the blocking estimate for switch pair k based on call i, α is a number between zero and 1 that represents the weight assigned to previous values of the blocking estimate, and r_i is defined to be 1 if call i is blocked and zero if it is not blocked. This blocking estimate is then used to trigger trunk reservation by applying thresholds such as those in Table 12.1. Several values of parameter α were tested, as were variations of the definition of r_i. One variation defined r_i as the average blocking over a short, fixed time interval such as 15 seconds. However, the results of many simulations show that automatic trunk reservation based on calls blocked is not as effective as that based on the 60-second update interval. The reason is that automatic trunk reservation based on calls blocked is overly sensitive in triggering trunk reservation to turn on and too slow to remove it, and consequently trunks are overreserved. Therefore, ATR-2 with 60-second switch-to-switch blocking update time appears most promising.

Table 12.14 illustrates the effect of total switch-blocking thresholds, corresponding to items A, B, C, E, and F in Table 12.5, on 65-switch network performance. As portrayed earlier in connection with Table 12.5, total switch blocking is used to dictate the load levels determined to carry the call. It is not obvious from the 65-switch network simulation results which threshold is better.

Table 12.15 shows results from a similar simulation run for the 103-switch network. It is clear that the total switch-blocking threshold of 3 percent performs better than that of 10 percent. Therefore, ATR-2 and 3 percent total switch-blocking thresholds with a 60-second switch-to-switch blocking update interval and total switch-blocking update interval are used for further studies.

TABLE 12.14 Effect of Total Switch-Blocking Thresholds on 65-Switch RTNR Network Blocking Performance (%)

Hour	10%	3%
	High day	
2	0.479	0.467
3	0.580	0.579
5	0.436	0.435
8	0.000	0.000
15	0.277	0.311
average	0.386	0.389
maximum	20.464	23.032
99%	7.393	7.203
90%	1.464	1.290
	Mother's Day	
20	16.949	16.842
maximum	97.496	97.935
90%	27.557	25.395

TABLE 12.15 Effect of Total Switch-Blocking Thresholds on 103-Switch RTNR Network Blocking Performance (High Day, %)

Hour	10%	3%
2	0.172	0.155
3	1.130	1.028
5	0.899	0.820
15	0.000	0.003
average	0.596	0.543
maximum	22.588	19.249
90%	1.873	1.569

Table 12.16 illustrates the effect of the number of states on 65-switch network performance. The results clearly show that the six-state model performs better in all aspects than the four-state model. Consequently, the combination of ATR-2, a 3 percent total switch-blocking threshold, a 60-second update interval, and a six-state model provides the most favorable RTNR threshold and parameter values. These design criteria are therefore used for further comparison studies of peak-day loads, focused overloads, transport failures, CCS network impact, and transition.

Next, we investigate the use of traffic demand for computation of trunk reservation, trunk reservation thresholds, and the selection of the most lightly loaded

**TABLE 12.16   Effect of the Number of States on
65-Switch RTNR Network Blocking Performance (%)**

	High day	
Hour	4 states	6 states
2	0.467	0.225
3	0.579	0.178
5	0.435	0.237
8	0.000	0.000
15	0.311	0.076
average	0.389	0.157
maximum	23.032	11.848
99%	7.203	2.869
90%	1.290	0.066
	Mother's Day	
20	16.842	16.476
maximum	97.935	97.578
90%	25.395	21.064

path from the candidate paths in a multiservice integrated network. The simulation results above use reservation based on the link size. However, use of estimated traffic demand for the computation of trunk reservation may be desirable because the traffic fluctuation with time can be taken into account. In addition, for multiservice integrated networks the link size may not be meaningful for an individual virtual network. Results from the high-day traffic simulation show that the network performance is comparable for the two reservation computation approaches. We furthermore use the peak-day traffic simulation to illustrate the network performance under the two different reservation computation approaches. The results, given in Table 12.17, reveal that the use of estimated traffic

**TABLE 12.17   Comparison of Traffic Demand and Link
Size for Reservation Computation for Christmas Day
Traffic Load (103-Switch, % Blocking)**

Hour	Link size	Traffic demand
10	5.0	4.9
11	8.9	8.7
12	12.4	11.7
average	9.0	8.6
maximum	64.5	75.8
99%	42.6	47.6
90%	22.0	18.7

TABLE 12.18   Effect of Use of the Most Lightly Loaded Candidate Path for
High-Day Traffic Load (% Blocking)

Hour	RTNR (next candidate path)	RTNR (most lightly loaded candidate path)
2	0.0	0.0
3	0.7	0.6
5	0.4	0.2
15	0.0	0.0
average	0.3	0.2
maximum	12.8	10.2
99%	3.2	2.5
90%	0.6	0.4
queries per call	0.10	0.08
delay call per call	0.012	0.008

demand for reservation provides better average blocking performance and distribution, although the maximum blocking is slightly higher. Because improvement in overall blocking is noticeable, the use of estimated traffic demand for trunk reservation is favored, especially because it is an appropriate quantity to use in a multiservice integrated network.

Figure 12.17 shows the RTNR via path determination involving the selection of the most lightly loaded path from the candidate routes. A simulation comparison study is conducted to investigate the impact of this approach on network performance. Table 12.18 shows the results. Clearly, the selection of the most lightly loaded via path is desirable in both average blocking and blocking distribution. Also, the number of query messages per call and the fraction of delayed calls are reduced. These benefits result from a more uniform link load distribution throughout the network.

Five percent of the link size is used for the determination of trunk reservation and link load state thresholds. Because trunk reservation is more effectively based on traffic estimate, it is desirable to test the effect of the percentage of trunk reservation on network performance. Table 12.19 shows the network performance for the various percentages tested for high-day traffic load. The comparison shows that the exact level of trunk reservation does not strongly influence network performance, although reservation is extremely important for the stabilization of network throughput, as shown in Section 12.3.2.

In order to further investigate the effect of the trunk reservation level, peak-day load is simulated. Table 12.20 shows the results for Christmas Day network performance. From these results, we conclude that trunk reservation in the range of 3 to 5 percent is acceptable.

The final sensitivity we investigate is the effect of the length of the periodic update interval. Studies described in Section 12.3.2 have shown that a one-

TABLE 12.19 Network Blocking Performance for Various Percentage Reservation for High-Day Traffic Load (%)

Hour	Reservation		
	3%	5%	7%
2	0.0	0.0	0.0
3	0.57	0.60	0.61
5	0.20	0.25	0.28
15	0.0	0.0	0.0
average	0.21	0.24	0.24
maximum	10.3	10.2	12.4
99%	2.2	2.5	2.6
90%	0.4	0.4	0.4

TABLE 12.20 Network Blocking Performance for Various Percentage Reservation for Christmas Day Traffic Load (%)

Hour	3% reservation	5% reservation	7% reservation
10	4.9	4.9	5.0
11	8.6	8.7	8.7
12	11.7	11.7	11.8
average	8.5	8.6	8.7
maximum	75.4	76.5	77.5
99%	48.0	49.4	49.8
90%	18.4	18.5	18.9

minute update interval provided appropriate responsiveness to load variations and changes in network blocking. However, the immediate triggering of trunk reservation, as described in Section 12.2, allows nearly instantaneous response to load variations and changes in network blocking and therefore allows the possibility of lengthening the periodic update interval.

In Table 12.21 we give the network blocking performance for high-day traffic loads for update intervals of reservation level and search depth parameters of one, two, three, four, and five minutes. In order to track rapid changes in the network load level (TL^i), updates of VT^i_{traf} are maintained at one-minute intervals. Table 12.22 gives similar results for Christmas Day traffic loads. There does not appear to be strong dependence in network blocking performance on the length of the update interval. It is concluded that an update interval for reservation level and search depth parameters in the range of one to three minutes provides acceptable performance.

TABLE 12.21   Network Blocking Performance for Various Periodic Update Intervals for High-Day Traffic Load (%)

Hour	Update interval				
	1 minute	2 minutes	3 minutes	4 minutes	5 minutes
2	0.00	0.00	0.01	0.01	0.01
3	0.59	0.58	0.60	0.60	0.58
5	0.24	0.23	0.25	0.24	0.23
15	0.00	0.00	0.00	0.00	0.00
average	0.23	0.22	0.24	0.23	0.22
maximum	10.1	11.0	10.2	11.8	12.1
99%	2.7	2.5	2.5	2.6	2.4
90%	0.37	0.40	0.40	0.39	0.39

TABLE 12.22   Network Blocking Performance for Various Periodic Update Intervals for Christmas Day Traffic Load (%)

Hour	Update interval				
	1 minute	2 minutes	3 minutes	4 minutes	5 minutes
10	4.9	4.9	5.0	4.9	4.9
11	8.7	8.6	8.7	8.6	8.7
12	11.7	11.7	11.7	11.7	11.7
average	8.6	8.6	8.6	8.6	8.6
maximum	75.6	76.3	76.5	76.5	76.9
99%	47.8	48.6	49.4	48.8	49.4
90%	18.9	18.7	18.5	18.7	18.8

12.3.4   Results for RTNR network performance

12.3.4.1   Results for high day.   Table 12.23 illustrates the performance of DNHR, DNHR plus NEMOS, and RTNR for a high-day network load pattern, in which it is clear that RTNR has the best overall performance.

12.3.4.2   Results for peak days and overload conditions.   Tables 12.24 and 12.25 give the simulation results for Mother's Day traffic in the 65-switch network model and Christmas Day traffic in the 103-switch network model, respectively. On these peak days the network is very heavily congested and network capacity is used to nearly its maximum extent with DNHR plus NEMOS. However, the RTNR method is still able to make some improvement in network performance for the same reasons cited above. We again see that without trunk reservation, as shown by the results with the three-state model in Table 12.24, the network exhibits instability and enters into a heavily congested state.

**TABLE 12.23   103-Switch Network Blocking Performance
for High-Day Traffic Load (%)**

Hour	DNHR	DNHR + NEMOS	RTNR
2	2.1	1.9	0.0
3	2.4	2.2	0.6
5	2.2	2.0	0.2
15	0.1	0.1	0.0
average	1.8	1.6	0.2
maximum	22.2		10.2
99%	8.8		2.5
90%	2.5		0.4

**TABLE 12.24   65-Switch Network Blocking Performance
for the Mother's Day Traffic Load (%)**

Hour 20	DNHR	DNHR + NEMOS	RTNR (3 states)	RTNR (4 states)	RTNR (6 states)
average	17.591	17.239	24.943	16.842	16.476
maximum	98.185		86.797	97.935	97.578
99%	77.276		75.145	70.981	75.537
90%	24.019		50.196	25.395	21.064

**TABLE 12.25   103-Switch Network Blocking Performance
for the Christmas Day Traffic Load (%)**

Hour	DNHR	DNHR + NEMOS	RTNR (4 states)	RTNR (5 states)	RTNR (6 states)
10	6.398	6.238	5.909	5.489	5.014
11	10.819	10.603	10.407	10.038	8.785
12	13.836	13.559	13.145	12.770	11.712
average	10.549	10.338	10.043	9.628	8.632
maximum	83.849		81.570	81.654	76.560
99%	57.165		56.211	55.536	49.459
90%	23.967		24.332	22.808	18.568

Tables 12.26–12.28 give the simulation results for focused overload conditions in the 103-switch network. Three simulations are reported in the tables: a partial focused overload in which 12 cities focus five times their normal load on Chicago, a five-times focused overload on Chicago, and a three-times focused overload on several switches, which include San Francisco, Sacramento, and Anaheim. Here again we see the performance improvement resulting from RTNR,

TABLE 12.26 103-Switch Network Blocking Performance for Partially Focused Overload to One Switch (CHCG7, %)

Hour	DNHR	DNHR + NEMOS	RTNR
2	0.091	0.082	0.000
3	0.097	0.087	0.000
5	0.050	0.045	0.000
15	0.004	0.004	0.000
average	0.063	0.057	0.000
maximum	27.801		0.000
99%	0.000		0.000

TABLE 12.27 103-Switch Network Blocking Performance for Focused Overload to One Switch (CHCG7, %)

Hour	DNHR	RTNR
2	5.824	5.762
3	6.328	6.288
5	6.288	6.260
15	4.820	4.761
average	5.878	5.832
maximum	72.072	75.726
99%	54.189	52.300
90%	0.000	0.000

TABLE 12.28 103-Switch Network Blocking Performance for Focused Overload to Multiple Switches (SNFC, SCRM, ANHM, %)

Hour	DNHR	RTNR
2	0.484	0.088
3	10.092	10.071
5	9.439	9.265
15	9.079	8.860
average	7.465	7.273
maximum	56.259	64.259
99%	44.098	44.286
90%	0.000	0.000

**TABLE 12.29   103-Switch Network Blocking Performance
Comparison for 11/18/88 Cable Cut in Sayreville, NJ (%)**

Hour	DNHR	RTNR
2	4.4	1.0
3	4.2	1.8
5	3.9	1.4
8	1.3	0.2
average	3.5	1.2
maximum	94.9	61.0
99%	50.9	16.4
90%	3.5	3.0

which outperforms DNHR plus NEMOS in practically every measure of blocking performance, except that the worst-case switch-to-switch blocking is sometimes worse for RTNR than for DNHR. However, as in the case of peak-day loads, the improvement is modest due to the already highly efficient use of network capacity under overload conditions with DNHR plus NEMOS, and much further improvement is not possible.

12.3.4.3   Results for network failures.   Table 12.29 gives simulation results for a large transport failure of about 33,000 trunks in the fiber cable connecting the Northeast Corridor, which occurred near Sayreville, New Jersey, on November 18, 1988. The performance of RTNR is markedly better than DNHR plus NEMOS. Under this failure, RTNR not only preserves a better overall blocking performance but also reduces switch isolation. Figure 12.24 shows the transient response of network attempts following the cable cut. In the figure, the cable cut in question occurs at minute 60 in the simulation. It is clear that RTNR greatly reduces the magnitude of the transient attempt rate and also measurably increases the network throughput under transport failure. This improvement is due in large measure to the RTNR immediate response under link failure. A bar chart comparison of the switch-pair blocking performance is given in Figure 12.25, which shows that RTNR eliminates isolations and significantly improves the distribution of blocking under failure.

Analysis of the network failure occurring on January 15, 1990, in which approximately one-half of the network capacity was unavailable for nine hours, shows that RTNR in comparison with DNHR would have eliminated isolations and significantly improved the distribution of blocking in the network. A significant reduction in CCS network load would also have been achieved.

12.3.4.4   Results for network design conditions.   Tables 12.30 and 12.31 illustrate the performance of RTNR for a network model that is ideally designed for DNHR; that is, the network is designed by the DNHR path-EFO model. We find that the performance of RTNR under average business day loads (Table 12.30) and high-day loads (Table 12.31) is equal to or better than DNHR and DNHR plus NEMOS

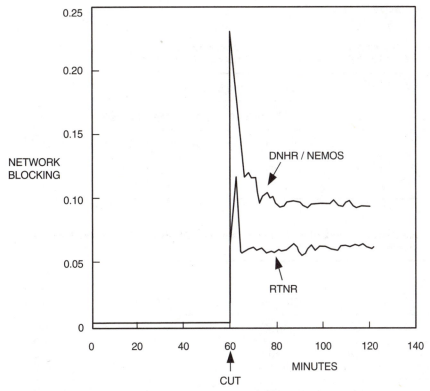

Figure 12.24 Network blocking performance for 11/18/88 cable cut in Sayreville, New Jersey

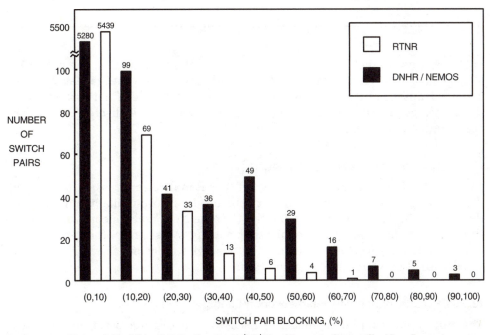

Figure 12.25 Network blocking distribution for 11/18/88 cable cut in Sayreville, New Jersey

TABLE 12.30   103-Switch Average Business Day Network Blocking Performance of DNHR and RTNR Using the DNHR Network Design (%)

Hour	Path-EFO model design		
	DNHR	DNHR + NEMOS	RTNR (4 states)
2	0.000	0.000	0.000
3	0.000	0.000	0.000
5	0.000	0.000	0.000
15	0.001	0.001	0.000

TABLE 12.31   103-Switch Network Blocking Performance for DNHR and RTNR Using the DNHR Network Design (High-Day, %)

Hour	Path-EFO model design			
	DNHR	DNHR + NEMOS	RTNR (4 states)	RTNR (6 states)
2	0.842	0.758	0.000	0.000
3	0.843	0.759	0.088	0.098
5	0.566	0.509	0.092	0.082
15	0.207	0.186	0.076	0.066
average	0.626	0.563	0.065	0.063
maximum	10.744		12.068	7.702
99%	3.302		2.725	2.418
90%	0.269		0.000	0.000

in all respects for the six-state model. These results indicate that the path-EFO model can be used for design of RTNR networks. However, other models can take better advantage of RTNR capability for more efficient use of network capacity; these include a transport flow optimization model and a discrete event flow optimization model discussed in Chapter 13.

12.3.5   Results for multiple ingress/egress routing

Table 12.32 illustrates the performance of RTNR multiple ingress/egress routing (MIER) in comparison with DNHR performance for international traffic between the United States and China during the Chinese New Year Celebration over the three-day period of heavy calling for that holiday. We see from these results the considerable improvement in the quality of service following the introduction of RTNR/MIER for Chinese New Year traffic.

12.3.6   Results for real-time internetwork routing

The call-by-call simulation model is used to measure the performance of the network under bilateral hierarchical (HIER) routing, which was the prior method

TABLE 12.32 Network Blocking Performance Comparison for Chinese New Year (%)

Day of Chinese New Year	DNHR (1991)	RTNR/MIER (1992)
day 1	73.2	31.6
day 2	44.6	6.0
day 3	6.0	0.0
Average	53.6	19.3

of operation, and RINR, which replaced bilateral hierarchical routing. A full-scale network model of the Pacific Rim network is used for the study. The model is simulated for a variety of traffic conditions to include average-day traffic, Chinese New Year and Christmas Day peak traffic, and network failure scenarios. Under the Chinese New Year peak traffic condition, the network is very heavily congested and network capacity is used to nearly its maximum extent. As illustrated in Table 12.33, the RINR strategy makes a substantial improvement in network performance across all service categories, and across a wide variation in key service traffic, compared with the previous HIER network.

The actual performance of the international RINR network during Chinese New Year 1996 agreed well with these simulation model predictions and confirmed the improved performance of RINR. Table 12.34 compares the performance of

TABLE 12.33 Chinese New Year Estimated Blocking Performance (%)

	Virtual network	
	HIER + RTNR/MIER	RINR + RTNR/MIER
out voice	34.5	0.3
in voice	35.5	3.2
out key	35.5	0.0
overall	35.2	2.5

TABLE 12.34 Network Blocking Performance Comparison for Chinese New Year (%)

	Virtual network		
	HIER (1991)	HIER + RTNR/MIER (1992)	RINR + RTNR/MIER (1996)
voice	73.2	31.6	0.7
key voice	—	—	0.0
average	73.2	31.6	0.7

international traffic between the United States and China during the Chinese New Year celebration for (1) bilateral HIER international routing plus HIER domestic routing without MIER in 1991, (2) HIER plus RTNR/MIER in 1992, and (3) RINR plus RTNR/MIER in 1996. The results represent actual performance measurements over the period during which the international routing methods were upgraded to RTNR/MIER and then to RINR. Over this period, traffic levels generally increased. We see from these results the considerable improvement in the quality of service following the introduction of RTNR/MIER and then RINR for Chinese New Year traffic. The predicted performance for HIER plus RTNR/MIER in 1996, as given in Table 12.33, is comparable to actual HIER plus RTNR/MIER performance in 1992, as shown in Table 12.34. It is also clear that key service protection provides essentially zero blocking, even under the most severe overload conditions in the network. Hence, from 1991 to 1996 as routing flexibility increased, quality of service substantially improved for the benefit of customers and increased profitability of the network.

12.3.7 Results for network service problems

RTNR capabilities allow adjustment of the D_{hl}^i and D_r^i factors, on a switch-to-switch basis, when excessive blocking is encountered on specific switch pairs. This adjustment of the depth factors is accomplished by setting appropriately higher values for the lower bounds on the depth factors, D_{HL}^i and D_R^i. An illustration of this capability is given in this section. A service problem is created in the 103-switch simulation network by inflating the switch-to-switch load on one switch pair. The thresholds for path-sequence selection are adjusted for this switch pair, as summarized in Table 12.35.

Results from the simulation of several scenarios of problem switch pairs are given in Table 12.36. For this simulation, specific problem switch pairs are created by inflating their loads to varying degrees for five scenarios. We see that

TABLE 12.35 Path-Sequence-Selection Thresholds for Problem Switch Pairs

Item	VT$_{traf}^i$	VT$_{eng}^i$	Total switch blocking, %	Switch–switch blocking, %	Path sequence	D_{hl}^i	D_r^i
A	>15	>0	[0, 3]	[0, 1]	1	8.0	8.0
				(1, 100]	2	8.0	8.0
B	≤15	>0	[0, 3]	[0, 1]	2	8.0	8.0
				(1, 100]	2	8.0	8.0
C	all	>0	(3, 10]	[0, 50]	1	8.0	8.0
				(50, 100]	2	8.0	8.0
D	all	>0	(10, 100]	[0, 50]	1	8.0	0
				(50, 100]	2	8.0	0

TABLE 12.36 Effect of Increased Alternate Routing Depth on Problem Switch-Pair Blocking (103-Switch)

Scenario number	RTNR (normal depth)		RTNR (increased depth)	
	Network blocking (%)	Problem switch-pair blocking (%)	Network blocking (%)	Problem switch-pair blocking (%)
1	0.002	0.2	0.0	0.0
2	0.011	1.1	0.0	0.0
3	0.055	4.9	0.0	0.0
4	0.114	9.2	0.0	0.0
5	0.191	15.4	0.0	0.0

the switch-to-switch control helps to alleviate the problem switch-pair blocking, and that this feature serves to mitigate blocking problems until additional trunks can be added to the network to eliminate the root cause of the blocking problems.

Note that the D_{hl}^i and D_r^i factors are set using the values in Table 12.5 for normal switch pairs. That is, heavily loaded and possibly reserved paths are used to route traffic for a switch pair only when the switch pair experiences blocking during the last periodic update interval, during which time several calls could be blocked. When the D_{hl}^i and D_r^i factors are increased, as is the case for the problem switch pair that uses the values in Table 12.36, calls for that switch pair can route on heavily loaded and reserved paths, regardless of its switch-to-switch blocking level. Therefore, for that switch pair, blocking performance is improved significantly, because some direct capacity of other switch pairs is used to complete calls.

12.3.8 Results for key services

RTNR implements a key service protection logic, described in Section 12.2. We investigate the relative performance of key services and normal services under network failure scenarios. Table 12.37 presents simulation results for a cable-cut scenario, in which a variable fraction of key service traffic is used. Before the network failure, there is no performance difference between key service

TABLE 12.37 103-Switch Key Voice and Normal Voice Service Blocking Performance for a Cable-Cut Scenario (%)

Percent of key service traffic	Total network		Cut-affected switch pairs	
	Normal voice	Key voice	Normal voice	Key voice
0	4.6	0.0	26.2	0.0
10	5.1	0.0	27.4	0.02
30	6.4	0.4	30.4	2.1

**TABLE 12.38   103-Switch Key Data and Normal Voice
Service Blocking Performance for a Cable-Cut Scenario (%)**

Percent of key service traffic	Normal voice (total network)	Key data (total network)
3	5.7	0.0
6	6.9	0.0

and normal service, which describes network behavior under normal operating conditions. After the network failure, the performance of both services is a function of the fraction of key service traffic.

As indicated in Table 12.37, the key service protection logic is effective in protecting the key service traffic not only in average network blocking but also in average blocking of the cut-affected switch pairs. This is due to the fact that key services can effectively reserve direct link capacity and use additional alternate routing under failure situations as compared with normal services. We observe increases in blocking for both streams of traffic as the fraction of key service traffic increases. This increase is attributed to the use of alternate routes for the key service calls, which tends to displace a slightly larger number of normal service calls. This effect is kept to a minimum, however, by the choking mechanisms for key service traffic described in Section 12.2. In order to maintain maximum network throughput, the fraction of key service traffic in the network should be choked at about 20 percent of total traffic.

Simulation studies are also performed to compare the performance of key data service and normal voice service. For these studies, the network is designed to meet availability criteria for key services under network failure scenarios. It is assumed that all trunks are capable of carrying both data traffic and voice traffic, and that data traffic demand varies from 3 percent to 6 percent of voice traffic demand. Table 12.38 gives the results for the simulation studies. After the cable cut, key data traffic is protected by the preferential key service routing treatment and by the network design. Key service protection allows key services to reserve direct capacity and route traffic onto heavily loaded and reserved paths. Because some data traffic is carried using the voice network capacity, the voice network performance is slightly degraded.

Hence, the RTNR key service protection mechanism provides an effective network capability for service protection. A constraint is that key service traffic should be a relatively small fraction (preferably less than 20 percent) of total network traffic.

12.3.9   Results for call setup delay performance

For a two-link call, RTNR may cause additional call setup delay because a query is needed for the determination of the via switch before the call is routed. However, this delay is comparable to the delay associated with crankback messages. And maximum call setup delay is reduced in the RTNR network as compared with

TABLE 12.39 Effect of Routing Options on Percentage of Delayed Traffic (Average Delays per Call)

Traffic load	DNHR	RTNR (no delay reduction)	RTNR (delay reduction)	RTNR (delay and query reduction)
average business day	0.004	0.064	0.0005	0.0019
high day	0.084	0.074	0.008	0.012
Christmas Day	0.35	0.221	0.120	0.123

the DNHR network, because DNHR can generate several crankbacks per call, whereas RTNR can generate at most one query message per call. As discussed in Section 12.2, the fraction of traffic that may experience additional call setup delay can be reduced with the RTNR delay reduction method.

Simulation results for the relative fraction of delayed calls are given in Table 12.39, across a representative set of network conditions including average business day, high day, and Christmas Day. We see that the delay performance is significantly improved with the RTNR delay reduction option in comparison with DNHR, across the whole spectrum of network conditions. Also, the RTNR delay reduction option performs comparably with RTNR without delay reduction in network blocking. These results, in combination with the improved network response to failures as illustrated in the above results, lead to the conclusion that RTNR improves service quality for customers in regard to blocking performance, delay performance, transmission performance, and other aspects.

12.3.10 Results for CCS network performance

This section addresses query message reduction. Table 12.40 shows that the relative number of crankback messages for DNHR is greater than the number of status query messages for RTNR without the delay reduction as the network gets busier. This is because the second link becomes busy more frequently as network load increases. With the RTNR delay reduction option, the relative number of request messages is approximately the same as without the delay reduction option. Analysis of CCS network performance with RTNR indicates that RTNR does not represent a significant impact on the CCS network, but for additional

TABLE 12.40 Impact of CCS Network Load for Various Traffic Conditions

Scenario	DNHR (crankbacks per call)	RTNR (no delay reduction) (queries per call)	RTNR (delay reduction) (queries per call)	RTNR (query reduction) (queries per call)
average business day (4 busy hours)	0.004	0.06	0.06	0.002
high day (4 busy hours)	0.08	0.07	0.08	0.01
Christmas Day (3 busy hours)	0.35	0.22	0.24	0.12

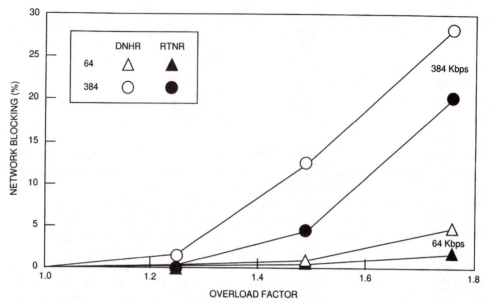

Figure 12.26 Network blocking for 64-kbps and 384-kbps services

protection under peak traffic and failure conditions it is desirable to reduce the number of queries when CCS congestion is detected. The reduction of queries is achieved by stopping the queries on via calls, for a fixed time period of one minute, to a terminating switch from which a CCS congestion message, the transfer control message, is received. The stored via switch is used until the path through it becomes busy, at which time a query is initiated to the terminating switch. We can see in Tables 12.39 and 12.40 that for the limiting case of query reduction activated at all times, the fraction of delayed traffic is increased, but the number of query messages is significantly reduced.

12.3.11   Results for multiservice integrated networks

A path-EFO model network design is obtained for two services, 64-kbps traffic and 384-kbps traffic, and a DNHR and RTNR comparison is performed with load inflation factors of 1.25, 1.5, and 1.75. Figure 12.26 shows the blocking performance for the 64-kbps service and the 384-kbps service for both DNHR and RTNR. Notice that RTNR performs uniformly better than DNHR in the study region for both 64- and 384-kbps services. The difference in performance between the two rates is due to the fact that a 384-kbps call requires six time slots and therefore has a higher blocking rate than a 64-kbps call during periods of network congestion.

12.3.12   RTNR transition analysis

Table 12.41 illustrates the performance of a mixed RTNR and DNHR network that would be necessary during transition from one method to the other. There

TABLE 12.41   Network Blocking Performance of a 103-Switch Mixed RTNR and DNHR Network During Transition (%)

Hour	100% DNHR 0% RTNR	75% DNHR 25% RTNR	50% DNHR 50% RTNR	25% DNHR 75% RTNR	0% DNHR 100% RTNR
2	2.138	1.904	1.144	0.410	0.017
3	2.407	2.126	1.572	1.247	0.763
5	2.199	1.990	1.282	0.964	0.453
15	0.061	0.036	0.017	0.017	0.000
average	1.793	1.597	1.063	0.708	0.337
maximum	22.172	22.350	14.210	13.285	15.015
99%	8.844	8.157	6.465	5.848	4.731
90%	2.463	2.317	1.507	0.933	0.517

are five columns in the table, representing 0 percent, 25 percent, 50 percent, 75 percent, and 100 percent deployment of RTNR, respectively. Calls between RTNR switches use only RTNR routing, whereas calls between all other switches use the DNHR method. As is shown by the results, the performance of the network improves monotonically from an all-DNHR network to an all-RTNR network. The results also show that this aspect of transition can be accommodated by the two strategies in coexistence.

12.3.13   Results for event-dependent real-time routing networks

In Chapter 1, we compared traffic routing methods that include fixed hierarchical routing and various event-dependent routing methods including learning with random routing (LRR) and success-to-the-top (STT) routing. In particular, the LRR model captures distributed real-time event-dependent routing methods currently in use, such as DAR and STR. We characterized network performance improvements that these approaches can achieve at the time of network overloads and failures. We also investigated a mix of dynamic routing methods (MXDR) in which STT, TSMR, and RTNR are implemented together within the same network. In MXDR, each switch uses one of these three methods.

Results for these distributed real-time event-dependent dynamic routing methods are presented in Tables 1.4, 1.6–1.9, and 1.11–1.15. These results illustrate the performance gains and design efficiencies potentially achievable with event-dependent dynamic routing methods in comparison with fixed hierarchical routing across all network configurations. However, the results show a large range of design and performance efficiencies. Some general observations and comparisons from these results are as follows. LRR, STT, and MXDR generally outperform the fixed hierarchical routing methods by a large margin under all scenarios. LRR is the least efficient of the real-time event-dependent routing methods. Because of the random selection of paths with LRR and possible loss of calls at a via switch, the blocking performance is not as good as with other methods. STT performs significantly better than LRR because of the use of crankback in STT via path selection, which avoids the loss of a call blocked at a via switch if another via path has idle capacity. In small networks, STT does quite well in comparison with more sophisticated state-dependent routing methods such as TSMR and RTNR. However, in larger networks the performance gap increases, which demonstrates the advantage of using state information in selecting paths, especially when there is a large number of paths to choose from. Results for MXDR reflect a weighted average of the individual performance of each dynamic routing strategy. These results show that it is unnecessary to standardize a single dynamic routing method to be used uniformly in a given network and that a mix of dynamic routing methods can achieve synergy and near-maximum throughput performance.

12.4 Conclusion

We have presented the results of analysis, design, and simulation studies that have formed the basis for routing design of distributed real-time dynamic routing for multiservice integrated networks, and we have examined the switch processing logic and traffic control methods for such multiservice integrated networks. The integrated network routing method allows a multiplicity of services to be provided on an integrated transport network and provides (1) partitioning of trunk capacity among virtual networks, each allocated a specific set of services, based on the traffic patterns for various times of the day; (2) dynamic trunk reservation to ensure that each service gets its allocated capacity while making unneeded capacity available to other services; (3) key service protection; (4) independent control of incoming and outgoing traffic for internetwork routing; (5) automatic selection of alternate paths transiting other countries to increase call completion in times of congestion for internetwork routing; (6) circuit capability selection for individual services such as fiber versus satellite transmission; and (7) automatic provisioning of route lists and other routing information.

A call-by-call simulation model is used to measure the performance of the distributed real-time dynamic routing networks in comparison with preplanned dynamic routing networks, under a variety of network conditions including average business day traffic, high-day traffic, Christmas Day traffic, Mother's Day traffic, link failure scenarios, and focused overload scenarios. We find that distributed real-time dynamic routing improves network performance in comparison with preplanned dynamic routing for the complete spectrum of network conditions simulated and provides significantly improved service quality, higher network throughput, and enhanced revenue.

Analysis of distributed real-time dynamic routing networks with RTNR, RINR, LRR, and STT as representative routing methods shows that such networks provide advantages for introduction into telecommunications networks. In particular, such networks can

- Implement a class-of-service routing feature for providing new service capabilities and extending dynamic routing features to emerging services for standardized fast feature introduction on a multiservice integrated network.

- Provide a self-healing-network capability that ensures networkwide path selection and immediate adaptation to failure.

- Improve network performance quality by lowering blocking and reducing call setup delay under all network load and failure conditions.

- Simplify and enhance the integration of voice and data networks, which allows dedicated network cost to be minimized and provides competitively priced services.

- Simplify the operation environment and lower operation costs.

- Simplify routing administration by eliminating load set periods, routing table updates from external systems, and other simplifications, thus minimizing routing errors and associated costs.

- Reduce load and stress on the CCS network.

- Provide the opportunity to further simplify and improve network designs and to achieve network capital cost reduction, as discussed in Chapter 13.

Overall, distributed real-time dynamic routing methods can provide service flexibility, fast feature introduction, and improved performance quality and enable service providers to achieve a better competitive position.

Chapter

13

Network Design for Real-Time Dynamic Routing Networks

13.1   Introduction

In this chapter we discuss network management and design models for real-time dynamic routing networks. Such models are applied to routing and capacity design of a wide class of real-time dynamic routing networks. As discussed in Chapter 1, Figure 1.1 illustrates the roles of network routing and network management and design. The central box represents the network, which can have various configurations, and the traffic and transport routing tables within the network. Routing tables describe the route choices from an originating switch to a terminating switch for a connection request for a particular service. Various implementations of real-time dynamic routing tables are reviewed briefly in Section 13.2.

Network management and design functions include real-time traffic management, capacity management, and network planning. A particular focus of this chapter is on capacity management, which ensures that network designs meet performance objectives at minimum cost, and on network design for real-time dynamic routing networks. Capacity management operates over a multiyear forecast interval and drives network capacity expansion. Under exceptional circumstances, capacity can be added on a short-term basis to alleviate service problems, but it is normally planned, scheduled, and managed over a period of one year or more. Network design includes routing design and capacity design, in which we investigate erlang flow optimization (EFO) models, transport flow optimization (TFO) models, and discrete event flow optimization (DEFO) models.

Our studies in Chapters 11 and 12 show that real-time dynamic routing leads to an improvement in network performance in comparison with preplanned dynamic routing networks. There is an opportunity to translate this improved performance into capacity design efficiencies with appropriate design models that reflect the efficiencies of real-time dynamic routing. In this chapter we analyze the impacts of real-time dynamic routing on the erlang flow optimization models discussed in Chapters 4–10 and introduce transport flow optimization models and discrete event flow optimization models.

The efficiency of handling uncertain load variations is the primary advantage of real-time dynamic routing strategies such as trunk status map routing (TSMR), discussed in Chapter 11, and real-time network routing (RTNR), discussed in Chapter 12. There are various categories of load variation that are accommodated in the network design process. These load variations can be categorized with respect to the time constant of the variation, as follows:

1. *Minute-to-minute variations*—The network is designed for the mean and variation of the load variations—that is, the random call arrivals and the traffic peakedness—within the hour. The minute-to-minute variation is modeled as a random process with a given mean and variance-to-mean ratio or peakedness in the network design procedure.

2. *Hour-to-hour variations*—The network is designed for the (typically 20-day) average hourly loads. The hour-to-hour variation of the load is modeled as a time-varying deterministic process in the network design procedure.

3. *Day-to-day variations*—The network is designed for the day-to-day randomness of the loads about their mean values. The day-to-day variation of the load is modeled as a random process with a gamma distribution and a specified variance level in the network design procedure.

4. *Week-to-week variations*—The network is designed for the forecasted average loads. The week-to-week or seasonal variation of the forecasted load is modeled

as a time-varying deterministic process in the design procedure as the loads are actually realized. The random component of the realized week-to-week loads is the forecast error, which is equal to the forecast load minus the realized load, and is modeled with a Gaussian distribution—for example, as an outcome of a Kalman filter forecast model [PaW82]. Forecast error impacts on capacity are accounted for in a separate, fast-acting design step called short-term capacity management.

The efficient handling of load variations from minute to minute, hour to hour, day to day, and week to week is therefore the principal area in which real-time dynamic routing might lead to additional capacity design efficiencies. Each of these load variations will be discussed in terms of its network design implications.

Studies discussed in Chapters 1, 11, and 12 show that real-time dynamic routing handles minute-to-minute or within-the-hour variations more efficiently than preplanned dynamic routing networks. This is evidenced by the relative performance of these routing methods under essentially all network scenarios for a given network load and capacity. The improved performance is observed when real-time dynamic routing is used on an ideal network design for preplanned dynamic routing—that is, a design in which there is zero reserve capacity for the case of preplanned dynamic routing. This example demonstrates that real-time dynamic routing is more efficient in handling the minute-to-minute variations of peaked overflow traffic. Therefore, these within-the-hour, minute-to-minute traffic peakedness impacts on network design capacity might be adjusted to capture this improved efficiency of real-time dynamic routing. To examine the sensitivity of peakedness design for real-time dynamic routing, we modeled the limiting case in which a real-time dynamic routing (TSMR) network is designed by the path-erlang flow optimization (path-EFO) model described in Chapter 6, under the assumption that all overflow traffic is Poisson—that is, the variance-to-mean ratio, or peakedness, is 1. This design yields a reduction in network cost of about 2 percent in comparison with the design that uses the peakedness methods presented in Chapter 6. However, although the simulation shows that the average network blocking performance is satisfactory, the average switch-to-switch blocking performance does not meet objectives in several instances. This unacceptable behavior results from the fact that the peaked overflow traffic, which still exists in the real-time dynamic routing network, is carried less efficiently than Poisson traffic, and the design methods must account for this.

The underlying preplanned dynamic routing model embedded in the erlang flow optimization model accounts for the hour-to-hour variations of the load. This model captures the hourly flow changes for a particular preplanned routing method such as two-link sequential path routing. More flexible flow realization achieves better network design, as demonstrated, for example, with the CGH method [CGH81] discussed in Chapter 6, which shows the possible reductions in hour-to-hour network design capacity with more flexible preplanned dynamic routing strategies. Real-time dynamic routing achieves the upper limit of flexible flow realization, and therefore new models are needed. For real-time dynamic routing hour-to-hour variation design, we present three models in Section 13.3.

One is a real-time erlang flow optimization model that captures the state-dependent nature of real-time dynamic routing through a fixed-point erlang flow optimization model. We also present a transport flow optimization model, which uses Truitt's ECCS method to convert hourly erlang load demand to virtual transport demands, and then these demands are routed using a linear programming flow and capacity optimization. A third multihour design approach is the discrete event flow optimization model.

As discussed in Chapter 6, for preplanned dynamic networks day-to-day variations are modeled in the path-EFO model by an equivalent load method. This approach models for each switch pair in the network an equivalent link, and Neal-Wilkinson methods are used to determine the equivalent Poisson load without day-to-day variation that gives the same equivalent link capacity as the load with day-to-day variation. If applied to real-time dynamic networks, this equivalent load method provides designs with network performance that more than meet the blocking objectives. The better-than-objective performance is achieved because the sizing method provides sufficient capacity for each traffic load by itself to meet the blocking objective, and it does not consider the fact that on any particular day at any particular time some traffic loads will be above their average load and some below. Application of the equivalent load method for preplanned dynamic networks increases the network cost by about 4–5 percent. This cost can be reduced for real-time dynamic networks by the modified equivalent load approach presented in Section 13.3, which better captures the effects of real-time dynamic routing.

The process of capacity management corrects for the effects of traffic forecast errors or, equivalently, for the effects of week-to-week variations not foreseen in the forecasted network design. When some links are found to be overloaded as a result of the realized traffic loads being larger than their forecasted values, additional trunks are provided to restore the network blocking performance to the objective level. As shown in Chapter 9, trunks are not usually disconnected in capacity management, and, as a result, the process leaves the network with a certain amount of reserve or idle capacity even when the forecast error is unbiased [FHH79]. Efficient routing techniques such as real-time dynamic routing lead to a lower level of reserve capacity because fewer capacity additions are necessary, as shown in Chapter 9 for preplanned dynamic routing networks. A reduction in reserve capacity of 5 percent or more is estimated for preplanned dynamic networks. Through the more efficient real-time routing table update mechanisms, real-time dynamic networks achieve equal or greater reserve capacity reduction because all available reserve capacity is accessed more efficiently with real-time dynamic routing.

13.2   Review of Real-Time Dynamic Routing Methods

Below we elaborate on the real-time dynamic routing methods addressed in this chapter. Dynamic traffic routing allows switch routing tables to be changed dynamically, either in a preplanned time-varying manner or on line in real time. Real-time dynamic traffic routing does not depend exclusively on precalculated routing tables. Rather, the switch senses the immediate traffic load and, if necessary, searches out new paths through the network—possibly, as in RTNR, on a call-by-call basis. With real-time dynamic routing strategies, routing tables change with a time constant less than a call-holding time.

We now briefly summarize various real-time dynamic routing methods. A number of real-time dynamic traffic routing strategies have been studied. These routing methods can be event dependent or state dependent.

13.2.1   Real-time event-dependent routing

Event-dependent real-time dynamic routing strategies may use learning models such as learning with random routing (LRR), success-to-the-top (STT) routing, dynamic alternate routing (DAR), and state- and time-dependent routing (STR). LRR is a decentralized call-by-call routing method with update based on random routing. LRR uses a simplified, decentralized learning method to achieve flexible real-time routing. The direct link is used first if available, and a fixed alternate path is used until it is blocked, in which case a new alternate path is selected at random as the alternate route choice for the next call overflow from the direct link. No crankback is used at a via switch or egress switch in LRR, so a call blocked at such a switch will be lost. Dynamically activated trunk reservation is used under call-blocking conditions. STT routing is an extension of the LRR method, in which crankback is allowed when a via path is blocked at the via switch and the call advances to a new random path choice. In a limiting case of STT, all possible one- and two-link path choices can be tried by a given call before the call is blocked.

13.2.2   Real-time state-dependent routing

State-dependent real-time dynamic routing strategies may change routing patterns in one of the following ways:

1. Every few minutes, as in worldwide international network (WIN) dynamic routing, STAR (system to test adaptive routing), and DR-5. These strategies recompute alternate-routing paths every five minutes based on traffic data.

2. Every few seconds, as in dynamically controlled routing (DCR) and trunk status map routing (TSMR).

3. On every call, as in RTNR.

In DCR the selection of candidate paths at each switch is recalculated every 10 seconds. The path selection is done by a central routing processor, based on the busy/idle status of all trunks in the network, which is reported to the central processor every 10 seconds. TSMR is a centralized real-time routing method with periodic routing table updates based on periodic network status. The TSMR routing method involves having an update of the number of idle trunks in each link sent to a network database every few seconds. Routing table updates are determined from analysis of the trunk status data using the TSMR dynamic routing method, which provides that the first path choice determined by the routing design be used if a circuit is available.

In RTNR, the routing computations are distributed among all the switch processors in the network. RTNR uses real-time exchange of network status information, with CCS query and status messages, to determine an optimal path from all possible one- or two-link path choices. With RTNR, the originating switch first tries the direct path and, if it is not available, finds an available least-loaded two-link path by querying the terminating switch through the CCS network for the busy/idle status of all links connected to the terminating switch. The originating switch compares its own link busy/idle status with that received from the terminating switch and finds the least-loaded two-link path on which to route the call.

13.2.3   Mixed dynamic routing

In a mixed dynamic routing (MXDR) network, many different methods of dynamic routing are used simultaneously. Calls originating at a given switch use the particular dynamic routing method implemented at that switch, and that switch could, for example, be a mix of STT, TSMR, and RTNR. Studies presented in Chapter 1 show that a mix of dynamic routing methods achieves good throughput performance in comparison with the performance of individual dynamic routing strategies.

13.2.4   Multiservice integrated dynamic routing

As discussed in Chapter 12, multiservice integrated dynamic routing allows independent control, by virtual network, of service-specific performance objectives, routing rules and constraints, and traffic data collection. Multiservice integrated dynamic routing allows the definition of virtual networks to carry assigned service types, where each virtual network is allotted a predetermined amount of the total network bandwidth. This bandwidth can be shared with other virtual networks but is protected under conditions of network overload or stress. Multiservice integrated dynamic routing is able to meet different blocking objectives and load levels for individual virtual networks. For example, virtual networks with different transmission capabilities, such as voice, 64-kbps, 384-kbps, and 1,536-kbps switched digital data services, can have different blocking objectives, as can virtual networks that require the same transmission capabilities, for example, domestic voice and international voice services.

Figure 1.25 illustrates the multiservice integrated dynamic routing method, which includes the following steps for call establishment:

1. At the originating switch, the class-of-service and terminating switch are identified.

2. The class-of-service and terminating switch information are used to select the corresponding virtual network routing table data, which include path sequence selection, reservation levels, load state thresholds, and traffic measurements.

3. An appropriate path is selected through execution of the path selection logic, and the call is established on the selected path.

As illustrated in Figures 1.25 and 1.26, the various virtual networks share bandwidth on the network links. For this purpose the multiservice integrated dynamic routing network is designed to handle the combined forecasted call loads for many classes-of-service. The network design allots a certain number of direct virtual trunks between switches to each virtual network, which is referred to as VT_{eng}. VT_{eng} is the minimum guaranteed bandwidth for a virtual network, but if the virtual network is meeting its blocking objective, other virtual networks are free to share the VT_{eng} bandwidth. The quantities VT_{eng} are chosen in the network design process such that their sum over all virtual networks sharing the bandwidth of the direct link is equal to or less than the total bandwidth on the link, as with the three VT_{eng} segments illustrated in Figures 1.25 and 1.26.

Each switch uses its current estimates of virtual network traffic volume and blocking performance to dynamically allocate link bandwidth to each virtual network. Under normal nonblocking network conditions, all virtual networks fully share all available capacity. Because of this, the network has the flexibility to carry a call overload between two switches for one virtual network if the traffic loads for other virtual networks are sufficiently below their design levels. An extreme call overload between two switches for one virtual network may cause calls for other virtual networks to be blocked, in which case trunks are reserved to ensure that each virtual network gets the amount of bandwidth allotted by the network design process. This reservation during times of overload results in network performance that is analogous to having a number of trunks between the two switches dedicated for each virtual network. With respect to trunk selection, as shown in Figure 12.21, some trunks are shareable by different services. The hunting of individual trunks follows an order from trunks dedicated to an individual service to the trunks shareable by two or more services. Hence, the design of multiservice integrated networks must account for the various complex interactions of bandwidth allocation and reservation, dynamic routing strategies, and capabilities of various subsets of trunk capacity for all classes-of-service on the integrated network.

We summarize the classifications of real-time dynamic routing routing methods discussed in this chapter in Table 13.1. In the following sections, we discuss methods for designing real-time dynamic routing networks. These methods include

- A multihour fixed-point erlang flow optimization (EFO) model, which extends the route- and path-EFO models discussed in Chapters 4–10 to real-time dynamic routing networks. In Section 12.4, this model is illustrated for the design of RTNR networks.

- A multihour transport flow optimization (TFO) model, which is applicable to the design of multiservice integrated networks with real-time dynamic routing. In Section 12.4, this model is illustrated for the design of multiservice integrated RTNR networks.

TABLE 13.1   Real-Time Dynamic Routing Methods

Routing method	Routing Table Update		Examples
	Frequency	Control point	
real-time event-dependent	call by call	ESS/CCS network	dynamic alternate routing (DAR); state- and time-dependent routing (STR); learning with random routing (LRR); success-to-the-top (STT) routing
real-time state-dependent	minutes	network management system	worldwide international network (WIN) dynamic routing; system to test adaptive routing (STAR); dynamic routing—five-minute (DR-5)
	seconds	network management system	dynamically controlled routing (DCR); trunk status map routing (TSMR)
	call by call	ESS/CCS network	real-time network routing (RTNR); mixed dynamic routing (MXDR); multiservice integrated dynamic routing

- A multihour discrete event flow optimization (DEFO) model, which is applicable to the design of multiservice integrated networks with real-time dynamic routing. In Section 12.4, results are discussed for DEFO model application to all categories of multiservice integrated real-time dynamic routing networks discussed in the book. As such the DEFO model is shown to have the widest range of application and flexibility for dynamic routing network design.

- An extension of the path-EFO model applicable to real-time dynamic networks, which includes an alternative day-to-day variation model. In Section 12.4, this model is illustrated for the design of TSMR networks.

- A network bandwidth allocation procedure for real-time dynamic networks, which allocates traffic to two sets of routing paths according to traffic patterns. In Section 12.4, this model is shown to improve network performance for example traffic patterns in a real-time dynamic routing network.

13.3   Network Management and Design

13.3.1   Multihour fixed-point erlang flow optimization model

A fixed-point erlang flow optimization model for real-time dynamic network design combines several techniques for achieving network design savings into a single approach. The steps of the design model and the techniques employed are briefly explained as follows:

1. Initial trunk requirements are generated based on the traffic load matrix input to the model. Trunks are allocated to individual links in accordance with the Kruithof allocation method [Kru37], which distributes trunks in proportion to the overall demand between switches.

2. Erlang flow is routed on the real-time path choice according to the real-time routing method. The real-time path choices for real-time event-dependent routing and real-time state-dependent routing are described in Section 13.2. Link, path, and route state probability estimates are used to estimate the erlang flows.

3. Link state probabilities are derived from link arrival rates determined in step 2, based on a birth–death model. Path and route state probabilities are then estimated from the link state probabilities.

4. Traffic completion performance is determined for each switch based on the converged link, path, and route state probabilities. The model uses the fixed-point method [Kat72, Kel86, Whi85, AsH93, WOY90, MGH91] to iterate between steps 2 and 3 in order to estimate erlang flow on the real-time path choices, as selected in each routing method.

5. From the estimate of blocked traffic at each switch in each time period, additional switch trunk requirements are determined for an updated trunk estimate. Trunks are allocated to individual links in accordance with the Kruithof allocation method, which distributes trunks in proportion to the overall demand between switches, as in step 1, and in accordance with link cost so that overall network cost is minimized. Sizing of individual links in this way ensures an efficient level of blocking on each link to optimally divide the load between the direct link and the overflow network.

The fixed-point erlang flow optimization model for real-time dynamic routing design is an iterative technique consisting of routing design steps 2 to 4 and capacity design steps 1 and 5. The routing design steps find the real-time paths between points in the network and flow the erlang load traffic onto the network capacity using the fixed-point method and birth–death model. The output from the routing design steps is the routed erlang flow and the fraction of traffic completed in each time period. This traffic completion performance is provided to capacity design step 5, which determines the new trunk capacity requirements of each switch and then each link to meet the design level of blocking. Once the links have been sized, the network is reevaluated to see if the blocking probability grade-of-service is met.

We now describe the fixed-point erlang flow optimization model routing design steps in more detail for an example RTNR network design.

13.3.1.1 Example of the fixed-point erlang flow optimization model. The fixed-point erlang flow optimization model computes network flows, as well as switch-to-switch blocking probabilities, for networks with real-time dynamic routing. Here we consider RTNR as an example of the model. An assumption of the fixed-point erlang flow model is that the stationary behavior of real-time dynamic routing networks can be computed based on the stationary, independent link state probabilities and on via switch selection probabilities, which are based on the link state probabilities and switch-to-switch blocking probabilities. An iterative fixed-point approach is used, and all overflow traffic is assumed to be Poisson.

Three steps are performed iteratively by the fixed-point erlang flow optimization model:

1. The link state probability model derives the link aggregate state probabilities from the link arrival rates, which are the outcome of the erlang flow model.

2. The erlang flow model uses the link aggregate state probabilities to derive the path state probabilities and the route state probabilities. With these state probabilities, the flow model determines the route flows, path flows, path arrival rates, and link arrival rates, where the latter are used in the link state probability model to compute the link state probabilities.

3. Finally, the automatic trunk reservation and path selection depth model computes the switch-to-switch blocking probability distribution, and from that models the path selection depth and the link reservation levels. A model of calls in progress is used to determine the probability that reservation and path selection depth are enabled.

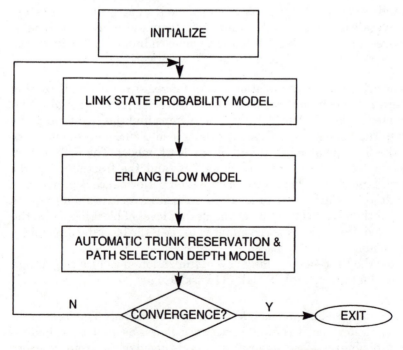

Figure 13.1 Fixed-point erlang flow optimization model for RTNR networks

Figure 13.1 illustrates the above three steps in the fixed-point erlang flow optimization model for RTNR networks. The convergence criterion is the sum of the squared differences in link state probability levels from the previous to the current iteration, summed over all links in the network. Note that the fixed-point method guarantees the existence of a solution but may not converge to a unique solution if trunk reservation is not sufficient [Aki84, Kru82]. Our numerical experience shows that, when RTNR trunk reservation is applied, the approximation method converges quickly to the fixed-point solution.

Link state probability model. As discussed in Chapter 12, RTNR uses six discrete load states for links: lightly loaded 1 (LL1), LL2, LL3, heavily loaded (HL), reserved (R), and all busy (B). The number of idle virtual trunks on the link is compared with the load state thresholds for the link to determine the load condition of the link. This determination is made every time a virtual trunk on the link is either seized or released.

The boundary of the six states on each link is defined in Table 12.2 and is illustrated in Figure 12.10. As discussed in Chapter 12, the link state thresholds depend on the following:

- Switch-to-switch blocking level NN_k for switch pair k, which determines the reservation level

- Switch-to-switch traffic load estimate TL_k and estimated virtual trunk requirements $VTtraf_k$ to meet the blocking probability grade-of-service objective requirement for switch pair k

The link aggregate state probability for state S is defined as

$$P_k(S) = \sum_{i=L(S)_k}^{U(S)_k} P_{ki} \qquad (13.1)$$

where $S = 0, 1, 2, 3, 4, 5$ and $L(S)_k$ and $U(S)_k$ are the lower and upper bounds of the aggregate link state S, for link k, as determined from Table 12.2. As illustrated in Figure 13.2, $S = 0$ denotes the busy state B, $S = 1$ denotes the reserved state R, $S = 2$ denotes the heavily loaded state HL, $S = 3$ denotes the lightly loaded 3 state LL3, $S = 4$ denotes the lightly loaded 2 state LL2, and $S = 5$ denotes the lightly loaded 1 state LL1. The quantity P_{ki} denotes the probability of i busy circuits on link k, which is computed from the birth–death equations for the link state. These equations are given as follows. The call termination rate on link k is

$$d_{ki} = i\mu$$

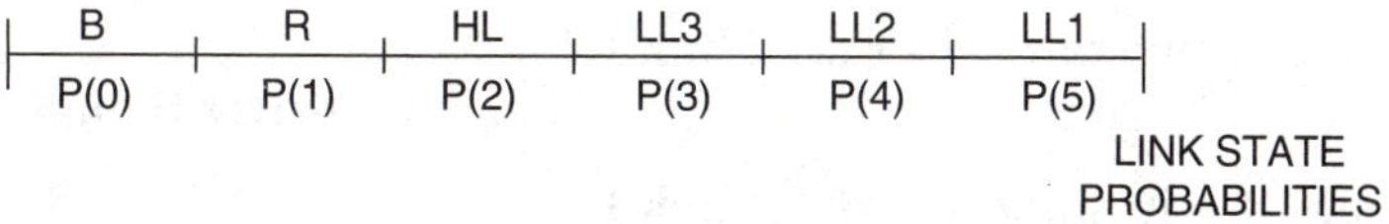

B	R	HL	LL3	LL2	LL1	
P(0)	P(1)	P(2)	P(3)	P(4)	P(5)	

LINK STATE
PROBABILITIES

Figure 13.2 Definition of link state probabilities

for $i = 1, 2, \ldots, C_k$, where C_k is the number of virtual trunks on link k and the call holding time is exponentially distributed with mean $1/\mu$. The call arrival rate is

$$b_{ki} = (D_k + A_{ki})\mu$$

for $i = 0, 1, \ldots, C_k - 1$, where D_k is the direct offered load for link k and A_{ki} is the overflow load offered to link k when it has i busy circuits. Then the balance equation

$$b_{ki}P_{ki} = d_{ki+1}P_{ki+1}$$

is solved for each link k.

To avoid numerical instability, the following procedure is used to compute the link state probability. This approach has been used by Wong and Yum [WoY90] and Mitra [MGH91] in studies of symmetric real-time routing networks. First, the initial point n^* is set to the integer part of $\min[D_k, C_k]$. Here, the choice of n^* will ensure that the probability density function for the link state is near the maximum value, and underflow problems are avoided. With the initial point n^* for forward and backward computation, the following method provides the P_{ki} values:

1. Initial step: Let $P_{kn^*} = 1$ and SUM $= 1$ be the initial values for computing link state probabilities.

2. Forward computation: For $i = n^* + 1, \ldots, C_k$, let $P_{ki} = P_{ki-1} \times b_{ki-1}/d_{ki}$ and SUM $=$ SUM $+ P_{ki}$.

3. Backward computation: For $i = n^* - 1, \ldots, 0$, let $P_{ki} = P_{ki+1} \times d_{ki+1}/b_{ki}$ and SUM $=$ SUM $+ P_{ki}$.

4. Normalization computation: For $i = 0, \ldots, C_k, P_{ki} = P_{ki}/$SUM.

The aggregate state probabilities, $P_k(S)$, are then computed from the P_{ki} values, as given in the Equation 13.1

Erlang flow model. This section describes the erlang flow model step in Figure 13.1. We first describe the procedures to compute the path state probabilities and the route state probabilities. We then compute link offered loads for each state S due to the overflow traffic. For a path consisting of two links, by the independence assumption, the path state probabilities $P_{PH}(S)$ are given by

$P_{PH}(S) = $ Pr (path is in state S)

$\qquad = $ Pr (both links are in state S)

$\qquad\quad + $ Pr (1st link L1 is in state S) Pr (2nd link L2

$\qquad\qquad\qquad\qquad\qquad\qquad\qquad\qquad$ is in state better than S)

$\qquad\quad + $ Pr (2nd link L2 is in state S) Pr (1st link L1

$\qquad\qquad\qquad\qquad\qquad\qquad\qquad\qquad$ is in state better than S)

$$= \Pr(S_{L1} = S)\Pr(S_{L2} = S) + \Pr(S_{L1} = S)\Pr(S_{L2} > S)$$
$$+ \Pr(S_{L1} > S)\Pr(S_{L2} = S)$$
$$= \Pr(S_{L1} = S)\Pr(S_{L2} \geq S) + \Pr(S_{L1} > S)\Pr(S_{L2} = S)$$

$$S = 0, 1, 2, 3, 4, 5$$

Here "state better than S" means a state with greater idle capacity than state S; for example, state LL1 has more idle capacity, or is "better," than state LL2. Figure 13.3 illustrates the computation of the path state probabilities from the link state probabilities. With the path probabilities defined, the route state probability is defined as

$$P_{RT}(S) = \Pr \text{ (route is in state } S)$$

$$= \Pr \text{ (at least one of its paths is in state } S \text{ and no path}$$
$$\text{is in a state better than } S \text{)}$$

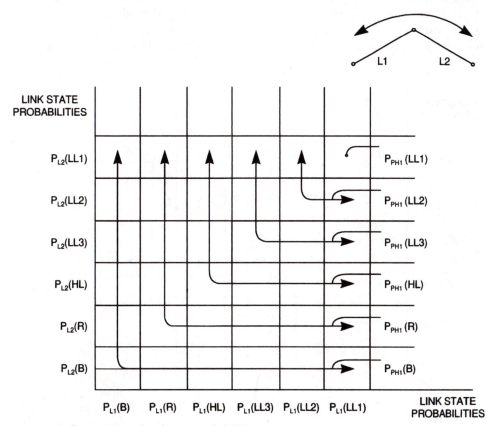

Figure 13.3   Computation of path state probabilities

To derive the route state probability, let $G_{PH}(S)$ be the cumulative distribution function of the path state probabilities—that is, for path p (PHp),

$$G_{PH_p}(S) = Pr(S_{PH_p} \le S)$$

Then the route state probabilities $R_{RT}(S)$ are given as follows:

$$P_{RT}(S) = Pr(\text{route is in state } S)$$

$$= \begin{cases} \prod_{PH_p} Pr(S_{PH_p} = 0), & S = 0 \\ \prod_{PH_p} G_{PH_p}(S) - \prod_{PH_p} G_{PH_p}(S - 1), & S = 1, 2, 3, 4, 5 \end{cases}$$

where the product is over all paths in the route for switch pair k. Figure 13.4 illustrates the computation of $P_{RT}(S)$.

In Chapter 12 we described the real-time routing procedure for RTNR. Here, in the fixed-point erlang flow optimization model, the steady-state behavior of RTNR is modeled to capture this real-time routing procedure, in which the direct overflow is allocated to via paths according to the route state probabilities, path state probabilities, and link state probabilities. That is, the

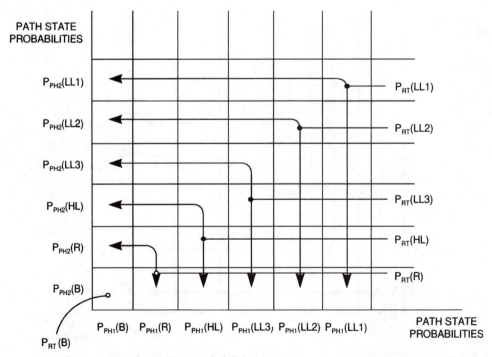

Figure 13.4 Computation of route state probabilities

via path flow is based on

- Overflow load α_k from link k
- Route state probabilities $P_{\mathrm{RT}}(S)$
- Path state probabilities $P_{\mathrm{PH}}(S)$
- Path sequence depth $\mathrm{DEPTH}_k(S)$ for the route for switch pair k in state S

The overflow load α_k from link k is given by

$$\alpha_k = D_k \times P_k \qquad (S = 0)$$

where D_k is the direct offered load for link k and P_k $(S = 0)$ is the probability of the busy state for link k. Let $\mathrm{DEPTH}_{\mathrm{RT}k}(S)$ equal the probability that the route for switch pair k is allowed, according to Tables 12.4 and 12.5, to overflow load to paths in state S. This quantity is computed in the adaptive trunk reservation and adaptive path selection depth model described in the next section. Then the flow carried by the route for switch pair k, in state S, is given by

$$f_{\mathrm{RT}k}(S) = \alpha_k \times P_{\mathrm{RT}k}(S) \times \mathrm{DEPTH}_{\mathrm{RT}k}(S)$$

Now define the probability sum for state S for paths in route k to be

$$PSUM_{\mathrm{RT}k}(S) = \sum_{p=1}^{P_k} P_{\mathrm{PH}p}(S)$$

where the summation is over all paths P_k in the route for switch pair k. Then the proportion of this flow carried on path p is approximated by the ratio of the probability that path p is in state S relative to the probability sum that the other paths in route k are in state S. That is,

$$f_{\mathrm{PH}p}(S) = f_{\mathrm{RT}k}(S) \times P_{\mathrm{PH}p}(S)/PSUM_{\mathrm{RT}k}(S)$$

Then the arrival rate to path p on the route for switch pair k when path p is in state S is given by the path-carried flow divided by the probability that the path is in state S. That is,

$$y_{\mathrm{PH}p}(S) = f_{\mathrm{PH}p}(S)/P_{\mathrm{PH}p}(S)$$

This expression is similar to that obtained by Wong and Yum [WoY90] for a symmetric network model. Following Wong's and Yum's approach, we let link L1 of path PHp be in state S_{L1} and link L2 be in state S_{L2}. Then the state of path PHp is

$$S_{\mathrm{PH}p} = \min(S_{\mathrm{L1}}, S_{\mathrm{L2}})$$

and the load offered to link L1 on path PHp, given that link L1 is in state S_{L1} and link L2 is in state S_{L2}, is given by

$$a_{L1}^{PHp}(S_{L1}) = \sum_{S_{L2}=0}^{5} y_{PHp}[\min(S_{L1}, S_{L2})] \times P_{L2}(S_{L2})$$

Then the arrival rate to link L1 from all overflow loads in the network, given that link L1 is in state S_{L1}, is given by

$$A_{L1}(S_{L1}) = \sum_{PHp} a_{L1}^{PHp}(S_{L1})$$

where the summation is over all paths PHp that contain link L1. These link arrival rates are used in the link state probability model to recompute the link state probabilities, as described above.

Automatic trunk reservation and path selection depth model. As discussed in Chapter 12, RTNR uses automatic trunk reservation, in which the number of reserved trunks depends on the switch-to-switch blocking detected over a periodic update interval of three minutes. In this manner, traffic attempting to alternate-route over the direct link of a triggered switch pair is subject to trunk reservation, and the direct traffic is favored for the triggered switch pair.

A binomial distribution is used to approximate the switch-to-switch blocking in the three-minute periodic update interval. That is, the expected three-minute peg counts for switch pair k are given by

$$\overline{PC}_k = TL_k \times 180 \times \mu$$

where $1/\mu$ is the mean holding time, in seconds. If the switch-to-switch blocking probability is NN_k for switch pair k, then the overflow counts are approximately binomially distributed in the three-minute update interval, with mean overflow count equal to

$$\overline{OV}_k = \overline{PC}_k \times NN_k$$

Here the stationary value of NN_k is computed from the results of the erlang flow model. Hence, with this estimate for $\overline{PC}_k$ and the binomial distribution of OV_k, the probability of each reservation level $i - P_k(RLi)$ for each switch pair k, as given in Table 12.1—is computed.

Recall that reservation is turned off when the calls in progress for switch pair k, CIP_k, exceed $VTtraf_k$. These dynamics of reservation turn-off are modeled by computing the probability that reservation is turned on for switch pair k, Pon_k. Pon_k is computed as the total probability for the condition that the number of calls in progress is less than or equal to $VTtraf_k$—that is, when $CIP \le VTtraf_k$. Pon_k is computed in this way under the assumptions that the number of calls in progress is Poisson distributed, with variance equal to the mean, and that the reservation turn-off process is independent of the reservation level triggering process.

With these estimates for $P_k(\text{RL}i)$ and Pon_k, we compute the mean reservation value according to

$$\overline{R}_k = \sum_{i=0}^{4} P_k(\text{RL}i) \times i \times 0.05 \times \text{VTtraf}_k \times Pon_k$$

We next compute the path sequence depth probabilities, $\text{DEPTH}_{\text{RT}k}(S)$, for the route associated with switch pair k, according to Tables 12.4 and 12.5. We compute the total switch blocking for switch j, TSB_j, as

$$\text{TSB}_j = \sum \text{OV}_k \Big/ \sum \text{PC}_k$$

where the summations are over all switch pairs k originating at switch j. Then, with the probability distribution of each switch-to-switch blocking and reservation level determined as described above, the probability is computed that the route uses the depth parameters given in Items A, B, C, D, E, and F of Table 12.5. Here again, because access to paths in the heavily loaded state and reserved state is regulated by a turn-off procedure when $\text{CIP}_k > \text{VTtraf}_k$, as described in Chapter 12, the depth probabilities for the reserved state $\text{DEPTH}_{\text{RT}k}$ ($S = 1$) and for the heavily loaded state $\text{DEPTH}_{\text{RT}k}$ ($S = 2$), as obtained from Table 12.5, are multiplied by Pon_k. Here, the computation of Pon_k is the same as described earlier.

13.3.2 Multihour transport flow optimization models

Our simulation studies of RTNR, discussed in Chapter 12, revealed that RTNR single-service voice networks can be designed adequately using the path-EFO model design for preplanned dynamic routing networks. However, this design does not model a multiservice integrated voice and data network, nor does it model the real-time routing method present in RTNR. Furthermore, it does not model new service requirements such as switched digital 64-kbps, 384-kbps, and 1,536-kbps data services. Hence, there is a need to model multiservice integrated network design, and the real-time routing method itself for proper capacity design.

A virtual trunk (VT) method lends itself to the design of reliable fixed transport routing networks, as discussed in Chapter 14, and also for the design of multiservice integrated real-time dynamic routing networks. The steps are illustrated in Figure 13.5. In the transport flow optimization (TFO) model design method, we first convert the input traffic demands with their associated load variations to equivalent traffic demands, then map these equivalent traffic demands to virtual trunk capacity demands, and finally optimally route the virtual trunk capacity demands on the traffic and transport network. This approach results in a simplified network design model for dynamic routing networks, in comparison with the path-EFO model for preplanned dynamic routing networks, that includes

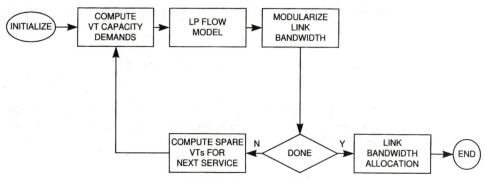

Figure 13.5 Transport flow optimization model

capability to design for multiple services and reflects real-time dynamic routing in the model. This TFO model also achieves significant reduction in processing run time and code size, enhanced robustness to network failures, and 2 to 4 percent capital cost savings in comparison with preplanned dynamic routing design.

Accordingly, the following methods are used to size the multiservice integrated real-time dynamic routing network. Each step of the block diagram in Figure 13.5 of the TFO model design procedure is described in the following sections.

13.3.2.1 Step 1—Computation of equivalent switch-to-switch loads and virtual trunk capacity demands. We begin with the computation of equivalent switch-to-switch loads that model the effects of within-the-hour load variations (peakedness), hour-to-hour load variations, and day-to-day load variations. In the TFO model design procedure, we model the total direct load plus aggregated background switch-to-switch loads of the shared network as an equivalent switch-to-switch load offered to an equivalent link. This equivalent load model captures the real-time dynamic routing efficiency of handling load variations and surges through sharing of capacity among many traffic loads on the shared network. The efficient handling of uncertain load variations and surges is the primary advantage of a flexible, real-time dynamic routing method, and the TFO model reflects this efficiency for the various categories of input load variation.

Real-time dynamic routing effectively gives each switch pair access to the capacity of a large number of parallel paths and, through use of real-time dynamic routing, makes the combination of parallel paths behave as a single large link. This equivalent larger link has, in turn, an equivalent larger switch-to-switch offered load. This equivalent switch-to-switch load R is approximated as the direct load A plus the aggregated background load on the shared network, BGL_k^i, and is evaluated for each switch pair k and service i as follows:

$$\mathrm{BGL}_k^i = \sum_{j=1}^{J_k^i} \min\{\max_h[A_j^{hi}(1)], \max_h[A_j^{hi}(2)]\}$$

where

J_k^i = the number of allowed paths for switch pair k for virtual network i

$A_j^{hi}(1)$ = the switch-to-switch load for the switch pair associated with link 1 of the two-link path j in hour h for virtual network i

$A_j^{hi}(2)$ = the switch-to-switch load for the switch pair associated with link 2 of the two-link path j in hour h for virtual network i

An equivalent link is sized for the equivalent switch-to-switch load, $R = A + \text{BGL}$, on the basis of Neal-Wilkinson engineering, which, as discussed in Chapter 2, models day-to-day load variations. The number of trunks N is calculated that is required in an equivalent link to meet the required blocking probability grade-of-service for the equivalent load R, with its specified peakedness Z and specified level of day-to-day load variation ϕ. We omit here, for simplicity, the subscripts k and the superscripts h and i. Based on Neal-Wilkinson theory, the equivalent link would require the following number of trunks N to satisfy the equivalent load R:

$$N = \overline{\text{NB}}(R, \phi, Z, B)$$

N = trunk requirement determined by the Neal-Wilkinson $\overline{\text{NB}}$ function

R = $A + \text{BGL}$

A = switch-to-switch load

BGL = background load of the shared network

ϕ = level of day-to-day variation of R

Z = peakedness of R

B = switch-to-switch blocking probability objective

Holding fixed the specified peakedness Z and the calculated number N of equivalent trunks, we next calculate what larger equivalent load R_e has required N trunks to meet the blocking objective B if the equivalent load is Poisson and has no day-to-day variation instead of the actual peakedness Z and level ϕ of day-to-day variation. R_e must then satisfy the following equation:

$$N = \overline{\text{NB}}(R_e, 0, 1, B)$$

where we have set $\phi = 0$, signifying no day-to-day variation, and $Z = 1$, signifying Poisson (random) traffic, or unity peakedness level. If we then multiply the switch-to-switch load A by the equivalent load factor $f_e = R_e/R$, the resulting equivalent switch-to-switch load denoted $f_e A$ then produces the required number of trunks when engineered for the same blocking objective, but with zero day-to-day variation and unity peakedness level. In summary, for purposes of determining the equivalent switch-to-switch loads for real-time dynamic routing network design, we replace the triple (A^h, ϕ^h, Z^h), characterizing the switch-to-switch load in hour h, by the triple $(f_e A^h, 0, 1)$.

We now turn to the calculation of the virtual trunk capacity demands given the equivalent switch-to-switch load $f_e A$. The virtual trunk capacity demands are

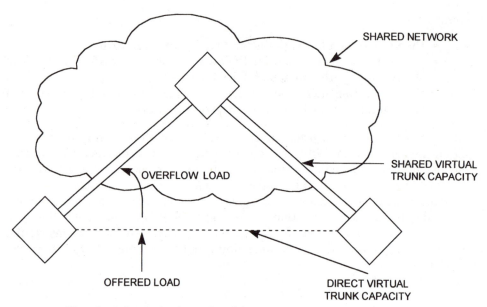

Figure 13.6 Virtual trunk capacity demand model

envisioned to be divided between "direct virtual trunk demands" and "shared virtual trunk demands," as illustrated in Figure 13.6. This model reflects the dynamics of real-time dynamic routing in the following way: Through the use of integrated trunk reservation techniques implemented by real-time dynamic routing methods, the direct virtual trunk capacity is reserved for the primary use of the direct traffic loads. The direct switch-to-switch traffic is routed over the direct virtual trunk capacity, and the shared virtual trunk demands are used as a "network pool" of shared capacity for all the overflow traffic loads, which are in effect routed over a shared network as illustrated in Figure 13.6. The direct capacity is used efficiently by maintaining a high occupancy; the shared capacity is used efficiently by sharing among a large number of traffic items. Hence, the direct and shared virtual trunk capacity demands correspond to the real-time dynamic routing method of efficiently sharing the available capacity among the traffic loads not routed over the direct link.

We first calculate the direct virtual trunks according to the economic sizing, or ECCS, procedure of Truitt [Tru54], which, as we have seen in Chapters 2 to 8, is used frequently in the design of traffic networks. The ECCS procedure is used to calculate the economic sizing for the direct virtual trunk capacity such that the load on the last virtual trunk (LLVT) equals the marginal capacity γ (equal to 0.78 erlang) divided by the cost ratio (CR) of the alternate-path cost to the direct-path cost—that is, LLVT $= \gamma/\text{CR}$. This procedure yields an efficient switch-to-switch blocking level B_k^{hi} such that we may calculate the direct virtual trunk demands for switch pair k in hour h for virtual network i as follows:

$$d_k^{hi} = \overline{\text{NB}}(f_e A_k^{hi}, 0, 1, B_k^{hi})$$

The shared virtual trunks for switch pair k for virtual network i in hour h, s_k^{hi}, are determined according to the overflow from the direct virtual trunks as follows:

$$s_k^{hi} = f_e A_k^{hi} \times \frac{B(f_e A_k^{hi}, d_k^{hi}) - \text{GOS}_k^{hi}}{\gamma}$$

Here, B denotes the Erlang B formula for link blocking given its offered load and trunk capacity, and GOS_k^{hi} denotes the blocking probability grade-of-service objective for switch pair k in hour h for virtual network i. If no alternate paths are available, then B_k^{hi} is set to the blocking probability grade-of-service objective because no shared virtual trunks can be used and all demand must be satisfied on the direct virtual trunks.

Table 13.2 summarizes the approximation of the equivalent load factor, f_e, and the blocking objective, B, based on the ECCS calculation, for the calculation of d_k^{hi} and s_k^{hi} for the case $r_i = 64$ kbps. The table is for the case $Z = 1$; other similar tables are generated for other values of Z and also for the case $r_i = 384$ kbps and $r_i = 1{,}536$ kbps in Tables 13.3 and 13.4. Larger

TABLE 13.2 Virtual Trunk Calculation Parameters ($r_i = 64$ kbps)

Offered load (erlangs) A_k^{hi}	Direct virtual trunk (d_k) blocking objective B_k^{hi}	Equivalent load factor f_e
[0,10]	1.0	1.04
(10,100]	$0.4 - [(A_k^{hi} - 10)/90] \times 0.2$	$1.04 - [(A_k^{hi} - 10)/90] \times 0.01$
(100,500]	$0.2 - [(A_k^{hi} - 100)/400] \times 0.15$	1.03
>500	0.05	1.03

TABLE 13.3 Virtual Trunk Calculation Parameters ($r_i = 384$ kbps)

Offered load (erlangs) A_k^{hi}	Direct virtual trunk (d_k) blocking objective B_k^{hi}	Equivalent load factor f_e
[0,10]	1.0	1.5
(10,100]	$0.4 - [(A_k^{hi} - 10)/90] \times 0.2$	$1.5 - [(A_k^{hi} - 10)/90] \times 0.4$
(100,500]	$0.2 - [(A_k^{hi} - 100)/400] \times 0.15$	1.1
>500	0.05	1.1

TABLE 13.4 Virtual Trunk Calculation Parameters ($r_i = 1{,}536$ kbps)

Offered load (erlangs) A_k^{hi}	Direct virtual trunk (d_k) blocking objective B_k^{hi}	Equivalent load factor f_e
[0,10]	1.0	2.0
(10,100]	$0.4 - [(A_k^{hi} - 10)/90] \times 0.2$	$2.0 - [(A_k^{hi} - 10)/90] \times 0.8$
(100,500]	$0.2 - [(A_k^{hi} - 100)/400] \times 0.15$	1.2
>500	0.05	1.2

values of f_e are used in Tables 13.3 and 13.4 than in Table 13.2 because wideband services occupying several 64-kbps (DS0) time slots attain a lower bandwidth efficiency per erlang of traffic than single-time-slot calls. These larger values are derived through computation of the relative bandwidth occupancy achieved for multislot calls versus single-slot calls as a function of traffic level.

13.3.2.2 Step 2—Initial design of direct link, computation of initial path flows. The design accommodates initial direct link sizing procedures, which include three features: (1) sizing for a minimum direct link capacity in the switch busy hour, (2) prove-in rules for narrowband and wideband terminations for data services, and (3) sizing for a traffic restoration level objective for robust network design.

Sizing for a minimum direct link capacity in the switch busy hour considers the capacity requirements of each direct link in the A-switch busy hour and the Z-switch busy hour, bh(A) and bh(Z), which are the hours of maximum total switch-to-switch load, and sizes the direct link for the maximum requirement in these two hours. That is,

$$\text{bh(A)} = \left\{ h : \sum_{k(\text{A})} A_k^{hi} \text{ is maximum} \right\}$$

$$\text{bh(Z)} = \left\{ h : \sum_{k(\text{Z})} A_k^{hi} \text{ is maximum} \right\}$$

$$T_k = \max[d_k^{\text{bh(A)}i}, d_k^{\text{bh(Z)}i}]$$

where $k(\text{A})$ and $k(\text{Z})$ are the switch pairs that include the A and Z switches and T_k is the number of trunks on the direct link. Sizing for this capacity level can ensure that minimum direct capacity is available in the network consistent with switch capacity requirements.

In the real-time dynamic routing network, design capacity for data services is deployed such that both voice and data services can share the capacity. An economic decision is made for this data service capacity to the effect that above a certain minimum traffic level it is justified to prove in narrowband and wideband data switch terminations on the direct link. That is, above this threshold it is less expensive to spend the incremental costs to build per-call control equipment or wideband terminals than to pay the cost of carrying the data traffic on an alternate path. On the direct link there is a modularity requirement of one 1,544-kbps (DS1) module of capacity for narrowband and wideband data terminations, but due to sharing on the alternate path, incremental costs can be paid proportional to the traffic level and no modularization penalty is incurred. In this manner the prove-in traffic thresholds are then calculated as shown in Table 13.5.

Sizing for a traffic restoration level objective is used to achieve a reliable transport network design, which responds to failure by continuing to provide connections to customers with little or no perceived interruption of service. Reliable design is achieved through the attainment of network transport reliability level objectives, or traffic restoration level objectives, that are

TABLE 13.5 Prove-In Thresholds for Direct Link Data Equipment

Data equipment	Prove-in threshold (erlangs)
narrowband 64-kbps switched digital data termination	1.16 (r_i = 64 kbps)
wideband 384-kbps or 1,536-kbps switched digital data termination	0.23 (r_i = 384 kbps) 0.06 (r_i = 1,536 kbps)

designed to eliminate network isolations. TFO model design helps achieve this reliable network design through appropriate trunk transport diversity design, which is one element of an integrated traffic and transport reliability design method.

As discussed in Chapter 14, the traffic restoration level objective specifies that for single transport link or switch failures, the traffic capacity of the network is designed to carry at least a certain minimum percentage of the design traffic load, denoted as the traffic restoration level objective. For example, if the traffic restoration level objective is 0.5, this means that following any single cable cut in the transport network at least 50 percent of the design traffic load will still be carried after the failure.

The model used for the traffic restoration level design of the direct link is shown in Figure 13.7, where we define the following:

P_k = number of fully diverse paths for switch pair k (switch pair A–B in Figure 13.7)

TRL = traffic restoration level objective

T_k = minimum number of trunks required for switch pair k to meet the traffic restoration level objective

Then for the case P_k = 2, the required number of surviving direct trunks after a transport failure is given by

$$T_k/2 = \overline{\text{NB}}\left(\max_h f_e A_k^h, 0, 1, 1 - \text{TRL}\right)$$

or, for general P_k, we have

$$T_k \times (P_k - 1)/P_k = \overline{\text{NB}}\left(\max_h f_e A_k^h, 0, 1, 1 - \text{TRL}\right)$$

Hence, the minimum number of required trunks T_k on the direct link can be determined from the above expressions. This minimum number of direct trunks is used to initialize link capacity. For switch pairs with zero switch-to-switch traffic demand—that is, when A_k^{hi} is zero for all hours h—the direct link is eliminated to ensure efficient capacity utilization.

The total of the direct and shared virtual trunk demands, $d_k^{hi} + s_k^{hi}$, is used in Step 2 to find an initial feasible flow on the network. This initial flow is found by flowing this total virtual trunk demand in each hour first on the direct link up to the maximum bound, as given in Step 3 below, and then onto the shortest

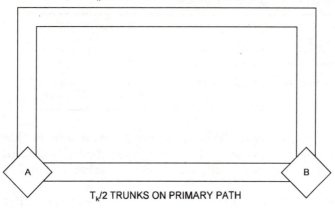

T_k/2 TRUNKS ON DIVERSE PATH

$T_k/2 = \overline{NB}$ (A, O, 1, 1-TRL)

TRL = TRAFFIC RESTORATION LEVEL OBJECTIVE

Figure 13.7 Traffic restoration level design model

(least-cost) two-link path. The feasible flow for switch pairs with no direct link capacity is initialized on the shortest (least-cost) two-link path.

13.3.2.3 Step 3—Linear programming flow model, setting upper bounds on path flows.
In Step 3 the virtual trunk demands $d_k^{hi} + s_k^{hi}$ are routed on the network using a linear programming flow model, as follows:

$$\text{Minimize} \sum_{l=1}^{L} \Delta a_l^i \times C_l$$

subject to

$$\sum_{k=1}^{K} \sum_{j=1}^{J_k^i} P_{jkl}^{hi} y_{jk}^{hi} \le a_l + \Delta a_l^i$$

$$\sum_{j=1}^{J_k^i} y_{jk}^{hi} = d_k^{hi} + s_k^{hi}$$

$$\text{Upper bound on direct path flow} = \max_h d_k^{hi}$$

where

Δa_l^i = added capacity on link l for virtual network i
C_l = cost per unit of capacity on link l
J_k^i = number of allowed paths (including the direct path) for switch pair k for virtual network i

P_{jkl}^{hi} = 1 if path j of switch pair k in hour h for virtual network i routes over link
   l ; 0 otherwise
y_{jk}^{hi} = flow assigned to path j of switch pair k in hour h for virtual network i
a_l = existing spare capacity of link l

In the above model, links l are numbered 1 to L, paths j are numbered 1 to J_k, switch pairs k are numbered 1 to K, hours h are numbered 1 to H, and virtual networks are numbered 1 to M. To reflect the maximum feasible carried load on a path due to path blocking, upper bounds are defined on the direct path flows equal to

$$\max_h d_k^{hi}$$

As in the path-EFO model discussed in Chapter 6, these bounds ensure that a feasible flow is assigned to the real-time dynamic routing network by the linear program, and here again the dynamics of real-time dynamic routing are reflected in the network design.

Either Karmarkar's algorithm or the linear programming heuristic technique used in path-EFO model network design for preplanned dynamic routing networks, as discussed in Chapters 6 and 7, can be used to solve the linear programming flow model. The heuristic method does not give rise to an optimal solution, but it helps us to reduce the size of the linear programming problem so that it can be more easily solved optimally by a method such as the Karmarkar algorithm [Kar84]. The size of the linear program is dominated by the number of variables y_{jk}^h, and when we examine the solutions by the heuristic method, we always find that only a small number of the path flow variables y_{jk}^h assume positive values. (Here, for simplicity, we drop the superscript i for the virtual network.) This suggests that we can drop many of the path candidates in the linear programming formulation without affecting the objective function value of the final solution.

To select the path routing variables for the reduced linear programming problem, we again make use of the heuristic solution in which we retain all those routing variables y_{jk}^h that assume positive values in the heuristic solution. In addition, we also want to include a few routing variables that take zero value in the heuristic solution. This is done as follows.

For each switch pair k and time period h, let j be a routing path in J_k. Suppose link l is included in path j. We define the spare bandwidth BWS_l^h of link l in time period h by

$$\mathrm{BWS}_l^h = \max_{t = 1, 2, ..., H} g_l^t - g_l^h$$

where

$$g_l^h = \sum_{k=1}^{K} \sum_{j=1}^{J_k} P_{jkl}^h y_{jk}^h - a_l$$

is the excess flow on link l in hour h.

We then define a metric

$$m_j = \sum_{l \varepsilon j} C_l \max_{h=1,2,\ldots,H} (0, d_l^h + s_l^h - \text{BWS}_l^h)$$

that measures the idle capacity of path j relative to the demand $d_l^h + s_l^h$. The routing paths j in J_k with zero path flow can be sorted in ascending order of m_j. We can then choose a fixed number of paths from the top of this sorted list as candidate paths for switch pair k and hour h. When we have selected candidate paths for all switch pairs and all time periods, we can discard the remaining routing paths. We can then drop from the linear programming formulation all those y_{jk} variables corresponding to the discarded routing paths j.

A further reduction of the linear programming problem can be made by dropping a certain number of the variables Δa_l. This can be done, again, by making reference to the heuristic solution. Those variables Δa_l that assume zero value in the heuristic solution are the obvious choices. But, as a conservative measure, we only discard a variable Δa_l if

$$\max_{h=1,2,\ldots,H} \frac{g_l^h}{a_l} \leq -0.02$$

That is, there is at least 2 percent spare capacity on the link after the heuristic solution. We present results in Section 13.4.2 that illustrate the use of this reduced linear programming solution method.

13.3.2.4 Step 4—Modularize link bandwidth.

In Step 4 each link is modularized to a modular value of capacity (for example, DS1 or DS3 modules) through the use of the following procedure. In this discussion we assume that there are M virtual networks i_m ordered with m ranging from 1 to M, in which the virtual network requiring the most total network bandwidth is virtual network i_1 and the virtual network requiring the least total network bandwidth is virtual network i_M.

First we define the total flow on link l in hour h for all virtual networks up to and including virtual network i_m as

$$\text{FLOWT}_l^{hi_m} = \sum_{n=1}^{m} \sum_{k=1}^{K} \sum_{j=1}^{J_k^{in}} P_{jkl}^{hi_n} y_{jk}^{hi_n} r_{i_n}$$

We define the background flow for link l corresponding to switch pair l in hour h for virtual network i as

$$\text{BGF}_l^{hi_m} = \sum_{j=1}^{J_l^{im}} \min[\text{FLOWT}_j^{hi_m}(1), \text{FLOWT}_j^{hi_m}(2)]$$

and the modular background flow for link l corresponding to switch pair l for virtual network i as

$$\mathrm{BGFmod}_l^{hi_m} = \sum_{j=1}^{J_l^{i_m}} \min[\mathrm{FLWrnd}_j^{hi_m}(1), \mathrm{FLWrnd}_j^{hi_m}(2)]$$

$\mathrm{FLOWT}_j^{hi_m}(1) = $ flow on link 1 of the two-link path j in hour h for virtual network i_m

$\mathrm{FLOWT}_j^{hi_m}(2) = $ flow on link 2 of the two-link path j in hour h for virtual network i_m

and where $\mathrm{FLWrnd}_j^{hi_m}$ is the rounded value of $\mathrm{FLOWT}_j^{hi_m}$, to the nearest modular bandwidth increment.

The modularization of each link is performed in two steps as follows. First, the added capacity of each link, $\Delta a_l^{i_m}$, is rounded to the nearest modular bandwidth. Then, for those links that are rounded *down,* we check whether the sum of the direct link flow plus background flow exceeds the modular direct link flow plus modular background flow, as follows: If

$$\max_h(\mathrm{FLOWT}_l^{hi_m} + \mathrm{BGF}_l^{hi_m} - \mathrm{FLWrnd}_l^{hi_m} - \mathrm{BGFmod}_l^{hi_m}) > 0$$

then we add a module to link l (corresponding to switch pair l); otherwise, we use the nearest rounding modular size. The result is the total modular link bandwidth:

$$\mathrm{TLBW}_l^{i_m} = \text{total modular link bandwidth}$$

On links that support wideband services—that is, where $r_i = 384$ kbps or $r_i = 1,536$ kbps—a minimum reserved bandwidth capacity for wideband services equal to $\mathrm{RWB}_l^{i_m}$ is ensured by the modularization process for these services. This reserved bandwidth capacity is necessary to ensure that the blocking probability grade-of-service objective is met for the wideband services. Therefore, we require that the direct link bandwidth available on link l for the wideband virtual network be greater than a minimum of two virtual trunks; that is,

$$(\mathrm{TLBW}_l^{i_m} - \max_h \mathrm{FLOWT}_l^{hi_{m-1}})/r_{i_m} \geq 2$$

for wideband virtual network i_m. If this minimum available bandwidth requirement is not met, then an additional module(s) is added to $\mathrm{TLBW}_l^{i_m}$ so that the requirement is met.

Illustrative values of $\mathrm{RWB}_l^{i_m}$, as a function of the added capacity for virtual network i, are shown in Table 13.6.

13.3.2.5 Step 5—Computation of spare virtual trunks for next virtual network.

Here, again, we assume that there are M virtual networks i_m, ordered with m ranging from 1 to M, in which the virtual network requiring the most total network bandwidth is virtual network i_1 and the virtual network requiring the least total network bandwidth is virtual network i_M. The spare bandwidth capacity $\mathrm{BWS}_l^{hi_m}$

TABLE 13.6   Minimum Reserved Bandwidth for Wideband Virtual Networks

Added direct link capacity for wideband virtual networks $(\text{TLBW}_l^{i_m} - \text{TLBW}_l^{i_{m-1}})$ (DS1s)	Minimum reserved bandwidth capacity $(\text{RWB}_l^{i_m})$	
	$(r_i = 384 \text{ kbps})$ (DS1s)	$(r_i = 1{,}536 \text{ kbps})$ (DS1s)
0	0	—
1	1	—
2–3	1	2
4–7	2	2
8–12	2	4
>12	2	6

added in Step 4 is computed as follows:

$$\text{BWS}_l^{h i_m} = \text{TLBW}_l^{i_m} - \max_h \text{FLOWT}_l^{h i_m} - \text{RWB}_l^{i_m}$$

This spare bandwidth capacity is converted to an equivalent virtual trunk capacity by dividing $\text{BWS}_l^{h i_m}$ by $r_{i_{m+1}}$. The resulting virtual trunk capacity $\text{VTS}_l^{h i_{m+1}}$ is used as the existing spare link capacity a_l in Step 2 for virtual network $m+1$, and the process repeats for this virtual network.

13.3.2.6   Step 6—Computation of VTeng.   The bandwidth allocations $\text{VTeng}_l^{i_m}$, discussed in Section 13.2 and in Chapter 12, are calculated for virtual network i_m on link l as follows:

$$\text{VTeng}_l^{h i_m} = \text{TLBW}_l^{i_M} \times \left\{ (d_l^{h i_m} + s_l^{h i_m}) \times r_{i_m} \Big/ \left[\sum_{n=1}^{M} (d_l^{h i_n} + s_l^{h i_n}) \times r_{i_n} \right] \right\} \Big/ r_{i_m}$$

and

$$\text{VTeng}_l^{i_m} = \max_h \text{VTeng}_l^{h i_m}$$

13.3.3   Multihour discrete event flow optimization model

We now illustrate the discrete event flow optimization (DEFO) model for dynamic routing network design. This model can be applied to successful design of a wide variety of dynamic routing networks, including all the various routing strategies identified in Table 13.1. DEFO models are used for fixed and dynamic traffic network design. These models optimize the routing of discrete event flows, as measured in units of individual calls, and the associated link capacities. Figure 1.46 illustrates the steps of the DEFO model. The event generator converts erlang traffic demands to discrete call events. The DEFO model provides routing logic according to the particular real-time dynamic routing method and routes the call events according to the dynamic routing logic. Discrete event flow optimization models use simulation models for routing table design to route discrete event demands on the link capacities, and the link capacities are then optimized to meet the required flow. In the DEFO models for real-time dynamic routing networks, we generate initial link capacity requirements based on the traffic load matrix input to the model. Based on experience with the model, an initial total link capacity demand for each switch is estimated based on a maximum design occupancy in the switch busy hour of 0.93. Then the occupancy of the total network link capacity in the network busy hour is adjusted to fall within the range of 0.84 to 0.89. Blocking performance is evaluated as an output of the discrete event model, and any necessary link capacity adjustments are determined. Trunks are allocated to individual links in accordance with the Kruithof allocation method [Kru37], which distributes link capacity in proportion to the overall demand between switches.

Kruithof's technique is used to estimate the switch-to-switch requirements p_{ij} from the originating switch i to the terminating switch j under the condition that the total switch link capacity requirements may be established by adding the entries in the matrix $\mathbf{p} = [p_{ij}]$. Assume that a matrix $\mathbf{q} = [q_{ij}]$, representing the switch-to-switch link capacity requirements for a previous iteration, is known. Also, the total link capacity requirements b_i at each switch i and the total link capacity requirements d_j at each switch j are estimated as follows:

$$b_i = \frac{a_i}{\gamma}$$

$$d_j = \frac{a_j}{\gamma}$$

where a_i erlangs is the total traffic at switch i, a_j erlangs is the total traffic at switch j, and γ is the average erlang-carrying capacity per trunk or switch design occupancy, as given above.

The terms p_{ij} can be obtained as follows:

$$\text{fac}_i = \frac{b_i}{\sum_j q_{ij}}$$

$$\text{fac}_j = \frac{d_j}{\sum_i q_{ij}}$$

$$E_{ij} = \frac{\text{fac}_i + \text{fac}_j}{2}$$

$$p_{ij} = q_{ij} E_{ij}$$

After the above equations are solved iteratively, the converged steady-state values of p_{ij} are obtained.

The DEFO model generates traffic call events according to a Poisson arrival distribution with a settable average holding time for exponentially distributed holding times. However, more general arrival streams can easily be used, such as peaked traffic arrivals and nonexponentially distributed holding times, because such models can readily be implemented in the DEFO routing simulation model. Traffic call events are generated in accordance with the traffic load matrix input to the model. These traffic call events are routed on the real-time dynamic routing path choice according to the real-time dynamic routing method, as modeled by a set of routing simulation modules that implement the real-time dynamic routing logic for each routing method. The routing design finds the real-time paths between switches in the network for each call event and flows the event onto the network capacity. Each real-time dynamic routing method attempts to share link capacity to the greatest extent possible in accordance with the distribution of loads in the network, with the objective of maximizing the utilization of network resources throughout the busy periods of the network.

The output from the routing design is the fraction of traffic completed in each time period. From this traffic completion performance, the capacity design determines the new link capacity requirements of each switch and each link to meet the design level of blocking. From the estimate of blocked traffic at each switch in each time period, an occupancy calculation determines additional switch link capacity requirements for an updated link capacity estimate. Such a link capacity determination is made based on the amount of blocked traffic. The total blocked traffic, Δa erlangs, is estimated at each of the switches, and an estimated link capacity increase ΔT for each switch is calculated by the relationship

$$\Delta T = \frac{\Delta a}{\gamma}$$

where, again, γ is the average erlang-carrying capacity per trunk. Thus, the ΔT for each switch is distributed to each link according to the Kruithof estimation method described above. The Kruithof allocation method [Kru37] distributes link capacity in proportion to the overall demand between switches and in accordance with link cost, so that overall network cost is minimized. Sizing individual links in this way ensures an efficient level of blocking on each link in the network, to optimally divide the load between the direct link and the overflow network. Once the links have been resized, the network is reevaluated to see if the blocking probability grade-of-service objective has been met, and, if not, another iteration of the model is performed.

We evaluate in the model the confidence interval of the engineered blocking. For this analysis, we evaluate the binomial distribution for the 90th percentile confidence interval. Suppose that for a traffic load of A erlangs in which calls arrive over the designated time period of stationary traffic behavior, there are on average m blocked calls out of n attempts. This means that there is an average observed blocking probability of

$$p1 = m/n$$

where, for example, $p1 = .01$ for a 1 percent average blocking probability. Now, we want to find the value of the 90th percentile blocking probability p such that

$$E(n, m, p) = \sum_{r=m}^{n} C_r^n p^r q^{n-r} \geq .90$$

where

$$C_r^n = \frac{n!}{(n-r)!r!}$$

is the binomial coefficient, and

$$q = 1 - p$$

Then the value p represents the 90th percentile blocking probability confidence interval. That is, there is a 90 percent chance that the observed blocking will be less than or equal to the value p. Methods given in [Wei63] are used to numerically evaluate the above expressions.

As an example application of the above method to the DEFO model, suppose that network traffic is such that 1 million calls arrive in a single busy-hour period, and we wish to design the network to achieve 1 percent average blocking or less. If the network is designed in the DEFO model to yield at most .00995 probability of blocking—that is, at most 9,950 calls are blocked out of 1 million calls in the DEFO model—then we can be more than 90 percent sure that the network has a maximum blocking probability of .01. For a specific switch pair where 2,000 calls arrive in a single busy-hour period, suppose we wish to design the switch pair to achieve 1 percent average blocking probability or less. If the network capacity is designed in the DEFO model to yield at most .0075 probability of blocking for the switch pair—that is, at most 15 calls are blocked out of 2,000 calls in the DEFO model—then we can be more than 90 percent sure that the switch pair has a maximum blocking probability of .01. These methods are used to ensure that the blocking probability design objectives are met, taking into consideration the sampling errors of the discrete event model.

The greatest advantage of the DEFO model is its ability to capture very complex routing behavior through the equivalent of a simulation model provided in software in the routing design module. By this means, very complex routing networks have been designed by the model, which include all of the routing methods discussed in Section 13.2, Chapter 12, and Table 13.1. In particular, these examples include MXDR, which combines LRR, TSMR, and RTNR in the same network, and multiservice integrated dynamic routing, described in Chapter 12 and Section 13.2. A flow diagram of the DEFO model, in which RTNR logical blocks described in Chapter 12 are implemented, is illustrated in Figure 13.8. The DEFO model is general enough to include all variations of these models and any real-time dynamic routing models yet to be determined. The DEFO model is therefore the basis for design of new dynamic routing methods, which are easily compared to other alternative implementations with the model. Design examples are presented in the next section.

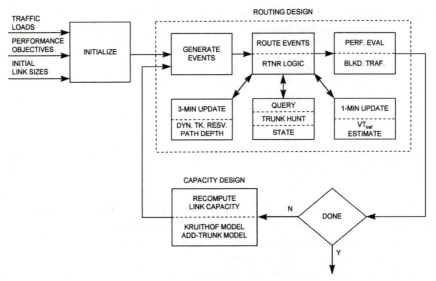

Figure 13.8 Discrete event flow optimization model with RTNR logic blocks

13.3.4 Day-to-day load variation design models

As discussed in Section 12.3.2 and in Chapter 6, an equivalent load method is used in the TFO model for real-time dynamic routing network design, and in the path-EFO model for preplanned dynamic routing network design to account for day-to-day variations. The method adjusts each traffic load A by a factor A_e/A, such that the following equations are satisfied.

$$N = \overline{\mathrm{NB}}(A, \phi, Z, B)$$

N = trunk requirement
A = switch-to-switch load
ϕ = level of day-to-day variation of A
Z = peakedness of A
B = switch-to-switch blocking probability objective
A_e = equivalent switch-to-switch load

and where $\overline{\mathrm{NB}}$ is a function mapping A, ϕ, Z, and B into the link requirement N.

Holding fixed the specified peakedness Z and the calculated circuits N, we calculate what larger equivalent load A_e requires N circuits to meet the blocking probability objective if the equivalent load has had no day-to-day variation:

$$N = \overline{\mathrm{NB}}(A_e, 0, Z, B)$$

where $\phi = 0$ signifies no day-to-day variation. The equivalent load A_e then produces the same equivalent number of circuits N when designed for the same

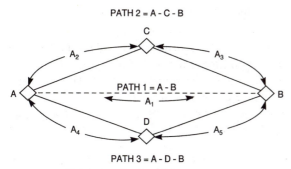

Figure 13.9 Equivalent load model

peakedness level but in the absence of day-to-day variation. The ratio A_e/A is referred to as the equivalent load factor.

As pointed out earlier, this equivalent load method provides sufficient capacity for each traffic load by itself to meet the blocking objective. It does not consider the fact that on any particular day at any particular time some traffic loads are above their average load and some below. Here we suggest a straightforward extension of the equivalent load method as applied to path-EFO model design to account for the way in which real-time dynamic routing effectively gives a traffic load access to the capacity of the large number of parallel paths between a pair of switches and makes the combination of parallel paths behave as a single large link. As shown in Figure 13.9, the route, or link bundle, between switches A and B consists of paths 1, 2, and 3. By means of real-time dynamic routing, traffic load A–B has access to the capacity normally used by traffic loads A_2 and A_3 on path 2 and by traffic loads A_4 and A_5 on path 3. Traffic load A–B therefore has access to a much larger number of circuits than those that would be required for just A_1 erlangs, and the operation of real-time dynamic routing is such that the capacity of the link bundle between switches A and B is used as if it were a single large link. The completion probability of traffic load A_1 calls, therefore, is more related to this larger bundle of circuits than to the circuits required for just A_1 erlangs.

Hence, real-time dynamic routing tends to make the link bundle between switches A and B behave as if it were an equivalent larger link; this equivalent larger link would, in turn, have an associated equivalent larger offered traffic load. This equivalent larger offered traffic load is approximately equal to $A_1 + \min[A_2, A_3] + \min[A_4, A_5]$. The minimum function is used for two-link paths because their contribution toward the equivalent larger link, and hence the equivalent larger traffic load, can be no larger than the smallest link capacity on the path. Here the added terms $\min[A_2, A_3] + \min[A_4, A_5]$ are referred to as the background load, BGL. Therefore, we adjust traffic load A_1 by the equivalent load factor associated with the equivalent larger offered traffic load, $A_1 + \text{BGL}$. This factor for $A_1 + \text{BGL}$ is smaller than the factor associated with just A_1.

13.3.5   Bandwidth allocation for real-time dynamic
routing networks

A bandwidth allocation technique is described here, which is based on the optimal solution of a network bandwidth allocation model for real-time dynamic routing networks. The model achieves significant improvement in both the average network blocking and switch-pair blocking distribution when the network is in a congested state, such as under peak-day loads. For link bandwidth allocation, the quantity $\text{VTeng}_l^{i_m}$ is calculated for each virtual network i_m on link l as described in Section 13.3.2.

For network bandwidth allocation, we consider a method to take advantage of predictable traffic demands. The bandwidth allocation is achieved by way of a prespecified set of alternate paths for each switch pair in the network. Thus, when a call enters the real-time dynamic network, the originating switch first tries to set up the call on the direct path. A further capability assumed for the real-time dynamic network is to segment alternate paths into two sets. If a direct trunk is not available, then the call can be connected through one of the first set of prespecified paths selected according to the idle trunk status. To determine those candidate paths, we present a model in which a linear programming problem is set up and solved. The candidate paths in the first set are then obtained from the solution to this problem. For each switch pair, the sets of candidate paths determined by the solution to the throughput-maximization problem are taken as the first-choice via paths. They will be used as the candidates for alternate paths when a call is blocked from the direct path. However, when the switch-to-switch blocking of a switch pair over a certain update period exceeds a specified level, the search for alternate paths can be extended into the second set of via paths, as illustrated in Figure 13.10. The latter paths are the remaining alternate paths that are not included in the first set of first-choice via paths. The reason for

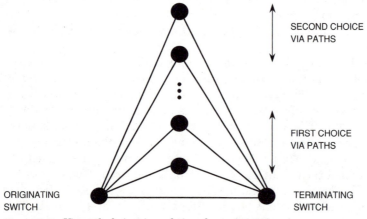

Figure 13.10 Via path choices in real-time dynamic routing

the second-choice via paths is to promote higher fairness of switch-to-switch blocking performance. This is especially important in situations such as (1) transport link failures, which can cause the loss of a significant amount of capacity in the network, possibly including the loss of direct paths for certain switch pairs to make their protection by trunk reservation impossible; (2) switch failures, which can cause the loss of alternate paths for other switch pairs; and (3) special events that can cause a significant increase of traffic demands for certain switch pairs.

The in-depth search for alternate paths is good for switch pairs that have excessively high blocking; these switch pairs can improve performance somewhat by using capacity of other switch pairs whose performance is less degraded. However, when high blocking occurs throughout the whole network, such a method could cause a drastic drop in call throughput. This situation can be detected by the fact that the total blocking at the originating switch is above a very high level. In such a situation, the search depth for each call originating from this switch should be reduced to the set of first-choice via paths or even a smaller set of paths, as described in Chapter 12, to prevent network instability.

Simulation results presented in Section 12.4 demonstrate that this network bandwidth allocation procedure reduces both the average network blocking and switch-pair blocking distribution when the traffic demands are higher than the design level. This suggests that this implementation of real-time dynamic routing can be profitable when the network is foreseen to face some excessive loads over a period of time, such as under peak-day loads.

In the network bandwidth allocation model, the network capacity and offered loads are given, and a throughput maximization problem is formulated and solved. Then, we obtain from the solution a set of candidate paths for each switch pair in the network. We first convert the traffic demands into virtual trunk capacity demands and then flow virtual trunk demands, rather than the erlang demands, when solving the linear programming problem.

When the equivalent virtual trunk demand for each switch pair is determined, the linear programming can now be set up. Note that the model is constructed for each given virtual network i and hour h. For simplicity, we omit the superscripts i and h from the model formulation. The objective of the linear program is to maximize total network throughput:

$$\text{Maximize} \sum_{k=1}^{K} \sum_{j=1}^{J_k} y_{jk}$$

subject to

$$\sum_{k=1}^{K} \sum_{j=1}^{J_k} P_{jkl} y_{jk} \leq a_l$$

$$\sum_{j=1}^{J_k} y_{jk} \leq \text{VT}_k$$

where

J_k = number of allowed paths (including the direct path) for switch pair k

P_{jkl} = 1 if path j of switch pair k routes over link l; 0 otherwise

y_{jk} = flow assigned to path j of switch pair k

a_l = existing trunk capacity on link l available for the given virtual network

VT_k = total VT demand for switch pair k ($= d_k + s_k$)

In this model, links l are numbered 1 to L, paths j are numbered 1 to J_k, and switch pairs k are numbered 1 to K. The first set of constraints requires that the sum of the flows on all the paths containing link l cannot exceed the total capacity of link l. The second set of constraints requires that the sum of flows on all the paths connecting switch pair k cannot exceed the virtual trunk demand of switch pair k. The Karmarkar algorithm [Kar84, KaR88] can be used to solve the linear programming flow model.

In a mesh network, the total number of links is almost the same as the total number of switch pairs. Thus, the number of constraints in the linear program is approximately 2 times the number of switch pairs, or $2K$. The number of variables is the same as that of paths, which is K times the average number of paths per switch pair. Assuming each alternate path has two links only, the total number of variables is of the order K^3, which is the dominating factor for the convergence speed of the Karmarkar algorithm. On the other hand, the constraint matrix of the linear programming problem is very sparse, because each path belongs to only one switch pair, and it contains only a few links. Thus, the particular variant of the Karmarkar algorithm given in [KaR88] is suitable for solving the problem. However, the excessive number of variables in this problem still affects the convergence speed of the Karmarkar algorithm. We found that a way of curtailing the computation time is to run the Karmarkar algorithm up to a certain level of convergence and then, as discussed in Section 13.3.2.3, to delete those paths with zero or very minimal path flows. A significant number of paths can be eliminated in this way. We then use the same Karmarkar algorithm method to solve the problem with the reduced sets of paths. The shortening of the path lists not only reduces the total computation time but also helps to force the flows of virtual trunks to be more concentrated on fewer paths. When we have obtained the optimal solution to the smaller problem, we can then select the candidate paths as the paths that have flows exceeding a threshold in the linear programming solution.

13.4   Modeling Results

13.4.1   Results for fixed-point erlang flow optimization model design

In this section we give numerical examples of the use of the fixed-point erlang flow optimization model for RTNR networks. Table 13.7 gives the trunk and

TABLE 13.7   Six-Switch Network Data

Originating switch	Terminating switch	Trunks	Load (erlangs)
1	2	36	27.5
1	3	24	7.0
1	4	324	257.8
1	5	48	20.5
1	6	48	29.1
2	3	96	25.1
2	4	96	101.6
2	5	108	76.8
2	6	96	82.6
3	4	12	11.9
3	5	48	6.9
3	6	24	13.2
4	5	192	79.4
4	6	84	83.0
5	6	336	127.1

TABLE 13.8   Total Network Blocking Comparison for Six-Switch Network

Overload factor	Simulation model	Fixed-point erlang flow optimization model
1.0	0.000	0.000
1.1	0.001	0.000
1.2	0.005	0.004
1.3	0.025	0.025
1.4	0.058	0.059

load information for the six-switch model [MiS91] used in the study. Results are given in Table 13.8 for the design load and for general overloads of 10, 20, 30, and 40 percent, respectively, in which we compare total network blocking for the simulation model and for the fixed-point erlang flow optimization model. The switch-to-switch blocking comparisons for the 40 percent overload case are given in Table 13.9 and those for the 30 percent overload case in Table 13.10.

It is noted that good agreement is achieved between the fixed-point erlang flow optimization model and the simulation model, and that switch-to-switch blocking estimates are usually within 1 percent of each other. In addition, the estimated blocking in the fixed-point erlang flow optimization model is almost always in the range of variability of the simulated blocking. We note also that good agreement is reported in [MGH91] between analytical methods and simulation for the case of symmetric least-busy-alternative routing networks with fixed trunk reservation.

We have used the fixed-point erlang flow optimization model to examine the performance of the six-switch network under variation of some of the RTNR parameters. First, we examined the reservation level threshold factor,

TABLE 13.9   Switch-to-Switch Blocking Comparison for Six-Switch Network (40% Overload)

Originating switch	Terminating switch	Trunks	Load (erlangs)	Simulation model	Fixed-point erlang flow optimization model
1	2	36	27.5	0.046	0.042
1	3	24	7.0	0.016	0.008
1	4	324	257.8	0.066	0.071
1	5	48	20.5	0.013	0.012
1	6	48	29.1	0.033	0.042
2	3	96	25.1	0.001	0.000
2	4	96	101.6	0.222	0.229
2	5	108	76.8	0.021	0.009
2	6	96	82.6	0.091	0.093
3	4	12	11.9	0.067	0.046
3	5	48	6.9	0.003	0.000
3	6	24	13.2	0.012	0.004
4	5	192	79.4	0.010	0.008
4	6	84	83.0	0.031	0.032
5	6	336	127.1	0.000	0.000

TABLE 13.10   Switch-to-Switch Blocking Comparison for Six-Switch Network (30% Overload)

Originating switch	Terminating switch	Trunks	Load (erlangs)	Simulation model	Fixed-point erlang flow optimization model
1	2	36	27.5	0.012	0.017
1	3	24	7.0	0.002	0.001
1	4	324	257.8	0.027	0.017
1	5	48	20.5	0.003	0.001
1	6	48	29.1	0.013	0.020
2	3	96	25.1	0.000	0.000
2	4	96	101.6	0.114	0.137
2	5	108	76.8	0.003	0.000
2	6	96	82.6	0.036	0.044
3	4	12	11.9	0.017	0.001
3	5	48	6.9	0.001	0.000
3	6	24	13.2	0.001	0.000
4	5	192	79.4	0.008	0.000
4	6	84	83.0	0.014	0.001
5	6	336	127.1	0.000	0.000

which multiplies $VTtraf_k$ to yield the reservation level threshold and which is given as 0.05 for reservation level 1 in Table 12.1. We examined a range of values for the cases of 20, 30, and 40 percent overload in the six-switch network. The results, given in Table 13.11, indicate that the exact level of trunk reservation does not strongly influence network performance, although reservation is extremely important for the stabilization of network throughput. These results from the fixed-point erlang flow optimization model are consistent with the results of the simulation model for various reservation level factors.

TABLE 13.11   Total Network Blocking vs. Reservation Level Threshold R_k Factor

Reservation level threshold factor (R_k = factor × VTtraf$_k$)	20% overload	30% overload	40% overload
0.02	0.0038	0.0250	0.0606
0.03	0.0038	0.0250	0.0602
0.04	0.0038	0.0250	0.0598
0.05	0.0038	0.0248	0.0595
0.06	0.0038	0.0245	0.0594
0.07	0.0038	0.0245	0.0593
0.08	0.0038	0.0245	0.0592
2.0	0.0038	0.0242	0.0603

TABLE 13.12   Total Network Blocking vs. Load State Threshold TK1$_k$ factor

Load state threshold factor (TK1$_k$ = factor × VTtraf$_k$)	20% overload	30% overload	40% overload
0.02	0.0038	0.0253	0.0599
0.03	0.0039	0.0252	0.0597
0.04	0.0039	0.0250	0.0595
0.05	0.0038	0.0248	0.0595
0.06	0.0048	0.0263	0.0605
0.07	0.0048	0.0264	0.0615
0.08	0.0055	0.0277	0.0615
2.0	0.0226	0.0456	0.0783

We also examined the link load state threshold factor, which multiplies VTtraf$_k$ to yield the load state threshold, and which is given as 0.05 for TK1$_k$ in Table 12.2. We examined a range of values for the cases of 20, 30, and 40 percent overload in the six-switch network. The results, given in Table 13.12, indicate that the load state threshold factor influences network performance rather strongly. They suggest that a threshold factor of 0.05 is a good choice across a range of conditions, which is the value implemented for RTNR operation.

Finally we examined different VTtraf$_k$ factors, which multiplies the switch-to-switch traffic load estimate, TL$_k$, to yield VTtraf$_k$ and which is given as 1.1 in Chapter 12. We examined a range of values for the cases of 20, 30, and 40 percent overload in the six-switch model. The results are given in Table 13.13, and they indicate that the VTtraf$_k$ factor has a substantial influence on network performance. The results suggest that a factor of 1.1 is a good choice across a range of conditions, which is the value implemented for RTNR operation. Hence the fixed-point erlang flow optimization model for real-time dynamic routing networks provides validation of some of the parameter values selected for RTNR implementation as determined through simulation models presented in Chapter 12.

TABLE 13.13   Total Network Blocking vs. Virtual Trunk Requirements VTtraf$_k$ factor

VTtraf$_k$ factor (VTtraf$_k$ = factor × TL$_k$)	20% overload	30% overload	40% overload
0.8	0.0050	0.0270	0.0646
0.9	0.0045	0.0265	0.0615
1.0	0.0041	0.0252	0.0602
1.1	0.0038	0.0248	0.0595
1.2	0.0036	0.0247	0.0601
1.3	0.0035	0.0258	0.0601
1.4	0.0041	0.0258	0.0601
2.0	0.0078	0.0274	0.0620

13.4.2   Results for transport flow optimization model design

Transport flow optimization (TFO) model capacity design procedures are tested on RTNR network models, and we find that these designs provide network performance within the design objectives, at a lower network cost than the path-EFO model design for preplanned dynamic routing networks with the same level of demand. These results are reported in this section.

A 103-switch model is used to test the path-EFO model and the TFO model design methods. Designs are made from zero initial capacity for a single voice service and for multiple services that include voice, 64-kbps switched digital data, and 384-kbps switched digital data. We first report initial tests with the single voice service. The results for the path-EFO model design for DNHR networks and the TFO model design for RTNR networks are given in Table 13.14. They indicate that the TFO model method is able to achieve about a 2 percent saving over the path-EFO model DNHR design. Using the results of the TFO model design for RTNR and DNHR path-EFO model designs, we observe that for the design load, both designs met the network performance objective—that is, zero blocking is obtained for both cases, based on network simulations. Tables 13.15 and 13.16 show the network performance comparison under a uniform inflation of network load of 10 and 20 percent, respectively. The results show that the RTNR TFO model design compares favorably with the DNHR path-EFO model design, except for the network blocking distribution for the 20 percent overload case, in which the RTNR design yields a somewhat higher 99th and 90th percentile blocking. Table 13.17 shows the network performance for Christmas Day traffic load.

TABLE 13.14   Network Design Comparison for Single-Service, Voice (103-Switch Model)

Design method	Trunks	Cost
DNHR path-EFO model	577K	639M
RTNR TFO model	570K	627M

TABLE 13.15 Network Blocking Performance Comparison for Single-Service RTNR TFO Model Design for 10% Traffic Overload (103-Switch Model)

Hour of day	DNHR path-EFO model design (% blocking)	RTNR TFO model design (% blocking)
9 to 10 A.M.	0.5	0.0
10 to 11 A.M.	0.7	0.1
2 to 3 P.M.	0.4	0.1
Sunday evening	0.0	0.0
Average	0.4	0.05
Switch-pair maximum	5.6	3.7
99% highest	1.5	1.0
90% highest	0.1	0.1

TABLE 13.16 Network Blocking Performance Comparison for Single-Service RTNR TFO Model Design for 20% Traffic Overload (103-Switch Model)

Hour of day	DNHR path-EFO model design (% blocking)	RTNR TFO model design (% blocking)
9 to 10 A.M.	4.2	1.6
10 to 11 A.M.	6.3	5.9
2 to 3 P.M.	5.7	5.0
Sunday evening	0.4	0.1
Average	4.4	3.4
Switch-pair maximum	35.4	25.9
99% highest	11.1	18.0
90% highest	5.1	10.1

TABLE 13.17 Network Blocking Performance Comparison for Single-Service RTNR TFO Model Design for Christmas Day Traffic Load (103-Switch Model)

Hour of day	DNHR path-EFO model design (% blocking)	RTNR TFO model design (% blocking)
10 A.M.	11.9	11.0
11 A.M.	18.9	18.7
12 noon	21.6	21.0
Average	17.8	17.3
Switch-pair maximum	96.9	75.0
99% highest	84.3	61.3
90% highest	48.2	40.6

TABLE 13.18 Network Design Comparison for Multiservice Integrated Network (50-Switch Model)

Design method	Trunks	Cost
DNHR path-EFO model	144K	156M
RTNR TFO model	143K	152M

TABLE 13.19 Network Blocking Performance for Multiservice RTNR TFO Model Design for 10% Traffic Overload (50-Switch Model)

Hour of day	Voice traffic (% blocking)	64-kbps traffic (% blocking)	384-kbps traffic (% blocking)
9 to 10 A.M.	0.0	0.0	0.0
10 to 11 A.M.	0.0	0.5	0.3
2 to 3 P.M.	0.0	0.9	0.0
Sunday evening	0.0	0.0	0.0
Average	0.0	0.4	0.1
Switch-pair maximum	0.6	9.7	2.6
99% highest	0.0	2.0	0.0
90% highest	0.0	0.0	0.0

TABLE 13.20 Network Blocking Performance for Multiservice RTNR TFO Model Design for 20% Traffic Overload (50-Switch Model)

Hour of day	Voice traffic (% blocking)	64-kbps traffic (% blocking)	384-kbps traffic (% blocking)
9 to 10 A.M.	0.1	0.0	2.2
10 to 11 A.M.	3.2	5.3	11.0
2 to 3 P.M.	3.0	4.6	11.2
Sunday evening	0.0	0.1	0.7
Average	1.8	2.8	6.6
Switch-pair maximum	21.6	35.1	32.7
99% highest	12.8	24.8	14.9
90% highest	5.9	12.9	9.5

The Christmas Day simulation results show that the RTNR TFO model designs compare favorably with the DNHR path-EFO model design, in both network throughput and switch-pair blocking distribution.

The results for the RTNR TFO model design for multiple services, voice, 64 kbps, and 384 kbps, are given in Table 13.18. These results are obtained for a 50-switch model and show a 2.6 percent saving with the RTNR TFO model design over the DNHR path-EFO model design. The DNHR path-EFO design is based on the multiservice integrated network design model presented in

TABLE 13.21 Network Blocking Performance for Multiservice RTNR TFO Model Design for Christmas Day Traffic Load (50-Switch Model)

Hour of day	Voice traffic (RWB_l^{im} not allocated to voice traffic, % blocking)	Voice traffic (RWB_l^{im} allocated to voice traffic, % blocking)
10 A.M.	5.1	4.6
11 A.M.	16.5	15.6
12 Noon	23.3	22.5
Average	15.9	15.1
Switch-pair maximum	73.6	73.8
99% highest	63.7	63.9
90% highest	45.0	44.1

Chapter 10. Simulation results for the RTNR TFO model design are given in Tables 13.19, 13.20, and 13.21 for 10 percent traffic overload, 20 percent traffic overload, and Christmas Day traffic load, respectively. The results show that the RTNR TFO model design for multiple services performs comparably to or even better than the single-service designs discussed above, but now for all services at once. Therefore, we can conclude that these multiservice designs are robust to traffic load variations and network disruptions, as are the single-service designs. Note that wideband traffic has somewhat higher blocking due to the greater difficulty in obtaining multiple contiguous time slots versus single-time-slot traffic. Note in Table 13.21 that no data traffic is assumed in the model for Christmas Day traffic load. Therefore, voice traffic can use data capacity for significant improvement in performance, especially if the minimum reserved bandwidth capacity for wideband services, RWB_l^{im}, is allocated to voice services on peak days.

Our experience with the TFO model designs shows that the heuristic method is very effective in obtaining a solution that is within 3.5 to 5 percent of the optimal solution generated by the Karmarkar algorithm [KaR88]. As described in Section 13.3.2.3, the heuristic method can also help to reduce drastically the size of the linear programming flow model. For example, for each switch pair k and time period h, the number of all routing paths is of the order K, where K is the total number of switches. After the reduction method is applied, the average number of paths per switch pair per time period is decreased from 50–100 paths to approximately four to six paths, depending on the characteristics of the input data. Thus, the reduction of routing variables is by an order of magnitude. On the other hand, the number of reducible a_l variables depends on the amount of existing capacity and is thus variable.

The reduction of variables in the linear programming flow model results in large savings in computation time. Table 13.22 illustrates the total run time for the design of a network consisting of 50 switches. The run time of the Karmarkar algorithm for the full-sized linear programming flow model takes 18 hours of CPU time on a superminicomputer system. This problem has approximately 25,000 constraints and 3 million variables. The heuristic solution,

TABLE 13.22   Comparison of Karmarkar Algorithm and Heuristic Design Methods

Method	Problem size	Optimal level	CPU time	Relative CPU time (% of full solution time)
Karmarkar	full	100%	18 hours	100%
heuristic	full	96%	0.25 hour	1.4%
Karmarkar	reduced	99.9%	1.25 hours	7.0%
Karmarkar	full	96%	3.6 hours	20%

on the other hand, takes only 0.25 CPU hour, and the Karmarkar algorithm applied to the reduced linear programming problem takes 1.25 CPU hours. Thus, the total computation time as taken by the size reduction approach consumes only 7.0 percent of the time taken by the Karmarkar algorithm alone in solving the full-sized problem. In this particular case, we assume that there is no existing capacity in the network. Thus, there is no reduction of the variables a_l. The final solution reached by the reduced Karmarkar algorithm approach is found to be within 0.1 percent of the optimal solution to the same problem. We consistently observed the same level of near optimality for all problems tested.

In the above test case, if we use the Karmarkar algorithm instead of the heuristic solution method to reach the same level of objective function value, it would have to run for 3.6 hours, which is 20 percent of the total time taken for the full-sized problem. Thus, we have observed several advantages of the heuristic solution method: (1) It achieves a feasible solution that is reasonably close to the optimal solution; (2) if further cost reduction is not needed, the heuristic solution can be used as the final solution; and (3) it serves as a good preprocessor by which the linear programming problem can be reduced and solved to obtain a near-optimal solution.

Hence, the optimal designs for real-time dynamic routing networks, using the Karmarkar algorithm approach for the solution of the TFO linear programming flow model, achieve a total of an approximately 5 to 8 percent reduction in network design cost in comparison with the path-EFO designs of preplanned dynamic networks. Here, the 5 to 8 percent reduction consists of a 2 to 3 percent cost reduction due to the TFO network design model, as discussed above, and a 4 percent cost reduction due to a Karmarkar algorithm optimal solution of the linear programming flow model, as discussed in this section.

13.4.3   Results for discrete event flow optimization model design

In Chapter 1 we illustrated extensive network design examples using DEFO models for the various real-time dynamic routing alternatives discussed in this chapter. In general, we find that most of the real-time dynamic routing alternatives have network design advantages in making them more highly utilized and better performing than fixed hierarchical networks. With the use of the DEFO model, comparisons are made between the real-time dynamic routing techniques described in Section 13.2.

It is shown in Chapter 1 in Tables 1.4 to 1.9 that the DEFO model design results for larger network models are in general agreement with results for smaller models. These results illustrate the cost savings potentially achievable with designs for metropolitan area networks, national intercity networks, and global international networks for real-time dynamic routing design in comparison with fixed hierarchical routing design. These design results show a large range of design efficiencies. LRR is the least efficient, and because of the random selection of paths and possible loss of calls at a via switch, the individual switch-pair blocking performance is difficult to control. This leads to less efficient network design, unless one is willing to relax the maximum switch-to-switch blocking objective. The NTT Japan implementation of event-dependent routing, STR, limits the choice of paths to a subset of total choices, which helps limit the negative effects of random path selection. STT achieves a significant improvement in design efficiency with the use of crankback to search for an available via path. RTNR achieves the most efficient design, with TSMR a close second, and MXDR reflects a weighted average of the individual performances of all the dynamic routing methods.

The DEFO model is also successfully applied to the design of integrated class-of-service dynamic routing networks, and it achieves much better design efficiency than other models. In typical cases, DEFO model designs achieve a 3–5 percent cost reduction for multiservice integrated networks in comparison with the TFO model designs. It is the basis for refinement of the real-time dynamic routing detailed implementation, in which alternative dynamic routing approaches are compared with the DEFO model. The successful design of such complex networks, both in terms of routing method design and capacity design, illustrates the advantage and flexibility of the DEFO model design approach.

13.4.4　Results for day-to-day load variation design

To apply the day-to-day variation model discussed in Section 13.3.4 for path-EFO model designs, we first find the appropriate background load to add to each traffic load. To estimate the background load, we considered the average background load as a function of distance for the 25-switch network, shown in Figure 11.5. The results are plotted in Figure 13.11, in which the traffic loads are grouped into 200-mile bands. We have also provided an approximate fit to the data for purposes of path-EFO model network design. For each switch pair, we add this background load (BGL) value to the switch-to-switch load and apply the equivalent load method to the sum. The equivalent load factor obtained through the use of this technique is then applied to each traffic load, and the network is designed for the adjusted loads by the path-EFO model for real-time dynamic network design. The result is a network that is 2–3 percent less costly than the path-EFO model capacity design for preplanned dynamic routing networks. The network performance results for a simulation of this alternative network design over 10 simulated daily load patterns are shown in Figures 13.12 and 13.13. We observe that the modified network performance under TSMR compares favorably with the DNHR network performance and meets the switch-to-switch blocking objectives.

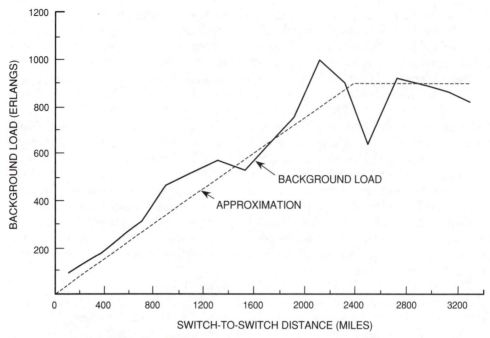

Figure 13.11   Background load vs. distance

These results demonstrate that additional network efficiencies are achievable with real-time dynamic routing with such alternative design procedures for day-to-day load variations.

Overall, applying the equivalent load technique increased the number of trunks in the 25-switch network by about 5.25 percent. This compares closely with the 5.2 percent increase in the number of trunks in the 10-switch model studied in [Kri82], which uses near-optimal techniques for computing day-to-day capacity augmentation in a real-time dynamic routing network. It is interesting to observe that the 5 percent inflation of network loads resulting from this approach corresponds closely to the expected 5 percent increase in network load on an average Monday (Figure 8.15). Hence, we should expect a similar increase in network capacity if we had designed the network for average Monday loads. This suggests that an interpretation for the equivalent loads might be an approximate average Monday load pattern.

13.4.5   Results for bandwidth allocation design

We now turn to the network bandwidth allocation model given in Section 13.3.5, and we use simulation to test the solution for paths that maximize throughput for a real-time dynamic network, based on solving a linear programming flow maximization problem with the traffic demands and network capacity given. The network bandwidth allocation model can be used to prespecify a set of routing paths for each switch pair so as to improve the blocking performance of a network using real-time dynamic routing.

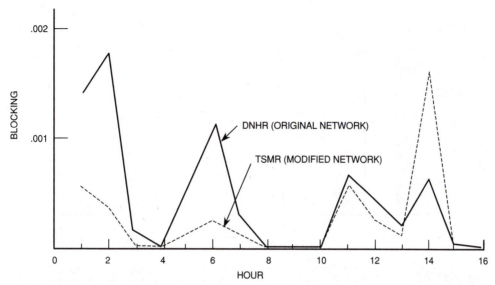

Figure 13.12 Average network blocking for DNHR (original network) and TSMR (modified network)

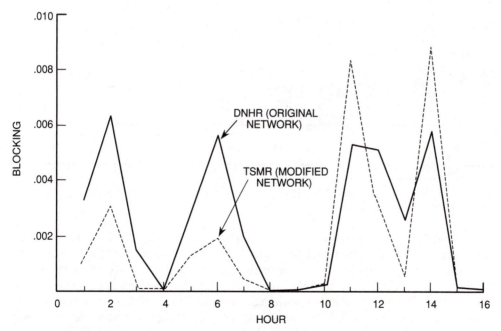

Figure 13.13 99th percentile switch-pair blocking for DNHR (original network) and TSMR (modified network)

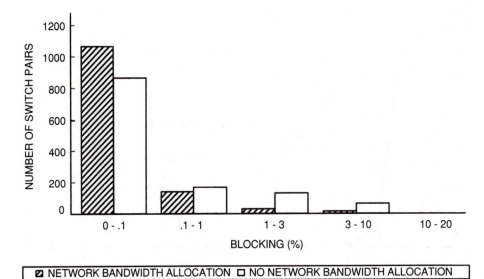

Figure 13.14 Network bandwidth allocation blocking distribution (50-switch model)

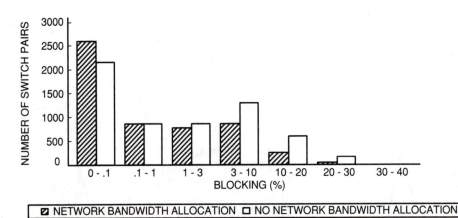

Figure 13.15 Network bandwidth allocation blocking distribution (100-switch model, 20% overload)

Two results are shown in Figures 13.14 and 13.15. The first result is based on a network of 50 switches. The network is first designed for the given traffic demands, using the TFO model described in Section 13.3. We then solve the throughput-maximization problem with each switch-to-switch load increased by 10 percent. The first-choice via paths can then be obtained from the solution as described in Section 13.3.5. In the simulation, the original loads are increased randomly by a factor uniformly varying between 8 and 12 percent. We find that the network average blocking of the network bandwidth allocation design is 0.27 percent, versus 0.5 percent without network bandwidth allocation. The blocking distribution is shown in Figure 13.14. The second result is based on a network of 100 switches. A 20 percent increase of switch-to-switch loads is assumed in the linear programming flow maximization problem, and an inflation factor uniformly varying between 15 and 25 percent is applied to each switch-to-switch load in the simulation. The network average blocking is seen to decrease from the 3.3 percent level to the 2.2 percent level. The switch-pair distribution is displayed in Figure 13.15. Both figures show a clear gain in the number of low-blocking switch pairs by the real-time dynamic network bandwidth allocation design.

We have thus seen that the network bandwidth allocation model for real-time dynamic networks achieves significant improvement in both the average network blocking and the switch-pair blocking distribution when the network is in a congested state. This suggests that the implementation of this network bandwidth allocation method can be profitable when the network is expected to face excessive loads over a period of time.

13.5 Conclusion

We have described the design and management of real-time dynamic routing networks. In Chapters 11 and 12 we showed that real-time dynamic routing networks allow improved performance and reduced network management and design costs because with real-time dynamic routing a number of network management and design operations have been simplified or eliminated, leading to savings in operations costs and expenses. In this chapter we address network design models for the design of real-time dynamic routing networks, some of which achieve near-optimal design. We have presented the results of design and performance studies for real-time dynamic routing networks, which include real-time event-dependent routing, centralized real-time state-dependent routing, and distributed real-time state-dependent routing networks. Analysis in this chapter shows that real-time dynamic routing provides key advantages for introduction into the network. In particular,

- It provides the opportunity to simplify and improve network designs and to achieve significant additional network capital cost reduction in comparison with preplanned dynamic routing.

- It provides a reliable network design capability that ensures networkwide path selection and immediate adaptation to failure.

- It improves network performance by lowering blocking and reducing call setup delay under all network load and failure conditions.

- It simplifies and enhances the integration of voice and data networks, which minimizes dedicated network cost and helps provide competitively priced services.

A fixed-point erlang flow optimization model is used to solve the nonlinear flow equations describing real-time dynamic network dynamic behavior. The link-state model provides the aggregate link-state probabilities, and the link-state probability computation solves the birth–death equations and models the adaptive nature of trunk reservation and path selection depth. The fixed-point erlang flow optimization model provides a method to calculate the traffic flow using the least-busy concept employed in real-time state-dependent routing networks. The model output for nonsymmetrical networks is provided, and the differences from a simulation model are presented. Good agreement is found between the fixed-point flow optimization model and the simulation model. We also used the fixed-point model to examine key RTNR parameters over a range of values, and the model provides validation of some of the parameter values selected for RTNR implementation.

A transport flow optimization (TFO) model is presented for real-time dynamic networks, in which a 2–3 percent network cost reduction is achieved in comparison with preplanned dynamic routing design with the path-EFO model. In addition, a Karmarkar algorithm optimal solution to the TFO linear programming model achieves an additional 5 or more percent reduction in network design cost in comparison with the designs of preplanned dynamic networks solved with heuristic design methods. A discrete event flow optimization model is shown to design the most complex real-time dynamic networks and to provide comparisons among the various approaches to real-time dynamic routing. The distributed real-time state-dependent routing approach, such as in RTNR, is shown to be most efficient, and learning models such as LRR among the least efficient. Day-to-day load variation design models for real-time dynamic networks show improved capacity design efficiency while meeting network performance objectives. Network bandwidth allocation models for real-time dynamic networks are shown to improve performance by accounting for optimal flows with given load patterns.

These optimization techniques attain significant capital cost reductions and network performance improvements by properly modeling the more efficient operation of real-time dynamic routing networks.

Reliable Traffic and Transport Routing Networks

14.1 Introduction

A telecommunications traffic network is normally designed for a performance objective aimed at a network that is fully intact. For example, the erlang flow optimization models and discrete event flow optimization models discussed in Chapters 4–13 give design methods for dynamic traffic routing networks for some desired blocking probability grade-of-service objective under normal network conditions. However, if there is a network failure this design may not provide sufficient surviving capacity to meet the required performance levels. For example, if a major fiber link fails, it could have a catastrophic effect on the network because traffic for many switch pairs could not use the failed link. Similarly, if one of the switches fails, it could isolate a whole geographic area until the switch is restored to service.

 With these two kinds of major failures in mind, we present here traffic and transport routing models to achieve reliable network design, or a self-healing network, so as to provide service for predefined restoration objectives for any transport link or switch failure in the network and continue to provide connections to customers with essentially no perceived interruption of service. This approach tries to integrate capabilities in both the traffic and transport networks to make the network robust or insensitive to failure. The basic aims of these models are to provide link diversity and protective trunk augmentation where needed so that specific "network robustness" objectives, such as *traffic restoration level* objectives, are met under failure events. This means that the network is designed so that it carries at least the fraction of traffic known as the traffic restoration level (TRL) under the failure event. For example, a traffic restoration level objective of 70 percent means that under any single transport link failure in the transport network, at least 70 percent of the original traffic for any affected switch pair is still carried after the failure; for the unaffected switch pairs, the traffic is carried at the normal blocking probability grade-of-service objective. These design models provide designs that address the network response immediately after a network event. It is also desirable to have transport restoration respond after the occurrence of the network event to bring service back to normal. Transport restoration is also addressed in this chapter.

Reliable network performance objectives may require, for example, the network to carry 50 percent of its busy-hour load on each link within five minutes after a major network failure, in order to eliminate isolations among switch pairs. Such performance may be provided through traffic restoration techniques, which include link diversity, traffic restoration capacity, and dynamic traffic routing. Reliable network performance objectives might also require a further reduction of the network blocking to less than 5 percent within, say, 30 minutes to limit the duration of degraded service. This is possible through transport restoration methods that utilize transport switches along with centralized transport restoration control. A further objective may be to restore at least 50 percent of severed trunks in affected links within this time period.

Traffic and transport routing models are studied with simulations of preplanned dynamic traffic routing (DNHR) networks and real-time dynamic traffic routing (RTNR) networks to examine the network response to a transport link or switch failure and observe whether the network satisfies the network robustness objectives for which it is designed. We investigate the capability for the cost-effective integration of traffic and transport protection for reliable network design. The approach combines traffic restoration capacity design, dynamic traffic routing, link transport diversity, and transport restoration capabilities into an overall approach for the design of a self-healing network. From the simulation studies we find that these traffic and transport restoration designs meet the restoration objective for the network and provide a robust network at a reasonable incremental cost. In particular, a load diversity model for traffic restoration design is shown to provide the required diversity level and augmented trunk requirements to meet the traffic restoration level objective. With link transport diversity and minimal traffic restoration capacity augmentation, we obtain designs to remove all network isolations and to meet a 50 percent traffic restoration level objective during transport failures. Traffic restoration level design allows for varying levels of diversity on different links in order to ensure a certain level of performance; for example, at least 50 percent of the offered load may be carried in the event of a failure.

The transport restoration process restores capacity for switched, as well as private-line, services in the event of link failures, and we model the interaction of the restoration of switched and private-line services. The restoration process is conducted via a centralized system that we assume restores the affected DS3s until all available restoration capacity is exhausted. Here, a DS3 transmission channel consists of 28 DS1 channels, where a DS1 channel consists of twenty-four 64-kbps DS0 channels. DS3s are restored according to a priority method, where individual DS1s are assigned weights corresponding to the type of service that is carried (such as common-channel signaling, dedicated 1.5-mbps services, or switched services). DS3 priorities are then determined by summing the weights of DS1 tributaries. These models reflect the higher-priority ranking of private-line services and restore these services first.

Optimization of the total cost of transport restoration capacity is possible through a design that increases sharing opportunities of the restoration capacity among different failure scenarios. Real-time transport restoration may also require the use of dedicated restoration capacity for each link and, thus, a lesser opportunity for sharing the restoration capacity. For the purpose of this analysis, we assume that all network transport may be protected with an objective level of restoration capacity. *Transport restoration level* (TPRL) is the term used to specify the minimal percentage of capacity on each transport link that is restorable. A transport restoration level is implemented in the model by restoring each affected link in a failure to a specified level. Here, we further distinguish between transport restoration level for switched circuits and private-line circuits and designate them by $TPRL_s$ and $TPRL_p$, respectively.

14.2   Reliable Traffic and Transport Routing Methods

In this section, we describe traffic restoration design models for survivable networks. Before we describe the models, we discuss the distinction between the traffic and transport networks and the concept of link diversity.

To distinguish between the traffic and transport networks, consider the example of a three-switch network in Figure 1.30. The transport (physical) network is depicted at the bottom of the figure, and the corresponding traffic (logical) network at the top. For example, the direct trunks for the traffic link connecting switches A and B may ride the path A–C–D–B in the physical network. There is not a traffic link between switches B and C, which means there are no direct trunks from switch B to C. A single link failure in the transport network may affect more than one traffic link. For example, in Figure 1.30, the failure of the transport link C–D affects traffic links A–D, C–D, and A–B. Link diversity refers to a trunk design in which direct trunks for a traffic link are split on two or more different physical paths. For example, in Figure 14.1, the direct

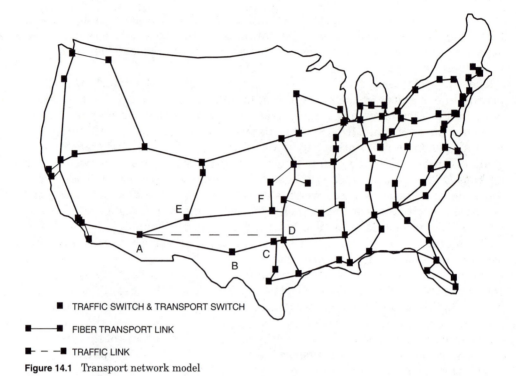

■ TRAFFIC SWITCH & TRANSPORT SWITCH

■———■ FIBER TRANSPORT LINK

■– – –■ TRAFFIC LINK

Figure 14.1 Transport network model

trunks for the traffic link A–D may be split on to the two physical paths A–B–C–
D and a physically diverse path A–E–F–D. A link diversity policy, say, of 70/30
corresponds to the fact that no more than 70 percent of the direct trunks for a
traffic link are routed on a single transport link for the different transport paths
for that link. The advantage of link diversity is that if a physical transport link
fails, the traffic for a particular switch pair can still use the direct trunks that
survived on the physical path not on the failed link.

We now present models for traffic restoration design. The first three models
correspond to design for transport link failure, and the fourth model to design for
switch failure.

14.2.1 Link failure model 1

Traffic restoration capacity design for a link failure has the objective of providing
sufficient diversity and trunk capacity so that for any single transport link
failure, the network meets a given traffic restoration level objective under the
failure event. That means that the network is designed to carry at least the
fraction TRL of the traffic load for a given failure event. For example, a TRL
objective of 50 percent means that under any single transport link failure
in the transport network, at least 50 percent of the traffic for any affected

switch pairs is still carried. Our link failure model 1 is based on a linear programming routing design model.

For a given number of switches, we assume that the traffic loads and the trunks between each switch pair are given. These initial trunks may be the ones that are designed for the normal blocking probability grade-of-service objective, such as in the path-erlang flow optimization model discussed in Chapters 6 and 7 or in the discrete event flow optimization model discussed in Chapter 13. We are given the transport path where the trunks for each switch pair are routed in the transport network and also the candidate traffic paths. The linear programming model selects the traffic paths to be used so that the total network cost is minimized, subject to the constraint that the design achieves a given traffic restoration level objective.

In the model, traffic demand is converted first to capacity demands, which we call virtual trunks (VTs), as discussed in Chapter 13, and then the virtual trunk demands are treated as equivalent traffic demands that are to be routed on the transport network. This is done for each switch pair in the network. The model assumes an underlying dynamic routing method that allows traffic loads to be divided between direct virtual trunk demands and shared virtual trunk demands. One can think of the virtual trunk concept in terms of the picture illustrated in Figure 13.6. Through the use of dynamic routing, direct traffic loads are routed over the direct virtual trunks, and the shared virtual trunk demands form a "network pool" of shared capacity that is used by the combined overflow traffic loads. The direct virtual trunk demand, d_k, is computed as given in Table 14.1. The shared virtual trunk demand, s_k, is computed according to the following formula:

$$s_k = \frac{\text{overflow load (erlangs)}}{0.87}$$

where we have used $\gamma = 0.87$ as the average erlang-carrying capacity of the shared overflow VTs, as defined in Chapter 13. Thus, the total virtual trunk demand is $d_k + s_k$.

Using the virtual trunk demands on each of the transport links in the transport path, we calculate the total number of virtual trunks going through each transport link and order the links in descending order of the total number of virtual trunks carried. This is the order of the "transport link scenarios" in which transport links are failed. With the scenarios determined, we perform the following three steps for each transport link failure scenario.

TABLE 14.1   Direct Virtual Trunk Calculation

Offered load (erlangs)	d_k design criterion
<10	$d_k = 0$
10–60	50% blocking
>60	10% blocking

14.2.1.1 Step 1—Survived traffic capacity and traffic restoration level objectives. For each transport link failure scenario, the number of trunks that survive on each of the traffic links in the traffic network is calculated using the transport path routing. We denote this survived trunk capacity on each traffic link i to be a_i. Then, for the affected switch pairs—that is, the ones that lost some trunks due to this transport link failure scenario—we set the traffic restoration level to be the objective virtual trunk demand to be carried. For the other switch pairs, the virtual trunk demand to be carried is set to the normal blocking probability grade-of-service objective. If the traffic restoration level objective for each switch pair k is t_k, then t_k = TRL if k is an affected switch pair, and t_k = the normal blocking probability grade-of-service objective otherwise.

14.2.1.2 Step 2—Linear programming model. The linear programming model minimizes trunk augmentation cost so as to carry the given traffic restoration level objective of traffic load. First, we introduce the following notation:

N = total number of switches
K = total number of switch pairs
L = number of traffic links
J_k = number of traffic paths for switch pair k
a_i = survived trunk capacity of traffic link i
y_i = augmented trunk capacity of traffic link i
r_{jk} = virtual trunk demand routed on path j of switch pair k
P_{jk}^i = 0/1 parameter, equal to 1 if traffic path j of switch pair k routes over traffic link i
v_k = total virtual trunk demand for switch pair k ($= d_k + s_k$)
M_i = incremental cost per trunk on traffic link i on existing transport network
t_k = traffic restoration level for switch pair k: TRL if affected switch pair, normal blocking probability grade-of-service objective otherwise

Now, the following linear programming model minimizes the incremental augmentation cost:

$$\text{Minimize} \sum_{i=1}^{L} M_i y_i \tag{14.1a}$$

subject to

$$\sum_{k=1}^{K} \sum_{j=1}^{J_k} P_{jk}^i r_{jk} \leq a_i + y_i, \qquad i = 1, \ldots, L \tag{14.1b}$$

$$\sum_{j=1}^{J_k} r_{jk} \geq t_k v_k, \qquad k = 1, \ldots, K \tag{14.1c}$$

$$r_{jk} \geq 0, \qquad j = 1, \ldots, J_k, \quad k = 1, \ldots, K \tag{14.1d}$$

$$y_i \geq 0, \qquad i = 1, \ldots, L \qquad\qquad (14.1e)$$

The constraints (14.1c) of this model say that for switch pair k, we must carry at least $t_k v_k$ of the total virtual trunk demand on all the possible candidate paths r_{jk}. The constraints (14.1b) of the model say that the total traffic flow on each traffic link i due to the flows for different switch-pair demands must equal the quantity on the left of the inequality, and that if this link has enough capacity (a_i) to carry the total flow, there is no need to augment any y_i virtual trunk capacity on this link. Otherwise, y_i of virtual trunk capacity must be added. These two sets of constraints must be satisfied so as to minimize the total augmentation cost.

14.2.1.3 Step 3—Add link augmentation capacity to transport path. Two different possible rules are used to add the augmented trunks on the transport network:

1. If trunks are added on an unaffected switch pair by the linear programming model, these trunks are equally divided among the number of different physical transport paths for that switch pair and added on them. If trunks are added on an affected switch pair, then the augmentation trunks, if any, are equally divided among the number of unaffected different physical paths for that switch pair and added on them.

2. If trunks are added on an unaffected switch pair by the linear programming model, these trunks are added on the shortest physical transport path for that switch pair. If it is an affected switch pair, the augmentation trunks, if any, are added on the shortest physical transport path for that switch pair as provided by the transport path routing if this path is not affected; otherwise, they are added on the next-shortest physical path for that switch pair.

After Step 3 is completed, we move to the next transport failure scenario and repeat Steps 1 to 3. One key point here is that once some trunks are added in a particular scenario, they are assumed to be there for the subsequent scenarios. When all the failure scenarios are completed, we obtain the updated number of trunks for each switch pair and, also, how the trunks are routed in the transport network, which in turn gives us the required diversity level. In this manner we obtain both the total trunk requirements and the link diversity level.

One variation of this design procedure is that instead of doing the design for all failure scenarios, we can restrict the failure scenarios to those that are the most important or to those for which the load is bigger than a certain threshold. Another variation of this procedure is to use different traffic restoration level objectives for switch pairs under different failure scenarios.

14.2.2 Link failure model 2

This approach is similar to the previous approach for a transport link failure scenario. In this model, instead of doing the scenarios one at a time in a sequential manner, we consider them all at once. This approach takes advantage of the noncoincidence of traffic, as in the dynamic traffic routing network design models described in previous chapters. Thus, to the linear programming model (14.1), one more dimension is added to consider different transport link failure scenarios concurrently. We present below the linear programming model for this method:

N = total number of switches

K = total number of switch pairs

L = number of traffic links

S = total number of failure scenarios

J_k^s = number of traffic paths for switch pair k in scenario s

a_i^s = survived trunk capacity of traffic link i in scenario s

r_{jk}^s = virtual trunk demand routed on path j of switch pair k in scenario s

y_i = augmented trunk capacity of traffic link i

P_{jk}^{is} = 0/1 parameter, equal to 1 if traffic path j of switch pair k routes over traffic link i in scenario s

v_k = total virtual trunk demand for switch pair k ($= d_k + s_k$)

M_i = incremental cost per trunk on traffic link i on existing transport network

t_k^s = traffic restoration level for switch pair k in scenario s: TRL if affected switch pair in scenario s, normal blocking probability grade-of-service objective otherwise

Now, the following linear programming model minimizes the incremental augmentation cost:

$$\text{Minimize} \sum_{i=1}^{L} M_i y_i \tag{14.2a}$$

subject to

$$\sum_{k=1}^{K} \sum_{j=1}^{J_k} P_{jk}^{is} r_{jk}^s \le a_i^s + y_i \qquad i = 1, \ldots, L, \quad s = 1, \ldots, S \tag{14.2b}$$

$$\sum_{j=1}^{J_k} r_{jk}^s \ge t_k^s v_k, \qquad k = 1, \ldots, K, \quad s = 1, \ldots, S \tag{14.2c}$$

$$r_{jk}^s \ge 0, \qquad j = 1, \ldots, J_k^s, \quad k = 1, \ldots, K, \quad s = 1, \ldots, S \tag{14.2d}$$

$$y_i \ge 0, \qquad i = 1, \ldots, L \tag{14.2e}$$

Here, too, the survived virtual trunk capacity is determined based on the initial transport path routing. As in the previous model, the same trunk augmentation procedure used in Step 3 is used here.

14.2.3   Link failure model 3 (load diversity design)

The third link failure model is quite different from the linear programming models presented in the previous two sections. However, it also uses the concepts of virtual trunks and link diversity. We call it load diversity design.

We assume that we have two distinct transport paths for direct trunks for each switch pair. Let v be the virtual trunk requirement for the traffic demand for a particular switch pair. Let d be the number of trunks to be put on the direct transport path and s be the number of trunks to be put on the other transport paths for the direct traffic link of the given switch pair. Let b be the number of trunks for this traffic link that are designed by the network design model. Let t be the traffic restoration level (TRL) objective under a link failure scenario. Let δb be the trunk augmentation that may be needed for this traffic link.

What we would like in a failure event is to carry a portion tv of the total virtual trunk demand for the affected switch pairs. Thus, if $tv \leq b/2$, we set $\delta b = 0$ (no augmentation) with $d = b - tv$ and $s = tv$. In this way, if either transport path fails, we can carry at least tv of the virtual trunk demand. On the other hand, if $tv > b/2$, then we want

$$(b + \delta b)/2 = tv$$

which implies

$$\delta b = 2tv - b$$

In this case, we set $d = s = (b + \delta b)/2$. The above procedure is repeated for every demand pair in the network. The incremental cost of the network is the cost of trunk augmentation and routing the direct trunks for each switch pair, if any, on two transport paths.

The above procedure can be extended to the general case of k distinct transport paths. So, for k distinct transport paths, if

$$tv \leq (k - 1)b/k$$

then $\delta b = 0$ with $d = b - tv$ on the first transport path, and $tv/(k - 1)$ on each of the other $(k - 1)$ transport paths. If

$$tv > (k - 1)b/k$$

then

$$\delta b = k\,tv - (k - 1)b$$

with each of the k transport paths having $(b + \delta b)/k$ trunks.

14.2.4   Switch failure model

The switch failure traffic restoration design incorporates the concept of dual homing, as discussed in Chapter 12, along with multiple ingress/egress routing. With single homing, the traffic from a particular geographical area normally goes to

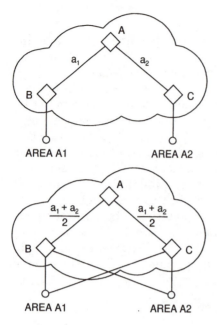

Figure 14.2 Illustration of dual homing

the single switch nearest to it in order to carry the ingress and egress traffic. For example, in Figure 14.2 the traffic from the served area A1 enters the network through switch B, and, similarly, the served area A2 traffic enters the network through switch C. Now, if switch B fails, then area A1 gets isolated. To protect against such an event, areas A1 and A2 are homed to more than one switch—to switches A and B in this case (Figure 14.2, bottom). This is the concept of dual homing, in which we address the issue of designing a reliable network when one of the switches may fail.

For every switch to be protected, we assign a dual-homed switch. Before failure, we assume that any load from a switch to the protected switch and its dual-homed switch is equally divided; that is, if the original load between area A1 and switch A is a_1, and between A and the dual-homed switch, B, is a_2, then we assume that under normal network conditions, the load between switches A and B is $(a_1 + a_2)/2$, and the same for the load between switches A and C. We refer to this concept as *balanced load*. Then, under a failure event such as a switch B failure, we carry load equal to $(a_1 + a_2)/2$ between switches A and C. (See Figure 14.2, bottom.) We call this design objective a 50 percent traffic restoration level objective in a manner very analogous to the link failure event. As we can see from the bottom of Figure 14.2, this restoration level of traffic from or to area A1 is then still carried.

The method used here is similar to the linear programming model (14.1) for link failure model 1. First the switch pairs that are going to be considered for the switch failure design scenarios are determined. Next, the dual-homing switches are determined for each switch and the balanced-load traffic routed accordingly. Then the v_k virtual trunks are computed for the balanced loads. For each switch pair, one of the switches is assumed to fail (say, switch B).

Then this switch cannot have any incoming or outgoing traffic and, also, cannot be a via switch for any two-link traffic between other switch pairs. Using these constraints, we solve the linear programming model (14.1). Then, we reverse the roles of the switches for this pair and solve the linear programming model (14.1) again with the above-mentioned constraints. This design process is repeated for every pair of candidate switches for each switch failure scenario.

14.2.5 Combined traffic and transport restoration models

Various traffic and transport designs are evaluated through the models discussed in the above sections. Here we introduce a transport link model, illustrated in Figure 14.3, where each transport cross section is assumed on the average to contain certain fractions of switched (N) and private-line (M) circuits. A portion of the switched and private-line circuits is further presumed to be restorable in real time, such as with ring or dual-feed transmission arrangements (lowercase n and m values).

Circuits that are not restored in real time are restored with transport restoration to a specified transport restoration level (TPRL) value. The lower part of Figure 14.3 demonstrates the interaction between the switched and private-line

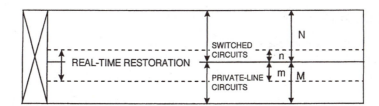

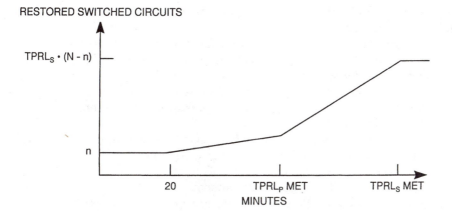

TPRL$_S$ = TRANSPORT RESTORATION LEVEL FOR SWITCHED CIRCUITS
TPRL$_P$ = TRANSPORT RESTORATION LEVEL FOR PRIVATE-LINE CIRCUITS

Figure 14.3 Transport restoration model

circuits in the restoration process. The restoration process is assumed to restore the first DS3 after 20 minutes, with one minute to restore each additional DS3. The restoration times are illustrative and are not critical to the reliable network design principles being discussed. In actuality, faster restoration times can be achieved, as illustrated in the example of transport restoration using FASTAR, discussed in Chapter 1 (Figure 1.4). SONET ring restoration can occur in about 50–200 milliseconds, and such real-time restoration is included in the model. A prioritization method is assumed, whereby DS3s that carry higher-priority private-line services are restored first. Because switched and private-line circuits are mixed at the DS3 level, some switched circuits are also restored.

Different levels of transport restoration may also be assigned for the private-line ($TPRL_p$) and switched ($TPRL_s$) networks. Each type of circuit demand is then restored to the corresponding level of restoration. Figure 14.3 also shows how the restoration level for switched circuits varies as a function of time. Some level of traffic circuits is restored in real time (n). After 20 minutes, transport restoration is initiated with one DS3 being restored in each minute, and with a smaller fraction of each DS3 being message traffic. The message portion in each DS3 subsequently increases to a larger fraction after private-line traffic is restored to its designated level $TPRL_p$. Transport restoration stops after both the $TPRL_p$ and $TPRL_s$ objectives are met.

14.3   Modeling Results

Here we discuss modeling results for the traffic restoration design models and the combined traffic and transport restoration design models described in Section 14.2.

14.3.1   Link failure model results

First, we discuss the link failure design results for link failure model 1 and link failure model 3, as described in Section 14.2. We use a network model consisting of 103 traffic switches in a national network model, in which traffic loads and link capacities are extracted from actual network data. Also, we use a fiber transport network model consisting of 113 transport switches, including 103 required for the traffic switches, and 148 transport links, as illustrated in Figure 14.1. This fiber network model is based on realistic network data from a national network model for the 1990 time frame.

For link failure model 1, we use the transport model to generate the transport path routing. For example, for the base network, we assume that all the trunks for a particular switch pair are routed on one physical path (a "100/0" link diversity policy), normally on the shortest physical transport path. Similarly, for a 50/50 link diversity policy, the trunks for a switch pair are split 50/50 over two different physical transport paths (normally, the shortest path and the second-shortest path).

We develop the failure scenarios as follows: Depending on the initial transport path routing, the number of circuits riding on different transport links for all

TABLE 14.2   Illustrative Self-Healing Network Design Objectives

Disruption	Base case	More stringent
ingress/egress	blocking to end-offices <50% after 5 mins. blocking to end-offices <5% after 120 mins.	blocking to end-offices <30% after 2 mins. blocking to end-offices <3% after 30 mins.
switch-to-switch	switch-pair blocking <50% after 5 mins. switch-pair blocking <5% after 120 mins.	switch-pair blocking <30% after 2 mins. switch-pair blocking <3% after 30 mins.
service	call failure per service per customer <50% after 5 mins. call failure per service per customer <5% after 20 mins.	call failure per service per customer <30% after 2 mins. call failure per service per customer <3% after 30 mins.

demand pairs is calculated. Then the links are ordered in descending order of the number of circuits carried on each transport link, which is the order in which we perform the failure scenarios. We then fail each of these transport links and adjust the survived trunk capacity as described in Section 14.2.

We assume the following costs: $275 for each traffic switch termination, including transport multiplexing on the transport switch; $130 per circuit-end for echo cancelers on circuits longer than 900 miles; and $0.22 per circuit mile of fiber transport. Traffic restoration level objectives used in the model are summarized in Table 14.2 along with other illustrative self-healing network design objectives.

Network costs for achieving several traffic restoration level objectives using link failure model 1 are shown in in Table 14.3, and those using link failure model 3 are shown in Table 14.4. The network costs are normalized to the cost of the base network designed for the normal blocking probability grade-of-service objectives with no link diversity. The network cost for each of the traffic restoration level objectives includes the cost for trunk augmentation and for link diversity. Most of the cost figures are computed for the hour of traffic load that is the largest total load. With the load diversity design model, we also computed the maximum load of all the load hours for each switch pair and then computed augmented trunks, if any. From these tables, we can see that for an incremental cost of about 5 percent of the base network cost, a 30 percent traffic restoration level objective can be met, and for an incremental cost of about 7 percent of the base network cost, a 50 percent traffic restoration level objective can be attained. However, the incremental cost grows by more than a factor of 2 in going from the 50 percent traffic restoration level objective to the 70 percent traffic restoration level objective. On the other hand, if we use the maximum of all load periods, the incremental cost increases more rapidly than the single busy-hour case, as shown in Table 14.4.

To study whether the design meets the traffic restoration level objectives, we did simulations of failure scenarios on the network model. All the simulations

TABLE 14.3 Incremental Cost for Different TRL
Objectives Using Link Failure Model 1

TRL objective	Incremental cost (LSP-3)
30%	5.4%
50%	9.1%

TABLE 14.4 Incremental Cost for Different TRL
Objectives Using Link Failure Model 3

	Incremental cost	
TRL objective	LSP-3	All LSPs
30%	4.1%	4.7%
50%	6.0%	9.3%
70%	14.6%	30.1%
100%	44.3%	77.2%

were done for two hours of simulation time. Of these two hours of simulation, some amount of time at the beginning of the simulation is spent to stabilize the network. We fail a transport link at the end of one hour of simulation. The second hour of the simulation runs with the trunks that survived in the network due to this failure. We assume a call retrial rate of 75 percent. For the simulation, we consider three link failure scenarios. These transport links are between New York, New York, and Newark, New Jersey; between Omaha, Nebraska, and Denver, Colorado; and between Sherman Oaks, California, and Stockton, California. For brevity, we will call them cut 1, cut 2, and cut 3, respectively.

We present network simulation results for the base network in Table 14.5. The base network does not have link diversity or traffic restoration trunk capacity to meet TRL objectives. In Tables 14.6, 14.7, and 14.8 we present network simulation results for the 50 percent traffic restoration level objectives. The results are for the second hour of simulation. It is interesting to note from Tables 14.5 and 14.6 that although more switch pairs are affected by the failures cut 1 and cut 2, the carried load and network blocking performances for the 50 percent traffic restoration level objective are observedly better than for the base network. However, for cut 3, performance with respect to average blocking and carried load for a 50 percent traffic restoration level is worse than the base network. This suggests that due to link diversity, the overall network performance might be worse, though this is not generally the case. This can occur because the total number of trunks routed on a given transport link is not constrained in either the base network design or the TRL design. It is therefore possible for a given transport link to have a larger number of trunks routed over it in the TRL design network than in the base design network, and if this link fails, the

TABLE 14.5   Network Performance with RTNR Base Network (No TRL Design)

Attribute	Cut 1	Cut 2	Cut 3
average network blocking	38.7	30.4	6.2
blocked calls	2,733,284	1,974,737	323,716
network carried load (erlangs)	359,759	374,718	407,436
dropped calls	64,634	51,558	20,995
affected trunks	106,512	88,152	31,320
affected switch pairs	828	1,101	231

TABLE 14.6   Network Performance with DNHR and RTNR (50% TRL Design)

Attribute	Cut 1	Cut 2	Cut 3
average network blocking:			
DNHR	23.9	17.7	10.2
RTNR	23.0	17.2	8.6
blocked calls:			
DNHR	1,455,465	1,017,866	549,302
RTNR	1,389,905	989,800	456,803
network carried load (erlangs):			
DNHR	384,813	393,554	402,763
RTNR	386,121	394,071	404,597
dropped calls:			
DNHR	50,943	39,796	33,962
RTNR	50,897	40,002	34,171
affected trunks	83,434	68,619	52,022
affected switch pairs	1,361	1,664	1,358

TABLE 14.7   Results from Cut 1 (RTNR Case)

		Number of switch pairs having blocking over	
	Average network blocking	50%	30%
base network	38.7%	715	751
50% TRL	23.0%	0	NA
70% TRL	8.6%	NA	0

TABLE 14.8   Results from Cut 2 (RTNR Case)

		Number of switch pairs having blocking over	
	Average network blocking	50%	30%
base network	30.4%	907	1,050
50% TRL	17.2%	0	NA
70% TRL	4.2%	NA	0

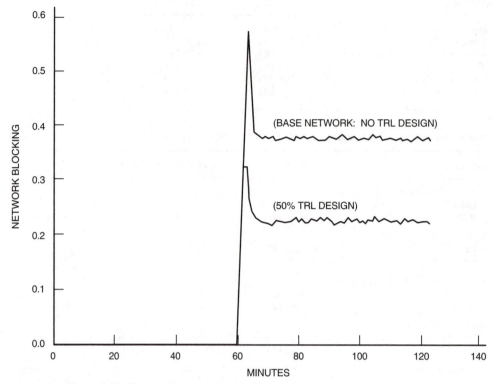

Figure 14.4 Network blocking performance for cut 1

TRL design network can have higher total blocking, even though all the individual switch pairs meet their TRL objectives.

We can also observe from Table 14.7 that with an increasing traffic restoration level objective the average network blocking goes down as expected. We see several hundred switch pairs having more than 50 percent (or 30 percent) blocking with the base network; however, with the traffic restoration level designed network there is no switch pair that could not carry the traffic at a level of, for example, at least 70 percent for the 70 percent traffic restoration level objective case (note that 30 percent maximum blocking can be equated with the 70 percent traffic restoration level objective). We present the same information for cut 2 in Table 14.6. We can make similar conclusions as with cut 1. To give insight on the behavior of traffic restoration, we present simulation results in Figures 14.4, 14.5, and 14.6 corresponding to these three cuts. In each of these figures, we plot average network blocking versus time from the starting point to the end of two hours of simulation. In these figures, we plot the base network design against the 50 percent traffic restoration level objective design to illustrate the transient effect of failure on the network blocking performance.

We conclude from the simulations that for a reasonable incremental cost and efficient use of link diversity and traffic restoration trunk capacity to meet

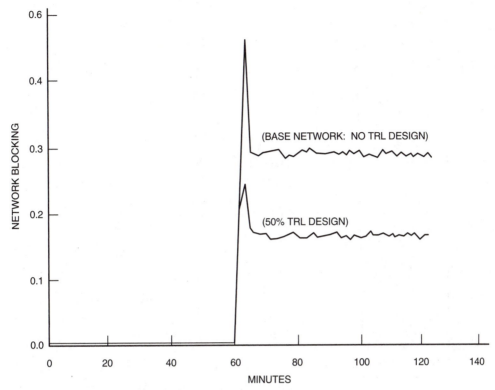

Figure 14.5 Network blocking performance for cut 2

TRL objectives, a robust network can be attained for link failure that meets self-healing-network design objectives.

14.3.2 Switch failure model results

For purposes of studying switch failure scenarios, we model three dual-home switch pairs in our network model. For each of these switch pairs, both switches are assumed to be in the same geographical vicinity. The switch failure design is done for a 50 percent traffic restoration level objective for each of these switch pairs, as described in Section 14.2. The traffic model is the same as the one used for the link failure design study.

Using the concept of balanced load as described in Section 14.2, we first determine the trunk requirements for the base network and then use the switch failure model to design the incremental trunk capacity required for the three switch pairs. The incremental trunk augmentation is 0.2 percent above the base network. Because our procedure involves evenly redistributing network loads between pairs of switches, the cost of augmentation to achieve the dual homing arrangement is insignificant.

To validate our design results, we again use a call-by-call simulation. We do three simulations, one for each pair of switches. In each case, one of the switches

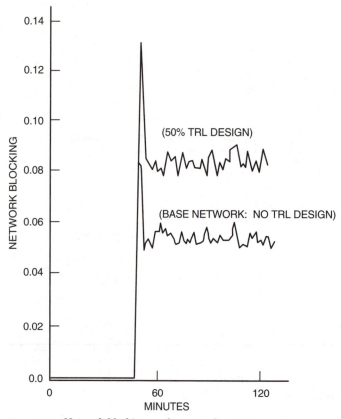

Figure 14.6   Network blocking performance for cut 3

is failed after one hour of simulation time, and the simulation then runs for another hour of simulation time.

Through the simulation model, we find that dual-homing ingress/egress traffic to two switches protects at least 50 percent of the total traffic from terminating on the failed switch. As another illustration of this point, from the simulation we observe that the difference between network blocking with traffic restoration level objective design and without traffic restoration level objective design is not significant. See, for example, the results shown in Figure 14.7. The simulation results also show that no unaffected switch pair experiences any increase in switch-pair blocking as the result of the switch failure. We ascribe this phenomenon to the robust and dynamic nature of the traffic routing.

14.3.3   Combined traffic and transport restoration design results

We study here several network designs with varying values of the traffic restoration level, transport restoration level, and diversity level and examine the network cost/benefit trade-offs for these designs. We use the load diversity model for TRL

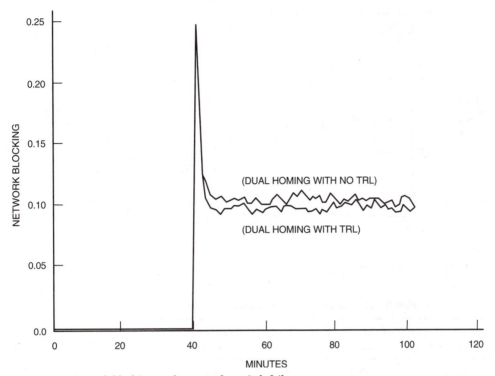

Figure 14.7 Network blocking performance for switch failure

link diversity design, as discussed in Section 14.2.3, and the transport restoration model illustrated in Figure 14.3. In the transport restoration model, the private line–to–switched circuit ratio in DS3s varies from an initial value of 65:35 to a final value of 20:80 as the restoration process proceeds. Also, the $N:M$ ratio is taken to be 7:3, the $n:m$ ratio as 1:2, and the $m:M$ ratio as 1:4.

Figure 14.8 demonstrates trunk augmentation requirements in the network as the traffic restoration level is varied from 0 to 100 percent for various levels of link diversity. The switch-pair blocking is shown for the worst-case switch pair. The added network cost (represented as a percentage of the base network cost) is an increasing nonlinear function of the traffic restoration level. Design results are summarized in Table 14.9. Average diversity level over all links is given, as are the additional trunks required to meet the traffic restoration level objective. The diversity level is calculated by taking the percentage of diversified trunks in various links over all network trunks. For a traffic restoration level of 50 percent with an average of 35 percent diversity, for example, individual links vary in their diversity levels, with anywhere from 0 percent to as much as 50 percent of their trunks diversified.

We see that the incremental cost of providing 30 percent traffic restoration level is about 32 percent lower than 50 percent traffic restoration level, and 70 percent traffic restoration level cost is 2.4 times the 50 percent traffic restoration level cost. Once again, the incremental cost represents a combination of

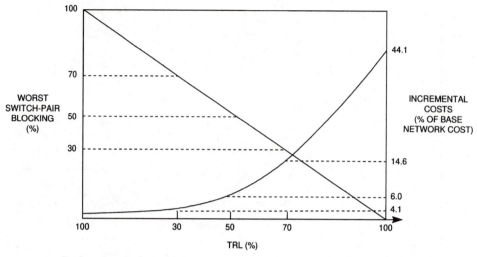

Figure 14.8   Performance and cost trade-off for TRL design

TABLE 14.9   Design Results for Various Traffic Restoration Values

Traffic restoration level objective	Augmented trunks (%)	Average diversity (%)	Incremental cost (%)	Incremental cost ($)
30	0.007	26	4.1	22M
50	0.2	35	6.0	33M
70	7.1	41	14.6	79M

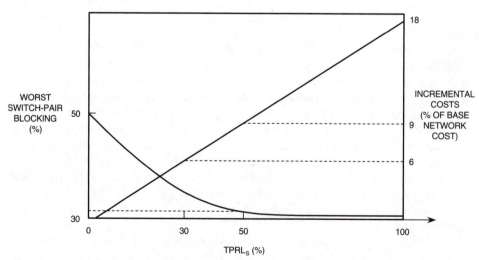

Figure 14.9   Performance and cost trade-off for TPRL$_s$ design (TRL = 50%)

augmented trunk capacity and additional fiber transport capacity needed by the TRL design to meet diversity objectives. The 50 percent traffic restoration level seems to provide a suitable middle ground for the costs. It is located at a point where the incremental benefits may still justify the incurred cost, whereas higher traffic restoration level values might be overly expensive. As stated above, traffic restoration level design results in varying levels of diversity in different links. The diversification of 50 percent of the load (50 percent traffic restoration level) results in a network average of 35 percent link diversity (65/35 policy) and few, if any, augmented trunks.

Figure 14.9 shows how varying levels of transport restoration can further enhance network performance for a given 50 percent traffic restoration level design. Here, the traffic restoration level is fixed at 50 percent and transport restoration level varies from 0 to 100 percent. The added transport miles and the worst switch-pair blocking are again drawn. Here, the cost curve is assumed linear, whereas the blocking curve is seen to be nonlinear. With a 50 percent traffic restoration level design, the cost of providing 50 percent transport restoration level for the network model is estimated at $49 million, or, equivalently, about 9 percent of the base network cost. This analysis also shows that 70 percent transport restoration capacity is required to restore 100 percent of the network under all transport link failure scenarios.

Table 14.10 demonstrates the performance of the combined 50 percent traffic restoration level and 50 percent transport restoration level design for three simulated link failures. These are (1) Newark–New York, (2) Omaha–Denver, and (3) Los Angeles–San Francisco. $TPRL_s$ is set at 50 percent and $TPRL_p$ at 100 percent. Transport restoration is initiated after 20 minutes for the first

TABLE 14.10   Network Performance for 50% TRL, 50% $TPRL_s$, and 100% $TPRL_p$

Attribute	Cut 1		Cut 2		Cut 3	
	1st hour	2nd hour	1st hour	2nd hour	1st hour	2nd hour
average network blocking:						
DNHR/NEMOS	0.146	0.011	0.083	0.000	0.034	0.001
RTNR	0.131	0.001	0.069	0.000	0.024	0.000
blocked calls:						
DNHR/NEMOS	815,579	56,242	439,870	1,586	176,024	3,489
RTNR	722,583	6,847	361,738	0	119,766	0
network carried load (erlangs):						
DNHR/NEMOS	395,993	412,597	403,917	414,043	410,009	413,990
RTNR	397,644	413,761	405,460	414,054	411,207	414,065
dropped calls:						
DNHR/NEMOS	50,943		39,796		33,962	
RTNR	50,897		40,002		34,171	
affected trunks	83,434		68,619		52,022	
affected switch pairs	1,361		1,664		1,358	

DS3, and each subsequent DS3 is restored in one minute. Two routing methods were used for the study: DNHR plus NEMOS real-time routing enhancement (discussed in Chapter 8) and RTNR. Performance is measured for the two hours following the outage. Some general observations may be made from these results. We see that cuts 1 and 2 have significantly larger impact on the amount of carried load in the network (about 3 percent of the network load) than cut 3. Traffic restoration level design provides relief in the minutes following an outage. By the second hour of each outage, significant performance improvements are observed with the partial restoration of failed facilities. These results generally confirm the validity of the design techniques presented in this chapter.

Figure 14.10 illustrates a typical instance of self-healing-network design with traffic restoration level objectives and transport restoration level objectives, as compared with the base network with no TRL or TPRL design objectives. In the example, a fiber link failure occurs in the model network 40 minutes after the beginning of the simulation, severing a large number of trunks in the network and cutting off thousands of calls. Therefore, in the simulation results shown in Figure 14.10, we see a large jump in the blocking at the instant of the cut. A transient flurry of reattempts follows as cut-off customers redial and reestablish their calls. This call restoration process is aided by the

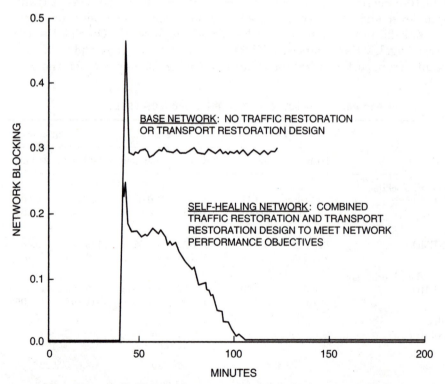

Figure 14.10 Network blocking performance comparison for base network design and traffic and transport restoration level design (transport link failure at minute 40)

traffic restoration level design, which provides link diversity and protective trunk capacity to meet the TRL objectives immediately following a failure. This TRL design, together with the ability of real-time dynamic routing to find surviving capacity wherever it exists, quickly reduces the transient blocking level, which then remains roughly constant for about 20 minutes until the transport restoration process begins. At 20 minutes after the link failure, the transport restoration process begins to restore capacity that was lost due to the failure. Blocking then continues to drop during that period when transport restoration takes place until it reaches essentially a level of zero blocking. Figure 14.10 illustrates the comparison between network performances with and without the traffic and transport restoration design techniques presented in this chapter.

14.4 Conclusion

In this chapter, we present traffic restoration design models for self-healing networks to attain a robust design for any transport link or switch failure. With these models we compute the cost for restoring traffic for different traffic restoration level objectives. From our simulation studies, we found that with efficient use of link diversity and at a reasonable incremental cost, one can obtain a robust network to respond to a network failure event such as a transport link or switch failure and still meet the network objectives. For a switch failure, it is also important to use the concept of dual homing in the network to avoid isolation of any geographical region.

We study integrating several reliable network design capabilities, which include dynamic traffic routing, multiple ingress/egress routing, link transport diversity routing, traffic restoration capacity design, transport restoration, and transport restoration capacity design. The study shows that an advantageous combination of capabilities is to do both of the following:

1. Implement traffic restoration capacity design utilizing link diversity and trunk augmentation to satisfy, for example, a traffic restoration level objective of 50 percent—that is, to carry at least 50 percent of offered load for each switch pair affected by a failure.
2. Provide a minimum transport restoration level of, for example, 50 percent for the failed trunks on affected links.

This traffic restoration level design allows for varying levels of diversity on different links to ensure a minimum level of performance; that is, 50 percent of the offered load may be carried in the event of a failure. Robust routing techniques such as dynamic traffic routing, multiple ingress/egress routing, and link transport diversity routing further improve response to switch or transport failures. Transport restoration is necessary to reduce network blocking to low levels. Given, for example, a 50 percent traffic restoration level design, it is observed that this combined with transport restoration of 50 percent of the failed trunks in affected links is sufficient to restore the traffic to low blocking levels. Therefore, the combination of traffic restoration level design and transport restoration level design is seen both to be cost-effective and to provide fast and reliable performance. The traffic restoration level design eliminates isolations between switch pairs, and transport restoration level is used to reduce the duration of poor service in the network. Traffic restoration techniques combined with transport restoration techniques provide the network with independent means to achieve reliability against multiple failures and other unexpected events and are perceived to be a valuable part of a reliable network design.

Dynamic Traffic and Transport Routing Networks

15.1 Introduction

This chapter describes and analyzes alternative traffic and transport network architectures in light of evolving technology for integrated broadband networks. We consider dynamic traffic routing and dynamic transport routing for such networks and the regimes in which dynamic routing and fixed routing are advantageous. We study alternative network architectures, which include dynamic traffic routing networks with fixed transport routing, fixed traffic routing networks with dynamic transport routing, and dynamic traffic routing networks with dynamic transport routing. Dynamic routing offers advantages of simplicity of design and robustness to load variations and network failures. These routing architectures are examined across a spectrum of network services including voice, data, video, and broadband services.

An important element of network architecture is the relationship between the transport network and the traffic network. An illustration of a transport network is shown in Figure 15.1, and Figure 1.30 illustrates the mapping of trunks in the traffic network onto the transport network of Figure 15.1. It is clear from Figures 15.1 and 1.30 that in a highly interconnected traffic network, many switch pairs will have a direct trunk connection where no direct physical path exists in the transport network. In this case a direct traffic trunk is obtained by cross-connecting through a transport switching location. This is distinct from the traffic situation known as alternative routing, in which a call is actually switched at an intermediate location. This distinction between cross-connecting and switching is a bit subtle, but it is fundamental to traffic routing of calls and transport routing of trunks. Referring to Figure 15.2, we illustrate one of the logical inconsistencies we encounter when we design the traffic network to be essentially separate from the transport network. On the alternative traffic route from switch B to switch D through A, the physical path is, in fact, up and back from B to A (a phenomenon known as "backhauling") and then across from B to D. The sharing of capacity by various traffic loads in this way actually increases the

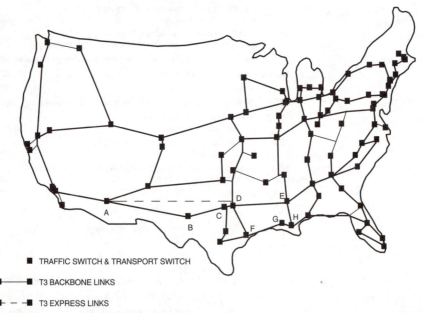

■ TRAFFIC SWITCH & TRANSPORT SWITCH

■——■ T3 BACKBONE LINKS

■– – –■ T3 EXPRESS LINKS

Figure 15.1 Transport network model

TRAFFIC NETWORK VIEW   TRANSPORT NETWORK VIEW

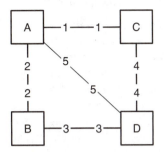
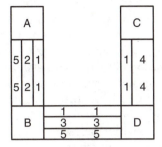

Figure 15.2   Fixed transport routing mesh network

efficiency of the network because the backhauled capacity to and from B and A is only used when no direct A-to-B or A-to-D traffic wants to use it. It is conceivable that under certain conditions, capacity could be put to more efficient use, and this is studied in this chapter.

15.2   Fixed and Dynamic Transport Routing Methods

In Chapters 3–13 we discussed a wide variety of dynamic traffic routing methods. In our studies in this chapter of dynamic traffic and transport routing, we incorporate these dynamic traffic routing methods into our traffic network models, such as the RTNR method described in Chapter 12. Dynamic traffic and transport routing networks can provide packet-switched services, as well as circuit-switched services. Packet services could include X.25, frame relay, internet, or asynchronous transfer mode (ATM) broadband ISDN-based services. The discussion of routing methods in this section applies equally well to both packet-switched services and circuit-switched services. Multiservice integrated dynamic routing methods used on such networks incorporate dynamic path selection, implemented by dynamic traffic and transport routing, which seeks out and uses idle network capacity by using frequent, perhaps call-by-call, traffic and transport routing table update decisions.

The trend in traffic and transport routing architecture is toward greater flexibility in resource allocation, which includes transmission, switching, and network management and design resource allocation. In this chapter we discuss future directions for traffic and transport network architectures with respect to a range of alternative directions, as follows:

1. *Density of link interconnection in the traffic network:* We mean by this the frequency with which switch pairs have "permanently" assigned logically connected direct links between them. We place *permanently* in quotes to indicate that this is a relative concept. Should a network provider provision direct links between switches on the transport network through the use of manual cross-connects and hold these links up for weeks or months at a time, then we consider these links permanent. On the other hand, dynamic link

rearrangement could change the assignment of transport capacity quickly and automatically and not result in a permanent link interconnection.

2. *Complexity of the alternative routing:* A network that uses many ways to get from one switch to another switch can be regarded as having complex alternative routing. One might imagine a simpler alternative routing pattern and high aggregation of traffic onto a small number of paths, perhaps only two paths, as in a ring network. One way to achieve this alternative of simpler routing patterns is to have fully dynamic assignment of bandwidth in the transport network. We investigate the effect of fast trunk rearrangement between different switch pairs and contrast this to networks in which the assignment of trunks is fixed and can be changed only slowly.

3. *Circuit- or packet-oriented transport network:* Different traffic and transport routing technologies lend themselves to different modes of operation. Broadband ISDN will incorporate the fixed-length packet or cell-based ATM protocol, as discussed in Chapter 1. Circuit mode, or cell mode, technology can have significant impact on the viability of different traffic and transport routing architectures.

Traffic and transport networks might incorporate a dense logical mesh transport routing architecture or a sparsely connected logical ring transport routing architecture. The fixed transport routing mesh architecture has dynamic traffic

TRAFFIC NETWORK VIEW 1

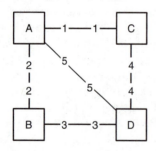

TRAFFIC NETWORK VIEW 2

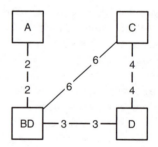

TRANSPORT NETWORK VIEW 1

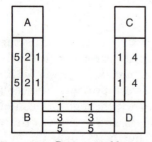

TRANSPORT NETWORK VIEW 2

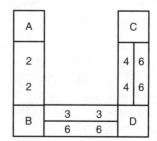

Figure 15.3 Rearrangeable transport routing mesh network

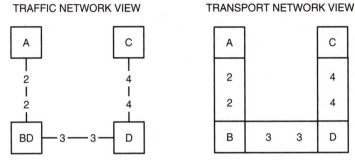

TRAFFIC NETWORK VIEW   TRANSPORT NETWORK VIEW

Figure 15.4 Real-time transport routing ring network

routing but fixed transport routing of trunk capacity. That is, traffic switching and transport network elements interconnect switches with logical transport link connections, forming a fixed yet dense logical mesh, as in Figure 15.2. The rearrangeable mesh architecture, shown in Figure 15.3, is like the fixed transport routing mesh architecture except that the logical trunk capacities can be rapidly rearranged mostly in a preplanned fashion—that is, they are not fixed. In the real-time transport routing ring architecture, shown in Figure 15.4, we use fixed traffic routing and real-time dynamic trunk capacity allocation, in which "virtual link" capacity can be changed essentially in real time, on a sparse logical and physical ring network. That is, the logical ring network has low logical (traffic network) connectivity in comparison with the mesh architecture, and it directly overlays the physical transport network. As will be discussed below, fixed traffic routing (or at most minimal alternative traffic routing) is used in the real-time transport routing ring network architecture—but flexible bandwidth allocation is employed, and calls may follow a second path on network element failure. Note that the rearrangeable transport routing mesh network is actually a hybrid of the mesh and real-time transport routing ring architectures; it has aspects of both. These integrated traffic and transport routing architectures could use either packet mode, circuit mode, or hybrid circuit/packet mode connection formats for the services supported.

In summary, three traffic and transport network architectures are examined:

1. Fixed transport routing mesh network (with dynamic traffic routing)

2. Rearrangeable transport routing mesh network (with dynamic traffic routing)

3. Real-time transport routing ring network (with fixed traffic routing)

A hybrid circuit-switched and high-speed packet-switched variation of the rearrangeable transport routing mesh network is also examined. We now describe details of the routing principles employed in these architectures.

15.2.1 Fixed transport routing mesh network

The fixed transport routing mesh architecture allows one- and two-link dynamic traffic routing between switches. The architecture provides the flexibility of dynamic traffic routing in responding quickly, without capacity augments, to variable and unforecasted traffic loads. Within the logical mesh network, separate routing is provided for each virtual network supported by the multiservice integrated dynamic routing network (for example, switched voice, switched digital 64-kbps B-channel data, switched digital 384-kbps H0-channel data, and switched digital 1,536-kbps H1-channel data). Bandwidth is shared among services on the logical trunk links and throughout the network through the use of class-of-service dynamic traffic routing techniques, as described in Chapter 12.

15.2.2   Dynamic transport routing

With dynamic transport routing, the logical transport bandwidth is shifted rapidly among switch pairs and services through the use of dynamic cross-connect devices. Figure 15.2 illustrates the basic difference between the physical transport network and the logical transport, or traffic, network. The figure indicates that a direct trunk is obtained by cross-connecting through a transport switching location, which is further illustrated in the network block of Figure 15.5. Thus, the traffic network is a logical network overlaid on a sparse physical one. Dynamic trunk capacity routing is necessary for the rearrangeable transport routing mesh network, and dynamic virtual link capacity allocation is required for the real-time transport routing ring network. The latter concept allows the logical transport network architecture to overlap the physical transport network architecture. Both concepts assume that circuit-oriented dynamic cross-connect or packet-oriented add/drop devices are traversed at each transport network cross-connect switch on a given traffic path, as illustrated in Figure 15.5. This is particularly promising when such a device has low cost.

Two alternatives for the transport network cross-connect switches are as follows:

1. *Digital cross-connects:* Circuit-oriented digital cross-connects reconfigure the link capacities at any instant under control of a network capacity and bandwidth allocation controller, as illustrated in Figure 15.5.

2. *High-speed packet add/drop:* High-speed packet add/drop devices provide simple routing of call connections and in the future may be implemented with photonic packet-switching technology.

High-speed packet network capability to share network capacity affords the most flexible use of the available network capacity. Such capabilities are emerging with ATM switching for B-ISDN applications [Min89]. ATM allows the possibility of simpler and far faster dynamic bandwidth switching capability at the fiber line rate.

One possible implementation is a fiber add/drop device such as pictured in Figure 15.6. This device works most naturally in a packet format in which no operations on the packets need be performed other than the placing and reading of headers. It only needs to decide what traffic is destined for a particular switching location and what traffic needs to be routed to the next switch. The fiber add/drop

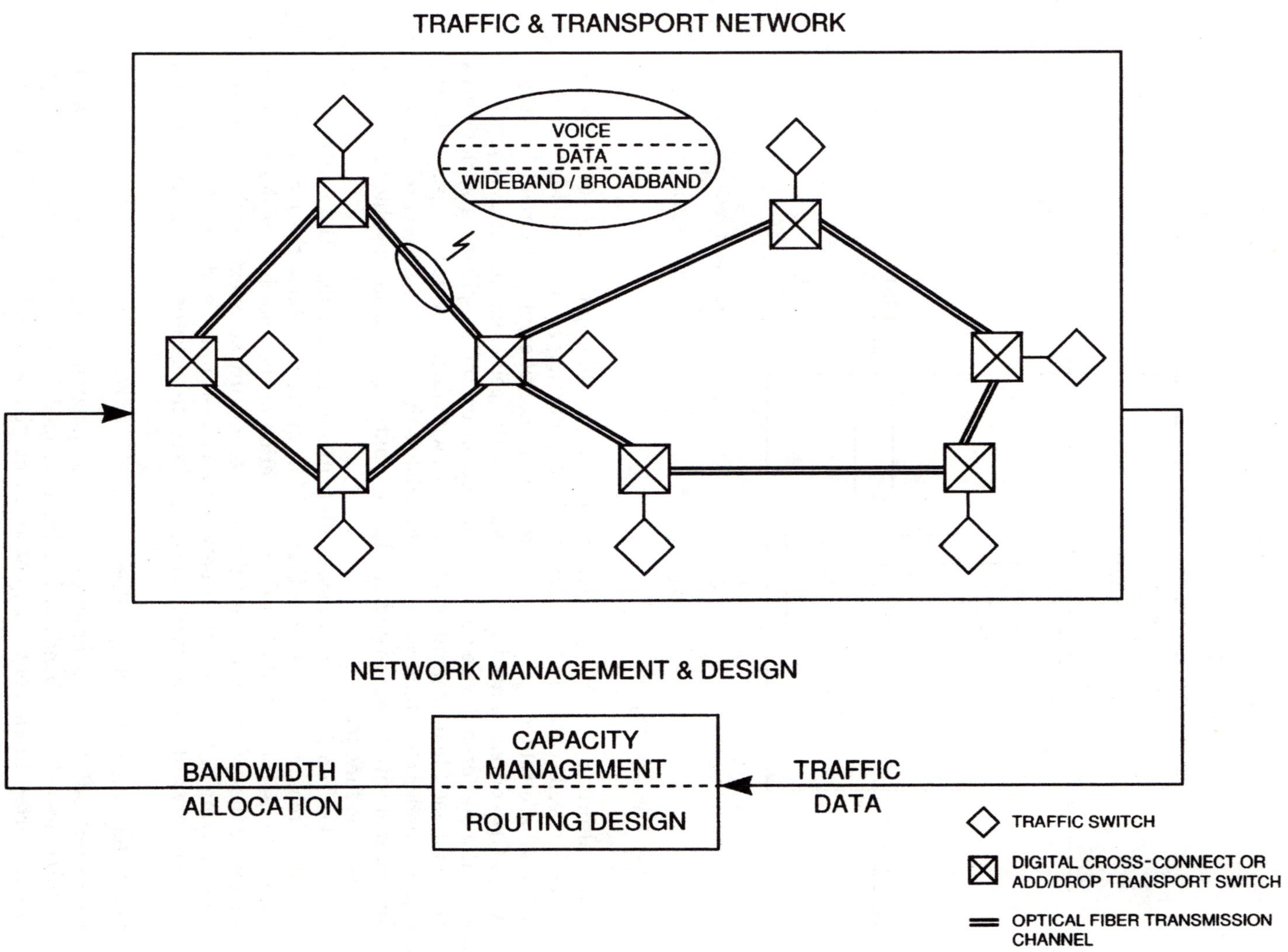

Figure 15.5 Capacity management and routing design for dynamic transport networks

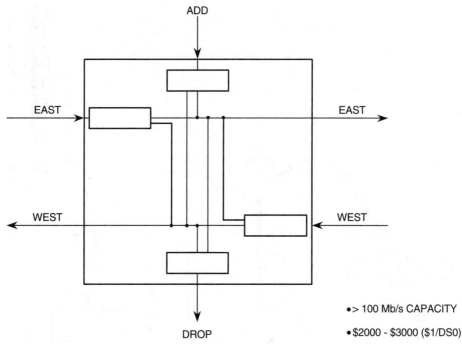

Figure 15.6 Fiber add/drop device

device also has the ability to write the routing information necessary on the headers of cells entering the network. ATM switching calls for fixed-length cells (packets) are routed by using a virtual path identifier/virtual circuit identifier (VPI/VCI) address placed in the header of the cell. This function is precisely what the add/drop device does, and in the future the simplicity of such devices may lend itself to incorporation of photonic switching, in which an optical device, possibly under electronic control, accomplishes the same task. Research has been undertaken with "mirror-window" devices that may someday be able to perform this routing function.

This add/drop device then passes all traffic it removes from the network to a traffic switch, as illustrated in Figure 15.5, in order to have the traffic broken down into its component pieces. The operation of the add/drop device might be viewed as that of a cross-connect that can reconfigure at an extremely rapid rate. This blurring of the distinction between a cross-connect and a switch is also a theme of forward-looking transmission and switching research and development.

15.2.3 Bandwidth allocation models

Bandwidth allocation is important to the functioning of any of the network architectures. In the bandwidth allocation model used here, traffic demands are converted to elements of bandwidth or virtual trunks (VTs). Figure 15.5 illustrates a network controller apportioning virtual link bandwidth to switches, then having the switches allocate this bandwidth to each call, and, in doing so, perform all necessary routing functions such as writing appropriate headers on cells. This allocation of bandwidth creates the equivalent of direct links, and we refer to these links as virtual links. Bandwidth is allocated to virtual links in accordance with traffic demands, and normally not all transport bandwidth is assigned; thus, there is a pool of unassigned bandwidth. In cases of traffic overload for a given switch pair, if possible, the switch first sets up calls on the virtual links that connect the switch pairs. If that is not possible the switch then sets up calls on the available pool of bandwidth. If there is available bandwidth, then at the completion of the call, the bandwidth is allocated to the virtual link to which it is assigned. In a similar manner, in the event that bandwidth is underutilized in a virtual link, excess bandwidth is released to the available pool of bandwidth and then becomes available for assignment to other switch pairs. The network capacity and bandwidth allocation controller reassigns network resources on a dynamic basis, through analysis of traffic data collected from the individual switches.

Figure 15.5 illustrates the bandwidth allocation model. We have suggested several virtual networks that are distinguished by their traffic characteristics, bandwidth requirements, and design performance objectives. Classes-of-service that may be carried on these virtual networks can include (1) narrowband 64 kbps B-channel circuit-mode services, (2) wideband 384 kbps H0-channel and 1,536 kbps H1-channel circuit-mode services, (3) broadband H2-channel and H3-channel services, (4) virtual-circuit packet-mode connections for constant-bit-rate and variable-bit-rate services, and (5) connectionless packet-mode services. There are two levels of bandwidth allocation that are accounted for in the bandwidth allocation method:

1. *Network bandwidth allocation:* When virtual trunk connections are requested, dynamic traffic routing and dynamic transport routing procedures are used to determine which network path has sufficient bandwidth for the service; if no such path exists the connection is blocked.

2. *Link bandwidth allocation:* A minimum guaranteed number of virtual trunk connections is allowed for each virtual network on each link; a link bandwidth allocation procedure is used to ensure that this minimum allocated bandwidth is provided for each virtual network. This procedure uses dynamic trunk reservation methods employed in dynamic traffic routing networks, as described in Chapters 8 and 11 to 13.

15.2.4 Rearrangeable transport routing mesh network

In the rearrangeable mesh architecture, we allow trunks between the various switches to be rearranged rapidly but in a preplanned fashion, such as by hour of the day. Rearrangeable transport routing capability enables rearrangement of the link capacities on demand. This capability appears most desirable for use in relatively slow rearrangement of capacity, such as for busy-hour traffic, weekend traffic, peak-day traffic, weekly redesign of link capacities, or for emergency restoration of capacity under switch or transport failure. At various times the demands

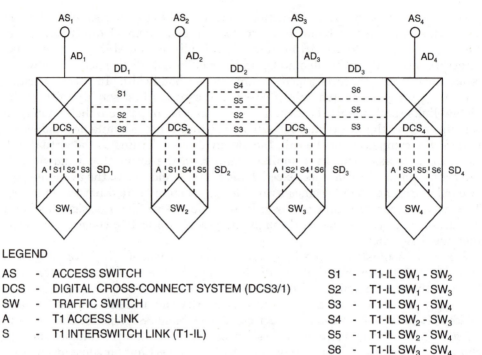

LEGEND

AS	-	ACCESS SWITCH	S1	-	T1-IL SW$_1$ - SW$_2$
DCS	-	DIGITAL CROSS-CONNECT SYSTEM (DCS3/1)	S2	-	T1-IL SW$_1$ - SW$_3$
SW	-	TRAFFIC SWITCH	S3	-	T1-IL SW$_1$ - SW$_4$
A	-	T1 ACCESS LINK	S4	-	T1-IL SW$_2$ - SW$_3$
S	-	T1 INTERSWITCH LINK (T1-IL)	S5	-	T1-IL SW$_2$ - SW$_4$
			S6	-	T1-IL SW$_3$ - SW$_4$

Figure 15.7 Rearrangeable transport routing mesh network

for switch and transport capacity by the various switch pairs and services that ride on the same optical fibers will differ. In this network, if a given demand for trunks between a certain switch pair decreases and a second goes up, we allow the trunks to be reassigned to the second switch pair. The ability to rearrange trunk capacity dynamically and automatically results in cost savings. Large segments of bandwidth can be provided on fiber routes, and then the transport capacity can be allocated at will with the rearrangement mechanism. This ability for simplified capacity management is discussed further in Section 15.3.

The rearrangeable transport routing mesh network is a dynamic transport routing concept for both circuit-switched and packet-switched virtual path (for example, ATM-based) networks. As illustrated in Figure 15.7, the rearrangeable transport routing mesh network concept includes traffic switches (SWs), digital cross-connect systems (DCSs), and access switches (ASs). Access switches include end-offices, access tandems, customer premises equipment, and overseas international switching centers. Here a T1 transmission channel consists of twenty-four 64-kbps DS0 channels and a T3 transmission channel consists of twenty-eight T1 channels. A DCS3/1 cross-connect system can switch (or "cross-connect") a T1 channel within one terminating T3 to a T1 channel within another terminating T3. A DCS3/3 cross-connect system can switch a T3 channel within one terminating fiber link to a T3 channel within another terminating fiber link. In the example illustrated, access switches connect to DCS3/1 cross-connects by means of access T1 links such as link AD$_1$. Switches connect to DCS3/1s by means of links such as SD_1. A number of T3 backbone links interconnect the DCS3/1 network elements, such as links DD_1 and DD_2. T3 backbone links are terminated at each end by DCS3/1s and are routed over fiber spans on the physical transport network on the shortest physical paths. T1 interswitch links (T1-ILs) are formed by cross-connecting T1 channels through DCS3/1s between a pair of switches. For example, the T1-IL S2 from SW_1 to SW_3 is formed by connecting T1 terminal equipment between SW_1 and SW_3 through links SD_1, SD_3, DD_1, and DD_2 by making appropriate cross-connects through DCS_1, DCS_2, and DCS_3. T1-ILs have variable T1 bandwidth capacity controlled by the rearrangeable transport routing mesh network design, as discussed in Section 15.3. Access T1 links are formed by cross-connecting T1 terminal equipment between access switches and switches— for example, access switch AS_1 connected on links AD_1 and SD_1 through DCS_1 to SW_1 or, alternatively, access switch AS_1 connected on links AD_1, DD_1, and SD_2 cross-connected through DCS_1 and DCS_2 to SW_2. For additional network reliability, switches and access switches are dual-homed to two DCS3/1s, possibly in different building locations.

Figure 15.1 illustrates an example set of T3 backbone links that overlays the physical fiber transport network. Some T3 backbone links, called T3 express links, overlay two or more T3 backbone links. Therefore, T3 express links traverse longer distances before terminating on DCS3/1s. These T3 express links are included in a T1-IL transport path if the T3 express link is fully traversed by the T1-IL transport path between network switches.

For example, in Figure 15.1, T3 express link AD is on the T1-IL transport path between switches A and E, which consists of T3 express link AD and T3 backbone link DE. Here, T3 backbone links AB, BC, and CD are traversed by T3 express link AD on the T1-IL transport path, which avoids demultiplexing at DCS3/1s B and C to the T1 level by only going through DCS3/3s at these locations. Hence, use of T3 express links leads to fewer DCS3/1 terminations and associated multiplexing and demultiplexing stages. Although only one express link is shown in Figure 15.1, the network design would have many express links.

Figure 1.31 illustrates the relationship of the DS0, T1, and T3 dynamic routing methods used in the rearrangeable transport routing mesh network. Dynamic traffic routing, such as class-of-service real-time network routing (RTNR), discussed in Chapter 12, is used at the DS0 level to route calls comprising the underlying traffic demand. DS0-level virtual link capacity allocations, denoted as VTeng, are made for each virtual network on the T1-IL capacity. For each call the originating switch analyzes the called number and determines the terminating switch, class-of-service, and virtual network. The originating switch tries to set up the call on the direct T1-IL, if one exists, to the terminating switch and, if unavailable, tries to find a two-link path using state-dependent routing logic. Preplanned rearrangeable transport routing is used at the T1 level to rearrange the T1-IL capacity as required to match the traffic demands and to achieve interswitch T1 diversity, access T1 diversity, and T1 restoration following switch, DCS, or fiber transport failures. The T1-IL capacities are allocated by the rearrangeable transport routing mesh network design module such that the bandwidth is efficiently used according to the level of traffic between the switches.

Preplanned rearrangeable transport routing is used at the T3 level to aggregate the interswitch and access T1-IL demands to the T3 backbone/express link level, and then to aggregate the backbone/express link T3 demands to the physical fiber link level. T3 restoration is used to restore T3 capacity through control of DCS3/3s in the event of transport or equipment failures.

Dynamic routing is thus used at each level in the transport hierarchy, and the switching node model of Figure 1.31 applies to both circuit-switched networks and packet-switched networks and to hybrids of the two technologies. However, with high-speed packet-switched networks, such as with ATM technology, there is greater flexibility due to lack of bandwidth modularity constraints, and there is the possibility of real-time bandwidth switching, due to the virtual bandwidth aspects of high-speed packet switching and routing. There are significant network design opportunities with dynamic transport routing, as illustrated in Figure 1.31. First, weekly design and rearrangement of T1 link capacity can approach zero reserve capacity designs. As discussed in Chapter 9, in-place capacity that exceeds the capacity required to exactly meet the design loads with the objective performance is called reserve capacity. Reserve capacity comes about because load uncertainties, such as forecast errors, tend to cause capacity buildup in excess of the network design that exactly matches the forecast loads. Reluctance to disconnect and rearrange link and transport capacity contributes to this reserve capacity buildup. Typical ranges for reserve capacity are from 15 to 25 percent or more of network cost. In Chapter 9 we found that dynamic traffic routing and fixed transport routing compared with fixed traffic routing and fixed transport routing provides a potential 5 percent reduction in reserve capacity

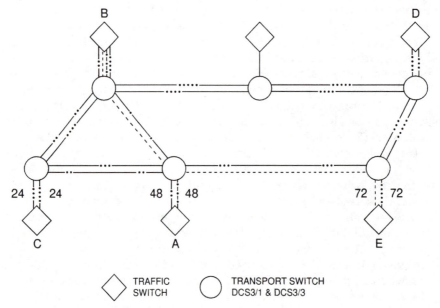

Figure 15.8 Rearrangeable transport routing network weekly arrangement (week 1 load pattern)

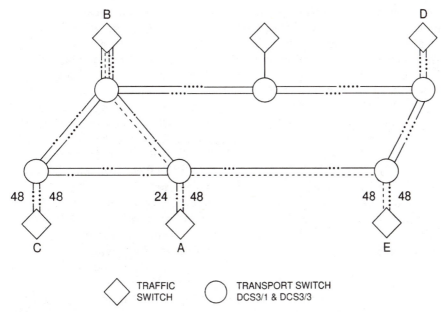

Figure 15.9 Rearrangeable transport routing network weekly arrangement (week 2 load pattern)

while retaining a low level of short-term capacity design. As illustrated in Section 15.4, dynamic traffic routing, together with weekly dynamic transport routing, achieves an additional 10 percent reduction in reserve capacity. With dynamic traffic routing together with real-time dynamic transport routing, we show in Section 15.4 further utilization and performance benefits if transport switching costs are sufficiently low.

Figures 15.8 and 15.9 illustrate the changing of routed transport capacity on a weekly basis between switch pairs A–B, C–D, and B–E, as demands between these switch pairs change on a weekly basis. These transport routing and capacity changes are made automatically in the dynamic transport network, in which diverse transport routing of trunks on T1-IL logical traffic links A–B and C–D is maintained by the dynamic transport routing network. As described in Chapter 14, link transport diversity achieves additional network reliability.

An illustration of how dynamic transport routing achieves reserve capacity reductions is given in Figure 1.32, which shows how transport demand is routed according to varying seasonal requirements. As seasonal demands shift, the dynamic transport network is better able to match demands to routed transport capacity, thus gaining efficiencies in transport requirements. The figure illustrates the variation of winter and summer trunk and transport capacity demands. With fixed transport routing, the maximum trunk capacity and transport capacity are provided across the seasonal variations, because in a manual environment without dynamic transport rearrangement it is not possible to disconnect and reconnect capacity on such short cycle times. When transport rearrangement is automated with dynamic transport routing, however, the trunk and transport design can be changed on a weekly, daily, or, with high-speed packet switching, real-time basis to exactly match the trunk and transport design with the actual network traffic and transport demands. Notice that in the fixed transport network there is unused trunk and transport capacity that cannot be used by any demands; sometimes this is called "trapped capacity," because it is available but cannot be accessed by any actual traffic demand. The dynamic transport network, in contrast, follows the traffic load with flexible transport routing, and together with transport network design it reduces the trapped capacity. Therefore, the variation of demands leads to capacity-sharing efficiencies, which in the example of Figure 1.32 reduce trunk termination capacity requirements by 50 trunk terminations, or approximately 10 percent compared with the fixed transport network, and by 50 DS0 transport capacity requirements, or approximately 14 percent.

Therefore, with dynamic traffic routing and dynamic transport routing design models, as illustrated in Figure 1.48, reserve capacity can be reduced in comparison with fixed transport routing, because with dynamic transport network design the link sizes can be matched to the network load. With dynamic transport routing, the link capacity disconnect policy becomes, in effect, one in which link capacity is always disconnected when not needed for the current traffic loads. Models given in [FHH79] predict reserve capacity reductions of 10 percent or more under this policy, and the results presented in Section 15.4 based on weekly dynamic transport design substantiate this conclusion.

Daily design and rearrangement of transport link capacity can achieve performance improvements for similar reasons, due to noncoincidence of transport capacity demands that can change daily. An example is given in Figures 15.10 and 15.11 for traffic noncoincidence experienced on peak days such as Christmas Day. In Figure 15.10, we illustrate the normal business-day routing of access demands and interswitch demands. On Christmas Day, however, there are many busy switches and many idle switches. For example, switch SW_2 may be relatively idle on Christmas Day (for example, if it were a downtown business switch), while SW_1 may be very busy. Therefore, on Christmas Day, SW_2 demands to everywhere else in the network are reduced, and through dynamic transport routing these transport capacity reductions can be made automatically. Similarly, SW_1 demands are increased on Christmas Day. Access demands such as those from AS_1 can be redirected to freed-up trunk termination capacity on SW_2, as illustrated in Figure 15.11, which also frees up trunk termination capacity on SW_1 to be used for interswitch demand increases. By this kind of access demand and interswitch demand rearrangement, based on noncoincident traffic shifts, more traffic to and from SW_1 can be completed because interswitch trunk capacity is increased, now using freed-up transport capacity from the reduction in the transport capacity needed by SW_2. On a peak day such as Christmas Day, the busy switches are often limited by interswitch trunk capacity; this rearrangement reduces or eliminates this bottleneck, as is illustrated in the Christmas Day dynamic transport network design example in Section 15.4.

The balancing of access and interswitch capacity throughout the network can lead to robustness to unexpected load surges. This load-balancing design is

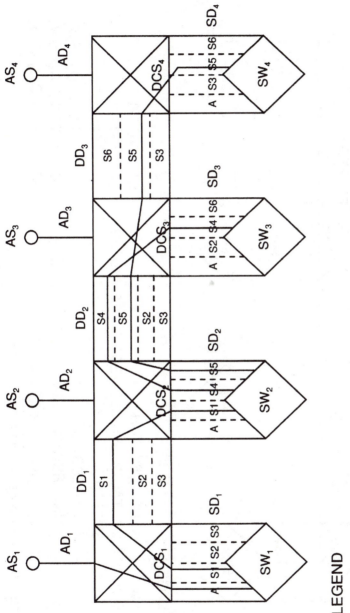

LEGEND

AS – ACCESS SWITCH
DCS – DIGITAL CROSS-CONNECT SYSTEM (DCS3/1)
SW – SWITCH
A   – T1 ACCESS LINK
S   – T1 INTERSWITCH LINK (T1-IL)

S1	–	T1-IL SW$_1$ - SW$_2$
S2	–	T1-IL SW$_1$ - SW$_3$
S3	–	T1-IL SW$_1$ - SW$_4$
S4	–	T1-IL SW$_2$ - SW$_3$
S5	–	T1-IL SW$_2$ - SW$_4$
S6	–	T1-IL SW$_3$ - SW$_4$

Figure 15.10 Rearrangeable transport routing peak-day design

559

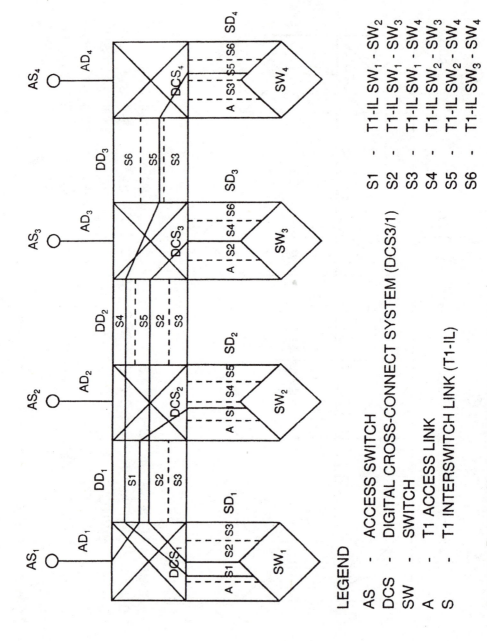

LEGEND

AS – ACCESS SWITCH
DCS – DIGITAL CROSS-CONNECT SYSTEM (DCS3/1)
SW – SWITCH
A   – T1 ACCESS LINK
S   – T1 INTERSWITCH LINK (T1-IL)

S1 – T1-IL SW_1 - SW_2
S2 – T1-IL SW_1 - SW_3
S3 – T1-IL SW_1 - SW_4
S4 – T1-IL SW_2 - SW_3
S5 – T1-IL SW_2 - SW_4
S6 – T1-IL SW_3 - SW_4

Figure 15.11 Rearrangeable transport routing peak-day design

illustrated in Section 15.4 with an example based on Hurricane Bob overload in the northeastern United States in August 1991. Capacity addition rearrangements based on instantaneous reaction to unforeseen events such as earthquakes could be made in the dynamic transport network.

Dynamic transport routing can provide dynamic restoration of failed capacity, such as that due to fiber cuts, onto spare or backup transport capacity. Dynamic transport routing provides a self-healing network capability such as discussed in Chapter 14 to ensure a networkwide path selection and immediate adaptation to failure. FASTAR [CED91], for example, implements central automatic control of DCS3/3 transport switching devices to quickly restore service following a transport failure. As illustrated in Figure 1.4, a fiber cut near Nashville, Tennessee, severed more than 129,000 DS0 circuits, of which about two-thirds were restored by FASTAR dynamic transport restoration in the first 11 minutes. Over the duration of this event, more than 12,000 calls were blocked in the AT&T switched network, almost all of them originating or terminating at the Nashville switch, and it is noteworthy that the blocking in the network returned to zero after the 37,000 trunks were restored, even though there were 30,000 trunks still out of service. That is, both FASTAR and RTNR were able to find available paths on which to restore the failed traffic. Hence, this example clearly illustrates how real-time dynamic traffic routing in combination with real-time dynamic transport routing can provide a self-healing network capability, and even if the cable is repaired two hours after the cut, degradation of service is minimal.

These examples illustrate that implementation of dynamic transport routing provides better network performance at reduced cost. These benefits are similar to those achieved by dynamic traffic routing, and, as shown, the combination of dynamic traffic and transport routing provides synergistic reinforcement to achieve these network improvements.

15.2.5 Hybrid circuit-switched and high-speed packet-switched transport networks

High-speed packet technology, such as ATM, has several possible applications including dedicated private-line replacement and efficient multimedia integration of voice, narrowband data, wideband data, image, and video traffic. In this section we discuss high-speed packet transport networks implemented with high-speed packet technology, such as the ATM technology discussed earlier in this chapter and in Chapter 1. High-speed packet transport technology components include a terminal adapter (TA), which serves as a gateway for high-speed packet transport networks, receives an incoming bit stream, converts the bit stream into 53-byte packets, called cells, and transmits the cells toward their destination on an outgoing link. Another hardware component is a high-speed packet-cross-connect system (PCS), which receives a cell on an incoming link and, on the basis of routing information in the cell header, routes the cell to an outgoing link. The PCS must be used in conjunction with a TA because the TA is needed to convert the bit stream to cells and back again.

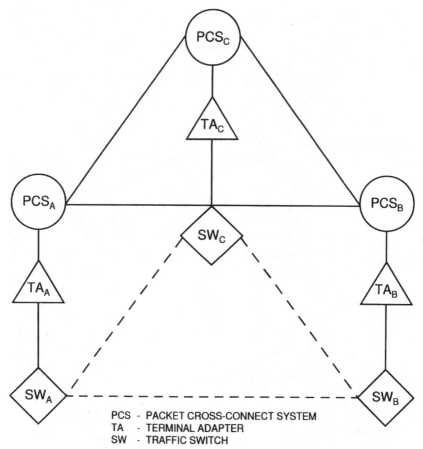

Figure 15.12 Hybrid circuit-switched and high-speed packet-switched ATM transport network

Because high-speed packet transport technology is efficient for multiplexing different bit streams and can be used to encode silence in voice transmissions or stationarity in video transmission, multiplexing efficiency gain and signal compression can be achieved on the transport network. For voice-compression applications, a five-to-one compression or more can be achieved for high-speed packet voice transmission in comparison with a full 64-kbps circuit-switched voice transmission. However, in compressing signals in this way, there is some chance of quality degradation during periods when the required instantaneous bit rate exceeds the channel capacity. In that case, some cells will be dropped and voice quality degraded.

In the hybrid circuit-switched and high-speed packet transport network design discussed in this chapter, each traffic switch is assumed to have a TA connected to a corresponding PCS, as is illustrated in Figure 15.12. A call from SW_A to SW_B, for example, could access (1) the direct circuit link from SW_A to SW_B, (2) the direct high-speed packet link from PCS_A to PCS_B, (3) the two-link circuit path through

SW_C, (4) the two-link high-speed packet path through PCS_C, (5) the two-link path that uses the circuit link from SW_A to SW_C and the high-speed packet link from SW_C to PCS_C to PCS_B to SW_B, or (6) the two-link path that uses the high-speed packet link to PCS_C and the circuit link from PCS_C to SW_C to SW_B. SW_A uses dynamic traffic routing to access the path choices.

Signaling for call setup can still use the common-channel signaling (CCS) network to exchange call setup information, status information, and call progress signals. The switch maintains the busy/idle status of circuit-switched links and high-speed packet links and thereby knows how many active virtual circuits (VCs) there are on each high-speed transport link connecting each PCS pair. If the number of active VCs on each link is less than a prescribed maximum number, the call is completed; otherwise, the CCS network signals the originating switch that there is no capacity available in the network.

In Section 15.4, results are presented on the design of the hybrid circuit-switched and high-speed packet transport network.

15.2.6 Real-time transport routing ring network

In the real-time transport routing ring network architecture, the logical transport network links exactly overlay the physical transport network links. In the simplest version of this network we use fixed traffic routing; that is, we allow no alternative traffic routing. We specify that the path taken between any two switches in the network will be the shortest path. This network is very different from the structure of current meshed networks, because the real-time transport routing ring network is sparse. For example, in the transport network model discussed in Section 15.4, the real-time transport routing ring network has only 168 logical trunk links out of a possible 2,145 in a 66-switch, fully connected network model. An immediate implication of this is that traffic paths in this network will require many more logical trunk links in a given connection than in the mesh network. This does not mean they are longer—as a matter of fact they are the same length or shorter because they follow the minimum distance transport route. It does mean that paths will pass through many more actual switches. The average is between four and five logical trunk links, and there are some paths that pass through as many as 12 logical links. Implementation of this architecture requires real-time reallocation of transport bandwidth to virtual links, or real-time dynamic transport routing, among the switch pairs and services that share the physical transport links. This rapid reallocation of bandwidth in real-time transport routing ring networks is especially amenable to high-speed packet technology such as ATM technology, which is the basis for B-ISDN. Methods to accomplish this are described in Sections 15.2 and 15.3.

15.3 Network Management and Design

In this section we review network management and design models for the various dynamic transport routing networks considered in Section 15.2.

15.3.1 Rearrangeable transport routing mesh network design model

In this model it is envisioned that the network design system obtains daily traffic data for the past 24-hour period, smooths and updates its current traffic estimates, allocates T1 bandwidth requirements to T1-ILs and access T1 links, allocates T3 bandwidth requirements to backbone/express links, routes T1-IL capacity on diverse routes through the backbone/express link network, computes a rearrangement sequence to minimize cross-connect activities in switches and DCSs, and populates the traffic routing, transport routing, and cross-connect data structures in the switches and DCSs to implement the next network rearrangement. Bandwidth allocation is controlled in the rearrangeable transport routing mesh network through (1) dynamic adjustment of T1-IL capacity based on traffic requirements and VTeng DS0 bandwidth allocation, (2) T1-IL and access T1 bandwidth allocation and routing on the T3 backbone/express link capacity, and (3) backbone/express link T3 bandwidth allocation and routing on the physical fiber links. Here, the design must meet network performance objectives for the estimated traffic loads with minimum cost and provide a robust design in the event of unforeseen load patterns and network failures.

The mathematical models given in this section are used to estimate traffic, size T1-IL capacity, reallocate access T1 capacity between overloaded and underloaded switches, compute diverse T1-IL capacity requirements, size backbone/express link T3 capacity, and rearrange network capacity. See Figure 15.13 for an illustration of the rearrangeable transport routing mesh network design model. Further details of the design model steps are now given.

15.3.1.1 Estimation of traffic. Based on traffic data received each day, the rearrangeable transport routing mesh network design model computes running estimates of (1) average daily loads, by day of week and peak day; (2) statistical day-to-day variation of the daily and peak-day load patterns; and (3) seasonal variation of load patterns. With these data the design model estimates the hourly load patterns and day-to-day variation parameters for the next several-day interval.

15.3.1.2 Sizing of T1-IL capacity. We note that T1 demand for private-line services can be included with the switched T1-IL demand. Switched T1-IL demand can be determined, for example, by the discrete event flow optimization model or a fixed-point erlang flow optimization model, as discussed in Chapter 13, which computes network flows, as well as switch-to-switch blocking probabilities, for networks with real-time dynamic traffic routing. An assumption of the fixed-point erlang flow optimization model is that the stationary behavior of real-time dynamic routing networks can be computed based on the stationary, independent link-state probabilities and on via switch selection probabilities, which are based on the link-state probabilities and switch-to-switch blocking probabilities. An iterative fixed-point approach is used, and all overflow traffic is assumed to be Poisson.

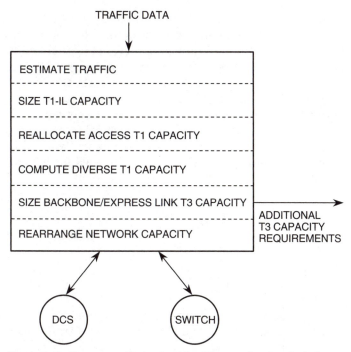

TRAFFIC DATA

ESTIMATE TRAFFIC

SIZE T1-IL CAPACITY

REALLOCATE ACCESS T1 CAPACITY

COMPUTE DIVERSE T1 CAPACITY

SIZE BACKBONE/EXPRESS LINK T3 CAPACITY

REARRANGE NETWORK CAPACITY

ADDITIONAL T3 CAPACITY REQUIREMENTS

DCS

SWITCH

Figure 15.13 Rearrangeable transport routing mesh network design model

Three steps are performed iteratively by the method. The link-state probability model derives the link aggregate state probabilities from the link arrival rates; the link arrival rates are the outcome of the erlang flow model. The erlang flow model uses the link aggregate state probabilities to derive the path state probabilities and the route state probabilities. With these state probabilities, the erlang flow model determines the route flows, path flows, path arrival rates, and link arrival rates, where the latter are used in the link-state probability model to compute the link-state probabilities. Finally, the automatic trunk reservation and path selection depth model computes the switch-to-switch blocking probability distribution, and from that it models the path selection depth and the link reservation levels. A model of calls in progress is used to determine the probability that reservation and path selection depth are enabled. The convergence criterion is the sum of the squared differences in link-state probability levels from the previous to the current iteration, summed over all links in the network. Our numerical experience shows that, when automatic trunk reservation is applied, the approximation method converges quickly to the fixed-point solution. At each iteration, the erlang flow model is used to compute blocked traffic for a given T1-IL capacity level, and then the T1-IL capacity is resized to meet the objective blocking performance. T1-IL capacity is modularized to the nearest modular T1 capacity.

Bandwidth is allocated for every virtual network on the T1-IL capacity. The bandwidth allocations for individual virtual networks, denoted as $VTeng^i$ for virtual network i, are shared among virtual networks under normal network conditions but are dedicated to virtual network i under congestion. The originating switch collects traffic data in real time and determines the level of virtual trunk demands to each destination in the network for each virtual network i. Based on these estimated demands and available capacity in the network, the switch allocates the DS0 capacity on the T1-IL links. VTeng^i is a minimum guaranteed bandwidth for a virtual network if there is blocking for the virtual network and sufficient traffic to use the VTeng^i bandwidth. If a virtual network is meeting its blocking objective, other virtual networks are free to share the VTeng^i bandwidth allocated to that virtual network. The quantity VTeng^i_k is calculated for virtual network i on link k as follows:

$$
\text{VTeng}^i_k = d_k \frac{\displaystyle\sum_{h=1}^{H} D^{hi}_k}{\displaystyle\sum_{n=1}^{M}\sum_{h=1}^{H} D^{hi_n}_k}
$$

Here, virtual networks n are numbered 1 to M, hours h are numbered 1 to H, d_k is the modular T1-IL capacity for switch pair k, and D^{hi}_k is the offered traffic load for switch pair k in hour h for virtual network i.

15.3.1.3 Reallocation of access T1 capacity between overloaded and underloaded switches. If the total interswitch termination capacity available on a switch is exceeded by the estimated interswitch T1 capacity demand, then additional interswitch termination capacity is generated for the switch by rehoming (rerouting) access T1 capacity to underloaded switches that have spare termination capacity. This access T1 rerouting frees terminations on the overloaded switches, which can then be used to satisfy the interswitch T1-IL demand. After terminations are rehomed to underloaded switches, switch-to-switch traffic loads are adjusted according to the number of access terminations moved and through use of a Kruithof iteration procedure [Kru37]. After the switch-to-switch traffic levels are adjusted, the T1-IL capacity requirements are recomputed. In this model, additional switch termination capacity requirements are determined to achieve sufficient capacity to meet network demands up to a specified level of overload.

15.3.1.4 Computation of diverse T1 capacity. The T1-IL sizing model provides for diverse sizing of the T1-IL capacity in the hour of maximum T1-IL traffic load, in order to achieve a traffic restoration level (TRL) objective under network failure. As discussed in Chapter 14, the traffic restoration level objective specifies that for single fiber transport link or switch failures, the T1-IL capacity of the network is sized to carry at least a certain minimum percentage of the design traffic load, denoted as the traffic restoration level objective. For example, if the traffic restoration level objective is 0.5, this means that following any single fiber cut in the transport network, at least 50 percent of the design traffic is still carried after the failure.

The model used for the traffic restoration level design of the T1-IL links is as follows:

TRL = traffic restoration level objective
T_k = minimum T1-IL capacity required for switch pair k to meet the traffic restoration level objective

The required number of surviving T1-IL capacity after a fiber transport failure is given by

$$T_k/2 = \overline{\text{NB}}\,(\max_h D_k^h, 0, 1, 1 - \text{TRL})$$

where the function $\overline{\text{NB}}$ gives the trunks required for a given offered load, level of day-to-day variation, peakedness, and blocking objective. Here, a minimum of $T_k/2$ trunks of the T1-IL capacity are routed, if possible, over each of two diverse T1 paths, while ensuring that T1 modularity conditions are met. A further constraint is that the total T1-IL capacity d_k must be greater than or equal to T_k. If that is possible it ensures that a failure of either diverse path allows enough T1 capacity to survive to meet the traffic restoration level objective. At times, however, because of modularity constraints, it is not possible to achieve this diversity condition on the T1-IL capacity (for example, if the T1-IL capacity only requires one T1, it is not possible to achieve the condition). Diverse T1 capacity is allocated, and, if possible, the $T_k/2$ diverse T1-IL capacity is realized on the diverse path. Similarly, the diverse access T1 capacity required to be split between network switches to achieve traffic restoration level objectives for switch failure is calculated in the same way. This same procedure is followed whether the access T1 capacity is brought into one building location or routed diversely into multiple DCS3/1 building locations.

Other related models for reliable network design are given in Chapter 14, in which linear programming models are formulated for diverse capacity design, in addition to the traffic restoration level method summarized here.

15.3.1.5 Sizing of backbone/express link T3 capacity. In this section we first describe the topological design of the rearrangeable transport routing mesh network in terms of backbone links, express links, and diverse paths. We then formulate the T3 capacity-sizing model for a rearrangeable transport routing mesh network. The rearrangeable transport routing mesh network is composed of DCS3/1s, DCS3/3s, and fiber links between them. Each DCS3/1 is co-located with a corresponding DCS3/3, although DCS3/3s may exist without corresponding DCS3/1s. To design candidate rearrangeable transport routing mesh network backbone links, we first find all the shortest paths between the DCS3/1s on the fiber network. We then break each shortest path into unique segments beginning and ending on a DCS3/1, and these segments are then the candidate backbone links. Diverse T1 paths through the T3 backbone link network are designed by first constructing a "violations matrix" for each T3 backbone link. This matrix describes the degree of physical overlap between any pair of T3 backbone links. The set of candidate backbone links defines a connectivity among the DCS3/1s. We also have the underlying fiber link topology for all the backbone links, and from physical routing information we can compute span violations between all backbone links. We use a span-diverse model to find a set of candidate span-diverse shortest paths between all DCS3/1s [Bha97]. This method allows for both switch and span violations with a weighted penalty function. Each T1-IL capacity requirement is split on two or more diverse T1 transport paths, so that typically the diverse T1 capacity $(T_k/2)$ is routed on the longer T1 path. For example, in Figure 15.1, the T1-IL capacity between switch F and switch H is split between T1 transport path FG–GH and T1 transport path FD–DE–EH, in which the diverse T1 capacity is routed on the FD–DE–EH path and the remaining T1-IL capacity is routed on the FG–GH path.

An express link is a "through T3," in that if there is more than one T3 of demand, for example, between DCS3/1 A and DCS3/1 D in Figure 15.1, the T3 express link AD may prove in. Then the T1-IL path between DCS3/1s A and E can route over express link AD and backbone link DE. With this transport routing we bypass DCS3/1s B and C by cross-connecting at the T3 level through the DCS3/3 at these switches, thus saving DCS3/1 terminations. We use a greedy heuristic to eliminate candidate express links that can never carry enough T1 demand to fill a T3. We do this by examining each candidate express link, one at a time, and flowing the T1-IL demand for the entire network. If the particular express link candidate carries less than one T3 of T1 flow, we discard it from the candidate list. We order the set of candidate express links in decreasing order by backbone link hop count. These are then inserted into the set of T1-IL paths starting with the one with the largest hop count, and thereby we replace the backbone links in the T1-IL paths with express links.

A minimum-cost linear programming (LP) model is solved for sizing backbone/express link T3 capacity, in which the T1-IL demands d_k are routed on the fiber link network as follows:

$$\text{Minimize} \sum_{f=1}^{F} M_f \, \Delta F_f$$

subject to

$$\sum_{k=1}^{K} \sum_{j=1}^{J_k} P_{jkb} y_{jk} \leq f_b \qquad b = 1, 2, \ldots, B$$

$$\sum_{b=1}^{B} Q_{bf} f_b \leq F_f + \Delta F_f \qquad f = 1, 2, \ldots, F$$

$$\sum_{j=1}^{J_k} y_{jk} = d_k \qquad k = 1, 2, \ldots, K$$

$$\sum_{j=1}^{J_k} \sum_{b=1}^{B} P_{jkb} Q_{bf} y_{jk} \le d_k - T_k/2 \qquad \begin{aligned} k &= 1, 2, \ldots, K \\ f &= 1, 2, \ldots, F \end{aligned}$$

$$y_{jk}, \quad f_b \ge 0 \qquad \begin{aligned} j &= 1, 2, \ldots, J_k \\ k &= 1, 2, \ldots, K \\ b &= 1, 2, \ldots, B \end{aligned}$$

Here, fiber links f are numbered 1 to F, backbone/express links b are numbered 1 to B, T1-IL paths j are numbered 1 to J_k, and switch pairs k are numbered 1 to K. In the above model,

P_{jkb} = 1 if path j of T1-IL k routes over backbone/express link b; 0 otherwise

Q_{bf} = 1 if backbone/express link b routes over fiber link f; 0 otherwise

y_{jk} = flow assigned to path j of switch pair k

d_k = T1-IL capacity for switch pair k

T_k = minimum T1-IL capacity required by switch pair k to meet the traffic restoration level objective

f_b = capacity assigned to backbone/express link b

F_f = existing transport capacity of fiber link f

ΔF_f = added transport capacity on fiber link f

M_f = incremental cost per T3 of added transport capacity on fiber link f

T3 transport capacity requirements are determined to achieve sufficient capacity to meet network demands up to a specified level of overload. Current fiber link T3 capacity is determined directly from network elements. The above model is solved for the required capacity additions and achieves efficient levels of transport fill rates. Because transport and equipment demands are aggregated to the fiber link level and total switch level, this provides a simpler and more robust network design.

Either Karmarkar's algorithm [Kar84] and/or a heuristic technique can be used to solve the LP multicommodity network flow model. One heuristic solution method that works well is as follows. First, the T1-IL demands are flowed over two diverse backbone/express link paths. We then find a minimum-cost alternative T3 path for each express link that minimizes DCS3/1 terminations. For a given express link, its minimum-cost alternate path is the concatenation of two or more shorter express or backbone links. Topologically, the minimum-cost alternate path comprises the same underlying fiber links as the shortest T3 backbone path. We then move traffic off partially filled express links onto its minimum-cost alternate path, starting with the longest express link and rounding the T1 flow down to a multiple of full T3s. The "overflow" T1 demand is split off and routed onto the corresponding minimum-cost alternate path. This procedure is repeated for each express link in the greedy ordered list and must terminate on backbone links. Backbone link capacity is then rounded up to the next T3 module of capacity.

15.3.1.6 Rearrangement of network capacity. Once the bandwidth allocation design is completed for the estimated traffic, a combinatoric optimization method computes the rearrangement method that minimizes cross-connect activities by first recognizing common route segments between the existing and target network arrangements and then finding the cross-connect actions that maximally reuse these common segments. In the model, we compute a sequence of T1-IL disconnect and connect orders so as to always maintain the maximum capacity connected throughout the network rearrangement. The rearrangeable transport routing mesh network design module communicates with the network elements to make the necessary rearrangement and cross-connect changes.

15.3.2 Hybrid circuit-switched and high-speed packet-switched transport network design models

The hybrid circuit-switched and high-speed packet-switched transport network model illustrated in Figure 15.12 is designed using the path-erlang flow optimization model described in Chapters 6 and 7. In the path-erlang flow optimization model each PCS is modeled as a separate switch in the network. Therefore, the 16-traffic-switch model used in the study, shown in Figure 15.14, becomes a 32-switch model when the 16 corresponding PCSs are added. In the model, which represents the initial DNHR implementation network in July 1984, there is only traffic-switch to traffic-switch load, and no traffic loads originate or terminate

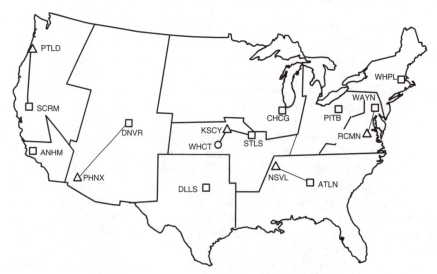

Figure 15.14 16-switch network model (initial DNHR implementation network)

on PCSs. The links between the traffic switches and their corresponding PCSs are sized to blocking probability objective of .001. Costs used in the study are illustrated in Figures 15.15 and 15.16, in which the five-to-one voice compression assumed in the model is reflected on the cost per mile of transport. In Figure 15.16, the trunk cost versus mileage is displayed, in which it is assumed that echo cancellation is required for trunks 1,165 miles and longer in the circuit-switched links, and echo cancellation cost is built into the terminal adapter (TA) function and included in the cost for all trunk mileages for the high-speed packet links. Hence, the cross-over in trunk cost occurs at 1,165 miles in the model. This cross-over leads to the result that links longer than 1,165 miles in the hybrid network are high-speed packet links, which leads to a lower network cost compared with the all circuit-switched design. These results are presented in Section 15.4.

15.3.3   Real-time transport routing ring
network design models

For the real-time transport routing ring network, because there is no alternate routing of traffic, the design procedure is relatively simple. The traffic demands of the various switch pairs are aggregated to the transport links, which overlay the logical links, and then each transport link is sized to carry the total traffic demand from all switch pairs that use the transport link for voice, data, and broadband traffic. As illustrated in Figure 15.17, one subtlety of the design procedure is deciding what blocking objective to use for sizing the transport links. The difficulty is that many switch pairs send traffic over the same transport link, and each of these switch pairs has a different number of transport links in its path. This means that for each traffic load, a different level of blocking on a given transport link is needed to ensure, say, a 1 percent level of blocking end to end. With many kinds of traffic present on the link, we are guaranteed an acceptable blocking probability grade-of-service objective if we identify the path through each transport link that requires the largest number of links, n, and size the link to a $1/n$ blocking objective. In Figure 15.17, link L_1 has a largest number n equal to 6, and link L_2 has a largest number n equal to 4. If the end-to-end blocking objective is 1 percent, then the link-blocking objectives are determined as given in the figure. We show that the real-time transport routing ring network sized in this simple manner still achieves significant efficiencies.

15.3.4   Capacity management

Capacity management is a large component of telecommunications network cost. Typically, large numbers of people are necessary to route and provision trunks over a transport network. Networks that have simplified trunk and traffic routing save on this administrative expense. Trunk provisioning incurs cost by putting trunks into the network and removing them when the demand shrinks or shifts. In the rearrangeable transport routing mesh network and real-time transport routing ring network, the provisioning of trunks takes place without the

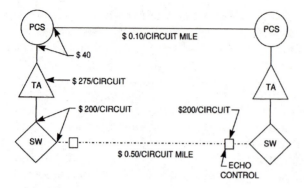

PCS = PACKET CROSS-CONNECT SYSTEM
TA = TERMINAL ADAPTER
SW = TRAFFIC SWITCH

Figure 15.15 Cost model for hybrid circuit-switched and high-speed packet-switched transport network

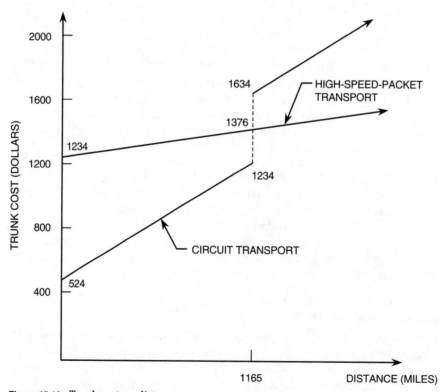

Figure 15.16 Trunk cost vs. distance

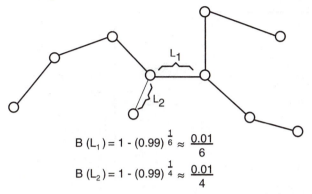

$$B\,(L_1) = 1 - (0.99)^{\frac{1}{6}} \approx \frac{0.01}{6}$$

$$B\,(L_2) = 1 - (0.99)^{\frac{1}{4}} \approx \frac{0.01}{4}$$

Figure 15.17 Real-time transport routing ring network design model

need for manual intervention, because the transport cross-connect and add/drop device can provision trunks automatically when needed. This automation saves the expense associated with manual operation. When capacity additions are required, they are aggregated to the T3 level of capacity on many fewer transport links in the network, as compared with managing a very large number of traffic links. For example, in a 100-switch network, there can be up to 4,950 traffic links, whereas transport links would number only a few hundred. This represents an order of magnitude reduction in the number of links on which capacity has to be provisioned.

15.4 Modeling Results

We now discuss design results that illustrate applications of the various architectures to multiservice and broadband ISDN networks. Using the above design models, we examine the advantages and disadvantages of the various architectures. We study network design efficiency and network management and design impacts. We also examine the robustness of the networks to overloads and failures. We study various-size network and traffic load models reflecting projections of service mix.

For cost assumptions, items in which large changes in technology could provide large changes in costs are bounded by two extremes. For switching cost per DS0 termination, we use at the high end $200 per termination, and with advances in VLSI technology, at the low end $20 per termination. Costs for transport switching, or cross-connecting, in the rearrangeable mesh architecture are based on forecasts for cross-connect products. As discussed in connection with Figure 15.6, the cost for the add/drop termination is less than $1 per DS0 termination, because almost all the cost comes from the control necessary to run the fabric. We chose a range of $5 to $20 because this device should cost no more than the lower bound for the switch termination and probably not significantly less than the least expensive cross-connect termination. We find that the conclusions are

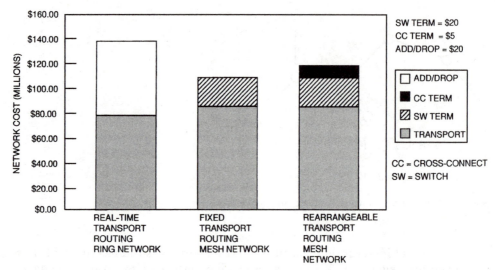

Figure 15.18 Cost comparisons of traffic/transport network architectures (single multihour voice traffic pattern)

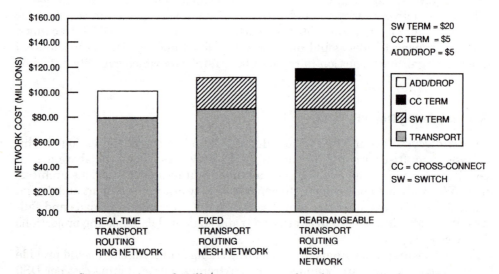

Figure 15.19 Cost comparisons of traffic/transport network architectures (single multihour voice traffic pattern)

not sensitive to small variations in these cost assumptions, if, say, an add/drop termination costs $30 per DS0.

15.4.1 Design comparisons across network architectures

Figures 15.18–15.21 present results for all traffic and transport network architectures with design results that include both voice and data services. These

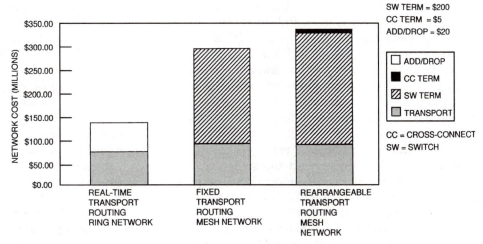

Figure 15.20 Cost comparisons of traffic/transport network architectures (single multihour voice traffic pattern)

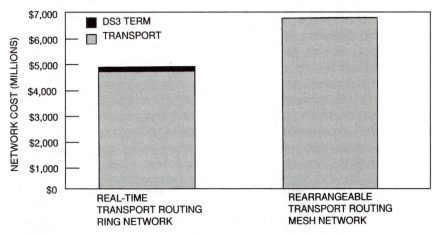

Figure 15.21 Cost comparisons of traffic/transport network architectures (integrated voice and broadband multihour traffic pattern)

designs consider a single multihour traffic pattern and do not include the effects of week-to-week traffic variation and reserve capacity, which are investigated in the next section. The only variation in costs between the figures is in the costs of the switch terminations and the terminations of the add/drop device. We use the term *add/drop* as shorthand for a termination on the real-time transport routing ring network transport switch. It is conceivable that other hardware could be used to run this network. In Figure 15.18, the switch termination and the add/drop termination each cost $20. The real-time transport routing ring network is quite efficient in its use of transport because of the aggregation of traffic loads to the transport links. We benefit here from the nonlinearity of the

Erlang blocking formula, in which large links have higher occupancy than small links for a given level of link blocking. The difficulty with the real-time transport routing ring network design in this instance is the large investment in transport switching/bandwidth allocation. In this network we traverse an average of approximately four via switches on an end-to-end path, and so we have to pay a great deal more for switching than in the other networks, in which we have to pass at most one via switch. Note that express links are not included in the designs presented in this section, which could reduce transport switching cost, as discussed in Section 15.3. The conclusion to draw from these results is that if the switching needed for a real-time transport routing ring network could not be made less expensive than the switching for a mesh network, then the real-time transport routing ring network will be prohibitively expensive.

The fixed mesh network has the lowest cost, but not by as much as one might expect. The extra costs incurred in the building of the rearrangeable transport routing mesh network are almost completely offset by the increased traffic efficiencies. This points to the possible use of new, more sophisticated cross-connects for uses other than restoration. The capability of moving bandwidth among different switch pairs can result in a more efficient network design; equivalently, for the same cost it can result in a more robust design. Efficiencies in network operations may well justify this type of network design, and further results for this architecture are presented in the next section, where the effect of reserve capacity is also taken into account.

Figure 15.19 shows results for the case in which the add/drop device is a factor of 4 less expensive than the switch termination. Efficiencies from using the real-time transport routing ring network in this case are fairly significant. The relative efficiencies of the two different mesh networks are constant for all of the figures—we change none of the costs or conditions for these networks in the different examples. For the real-time transport routing ring network it is significant that we can get higher efficiency for this architecture and also get a network with greater available bandwidth for improved performance to load variations and failures. In the next section this point is quantified. Finally, Figure 15.20 shows the results for a perhaps extreme case in which the real-time transport routing ring network terminations are a factor of 10 less expensive than switch terminations of the current technology. This figure points to a possible significant benefit associated with real-time transport routing ring networks if switching costs drop accordingly.

Figure 15.21 shows the comparison between a fixed mesh network design to carry the load of Figures 15.18–15.20, along with a broadband erlang load equal to 10 percent of the voice erlang load. We assume a cost of $600 for a DS3 termination for the broadband traffic for both the fixed mesh network and the real-time transport routing ring network and assume $0.50 per DS0 mile for all traffic. Switch terminations at the DS0 level are assumed to cost $20, and narrow bandwidth add/drop terminations, $5. In this case, because the cost of transport of such high-bandwidth services starts to become large, the significant savings in transport miles provided by the real-time transport routing ring network are economically significant. Figure 15.21 shows a real-time transport routing ring network that is 25 percent more efficient than the fixed mesh network. This result also points to one of the difficulties of planning for broadband networks. When high-bandwidth services materialize, they could dominate all other traffic carried by the network. At a level of broadband traffic equal to 10 percent of voice traffic, the broadband portion of the network has 70 times the bandwidth requirement of the narrow bandwidth voice and narrowband data traffic. Small uncertainties in the number of broadband calls therefore make for large uncertainties in the design. Again, this points to the need for the most flexible routing possible, so that bandwidth can be reassigned as needed.

15.4.2 Performance comparisons across network architectures

One goal of traffic network design is the lowest-cost design that carries the forecast traffic at a given blocking probability grade-of-service objective. As discussed in Chapter 9, forecasts are subject to error and need to be accounted for in the design procedure, which can include routing table design updates to address changes in load not accounted for in the forecast. Another approach, called the forecast risk analysis model [LyM88], is an example of a stochastic demand model in which forecast errors are accounted for in the network design. A further approach is to design networks that are more stable to variations in traffic loads, which is important when the network carries new services that have no history from which to extrapolate required capacity. When these new services have large bandwidth demands, the ability to transfer bandwidth flexibly between many switch pairs is of even greater value. We can overprovide for some small percentage of voice calls without paying a large penalty, but a networkwide overprovisioning of 45-Mbps trunks could be prohibitive. The other important question to be addressed in network design is the ability to withstand failures of various components of the network. As the transport and switching elements have ever-increasing capacity, this tends to concentrate more load on fewer network elements. Although this may be cost-effective, it leaves the network more vulnerable to failures of the network elements.

Figure 15.22 shows the ability of the traffic and transport architectures to withstand forecast variations. The bars show the number of switch pairs with blocking between the indicated levels. The total number of switch pairs in the 66-switch model network is 2,145. These results are for network designs for voice and data traffic. The loads in the networks are all perturbed by 10 percent about a zero mean. In other words, some switch pairs have demands higher and some lower than the traffic forecast. This figure shows the flexibility with which network bandwidth can be reconfigured to provide bandwidth where it is needed.

The first conclusion is that the real-time transport routing ring network is the most robust of the traffic and transport architectures. Almost all of the switch pairs have blocking performance lower than the design objective of 1 percent blocking. In essence, the traffic perturbation is not felt at all. The reason for this is that the switch-pair loads are perturbed about a zero mean, and so when these loads are aggregated to the links in the real-time transport routing ring network, the law of large numbers takes over. That is,

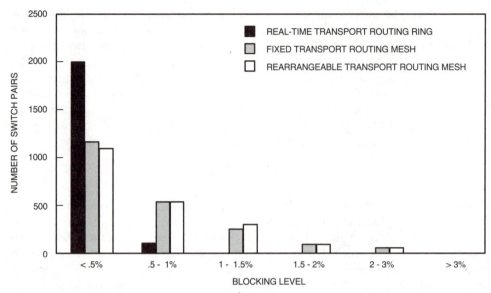

Figure 15.22 Performance comparisons of traffic/transport architectures under traffic load variations

there are many switch pairs routed over a given transport link in the real-time transport routing ring network, which implies that on average there is little or no perturbation in the total load on the link. It may at first seem surprising that a network that allows no alternative traffic routing is so able to accommodate uncertainties in traffic. However, alternative traffic routing is only alternate-routed in the logical network, and we need to consider how the traffic is routed over the physical transport network. The real-time transport routing ring network provides call-by-call reallocation of transport capacity, which has the same effect as dynamic traffic routing and provides the greatest flexibility in bandwidth allocation. That is, the ability to do real-time dynamic

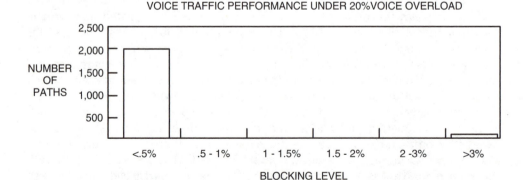

Figure 15.23 Integrated network performance (real-time transport routing ring network design for voice & broadband traffic), voice traffic above design load, broadband traffic below design load

transport bandwidth allocation is important, either through packet-oriented real-time dynamic transport routing ring technology or through the circuit-oriented rearrangeable mesh technology.

We also model the performance benefits of sharing capacity in a broadband network. Here, the broadband network is sized according to the principles of forecast risk analysis, and the virtual network bandwidth allocations are protected with dynamic trunk reservation. We assume a uniform overload of 30 percent on all voice traffic, although broadband traffic arrives at the expected rate. In Figure 15.23 we show the number of switch pairs experiencing blocking of voice calls in the range listed at the bottom of the chart. It is clear that integration of services in a real-time transport routing ring network has advantages for network performance.

15.4.3 Results for rearrangeable transport routing mesh networks

The implementation of a rearrangeable transport routing mesh network allows significant reductions in capital costs and network management and design expense with rearrangeable transport capacity design methods. Automated T1 provisioning and rearrangement lead to annual operations expense savings. Other network management and design impacts, leading to additional reduction in operations expense, are to simplify T1 provisioning systems; automate preservice trunk testing and simplify maintenance systems; integrate trunk forecasting, administration, and bandwidth allocation into T3-level capacity planning and delivery; simplify switch and transport planning; and automate inventory tracking.

15.4.3.1 Design for traffic loads with week-to-week traffic variation. Rearrangeable transport routing mesh network design allows more efficient use of switch capacity and transport capacity and can lead to a reduction of network reserve trunk capacity by about 10 percent, while improving network performance, as illustrated by modeling results presented in the following sections. Table 15.1 illustrates a comparative forecast of a national intercity network's trunking requirements for the base case without dynamic transport routing and the network requirements with rearrangeable transport routing mesh network design. When week-to-week

TABLE 15.1 Dynamic Transport Routing Capital Savings with Week-to-Week Traffic Variations

Design date	Number of trunks, fixed transport routing design	Number of trunks, rearrangeable transport routing mesh network design	Trunk savings (%)
4Q95	1,073,976	937,752	12.7
4Q96	1,125,696	987,120	12.3
4Q97	1,167,792	1,039,248	11.0
4Q98	1,222,128	1,095,024	10.4

traffic variations, which reflect seasonal variations, are taken into account, as in this analysis, the rearrangeable transport routing design can provide a reduction in network reserve capacity, as discussed in Chapters 1 and 9. As shown in the table, the trunk savings always exceed 10 percent, which translates into a significant reduction in capital expenditures.

Rearrangeable transport routing mesh network design for T3 transport capacity achieves 90 percent or more average T1-to-T3 fill rates, which further reduces transport costs. The rearrangeable transport routing network implements automated interswitch and access T1 diversity, T1 restoration, and switch backup restoration to enhance the network survivability over a wide range of network failure conditions. We now illustrate rearrangeable transport routing network performance under design for normal traffic loads, fiber transport failure events, unpredictable traffic load patterns, and peak-day traffic load patterns.

15.4.3.2 Performance for network failures. Simulations are performed for the fixed transport and rearrangeable transport network performance for the January 4, 1991, fiber cut in Newark, New Jersey, in which approximately 70,000 trunks were lost. The results are shown in Table 15.2. Here, a threshold of 50 percent or more switch-pair blocking is used to identify switch pairs that are essentially isolated; hence, the rearrangeable transport network design eliminates all isolations during this network failure event.

An analysis is also performed for the network performance after T3 transport restoration, in which the fixed and rearrangeable transport network designs are simulated after 29 percent of the lost trunks are restored. The results are shown in Table 15.3. Again, the rearrangeable transport network design eliminates all network isolations, some of which still exist in the base network after T3 restoration. From this analysis we conclude that the combination of dynamic traffic routing, T1-IL diversity design, and T3 transport restoration provides synergistic network survivability benefits. Rearrangeable transport network design

TABLE 15.2 Network Blocking Performance for 1/4/91 Fiber Cut in Newark, NJ

	Average network blocking (%)	Number of switch pairs with blocking >50%
fixed transport routing	14.4	963
rearrangeable transport routing	4.2	0

TABLE 15.3 Network Blocking Performance for 1/4/91 Fiber Cut in Newark, NJ (after T3 Restoration)

	Average network blocking (%)	Number of switch pairs with blocking >50%
fixed transport routing	7.0	106
rearrangeable transport routing	0.6	0

TABLE 15.4   Rearrangeable Transport Network Restoration Following Various Span Cuts

Span cut	T1s cut	T1s restored by T3 restoration	Additional T1s restored by T1 restoration
K4011	3,491	2,640	451
N2474	3,733	1,113	642
X0499	4,957	2,409	1,228
N1474	4,577	1,851	1,181

automates and maintains T1-IL diversity, as well as access network T1 diversity in an efficient manner, and provides automatic T1/T3 transport restoration after failure.

A related aspect of dynamic transport routing reliability design is the ability to perform T1, as well as T3, restoration. Some T1 transport channels are idle at times, and in the event of a transport failure they can be used for restoration of active T1 channels that are not otherwise restored by T3 restoration. Some analysis of the level of T1 restoration possible in addition to T3 restoration is made on various transport network fiber span cuts in the network model illustrated in Figure 15.1. Table 15.4 summarizes the results, illustrating that the dynamic transport routing capability to restore both T3 and T1 channels provides additional network robustness and resilience to transport failures.

A final network reliability example is given for dual-homing transport demands on various DCS transport switches. In one example, a DCS3/1 failure at one transport switch location at Littleton, Massachusetts, in the model illustrated in Figure 15.1 is analyzed, and results given in Table 15.5. Because transport demands are diversely routed between switches and dual-homed between access switches and DCS devices, this provides additional network robustness and resilience to traffic switch and transport switch failures. When the network is designed for load balancing between access and interswitch demands, as discussed in Section 15.3.1.3, and T3/T1 restoration is performed, the performance of the rearrangeable transport routing mesh network is further improved.

TABLE 15.5   Rearrangeable Transport Routing Performance under DCS3/1 Failure

	Average network blocking (%)	Number of switch pairs with blocking >50%
fixed transport routing network	4.1	231
rearrangeable transport routing network, dual-homing, diversity design	1.3	0
rearrangeable transport routing network, dual-homing, diversity design (after load balancing and T3/T1 restoration)	0.6	0

TABLE 15.6   Network Blocking Performance for 10% Traffic Overload (129-Switch Network Design for Normal Traffic Loads)

Hour of day	Fixed transport routing network design (% network blocking)	Rearrangeable transport routing network design (% network blocking)
9 to 10 A.M.	0.19	0
1 to 2 P.M.	0.30	0
8 to 9 P.M.	0	0
average	0.11	0
switch-pair maximum	17.3	0

15.4.3.3   Performance for general traffic overloads. A national network model of 129 switches is designed for normal engineered traffic loads with the methodology described in the above sections, and it results in a 15 percent savings in reserve trunk capacity over the fixed transport routing model. In addition to this large savings in network capacity, the network performance under a 10 percent overload results in the performance comparison illustrated in Table 15.6. Hence, rearrangeable transport routing mesh network designs achieve significant capital savings while also achieving superior network performance.

15.4.3.4   Performance for unexpected traffic overloads. Rearrangeable transport routing mesh network design provides load balancing of switch traffic load and T1-IL capacity so that sufficient reserve capacity is provided throughout the network to meet unexpected demands on the network. The advantage of such design is illustrated in Table 15.7, which compares the simulated network blocking for the fixed transport routing network design and rearrangeable transport routing mesh network design during the evening hours of August 19, 1991, when Hurricane Bob caused severe traffic overloads in the northeastern United States. Such unexpected focused overloads are not unusual in a switched network, and the additional robustness provided by rearrangeable transport routing mesh network design to the unexpected traffic overload patterns is clear from these results.

TABLE 15.7   Network Blocking Performance for Unexpected Traffic Overloads (8/19/91, Hurricane Bob)

Hour of day	Fixed transport routing network design (% network blocking)	Rearrangeable transport routing network design (% network blocking)
6 to 7 P.M.	0.01	0
7 to 8 P.M.	1.15	0.85
8 to 9 P.M.	0.44	0.21
average	0.43	0.28
switch-pair maximum	22.7	13.3

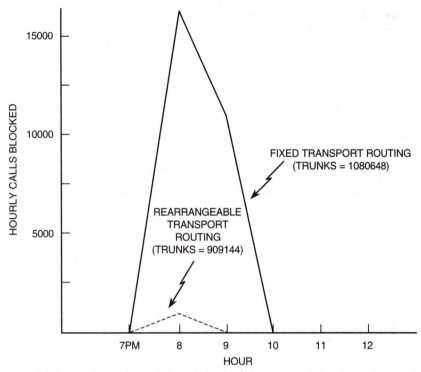

Figure 15.24 Blocking-triggered rearrangement in rearrangeable transport routing network (25% overload on Jackson, Mississippi, switch)

Another illustration of the benefits of load balancing is given in Figure 15.24, in which a 25 percent traffic overload is focused on a switch in Jackson, Mississippi. Because the rearrangeable transport network is load balanced between access demands and interswitch demands, this provides additional network robustness and resilience to unexpected traffic overloads, even though the dynamic transport routing network in this model has more than 15 percent fewer trunks than the fixed transport routing network. In this example, blocking-triggered rearrangement is allowed in the rearrangeable transport network. That is, as soon as switch-pair blocking is detected, additional T1-IL capacity is added to the affected links by cross-connecting spare switch-termination capacity and spare T1-IL capacity, which has been freed up as a result of the more efficient rearrangeable network design. As can be seen from the figure, this greatly improves the network response to the overload.

15.4.3.5 Performance for peak-day traffic loads. A rearrangeable transport network design is performed for the Christmas 1990 traffic loads, and simulations performed for the base network and rearrangeable transport network design for the Christmas Day traffic. Results for the interswitch blocking are summarized in Table 15.8. Clearly, the rearrangeable transport network design eliminates the interswitch network blocking, although the access network blocking may still

TABLE 15.8   Network Blocking Performance Comparison for Christmas 1990

Hour of day	Fixed transport routing network design (% network blocking)	Rearrangeable transport routing network design (% network blocking)
9 to 10 A.M.	17.2	0
10 to 11 A.M.	22.2	0
11 to 12 A.M.	29.7	0

TABLE 15.9   Estimated Peak-Day Revenue Recovered for Rearrangeable Transport Routing Network Designs

Peak day	Blocked calls (millions)	Average blocking (%)	Estimated revenue recovered ($ millions)
Christmas	5.9	6.1	2.6
Mother's Day	1.5	1.5	0.7
Thanksgiving	0.2	0.4	0.1
Father's Day	0.2	0.2	0.1
Total			3.5

exist but is not quantified in the model. Given this interswitch blocking reduction resulting from rearrangeable transport network peak-day design, we estimate annual revenue increases of $3.5 million in recovered lost revenue, as per Table 15.9. In addition to increased revenue, customer perception of network quality is also improved for these peak-day situations.

15.4.4   Results for hybrid circuit-switched and high-speed packet-switched networks

The hybrid circuit-switched and high-speed packet-switched network model illustrated in Figure 15.12 is designed using the path-erlang flow optimization model described in Chapters 6–7. Each PCS is modeled as a separate switch in the network, and therefore the 16-traffic-switch model used in the study, Figure 15.14, becomes a 32-switch model when the 16 corresponding PCSs are added. The cost comparison of the designs for the circuit-switched network and the hybrid circuit-switched and packet-switched network is given in Table 15.10. Notice that although there are slightly fewer trunks in the hybrid design, the hybrid network has more trunk miles and more trunks longer than 1,165 miles. Of the 120 possible links in the circuit-switched network, 68 have echo control. As shown in Table 15.11, only eight trunks in the hybrid network have echo control. There are 4,000 fewer traffic-switch terminations in the hybrid network. Note that an incoming DS0 to a TA can, on one call, support a virtual circuit that connects two switches A and B, and on another call, it can support a virtual circuit that connects switches A and C. In this sense, the "trunks" that provide access to the

TABLE 15.10 Comparison of the Circuit and Hybrid Circuit-Switched and Packet-Switched Networks

	Circuit	Hybrid
total trunks	23,711	23,316
total trunk miles	24,449,136	25,327,678
number of trunks longer than 1,165 miles	6,724	7,732
total trunk miles of trunks longer than 1,165 miles	13,592,469	15,267,274
number of echo-controlled links	68	
number of traffic-switch terminations	47,422	43,031
total cost	$30,069,792	$24,801,139
		(17.5% savings)

TABLE 15.11 A More Detailed Look at the Hybrid Network

	Traffic-switch-to-PCS trunks	Packet trunks	Circuit trunks
total number of trunks	17,943	10,772	12,544
total trunk miles		17,346,704	7,980,974
total cost	$10,357,223	$2,982,686	$11,461,220
total number of trunks longer than 1,165 miles		7,732	8
total trunk miles of trunks longer than 1,165 miles		15,267,274	20,286

high-speed packet-transport network are different from circuit-switched trunks, because they are not dedicated between two endpoints. We expect greater network flexibility as a consequence, which is reflected in the decreased number of traffic-switch terminations, and a hybrid network design cost that is 17.5 percent less expensive than the circuit network design cost.

More insight into the hybrid network design can be gained by comparing circuit-switched and packet-switched transport. In Table 15.11 we see that the .001 blocking probability objective for the traffic-switch-to-PCS links results in almost 18,000 trunks on these links that provide access to 10,772 high-speed packet trunks. Although there are almost 1,800 more circuit trunks than high-speed packet trunks in the network, more than two-thirds of the total trunk miles in the hybrid network are in high-speed packet transport. The cost allocated to the high-speed packet transport is 54 percent of the total network cost, of which the cost of the traffic-switch-to-PCS links accounts for 78 percent. Thus, access to the high-speed packet transport network requires the large majority of capital investment.

Table 15.12 shows the sensitivity of the hybrid network cost to the cost of the TA. We express this sensitivity as the relative difference between the costs of

TABLE 15.12   Sensitivity of Hybrid Network
Cost to TA Cost

TA cost ($/circuit)	% savings with hybrid network
0	41.7
100	30.6
200	22.3
300	16.1
350	13.1
400	9.9
450	6.5
500	4.1
550	1.1

the circuit network and the hybrid network. Thus, each entry in the right-hand column of the table is computed by subtracting from the circuit network cost the cost of the hybrid network designed with the corresponding per-circuit TA cost in the left-hand column, and then dividing that difference by the circuit network cost. As the results show, the hybrid network cost is very sensitive to the cost of the TA, and the cost increases as the per-circuit TA cost increases.

15.5   Conclusions

In this chapter, we present and analyze alternative traffic and transport network architectures, which include bandwidth allocation methods and traffic and routing control plans. These architectures extend dynamic routing concepts to integrated broadband networks and suggest architectures toward which broadband networks might evolve. The architecture alternatives provide to varying degrees the advantages of increased network efficiency, improved performance, and increased network flexibility. We find that networks benefit more in these measures as the ability to reassign transport bandwidth is increased, either through packet-oriented switching technology or circuit-oriented cross-connect technology. When there is capability for real-time dynamic transport bandwidth allocation, a significant decrease in alternate traffic routes does not degrade performance and can simplify network management and design. We present results of a number of analysis, design, and simulation studies related to dynamic transport network architectures including rearrangeable transport routing mesh networks and real-time transport routing ring networks. We also investigate hybrid circuit-switched and high-speed packet-switched network designs.

Rearrangeable transport routing is a routing and bandwidth allocation method, which combines dynamic traffic routing with dynamic transport routing and for which we provide associated network design methods. A call-by-call simulation model is used to measure the performance of the network for rearrangeable transport routing mesh network design in comparison with the base network design, under a variety of network conditions including normal daily load patterns, unpredictable traffic load patterns such as caused by Hurricane Bob, known traffic overload patterns such as occur on Christmas Day, and network failure conditions such as the January 4, 1991, Newark, New Jersey, fiber cut. We find that rearrangeable transport routing mesh network design improves network performance in comparison with the base network for all network conditions simulated. In particular, the ability of the rearrangeable transport routing mesh network to enhance network performance under abnormal and unpredictable traffic load patterns results from the improved robustness of the network design, which is achieved while reducing network costs. The ability of the rearrangeable transport routing mesh network design to enhance network performance under failure arises from automatic interswitch T1 and access T1 diversity in combination with the networkwide path selection and immediate adaptation to failure available with dynamic traffic routing combined with T1/T3 transport restoration. We show that higher network throughput and enhanced revenue should accrue from deployment of a rearrangeable transport routing mesh network, and at the same time capital savings should result. A rearrangeable transport routing mesh network architecture provides dynamic traffic and transport routing that meets customer-oriented goals of service flexibility and performance quality.

Results presented for meshed network designs of hybrid circuit-switched and high-speed packet-switched networks show that design cost savings can be achieved with such hybrid networks in comparison with all-circuit-switched networks. This, in part, is achieved by the lower costs of the packet-switched network technology, and also its ability to provide savings through advanced signal-processing capabilities such as voice compression. Under certain assumptions, such hybrid network design savings exceed 17 percent, along with a large reduction in traffic-switch terminations in the hybrid network design. Therefore, this high-speed packet-switching technology and the possible network efficiency and performance improvements it gains suggest that the technology may be attractive for future network evolution.

This conclusion concerning the value of high-speed packet-switching technology is further supported by the results presented for real-time transport routing ring networks. Real-time transport routing ring networks yield the greatest advantages and provide simplified network management and design along with maximum flexibility to apportion network resources, especially when implemented with high-speed packet technology, such as ATM technology introduced with broadband ISDN. In particular, we find that a real-time transport routing ring network (1) is the most robust of the traffic and transport architectures studied, (2) provides the greatest flexibility in bandwidth allocation, and (3) is the simplest network to design and manage. If the needed switching technologies become available at low enough costs, such a traffic architecture may provide a possible direction for integrated broadband networks.

Dynamic Routing Feasibility, Economics, and Implementation

16.1   Introduction

In Chapters 1 to 15, we discussed a wide range of dynamic traffic routing and dynamic transport routing alternatives and presented detailed analysis of many of these methods through a large variety of network models. We described possible future network architectures incorporating dynamic routing and high-lighted areas of potential network efficiency and performance gains. We have also presented network management and design methods associated with each routing strategy to possibly realize those potential benefits. However, we have not yet addressed in detail the overall technical feasibility of dynamic routing with regard to network impact, operational impact, business justification, and implementation requirements to actually put dynamic routing into operation in a network. In this chapter we address those technical feasibility issues, impact areas, and implementation aspects of dynamic routing networks.

First we briefly review what is required to implement a dynamic routing network. In Chapter 1 we outlined the fundamental impacts that a dynamic routing network has on (1) network configuration and architecture, (2) routing table design, and (3) network management and design. Let us briefly examine each of these impacts. To implement a national dynamic routing network, for example, a typical multiple-level fixed hierarchical routing network configuration would normally evolve to a two-level configuration. In the two-level structure, end-offices would continue to home on intercity switches just as they did in the hierarchical network. The primary technologies needed for dynamic routing implementation are ESS stored program control, common-channel signaling, and automated network management and design systems implementing dynamic network support functions. These capabilities are necessary to enable the migration of network configuration to a nonhierarchical structure and to enable the network architecture functional implementation to migrate to dynamic routing. Many viable dynamic routing methods are discussed in the book, and implemented examples are summarized in Chapter 1. The routing implementation complexity and network performance trade-off are important considerations in selecting a dynamic routing method, and each method has different impacts. For example, real-time event-dependent routing methods such as LRR are less complex to implement than most other methods, but other more costly methods, such as RTNR, have much better network performance. Moreover, within a given type of dynamic routing, such as preplanned dynamic routing, DNHR may be selected over CGH routing because DNHR is simpler to implement given that sequential path selection requires no randomization, and the difference in performance is insignificant.

In regard to network management and design implementation, dynamic routing networks have fundamental impact on real-time traffic management, capacity management and network design, and network planning. Real-time traffic management requires collection of new network surveillance data and implementation of new controls oriented to nonhierarchical dynamic networks. Capacity management and network design, which includes routing design and capacity design, can require large-scale optimization of network routing and capacity, such as for preplanned dynamic routing design with a path-EFO model, as discussed in Chapters 6 and 7. For real-time routing networks, explicit routing design might be avoided by using, for example, the transport flow optimization model discussed in Chapter 13. Large-scale optimization of network capacity may still be required, however. Routing design techniques discussed in Chapters 8 and 9 provide routing table updates for dynamic routing networks in order to access existing network capacity and to provide a minimum-cost augmentation to meet the forecasted and current network demands. Such methods tend to reduce network reserve capacity and thus further increase efficiency. Such network design techniques allow routing table updates to replace trunk augmentation whenever possible. Because dynamic routing implementation changes the basic network configuration and routing, these attributes need to be reflected in network planning models supporting switch planning, transport planning, and other planning functions.

Network management and design operations in a dynamic routing environment involve using the new design models incorporated into modified or new systems for performing forecasting, capacity management, and design, as described in detail in Chapter 17. Work centers that use these modified systems need to have people trained and knowledgeable on dynamic routing impacts on work center functions, which are also described in Chapter 17. In some cases, work center functions such as real-time traffic management and capacity management are more naturally centralized in a dynamic routing network, which may have a nonhierarchical structure and better operate as a single network rather than an assembly of regional networks, as in a fixed hierarchical network structure. Personnel in the centralized real-time traffic management work center, for example, manage the entire dynamic routing network rather than a single region, and, similarly, personnel in the centralized capacity management work center plan the entire dynamic routing network rather than a single region. Additional switch-to-switch traffic data are fundamental to performing dynamic routing real-time traffic management and capacity management. Such switch-to-switch data can be collected in the switching system as part of the software upgrade to implement dynamic routing. Data collection systems must be configured to collect such data from the switching elements and pass them on to the systems that support the work centers in performing their functions. Routing administration in the dynamic routing network requires a system to take the output of a routing table design, for example, and perform the routing table updates in the switch through the switch "recent change" interface to make routing table changes essentially automatic.

Many issues are addressed in this chapter relating to the feasibility and economic viability of dynamic routing network implementation, which include dynamic routing network impacts, operational feasibility, business justification, and implementation requirements. Study results reported in the chapter show that dynamic routing implementation is both technically feasible and economically justifiable. In Sections 16.2.1 to 16.2.3, we address dynamic routing network impacts with regard to network overload and failure performance, transmission performance, and processing load on the ESS and CCS network elements. In Sections 16.2.4 to 16.2.6, we address dynamic routing operational feasibility with regard to traffic data collection needs and capacity management and network design needs, including large-scale network optimization feasibility. Network management and design implementation for dynamic routing networks is addressed in detail in Chapter 17. In Section 16.3, we address the dynamic routing business justification with regard to dynamic routing economic benefits, switch and CCS network development requirements and costs, network management and design systems development requirements and costs, and transition and other costs and give an overall economic analysis of these costs and benefits. The preplanned sequential two-link routing method, DNHR, is used as an example for several of the feasibility and economic studies discussed in Sections 16.2 and 16.3. In Section 16.3.5, we give examples of economic benefits derived in a real-time state-dependent routing network, with RTNR class-of-service and multiple ingress/egress routing as an example implementation. In Section 16.4, we address dynamic routing implementation with regard to standardized interworking requirements and information exchange needs between dynamic routing switches and other network elements. Dynamic routing internetworking is especially important in a multivendor network implementation. In Section 16.5, we discuss common-channel signaling messaging requirements and give examples of call setup with these messages for preplanned dynamic routing (DNHR), distributed real-time state-dependent routing (RTNR), centralized real-time state-dependent routing (TSMR), and real-time event-dependent routing (LRR).

This chapter summarizes the results of the technical feasibility and impact studies discussed above and outlines implementation and transition plans required to implement a dynamic routing network. The overall conclusion is that dynamic routing is both economically and technically feasible to implement.

16.2 Feasibility Studies

In this section we address the impacts of dynamic routing with regard to network overload and failure performance, transmission performance, and processing load on the ESS and CCS network elements. Then we address dynamic routing operational feasibility with regard to traffic data collection needs and capacity management and network design needs, including large-scale network optimization feasibility.

16.2.1 Network overload and failure performance

A network designed for greater efficiency might be expected to be more susceptible to network overload and failure, because less network capacity possibly means fewer places to route calls. Evaluation of dynamic network designs under overload and failure is made using a simulation model that includes real-time automatic traffic management controls, as well as dynamic traffic routing capability. The automatic traffic management controls available in this simulator are selective dynamic overload control (SDOC) and selective trunk reservation (STR) [Mum76]. SDOC helps prevent switch congestion by limiting attempts coming to a switch from other connecting switches. Maximum control is imposed on hard-to-reach traffic that has a low probability of completion; hence, the control is selective. STR helps prevent trunk congestion by using trunk reservation techniques to give hard-to-reach traffic limited access to available trunks. Another control prohibits an excessive number of calls from queuing for a limited number of call registers in the ESS call processing. Such controls behave similarly to dynamic overload control in turning calls away from congested switches. The simulation model includes retrials for calls that fail to complete. The study compared fixed hierarchical routing and preplanned dynamic traffic routing (DNHR) in a 30-switch intercity network model (shown in Figure 6.12), and we summarize the results here. The difference in network cost between the hierarchical and dynamic networks exceeded 14 percent. General overloads, focused overloads, transport failures, and switch failures are studied.

The results for general overloads are shown in Figure 16.1. Each switch-to-switch load was increased above its engineered value as indicated on the horizontal axis. The vertical axis measures the average number of calls in progress and is a measure of how efficiently the network is completing calls. With ESS traffic management controls applied, as shown in the figure, both networks improve dramatically and have comparable performance, even at up to 250 percent of design traffic load. The implication is that switch congestion is the dominant effect; the large difference in the number of trunks in the DNHR and hierarchical networks has no visible effect on the results. The results for the situation without controls are encouraging because, as also shown in Figure 16.1,

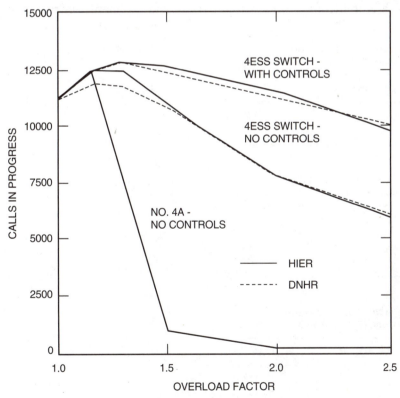

Figure 16.1 Network overload performance

the dynamic routing network does not degrade under very large general overloads nearly as badly as did the network of electromechanical No. 4A toll switches, where switch congestion in the absence of controls had a catastrophic effect on network performance. This behavior results from the automatic controls internal to the switch call-processing software, which were modeled in the simulation model.

What is concluded from these results is that the 14 percent difference in network cost and 9 percent difference in number of trunks has little effect on network performance. The results for focused overload are shown in Table 16.1, where all traffic destined for Oakland was increased by a factor of 8.

Again, with traffic management controls applied, the networks behave comparably well, with fixed hierarchical routing doing slightly better (some overcontrol in the dynamic routing network may account for this small difference). The results for transport failure are shown in Table 16.2, where all trunks terminating at Dallas, New Orleans, Birmingham, Denver, Salt Lake City, and Reno are out of service, which accounted for about 3 percent of the total trunks being unavailable.

The fact that the dynamic routing network carries more calls is indicative of its ability to route traffic around such failures. When traffic management controls are applied, both networks perform worse—this indicates overcontrol for this network condition. The switch failure in Table 16.3 exhibits similar properties to the transport failure, with the dynamic network again performing slightly better than the hierarchical network without traffic management controls, and

TABLE 16.1 Average Number of Calls in Progress for 8× Focused Overload on Oakland

	DNHR	HIER
no controls	1,533	3,967
controls	10,511	10,807

TABLE 16.2 Average Number of Calls in Progress for Transport Failure

	DNHR	HIER
no controls	10,747	10,494
controls	10,379	10,263

TABLE 16.3 Average Number of Calls in Progress for Southbend (SBND) Switch Failure

	DNHR	HIER
no controls	9,978	9,911
controls	9,907	9,998

slightly worse with traffic management controls. These results show that dynamic routing networks are robust under overloads and failures with automatic traffic management controls active, and they exhibit performance comparable to that of hierarchical routing networks either with or without traffic management controls applied.

16.2.2 Transmission performance

Network design techniques for dynamic networks achieve their greatest utilization efficiency and least network cost by allowing routing to be left unconstrained with regard to path lengths. Transmission constraints must be imposed, however, to force the network design to allow only those paths that satisfy transmission grade-of-service objectives. Also, the interaction of dynamic network design with transmission technologies such as compressed voice and satellite transmission must be considered.

A study is conducted using the 28-switch intercity model, in which the transmission performance of the fixed hierarchical routing design and the dynamic network design are compared. The results for analog transmission networks are displayed in Figure 16.2, in which the loss-noise-echo grade-of-service is displayed

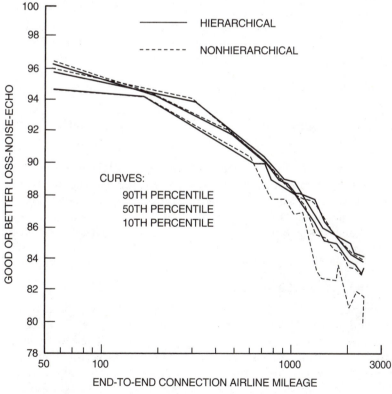

Figure 16.2 Comparison of transmission service quality for analog transmission networks

versus the direct (airline) distance of the connection. The grade-of-service is determined from customer opinion models and can be interpreted as the relative transmission quality of a connection. For a given connection distance, there is a distribution over the various traffic loads in that distance range of the percentage of customers perceiving connection quality as being good or better. This distribution is dependent on variations in traffic routing. The 10th, 50th, and 90th percentiles of this distribution are displayed.

It can be seen that for distances exceeding 1,000 miles, dynamic routing design suffers a 1 to 2 percent degradation in comparison with the hierarchical network design. This result stems from the greater likelihood of routing over two-link connections in the dynamic network. Below 1,000 miles, the transmission grade-of-service performances are entirely comparable. The results also show that the worst possible connection in a dynamic network was almost always better than the worst possible connection in a hierarchical network. This difference arises from the maximum of two links in the dynamic network as opposed to seven in the hierarchical network and leads to less contrast in possible connection qualities for dynamic network design. A study of metropolitan dynamic routing transmission performance used data from the 42-switch model of Chicago. Distributions of the ratios of path to direct air mileage are the same for the dynamic network design and the hierarchical design, and therefore no transmission problems arise.

The conclusion reached in this analysis is that transmission performance is substantially preserved in the dynamic network environment. However, to ensure this performance in an analog transmission network environment, the following constraints are imposed on the national intercity network:

- The total length of a two-link connection does not exceed twice the length of the direct path.

- Satellites and speech compression devices do not appear in tandem in any combination.

As networks with digital transmission are deployed, the differences in transmission performance of dynamic networks and hierarchical networks disappear, and the above transmission constraints can be relaxed. In fact, dynamic routing transmission performance is improved in an all-digital network when paths in the dynamic routing network are limited to two links versus, perhaps, seven in the hierarchical network.

16.2.3   Processing load on ESS/CCS network

The dynamic routing call processing consumes real-time resources both in the switching system and in the CCS network, and the acceptability and cost of the additional load requires assessment. Simulation models are used to measure impacts related to the switching and signaling load in both a hierarchical network and a dynamic routing network. These impacts include originating and terminating attempts, tandem load, and crankback traffic, which are then translated into real-time consumption and cost.

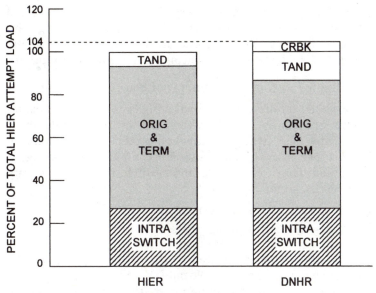

Figure 16.3 Network sum of peak-day switch loads

The 30-switch model is used for the simulation study. This model is designed for the average business day (ABD) load for both dynamic routing and hierarchical routing. Ten high Monday loads are generated using the high Monday average load distribution, which is shown in Figure 8.15. The individual switch-to-switch loads are further perturbed using a gamma distribution with a variance corresponding to very high day-to-day traffic load variations [Wil71]. We note that the average peak Monday traffic exceeds the ABD traffic by about 5–10 percent. We note also that the average volatility of the switch-to-switch loads, defined as the ratio of the peak-day switch load to the average of the 10 high-day loads, exceeds 1.08. If the switch capacity is designed for the high-day load, the comparison of interest is of the sum of the high-day loads in the dynamic routing network with the same sum in the hierarchical network. These sums should reflect the increased switch loads due to dynamic routing. The results are compared in Figure 16.3, and we find that the average dynamic network load is higher by about 4 percent as a result of increased tandem attempts and crankback attempts. The average volatility in the simulations proved to be slightly above 1.08 for the dynamic routing network. One additional result from these simulations is that the dynamic network completed an average of 0.4% more calls than the hierarchical network, because of greater routing flexibility. This effect equates into considerable additional revenue.

Translating the increased switch load into cost reflects the potential cost of earlier exhaust of dynamic routing switches versus hierarchical switches under this increased load. (Switch exhaust occurs when the switch load reaches its maximum capacity.) To do this, we compare the switch loads in the 140-switch network model. Through the use of hierarchical and dynamic routing designs for this model, the peak load on each switch is determined over the study period 1988–1997. The total load is broken down into its originating, terminating, tandem, and crankback load components. The exhaust date for both hierarchical and dynamic routing switches is then determined based on the following assumptions:

- The switch is attempt-limited at 550,000 attempts per hour.

- Intraswitch traffic completed totally within a switching center area (for example, traffic between two end-offices homed on the same switch) equals about 22 percent of the total switch load. In addition, about 5 percent of the switch load is destined to switches other than the large-switch dynamic routing switches and is hierarchically routed. The busy hours for intraswitch traffic and intercity traffic are assumed to be the same.

- The high-day load is 13 percent above the average business day load.

- The average call-holding time is 3.5 minutes.

The profiles of the cumulative number of switches exhausting, by year, over the 10-year study period are shown in Figure 16.4. We note that over the study period the cumulative number of exhausts for the dynamic routing network is consistently above the number for the hierarchical network, but not all switches exhaust

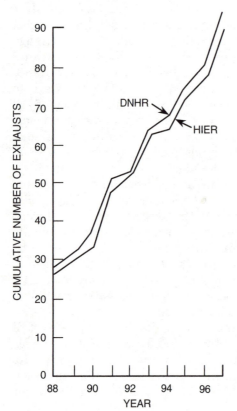

Figure 16.4 Cumulative 4ESS exhaust for DNHR and hierarchical network

earlier in the dynamic network. In particular, large switches, notably regional centers in the hierarchical network, are deloaded by the use of dynamic routing. The result of this analysis is that the model predicts a net cost of $19 million for dynamic routing design in terms of earlier switch exhausts.

A study of increased CCS load is based also on the 140-switch network results discussed above. Dynamic routing and hierarchical network designs are compared, and the increase in CCS signal transfer point (STP) loads is estimated. The results show that the 10-year cost of the load increase is negligible.

16.2.4 Traffic data collection

Adequate switch-to-switch traffic data are essential for supporting the capacity management, including capacity design and routing design, of the dynamic routing network. *Switch-to-switch* in the metropolitan network signifies end-office-originating to end-office-terminating traffic; *switch-to-switch* in the intercity network signifies intercity-switch-originating to intercity-switch-terminating traffic. As illustrated in Table 16.4, traffic data needs for capacity management include weekly availability of hourly switch-to-switch usage, attempt count, and blocked call count to provide switch-to-switch blocking and switch-to-switch carried traffic load measurements.

Studies conducted for both intercity networks and metropolitan networks related to data collection show that message-billing data satisfy the need for switch-to-switch load forecasting. The data requirements for intercity dynamic network capacity management can be met by collecting the data in the switch itself. Performance monitoring, short-term capacity management, and routing table design do not require the detailed NPA-NXX data needed in forecasting; hence the measurements are aggregated for each destination switch. The 10-year cost of collecting the intercity data is estimated to be $4 million, taking these factors into consideration:

- Real-time consumption in the switch processor

- Increased memory requirements

TABLE 16.4 Switch-to-Switch Traffic Data Requirements for Dynamic Routing Network

Data item	Needed for	Frequency of collection
switch-to-switch* usage (NNX to NNX)	forecasting	four (seasonal) 20-day studies each year
switch-to-switch usage	performance monitoring; short-term capacity management	daily
switch-to-switch attempt count	performance monitoring; short-term capacity management	daily
switch-to-switch blocked call count	performance monitoring; short-term capacity management	daily

*Intercity switch to intercity switch for national intercity networks; end-office to end-office for metropolitan area networks

- Traffic data collection, processing, and transmission
- Operations and administration

The traffic data collection alternatives for meeting the above needs are as follows:

- Billing data in conjunction with a routing database to determine the loads carried in an intercity dynamic routing network (not applicable to metropolitan networks that have no billing data). The ability to use billing data to meet the daily traffic data requirement places additional requirements on some billing systems, and billing data often do not provide switch-to-switch blocking measurements.

- Link-based measurements (usage, peg count, and overflow) plus a form of load disassembly to determine an "equivalent" set of switch-to-switch loads. Although the load disassembly procedure can be implemented for fixed routing, where the traffic overflow patterns are well structured, studies indicate that it becomes difficult in a nonhierarchical dynamic network that is not fully connected, has no "first routes," and has routing patterns that vary with time or state of the network. In addition, link-based measurements do not provide a direct measurement of switch-to-switch blocking.

- Direct measurement of the traffic by implementation of a switch-to-switch data collection system and provision of necessary enhancements to satisfy dynamic routing needs. This appears to be the only practical alternative for satisfying all the traffic data collection requirements stated above. Such a system serves both the intercity and the metropolitan dynamic network design applications. One approach involves measuring switch-to-switch peg count, blocked call count, and usage measurements in the switch. Another alternative uses billing data for estimating call-holding time and avoids online switch-to-switch usage measurements, which may be costly to obtain in some switching systems. In the intercity network, the holding time measurement would continue to be provided by billing data.

The latter alternative has functioned well in the actual implementation of dynamic routing traffic data collection discussed in Chapter 17 and, hence, is feasible for dynamic routing networks.

16.2.5 Capacity management

Preplanned dynamic routing capacity management envisions the following:

- The ongoing collection and processing of switch-to-switch traffic load and blocking data to identify network service problems in which the objective switch-to-switch blocking is exceeded.

- Frequent routing table redesign to reoptimize the network routing patterns and link capacities in response to service problems.

- Updating switch routing tables and implementing network link-size changes.

In many ways, dynamic routing capacity management in a dynamic routing environment is similar to control in a fixed routing environment, although because of the volume of changes for preplanned routing table updates there is greater dependence on mechanization to make the decisions necessary to control the network routing tables. Therefore, dynamic routing network design requires (1) more frequent routing table changes, which need very fast implementation and necessitate a highly mechanized process, and (2) that network management and design work centers be able to track routing table status automatically. Both of the above capabilities have been successfully demonstrated for dynamic routing networks, as discussed in Chapter 17.

16.2.6 Network design and large-scale optimization

Design of a dynamic routing network requires that all network switches be considered simultaneously in the design, and a significant question is whether such designs are feasible with presently available design models and reasonable computational resources. Advances in capacity management design and large-scale optimization techniques have made possible greatly reduced run times and storage requirements necessary for designing large-scale dynamic routing networks. Examples of these advances are the following:

- A path-erlang flow optimization model, described in detail in Chapters 6–7, affords great reductions in computer storage requirements and significant improvements in the run time of the routing design step in comparison with the route-erlang flow optimization model discussed in Chapters 4–6.

- Heuristic procedures developed to solve the routing design and optimization step run hundreds of times faster than the most efficient commercial linear programming package tested, MPSX-370, at the expense of a small decrease in optimization accuracy.

- Improved computational procedures to perform necessary traffic calculations—such as for the Erlang-B model, which calculates link blocking given the offered load and link size—improved by a factor of 20 the speed achieved with earlier procedures.

These advances apply also to the network design models presented in Chapter 13 for real-time dynamic routing networks. For example, the transport flow optimization (TFO) model requires no routing design, which represents a further computational simplification, and the heuristic procedure for the linear programming optimization of the capacity design also applies to the TFO model. A 190-switch model of the intercity network is used to test the path-erlang flow optimization model for a design of six daily load set periods. This model represents most major metropolitan areas within the continental United States, and is designed on a mainframe computer system in three hours of CPU time. We conclude from these results that the design of dynamic routing networks is feasible for full-scale network applications, with available design models and computer technology. As computer systems continue to evolve with faster processors and larger memory capacity, these results should continue to improve.

16.3 Implementation Requirements and Economic Analysis

In this section we quantify the costs and benefits of dynamic routing implementation in terms of (1) development costs of switch and network management and design capabilities; (2) capital savings in switching and transport costs with improved network design; (3) expense savings with automated and centralized network management and design, shared across all classes-of-service and with dynamic routing to mitigate trunk-churning activity; (4) improved call completion and service protection under network stress such as overload or failures and more uniform service with blocking levels designed on a switch-to-switch basis; (5) improved transmission grade-of-service with a maximum of two links versus more links in fixed hierarchical networks; and (6) additional revenue with improved call completion and new service opportunities.

16.3.1 ESS/CCS network development requirements

Electronic switching systems require software changes to implement dynamic routing capabilities in the network. CCS crankback capabilities, time-sensitive allocation of traffic to different routes, and real-time dynamic load allocation techniques are examples of features that might be implemented in the switch to provide dynamic routing capability.

Here we give an example of switch requirements for the implementation of DNHR, and generally these capabilities will have needs parallel to those for other routing systems [CDK83]. The call-processing software architecture of the switch is required to deal with both the existing hierarchical routing logic and data structures and the new DNHR logic and data structures. The identification of the originating switch (OS), via switch (VS), and terminating switch (TS), is accomplished through the use of a traveling class mark accompanying the CCS initial address message, as discussed in Section 16.4. Another aspect of the enhanced call-processing capability that affects the software logic is the concept of routing a call from the OS to the TS rather than on a link-by-link basis, which is the main thrust of the implementation of hierarchical (progressive) routing. To achieve the OS-to-TS routing requires that the OS maintain control of the call until assured of the acceptance of the call by the TS. A crankback signal is used by the VS to inform the OS that the call cannot be forwarded to the TS and that, therefore, the OS should try another path to deliver the call to the TS. The traveling class mark and crankback signals both use parameters in CCS messages, as discussed in Section 16.4.

In order to provide time-varying routing, both the existing and the new data structures are necessary. In order to not seriously affect the existing call-processing and routing logic, the existing routing data structures are maintained as much as possible. One approach to this is to build bridge data structures between the existing and new data structures, which achieve the time-varying aspects of the routing and can be returned to the existing routing structures to continue normal call processing. The administration of dynamic routing requires changes to the traffic data collection, network management, and routing administration functions, which are discussed later in this section. Interworking of dynamic routing and hierarchical routing requires a transition capability. This is because introduction of dynamic routing features must be done while the existing hierarchical routing continues to operate, and so that dynamically routed calls can be initiated in a transparent manner, with the ability to back out if necessary, without interrupting the ability to continue to process calls.

The affected call-processing functions include digit analysis, routing translation, route selection, and trunk hunting. Digit analysis determines that a valid and sufficient digit pattern has been received so that routing translations can take place. Digit analysis yields a routing classification and a pointer to the appropriate routing table. In the case of dynamic routing, digit analysis also yields the TS for the call. If the digit analysis determines that the routing classification is hierarchical, the pointer points to a hierarchical routing table. If the digit analysis determines that the routing classification is dynamic routing, the pointer points to a dynamic routing table. Therefore, dynamic routing requires a new routing classification and pointer to the new dynamic routing table structures. It is also necessary to be able to link the hierarchical routing table to a dynamic routing table, because a call can first use hierarchical routing and overflow to dynamic routing, as illustrated in Figure 1.10 and now further explained.

Merging dynamic routing switches and hierarchical routing switches creates the need to separate the load that is dynamically routed from the load that is hierarchically routed. For national intercity networks, metropolitan area networks, and global international networks, dynamic routing switches usually must be merged with nondynamic routing switches because not all switch types support the dynamic traffic routing software. As illustrated in Figure 1.10, the dynamic network handles traffic load that originates and terminates in the dynamic routing network but also must route various overflow and through-switched loads from the subtending hierarchical network. These overflow and through-switched loads are also regarded as switch-to-switch loads within the dynamic routing network. For example, in Figure 1.10, overflow from the traffic routed from E2 to E3 is an example of such an overflow load in the dynamic routing network. Another example is the through-switched load from E01 to E05. This load enters the dynamic routing network at switch T1 and terminates at switch T3. A dynamic routing originating switch must therefore be able to identify a call first entering the dynamic routing network and also determine its destination TS within the network. Once a call is identified from the routing table translations as being in the dynamic network, the dynamic routing software is employed to complete the call between the OS and TS.

The metropolitan area dynamic network has analogous but somewhat different traffic routing requirements. In Figure 1.6, switches E_2, E_3, E_4, E_5, T_1, and T_2 are in the dynamic routing network, but E_1 and E_6 are not. Calls from E_2 to E_1 are hierarchically routed (directly and through T_1), whereas calls from E_2 to E_3, E_4, and E_5 are dynamically routed. Link E_2–T_1 therefore carries both dynamically routed and nondynamically routed traffic.

Routing translation retrieves the applicable routing table for the received digits. Route and link selection match the needs of the call to the link attributes, such as

CCS signaling, in the routing table. Routing table data must distinguish among hierarchical paths, one-link dynamic routing paths, two-link dynamic routing paths, and two-link dynamic routing paths with trunk reservation (real-time paths). Routing table data must also distinguish the time-varying routing data as used in the OS function for dynamic routing and the time-fixed data used in the VS function for dynamic routing. Time-variable aspects include the time of day and the day of the week. Furthermore, each destination TS is represented by two routing tables: both an active and an inactive routing table. The appropriate active or inactive routing table is chosen via a pointer, which can be updated along with other dynamic routing table data. This capability allows routing table data to be updated and then switched simultaneously from the old data to the new data. The time of day and day of the week are used to determine the load set period (LSP) from an LSP data structure. The LSP and active or inactive routing table pointer are used to retrieve the corresponding path list from the dynamic routing table data.

Trunk hunting provides the mechanism to select an outgoing trunk from a link and forward the call toward the destination. If there is no idle trunk in the selected link, the search continues with the next link in the routing table. If the last link in the routing table is searched and no idle trunk is found, trunk hunting terminates and blocks the call. Dynamic routing trunk hunting must also reflect the restrictions of the different paths types: one-link, two-link, or real-time.

Routing data administration requires an automated data interface for a mechanized routing administration system to provide automatic updates of routing table data. Adequate switch-to-switch traffic data are essential for supporting the capacity management and real-time traffic management of the dynamic routing network. The data needs for dynamic routing are identified in Section 16.2.4 and include OS-to-TS attempt counts, usage, and blocked call counts. Switch software modifications are required to gather, consolidate, and output these data together with the existing traffic data to an automated data collection system, which in turn delivers the data to the dynamic routing capacity management system.

Modifications are needed in the real-time traffic management functions in the switch software to monitor the dynamic routing network and to control traffic on an OS-to-TS basis, rather than on a per-link basis. For network monitoring, the new OS–TS attempt counts, usage, and blocked call counts are provided to the automated real-time traffic management system on a five-minute basis. For network control, the switch software provides enhanced, OS–TS-oriented implementation of selective dynamic overload control (SDOC) and selective trunk reservation (STR). SDOC helps prevent switch congestion by limiting attempts coming to a switch from other connecting switches, and it exerts maximum control on hard-to-reach (HTR) traffic that has a low probability of completion; hence the control is selective. STR helps prevent trunk congestion by using trunk reservation techniques to give HTR traffic limited access to available trunks. Dynamic routing network real-time traffic management requires that SDOC and STR be fully automatic controls rather than maintaining the manual administration and activation of the hierarchically oriented implementation of SDOC and STR [Mum76]. For example, with STR two trunk-reservation thresholds are determined by the switch software based on link size rather than being manually inserted. SDOC and STR are OS–TS oriented for dynamic routing. For example, with SDOC a call offered to an overloaded VS is either canceled at the OS or advanced to another VS for a given TS destination for the call. STR differentiates between one-link and two-link traffic rather than between first-routed and alternate-routed, as in the hierarchical network. SDOC and STR use HTR control selectivity differently than in the hierarchical network, where HTR codes are detected on a per-link basis. With dynamic routing, HTR codes are detected by a TS and communicated by it to all the other switches in the network. An OS then treats the TS-detected HTR code as hard-to-reach on all dynamic routing links. SDOC and STR are automatically enabled for the dynamic routing network and automatically determine whether a controlled call is canceled or skipped to another path.

Dynamic routing transition capabilities must enable the creation and growth of the dynamic routing network from the existing hierarchical network, without service interruption. Hence, routing in the remnant hierarchical network must be retained, and many dynamic routing switches must simultaneously change from hierarchical routing to dynamic routing. Two cases of transition are conversion from an exclusively hierarchical network to include both hierarchical and dynamic routing and the growth of the dynamic routing network by the addition of new switches. The active-or-inactive routing table capability is used to create an active set of routing tables that continue to perform hierarchical routing and the inactive set that performs dynamic routing. After population of both sets of routing tables, digit translations are pointed to the active tables, which continue to perform hierarchical routing. The active-or-inactive pointer is set to change at the specific time and date of conversion to dynamic routing. This time of interchange is set the same at each switch, so that a synchronized creation of the initial dynamic routing network occurs. After the change, the previously active dynamic routing tables remain unchanged, which makes possible a back-out procedure to the old (hierarchical) routing tables.

16.3.2 Network management and design system development requirements

As illustrated in Figure 1.33, real-time traffic management provides monitoring of network performance through collection and display of real-time traffic and performance data and allows traffic management controls, such as code blocks, call gapping, and reroute controls, to be used by network managers when circumstances warrant. Monitoring of network performance is illustrated in Figures 1.34 and 1.35. Figure 1.34 illustrates network managers performing these functions in the network operations center located in Bedminster, New Jersey. Figure 1.35 illustrates real-time congestion performance data for the January 17, 1994, Los Angeles earthquake, which is displayed on video wallboards and on network managers' terminals. A number of real-time traffic management controls are used to control network congestion. Directionalization of a link by using trunk reservation can favor calls originating from a disaster area, for example. Cancellation of alternate routing removes alternate-routed traffic from a link and thereby reduces the load on the distant switching systems, as well as the average number of links per call. Reroute controls route overflow traffic to a link that is not in the normal routing pattern. Code-blocking controls block calls to a particular destination code. Call-gapping controls, as illustrated in Figure 1.36, allow one call for a controlled code or set of codes to be accepted into the network by each switch once every x seconds, and calls arriving after the accepted call are rejected for the next x seconds. In this way, call gapping throttles the calls and prevents the overload of the network to a particular focal point. As discussed in the previous section, SDOC and STR are automatic controls that sense congestion and take control action. SDOC senses congestion in a switch, for example, by measuring queue length of calls waiting for senders, and it sends a signal to connecting switches that skip or cancel some portion of traffic to the congested switch. SDOC uses completion statistics by code, and traffic control is performed on the hard-to-reach codes. STR helps prevent trunk congestion by using trunk reservation techniques to give hard-to-reach traffic limited access to available trunks.

Real-time traffic management systems and real-time traffic management practices require changes in order to accommodate dynamic routing and hierarchical routing in the same network. For instance, expansive controls in the hierarchical or dynamic routing part of the network can interact with each other and so can cancellation of alternate paths. As discussed above, real-time traffic management controls for a hierarchical fixed routing network are modified for a nonhierarchical dynamic network. For example, dynamic routing table design automatically designs alternate-path choices to maximize network flow and largely replaces equivalent manual functions in the hierarchical network. Because maximum flow is also the basic objective of real-time traffic management controls, dynamic routing automatically assists in attaining this objective. Restrictive controls that affect specific traffic loads, such as code blocks and the hard-to-reach control, are appropriately modified for the dynamic routing network, as discussed earlier. Such restrictive controls have a major role in managing a telecommunications network in both fixed and dynamic routing environments.

In the event of unusual circumstances, the traffic manager would have full network monitoring information on the status of the dynamic routing network and controls to properly implement real-time traffic management techniques. With the deployment of real-time traffic management systems, network managers have automated tools to diagnose problems and recommend control actions in the more complex dynamic routing environment. The real-time traffic management system software must be updated to achieve the required dynamic routing surveillance and control functions described in detail in Chapter 17.

As illustrated in Figure 1.37, capacity management provides for projection of demands, including adjustments for business forecasts and projected new service demands, and execution of the network design model to determine the capacity requirements in the forecast horizon. The updated capacity requirements are sent to switching and transport provisioning systems so that capacity expansion is implemented on a scheduled basis to meet the projected demands. As illustrated in

Figure 1.37, capacity management provides monitoring of network performance through collection and display of daily traffic and performance data and, if service problems are detected, allows capacity and routing redesign and implementation to alleviate the service problems. Under exceptional circumstances, capacity can be added on a short-term basis to alleviate service problems but is normally planned, scheduled, and managed over a period of several months to one year or more. Network design encompasses capacity and routing design.

Capacity management systems must be modified to allow the dynamic routing capabilities to influence network design. Several functions within these systems need to be modified, such as use of switch-to-switch traffic data, the ability to handle multiple load set periods, the treatment of the dynamic network design model as a separate program module, and the incorporation of dynamic routing design. Each of these required changes is described in detail in Chapter 17. Necessary changes in dynamic routing table design include the ability to make routing table changes introduced by the dynamic network design system and the need for switch-to-switch traffic data. The routing table design procedures described above imply frequent routing updates in the network. One essential requirement for this is that the switch be programmed to automatically accept routing table changes, as described in the previous section, versus a manual entry system that is often used in fixed hierarchical routing networks. A mechanized interface is necessary because of the large number of routing changes that may occur in the routing table design. Such a capability can be implemented for the dynamic routing network with an automated routing administration system. With this interface, the routing table updates output from the network design system are sent via data link to the automated routing administration system. These routing changes are then input to the switching systems.

Switch-planning systems must reflect the impact of dynamic routing on network design. These systems could use design models that approximate the network efficiency improvement with dynamic routing rather than compute exact designs, which could require more computational resources. Switch planning could be done separately for individual areas, or all dynamic routing switches could be considered together in the switch-planning tool. This latter approach represents a departure from practices in fixed hierarchical networks, but it is feasible to accomplish, especially if approximate design techniques are incorporated. Dynamic routing design models and impacts must be included in the switch-planning system software [DaF83].

16.3.3 Transition costs

Studies show that the distributions of link and switch loads shift when dynamic routing is introduced. The cost of evolving to this redistributed network is evaluated using the 30-switch intercity network model shown in Figure 6.12, to simulate the evolution to an intercity dynamic routing network. In the study it is assumed that there are 10 dynamic routing switches in year 1, 20 in year 2, and 30 in years 3–10. The evolution strategy uses the dynamic routing capacity management design techniques described in Section 16.2.5; no disconnects are

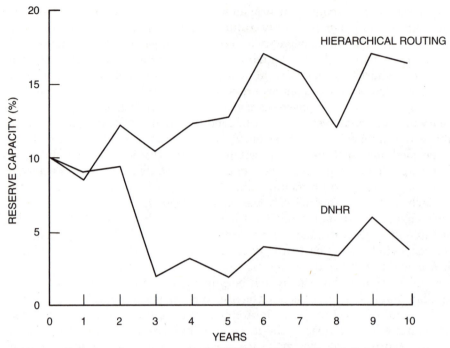

Figure 16.5 Evolution of reserve capacity for DNHR and hierarchical routing

allowed in either the dynamic routing network or the hierarchical network beyond year 3. During the three-year transition period, however, the dynamic routing designs are assumed to be implemented to allow disconnects from the existing link capacities.

Of particular interest is the evolution of network cost, especially the component due to reserve capacity. These results are shown in Figure 16.5. We can see the high level of reserve capacity as dynamic routing is introduced; this, in effect, delays the capital trunk savings until later years. The difference in reserve capacity between the hierarchical network and the dynamic routing network averages from 5 to 7 percent in the steady state and yields an overall dynamic routing network design savings on the order of 20 percent of the cost of the hierarchical network. It is found that the transition strategy does not favor leaving hierarchical trunks in place during the transition. Such a strategy leaves capacity in the wrong places in the dynamic routing network, and capital trunk savings are delayed. A policy of disconnecting trunks down to the required dynamic routing link size in the year that a link is converted to dynamic routing is a good strategy for achieving early savings.

The rearrangements required during the three-year transition are shown in Figure 16.6. The operational cost of this rearrangement in the actual DNHR transition is estimated to be $5–10 million. Loss of capital trunk savings due to the three-year transition amounts to $30 million. In other words, if a transition to

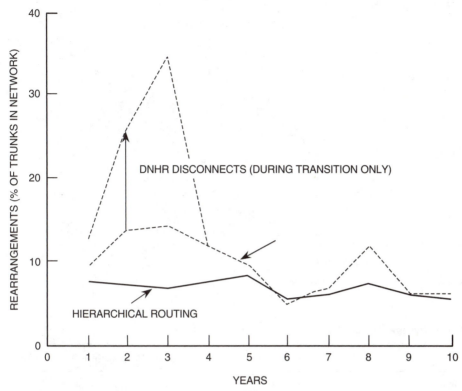

Figure 16.6   Rearrangements for DNHR and hierarchical routing

full DNHR deployment could be achieved a year sooner, the capital trunk savings would increase by about $30 million.

16.3.4   Economic analysis

Early predictions of dynamic routing economic benefits are based on designs for small (28–30-switch) network models, and there was a need to test dynamic network economics on a full-scale network model and to evaluate costs as completely as possible. A 215-switch intercity model, shown in Figure 16.7, is designed for the 1989 October loads. It represents 140 major city locations with large-switch capacity (140 switches identified in Figure 16.7), and 75 medium-switch locations serving large but relatively smaller metropolitan areas. Ten load set periods are used to design the dynamic routing network, and three load set periods are used to design the hierarchical fixed routing network.

Two dynamic network designs are made, in which the first design allows dynamic routing only among the 140 large switches, with the remaining 75 medium switches using fixed hierarchical routing. The second design allows dynamic routing among all 215 switches. The results are as shown in Tables 16.5 and 16.6, and the savings shown include the cost penalties listed in Table 16.7. The 15 percent design savings corresponds well to the predictions of smaller dynamic

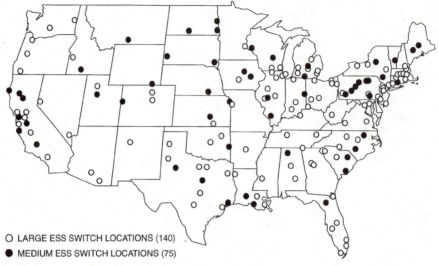

O LARGE ESS SWITCH LOCATIONS (140)
● MEDIUM ESS SWITCH LOCATIONS (75)

Figure 16.7 215-node intercity network model

TABLE 16.5 Dynamic Routing Design Results (140-Switch Model)

	With speech compression	Without speech compression
% savings	16%	14%
capital savings	$560M	$540M

TABLE 16.6 Dynamic Routing Design Results (140-Large-Switch Model and 75-Medium-Switch Model)

	With speech compression	Without speech compression
% savings	17%	15%
capital savings	$710M	$690M

TABLE 16.7 Dynamic Routing Implementation Costs

ESS real-time load	$19M
CCS network load	$1M
traffic data collection	$4M
added routing table update load	$3M
development	$42M
transition rearrangements	$10M
Total:	$79M

network design models, as discussed in Chapter 1. Additional savings on the order of 5 percent of network reserve capacity cost are achieved with dynamic network routing table update procedures, as described above and in Chapter 9. Typical inflation rates of 7–8 percent are assumed. Termination costs are about $230 per termination, and transport costs are about $0.70 per circuit mile, which includes intermediate multiplex equipment. Termination costs include terminal multiplex equipment, trunk termination equipment, switch processor, and common equipment. Speech compression equipment allows per-mile transport cost reduction of a factor of 2 on certain network links, and flow on such links increases by almost 30 percent in the dynamic network design relative to the hierarchical network design, because the dynamic network cost optimization recognizes the lower cost of speech compression links, especially on the long-mileage circuits. This increase in the use of speech compression links yields an overall increase in dynamic network savings, as shown in Tables 16.5 and 16.6.

Metropolitan area network capital savings resulting from the implementation of dynamic routing are estimated to be about 7–10 percent, based on studies of the Chicage metropolitan model and other models, as discussed in Chapter 1. Additional benefits are projected for deloading tandem switches using metropolitan area dynamic routing, because the tandem routing load is transferred to end-offices performing the dynamic routing via function.

16.3.5 Other benefits

As discussed in the previous section, dynamic routing reduces switching and transmission capital costs through advanced design and optimization techniques. Network capital savings are estimated to be 15–20 percent of total network capital cost, or more than $500 million in total network capital savings for the network under study. Dynamic routing networks achieve benefits not quantified in the economic analysis in the previous section, which include (1) customer revenue retention through improved network reliability; (2) additional revenue through improved call completions; (3) new service revenue based on priority routing technology such as 800 Gold Service; (4) ISDN service revenues through integration of voice and ISDN data services, allowing competitive price reduction of ISDN services; (5) operational expense savings; (6) switch development cost avoidance and fast feature introduction through standardized class-of-service routing functions and capacity sharing among services; and (7) network cost avoidance through circuit selection capability routing to provide specific service characteristics on a shared network versus a dedicated overlay network. We briefly illustrate each of these benefits.

Dynamic routing allows a marked improvement in network reliability, as discussed in Chapter 14, by providing for the selection of many more possible routes between every pair of cities than is possible with hierarchical routing, for every call. Dynamic routing improves service quality through improved robustness to failure and load surges by responding in real time to such failures and overloads, and it continues to provide connections to customers with essentially no perceived interruption of service. Dynamic routing provides transport routing diversity, as well as multiple ingress/egress routing, to help respond to transmission and switch failures. Transport-diverse routing allows selection of diverse traffic paths through the transport network, and multiple ingress/egress routing allows selection of diverse traffic paths into and out of the switching network. Network reliability provided by dynamic routing is a strategic network advantage for a service provider, allowing attraction and retention of customers. For example, dynamic routing service quality allowed AT&T to make a "never miss a call" guarantee, which was promoted, for example, during the 800 portability transition. According to a February 10, 1997, *Wall Street Journal* article quoting AT&T Vice President Gail McGovern, this guarantee was responsible for AT&T losing only 2 percent rather than 10 percent of market share. Hence, in a $10 billion annual 800 services market, this retention of 8 percent represents an annual revenue retention of $800 million in 800 services.

Dynamic routing provides additional revenue through improved call completions, as discussed throughout the book. Estimated switched-services revenue increases as a result of the introduction of dynamic routing technology exceed $100 million annually.

Dynamic routing technology allows the introduction of new "key" services with priority routing, as discussed in Chapter 12, which include 800 Gold Service, software-defined network key service, international priority routing service, and others. Revenues from new dynamic routing priority routing services exceed $200 million annually.

Integration of voice and switched digital ISDN data services, as discussed in Chapters 10, 12, and 13, has allowed cost reduction and thereby competitive price reduction of ISDN services, attracting significant ISDN service revenues exceeding $100 million annually.

Dynamic routing simplifies the operational environment and reduces operations costs by centralizing and automating some operations functions, such as real-time traffic management as performed by NEMOS and the network operations center, as further discussed in Chapter 17. In all, a reduction of approximately 100 or more operations personnel positions has been achieved with the introduction of dynamic routing technology, through operations automation, simplification, and centralization. The reduction of 100 network management personnel represents an annual savings of $15 million.

Dynamic class-of-service routing, as discussed in Chapter 12, increases service flexibility for fast feature introduction through standardized routing functions and capacity sharing among services. Class-of-service routing can be used for introduction of all new services using dynamic routing on an integrated shared network, and it also, as discussed in Chapter 12, allows individual voice and switched digital ISDN data services to share network bandwidth on an integrated transport network. Avoidance of special switch development to employ dynamic routing for all new services, through use of dynamic routing class-of-service capability, is estimated to save in excess of $0.5 million to $1 million in new switch development cost for every newly introduced service.

Dynamic routing circuit selection capability, as discussed in Chapter 12, allows specific hardware characteristics, such as high-fidelity voice transmission, to be provided on an integrated shared network as opposed to a dedicated overlay network (such as on a dedicated switch with dedicated trunks having the desired characteristic to all other switches). Capacity sharing versus dedicated overlay capacity on a dedicated overlay switch represents a capital avoidance of approximately $40 million.

16.4   Dynamic Routing Interworking Requirements

The studies discussed in Section 16.3 show that large economic and service benefits will accrue from implementing dynamic routing methods. However, to fully achieve these benefits, standardized CCS messages are required so that switching equipment from different vendors interacts to implement dynamic routing methods in a coordinated fashion. This is important because most networks are multiple-vendor switching networks. It is desirable that a maximal set of dynamic routing techniques be enabled through such standardization, which should include most dynamic routing methods in use in networks today. With four CCS messages, all methods described here plus others could be implemented by service providers across different-vendor switching technologies in their networks. They could also allow interworking with switches in other service providers' networks, which may in fact use other types of dynamic routing. All of these messages are used during the call setup procedure to select a path for the call to be routed on and are defined as follows:

1. *Traveling class mark*—controls the route choice at an originating switch, via switch, or terminating switch

2. *Crankback*—returns call control to the originating switch from a via switch or terminating switch

3. *Query*—requests link status from a switch

4. *Status*—contains link status information

We now describe each of these messages and their functions in more detail.

When a call is alternate-routed to a switch, the routing technique must ensure proper handling at the alternate switch, which could mean that the call is not again routed to an alternate path or routed to a previously visited switch. To help the switch select an appropriate path to forward a call, an indicator (traveling class mark) can be put in the initial call setup message. This explicitly informs the switch whether alternate routing has or has not been done. When alternate routing has been done, the switch may, for example, then restrict the call to the direct link to the destination. A call may be routed by a switch to a path on which, somewhere downstream, the call cannot be supported. In such a case, the call must revert to a more appropriate path or inform the switches that selected it that the path should be avoided for subsequent calls. Information in the form of a crankback message needs to be returned to the switches that selected it. Alternate-route selection may require information to be obtained from switches.

Such information can be queried directly by the switch when it implements the alternate-route selection process, or it can be queried by a database when the process is implemented outside the switch. Examples of the types of information to support the route selection process could include (1) the number of idle trunks on all links for a given switch; (2) the number of calls that overflowed their direct link since the last query, on all links for a given switch; and (3) an alternate-route recommendation, which identifies the switch to use if the direct link is busy for all destinations on a given switch.

All of these messages may be used during the call setup procedure to select a path for the call to be routed on. However, communication of this information does not need to be tied to call setup. We now illustrate each of these messages and their functions in more detail. The description applies equally well to global international network dynamic routing applications, national intercity network dynamic routing applications, metropolitan area network applications, and private network dynamic routing applications. Here, the emphasis is on how the CCS messages are interpreted rather than on how or when these messages are generated or what actions the switch takes based on the messages.

16.4.1   Traveling class mark

Traveling class marks (TCMs) can be used within the dynamic routing network to control the routing choices at various switches. This section presents both the routing choices available at a switch in the dynamic network and the use of the TCMs and other CCS messages to enforce the nonhierarchical routing constraints with crankback within the dynamic network.

TCMs can be derived on a call-by-call basis during call routing, and they accompany an initial address message (IAM) to help differentiate the various routing treatments that are provided for a call that arrives at a switch in the dynamic network.

Table 16.8 indicates possible parameter values for the CCS TCM message, and all of the valid choices of routing treatments for TCM values are summarized in Table 16.9. Here, H denotes hierarchical routing translation, D denotes dynamic routing translation, H $\rightarrow$ D denotes H overflowing to D translation, OS denotes the originating switch, VS denotes a via switch, and TS denotes the terminating switch. The table indicates that the TCM values together with the internal translations may determine the routing treatment. The internal translations include digit analyses and routing translation, as discussed in Section 16.3.1. The significance of the TCM values is now discussed.

TABLE 16.8   Traveling Class Mark Message Components

message type	traveling class mark
operation	control routing treatment at a switch
parameters	TCM value

TABLE 16.9 Traveling Class Mark Values and Routing Treatments

Dynamic network TCM values	Internal translations	Routing treatment
enter	H	hierarchical
enter	H → D	hierarchical overflowing to OS
enter	D	OS
via 1	D	VS1, VS2, VS3
via 2	D	VS1, VS2, VS3
exit 1	H	TS
exit 2	H	TS

The ENTER value in the TCM field can be used at the interface between the hierarchical connecting network and the dynamic network. This value can be chosen to be consistent with current (hierarchical) CCS use for an IAM. The ENTER value means that the receiving switch is free to advance a call as its routing translation dictates. The use of the ENTER value in the TCM is limited to calls

- Entering the dynamic network

- Entering the hierarchical network

- Remaining within the hierarchical network

The VIA 1 value in the TCM field restricts the receiving switch to perform only the VS routing treatment; that is, the VS interprets the message to route the call on possibly one or two transit links from the entry transit switch to the exit transit switch in the transit (via) country. A parameter of the via 1 TCM can indicate whether the transit switch is the first (entry) transit switch, VS1, second (via) transit switch, VS2, or third (exit) transit switch, VS3, in the transit country. That is, if two transit links are used in the transit country, the entry transit switch (VS1) can send the via transit switch (VS2) a via 1 TCM with the parameter set to 2, and the via transit switch (VS2) can send to the exit transit switch (VS3) a via 1 TCM with this parameter set to 3. It is possible for VS1 to send the call directly to the TS if transit links are not used in the routing implementation. If the call cannot be completed to the TS, the entry transit switch VS1 can then send a crankback message to the OS.

The VIA 2 value in the TCM field indicates to the receiving switch that it is a VS and that trunk reservation should be applied in routing the call on possibly one or two transit links from the entry transit switch to the exit transit switch in the transit (via) country. A parameter of the via 2 TCM indicates whether the transit switch is the first (entry) transit switch, VS1, second (via) transit switch, VS2, or third (exit) transit switch, VS3, in the transit country. That is, if two transit links are used in the transit country, the entry transit switch (VS1) sends the via transit switch (VS2) a via 2 TCM with the parameter set to 2, and the via transit switch (VS2) sends the exit transit switch (VS3) a via 2 TCM with this parameter set to 3, where trunk reservation is used on all transit links. It is possible that VS1 sends the call directly to the TS if transit links are not used in the routing implementation. If the call cannot be completed to the TS, the entry transit switch VS1 sends a crankback message to the OS.

Either the EXIT 1 or the EXIT 2 value in the TCM field is interpreted to restrict the receiving switch to perform only the TS routing treatment. The two exit values can be used to distinguish between one-link and multiple-link paths and can then be used for trouble analysis.

16.4.2 Crankback

In a dynamic routing network, the OS in some dynamic routing implementations maintains control of a call until one of the available routing paths for it can be accessed to the designated TS, or the call is blocked from the network. As each routing path is attempted, the VS in some dynamic routing implementations returns control of the call to the OS if no direct connection to the TS can be established. Generally, a crankback message can be sent to the OS from either the VS or the TS. Hence, a CCS crankback message returned to the OS from a VS would, in general, be interpreted to cause the OS to route advance to the next available path in the routing table. If the crankback message is received from the TS, the OS, in general, will interpret this message to mean it should select another available TS to which to route the call.

Table 16.10 indicates possible parameter values for the CCS crankback message. Here, the world switch number represents a unique numerical identifier for each switch in the dynamic routing network. In this illustration, the VS/TS parameter is set to 1 for a VS crankback message and set to 2 for a TS crankback message.

16.4.3 Query

As part of the call setup procedure to select a routing path for a given call, the OS can initiate a link load status query to another switch in the dynamic network. This query may, for example, be directed to the TS or perhaps to a VS and can be used by the OS to derive a best VS to connect a call. For purposes of illustration here, we assume that the query is sent by the OS to the TS; however, there are many other cases for use of these messages.

TABLE 16.10 Crankback Message Components

message type	crankback
operation	return call control to originating switch
parameters	transport capability
	service identity
	world switch number indicator
	VS/TS

TABLE 16.11   Query Message Components

message type	query
component type	invoke
operation	request dynamic routing information
parameters	transport capability
	service identity
	load level indicator
	world switch number indicator
	information type

Table 16.11 indicates possible parameter values for the CCS Query message. *Information type* is defined in the next section under the discussion of the status message components.

16.4.4   Status

As explained above, as part of the call setup procedure to select a routing path for a given call, the OS can initiate a link load status query to another switch in the dynamic network. This query may, for example, be directed to the TS or perhaps to a VS and can be used by the OS to derive a best VS to connect a call.

After receiving a query message from the OS, the TS sends a status message back to the OS, which can be used by the OS as part of the call setup procedure to select a routing path for a given call. The TS can encode the trunks available on each link connected from the TS to every other switch in the dynamic network into a given number of discrete levels. For example, the encodings could denote greater than 10 percent idle trunks (lightly loaded state), less than 10 percent idle trunks (heavily loaded state), fewer than the number of reserved trunks available (reserved state), or all trunks busy (busy state). These four states can be mapped into 2 bits of information corresponding to each link from the TS to every other switch in the dynamic network, and in this example these status bits are sent in the status message from the TS to the OS.

The OS can use the status information received from the TS or from candidate VSs to make a routing path decision, which can be based on economics, load status derived from query-status exchanges, its own estimates of traffic conditions, and other data parameters. The OS can then attempt to set up the call on the selected path using the TCM control messages and crankback messages described above.

Table 16.12 indicates possible parameter values for the CCS status message. Some of the types of information that could be transmitted by this message are given in the following table:

message type	status
information type	contents of information table
status	link states
RECOM1	first-choice routing recommendations
RECOM2	second-choice routing recommendations

TABLE 16.12   Status Message Components

message type	status
component type	return result — last
operation	set dynamic routing information
parameters	transport capability
	service identity
	world switch number indicator
	information type
	number of bits per entry
	information table

Other information types could be added to enable the transmission of other useful information, such as the number of arriving calls and overflow calls. For centralized dynamic routing systems, one recommendation can be sent for each destination with the "information type" field set to RECOM#, and in distributed dynamic routing systems, an encoded message of all link statuses can be returned with the "information type" field set to status. The corresponding query message can use the information type to specify status, RECOM#, or other requested status information. Other parameters can be defined for needed dynamic routing information exchange.

16.5   Examples of Dynamic Routing Use of CCS Messages

We now illustrate the use of the four CCS messages described in the previous section, with typical examples of dynamic routing methods. These examples do not always reflect the precise way some dynamic routing techniques are implemented and only serve to illustrate the use of the messages across a variety of different methods.

16.5.1   Dynamic nonhierarchical routing (DNHR)

As discussed in Chapters 6–10, DNHR uses a time-varying routing system to respond to network load variations and incorporates one- and two-link path routing between originating and terminating switches. A sequence of engineered two-link paths is designed by the network design model and is supplemented with a sequence of additional two-link (real-time) paths to be used only when idle capacity is available above a trunk reservation level. DNHR uses the traveling class mark and crankback CCS messages in call setup. CCS crankback can be used when the second link of a two-link path is blocked at a via switch to allow the originating switch to advance to the next path, and it also can be used to implement multiple ingress/egress routing.

Figure 16.8 illustrates the use of the CCS messages in setting up a DNHR call, as follows:

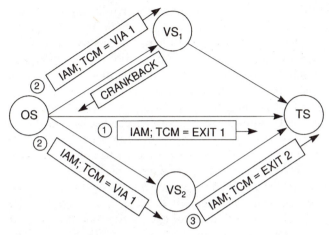

Figure 16.8 Use of CCS messages in setting up a DNHR call

Step 1—The originating switch (OS) first tries the direct link for a call destined for the terminating switch (TS). If there is a trunk available, the OS selects a trunk and sends a CCS IAM with the TCM set to EXIT1. Steps 2 and 3 are skipped unless there is no direct trunk available, in which case the OS proceeds to Step 2.

Step 2—The OS finds the next (two-link) path in its routing table. If there are no more paths to try in the routing table, the OS proceeds to Step 5. If there is another two-link path in the routing table, the OS searches for an idle trunk on the first link. If there is a trunk available, the OS selects the trunk and sends a CCS IAM with the TCM set to VIA1 to the via switch (VS). If there is no free trunk on the first link, the OS returns to the beginning of Step 2. Otherwise, the process continues with Step 3.

Step 3—The VS receives the call and determines the TS through the normal number translation procedures. The VS searches for an idle trunk on the link to the TS. If there is a trunk available, the VS selects the trunk and sends a CCS IAM with the TCM set to EXIT2, and the process continues with Step 4. If there is no free trunk available on the link to the TS, the VS sends a CCS crankback message to the OS, which proceeds again with Step 2.

Step 4—The TS receives the call and TCM and routes the call to the next downstream switch and thus toward the final destination in the normal fashion. This ends the call flow steps.

Step 5—The OS blocks the call and returns the CCS national trunk congestion (NTC) message to the upstream switch from which the call was received. This ends the call flow steps.

16.5.2 Real-time network routing (RTNR)

As discussed in Chapter 12, RTNR is a decentralized routing method with call-by-call updates based on real-time network status. RTNR routing first selects the direct link between the OS and the TS. When no direct trunks are available, the OS checks the availability and load conditions of all of the two-link paths to the TS on a per-call basis. If any of these two-link paths are available, the call is set up over the least-loaded two-link path. An available two-link path is considered to be lightly loaded if the number of idle trunks on each link exceeds a threshold level. In order to determine all of the switches in the network that satisfy this criterion, the OS sends a query message to the TS over the CCS network, requesting the TS to send a list of the switches to which it has lightly loaded links. Upon receiving this list of switches in the return CCS status message from the TS, the OS compares this list with its own list of switches to which it has lightly loaded links. Any switch that appears in both lists currently has lightly loaded links to both the OS and the TS and therefore can be used as the via switch for a two-link connection for this call. The OS then simply ANDs the bit map it receives from the TS, which lists all of the lightly loaded links out of the TS, with its own bit map to produce a new bit map that identifies all the via switches with lightly loaded links to both the OS and the TS. RTNR uses the query, network status, traveling class mark, and crankback CCS messages in call setup. CCS crankback messages are used in RTNR to implement multiple ingress/egress routing.

Figure 16.9 illustrates the use of the CCS messages in setting up an RTNR call, as follows:

Step 1—The OS first tries the direct link for a call destined for the TS. If there is a trunk available, the OS selects a trunk and sends a CCS IAM with the TCM set to EXIT1, and the process skips to Step 4. If there is no direct trunk available, the OS proceeds to Step 2.

Step 2—The OS sends a CCS query message to the TS requesting the list of lightly loaded links. The TS returns this information to the OS in a CCS status message. By comparing its own link status information with the information in the status message received from the TS, the OS finds the most lightly loaded two-link path to the TS. If no such path is found, the OS proceeds to Step 5. Otherwise, the OS searches for an idle trunk on the first link in the selected two-link path, selects the trunk, and sends a CCS IAM with the TCM set to VIA1 to the VS.

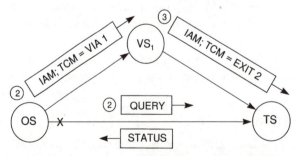

Figure 16.9 Use of CCS messages in setting up an RTNR call

Step 3—The VS receives the call and determines the TS through the normal number translation procedures. The VS searches for an idle trunk on the link to the TS, selects the trunk, and sends a CCS IAM with the TCM set to EXIT2, and the process continues with Step 4.

Step 4—The TS receives the call and TCM, and routes the call to the next downstream switch toward the final destination in the normal fashion. This ends the call flow steps.

Step 5—The OS blocks the call and returns the CCS NTC message to the upstream switch from which the call was received. This ends the call flow steps.

16.5.3 Trunk status map routing (TSMR)

As discussed in Chapter 11, TSMR is a centralized real-time dynamic routing method with periodic updates based on periodic network status. TSMR provides periodic real-time routing decisions in the dynamic routing network. The TSMR method involves having an update of the number of idle trunks in each link sent via the CCS status message to a network database every T (for example, 10) seconds. Routing tables are determined from analysis of the trunk status data using the TSMR dynamic routing method, which provides that the first path choice be the first design path (usually the direct path) if it exists and is available. If the first path is busy, the second path is selected from the list of feasible paths on the basis of having the greatest number of idle circuits at the time; this path update is performed every T seconds. In this example, TSMR uses the network status, traveling class mark, and crankback CCS messages during call setup. CCS crankback can be used when the second link of a two-link path is blocked at a via switch, to allow the originating switch to advance to the next path, and is also used to implement multiple ingress/egress routing. DCR is a TSMR-like dynamic routing method that could also be implemented through use of the standard CCS messages.

Figure 16.10 illustrates the use of the CCS messages in setting up a TSMR call, as follows:

Step 1—All switches send periodic CCS status messages to a central database containing the status of all connected links. The OS first tries the direct link for a call destined for the TS. If there is a trunk available, the OS selects a trunk and sends a CCS IAM with the TCM set to EXIT1, and the process continues with Step 4. If there is no direct trunk available, the OS proceeds to Step 2.

Step 2—The OS sends a CCS query message to the central database requesting the list of lightly loaded links connected to the TS. The central database returns this information to the OS in a CCS status message. By comparing its own link status information with the information in the status message received from the central database, the OS finds the most lightly loaded two-link path to the TS. If no such path is found, the OS proceeds to Step 5. Otherwise, the OS searches for an idle trunk on the first link in the selected two-link path, selects

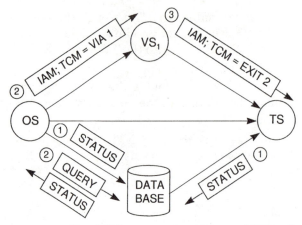

Figure 16.10 Use of CCS messages in setting up a TSMR call

the trunk, and sends a CCS IAM with the TCM set to VIA1 to the VS. If the first link is blocked, the OS proceeds to Step 5.

Step 3—The VS receives the call and determines the TS through the normal number translation procedures. The VS searches for an idle trunk on the link to the TS, selects the trunk, and sends a CCS IAM with the TCM set to EXIT2, and the process continues with Step 4. If the link to the TS is blocked, the process continues with Step 5.

Step 4—The TS receives the call and TCM and routes the call to the next downstream switch toward the final destination in the normal fashion. This ends the call flow steps.

Step 5—The OS blocks the call and returns the CCS NTC message to the upstream switch from which the call was received. This ends the call flow steps.

16.5.4   Learning with random routing (LRR)

As discussed in Chapter 12, LRR is a decentralized call-by-call method with update based on random routing. LRR uses a simplified decentralized learning method to achieve real-time dynamic routing. The direct link is used first if available, and a fixed alternate path is used until it is blocked. In this case a new alternate path is selected at random as the alternate path choice for the next call overflow from the direct link. No crankback is used at a via switch or egress switch in LRR, so a call blocked at a via switch or egress switch will be lost. In this example, LRR uses the traveling class mark message to control call setup on two-link paths. DAR is an LRR-like dynamic routing method that could also be implemented through use of the standard CCS messages.

Figure 16.11 illustrates the use of the CCS messages in setting up an LRR call, as follows:

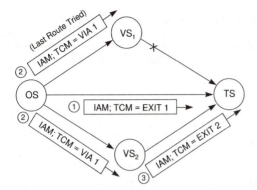

Figure 16.11 Use of CCS messages in setting up an LRR call

Step 1— The OS first tries the direct link for a call destined for the TS. If there is a trunk available, the OS selects the trunk and sends a CCS IAM with the TCM set to EXIT1, and the process continues with Step 4. If there is no direct trunk available, the OS proceeds to Step 2.

*Step 2—*The OS tries the two-link path that was tried successfully on the previous two-link call to the TS. If this path does not exist the OS finds the next (two-link) path in its route list. The OS searches for an idle trunk on the first link. If there is a trunk available, the OS selects the trunk and sends a CCS IAM with the TCM set to VIA1 to the via switch (VS), and the process continues with Step 3. If there is no free trunk on the first link, the OS proceeds to Step 5.

*Step 3—*The VS receives the call and determines the TS through the normal number translation procedures. The VS searches for an idle trunk on the link to the TS. If there is a trunk available, the VS selects the trunk and sends a CCS IAM with the TCM set to EXIT2, and the TS proceeds with Step 4. If there is no free trunk available on the link to the TS, the VS blocks the call, and the process continues with Step 5.

*Step 4—*The TS receives the call and TCM and routes the call to the next downstream switch, and thus toward the final destination, in the normal fashion. Having completed the call to the TS, the OS stores the successful two-link route for use by the next two-link call to the TS. This ends the call flow steps.

*Step 5—*The OS blocks the call and returns the CCS NTC message to the upstream switch from which the call was received. This ends the call flow steps.

Hence, with four CCS messages, all of the above dynamic routing methods, plus all others discussed in Chapter 1 and throughout the book, plus others yet to be devised, could be implemented within a multivendor switching environment.

In order to implement any dynamic routing method between equipment from different switch vendors, standardized CCS messages of the type discussed in this section are required [AKA94, AsH94]. At the same time, switch vendors need

TABLE 16.13 CCS Message Volumes for Dynamic Routing (50-Switch Global International Network Model, Busy-Hour Period, 69K Erlangs)

	LRR	RTNR	D-TSMR	TSMR	DNHR	HIER
traveling class mark/IAM	910,800	910,800	910,800	910,800	935,600	910,800
crankback	—	—	—	—	24,800	—
query	—	82,800	—	36,000	—	—
status	—	82,800	1,764,000	72,000	—	—
total:	910,800	1,076,400	2,674,800	1,018,800	960,400	910,800

to implement features in their switching software to send and respond to the standard CCS messages, as follows:

1. If a query message is received, the switch returns a status message.

2. If a traveling class mark is received, the switch routes the call to its destination or sends a crankback message.

3. The switch sends periodic status messages to a specified list of CCS addresses.

ITU-T is studying the standardization of information exchange and switch capabilities, so that each carrier could implement its desired routing method and thereby reap the economic and performance benefits of dynamic routing networks.

Table 16.13 compares signaling message volumes for the four routing methods discussed in this section, assuming a busy-hour traffic load of 69,000 erlangs in a 50-switch global international network model. Two implementations of TSMR are considered: a distributed (D-TSMR) and a centralized implementation. Status messages are sent every five seconds either to every other switch in the network for the distributed implementation or to the central database in the centralized implementation. Distributed TSMR requires by far the most signaling messages, whereas LRR requires the least.

16.6 Conclusion

The dynamic routing network impacts, operational feasibility, business justification, and implementation requirements reported in this chapter show the following results:

■ Studies of dynamic routing networks subjected to overload and failure conditions demonstrate a performance comparable to that of a fixed hierarchical network and in some cases, such as those of transport failures, better than that of the hierarchical network.

■ With a digital network transmission technology, dynamic routing implementation does not degrade transmission performance and in fact improves it in comparison with hierarchical routing by limiting connections to at most two links. With an analog network transmission technology, dynamic routing transmission grade-of-service with respect to the percentage of customers perceiving transmission quality as good or better is degraded by about 1–2 percent relative to hierarchical network at longer mileage bands. This change arises from greater use of two-link routing at longer distances but is mitigated by imposing design constraints such as limiting total path distance, prohibiting tandem satellite links, and prohibiting tandem voice compression links.

- Dynamic routing switch real-time processing load and common-channel signaling impacts lead to additional loading of these elements and an economic penalty of earlier switch exhausts relative to the hierarchical network, which is included in the economic analysis. However, the routing flexibility inherent in dynamic routing can more efficiently distribute tandem load among the dynamic routing switches and reduce the switch network penalty.

- Dynamic routing operational requirements such as switch-to-switch data collection and routing administration are shown to be feasible. In regard to dynamic network design requiring large-scale optimization, such as with the path-EFO model, for example, advances in dynamic routing large-scale-optimization techniques make it computationally feasible to design and service 200-switch and larger dynamic routing networks using presently available computer technology.

- Business justification analysis shows a highly positive result, in that the dynamic routing benefits far outweigh the costs. We review capital savings for intercity network examples and metropolitan network examples, and based on full-scale intercity network models, design savings in the range of 14–16 percent and more have been verified. Metropolitan area network savings are estimated to be 7–10 percent and more, based on different examples of metropolitan area network design. Additional benefits are projected for deloading tandem switches using metropolitan area dynamic routing. Capital savings in an intercity network example are estimated to be $500 million over a 10-year period. A reduction in network reserve capacity cost in the range of 0–5 percent is achieved by routing table update flexibility inherent in dynamic routing networks to access existing network capacity.

- Dynamic routing flexibility results in fewer network blockages compared with hierarchical routing, which allows significant additional revenues from fewer abandoned calls. Achievement of near-perfect network reliability is extremely important to customers, especially to business customers such as 800 service providers, whose revenue in turn depends on network availability. Retention of these customers, especially in a competitive environment where customers have a choice of service providers, is enhanced by dynamic routing implementation. The additional call-completion and customer-retention revenue are shown to have impacts of several hundred million dollars annually.

- Dynamic routing implementation requirements are described in the chapter, which include switch development, real-time traffic management system development, capacity design system development, routing administration system development, switch-to-switch data collection system development, and network planning systems development.

- Standardized information exchange is necessary so that switching equipment from different vendors can interact to implement dynamic routing methods in a coordinated fashion and thus allow service providers who have multivendor networks to reap the efficiency and performance benefits of dynamic routing in their networks. This information exchange is illustrated with the use of standardized common-channel signaling messages for various dynamic routing methods.

Network Management and Design Implementation for Dynamic Routing Networks

17.1   Introduction

As discussed in Chapter 1, Figure 1.1 illustrates a model for network routing and network management and design. The central box represents the network, which can have various configurations, and the traffic routing tables and transport routing tables within the network. Routing tables describe the route choices from an originating switch to a terminating switch for a connection request for a particular service. Hierarchical, nonhierarchical, fixed, and dynamic routing tables have all been discussed in the book. Routing tables are used for a multiplicity of services on the telecommunications network.

Network management functions include real-time traffic management, capacity management, and network planning. Figure 1.1 illustrates these functions as interacting feedback loops around the network. The input driving the network is a noisy traffic load, consisting of predictable average demand components added to unknown forecast error and other load variation components. The feedback controls function to regulate the service provided by the network through real-time traffic management controls, capacity adjustments, and routing adjustments. Real-time traffic management provides monitoring of network performance through collection and display of real-time traffic and performance data and allows traffic management controls such as code blocks, call gapping, and reroute controls to be inserted when circumstances warrant. Capacity management includes capacity forecasting, daily and weekly performance monitoring, and short-term network adjustment. Forecasting operates over a multiyear forecast interval and drives network capacity expansion. Daily and weekly performance monitoring identify any service problems in the network. If service problems are detected, short-term network adjustment can include routing table updates and, if necessary, short-term capacity additions to alleviate service problems. Updated routing tables are sent to the switching systems either directly or via an automated routing update system. Short-term capacity additions are the exception, and most capacity changes are normally forecasted, planned, scheduled, and managed over a period of months or a year or more. Network design embedded in capacity management includes routing design and capacity design. Network planning includes longer-term switch planning and transport network planning, which operates over a horizon of months to years to plan and implement new switch and transport capacity.

Network management and design implementation is the subject of this chapter. We focus on the dynamic nonhierarchical routing (DNHR) network in operation in the AT&T network from 1984 to 1991, which implemented the preplanned dynamic two-link sequential path routing method discussed in Chapter 6. The initial 16-switch implementation of DNHR, illustrated in Figure 15.14, was cut over into network operation on July 14, 1984. Because this network is the first such dynamic routing network in actual use, it presents a very interesting case study of how a dynamic routing network can be managed, designed, and operated.

In Sections 17.2 to 17.5, we focus on the steps involved in

- Real-time traffic management of the dynamic routing network (Section 17.2)

- Capacity forecasting in the dynamic routing network (Section 17.3)
- Daily and weekly performance monitoring (Section 17.4)
- Short-term network adjustment in the dynamic routing network (Section 17.5)

This material is the result of interviews with J. S. Dudash from the AT&T Network Management District, T. R. Brown from the AT&T Network Forecasting District, and G. R. McCurdy from the AT&T Network Servicing District. For each of these three topics, we illustrate the steps involved with actual data taken from the 16-switch operational dynamic routing network.

Finally, in Section 17.6 we illustrate how each of these network management and design functions changes with the implementation of real-time dynamic routing. We illustrate this with the implementation of real-time network routing (RTNR), which replaced DNHR starting in 1991. With RTNR, several changes and improvements occurred in real-time traffic management and capacity management, and these are discussed.

17.2   Real-Time Traffic Management

In this section we concentrate on the surveillance and control of the dynamic routing network. We also discuss the interactions of traffic managers with other work centers responsible for dynamic routing network operation and contrast dynamic routing real-time traffic management with hierarchical routing real-time traffic management. Dynamic routing real-time traffic management functions are performed from the Network Operations Center located at Bedminster, New Jersey, and are supported by the network management operations system (NEMOS). A functional block diagram of NEMOS is illustrated in Figure 1.33.

17.2.1   Real-time performance monitoring

The surveillance of the dynamic routing network is performed through use of the highest true-overflow-count (TOC) pair display, which is monitored at all times (an example is shown in Figure 17.6). This display is used in the auto-update mode, which means that every five minutes NEMOS automatically updates the exceptions shown on the map itself and displays the eight switch pairs with the highest true overflow count. NEMOS will display more than eight switch pairs when the low-blocking switch pairs have the same blocking. For example, if the eighth-, ninth-, and tenth-highest switch pairs have equal true overflow counts, all ten pairs are displayed.

NEMOS also has displays that show the high true-overflow-percent (TOP) pairs within threshold values (Figure 17.1). This display, which reflects the total network, has the manual true-overflow-percent thresholds set with the low threshold at 0 and the high threshold at 100. This particular display shows any dynamic routing network switch pair with any overflow between those switches, with the thresholds set as they are. Therefore, it shows every switch pair for which data are being received and recognized by NEMOS.

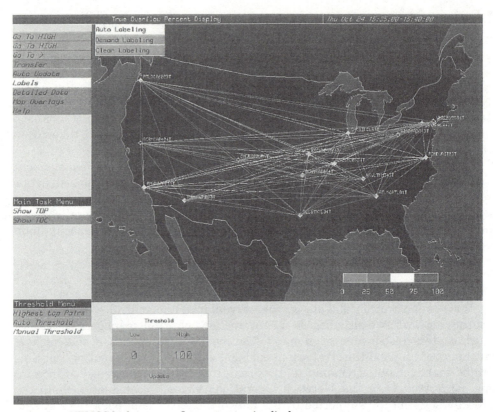

Figure 17.1 NEMOS high true-overflow-percent pairs display

Traffic managers are most concerned with what calls can be rerouted and therefore want to know the location of the heaviest concentrations of blocked calls. For that purpose, overflow percentages can be misleading at times. From the revenue standpoint, the difference between 1 percent and 10 percent blocking on a switch pair may favor concentration on the 1 percent blocking situation, because there are more calls to reroute. NEMOS can also display all the exceptions that there are with the auto threshold display, which displays everything exceeding the present threshold—normally either 1 percent true overflow percent or 1 true overflow count. This display then shows the total blocked calls and not just the highest pairs.

For peak-day operation, or operation on a high Monday, traffic managers work back and forth between the auto threshold display and the highest true-overflow-count pair display. They spend most of their time with the auto threshold display, where they see everything that is being blocked. Then, when traffic managers want to concentrate on clearing out some particular problem, they look at the highest true-overflow-count pair display, an additional feature of which is that it allows the traffic manager to see the effectiveness of controls. NEMOS subtracts the reroute successes from the dynamic routing network real-time overflow count,

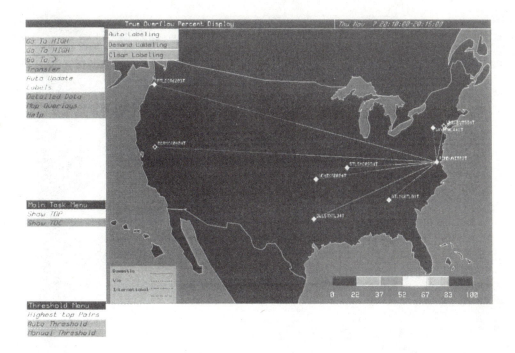

Figure 17.2 Focused overload on Richmond, Virginia

so as the reroute controls are successful in completing calls that would otherwise be blocked, the traffic manager sees this in near real time using this display.

The traffic manager can recognize certain patterns from the surveillance data. For example, Figure 17.2 illustrates a focused overload on Richmond, Virginia, due to the November 5, 1985, flooding situation discussed further in Sections 17.3, 17.4, and 17.5. This situation caused heavy calling into the Richmond area that went on for three evenings. The traffic pattern there was that most locations showed heavy overflow into and out of Richmond. That particular focused overload going into Richmond had no focused completion problems below Richmond; it was just caused by uniformly heavy calling because of the flooding. Figure 17.2 shows the heavy calling on the evening of November 7, 1985. The display shows the true overflow percent; the manual threshold in this case has the low setting at 1 percent and the high setting at 100 percent. This display shows any switch pair in the dynamic routing network that exceeds 1 percent true overflow percent. The display in Figure 17.3, which again illustrates the Richmond flooding problem, is an example of the highest true-overflow-percent pairs. This particular threshold selection displays the eight switch pairs with the highest true overflow percent, even though there might be additional switch pairs with blocking in the network.

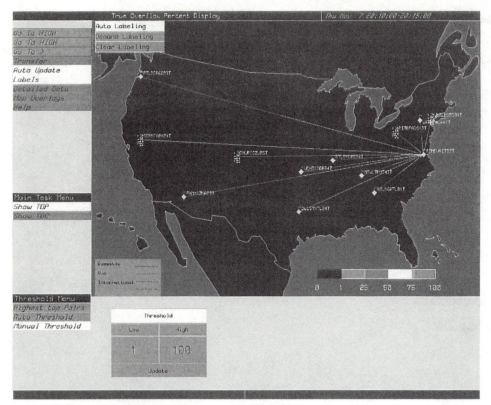

Figure 17.3 Highest true-overflow-percent pairs during the Richmond flood

On this particular display the bar chart with a color-coded representation indicates that the lowest percent overflow is 22 percent and the highest is 83 percent.

One of the other things traffic managers can see with NEMOS using the highest true-overflow-count pair display is a switch failure. This is because the true overflow count includes calls affected by the failure, which will identify the failed switch in the next five-minute update. In one instance, a machine congestion level 3 (MC3) condition (switch failure) occurred at 49 minutes after the hour; the next NEMOS update 50 minutes after the hour showed the effects of the MC3. Transport failures also show on the displays, but the resulting display pattern depends on the failure itself.

17.2.2 Network control

As discussed in Chapters 1 and 16, the dynamic routing network has automatic controls built into the switch and also has automatic and manual controls that can be activated from NEMOS. We first describe the controls and what they do, and then we discuss how the dynamic routing traffic managers work with these controls. Two protective automatic real-time traffic management controls are used in the dynamic routing network: selective dynamic overload control (SDOC), which responds to switching system congestion, and selective trunk reservation (STR), which responds to trunk congestion. SDOC and STR are selective in the sense that they control traffic destined for hard-to-reach points more stringently than other traffic. As discussed in Chapter 16, SDOC and STR in this case are the enhanced implementations for dynamic routing networks.

The complexity of dynamic routing makes it necessary to place more emphasis on fully automatic controls that are reliable and robust and do not depend on manual administration. SDOC and STR respond automatically within the switching system software program. For STR, the automatic response is coupled with two trunk reservation threshold levels, represented by the number of idle trunks in a dynamic routing link. STR trunk reservation levels are automatic functions of the link size.

SDOC and STR are not strictly link-dependent but also depend on the switch pair to which a controlled call belongs. A call offered to an overloaded via switch will either be canceled at the originating switch or advanced to an alternate via switch, depending on the destination of the call. STR differentiates between one- and two-link calls and first- and alternate-routed calls.

SDOC and STR also use a simplified method of obtaining hard-to-reach control selectivity. In the dynamic routing network, hard-to-reach codes are detected by the terminating switch, which communicates them to the originating switches and via switches. Because the terminating switch is the only exit point from the dynamic routing network, the originating switch treats a hard-to-reach code detected by a terminating switch as hard to reach on all dynamic routing links.

SDOC is permanently enabled on all dynamic routing links. STR is automatically enabled by an originating switch on all dynamic routing links when that originating switch senses general network congestion. STR is particularly important in the dynamic routing network because it minimizes the use of two-link connections and maximizes useful network throughput during overloads. The automatic enabling mechanism for STR ensures its proper activation without manual intervention. SDOC and STR automatically determine whether to subject a controlled call to a cancel or skip control. In the cancel mode, affected calls are blocked from the network, whereas in the skip mode such calls skip over the controlled link to an alternate link. SDOC and STR are completely automatic controls. Capabilities such as automatic enabling of STR, the automatic skip/cancel mechanism, and the STR one-link/two-link traffic differentiation adapt these controls to the dynamic routing network and make them robust and powerful automatic controls.

Code-blocking controls block calls to a particular destination code. These controls are particularly useful in the case of focused overloads, especially if the calls are blocked at or near their origination. Code blocking controls need not block all calls, unless the destination switch is completely disabled through natural disaster or equipment failure. Switches equipped with code-blocking controls can typically control a percentage of the calls to a particular code. The controlled code may be NPA, NXX, NPA-NXX, or NPA-NXX-XXXX, when in the latter case one specific customer is the target of a focused overload.

Figure 1.36 illustrates a call-gapping control typically used by network managers in a focused call overload, such as sometimes occurs with radio call-in give-away contests. Call gapping allows one call for a controlled code or set of codes to be accepted into the network, by each switch, once every x seconds, and calls arriving after the accepted call are rejected for the next x seconds. In this way, call gapping throttles the calls and prevents the overload of the network to a particular focal point.

An expansive control is also available in dynamic routing. Dynamic routing reroute is able to modify routes by inserting additional paths at the beginning, middle, or end of a route sequence. Such reroutes can be inserted manually or automatically through NEMOS. When a reroute is active on a switch pair, STR is prevented on that switch pair from going into the cancel mode, even if the overflow is heavy enough on a particular switch pair to trigger the STR cancel mode. Hence, if a reroute is active, calls do have a chance to use the reroute paths and are not blocked prematurely by the STR cancel mode.

In the dynamic routing network, a display is used to graphically represent the controls in effect. Depending on the control in place, either a certain shape or a certain color will tell traffic managers which control is implemented. Traffic managers are able to tell if a particular control at a switch is the only control on that switch. Different symbols are used for the switch depending on the controls that are in effect.

17.2.3   Work center functions

17.2.3.1   Automatic controls.   The dynamic routing network has automatic controls, as described above, and if there is spare capacity, traffic managers can decide to reroute. In the November 5, 1985, Richmond flooding situation, the links were occupied sufficiently, and there was no network capacity available for reroutes until shortly before the problem went away. The STR control was active at the time. In order to get calls out of Richmond, traffic managers went in manually and disabled the STR control at Richmond. This gave preference to calls going out of Richmond. Therefore, Richmond was doing much better at completing outgoing calls than were the other switches at completing incoming calls. This control resulted in using the link capacity a little more efficiently.

Traffic managers can manually enable or inhibit STR and also inhibit the skip/cancel mechanism for both STR and SDOC. Traffic managers monitor SDOC controls very closely because they indicate switching congestion or failure. Therefore, SDOC activations are investigated much more thoroughly and more quickly than STR activations, which are frequently triggered by normal heavy traffic. SDOC did not trigger, for example, in the Richmond flooding situation. One of the items that has changed real-time traffic management strategy is the effective automatic control capabilities of the switch. With dynamic routing and automatic controls, traffic managers are no longer as concerned about switching system overload due to heavy calling as they were with hierarchical routing and electromechanical switching technology. Traffic managers are, of course, very concerned with switch failures, which are sometimes accompanied by SDOC controls.

17.2.3.2 Code controls. Code controls are used to cancel calls for very hard-to-reach codes, but they were not needed in the Richmond situation because there were no cases of very hard-to-reach terminating points below the Richmond switch. Code control is used when calls cannot complete to a point in the network or there is isolation. For example, traffic managers used code controls extensively for the September 19, 1985, Mexico City earthquake situation, in which there was isolation. Normal hard-to-reach traffic caused by heavy calling volumes will be blocked by the STR control, as described above.

Traffic managers use data on hard-to-reach codes in certain situations for problem analysis. For example, if there is a problem in a particular area, one of the early things traffic managers look at is the hard-to-reach data to see if they can identify one code or many codes that are hard to reach and if they are from one location or several locations. For example, there is a customer that has three answering locations that have calling volumes in excess of what they can handle. With the use of Advanced 800 Service, the customer is able to control the distribution of traffic to each answering location. Traffic managers can use the hard-to-reach capability by looking at the traffic patterns to see how the customer's load is distributed over those three locations. Then, control can be placed on terminating traffic using a call gap control at the switch nearest to the answering location.

Management of these customer-controlled networks also entails going back to work with the customer not only to put in more lines but also to help them in distributing the traffic more evenly. Control of these networks used by customers can be very complex, and destination-oriented control capabilities help control them. Traffic managers use such controls in addition to manual intervention to help customers control their networks at the terminating locations.

17.2.3.3 Reroute controls. Reroutes in the dynamic routing network depend primarily on the automatic reroute capability in NEMOS to initiate reroutes, as described in Chapter 8. One exception to that is when the traffic manager knows of some condition in the network that NEMOS would not be aware of, just from communicating with various locations beyond the view of NEMOS. Another exception to that could be knowing of conditions that are recurring periodically, where traffic managers can go in manually and apply a reroute prior to the problem occurring. There are enough of these events that traffic managers still use the manual reroute even though the automatic reroute capability is there. This allows the traffic manager to anticipate the blocking exceptions before the NEMOS automatic reroute reacts to them.

Reroutes are used primarily for transport failures or heavy traffic surges, such as traffic on heavier than normal days, where the surge is above the normal capabilities of the network to handle the load. Those are the two prime reasons for rerouting. Traffic managers do not *usually* reroute into a disaster area. There are times, however, when things are quiet and there are only one or two links overflowing; in that case traffic managers might reroute into a disaster area if they cannot complete the calls in any other way. However, in peak periods they definitely do not want to do that.

17.2.3.4 Peak-day control. Peak-day routing in the dynamic routing network involves using the direct path as the only engineered path and then the remaining available paths as real-time paths (engineered and real-time paths are discussed in Chapter 8). This routing method is very effective because it favors single-link traffic when it is needed. The effectiveness of the additional real-time paths and reroute capabilities depends very much on the peak day itself. The greater the peak-day traffic, the less effective the real-time paths are. That is, on the higher peak days, such as Christmas and Mother's Day, the network is filled with single-link connections and is completing more calls. On lower peak days, such as Easter or Father's Day, the real-time paths and rerouting capabilities are much more effective. This is because the peaks, although they are high and have an abnormal traffic pattern, are not quite as high as on Christmas or Mother's Day. So on these days there is additional capacity to complete calls on the real-time paths and NEMOS reroute paths.

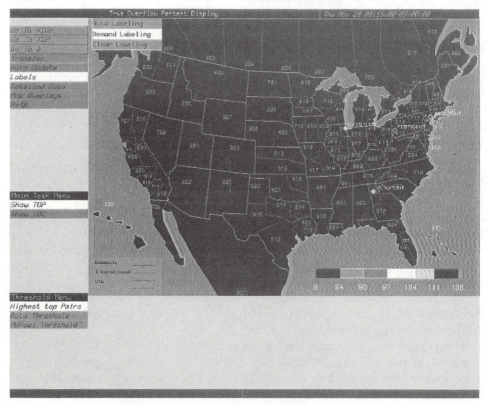

Figure 17.4 Traffic load on Thanksgiving Day 1985, 8:50 A.M. to 9:00 A.M.

Reroute paths are used on the peak days in addition to the real-time paths. Reroute paths are particularly available in the early morning and late evening. Depending on the peak day, at times there is also a lull in the afternoon, and NEMOS can find reroute paths that are available. The performance of the NEMOS automatic reroute is illustrated in Chapter 18 for Easter 1988 traffic (Figure 18.4).

17.2.4 Real-time traffic management experience on Thanksgiving Day, 1985

Thanksgiving Day is not one of the very highest peak days—it does not have nearly the traffic of a Christmas or a Mother's Day. In the Thanksgiving traffic pattern, the morning normally starts off lighter than on the higher peak days. It starts to build more gradually and has two heavy peaks. One peak occurs around lunchtime, near the starting time of the football games, and the other peak is in the evening. The other times are heavy but not the constant 99–100 percent occupancy that occurs on Christmas and Mother's Day.

The load patterns exhibited on Thanksgiving Day 1985 are illustrated in Figures 17.4–17.7. Figure 17.4 shows the time interval 8:50 A.M. to 9:00 A.M. It

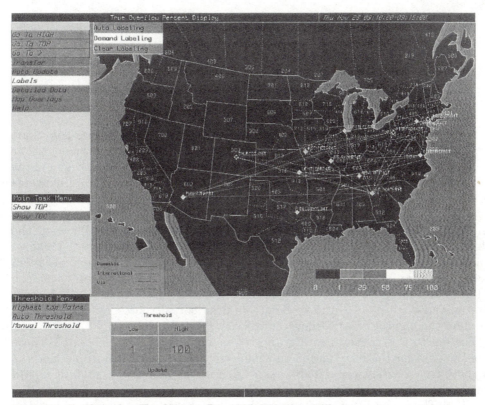

Figure 17.5 Traffic load on Thanksgiving Day 1985, 9:10 A.M. to 9:15 A.M.

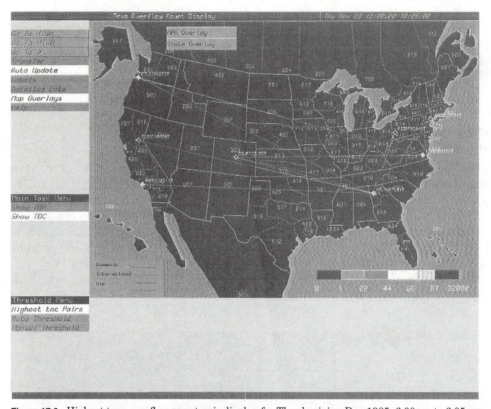

Figure 17.6 Highest true-overflow-count pair display for Thanksgiving Day 1985, 6:00 P.M. to 6:05 P.M.

reflects the highest true-overflow-percent pairs early in the morning. This chart indicates that Chicago, Pittsburgh, and White Plains had a high true overflow percent to Atlanta. This was shortly after heavy overflow conditions began in the dynamic routing network. The second chart, shown in Figure 17.5, illustrates 9:10 A.M. to 9:15 A.M., and it reflects dynamic routing switch pairs that have a true overflow percent between 1 and 100. As can be seen on this chart, Atlanta is still the primary-focus location, with traffic building to both Dallas and Phoenix. The third chart, shown in Figure 17.6, is labeled 18:00–18:05 and reflects the highest true-overflow-count pairs for the early evening on Thanksgiving Day. This again reflects Atlanta being the focal point, although in this case the traffic between Atlanta and the West Coast is very much in evidence. Note that although there are many switch pairs with overflow, this time period represents a low period because the highest true overflow count for any switch pair is only 87.

The fourth chart, shown in Figure 17.7, is time-stamped 20:05–20:10. These are the data for the busy period on Thanksgiving evening, again reflecting the highest true-overflow-count pairs. As evidenced on this chart, Atlanta is still one of the busiest switches, and the pattern is similar to what is shown in Figure 17.6. The major difference between these two charts is the number of overflow

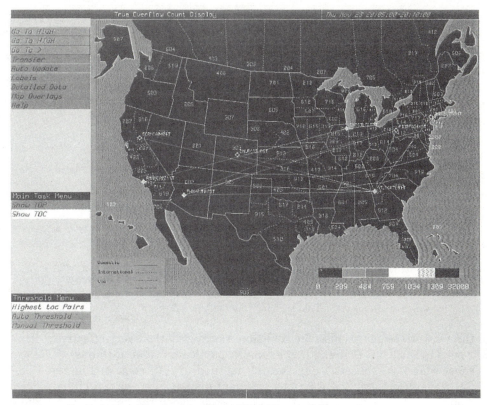

Figure 17.7   Traffic load on Thanksgiving Day 1985, 8:05 P.M. to 8:10 P.M.

counts; although the 18:00–18:05 time period had a high of 87 counts for any switch pair in the five-minute interval, the 20:05–20:10 chart has a high of 1,309 counts on one switch pair. The latter chart also has a much higher count for the other switch pairs that are displayed. The lowest overflow count displayed in the busy period is 209 versus only one overflow count for the 18:00–18:05 time period.

The peak-day routing method was employed on Thanksgiving Day. As described above, this method uses the direct path between switch pairs as the only engineered path, followed by 13 real-time via paths. This method had been very effective on the peak days that it had been used. Traffic managers' experience is that the peak-day routing is much more effective during the lighter peak days such as Thanksgiving, Easter, and Father's Day. With the lighter loads, when the network is not fully saturated, there is a much better chance of using the real-time paths. However, when we enter the network busy hour or combination of busy hours, with a peak load over most of the network, the routing method at that point drops back to one-link routing because of the effect of trunk reservation. At other times the real-time paths appear to be very effective as far as completing calls. Even in the busy hour, some switch pairs that do not have high overflow

```
ONE WAY MEASUREMENTS                          (OMN SAOWM)              PAGE = 2 MORE
SELECTION:_ _                                                         12/03/85 08:36:05
OTS CLLI# NSVLTNMT43T              TTS CLLI# SCRMCA0404T
DATE: 11 28 85    {11/24/85-12/02/85}                                DOM/INT CODE: D

                                  CARRIED      ENG         RT          NM CALLS
SELN        HR        PEG CT      LOAD         OVFL        OVFL        SAVED

   9        08            25        190          0           0             0
  10        09           273        432        217         188             0
  11        10          1252        287       1171        1171             0
  12        11          2074        355       2005        2004             0

  13        12          2182        217       2129        2128             0
  14        13          2194        260       2105        2096             0
  15        14          1871        483       1747        1686             0
  16        15          1339       1000       1156         976             0

                                                          COMMAND:_____

1-SADBS - DAILY BLKNG SUM      2-SADBD - DAILY BLKNG DET      3-SRRTC - RTE CONFIGURATION
4-SAWBD - WKLY BLKNG DETL      5-SAWBS - WKLY BLKNG SUM       6-SGHDX - HR/DY-EXCLUSION
```

Figure 17.8   One-way measurements on Thanksgiving Day 1985, part 1

in the network appear to use the real-time routing method very effectively. This was seen in the Thanksgiving Day example, particularly out of Dallas.

These effects of the peak-day routing method are all seen in Figures 17.8–17.11, which represent one-way measurement displays for two switch pairs over 16 hours on Thanksgiving Day. The difference between the engineered overflow (from the direct path) and the real-time overflow (from all 14 paths) represents the number of calls completed on the real-time paths. During the daytime hours, few calls get through on the real-time paths because the peak loads and the effects of STR cancel. In the evening hours, however, when there is more capacity in the network, the real-time paths are very effective in completing additional calls that might not otherwise have been completed.

Control of the network with automatic controls was predominant in the dynamic routing network on Thanksgiving Day. The primary control that was active was STR, which during the busy hours was not only in skip mode but also in cancel mode. This resulted in large quantities of calls being canceled. When STR triggered the cancel mode for a switch, the actual calls affected were only between those switches that had high traffic loads between them; hence STR did not affect traffic to all switch pairs. There were several occasions of SDOC being activated during the day. This was triggered by machine congestion signals MC1 and MC2, which occur during switch overload. A lesser quantity of traffic is affected with SDOC than with STR. The effect of SDOC being triggered depends on whether the calling to that particular switch is for terminating traffic or via traffic. Terminating traffic will be canceled on a percentage basis. Via traffic, however, could be skipped rather than canceled, dependent on the amount of total real-time path overflow. If the overflow load is high enough,

```
ONE WAY MEASUREMENTS                  (OMN SAOWM)              PAGE = 3 LAST
SELECTION:_ _                                                 12/03/85 08:36:17
OTS CLLI# NSVLTNMT43T          TTS CLLI# SCRMCA0404T
DATE: 11 28 85  {11/24/85-12/02/85}                          DOM/INT CODE: D
```

SELN	HR	PEG CT	CARRIED LOAD	ENG OVFL	RT OVFL	NM CALLS SAVED
17	16	830	1534	593	0	0
18	17	826	1809	618	0	0
19	18	865	2223	720	1	0
20	19	751	2096	620	66	0
21	20	795	1709	634	166	0
22	21	393	980	273	90	0
23	22	255	841	134	0	0
24	23	60	444	5	0	0

COMMAND:_____

```
1-SADBS - DAILY BLKNG SUM    2-SADBD - DAILY BLKNG DET    3-SRRTC - RTE CONFIGURATION
4-SAWBD - WKLY BLKNG DETL    5-SAWBS - WKLY BLKNG SUM     6-SGHDX - HR/DY-EXCLUSION
```

Figure 17.9   One-way measurements on Thanksgiving Day 1985, part 2

```
ONE WAY MEASUREMENTS                  (OMN SAOWM)              PAGE = 2 MORE
SELECTION:_ _                                                 12/03/85 08:37:37
OTS CLL1# PHNXAZMA03T          TTS CLLI# CHCGILCL57T
DATE: 11 28 85    {11/24/85-12/02/85}                        DOM/INT CODE: D
```

SELN	HR	PEG CT	CARRIED LOAD	ENG OVFL	RT OVFL	NM CALLS SAVED
9	08	1037	3444	168	0	0
10	09	10117	12674	8436	8002	0
11	10	21401	8879	20044	20044	0
12	11	20177	8626	18768	18725	0
13	12	14799	9942	13407	12987	0
14	13	5666	12855	4120	3127	0
15	14	1945	9598	243	0	0
16	15	972	5181	11	0	0

COMMAND:_____

```
1-SADBS - DAILY BLKNG SUM    2-SADBD - DAILY BLKNG DET    3-SRRTC - RTE CONFIGURATION
4-SAWBD - WKLY BLKNG DETL    5-SAWBS - WKLY BLKNG SUM     6-SGHDX - HR/DY-EXCLUSION
```

Figure 17.10   One-way measurements on Thanksgiving Day 1985, part 3

```
ONE WAY MEASUREMENTS                    (OMN SAOWM)                PAGE =3 LAST
SELECTION:_ _                                                     12/03/85 08:37:53
OTS CLLI# PHNXAZMA03T          TTS CLLI# CHCGILCL57T
DATE: 11 28 85  {11/24/85-12/02/85}                              DOM/INT CODE: D
```

SELN	HR	PEG CT	CARRIED LOAD	ENG OVFL	RT OVFL	NM CALLS SAVED
· 17	16	980	4360	0	0	0
18	17	996	4465	0	0	0
19	18	1673	7369	191	26	0
20	19	3452	11536	1794	1233	0
21	20	4458	11434	3030	2523	0
22	21	3736	10814	2433	1800	0
23	22	686	5921	108	0	0
24	23	153	1304	0	0	0

```
                                                   COMMAND:_____
```

```
1-SADBS - DAILY BLKNG SUM    2-SADBD - DAILY BLKNG DET    3-SRRTC - RTE CONFIGURATION
4-SAWBD - WKLY BLKNG DETL    5-SAWBS - WKLY BLKNG SUM     6-SGHDX - HR/DY-EXCLUSION
```

Figure 17.11 One-way measurements on Thanksgiving Day 1985, part 4

the traffic could be canceled by the automatic skip-to-cancel mechanism in SDOC. SDOC operates in the exact same fashion as the skip-to-cancel mechanism for STR. The machine congestion levels 1 and 2 that we saw on Thanksgiving Day were load related. This is the switch's method of shedding load to prevent it from running into severe congestion.

As the load increases and all via routes become congested, the traffic managers begin to look in the hierarchical part of the network for capacity. They have been successful in finding capacity to various switches through switches that normally carry primarily business traffic and other switches with large links that have capacity to the affected switches. Some of the hierarchical routes that did have capacity were through the New York (Broadway) switch; the Houston, Texas, switch, for certain traffic loads; and the Los Angeles switch, in some cases. These are primarily the switches that are very heavily used by business loads on normal business days. The caution that traffic managers have to use when employing these reroutes between dynamic routing switches via a hierarchical switch is that they must use the cancel reroute overflow control to prevent any rerouted traffic from coming back up into the dynamic routing network. The other part of the rerouting strategy is that before the dynamic routing network becomes congested the traffic managers try to find reroute capacity that can be used to complete traffic before it gets up into the dynamic routing network.

17.2.5 Interfaces to other work centers

The main interaction traffic managers have is with the capacity managers. Traffic managers notify capacity managers of conditions in the network that are affecting the data that they use in making decisions as to whether or not to add capacity.

Examples are transport failures and switch failures that would distort traffic data. A machine congestion level 3 (MC3) triggers SDOC; SDOC cancels all traffic destined to a switch while the MC3 is active. All calls to the failed switch are reflected as real-time overflow calls for the duration of the MC3 condition. This can be a considerable amount of canceled traffic. The capacity manager notifies traffic managers of the new link capacity requirements that they are trying to get installed but that are delayed. Traffic managers can then expect to see blocking on a daily basis or several times a week. This type of information is passed back and forth on a weekly, and in many cases, daily, basis.

17.2.6 Comparison of dynamic routing and hierarchical routing real-time traffic management

Dynamic routing uses more fully automated real-time traffic management than is used in the hierarchical network. The NEMOS automatic reroute capability is an example. Another example is the capability for NEMOS to recommend via paths for reroutes that could then be put in manually. This feature avoids the manual effort of finding a reroute path from the data available to traffic managers. It makes manual rerouting much simpler from that standpoint.

Dynamic routing has also complicated real-time traffic management. For example, if traffic managers run into a routing type problem in the network, identifying the path that is taken by a particular call associated with the problem becomes more complex. For example, there was a trouble report from Nashville when customers there could not complete international calls through the Sacramento international switching center during a particular load set period. The basic problem was incorrect routing in Atlanta. It took a while to resolve this problem, primarily because it involved international traffic routing. A related problem was that the Nashville–Sacramento direct path was not used during the problem load set period, and the first path was a via path through Atlanta. Because there was incorrect routing data in Atlanta that only affected international calls, the problem only occurred during this load set period. It took some time to piece all these clues together because of the complexity of dynamic routing tables. Traffic managers can become misled unless they are thoroughly familiar with the routing in place at the time, with the dynamic routing network, and with troubleshooting such problems.

The code- and switch-routing display in NEMOS helps greatly in resolving such problems. By making a single request, NEMOS pulls the appropriate routing data from the routing database and then pulls the corresponding data from the ESS switch and displays the two on the same display, highlighting differences. Hence, this display has real advantages in determining what the routing should be when traffic managers encounter a problem of the type described above.

Dynamic routing crankbacks can complicate traffic data and cause them to be misleading. The reason is that link overflow counts are not scored due to crankbacks; that is, when there is a crankback on a particular link, an overflow count is not scored. For example, a problem on Christmas Day was related to a lack of sufficient signaling transceivers in White Plains. Associated with this problem, the White Plains–to-Wayne link was being used very heavily for via traffic from White Plains to all other switches. This link normally has very little direct-routed traffic on it, so the percent occupancy was very low. Once a call got to Wayne from White Plains, Wayne could not complete the call to almost any location. As a result of this, the Wayne switch cranked the call back to White Plains. White Plains had a high real-time overflow percentage and had proceeded from the skip to the cancel mode of STR.

When this particular condition occurred, traffic managers observed a very short holding time and very low overflow from White Plains to many locations. The cause of this pointed to a link problem and was indicated by no overflow and short holding time on the White Plains–to-Wayne link. In essence, the problem was really a lack of transceivers in the White Plains switch, and the scoring or not scoring of the link overflow when the switch received a crankback tied into the current coding, which does not score link overflow when we cancel the call. The call does not proceed any farther, and it is very misleading in that situation.

On a peak day this occurs on four links: Wayne–White Plains, Anaheim–Phoenix, Chicago–St. Louis, and Chicago–Pittsburgh. These links are used for business-day traffic. On peak days, these links have very little usage and a lot of capacity, but when a call reaches the via switch, there are many links congested to most destinations. This leads to many crankbacks.

Traffic managers need to be intimately familiar with the details of the dynamic routing implementation and the exact means by which traffic data are scored in order to analyze problems. In finding solutions, traffic managers must be fully trained on the implementation and use of controls applicable to the dynamic routing network. Although problems occur less frequently, because of the efficiency and automation introduced with dynamic routing, analyzing problems and finding solutions when problems do occur requires new and deeper knowledge on the part of traffic managers when faced with managing the dynamic routing network.

17.3   Capacity Management—Forecasting

In this section we concentrate on the forecasting of dynamic routing switch-to-switch loads and the sizing of the dynamic routing network. We also discuss the interactions of network forecasters with other work centers responsible for dynamic routing network operations, and we contrast dynamic routing network forecasting with hierarchical network forecasting.

Dynamic routing network forecasting functions are performed from the Capacity Administration Center and are supported by the Network Forecasting System (NFS). A functional block diagram of NFS is illustrated in Figure 1.37. In the following two sections we discuss the steps involved in each functional block.

17.3.1   Load forecasting

17.3.1.1   Common master file functions.   The first step in the forecasting process is the planning of the dynamic routing switches themselves. Forecasters in collaboration with switch planners need to plan when and how each dynamic routing switch will come into service. Once that has been determined, forecasters put the switch information into the planning databases called the common master files. Within the common master files there are a number of databases that define various specific components of the network itself: intercity switches, end-offices, access tandems, transport points of presence, buildings, manholes, microwave towers, and so forth. The database is large, and it has a wide range of information in it.

The files that forecasters are primarily interested in for designing and forecasting the dynamic routing network are the intercity switch identification files, an example of which is shown in Figure 17.12. This figure illustrates the case of the Richmond, Virginia, switch, which we use as an example throughout this section. Forecasters enter into this file the common language code or name of the switch, the date the switch goes into service, the date the switch goes out of service (this often is blank), where the switch homes, whether or not the switch can do 6-digit translation, the dynamic routing exclusion indicator (indicating whether to exclude this switch as a via switch or not), the equipment type (indicating 4ESS or other switch type), the types of trunks, the type of signaling, and the rank or class of the switch.

For a dynamic routing switch, there is the dynamic routing indicator. Forecasters enter a Y in this field when they want to bring this switch into the dynamic routing network. Once the dynamic routing indicator is turned on, this triggers all the downstream processes involved in forecasting the dynamic routing network. Forecasters also indicate whether the switch is a gateway switch

```
TOLL SWITCH IDENTIFICATION                    (EP1 TS1)          PAGE = 1 LAST
ACTION: I                      CMF DATA BASE              11/08/85 14:50:59

TS CLLI: { RCMD VA IT03T }         FROM DATE: { 10 85 }          TO DATE: 06 86

HOME CLLI(S) ->  2WAY/IN:  WAYN PA LA42T          ROUTING EXCEPTIONS: N
                     OUT:  _ _ _ _ _ _ _ _ _ _
DNHR EXCL IND: N
EQUIP TYPE: 4E_        GENERIC: 4E9_ _ _ _ _ _      DIR: 2WAY       2 WIRE: _

SIGNALING TRANSMIT           DP:  Y  MF:  Y  CCIS:  Y  ITT5:  -  ITT6:  -  ITT7:  -
            RECEIVE          DP:  Y  MF:  Y  CCIS:  Y  ITT5:  -  ITT6:  -  ITT7:  -

CLASS: 2        DNHR: Y    GATEWAY: N    900 SWITCH: N    OSO: Y    CAMA: Y

ACTION POINT: Y    NSC: _    NETWORK REGION: C    OWNING COMPANY: ATTC

RESPONSIBILITY ->             PEA07     LEA07         UPDATE: FEA07

                                          COMMAND:_ _ _ _ _ _ _ _ _ _

1-TS2              2-TS4              3-TS5
4-TS9              5-QNP1             6-QTS3
```

Figure 17.12 Intercity switch identification file

for international switching, whether the switch carries 900-service traffic, whether it is an originating screening switch for 800-traffic, whether it can perform centralized automatic message accounting, whether it is an access point or an action point for some of the enhanced services offered, who the owning companies are, whether it is an AT&T switch or whether it is owned by an independent entity, and the responsibility codes (identifying the planning engineer, access engineer, and person who has responsibility for updating this display).

Another important part of the common master file database, from a forecasting standpoint, is the files that describe the end-offices that actually contribute the load to the dynamic routing switches. These files are called end-office identification files, and one is shown in Figure 17.13. They are similar to the intercity switch identification files in many ways. They show the date periods in which the record is in effect and the common language code of the end-office. For example, Figure 17.13 shows "RCMD VA ITCG0," which means it is a 1ESS switch in Richmond, Virginia. The figure gives the homing arrangement, which shows that this end-office is homed for two-way traffic to the Richmond switch, identified in Figure 17.12 as a dynamic routing switch.

Hence, this end-office contributes traffic to the Richmond switch. Within the end-office identification file there are approximately 200–300 end-offices homed on the Richmond 4ESS switch. Other information in this file includes the homing exceptions, if any; whether or not high-usage links are built out of the switch (since divestiture in 1984, high-usage links are not built out of end-offices); whether or not the switch can handle all the digits for international direct distance dialing; where the operator traffic is routed that originates from customers served by

```
END OFFICE IDENTIFICATION                    (EP1 EO1)              PAGE = 1 LAST
ACTION: Q                           TTN DATABASE              11/08/85 15:04:55
EO CLLI: { RCMD VA ITCG0 }          FROM DATE: { 10 85 }           TO DATE: 12 04

HOMING  2W/IN:  RCMD VA IT03T       ->AT: _  TS: Y  IF VIA FPOP: _ _ _ _ _ _ _ _ _ _
          OUT:  _ _ _ _ _ _ _ _ _ _ ->AT: _  TS: _  IF VIA FPOP: _ _ _ _ _ _ _ _ _ _

HOMING EXCEPTIONS: N                     HIGH USAGE CAPABILITY: N

EQUIP TYPE: IE_      IDDD: Y       EQUAL ACCESS: _

OPERATOR SVC LOCATION -> COIN:  RCMD  VA  GR1UD   NON COIN:  RCND  VA  GR1UD

SIGNALING TRANSMIT    DP: Y  MF: Y  CCIS: _
          RECEIVE     DP: Y  MF: Y  CCIS: _

LAMA: _        ANI: Y        NETWORK REGION: C        OWNING COMPANY:  CV
RESPONSIBILITY ->      FEA07      PEA07                    UPDATE: LEA07

Z005 INQ FORCED: 'C' ACTN WITH KEY CHG OR TRANFER        COMMAND:_ _ _ _ _ _ _ _ _ _ _ _ _ _ _

1-EO2                 2-EO4                 3-EO5
4-AT1                 5-REOI                6-TEO1
```

Figure 17.13 End-office identification file

this end-office; whether or not the switch can transmit or receive dial pulse, multifrequency, or CCS; whether or not the switch can perform local automatic message accounting or automatic number identification; who the owning company is (CV indicates the C&P Company of Virginia); and the responsibility codes.

When a forecast cycle is started, which forecasters do twice a year, the first step is to extract the relevant pieces of information from the common master file database that are necessary to drive NFS. One of the files generated in this process is the intercity switch master file, shown in Figure 17.14. This file is readily available to all forecasters to check the different switches in the network that will be included in this particular forecast period. It consolidates some of the information that is in the intercity switch identification file, and it also incorporates information from several other files that are in the common master file database. Figure 17.14 illustrates one page of the intercity switch master file. There is administrative information at the top that indicates that this is the December 1985 forecast view; this is when these files were frozen, which means that this is the snapshot of the common master file database for that forecast view. This is the network structure that this forecast is based on, which includes end-office information, intercity switch information, and homing arrangements.

We illustrate this report with reference once again to the Richmond, Virginia, switch. On the first line entry for Richmond (there are five), we see that it indicates 4P85, which refers to the fourth period of 1985. Because this copy is taken from the December 1985 forecast, 4P85 is the base year, or base period. There is also a direction indicator (3 means two-way) and the home intercity switch (Wayne, Pennsylvania). Next are listed the vertical and horizontal (V and H) coordinates; these are five numeric characters used for mileage calculations. This information is taken from the building file in the common master files, and forecasters key off the building code in the seventh and eighth characters of the common language code. There is also a two-character code used to identify cutovers. If there is a major rearrangement that is associated with the switch, a cutover is used to flag all of the links associated with that switch for special attention during that particular period. There is a code for the signaling and switch information; in this case, it is 77. Looking at the cross-reference tables, we find that 77 means that the switch is a 4ESS switch with CCS signaling capability. The rank of the switch in this case is 2. In the outlet field there is an M, for multioutlet, which means the switch can support high-usage links anywhere that the mechanized process can justify a link (S means single outlet with no high-usage links allowed). There is also a field that could contain a discontinue indicator as signified by a D; this would indicate contract elimination with divestiture.

The next field is the dynamic routing indicator, and for Richmond the dynamic routing indicator field indicates "yes" in all time periods. If we look at Rochelle Park, New Jersey (RCPKNJ), however, we see that in the second period of 1987, as opposed to 4P85, the dynamic routing field is marked with a Y, which says that in this forecast period the switch will become part of the dynamic routing network. Other fields include the originating screening switch function (Y means Richmond can perform the 800 originating screening switch functions); the priority code (used to sort links for engineering purposes); the engineering field

AT&T COMMUNICATIONS
DEC 85 - FROZEN
SORT ORDER (TOLL SWITCH CLLI. FP. DIR)

TOLL SWITCH MASTER

FORM NO. M1077
RET. CODE 02P-10000

TS CLLI	P&YR	DIR	HOMED ON TS	TS V/H	COORD	CUT	MOP	RANK	OUT	DISC	DNHR	OSO	SATL	PRI	ENG	ACID	NPA	REGION/SECTION TF	REGION/SECTION AE
QUBCPQ1402T	4P86	3	QUBCPQ1403T	03682	01896		64	4	S			N		840	A	CAXJ	418	NE12	NE12
QUBCPQ1402T	1P88			00000	00000					D				000		CAXJ	418	NE12	NE12
QUBCPQ1403T	4P85	3	MTRLYQ0207T	03682	01896		S4	2	M			N		840	A	CAXJ	418	NE12	NE12
QUBCPQ1403T	4P86	3	SPFDMABR02T	03682	01896		S4	3	S			N		840	A	CAXJ	418	NE12	NE12
RBSNILXE06T	4P85	3	OLNYILXE50T	06544	03148		52	4	M			N		370	A	LB	618	CE09	CE12
RBSNILXE06T	3P86			00000	00000					D				000		LB	618	CE09	CE12
RCCYIAXC02T	4P85	3	SXCYIADT17T	06377	04496		51	4	M			N		330	A	NWIA	712	CE20	CE54
RCCYIAXC02T	1P88	3	OMAHNENW14T	06377	04496		51	4	S			N		330	A	NWIA	712	CE20	CE54
RCFRILRC55T	3P87	3	PEORILPJ5IT	06022	03675	RI	87	3	M			Y		390	L	LB	815	CE08	CE02
RCFRILRC55T	4P87	3	PEORILPJ5IT	06022	03675		87	3	M			Y		390	L	LB	815	CE08	CE02
RCFRILRT51T	4P85	3	PEORILPJ5IT	06021	03675		67	3	M			Y		390	L	LB	815	CE08	CE02
RCFRILRT51T	3P87			00000	00000					D				000		LB	815	CE08	CE02
RCHEILXA50T	4P85	3	RCFRILRT51T	06086	03636		84	4	M			N		390	A	LB	815	CE08	CE04
RCHEILXA50T	3P87	3	RCFRILRC51T	06086	03636		84	4	M			N		390	A	LB	815	CE08	CE04
RCHEILXA50T	3P88			00000	00000					D				000		LB	815	CE08	CE04
RCHLSCXB02T	4P85	3	CLMASCTL03T	06731	01692		54	4	M			N		530	A	SBSC	803	S011	SO22
RCHLSCXB02T	2P86	3	CHRLNCCA03T	06731	01692		54	4	M			N		530	A	SBSC	803	S011	SO22
RCISILRI01T	4P85	3	PEORILPJ51T	06276	03815		34	4	M			N		390	L	LB	309	CE08	CE14
RCISILRI01T	1P87	3	PEORILPJ51T	06276	03815		34	4	S			N		390	L	LB	309	CE08	CE14
RCISILRI01T	2P87			00000	00000					D				000		LB	309	CE08	CE14
RCLKWIXB31T	4P85	3	EUCLWIO184T	05579	04359		C4	4	M			N		360	A	WT	715	CE11	CE40
RCLKWIXB31T	1P88			00000	00000					D				000		WT	715	CE11	CE40
RCMDINXB03T	4P85	3	IPLSIN0102T	06157	02815		C4	5	S			N		410	A	NB	317	CE02	CE22
RCMDVAIT03T	4P85	3	WAYNPALA42T	05928	01476		77	2	M		Y	Y		580	L	CV	804	EA07	EA07
RCMDVAIT03T	1P87	3	WAYNPALA42T	05928	01476	RG	77	2	M		Y	Y		580	L	CV	804	EA07	EA07
RCMDVAIT03T	2P87	3	WAYNPALA42T	05928	01476		77	2	M		Y	Y		580	L	CV	804	EA07	EA07
RCMDVAIT03T	1P88	3	WAYNPALA42T	05928	01476	RH	77	2	M		Y	Y		580	L	CV	804	EA07	EA07
RCMDVAIT03T	2P88	3	WAYNPALA42T	05928	01476		77	2	M		Y	Y		580	L	CV	804	EA07	EA07
RCMTNCXA03T	4P85	3	GNBONCEU03T	06232	01329		77	3	M			Y		520	A	SNBC	919	SO04	SO12
RCMTNCXA03T	3P87	3	GNBONCEU03T	06232	01329	RC	77	3	M			Y		520	A	SNBC	919	SO04	SO12
RCMTNCXA03T	4P87	3	GNBONCEU03T	06232	01329		77	3	M			Y		520	A	SNBC	919	SO04	SO12
RCMTVAXA02T	4P85	3	RONKVALK02T	06241	01761		S4	4	M			N		580	A	CV	703	EA07	EA07
RCMTVAXA02T	4P86	3	RONKVALK03T	06241	01761		S4	4	M			N		580	A	CV	703	EA07	EA07
RCMTVAXA02T	3P87			00000	00000					D				000		CV	703	EA07	EA07
RCPKNJ0203T	4P85	3	NWRKNJ0208T	04975	01440		77	4	M			Y		690	L	NJ	201	EA10	EA10
RCPKNJ0203T	2P87	3	NWRKNJ0208T	04975	01440		77	4	M		Y	Y		690	L	NJ	201	EA10	EA10
RCPKNJ0203T	1P88	3	NWRKNJ0208T	04975	01440		77	3	M		Y	Y		690	L	NJ	201	EA10	EA10
RDCYCA0204T	4P85	3	SNJSCA0241T	08556	08682		47	3	M			Y		182	L	PTBS	415	WE31	WE31
RDCYCA0204T	2P87			00000	00000					D				000		PTBS	415	WE31	WE31
RDDRAB0102T	4P85	3	CLGRAB2104T	05168	07820		64	4	S			N		150	A	ABGT	403	WE81	WE81
RDDRAB0102T	4P86	3	CLGRAB2104T	05168	07820		64	4	M			N		150	A	ABGT	403	WE81	WE81
RDGCCAXF39T	4P85	3	ANHMCA0211T	08896	07825		14	4	S			N		171	A	PTSO	619	WE15	WE15
RDLDCAXF74T	4P85	3	ONTRCAXP80T	09181	07687		54	4	S			N		171	A	PTSO	714	WE15	WE15

TF01.KP.126P1 09/06/85

PAGE 39

Figure 17.14 Intercity switch master file

(indicates who owns the switch—L means AT&T and A means the Regional Bell Operating Companies); the associated company identification codes (used to sort trunk requirements into interstate and intrastate for regulatory purposes and administrative reports); the numbering plan area or NPA that the switch resides in (Richmond is in 804 in Virginia); and the region and section (identifying the trunk forecaster responsible for the switch and the access engineer responsible for the switch). All this information is needed to drive the forecasting process, for administrative purposes, routing purposes, mileage calculations, and so forth.

Another report derived from the common master files at the beginning of a forecast view is the subtending trunking entities file (shown in Figure 17.15), which identifies the end-offices that are homed on each intercity switch. As stated earlier, the two items of interest are the dynamic routing switch and the end-offices that are contributing traffic to that switch. This particular report lists by intercity switch all of the end-offices that are homed on it, and it also breaks the end-office down by the type of service, method of recording, and direction of traffic. Again, the information is listed by forecast period, by year. There is administrative information as to the region and section, which company owns this particular end-office (these are owned by C&P of Virginia), and the NPA.

The listing of the end-offices homed on the Richmond switch tells exactly which one is sending which type of traffic to the switch. Many of the end-offices are coded 003 for all types of traffic, all methods of recording, and both directions, so virtually all the traffic is going to the Richmond switch. There are normally about 200 to 300 of these end-offices homed on an intercity switch such as Richmond. Richmond serves the great majority of the entire south-central part of Virginia, and the intercity switch master file and the subtending trunking entities file are the files that actually drive the forecasting process with the information on the switches and the load information that will come from the end-offices themselves.

Another report generated from the common master files at the beginning of a forecast view is the dynamic routing switch report, shown in Figure 17.16. This report is a listing of all the dynamic routing switches in the forecast, which have been identified based on the dynamic routing indicator being turned on in the intercity switch identification file. They are listed alphabetically by common language identification code of the dynamic routing switch. The report also indicates the region, the section, and the period and year that this particular location becomes a dynamic routing switch. Forecasters use this report to verify all the dynamic routing switches in this particular forecast cycle.

17.3.1.2 Load aggregation, basing, and projection functions.

Once NFS starts processing data from the centralized message database, four study periods of information are extracted each year: from March, May, August, and November, each a 20-day study period. From the centralized message database, there is a 5 percent sample of billed messages, and that sample is pulled for 20 days that we designate in advance, plus a few spare days in case forecasters have to discard some days of data. Forecasters equate that 5 percent, 20-day sample to one average business day. They extract 16 hours of average business day load information and drop the hours between midnight and 8:00 A.M. The load information consists of messages

EASTERN REGION SECTION 07

TOLL SWITCHING SYSTEM SUBTENDING TRUNKING ENTITIES

SORT ORDER (TF REG/SEC, TS CLLI, TE CLLI)

TS CLLI

RCMD VA IT03T

TE CLLI	TOS	MOR	DIR	P&YR-P&YR	REGN	SECT	ACID	NPA	MH
BRWR VA XARLL	2	2	2	1P87-4P88	EA	07	CV	703	
	2	3	2	1P87-4P88					
	3	3	2	1P87-4P88					
	3	0	3	1P87-4P88					
BRWR VA XASGO	0	0	1	1P87-4P88	EA	07	CV	703	
	0	1	2	1P87-4P88					
	2	2	2	1P87-4P88					
	2	3	2	1P87-4P88					
	3	2	2	1P87-4P88					
	3	3	2	1P87-4P88					
	0	0	3	4P89-4P90					
BRWY VA XARLL	0	0	1	1P87-4P88	EA	07	CV	703	
	0	1	2	1P89-4P88					
	2	2	2	1P89-4P88					
	2	3	2	1P89-4P88					
	3	2	2	1P87-4P88					
	3	3	2	1P87-4P88					
	0	0	3	4P89-4P90					
BRWY VA XASGO	0	0	1	1P87-4P88	EA	07	CV	703	
	0	1	2	1P87-4P88					
	2	2	2	1P87-4P88					
	2	3	2	1P89-4P88					
	3	2	2	1P87-4P88					
	3	3	2	1P87-4P88					
	0	0	3	4P89-4P90					
BTHI VA BICGO	0	0	3	4P85-4P90	EA	07	CV	804	
BVRD VA XARLL	0	0	3	4P85-4P90	EA	07	CV	804	
BWLG VA XACGO	0	0	3	4P85-4P90	EA	07	CV	804	
BWLG VA XADGO	0	0	3	4P85-4P90	EA	07	CV	804	

TS CLLI

RCMD VA IT03T

TE CLLI	TOS	MOR	DIR	P&YR-P&YR	REGN	SECT	ACID	NPA	MH
BWLG VA XASGO	0	0	3	4P85-4P90	EA	07	CV	804	
BYTN VA XARLL	0	0	1	1P88-4P88					
	0	1	2	1P88-4P88					
	2	2	2	1P88-4P88					
	2	3	2	1P88-4P88					
	3	3	2	1P88-4P88					
	0	0	3	4P89-4P90					
BYTN VA XASGO	0	1	1	3P87-4P88	EA	07	CV	804	
	0	2	2	3P87-4P88					
	2	2	2	3P87-4P88					
	2	3	2	3P87-4P88					
	3	3	2	3P87-4P88					
	0	0	3	4P89-4P90					
CALL VA XARLL	0	0	1	4P86-4P89	EA	07	CV	804	
	0	1	2	4P86-4P89					
	2	2	2	4P86-4P89					
	2	3	2	4P86-4P89					
	3	3	2	4P86-4P89					
	0	0	3	4P90-4P90					
CALL VA XASGO	0	0	1	2P86-4P89	EA	07	CV	804	
	0	1	2	2P86-4P89					
	2	2	2	2P86-4P89					
	2	3	2	2P86-4P89					
	3	3	2	2P86-4P89					
	0	0	3	4P90-4P90					
CCHS VA XARLL	0	0	1	1P88-4P88	EA	07	CV	804	
	0	1	2	1P88-4P88					
	2	2	2	1P88-4P88					

Figure 17.15 Subtending trunking entities file

```
AT&T-C                                                         FORM NO. M3480
DEC 1985 - FROZEN
TRUNK FORECASTING        DYNAMIC ROUTING SWITCH REPORT    RETENTION CODE 02R-01000
```

DYN RTE SW	REGION	SECTION	PERIOD & YEAR
MOBLALAZ01T	SO	05	2P87
MPLSMNDT18T	CE	18	2P87
MTGMALMT01T	SO	05	2P87
NRFLVABS03T	EA	07	2P87
NSVLTNMT43T	SO	12	2P86
NWHNCT0205T	NE	06	2P87
NWORLAMA04T	SO	10	2P87
NWRKNJ0208T	EA	10	2P86
NYCMNYBW24T	NE	02	2P87
NYCMNYBW51T	NE	02	2P87
NYCMNY5450T	NE	02	2P87
NYCQNYRP08T	NE	03	2P87
OJUSFLTL03T	SO	07	2P87
OKBRIL0A52T	CE	07	4P87
OKCYOKCE04T	MW	02	4P87
OKLDCA0344T	WE	31	4P87
OMAHNENW14T	CE	16	2P86
ORLDFLMA03T	SO	07	2P86
PEORILPJ51T	CE	08	4P87
PHLAPASL42T	EA	01	2P87
PHNXAZMA03T	WE	90	2P86
PITBPADG43T	EA	04	2P86
PTLDOR6203T	WE	22	2P86
RCMDVAIT03T	EA	07	2P86
RCPKNJ0203T	EA	10	2P87
RENONVO344T	WE	53	4P87
SCRMCA0404T	WE	56	2P86
SHOKCA0296T	WE	15	4P87
SKTNCA0107T	WE	33	4P87
SLKCUTMA02T	WE	40	2P86
SNANTXCA02T	MW	06	4P87
SNDGCA0787T	WE	10	4P87
SNFCCA2147T	WE	31	4P87
SNJSCA0241T	WE	31	2P86
SPFDILSD51T	CE	09	4P87
SPFDMABR02T	NE	07	2P87
SPFDMOTL04T	MW	01	4P87
SPKNWA0102T	WE	24	2P87
STLSMO0934T	CE	09	2P86
STTLWA0604T	WE	23	2P87
SYRCNYSU13T	NE	05	2P87
TAMPFLCO02T	SO	06	2P87
TOLDOH2103T	CE	06	4P87
TULSOKTB04T	MW	02	4P87
WASHDCSW06T	EA	05	2P87

```
TFO1.KP175P     SEP 6, 1985
VERSION 01                                                          PAGE 2
```

Figure 17.16 Dynamic routing switch report

and CCS (units of 100 call seconds) by study period. It is identified by method of recording (operator handled, operator serviced, customer dialed), and by service (800, OUTWATS, overseas, or 900 service) and further broken down as business service or residential service. The data are also identified by direction.

Load data are also pulled for Saturday and Sunday for those same four study periods. NFS processes Saturday and Sunday data separately, up through the basing module of load forecasting, and from that point on it combines the Saturday and Sunday data together into a weekend load. NFS combines the data by comparing the Saturday and Sunday load for each switch pair for each individual hour and then selects the highest load for that switch pair and hour as the representative weekend load. In the load aggregation step, NFS applies nonconversation time factors to equate the erlang load obtained from billed erlang load to the actual holding time erlang load.

One of the first steps in load forecasting is to aggregate all of the end-office-to-end-office loads, up to the switch-pair level. This produces what we call the switch-to-switch traffic item sets. These switch-to-switch traffic item sets are then routed, using a fairly simple set of routing rules, to the candidate links. As discussed in Chapter 2, what the routing process does is to take a predetermined network structure, which includes candidate links, and determine the first-route path for the traffic item sets using a network structure function. The traffic item sets are then routed on these paths. In dynamic routing, candidate links exist between every pair of dynamic routing switches, and therefore all the dynamic routing switch pairs are fully interconnected with final links. Final links are low-blocking links that do not overflow. This procedure causes the hierarchical routing process to accumulate switch-to-switch loads for every dynamic routing switch pair. The use of final links for dynamic routing switch pairs is a key step in the load forecasting process and is also significant in the interface between hierarchical design and dynamic routing design, as discussed below.

NFS now projects those aggregated loads into the future. To do this, NFS uses a model called the sequential projection model [PaW82], which uses smoothing techniques to compare the current measured data with the previously projected data and determines an optimal estimate of the base and projected loads. The result is the initially projected loads that are ready for forecaster adjustments and business/econometric adjustments, which we now discuss.

17.3.1.3 Load adjustment cycle and view of business adjustment cycle. Once NFS smooths and projects the data, forecasters then enter a load adjustment cycle. This is an online process that has the capability to go into the projected load file for all the forecast periods for all the years and apply forecaster-established thresholds to those loads. For example, if the forecaster requests to see any projected load that has deviated more than 15 percent from what it was projected to be in the last forecast cycle, the load analysis module searches through all the switch pairs that the forecaster is responsible for, sorts out the ones that are beyond these thresholds, and prints them on a display. The forecaster then has the option to change the projected loads or accept them.

| AT&T COMMUNICATIONS | NFS "CURRENT VIEW LOAD - BASE/PROJECTED" ANALYSIS | FORM MXXXX |
| NFS | AAA I9XX VIEW | RETENTION CODE XXXXXXXXX |

AAAA AA AAAA   AAAA AA AAAA   XXXXXXX   CONTROL: AA99 A/B NON-CONTROL: AA99

YEAR	FORECAST PERIOD 1 (MAR)		FORECAST PERIOD 2 (MAY)		FORECAST PERIOD 3 (AUG)		FORECAST PERIOD 4 (NOV)	
	CCS LOAD		CCS LOAD		CCS LOAD		CCS LOAD	
LSP	BASE	PROJ	BASE	PROJ	BASE	PROJ	BASE	PROJ
19XX								
MORN	999999	999999	999999	999999	999999	999999	999999	999999
AFTN	999999	999999	999999	999999	999999	999999	999999	999999
EVNG	999999	999999	999999	999999	999999	999999	999999	999999
WKED	999999	999999	999999	999999	999999	999999	999999	999999
19XX								
MORN	999999	999999	999999	999999	999999	999999	999999	999999
AFTN	999999	999999	999999	999999	999999	999999	999999	999999
EVNG	999999	999999	999999	999999	999999	999999	999999	999999
WKED	999999	999999	999999	999999	999999	999999	999999	999999
19XX								
MORN	999999	999999	999999	999999	999999	999999	999999	999999
AFTN	999999	999999	999999	999999	999999	999999	999999	999999
EVNG	999999	999999	999999	999999	999999	999999	999999	999999
WKED	999999	999999	999999	999999	999999	999999	999999	999999
19XX								
MORN	999999	999999	999999	999999	999999	999999	999999	999999
AFTN	999999	999999	999999	999999	999999	999999	999999	999999
EVNG	999999	999999	999999	999999	999999	999999	999999	999999
WKED	999999	999999	999999	999999	999999	999999	999999	999999

REPORT BASED ON THE FOLLOWING INPUT PARAMETERS   (REG SEC:_ _ _ _ _   (cont: _ NON-CONT: _ BOTH_
(TS/TS:_ _)   (LATA/LATA:_ _ _ _ _ _ _ _ _ _)   (NPAINPA:_ _ _ _ _ _)

NFS.XXXX   (LAST) PAGE XXXX
VERSION NFSRPT.6

Figure 17.17   Base and projected load analysis

Figure 17.17 illustrates the format of one of the displays that comes out of the load analysis process, which the forecaster can obtain from the online system based on the threshold that the forecaster established. The report is for one switch pair for four quarterly periods: March, May, August, and November. As shown in Figure 17.17, NFS displays four years of information on one display with four forecast periods. The common language codes at the top represent the switch pair that is displayed, and the control information indicates who controls the report. The load information is the traffic load for a particular year and load set period, and it is given for both the base load and the projected load. Hence, with this display, the forecaster can see what the base load and the projected loads are at this point. The forecaster has the prerogative to change the projected loads, but not the base loads, in four representative load set periods: morning, afternoon, evening, and weekend. For example, if the forecaster adjusts the morning load from 1,000 to 1,200 CCS, the process takes all the morning hours between 7:00 A.M. and

AT&T COMMUNICATIONS NFS "VIEW/VIEW LOAD" ANALYSIS FORM MXXXX
NFS AAA I9XX VIEW RETENTION CODE XXXXXXXXX

AAAA AA AAAA   AAAA AA AAAA   XXXXXXX   CONTROL: AA99 A/B   NON-CONTROL: AA99

YEAR	FORECAST PERIOD 1 (MAR)			FORECAST PERIOD 2 (MAY)			FORECAST PERIOD 3 (AUG)			FORECAST PERIOD 4 (NOV)		
	PREVIOUS	PREVIOUS		PREVIOUS	PREVIOUS		PREVIOUS	PREVIOUS		PREVIOUS	PREVIOUS	
	UNADJ	UNADJ	%	UNADJ	UNADJ	%	UNADJ	UNADJ	%	UNADJ	UNADJ	%
LSP	CCS	CCS	CHG	CCS	CCS	CHG	CCS	CCS	CHG	CCS	CCS	CHG
19XX												
MORN	999999	999999	999	999999	999999	999	999999	999999	999	999999	999999	999
AFTN	999999	999999	999	999999	999999	999	999999	999999	999	999999	999999	999
EVNG	999999	999999	999	999999	999999	999	999999	999999	999	999999	999999	999
WKED	999999	999999	999	999999	999999	999	999999	999999	999	999999	999999	999
19XX												
MORN	999999	999999	999	999999	999999	999	999999	999999	999	999999	999999	999
AFTN	999999	999999	999	999999	999999	999	999999	999999	999	999999	999999	999
EVNG	999999	999999	999	999999	999999	999	999999	999999	999	999999	999999	999
WKED	999999	999999	999	999999	999999	999	999999	999999	999	999999	999999	999
19XX												
MORN	999999	999999	999	999999	999999	999	999999	999999	999	999999	999999	999
AFTN	999999	999999	999	999999	999999	999	999999	999999	999	999999	999999	999
EVNG	999999	999999	999	999999	999999	999	999999	999999	999	999999	999999	999
WKED	999999	999999	999	999999	999999	999	999999	999999	999	999999	999999	999
19XX												
MORN	999999	999999	999	999999	999999	999	999999	999999	999	999999	999999	999
AFTN	999999	999999	999	999999	999999	999	999999	999999	999	999999	999999	999
EVNG	999999	999999	999	999999	999999	999	999999	999999	999	999999	999999	999
WKED	999999	999999	999	999999	999999	999	999999	999999	999	999999	999999	999

REPORT BASED ON THE FOLLOWING INPUT PARAMETERS: (REG SEC:_ _ _ _)   (cont: _ NON-CONT: _ BOTH_   (TS CLLI:———)
(FROM FP:_ _ _ _ TO FP:_ _ _ _)   (CURRENT CCS VALUE:_ _ _ _ _ _)   (FROM % CHG:_ _ _ _ TO % CHG:_ _ _   POSITIVE:_ NEGATIVE:_ BOTH:_)

NFS.XXXX (LAST) PAGE XXXX
VERSION NFSRPT.6

Figure 17.18 Comparison of previous and current load projections

noon and prorates an adjustment in the same proportion as the 1,000-to-1,200 CCS adjustment; it will do likewise for any of the other load set periods. That adjustment is also made to all categories of service. If the forecaster wants to be specific and adjust only a specific category of service, there are displays that can be called up to adjust just the 800 service, just business service, just residence service, and so on.

Figure 17.18 illustrates the format of a display that gives a view-to-view load analysis. The display illustrated in Figure 17.17 compares the base load with the projected load; this display compares what a load was projected to be in the last forecast cycle with what it is projected to be in this forecast cycle. This display also shows all four quarterly periods at the same time and four years at one time by morning, afternoon, evening, and weekend load set periods. It shows the unadjusted traffic load projected in the previous forecast view. For instance, as in the previous example, if a switch-to-switch load is adjusted from 1,000 to 1,200

CCS for the fourth period of 1986, the current unadjusted CCS load might show up as 1,196 CCS, and the forecaster is able to see that the adjustment made in the last forecast view was appropriate. With this display the forecasters can track what they did in the last forecast view compared with what the current view looks like. Forecasters can also call up similar displays by category of service.

For the dynamic routing switch pairs, NFS assumes that there is a final link between all dynamic routing switch pairs, so as to properly aggregate traffic up to those pairs. When a dynamic routing switch pair is called up on these load adjustment displays, what the forecaster sees is the first-routed traffic that is used in the dynamic routing design step. NFS passes these switch-to-switch dynamic routing loads into the hierarchical design model, which sizes the final link. This occurs after NFS has gone through the lower-level links in the hierarchy, processed and sized them, and calculated their overflow. NFS routes the overflow up to the appropriate final links in the dynamic routing network. In this process, NFS captures the first-routed traffic, overflow traffic, and the characteristics of that overflow traffic in terms of peakedness and day-to-day variation.

After the adjustment cycle is complete and the forecasters have adjusted the loads to account for missing data, erroneous data, more accurate current data, or specifically planned events that cause a change in load, forecasters then apply the view of the business adjustments. Up to this point, the projection of loads has been based on mathematical formulas and network structure changes, as well as the base study period billing data. There has been no consideration yet regarding what the economy may do next year or the year after, what employment or unemployment may be like, what the gross national product will be, what disposable income may be like, and what the impact of competition or rate reductions or changes may be. The view of the business adjustment is intended to adjust the future network loads to compensate for the effects of competition, rate changes, and econometric factors on the growth rate.

The marketing analysis and forecasting process tries to encompass those factors in an adjustment that is applied to the NPA-to-NPA growth rates. One of the reports used in the business adjustment step is the NPA growth factor report, shown in Figure 17.19. This report is given by NPA pair and lists all 13 years of the forecast. There is a comparison of the market analysis adjustment to the growth rate with the growth rate estimated in the sequential projection model. The report shows the sequential projection algorithm (SPA) growth rate broken down by category of service for the average business day and for the weekend. It shows the growth rate for business, residence, OUTWATS, 800, and 900 service. There is a separate report for the overseas traffic that is being carried on the domestic network. We see the initial sequential projection model growth rate for this NPA pair and year and the market analysis adjustment factor to that growth rate. In this example the market adjustment tends to clip the growth rate down in all years for the business and residence average business day traffic. As we have said, the adjustment is applied by category of service. This report is broken down by year over the 13-year forecast; there is one of these reports for each of the March, May, August, and November study periods.

AT&T-TF
VIEW DEC 1985
DR/MK CYCLE BV
SPA CYCLE UN

NAP-NPA (N. AM.) GROWTH FACTOR REPORT

FORM NO. M3466
RET. CODE Q2R-01000

NPA-PAIR SYS LSP	215 804 COS	FORECAST PERIOD - NOV YEAR 0	YEAR 1	YEAR 2	YEAR 3	YEAR 4	YEAR 5	YEAR 6	YEAR 7	YEAR 8	YEAR 9	YEAR 10	YEAR 11	YEAR 12	YEAR 13
BR ABD SPA	MTS/BUS	0.9400 1.0526	0.9197 1.0499	0.9424 1.0473	1.0226 1.0451	0.9685 1.0431	0.9765 1.0413	0.9564 1.0396	0.9414 1.0397	0.9520 1.0398	0.9436 1.0398	0.9563 0.0399	0.9439 1.0400	0.9432 1.0400	0.9413 1.0401
BR ABD SPA	MTS/RES	0.9912 1.0411	0.9953 1.0395	1.2000 1.0382	1.0734 1.0367	1.1299 1.0354	1.1841 1.0342	0.9424 1.0330	1.0331 1.0331	1.0278 1.0332	1.0324 1.0332	1.0308 1.0333	1.0343 1.3334	1.0348 1.0335	1.0363 1.0335
BR WKE SPA	MTS/BUS	1.0100 1.0700	0.9053 1.0654	1.0828 1.0614	1.0816 1.0579	1.0581 1.0547	1.0470 1.0519	1.0007 1.0493	0.9826 1.0493	0.9794 1.0493	0.9831 1.0493	0.9770 1.0493	0.9793 1.0493	0.9761 1.0493	0.9738 1.0493
BR WKE SPA	MTS/RES	1.0559 1.0700	0.9341 1.0654	1.2000 1.0614	1.1588 1.0579	1.1491 1.0547	1.1226 1.0519	0.9348 1.0493	1.0526 1.0493	1.0454 1.0493	1.0562 1.0493	1.0521 1.0493	1.0541 1.0493	1.0527 1.0493	1.0527 1.0493
MK ABD SPA	OUTWATS	1.0538 1.0548	1.0366 1.0520	1.0254 1.0493	1.0010 1.0470	1.0608 1.0449	1.0677 1.0430	1.0119 1.0412	1.0254 1.0412	1.0359 1.0413	1.0360 1.0413	1.0334 1.0413	1.0340 1.0414	1.0354 1.0414	1.0419 1.0415
MK WKE SPA	OUTWATS	0.9742 1.0700	1.0499 1.0654	1.0631 1.0614	1.0759 1.0578	1.1412 1.0547	1.1481 1.0519	1.0778 1.0493	1.0847 1.0493	1.0958 1.0493	1.0834 1.0493	1.0890 1.0493	1.0896 1.0493	1.0902 1.0493	1.0962 1.0493
MK ABD SPA	800/SVC	1.1934 1.0555	1.0811 1.0525	1.0780 1.0501	1.0688 1.0477	1.0225 1.0455	1.0341 1.0435	1.0633 1.0417	1.0585 1.0417	1.0659 1.0418	1.0568 1.0418	1.0644 1.0419	1.0620 1.0419	1.0598 1.0420	1.0602 1.0420
MK WKE SPA	800/SVC	1.1952 1.0700	1.0738 1.0654	1.0326 1.0614	1.0539 1.0579	1.0074 1.0547	1.0034 1.0519	1.0426 1.0493	1.0314 1.0493	1.0381 1.0493	1.0233 1.0493	1.0304 1.0493	1.0292 1.0493	1.0264 1.0493	1.0261 1.0493
MK ABD SPA	900/SVC	NC 1.0387	NC 1.0376	NC 1.0344	NC 1.0333	NC 1.0322	NC 1.0312	NC 1.0302	NC 1.0306	NC 1.0303	NC 1.0303	NC 1.0306	NC 1.0305	NC 1.0307	NC 1.0302
MK WKE SPA	900/SVC	NA 1.0700	1.0654	1.0614	1.0578	1.0547	1.0518	1.0493	1.0493	1.0493	1.0493	1.0493	1.0493	1.0493	1.0493

Figure 17.19 NPA growth factor report

17.3.2 Network design

We now have our dynamic routing switch-pair loads, adjusted by the forecasters, and have applied the view of the business adjustments to those projected loads. Now we are ready to design the network for those loads. The first step is the hierarchical design. As discussed earlier, NFS takes all of the loads and all of the candidate links through the hierarchical design process. Out of that process, NFS generates a set of engineering data records, such as the link record illustrated in Figure 17.20, which is for the Richmond–White Plains link.

This report shows administrative information across the top: the retention periods, the codes, the form numbers, and the regional or sectional capacity manager who is responsible for this particular link. Then the link itself is identified: Richmond, Virginia, to White Plains, New York. Both of these switches are shown as dynamic routing switches. After the common language codes, the report gives the mileage of the link. The airline mileage is calculated from the V and H coordinates in the intercity switch master files and is multiplied by an adjustment factor to determine route miles. This link record is broken down in detail for the first five years of the forecast into 13 forecast periods. Because these data are from the December 1985 view, the first period to show up is the second period of 1986, then the third and fourth periods of 1986. In year 2, there are the first, second, third, and fourth periods of 1987; in year 3, we have the first, second, third, and fourth periods of 1988. Year 4 is represented by the last quarterly period of 1989, 4P89; year 5 is represented by 4P90. Effectively, there are five years of network structure, the first three years with detail down to individual quarterly period and the last two years with just the year-end network configuration. This report gives all of the engineering information relative to this link.

As stated earlier, all dynamic routing switch pairs in the hierarchical design are final links, which are sized in the hierarchical network design process just like final links in the real network. This report is also used for the actual hierarchical links. First, it gives the peakedness factor for the first-routed traffic as 1.0, and then the first-route load between Richmond and White Plains in the second period of 1986 as 2,493 CCS. Because this link has come through the hierarchical design process, it has accumulated 2,769 CCS of overflow load. Because it is a final link, it traps all overflow load from links beneath it but does not overflow to any other links. The total offered load, then, is 5,262 CCS, with a link peakedness factor of 2.35. Also included in the report is other information on these switch pairs from the switch master file. The type of signaling is CC, which means these two switches use common-channel signaling to communicate with one another. The report also lists the load set period that is used to size the link; E indicates that it is an evening load set period that is used to size the link and also determine the peakedness factor. If there is a cutover code involved, it is flagged in the appropriate forecast period and year. There are several cutover codes for this link, which indicates that there is a lot of cutover activity around Richmond and White Plains.

The modular indicator shows that this link is sized using modular engineering rules. The engineering schedule in this case indicates that the link is sized

AT&T-C MOST LIKELY
DEC 1985 VIEW EA07 PR
EASTERN REGION SECTION 07

PRELIMINARY ROUTING
LINK RECORD

FORM NO. MI 139
RET. CODE 02P-01000

LINK RCMD VA IT03T - WHPL NY 0504T MILEAGE: 412   SUB-GRP START 00 SUB-GRP PERCENT 50 25 25

	YEAR 1			YEAR 2				YEAR 3				YEAR 4	YEAR 5
	2P86	3P86	4P86	1P87	2P87	3P87	4P87	1P88	2P88	3P88	4P88	4P89	4P90
IRS PEAK CTR	1.00	1.00	1.00	1.00	1.00	1.00	1.00	1.00	1.00	1.00	1.00	1.00	1.00
LOAD CCS	2493	2397	2435	2768	3530	2890	2894	3063	3902	3185	3152	3511	4090
LL OFLO REC	2769	2485	2014	2995	19	104	28	24	35	2	39	63	126
AC OFLO REC	0	0	0	0	0	0	0	0	0	0	0	0	0
LINK ADJMT	0	0	0	0	0	0	0	0	0	0	0	0	0
TOT OFRD CCS	5262	4882	4449	5763	3549	2994	2922	3087	3937	3187	3191	3574	4216
LINK PEAK FCTR	2.35	2.39	2.32	2.23	1.01	1.04	1.02	1.01	1.02	1.00	1.02	1.03	1.05
M/O	CC	CC	CC	CC	CC	CC	CC	CC	CC	CC	CC	CC	CC
LSP	E	E	E	E	E	E	E	E	E	E	E	E	E
CUTOVER CODE:		WH		RG		WL		RH					
MOD IND	M	M	M	M	M	M	M	M	M	M	M	M	M
ENG SCH NON MCD	B.01M	B.01M	B.01M	B.01M	B.01L	B.01L	B.01L	B.01L	B.01L	B.01L	B.01L	B.01L	B.01L
MOD OR LTCCS													
MOD TRK CHANGE	5	6	-3	2	-2	3	5	0	-2	-3	-3	-3	1
TRUNK FU 2W	2204	192	168	216	120	108	108	108	132	108	108	120	144
REQMT 1W IN	0	0	0	0	0	0	0	0	0	0	0	0	0
1W OUT	0	0	0	0	0	0	0	0	0	0	0	0	0
TRUNK IF 2W	0	0	0	0	0	0	0	0	0	0	0	0	0
1W IN	0	0	0	0	0	0	0	0	0	0	0	0	0
1W OUT	0	0	0	0	0	0	0	0	0	0	0	0	0
TRUNK TR 2W	0	0	0	0	0	0	0	0	0	0	0	0	0
1W IN	0	0	0	0	0	0	0	0	0	0	0	0	0
1W OUT	0	0	0	0	0	0	0	0	0	0	0	0	0
TOTAL REQMT	204	192	168	216	120	108	108	108	132	108	108	120	144
CCS CARRIED	5210	4833	4406	5705	3515	2964	2894	3057	3898	3155	3159	3538	4174
CCS OVERVIEW	0	0	0	0	0	0	0	0	0	0	0	0	0
VARIANCE	0	0	0	0	0	0	0	0	0	0	0	0	0

Figure 17.20 Richmond–White Plains link record

as a final link to a B.01M blocking probability grade-of-service objective with medium (M) day-to-day variations. As discussed in Chapter 2, the B.01M objective indicates that the Neal-Wilkinson sizing model is used for an average blocking of 1 percent, with medium day-to-day variation considered. This is true for the second, third, and fourth periods of 1986 and the first period of 1987. Notice what happened in the second period of 1987. The day-to-day variation drops to L, which means low day-to-day variation. At the same time, the link peakedness factor drops from 2.23 down to 1.01. In fact, the link carries almost nothing but first-routed traffic. When we discuss the next report, we will see what caused this to happen. As NFS modularizes the link, the modular trunk change indicates that the change for 2P86 is five additional trunks, which is the final modular size compared with the exact size. The report shows that in some cases there is a shortage of up to three trunks. The next item gives the number of trunks, which are all two-way trunks. There is a capability here to break down the trunks to one-way in, one-way out, and so forth, if necessary. In dynamic routing, all trunks are two-way trunks.

The final detail on this report is the carried traffic load. Because this is a final link, the bulk of the traffic is carried; no traffic overflows to another link. For high-usage links there is traffic that overflows, and NFS also calculates the variance of that overflow. This information on the CCS overflow and variance of the overflow is carried forward in the hierarchical design process to the alternate path.

The next report in this series, shown in Figure 17.21, is the link-routed items report. This report shows all the items of traffic that are routed on this link in the load aggregation module, which aggregates traffic items to the candidate links. It shows each of the individual traffic items that were routed from subtending switches up to this link. It identifies those items by direction and gives the percent contribution to the total load in that particular direction. Again, there is the same type of administrative information: the link identifier (Richmond–White Plains) and the forecast periods for each of the five years. This report also gives the base-year configuration and shows that the base year is 4P85. Under the heading 2P86, the number 21 means that this load is representative of hour 21 (9 P.M., central time or network time). As indicated earlier for the link record, NFS uses the evening load set period to size the link; this report shows 21E, which means that hour 21 from the evening load set period is the sizing hour. The report shows the total traffic, in and out: 1,349 and 1,144 CCS, respectively, for a total first-route load of 2,493 CCS.

The next item of relevant information on the link-routed items report is the breakdown of the loads in the morning, afternoon, and evening load set periods. Here, the report shows that the evening load is 2,493 CCS. The morning load is actually higher—2,644 CCS—the afternoon load is 2,181 CCS, and the weekend load is 1,831 CCS. With the network cluster concept discussed in Chapter 2, we do not always have to use the highest load to size the link. There is a breakdown of the 2,493 CCS by category of service (information carried through from the beginning), which shows how much of this traffic is business, residence, OUTWATs, 800, and overseas service.

AT&T-C MOST LIKELY
DEC 1985 VIEW EA07 PR

PRELIMINARY ROUTING
LINK ROUTED ITEMS

FORM NO. M1133
RET. CODE 02P-01000

EASTERN REGION SECTION 07   CUTOFF OO CCS

LINK RCMD VA IT03T - WHPL NY 0504T LL

	BASE YR	YEAR 1			YEAR 2				YEAR 3				YEAR 4	YEAR 5
	4P85	2P86	3P86	4P86	1P87	2P87	3P87	4P87	1P88	2P88	3P88	4P88	4P89	4P90
	H21 LE	21 E	21 E	20 E	20 E	21 E	21 E	19 E	20 E	21 E	19 E	19 E	19 E	19 E
TOTAL IN TO A OFFICE	875	1349	1308	1290	1489	1917	1575	1583	1677	2119	1757	1727	1980	2257
TOTAL OUT OF A OFFICE	796	1144	1088	1144	1278	1613	1314	1310	1385	1781	1428	1424	1581	1832
GRAND FIRST ROUTE	1671	2493	2396	2434	2768	3530	2889	2893	3062	3901	3185	3152	3511	4090
TOTAL FIRST ROUTE BY TIME OF DAY														
MORNING	2017	2644	2255	2412	2608	2599	2773	2617	2804	2653	2898	2670	2842	3111
AFTERNOON	1842	2181	1866	2514	2188	2603	2563	2690	2486	2649	2614	2766	2939	3214
EVENING	1671	2493	2396	2434	2768	3530	2889	2893	3062	3901	3185	3152	3511	4090
WEEKEND	2846	1831	2992	3153	2395	3111	3548	3639	2828	2988	3954	4176	4623	5211
INTERSTATE FIRST ROUTE TRAFFIC BY CL. OF SVC.														
MTS-BUS	211	223	286	247	283	270	340	359	353	341	438	465	533	618
MTS-RES	1404	2079	1974	1974	2150	2949	2374	2198	2404	3271	2586	2369	2670	3168
CW	30	86	37	130	61	92	37	179	64	98	38	184	196	207
800 SVC	5	9	15	20	39	10	17	9	43	11	19	10	11	11
900 SVC														
OVS MTS	20	94	81	62	229	208	120	144	195	180	101	122	101	80
TOTAL INTERSTATE	1671	2493	2395	2435	2765	3532	2891	2892	3061	3903	3185	3152	3513	4085
INTERSTATE FIRST ROUTE TRAFFIC BY CL. OF SVC.														
MTS-BUS														
MTS-RES														
CW														
800 SVC														
900 SVC														
OVS MTS														
TOTAL INTERSTATE														
INTERSTATE FIRST ROUTE TRAFFIC BY CL. OF SVC.														
MTS-BUS	211	223	286	247	283	270	340	359	353	341	438	465	533	618
MTS-RES	1404	2079	1974	1974	2150	2949	2374	2198	2404	3271	2586	2369	2670	3168
CW	30	86	37	130	61	92	37	179	64	98	38	184	196	207
800 SVC	5	9	15	20	39	10	17	9	43	11	19	10	11	11
900 SVC														
OVS MTS	20	94	81	62	229	208	120	144	195	180	101	122	101	80
GRAND TOTAL	1671	2493	2395	2435	2765	3532	2891	2892	3061	3903	3185	3152	3513	4085

Figure 17.21 Link-routed items report

Figure 17.21 lists only the first-route items of traffic on a link. Figure 17.22, which is a continuation of Figure 17.21, illustrates the overflow items of traffic. This shows each individual item of overflow load from each subtending link beneath the Richmond–White Plains link. In the listing of overflow items, the Albany-to-Richmond item is at the top, which comes about because Albany subtends White Plains. The overflow is given by direction, and this top item is in direction 1, which means that 911 CCS of the load from the Albany switch to the Richmond switch overflows. In the other direction, 702 CCS overflow from the Richmond switch to the Albany switch. The report shows the subtending networks, which include Buffalo–Richmond, where Buffalo subtends White Plains; Cambridge–Richmond, where Cambridge subtends White Plains; Norfolk–White Plains, where Norfolk subtends Richmond; New Haven–Richmond, where New Haven subtends White Plains; New York–Richmond, where New York subtends White Plains; Richmond–Springfield, where Springfield subtends White Plains; and Roanoke–White Plains, where Roanoke subtends Richmond.

Recall from the link record (Figure 17.20) that in 2P87 there is a very significant change in the peakedness factor and in the day-to-day variation. In Figure 17.22, 2P87 indicates that virtually all of the overflow disappears. In that period all of the subtending switches except Roanoke—Albany, Buffalo, Cambridge, New Haven, Norfolk, New York City, and Springfield—become dynamic routing switches. Hence, they are all connected to Richmond with a final link and to White Plains with a final link, and they no longer overflow any traffic to the Richmond–White Plains final link.

This is the type of information that is output for all of the artificial, hierarchical final links that provide the interface to the dynamic routing network. This traps the switch-to-switch load and the overflow offered to the dynamic routing network and calculates the peakedness and the day-to-day variation factor, which is everything needed to run the path-erlang flow optimization (path-EFO) model, as far as loads and their traffic characteristics. The other information needed to run the path-EFO model is the set of control parameters shown in Figure 17.23. These parameters include switch and transport information necessary to drive the path-EFO model. The dynamic routing intertandem link file is generated starting from the dynamic routing flag in the intercity switch master file and the common master file.

A list of all the dynamic routing switch pairs is sent to the transport planning database, from which is extracted transport information relative to the transport network between the switch pairs on that list. Forecasters combine that information with other planning detail relative to the time period during which the switches are coming into the dynamic routing network, who the responsible forecasters are, what type of switch it is, and what type of termination is on the switch (digital or analog). With that transport information, forecasters use the incremental cost files to create a cost per trunk mile; the route cost multiplier represents the overall mileage of the transport path. Forecasters also provide termination cost per trunk, minimum and maximum link sizes, modular indicator, and satellite transport indicator. Another item of information that goes into this file is very relevant to driving the path-EFO model; that is the in-service

AT&T-C MOST LIKELY
DEC 1985 VIEW EA07 PR
EASTERN REGION SECTION 07   CUTOFF OO CCS
PRELIMINARY ROUTING
LINK RECORD
FORM NO. M1133
RET. CODE 02P-01000

NY 0504T - WHPL

LINK RCMD VA IT03T - WHPL	LL	BASE YR 4P85 H21 LE	2P86	3P86	4P86	1P87	2P87	3P87	4P87	1P88	2P88	3P88	4P88	4P89	4P90
			YEAR 1				YEAR 2				YEAR 3			YEAR 4	YEAR 5
OVERFLOW			21 E	21 E	20 E	20 E	21 E	21 E	19 E	20 E	21 E	19 E	19 E	19 E	19 E
ALBYNYSS05T - RCMDVAIT03T	1		911	747	456	953									
ALBYNYSS05T - RCMDVAIT03T	2		702	736	480	974									
BFLONYFRO5T - RCMDVAIT03T	1				29	125									
BFLONYFRO5T - RCMDVAIT03T	2				34	98									
CMBRMA0119T - RCMDVAIT03T	1			0	0	23									
CMBRMA0119T - RCMDVAIT03T	2			0	0	21									
NRFLVA0S03T - WHPKBY0504T	1		75	59	58	75									
NRFLVA0S03T - WHPLNY0504T	2		56	74	67	141									
NWHNCT0205T - RCMDVAIT03T	1		169	25	106	9									
NWHNCT0205T - RCMDVAIT03T	2		157	25	101	9									
NYCONYRP08T - RCMDVAIT03T	1		94	126	112	117									
NYCONYRP08T - RCMDVAIT03T	2		105	144	111	116									
RCMDVAIT03T - SPFDMABR02T	1		124	258	230	185									
RCMDVAIT03T - SPFDMABR02T	2		195	238	219	144									
RONKVALK02T - WHPLNY0504T	1		108	31											
RONKVALK02T - WHPLNY0504T	2		75	23											
RONKVALK03T - WHPLNY0504T	1				8	3	12	55	18	12	20	1	24	39	79
RONKVALK03T - WHPLNY0504T	2				5	4	8	49	11	13	15	1	15	24	47
TOTAL INTERSTATE OVERFLOW			2769	2485	2014	2995	19	104	28	24	35	2	39	63	126
TOTAL INTRASTATE OVERFLOW			2769	2485	2014	2995	19	104	28	24	35	2	39	63	126
GRAND TOTAL OVERFLOW			2769	2485	2014	2995	19	104	28	24	35	2	39	63	126

Figure 17.22 Overflow traffic from Figure 17.21

AT&T-COMMUNICATIONS
TRUNK FORECASTING
FROZEN
DNH INTERTANDEM LINK FILE
EASTERN REGION
SECTION 07
FORM M3479
RETENTION CODE 02R-01000

DNH TOLL SWITCH: RCMDVAIT03T   TYPE: D

DECEMBER 1984 VIEW

NEW REC	UPD REC	TS CLLI	TYPE	..PERIOD.. FROM	TO	COMP	ENG RESP	TRKS PEND	EFF DATE	IN SVC	TERM-TYPE TS-A	TS-B	COST PER TRK MILE	RC MULT	TERM COST PER TRUNK	MIN LINK	MAX LINK	MOD IND	FAX TYP	SAT IND
		WCHTKSBR24T	D	2P85	4P89	LL	MW03	0	/ /	24	D	D	1.50	1.55	0.00	3	9999	Y		N
		WHPLNY0504T	D	2P85	4P89	LL	EA07	0	/ /	120	D	D	1.50	1.22	0.00	3	9999	Y		N
		WKSHWI0231T	D	4P87	4P89	LL	CE11	0	/ /	60	D	D	1.50	1.45	0.00	3	9999	Y		N
		WPBHFLAN04T	D	2P87	4P89	LL	S007	0	/ /	36	D	D	1.50	1.71	0.00	3	9999	Y		N

TF50.KP150P 1
VERSION 1   EASTERN REGION   07   11-15-84   RCMDVA1T03T   PAGE 12

Figure 17.23 Dynamic routing intertandem link file

trunk quantity. If the dynamic routing switch pair is in service in the dynamic routing network, forecasters provide the actual in-service dynamic routing trunk quantities or those that are planned to be in service within the next 60–90 days. This file then supplies the physical transport information, cost information, and in-service trunk information to the path-EFO model.

Once this information is provided to the intertandem link file and all of the load information is available from hierarchical design, NFS has everything necessary to run the path-EFO model. Once the information has been processed in the path-EFO model, NFS outputs the dynamic routing trunk forecast report, shown in Figure 17.24. This report lists all the switch pairs in alphabetical order and the trunk requirements between those switch pairs as they evolve for each forecast period in the path-EFO model. For Richmond–White Plains the report shows that there are trunks in forecast period 2, which indicates that this switch pair is in the dynamic routing network, and also the trunk requirements as they are developed from the path-EFO model. Some links start to show up in forecast period 6, which is the second period of 1987, when there is a second cutover of dynamic routing switches; there are some other links that show up in forecast period 8, when another set of switches is cut over to dynamic routing.

We get more detailed information in the dynamic routing link summary report, shown in Figure 17.25. This report gives the routing tables by the 10 load set periods of the average business day and the five load set periods for the weekend. For the Richmond–White Plains switch pair, the routing tables for all 10 of the average business day load set periods are shown, as is the load that is offered to the switch pair; this is the load that we brought up from hierarchical design. There is the switch-to-switch blocking for this switch pair in that load set period. The routing paths are flagged to distinguish the engineered paths from the real-time paths. For load set period 1, which is early in the morning, the load between Richmond and White Plains is 3,130 CCS, with virtually no blocking at that time of day. The first path is the direct link, Richmond–White Plains or White Plains–Richmond, and this is also the only engineered path. Then there are the real-time paths that are available in this routing table: via New York City (Rego Park); Washington, D.C.; New York City (Broadway 24); Wayne; Baltimore; or Pittsburgh. As we progress through the day, load set periods 7, 8, and 9 have the biggest loads. Load set period 7 has 4,215 CCS, but there is still just one direct-link engineered path. In load period 3 there are only 3,144 CCS, but now there are four engineered paths needed to meet the switch-to-switch blocking criterion in load set period 3. This is expected, because load set period 3 is generally when the network as a whole loads up the most, and more paths are required to meet the blocking probability grade-of-service objective.

The output of the path-EFO model is put into the general forecast file, and it has all of the detailed routing information in it. This file contains all the paths, how much load is routed on each of the paths, how much is via traffic, and how much is first-routed traffic. Once the path-EFO model has run for a forecast cycle, the general forecast file and routing information are sent downstream to the provisioning systems, planning systems, and capacity management system, and the capacity manager takes over from there as far as implementing the routing

DYNAMIC ROUTING TRUNK FORECAST
DEC 1985 VIEW
PR CYCLE
ABD STUDY TYPE

FORM M3906
RETENTION CODE: 02P-13000

LOW TS	HIGH TS	FP02	FP03	FP04	FP05	FP06	FP07	FP08	FP09	FP10	FP11	FP12	FP13	FP14
PTLDOR6203T	SNDGCA0787T	*	*	*	*	*	*	180	180	180	180	180	228	324
PTLDOR6203T	SNFCCA2147T	*	*	*	*	*	*	312	336	336	336	336	336	360
PTLDOR6203T	SNJSCA0241T	240	360	360	396	396	444	444	444	444	444	444	444	444
PTLDOR6203T	SPFDILSD5IT	*	*	*	*	*	*	36	36	36	36	36	48	60
PTLDOR6203T	SPFDMABR02T	*	*	*	*	24	24	24	24	24	24	24	24	36
PTLDOR6203T	SPFDMOTL04T	*	*	*	*	*	*	12	12	12	12	12	12	24
PTLDOR6203T	SPKNWA0102T	*	*	*	*	324	348	348	348	360	480	480	504	540
PTLDOR6203T	STLSMO0934T	84	84	84	84	*	*	84	84	84	84	84	84	84
PTLDOR6203T	STTLWA0604T	*	*	*	*	2784	3264	3264	3384	3528	4152	4152	4392	4896
PTLDOR6203T	SYRCNYSU13T	*	*	*	*	12	12	12	12	12	12	12	12	12
PTLDOR6203T	TAMPFLCO02T	*	*	*	*	24	24	24	24	24	24	24	24	24
PTLDOR6203T	TOLDOH2103T	*	*	*	*	*	*	24	24	24	24	24	24	24
PTLDOR6203T	TULSOKTB04T	*	*	*	*	*	*	24	24	36	36	36	36	36
PTLDOR6203T	WASHDCSW06T	36	36	36	36	84	84	84	84	84	84	84	96	96
PTLDOR6203T	WAYNPALA42T	36	36	36	36	36	36	36	36	36	36	36	36	36
PTLDOR6203T	WCHTKSBR24T	36	36	36	36	36	36	36	36	36	36	36	36	36
PTLDOR6203T	WHPLNY0504T	60	60	60	60	60	60	60	60	60	60	60	60	60
PTLDOR6203T	WKSHWI0231T	*	*	*	*	*	*	72	72	72	72	72	72	72
PTLDOR6203T	WPBHFLAN04T	*	*	*	*	*	0	0	0	0	0	0	12	12
RCMDVAIT03T	RCPKNJ0203T	*	*	*	*	60	60	60	60	60	60	60	60	72
RCMDVAIT03T	RENONV0344T	*	*	*	*	*	*	12	12	12	12	12	12	12
RCMDVAIT03T	SCRMCA0404T	48	48	48	48	48	48	48	48	48	48	48	48	48
RCMDVAIT03T	SHOKCA0296T	*	*	*	*	*	*	48	48	48	48	48	48	48
RCMDVAIT03T	SKTNCA0107T	*	*	*	*	*	*	24	24	24	24	24	24	24
RCMDVAIT03T	SLKCUTMA02T	24	24	24	24	24	24	24	24	24	24	24	24	24
RCMDVAIT03T	SNANTXCA02T	*	*	*	*	*	*	36	36	36	36	36	36	36
RCMDVAIT03T	SNDGCA0787T	*	*	*	*	*	*	36	36	36	36	36	36	36
RCMDVAIT03T	SNFCCA2147T	*	*	*	*	*	*	24	24	24	24	24	24	24
RCMDVAIT03T	SNJSCA0241T	24	24	24	24	24	24	36	36	36	36	36	36	36
RCMDVAIT03T	SPFDILSD51T	*	*	*	*	*	*	0	0	0	0	0	0	0
RCMDVAIT03T	SPFDMABR02T	36	36	36	36	36	36	36	36	36	36	36	36	36
RCMDVAIT03T	SPFDMOTL04T	*	*	*	*	*	*	48	48	48	48	48	48	48
RCMDVAIT03T	SPKNWA0102T	*	*	*	*	*	*	36	48	48	48	48	48	48
RCMDVAIT03T	STLSMO0934T	36	36	36	36	36	36	36	36	36	36	36	36	36
RCMDVAIT03T	STTLWA0604T	24	24	24	24	24	24	24	24	24	24	24	24	24
RCMDVAIT03T	SYRCNYSU13T	*	*	*	*	*	*	24	24	24	24	24	24	24
RCMDVAIT03T	TAMPFLCO02T	*	*	*	*	*	*	24	24	24	24	24	24	24
RCMDVAIT03T	TOLDOH2103T	*	*	*	*	*	*	24	24	24	24	24	24	24
RCMDVAIT03T	TULSOKTB04T	*	*	*	*	*	*	24	24	24	24	24	24	24
RCMDVAIT03T	WASHDCSW06T	*	*	*	*	528	600	600	600	624	624	624	696	696
RCMDVAIT03T	WAYNPALA42T	264	264	264	252	252	252	252	264	264	264	264	264	264
RCMDVAIT03T	WCHTKSBR24T	24	24	24	24	24	24	24	24	24	24	24	24	24
RCMDVAIT03T	WHPLNY0504T	180	180	192	216	216	216	216	216	216	216	216	228	252
RCMDVAIT03T	WKSHWI0231T	*	*	*	*	*	*	60	60	60	60	60	60	60
RCMDVAIT03T	WPBHFLAN04T	*	*	*	*	*	36	36	36	36	36	36	36	36

Figure 17.24 Dynamic routing trunk forecast report

AT&T COMMUNICATIONS
TF
FORECAST PERIOD 14
PT-PT PAIR SUMMARY

DYNAMIC ROUTING LINK SUMMARY
DEC 1985 VIEW
PR CYCLE
ABD STUDY TYPE

FORM M3905
RETENTION CODE: 02P-13000

RCMDVAIT03T - WCHTKSBR241

	LSP1	LSP2	LSP3	LSP4	LSP5	LSP6	LSP7	LSP8	LSP9	LSP10
LOAD: 500 BLKG: .0000	500 .0000	633 .0028	552 .0017	357 .0005	475 .0050	474 .0044	541 .0000	551 .0003	546 .0008	389 .0011
01	WCHTKSBR24T	WCHTKSBR24T	WCHTKSBR24T	WCHTKSBR24T	OKCYOKCE04T	WCHTKSBR24T	WCHTKSBR24T	WCHTKSBR24T	WCHTKSBR24T	WCHTKSBR24T
02	DLLSTXTL341	TULSOK1B04T	WAYNPALA42T	HSTNTX0I44T*	PITBPADG43T	WAYNPALA42T*	DLLSTXTL34T	DLLSTXTL34T	WKSHWI0231T	NSVLTNMT43T*
03	CHCGILCL57T*	WAYNPALA42T	PITBPADG43T	CHCGILCL57T*	CLSPCOMA02T	NYCQNYRP08T*	WHPLNY0504T*	WAYNPALA42T*	WHPLNY0504T*	CHCGILCL57T*
04	DNVRCOZJ05T*	WHPLNY0504T	WHPLNY0504T	PITBPADG43T*	KSCYMO0904T	NYCMNYBW24T*	CHCGILCL57T*	CNCNOHWS14T*	KSCYMO0904T*	NWHNCT0205T*
05	FTWOTXED24T*	CHCGILCL57T*	GNBONCEU03T	NYCQNYRP08T*	NYCQNYRP08T	KSCYMO0904T*	ATLNGATL01T*	ATLNGAYL01T*	CHCGILCL57T*	NYCMNYBW24T*
06	HSTNTX01441T*	ATLNGATL01T*	ARTNVACK04T	LTRKARFRI5T*	GNBONCEU03T	OKCYOKCE04T*	NYCMNYBW24T*	OKBRILOA52T*	ATLNGATL01T*	OKBRILOA52T*
07	ATLNGATL01T*	CLSPCOMA02T*	CHCGILCL57T	ATLNGATL01T*	TULSOKTB04T*	CHCGILCL57T*	WKSHWI0231T*	CHCGILCL57T*	DLLSTXTL34T*	ATLNGATL01T*
08	NYCQNYPR08T*	KSCYMO0904T*	ATLNGATL01T*	OKBRILOA52T*	CLMBOH1103T*	PITBPADG43T*	OKBRILOA52T*	WHPLNY0504T*	PITBPADG43T*	ATLNGATL01T*
09	PITBPADG43T*	DNVRCOZJ05T*	CLMBOH1103T*	WKSHWI0231T*		ATLNGATL01T*		PITBPADG43T*	WAYNPALA42T*	PHLAPASL42T*
10	WAYNPALA42T*	DLLSTXTL34T*		OKCYOKCE04T*		DLLSTXTL34T*		NYCMNYBW24T*	NSVLTNMT43T*	PITBPADG43T*
										WAYNPALA42T*

RCMDVAIT03T-WHPLNY0504T

	LSP1	LSP2	LSP3	LSP4	LSP5	LSP6	LSP7	LSP8	LSP9	LSP10
LOAD: 3130 BLKG: .0000	3130 .0000	3328 .0024	3144 .0003	2562 .0000	3345 .0040	3023 .0000	4215 .0001	4062 .0017	3926 .0047	3296 .0000
01	WHPLNY0504T	WHPLNY0504T	WHPLNY0504T	WHPLNY0504T	WHPLNY0504T	WHPLNY0504T	WHPLNY0504T	WHPLNY0504T	WHPLNY0504T	WHPLNY0504T
02	NYCQNYPR08T*	WAYNPALA42T*	WASHDCSW06T*	ARTNVACK04T*	NYCQNYRP08T	NYCMNYBW24T*	PHLAPASL42T*	PHLAPASL42T	WAYNPALA42T	HRBGPAHA42T*
03	WASHDCSW06T*	NYCQNYRP08T*	BLTMMDCH01T	NYCQNYRP08T*	NYCMNYBW24T*	NYCQNYRP08T*	WAYNPALA42T*	BLTMMDCH01T*	RCPKNJ0203T*	WASHDCSW06T*
04	NYCMNYBW24T*	NYCMNYBW51T*	NYCMNYBW24T	WAYNPALA42T*	WAYNPALA42T*	NYCQNYRP08T*	NYCMNYBW51T*	NYCMNYBW24T*	PHLAPASL42T*	WAYNPALA42T*
05	WAYNPALA42T*	NYCMNY5450T*	NYCQNYRP08T*	NYCMNYBW24T*	NYCMNY5450T*	WASHDCSW06T*	NYCMNYBW24T*	WASHDCSW06T*	PITBPADG43T*	BLTMMDCH01T*
06	BLTMMDCH01T*	NRFLVABS03T*	NYCMNY5450T*	BLTMMDCH01T*	BLTMMDCH01T*	BLTMMDCH01T*	PITBPADG43T*	PITBPADG43T*	WASHDCSW06T*	NYCMNYBW24T*
07	PITBPADG43T*	NWRKNJ0208T*	NWRKNJ0208T*	PITBPADG43T*	RCPKNJ0203T*	PITBPADG43T*	WASHDCSW06T*	WAYNPALA42T*	ARTNVACK04T*	PITBPADG43T*

RCMDVAIT03T-WKSHWI0231T

	LSP1	LSP2	LSP3	LSP4	LSP5	LSP6	LSP7	LSP8	LSP9	LSP10
LOAD: 1654 BLKG: .0003	1654 .0003	1977 .0020	2032 .0015	1414 .0021	2464 .0037	1833 .0008	1224 .0008	1283 .0000	1582 .0001	1377 .0030
01	WKSHWI0231T	WKSHWI0231T	WKSHWI0231T	WKSHWI0231T	WKSHWI0231T	WKSHWI0231T	WKSHWI0231T	WKSHWI0231T	WKSHWI0231T	WKSHWI0231T
02	ARTNVACK04T	OKBRILOA52T*	GNBONCEU03T	NSVLTNMT43T*	OKBRILOA52T	OKBRILOA52T	WAYNPALA42T*	CHCGILCL57T	CLEVOH0203T	PHLAPASL42T*
03	AKRNOH2503T*	GDRPMIBL50T	NRFLVABS03T	DTRTMIBH50T*	PITBPADG43T	NYCQNYRP08T	WASHDCSW06T*	WAYNPALA42T*	OKBRILOA52T	CHRLNCCA03T*
04	ATLNGATL01T*	CHCGILCL57T	ARTNVACK04T	NYCQNYRP08T*	WAYNPALA42T	ARTNVACK04T*	GNBONCEU03T*	NYCMNYBW51T*	WHPLNY0504T*	OKBRILOA52T*
05	NYCQNYRP08T*	WASHDCSW06T*	CLMBOH1103T*	WAYNPALA42T*	ARTNVACK04T*	WHPLNY0504T*	ATLNGATL01T*	OKBRILOA52T*	HRBGPAHA42T*	NYCMNYBW24T*
06	CHCGILCL57T*	ARTNVACK04T*	GDRPMIBL50T*	PITBPADG43T*	NRFLVABS03T*	WAYNPALA42T*	NYCMNYBW24T*	NYCMNYBW24T*	WAYNPALA42T*	ATLNGATL01T*
07	PITBPADG43T*	MPLSMNDT18T*	CHCGILCL57T*	ATLNGATL01T*	BLTMMDCH01T*	PITBPADG43T*	WHPLNY0504T*	WHPLNY0504T*	ATLNGATL01T*	CHCGILCL57T*
08	WAYNPALA42T*	NRFLVABS03T*	BLTMMDCH01T*	WAYNPALA42T*	GDRPMIBL50T*	CHCGILCL57T*	NYCMNYBW24T*	PITBPADG43T*	PITBPADG43T*	PITBPADG43T*
09	NYCMNYBW24T*	WAYNPALA42T*	OKBRILOA52T*	NYCMNYBW24T*	CHCGILCL57T*	NYCMNYBW24T*	PITBPADG43T*	PHLAPASL42T*	CHCGILCL57T*	WHPLNY0504T*
10	WHPLNY0504T*	PITBPADG43T*	WASHDCSW06T*	WHPLNY0504T*	WASHDCSW06T*	ATLNGATL01T*	CHCGILCL57T*		PHLAPASL42T*	

NOTE: * INDICATES A REAL-TIME PATH

Figure 17.25 Dynamic routing link summary report

and the link capacity for the next six months. The forecast of dynamic routing trunk requirements is merged back with the hierarchical requirements and sent to all the forecasters.

17.3.3 Work center functions

Before dynamic routing, about 80 people in the regional work centers performed forecasting and capacity management for the hierarchical network. For dynamic routing, with automated forecasting processes, capacity management and forecasting operation have been fully centralized. The forecasting and capacity management of the full dynamic routing network requires a work center of about six first-level supervisors and one second-level manager. That work center has full responsibility for forecasting and capacity management of the dynamic routing network.

Work is divided among the first-level supervisors on a geographic basis so that the dynamic routing forecaster and capacity manager for a region can work with specific work centers within the region. These work centers are the switch planning and implementation organizations and the transport planning and implementation organizations. Their primary interface is with the system that is responsible for issuing the circuit orders to augment links. Another interface is with the routing organization that processes the routing information coming out of NFS, going through the capacity management system and then directly into the routing system, where it is loaded into the switches.

The bulk of the time is spent in capacity management of the network. The forecasting cycle runs for both the hierarchical and the dynamic routing parts of the network for eight weeks from the time the master file databases are frozen, and it runs twice a year. Because the capacity management function is an every-day, ongoing function, the majority of the time is spent in capacity management. With the highly mechanized interfaces with the circuit-provisioning and automated routing systems, forecasters spend little time communicating with those work centers.

NFS provides a considerable amount of automation and formats the data so that they are readily usable. The forecasting function for dynamic routing is therefore not the long, laborious process that it is in the hierarchical network. For example, the forecasting process at one time took $5\frac{1}{2}$ months to produce a cycle. In that process there were seven-plus weeks of manual intervention in each cycle, with various kinds of adjustments. In the past, much of the analysis of loads and switch-to-switch data was all manual, using microfiche, and required poring over reports such as the link-routed items report to see where traffic was routed. With NFS, the system runs in a total of eight weeks. NFS runs in almost less time than the forecasters took to intervene manually in the old process. The load analysis module of NFS is designed to flag the outliers, and indicate where something needs attention. If something does not need human attention, the mechanized process does all the looking, verifying, and checking. So, with a mechanized process made possible by NFS and the capacity management system, the forecasting work center does not need to staff as many people.

With these highly automated systems, people spend their time on more productive activities. By combining the forecasting job and the capacity management job into one centralized operation, additional efficiencies are achieved from a reduction in fragmentation. By centralizing the operations, this automatically does away with several job functions that are duplicated in the regional operation. And, with the automation, time need only be spent to clear a problem or analyze outliers, rather than to check and verify everything.

Dynamic routing operation requires people who are able to understand and deal with a more complex network, and the network complexity will continue to increase as new technology and services are introduced. The need, then, is a select work center of experts who have the capability and mental capacity to grapple with this complexity and understand all that is going on, and not to isolate that expertise in some small region of the country.

Other disciplines are going the same way as far as centralization is concerned. In the switch-planning system, the exhaust prediction module is run and used centrally. The centralized work center of switch planners comes up with an overall switch plan. In transport planning, there is a centralized transport-planning task force, which uses the NFS forecast to produce a long-range transport evolution plan. A centralized equipment-ordering and inventory control organization bulk-orders all equipment required for the network. This has led to a much more efficient use of inventory.

17.3.4   Interfaces to other work centers

Network forecasters work cooperatively with switch planners, transport planners, real-time traffic managers, and capacity managers. With dynamic routing, the centralized force expands and becomes more operative than it has been, and forecasting, capacity management, and real-time traffic management tie together closely. One way to develop those close relationships is by having centralized, compact work centers. The forecasting process essentially drives all the downstream construction and planning processes for the entire company.

17.3.5   Comparison of dynamic routing and hierarchical network forecasting

The most notable observation is that dynamic routing is more complex and is therefore more difficult for the forecasters to understand. At the same time, because of the highly automated process, there is no real need for people to interact closely with the system and analyze all the information. Hierarchical forecasting often does not have mechanized support that could do routing and other functions automatically: People then have to do it manually. That is a major difference from dynamic routing, and the concept that forecasters do not need to concern themselves with the final link requirements, but only with the correct switch-to-switch load that is given to the system, is a fundamental change.

17.4 Capacity Management—Daily and Weekly Performance Monitoring

In this section we concentrate on the analysis of switch-to-switch capacity management data and the design of the dynamic routing network. Dynamic routing capacity management becomes mandatory at times, as seen from the switch-to-switch traffic data, when significant blocking problems are extant in the network or when it is time to implement a new quarterly forecast. We discuss the interactions of capacity managers with other work centers responsible for dynamic routing network operation, and we contrast dynamic routing network capacity management with hierarchical network capacity management.

Dynamic routing capacity management functions are performed from the capacity administration center and are supported by the capacity management system. A functional block diagram of the capacity management system is illustrated within the lower three blocks of Figure 1.37. In the following sections we discuss the processes in each functional block.

17.4.1 Daily blocking analysis functions

The daily blocking summary display, illustrated in Figures 17.26 and 17.27, gives a breakdown of the highest to the lowest switch-pair blocking that occurred the preceding day. This is an exception-reporting system, in which there is an ability to change the display threshold. For example, the capacity manager can request to see only switch pairs whose blocking level is greater than 10 percent,

```
DAILY BLOCKING SUMMARY                        (OMN SADBS)            PAGE = 1 MORE
SELECTION: _ _ _                                                    11/08/85 09:15:51
DATE: 11 03 85  {11/03/85-11/07/85}                                 RESP CODE: ****
                                                                    DOM/INT CODE: D

SCREEN THRESHOLD: _ _ , _ _ %     SYSTEM THRESHOLD: 1.00 %

                                      NO. OF      HIGHEST       HIGH       SRVCR
  SELN    A SWITCH CLLI   Z SWITCH CLLI   OCCUR     % RT OVFL      HOUR      THRSH%

    1     ATLNGATL01T    SCRMCA0404T       2         30.11          18
    2     ANHMCA0211T    SCRMCA0404T       2         13.84          22
    3     ATLNGATL01T    DNVRCOZJ05T       2         11.72          20
    4     ATLNGATL01T    RCMDVAIT03T       1          8.03          15
    5     ANHMCA0211T    ATLNGATL01T       1          5.88          20
    6     CHCGILCL57T    SCRMCA0404T       1          4.05          20
    7     ATLNGATL01T    PHNXAZMA03T       2          4.02          20
    8     ANHMCA0211T    PTLDOR6203T       1          3.07          22
    9     CHCGILCL57T    PHNXAZMA03T       1          2.17          20
   10     RCMDVAIT03T    WHPLNY0504T       1          1.23          15

                                                      COMMAND:_____

1-SADBS - DAILY BLKNG SUM    2-SAWBS - WKLY BLKING SUM   3-SASPI - SP INFORMATION
4-SRRTC - RTE CONFIGURATN    5-SGHDX - HR/DY-EXCLUSION   6-SAOWM - ONE WAY MEASMTS
```

Figure 17.26 Daily blocking summary display, part 1

```
DAILY BLOCKING SUMMARY              (OMN SADBS)              PAGE = 2 LAST
SELECTION: _ _ _                                            11/08/85 09:16:20
DATE: 11 03 85   {11/03/85-11/07/85}                        RESP CODE: ****
                                                            DOM/INT CODE: D

SCREEN THRESHOLD: _ _ , _ _ %    SYSTEM THRESHOLD: 1.00 %
```

SELN	A SWITCH CLLI	Z SWITCH CLLI	NO. OF OCCUR	HIGHEST % RT OVFL	HIGH HOUR	SRVCR THRSH%
11	ANHMCA0211T	CHCGILCL57T	1	1.12	20	
12	PHNXAZMA03T	SCRMCA0404T	1	1.11	22	
13	ATLNGATL01T	PTLDOR6203T	1	1.03	20	

```
                                              COMMAND:_____

1-SADBD - DAILY BLKNG DET    2-SAWBS - WKLY BLKING SUM    3-SASPI - SP INFORMATION
4-SRRTC - RTE CONFIGURATN    5-SGHDX - HR/DY-EXCLUSION    6-SAOWM - ONE WAY MEASMTS
```

Figure 17.27 Daily blocking summary display, part 2

```
DAILY BLOCKING SUMMARY              (OMN SADBS)              PAGE = 1 LAST
SELECTION: _ _ _                                            11/08/85 09:16:52
DATE: 11 03 85   {11/03/85-11/07/85}                        RESP CODE: ****
                                                            DOM/INT CODE: D

SCREEN THRESHOLD: _ _ , _ _ %    SYSTEM THRESHOLD: 1.00 %
```

SELN	A SWITCH CLLI	Z SWITCH CLLI	NO. OF OCCUR	HIGHEST % RT OVFL	HIGH HOUR	SRVCR THRSH%
1	ATLNGATL01T	SCRMCA0404T	2	30.11	18	
2	ANHMCA0211T	SCRMCA0404T	1	13.84	22	
3	ATLNGATL01T	DNVRC0ZJ05T	1	11.72	20	

```
                                              COMMAND:_____

1-SADBD - DAILY BLKNG DET    2-SAWBS - WKLY BLKING SUM    3-SASPI - SP INFORMATION
4-SRRTC - RTE CONFIGURATN    5-SGHDX - HR/DY-EXCLUSION    6-SAOWM - ONE WAY MEASMTS
```

Figure 17.28 Daily blocking summary with 10 percent switch-blocking display threshold

as illustrated in Figure 17.28. If the threshold is blank, the report displays all the switch pairs that had blocking greater than the system threshold, which is currently 1 percent, as shown in Figures 17.26 and 17.27.

This display therefore gives a snapshot of what went on the day before in the dynamic routing network. As an example, take the situation on November 5, 1985. Figures 17.29 and 17.30 show an exceptionally high-percentage blocking for hours 18 and 19. This caused the capacity managers to look farther, and they found that switch pairs that include Richmond, Virginia, were predominant in the display. In fact, Richmond came up as a switch in all the switch pairs.

```
DAILY BLOCKING SUMMARY                    (OMN SADBS)          PAGE = 1 MORE
SELECTION: ___                                                11/08/85 08:36:04
DATE: 11 05 85  {11/03/85-11/07/85}                           RESP CODE: ****
                                                              DOM/INT CODE: D

SCREEN THRESHOLD: __ , __ %      SYSTEM THRESHOLD: 1.00 %
```

SELN	A SWITCH CLLI	Z SWITCH CLLI	NO. OF OCCUR	HIGHEST % RT OVFL	HIGH HOUR	SRVCR THRSH%
1	RCMDVAIT03T	WAYNPALA42T	4	51.15	19	
2	ATLNGATL03T	RCMDVAIT03T	4	40.12	19	
3	RCMDVAIT03T	WHPLNY0504T	3	26.17	18	
4	PITBPADG43T	RCMDVAIT03T	3	17.21	18	
5	RCMDVAIT03T	STLSMO0934T	3	13.95	19	
6	CHCGILCL57T	RCMDVAIT03T	3	10.42	18	
7	RCMDVAIT03T	WCHTKSBR24T	2	9.94	19	
8	ANHMCA0211T	RCMDVAIT03T	2	9.43	19	
9	DLLSTXTL34T	RCMDVAIT03T	1	8.58	19	
10	NSVLTNMT43T	RCMDVAIT03T	2	3.88	18	

```
                                              COMMAND:_____


1-SADBD - DAILY BLKNG DET     2-SAWBS - WKLY BLKING SUM     3-SASPI - SP INFORMATION
4-SRRTC - RTE CONFIGURATN     5-SGHDX - HR/DY-EXCLUSION     6-SAOWM - ONE WAY MEASMTS
```

Figure 17.29 Daily blocking summary on November 5, 1985, part 1

```
DAILY BLOCKING SUMMARY                    (OMN SADBS)          PAGE = 2 LAST
SELECTION: ___                                                11/08/85 08:37:15
DATE: 11 05 85  {11/03/85-11/07/85}                           RESP CODE: ****
                                                              DOM/INT CODE: D

SCREEN THRESHOLD: __ , __ %      SYSTEM THRESHOLD: 1.00 %
```

SELN	A SWITCH CLLI	Z SWITCH CLLI	NO. OF OCCUR	HIGHEST % RT OVFL	HIGH HOUR	SRVCR THRSH%
11	DNVRCOZJ05T	RCMDVAIT03T	2	3.29	20	
12	PHNXAZMA03T	RCMDVAIT03T	2	3.06	20	
13	KSCYMO0904T	RCMDVAIT03T	3	2.96	20	
14	RCMDVAIT03T	SCRMCA0404T	1	2.42	20	
15	PTLDOR6203T	RCMDVAIT03T	2	1.39	19	

```
                                              COMMAND:_____


1-SADBD - DAILY BLKNG DET     2-SAWBS - WKLY BLKING SUM     3-SASPI - SP INFORMATION
4-SRRTC - RTE CONFIGURATN     5-SGHDX - HR/DY-EXCLUSION     6-SAOWM - ONE WAY MEASMTS
```

Figure 17.30 Daily blocking summary on November 5, 1985, part 2

```
┌──────────────────────────────────────────────────────────────────────────┐
│ HOUR/DAY EXCLUSION              (OMN SGHDX)              PAGE = 1 LAST      │
│ ACTION: I                                               11/08/85 08:42:19  │
│                            RESP CODE: AE10                                 │
│                                                                            │
│    A SWITCH#     RCMDVAIT03T    Z SWITCH# WAYNPALAF2T                       │
│                                                                            │
│         DATE:    11 05 85            DAY: TUE                               │
│                                                                            │
│         HOUR:    19                                                        │
│                                                                            │
│  EXCLUDE HOUR:   Y OR DAY: _                                               │
│                                                                            │
│ REINSTATE HOUR:  _ OR DAY: _                                               │
│                                                                            │
│ EXCLUSION CODE:  S100                                                      │
│                                                                            │
│    EXPLANATION:  FLOODING_-_VIRGINIA_____  │
│                  _____ │
│                                                                            │
│ Z015   TRANSFER INQUIRY COMPLETED                    COMMAND:_____   │
│                                                                            │
│ 1-CLEAR DATA FROM SCREEN  2-SADBD - DAILY BLKING DET   3-SADBS - DAILY BLKNG SUM │
│ 4-SRRTC - RTE CONFIGURATN  5-SOUSC - STATUS CODE UPD                       │
└──────────────────────────────────────────────────────────────────────────┘
```

Figure 17.31   Hour/day exclusion display

Next, capacity managers investigated to find out whether they should exclude these data and, if so, for what reason. One reason for excluding data is to keep them from downstream processing if they are associated with an abnormal network condition. This would prevent designing the network for this type of nonrecurrent network condition. In order to find out what the network condition was, capacity managers first call the traffic managers. In this case, traffic managers indicated that Richmond had experienced flooding the night before, which caused an abnormal network condition. Therefore, capacity managers made the decision to exclude the data.

To exclude the data, capacity managers pick out on the display the switch pair that should be excluded—for example, Richmond–Wayne. They then call up the hour/day exclusion display, illustrated in Figure 17.31, to make the exclusion transaction. On the hour/day exclusion display, capacity managers indicate that for the 5th of November, hour 19, there was a flooding condition in Richmond, Virginia, and the data are excluded for that hour. This is done for all switch pairs that include Richmond that appear to be abnormal.

If there is a need for further investigation associated with the daily blocking summary data, or for another level of detail, capacity managers can use the daily blocking detail display illustrated in Figures 17.32, 17.33, and 17.34. For example, they may wish to investigate the four occurrences of blocking for Richmond–Wayne in hour 19, shown in Figure 17.29. The daily blocking detail is a series of displays that gives data for 24 hours. Capacity managers look at the first display to come up (Figure 17.32), which displays the hours from midnight to 7:00 A.M.; there was no blocking for those hours. The next display (Figure 17.33) is for 8:00 A.M. to 3:00 P.M., and the last display (Figure 17.34) gives the hours from 4 P.M. to 12 midnight. On this latter display appear the four occurrences of

```
DAILY BLOCKING DETAIL                         (OMN SADBD)           PAGE  = 1 MORE
SELECTION: _ _                                                      11/08/85 08:49:14
A SWITCH# RCMDVAIT03T           Z SWITCH# WAYNPALA42T               RESPCODE: AE10
DATE: 11 05 85  {11/03/85-11/07/85}                                DOM/INT CODE: D
SYSTEM THRESHOLD:   1.00 %      SERVICER THRESHOLD:      %
```

SELN	HR	PEG CT	RT OVFL	% RT OVFL	NM CALLS SAVED	% NC	CARRIED LOAD	STAT CODE	EXC IND
1	00	30	0	00	0		112	D004	
2	01	17	0	00	0		87	D004	
3	02	8	0	00	0		42	D004	
4	03	5	0	00	0		64	D004	
5	04	8	0	00	0		45	D004	
6	05	40	0	00	0		104	D004	
7	06	157	0	00	0		404	D004	
8	07	462	0	00	0		938	D004	

```
Z015   TRANSFER INQUIRY COMPLETE                          COMMAND:_____

1-SADBS - DAILY BLKNG SUM      2-SAWBS - WKLY BLKING SUM   3-SASPI - SP INFORMATION
4-SRRTC - RTE CONFIGURATN      5-SGHDX - HR/DY-EXCLUSION   6-SAOWM - ONE WAY MEASMTS
```

Figure 17.32 Daily blocking detail display for November 5, 1985, part 1

```
DAILY BLOCKING DETAIL                         (OMN SADBD)           PAGE  = 2 MORE
SELECTION: _ _                                                      11/08/85 08:49:53
A SWITCH# RCMDVAIT03T           Z SWITCH# WAYNPALA42T               RESPCODE: AE10
DATE: 11 05 85  {11/03/85-11/07/85}                                DOM/INT CODE: D
SYSTEM THRESHOLD:   1.00 %      SERVICER THRESHOLD:      %
```

SELN	HR	PEG CT	RT OVFL	% RT OVFL	NM CALLS SAVED	% NC	CARRIED LOAD	STAT CODE	EXC IND
9	08	1088	0	0.00	0		1894	D0.004	
10	09	1272	0	0.00	0		2125	D0.004	
11	10	1259	0	0.00	0		2163	D0.004	
12	11	642	0	0.00	0		1053	D0.004	
13	12	799	0	0.00	0		1329	D0.004	
14	13	1241	0	0.00	0		2052	D0.004	
15	14	1316	0	0.00	0		2304	D0.004	
16	15	926	0	0.00	0		1722	D0.004	

```
Z015 TRANSFER INQUIRY COMPLETE                            COMMAND:_____

1-SADBS - DAILY BLKNG SUM      2-SAWBS - WKLY BLKING SUM   3-SASPI - SP INFORMATION
4-SRRTC - RTE CONFIGURATN      5-SGHDX - HR/DY-EXCLUSION   6-SAOWM - ONE WAY MEASMTS
```

Figure 17.33 Daily blocking detail display for November 5, 1985, part 2

```
DAILY BLOCKING DETAIL                 (OMN SADBD)            PAGE = 3 LAST
SELECTION: _ _                                              11/08/85 08:50:01
A SWITCH# RCMDVAIT03T        Z SWITCH# WAYNPALA42T           RESPCODE: AE10
DATE: 11 05 85  {11/03/85-11/07/85}                         DOM/INT CODE: D
SYSTEM THRESHOLD:  1.00 %     SERVICER THRESHOLD:     %
```

SELN	HR	PEG CT	RT OVFL	% RT OVFL	NM CALLS SAVED	% NC	CARRIED LOAD	STAT CODE	EXC IND
17	16	463	0	00	0		1111	D004	
18	17	1005	36	3.58	0		2182	D004	E
19	18	3102	1018	32.81	0		4387	D004	E
20	19	3089	1580	51.15	0		4433	D004	E
21	20	822	184	22.34	0		2758	D004	E
22	21	294	0	00	0		1322	D004	
23	22	347	0	00	0		1390	D004	
24	23	72	0	00	0		374	D004	

```
Z015 TRANSFER INQUIRY COMPLETE                    COMMAND:_____

1-SADBS - DAILY BLKNG SUM     2-SAWBS - WKLY BLKING SUM    3-SASPI - SP INFORMATION
4-SRRTC - RTE CONFIGURATN     5-SGHDX - HR/DY-EXCLUSION    6-SAOWM - ONE WAY MEASMTS
```

Figure 17.34 Daily blocking detail display for November 5, 1985, part 3

blocking first identified in Figure 17.29. There are three hours that have rather high blocking, and capacity managers would exclude those three hours. However, in hour 17 there is only 3.58 percent blocking, which will probably not affect the database, so based on this information, capacity managers can make the decision to exclude the data or not exclude them. This is another level of detail used in daily blocking analysis.

Another investigation that capacity managers may conduct in the case that traffic managers are not able to help determine the network condition is to look at further detail. Another level of detail is contained in the one-way measurements display, illustrated in Figures 17.35 to 17.40. These displays show the level of traffic pressure being applied in each direction. They are good examples of the Richmond–Wayne condition, which show that in the direction from Richmond to Wayne, the pressure was higher than in the other direction. This conclusion is fairly consistent for the midnight-to-7:00-A.M. time frame and consistent again in the 8:00-A.M.-to-3-P.M. time frame. In the night-time hours, when there was a blocking condition, the pressure was greater in the direction from Wayne to Richmond. Comparison of Figures 17.38, 17.39, and 17.40 with Figures 17.41, 17.42, and 17.43 shows these differences in the Wayne-to-Richmond traffic. Apparently, in the night-time hours, people outside of the Richmond area tried to call people in the Richmond area to find out if they are all right, and that effect shows up on the one-way measurements display, which also gives an indication of the real-time traffic management calls saved by automatic NEMOS reroutes, as discussed in Section 17.2.

```
ONE WAY MEASUREMENTS                        (OMN SAOWM)          PAGE  =1 MORE
SELECTION:_ _                                                   11/08/85 08:51:55
OTS CLLI# RCMDVAIT03T          TTS CLLI# WAYNPALA42T
DATE: 11 05 85    {11/03/85-11/07/85}                          DOM/INT CODE: D

                                CARRIED        ENG         RT         NM CALLS
SELN        HR        PEG CT     LOAD          OVFL        OVFL       SAVED
----        --        ------     -------       ----        ----       -------
  1         00          21         77            0           0           0
  2         01          26         72            0           0           0
  3         02          10         17            0           0           0
  4         03           3          6            0           0           0

  5         04          10         24            0           0           0
  6         05          41         82            0           0           0
  7         06         155        379            0           0           0
  8         07         466        863            0           0           0

Z015 TRANSFER INQUIRY COMPLETE                            COMMAND:_____

1-SADBS - DAILY BLKNG SUM      2-SADBD - DAILY BLKNG DET      3-SRRTC - RTE CONFIGURATN
4-SAWBD - WKLY BLKNG DETL      5-SAWBS - WKLY BLKNG SUM       6-SGHDX - HR/DY-EXCLUSION
```

Figure 17.35 One-way measurements display for November 5, 1985, part 1a

```
ONE WAY MEASUREMENTS                        (OMN SAOWM)          PAGE  =2 MORE
SELECTION:_ _                                                   11/08/85 08:52:07
OTS CLLI# RCMDVAIT03T          TTS CLLI# WAYNPALA42T
DATE: 11 05 85    {11/03/85-11/07/85}                          DOM/INT CODE: D

                                CARRIED        ENG         RT         NM CALLS
SELN        HR        PEG CT     LOAD          OVFL        OVFL       SAVED
----        --        ------     -------       ----        ----       -------
  9         08        1141       1940            0           0           0
 10         09        1203       2042            0           0           0
 11         10        1329       2320            0           0           0
 12         11         621       1108            0           0           0

 13         12         743       1320            0           0           0
 14         13        1303       2280            0           0           0
 15         14        1420       2385            0           0           0
 16         15         964       1917            0           0           0

Z015 TRANSFER INQUIRY COMPLETE                            COMMAND:_____

1-SADBS - DAILY BLKNG SUM      2-SADBD - DAILY BLKNG DET      3-SRRTC - RTE CONFIGURATN
4-SAWBD - WKLY BLKNG DETL      5-SAWBS - WKLY BLKNG SUM       6-SGHDX - HR/DY-EXCLUSION
```

Figure 17.36 One-way measurements display for November 5, 1985, part 1b

```
ONE WAY MEASUREMENTS                    (OMN SAOWM)              PAGE = 3 LAST
SELECTION:_ _                                                   11/08/85 08:52:47
OTS CLLI# RCMDVAIT03T          TTS CLLI# WAYNPALA42T
DATE: 11 05 85     {11/03/85-11/07/85}                          DOM/INT CODE: D
```

SELN	HR	PEG CT	CARRIED LOAD	ENG OVFL	RT OVFL	NM CALLS SAVED
17	16	434	932	0	0	0
18	17	731	1885	32	32	0
19	18	2454	4018	539	503	0
20	19	2514	4411	1040	977	0
21	20	693	2804	58	58	0
22	21	201	1157	0	0	0
23	22	253	1292	0	0	0
24	23	70	348	0	0	0

```
                                                    COMMAND:_____
```

```
1-SADBS - DAILY BLKNG SUM    2-SADBD - DAILY BLKNG DET    3-SRRTC - RTE CONFIGURATN
4-SAWBD - WKLY BLKNG DETL    5-SAWBS - WKLY BLKNG SUM     6-SGHDX - HR/DY-EXCLUSION
```

Figure 17.37   One-way measurements display for November 5, 1985, part 1c

```
ONE WAY MEASUREMENTS                    (OMN SAOWM)              PAGE = 1 MORE
SELECTION:_ _                                                   11/08/85 08:53:13
OTS CLLI# WAYNPALA42T          TTS CLLI# RCMDVAIT03T
DATE: 11 05 85     {11/03/85-11/07/85}                          DOM/INT CODE: D
```

SELN	HR	PEG CT	CARRIED LOAD	ENG OVFL	RT OVFL	NM CALLS SAVED
1	00	38	147	0	0	0
2	01	7	101	0	0	0
3	02	5	66	0	0	0
4	03	7	121	0	0	0
5	04	6	65	0	0	0
6	05	38	125	0	0	0
7	06	158	429	0	0	0
8	07	457	1012	0	0	0

```
                                                    COMMAND:_____
```

```
1-SADBS - DAILY BLKNG SUM    2-SADBD - DAILY BLKNG DET    3-SRRTC - RTE CONFIGURATN
4-SAWBD - WKLY BLKNG DETL    5-SAWBS - WKLY BLKNG SUM     6-SGHDX - HR/DY-EXCLUSION
```

Figure 17.38   One-way measurements display for November 5, 1985, part 2a

```
ONE WAY MEASUREMENTS                          (OMN SAOWM)            PAGE = 2 MORE
SELECTION:_ _                                                        11/08/85 08:53:23
OTS CLLI# WAYNPALA42T            TTS CLLI# RCMDVAIT03T
DATE: 11 05 85    {11/03/85-11/07/85}                               DOM/INT CODE: D

                                   CARRIED      ENG       RT        NM CALLS
     SELN        HR      PEG CT     LOAD         OVFL      OVFL      SAVED
     ────        ──      ──────     ──────       ────      ────      ──────
       9         08       1035       1847         0         0           0
      10         09       1340       2208         0         0           0
      11         10       1189       2005         0         0           0
      12         11        663        997         0         0           0

      13         12        855       1337         0         0           0
      14         13       1178       1823         0         0           0
      15         14       1211       2223         0         0           0
      16         15        888       1527         0         0           0

                                                        COMMAND:_____

  1-SADBS - DAILY BLKNG SUM     2-SADBD - DAILY BLKNG DET    3-SRRTC - RTE CONFIGURATN
  4-SAWBD - WKLY BLKNG DETL     5-SAWBS - WKLY BLKNG SUM     6-SGHDX - HR/DY-EXCLUSION
```

Figure 17.39   One-way measurements display for November 5, 1985, part 2b

```
ONE WAY MEASUREMENTS                          (OMN SAOWM)            PAGE = 3 LAST
SELECTION:_ _                                                        11/08/85 08:53:33
OTS CLLI# WAYNPALA42T            TTS CLLI# RCMDVAIT03T
DATE: 11 05 85    {11/03/85-11/07/85}                               DOM/INT CODE: D

                                   CARRIED      ENG       RT        NM CALLS
     SELN        HR      PEG CT     LOAD         OVFL      OVFL      SAVED
     ────        ──      ──────     ──────       ────      ────      ──────
      17         16        492       1289         0         0           0
      18         17       1278       2479        42        40           0
      19         18       3749       4756      1610      1532           0
      20         19       3664       4455      2228      2183           0

      21         20        950       2712       309       309           0
      22         21        387       1486         0         0           0
      23         22        440       1487         0         0           0
      24         23         73        399         0         0           0

                                                        COMMAND:_____

  1-SADBS - DAILY BLKNG SUM     2-SADBD - DAILY BLKNG DET    3-SRRTC - RTE CONFIGURATN
  4-SAWBD - WKLY BLKNG DETL     5-SAWBS - WKLY BLKNG SUM     6-SGHDX - HR/DY-EXCLUSION
```

Figure 17.40   One-way measurements display for November 5, 1985, part 2c

17.4.2 Study-week blocking analysis functions

The capacity management system can move from daily to weekly blocking analysis. This normally occurs after capacity managers form the weekly average using the previous week's data. The study-week data are used later in the downstream processing to develop the study-period average. The weekly blocking data are set up basically the same way as the daily blocking data. They give the switch pairs that had blocking for the week, as illustrated in Figure 17.41. This study-week blocking analysis function gives another opportunity to review the data to see if there is a need to exclude any weekly data.

Normally, blocking exclusions are made on a daily basis because the condition of the network is current. It is easier to remember overnight what happened the day before than to go back to reconstruct major failures or network conditions at the end of the week. The other advantage of excluding data on a daily basis is that, at any time before the weekly averages are formed, capacity managers have the ability to go back and remove that exclusion. So they may exclude data but find out before the weekly average is formed that they do not want to exclude them after all. If capacity managers make the decision not to exclude the data, they can leave them in and let them be used to form the weekly averages. Once the system forms the weekly averages, there is another opportunity to exclude data from the study-period average.

There are 24 hours of data available in the weekly blocking detail display, as illustrated in Figures 17.42 to 17.44. Figure 17.42 illustrates excluded data for the Anaheim–Phoenix switch pair on October 27 in hour 2, which is 2:00 A.M. in the morning network time. The weekly exclusion display can be called from the weekly blocking detail display by selecting key 5 (SGWKX) at the bottom of the display.

```
WEEKLY BLOCKING SUMMARY                    (OMN SAWBS)        PAGE =1 LAST
SELECTION:__                                                 11/08/85 08:56:25
DATE: 10 27 85 {10/06/85-11/03/85}                          RESP CODE: ****
TYPE OF AVERAGE: ABD                                        DOM/INT CODE: D
SCREEN THRESHOLD: __,__%    SYSTEM THRESHOLD: 1.00 %

                                      NO. OF   HIGHEST                SRVCR
 SELN   A SWITCH CLLI   Z SWITCH CLLI OCCUR   % RT OVFL   HOUR       THRSH%

   1   ANHMCCA0211T    PHNXAZMA03T       1      5.46       11
   2   ANHMCCA0211T    DLLSTXTL34T       1      3.52       11
   3   ATLNGATL01T     DLLSTXTL34T       3      2.23        9
   4   DLLSTXTL34T     SCRMCA0404T       1      1.50       11


DO38 * WARNING * CURRENT WORK-WEEK DATE WAS USED         COMMAND:_____

1-SAWBD - WKLY BLKNG DETL   2-SADBS - DAILY BLKING SUM   3-SASPI - SP INFORMATION
4-SRRTC - RTE CONFIGURATN   5-SGHKX - WEEK EXCLUSIONS    6-SGHDX - HR/DY-EXCLUSION
```

Figure 17.41 Weekly blocking summary

```
WEEKLY BLOCKING DETAIL                      (OMN SAWBD)        PAGE = 1 MORE
SELECTION:_ _                                                 11/08/85 09:00:03
DATE: 10 27 85   {10/06/85-11/03/85}                          RESP CODE: AW10
TYPE OF AVERAGE: ABD                                          DOM/INT CODE: D
A SWITCH# ANHMCA0211T            Z SWITCH# PHNXAZMA03T
SYSTEM THRESHOLD:  1.00 %     SERVICER TRESHOLD:   %
```

SELN	HR	PEG CT	RT OVFL	% RT OVFL	NM CALLS SAVED	% NC	CARRIED LOAD	STAT CODE	EXC IND
1	00	331	0	0.00	0	0.00	1605	5	
2	01	128	0	0.00	0	0.00	689	5	
3	02	66	0	0.00	0	0.00	318	4	E
4	03	48	0	0.00	0	0.00	142	5	
5	04	24	0	0.00	0	0.00	79	5	
6	05	24	0	0.00	0	0.00	76	5	
7	06	33	0	0.00	0	0.00	89	5	
8	07	101	0	0.00	0	0.00	222	5	

```
Z015 TRANSFER INQUIRY COMPLETE                    COMMAND:_____

1-SAWBS - WKLY BLKNG SUM     2-SADBS - DAILY BLKING SUM    3-SASPI - SP INFORMATION
4-SRRTC - RTE CONFIGURATN    5-SGWKX - WEEK EXCLUSIONS     6-SGHDX - HR/DY EXCLUSION
```

Figure 17.42 Weekly blocking detail, part 1

```
WEEKLY BLOCKING DETAIL                      (OMN SAWBD)        PAGE = 2 MORE
SELECTION:_ _                                                 11/08/85 09:00:13
DATE: 10 27 85   {10/06/85-11/03/85}                          RESP CODE: AW10
TYPE OF AVERAGE: ABD                                          DOM/INT CODE: D
A SWITCH# ANHMCA0211T            Z SWITCH# PHNXAZMA03T
SYSTEM THRESHOLD:  1.00 %     SERVICER THRESHOLD    %
```

SELN	HR	PEG CT	RT OVFL	% RT OVFL	NM CALLS SAVED	% NC	CARRIED LOAD	STAT CODE	EXC IND
9	08	492	0	0.00	0	0.00	1120	5	
10	09	1449	0	0.00	0	0.00	2806	5	
11	10	3054	0	0.00	0	0.00	5654	5	
12	11	4483	316	5.46	0	7.04	7467	5	
13	12	3595	0	0.00	0	0.00	6482	5	
14	13	2658	0	0.00	0	0.00	4558	5	
15	14	2605	0	0.00	0	0.00	4241	5	
16	15	3304	0	0.00	0	0.00	5656	5	

```
Z015 TRANSFER INQUIRY COMPLETE                    COMMAND:_____

1-SAWBS - WKLY BLKNG SUM     2-SADBS - DAILY BLKING SUM    3-SASPI - SP INFORMATION
4-SRRTC - RTE CONFIGURATN    5-SGWKX - WEEK EXCLUSIONS     6-SGHDX - HR/DY EXCLUSION
```

Figure 17.43 Weekly blocking detail, part 2

```
WEEKLY BLOCKING DETAIL                    (OMN SAWBD)          PAGE = 3 LAST
SELECTION:_ _                                                 11/08/85 09:00:30
DATE: 10 27 85   {10/06/85-11/03/85}                          RESP CODE: AW10
TYPE OF AVERAGE: ABD                                          DOM/INT CODE: D
A SWITCH# ANHMCA0211T            Z SWITCH# PHNXAZMA03T
SYSTEM THRESHOLD: 1.00 %     SERVICER THRESHOLD      %
```

SELN	HR	PEG CT	RT OVFL	% RT OVFL	NM CALLS SAVED	% NC	CARRIED LOAD	STAT CODE	EXC IND
17	16	3472	0	0.00	0	0.00	6216	5	
18	17	2883	0	0.00	0	0.00	5376	5	
19	18	1944	0	0.00	0	0.00	3883	5	
20	19	1644	0	0.00	0	0.00	4020	5	
21	20	1805	0	0.00	0	0.00	5139	5	
22	21	1653	0	0.00	0	0.00	5692	5	
23	22	1193	0	0.00	0	0.00	4647	5	
24	23	665	0	0.00	0	0.00	2915	5	

```
Z015 TRANSFER INQUIRY COMPLETE                    COMMAND:_____

1-SAWBS - WKLY BLKNG SUM      2-SADBS - DAILY BLKING SUM    3-SASPI - SP INFORMATION
4-SRRTC - RTE CONFIGURATN    5-SGWKX - WEEK EXCLUSIONS      6-SGHDX - HR/DY EXCLUSION
```

Figure 17.44   Weekly blocking detail, part 3

From the weekly exclusion display, the hour/day exclusion display for Anaheim–Phoenix can be obtained, as illustrated in Figure 17.45, which shows that there was an MC3 condition (switch failure) at Anaheim. That represented an abnormal traffic condition, and the MC3 at Anaheim drove the blocking up. Because capacity managers did not want those data to be included in a problem hour later on in the study-period average, they excluded them.

```
HOUR/DAY EXCLUSION                       (OMN SGHDX)          PAGE = LAST
ACTION: I                                                    11/08/85 09:03:21
                                   RESP CODE: AE10

        A SWITCH#     ANHMCA0211T    Z SWITCH# PHNXAZMAO03T

             DATE:    10 27 85       DAY: SUN

             HOUR:    02

   EXCLUDE HOUR:      Y OR DAY: _

  REINSTATE HOUR:     _ OR DAY: _

  EXCLUSION CODE:     S300

    EXPLANATION:      MC3_ANHM_____
                      _____

                                                 COMMAND:_____

1-CLEAR DATA FROM SCREEN  2-SADBD - DAILY BLKING DET   3-SADBS - DAILY BLKNG SUM
4-SRRTC - RTE CONFIGURATN  5-SOUSC - STATUS CODE UPD
```

Figure 17.45   Hour/day exclusion display for the Anaheim–Phoenix switch pair

17.4.3 Study-period blocking analysis functions

Once each week, the study-period average is formed using the most current four weeks of data in the last nine weeks. The study-period blocking summary, illustrated in Figure 17.46, gives an idea of the blocking during the most current study period. In the week of October 27, there are four switch pairs that experienced average business day average blocking greater than 1 percent, which is the system threshold. Of these four exceptions, the Dallas–Sacramento switch pair would be a prime candidate to possibly run the path-EFO model to solve the blocking problem. It is also the only switch pair over the 3 percent threshold that triggers a network design analysis.

In order to determine whether they should run the path-EFO model for that problem hour, capacity managers first look at the study-period blocking detail display, illustrated in Figure 17.47. For the Dallas–Sacramento switch pair they get a snapshot of the 24 hours of data to see if there are any other hours for that switch pair that should be investigated. In this particular week, there is only one hour, hour 11, that should be investigated. Next, capacity managers go to the study-period information display, shown in Figure 17.48, for hour 11 to see more detail for this problem hour in the current study period. On this display they see that during the current study period there is an average of 3.19 percent real-time overflow for the four weeks. This study-period average is made up of four study weeks, and the week of October 13 is when there is a high blocking condition for the Dallas–Sacramento switch pair.

At about that same time, capacity managers placed trunks on order for the Dallas–Sacramento link. They can find out what date the circuits are due and other relevant information on the circuit activity by calling up the link information

```
STUDY PERIOD BLOCKING SUMMARY                (OMN SASPS)        PAGE = 1 LAST
SELECTION:_ _                                                   11/08/85 09:05:15
STUDY PERIOD DATE: 10 27 85   {09/29/85-10/27/85}               RESP CODE: ****
TYPE OF AVERAGE: ABD
SCREEN THRESHOLD: _ _ , _ _ %     SYSTEM THRESHOLD: 1.00 %
```

SELN	A SWITCH CLLI	Z SWITCH CLLI	NO. OF OCCUR	HIGHEST % RT OVFL	HR	SRVCR THRSH%
1	DLLSTXTL34T	SCRMCA0404T	1	3.19	11	
2	ANHMCA0211T	DLLSTXTL34T	1	1.92	11	
3	ANHMCA0211T	PTLDOR6203T	1	1.91	22	
4	ANHMCA0211T	PHNXAZMA03T	1	1.37	11	

```
D022 DEFAULTS HAVE BEEN APPLIED ON INPUT                  COMMAND:_____

1-SASPD - SP BLKNG DETAIL     2-SASPI - SP INFORMATION      3-SAWBS - WKLY BLKING SUM
4-SRRTC - RTE CONFIGURATN     5-SGHDX - HR/DY-EXCLUSION     6-SGWKX - WEEK EXCLUSIONS
```

Figure 17.46 Study-period blocking summary

```
STUDY PERIOD BLOCKING DETAIL                    (OMN SASPD)        PAGE 1 LAST
SELECTION:_ _ _ _                                                 11/08/85 09:08:47
A SWITCH CLLI# DLLSTXTL34T        Z SWITCH CLLI# SCRMCA0404T      RESP CODE: AW10
STUDY PERIOD DATE: 10 27 85   {09/29/85-10/27/85}
TYPE OF AVERAGE: ABD
SCREEN THRESHOLD: 1.00 %      SYSTEM THRESHOLD: 0.00 %
```

SELN	HOUR	% RT OVFL	SELN	HOUR	% RT OVFL	SELN	HOUR	% RT OVFL
1A	00	0.0	IB	08	0.00	1C	16	0.00
2A	01	0.0	2B	09	0.00	2C	17	0.00
3A	02	0.0	3B	10	0.00	3C	18	0.00
4A	03	0.0	4B	11	3.19	4C	19	0.00
5A	04	0.0	5B	12	0.00	5C	20	0.00
6A	05	0.0	6B	13	0.00	6C	21	0.00
7A	06	0.0	7B	14	0.00	7C	22	0.00
8A	07	0.0	8B	15	0.00	8C	23	0.00

```
                                               COMMAND:_____

1-SASPS - SP BLKNG SUM      2-SASPI - SP INFORMATION     3-SAWBS - WKLY BLKING SUM
4-SRRTC - RTE CONFIGURATN   5-SGHDX - HR/DY-EXCLUSION    6-SGWKX - WEEK EXCLUSIONS
```

Figure 17.47 Study-period blocking detail

```
STUDY PERIOD INFORMATION                        (OMN SASPI)        PAGE 1 LAST
A SWITCH CLLI# DLLSTXTL34T        Z SWITCH CLLI# SCRMCA0404T       11/08/85 9:09:41
HOUR# 11          TYPE OF AVERAGE: ABD                            RESP CODE: AW10
STUDY PERIOD DATE: 10 27 85              {09/29/85-10/27/85}
```

DATE	PEG CT	OVFL	% RT OVFL	NM SAVED	ENG BLOCKED	% NC	NO. VALID DAYS-BLKG	DATA SUMT
CURRENT SP	1391	81	3.19	0	81	5.82	20	0
SW 10/27/85	1305	25	1.50	0	25	1.91	5	
SW 10/20/85	1389	26	1.54	0	26	1.87	5	
SW 10/13/85	1480	272	9.70	0	272	18.37	5	
SW 10/06/85	1389	0	0.00	0	0	0.00	5	
SP 09/22/85								

	PEAKEDNESS	DAYDAY VARIATION	OFFERED LOAD	NO. VALID DAYS-LOAD
CURRENT SP LOADS	28.37	H	5034	20
PREVIOUS SP LOADS				

```
                                               COMMAND:_____

1-SASPS - SP BLKNG SUM      2-SASPD - SP BLKING DETAIL   3-SAWBS - WKLY BLKING SUM
4-SRRTC - RTE CONFIGURATN   5-SGHDX - HR/DY-EXCLUSION    6-SGWKX - WEEK EXCLUSIONS
```

Figure 17.48 Study-period information display

```
LINK INFORMATION                              (OMN SGLKI)          PAGE 1 LAST
ACTION: I                                                         11/08/85 09:11:38

LINK# DLLSTXTL34T    SCRMCA0404T

IN-EFFECT:

               TRUNKS      SAT     FBR OPT
               IN SVC      (Y/N)    (Y/N)        TRUNK TYPE
               _____     _____    _____      _____

               00336       N        N           DN11ITCC    -------

PENDING:

               TRUNKS      SAT     FBR OPT                          STATUS
EFF DATE       PENDING     (Y/N)    (Y/N)        TRUNK TYPE         (P/D)
_____       _____     _____    _____      _____        _____

11/15/85       00420       N        N           DN11ITCC    -------    P
--------       -----       -        -           -------     -------    -

                                                       COMMAND:_____

1-CLEAR DATA FROM SCREEN    2-SRRTC - RTE CONFIGURATN   3-SADBD - DAILY BLKING DET
4-SADBS - DAILY BLKNG SUM   5-SASPI - SP INFORMATION
```

Figure 17.49 Link information display

display, illustrated in Figure 17.49. The link information display tells the capacity managers the number of trunks in service and gives an indication of the type of transport being used. The "pending" columns indicate that on effective date November 15 the trunk size should increase from 336 to 420. Because there are trunks coming up on November 15, and these trunks were ordered on October 13, capacity managers made the decision not to run the path-EFO model.

17.5 Capacity Management—Short-Term Network Adjustment

17.5.1 Network design functions

There are several features available in the path-EFO model. On the path-EFO model parameter display illustrated in Figure 17.50, capacity mangers can set the parameters on the type of design they wish to run. First, there is the capability of running the path-EFO model for average business day or weekend traffic loads. At the same time, they select the blocking probability grade-of-service objective to which they want to design the network.

With the run options field, capacity managers can select the routing change option and the routing change effective date. This routing change feature is the one used in almost all cases. With this option, the path-EFO model makes routing table changes to utilize the network capacity that is in place to minimize blocking. The path-EFO model also designs the network to the specified blocking

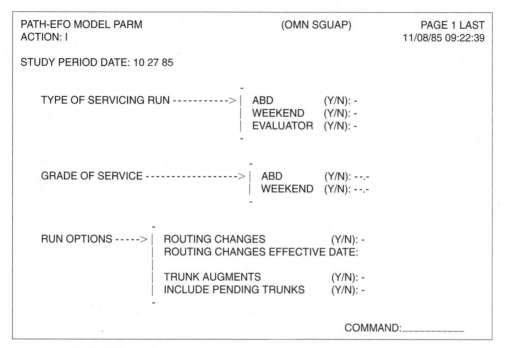

Figure 17.50   Path-EFO model parameter display

probability grade-of-service objective. If it cannot meet the objective with the network capacity in place, it specifies how many trunks to add to which links in order to meet it. So the routing changes option gives both the routing table changes to achieve the best blocking probability grade-of-service objective and, in addition, the trunks required in order to meet that objective.

Because the path-EFO model runs on Wednesday, this means that on Monday and Tuesday the capacity manager has to review all of the weekend activity, including the previous Friday's activity, and make the decision to exclude data. The path-EFO model is run every Wednesday morning, which gives the capacity manager the remaining week to administer those changes and to take care of circuit-order activity that has been issued. The routing table update implementation is a hands-off operation—it is automatic from the capacity management system all the way through to the ESS switches.

The evaluator option is used to determine the carried traffic per trunk, or network efficiency, for every link in the network for the busiest load set period. This report is illustrated in Figures 17.51 to 17.54. For example, as illustrated in Figure 17.51, the Anaheim-to-Chicago switch pair has an average of 34.9 CCS per trunk. A number of weekend load set periods have the highest CCS per trunk. With this report, capacity managers can also sometimes find trouble spots. For example, the Wichita–Kansas City switch pair has only 16.9 CCS per trunk, which is considerably lower than some of the other switch pairs. This occurred in load set period 3, which is often the highest hour on an average business day.

STUDY PERIOD DATE: 10/20/85

FORECAST PERIOD: 4P85

NETWORK AVERAGE CCS/TRK: 30.7

A SWITCH	Z SWITCH	ENG LSP	CCS/TRK
ANAHEIM, CA	ATLANTA, GA	15	32.6
	CHICAGO, IL	15	34.9
	DALLAS, TX	5	33.8
	DENVER, CO	15	34.7
	KANSAS CITY, MO	12	31.0
	NASHVILLE, TN	6	33.0
	PHOENIX, AZ	15	24.4
	PITTSBURG, PA	12	29.6
	PORTLAND, OR	12	34.9
	RICHMOND, VA	15	30.1
	SACRAMENTO, CA	15	33.2
	ST. LOUIS, MO	12	34.9
	WAYNE, PA	4	33.9
	WICHITA, KS	15	19.6
	WHITE PLAINS, NY	5	34.4
ATLANTA, GA	CHICAGO, IL	15	34.5
	DALLAS, TX	6	34.3
	DENVER, CO	15	35.2
	KANSAS CITY, MO	5	34.6
	NASHVILLE, TN	2	18.3
	PHOENIX, AZ	15	32.9
	PITTSBURG, PA	15	35.6
	PORTLAND, OR	3	33.9
	RICHMOND, VA	15	32.9
	SACRAMENTO, CA	15	34.9
	ST. LOUIS, MO	5	30.6
	WAYNE, PA	15	31.8
	WICHITA, KS	15	32.2
	WHITE PLAINS, NY	14	35.3
CHICAGO, IL	DALLAS, TX	3	35.1
	DENVER, CO	15	35.1
	KANSAS CITY, MO	3	31.9
	NASHVILLE, TN	3	28.7
	PHOENIX, AZ	15	34.4
	PITTSBURG, PA	3	25.0
	PORTLAND, OR	3	33.6
	RICHMOND, VA	14	33.4
	SACRAMENTO, CA	15	31.3

PAGE 1

Figure 17.51   Network efficiency report, part 1

STUDY PERIOD DATE: 10/20/85

FORECAST PERIOD: 4P85

NETWORK AVERAGE CCS/TRK: 30.7

A SWITCH	Z SWITCH	ENG LSP	CCS/TRK
CHICAGO, IL	ST. LOUIS, MO	3	22.3
	WAYNE, PA	3	34.1
	WICHITA, KS	3	33.5
	WHITE PLAINS, NY	3	33.2
DALLAS, TX	DENVER, CO	3	33.4
	KANSAS CITY, MO	6	30.7
	NASHVILLE, TN	3	33.6
	PHOENIX, AZ	15	31.1
	PITTSBURG, PA	2	28.0
	PORTLAND, OR	3	34.0
	RICHMOND, VA	15	33.8
	SACRAMENTO, CA	15	29.6
	ST. LOUIS, MO	5	21.9
	WAYNE, PA	3	34.5
	WICHITA, KS	3	31.4
	WHITE PLAINS, NY	3	34.7
DENVER, CO	KANSAS CITY, MO	15	29.7
	NASHVILLE, TN	15	33.3
	PHOENIX, AZ	15	27.8
	PITTSBURG, PA	15	33.4
	PORTLAND, OR	3	33.4
	RICHMOND, VA	15	34.6
	SACRAMENTO, CA	12	31.1
	ST. LOUIS, MO	15	27.4
	WAYNE, PA	15	33.8
	WICHITA, KS	15	26.5
	WHITE PLAINS, NY	15	34.2
KANSAS CITY, MO	NASHVILLE, TN	3	31.5
	PHOENIX, AZ	11	29.9
	PITTSBURG, PA	3	33.3
	PORTLAND, OR	12	33.0
	RICHMOND, VA	2	25.6
	SACRAMENTO, CA	12	35.2
	ST. LOUIS, MO	3	24.3
	WAYNE, PA	3	21.9
	WICHITA, KS	3	16.9
	WHITE PLAINS, NY	3	29.4

PAGE 2

Figure 17.52   Network efficiency report, part 2

STUDY PERIOD DATE: 10/20/85

FORECAST PERIOD: 4P85

NETWORK AVERAGE CCS/TRK: 30.7

A SWITCH	Z SWITCH	ENG LSP	CCS/TRK
NASHVILLE, TN	PHOENIX, AZ	15	22.9
	PITTSBURG, PA	3	24.0
	PORTLAND, OR	3	31.8
	RICHMOND, VA	15	30.9
	SACRAMENTO, CA	15	35.1
	ST. LOUIS, MO	3	16.1
	WAYNE, PA	3	32.2
	WICHITA, KS	3	32.1
	WHITE PLAINS, NY	3	32.6
PHOENIX, AZ	PITTSBURG, PA	15	34.4
	PORTLAND, OR	12	35.6
	RICHMOND, VA	15	30.8
	SACRAMENTO, CA	15	34.7
	ST. LOUIS, MO	11	26.2
	WAYNE, PA	14	31.4
	WICHITA, KS	15	29.4
	WHITE PLAINS, NY	2	32.2
PITTSBURG, PA	PORTLAND, OR	3	26.6
	RICHMOND, VA	15	25.9
	SACRAMENTO, CA	15	27.7
	ST. LOUIS, MO	2	16.9
	WAYNE, PA	2	30.3
	WICHITA, KS	2	29.5
	WHITE PLAINS, NY	15	22.5
PORTLAND, OR	RICHMOND, VA	15	32.1
	SACRAMENTO, CA	15	28.2
	ST. LOUIS, MO	3	34.4
	WAYNE, PA	5	33.7
	WICHITA, KS	5	32.1
	WHITE PLAINS, NY	5	34.0
RICHMOND, VA	SACRAMENTO, CA	15	34.5
	ST. LOUIS, MO	2	34.7
	WAYNE, PA	15	24.7
	WICHITA, KS	13	32.0
	WHITE PLAINS, NY	15	29.2

PAGE 3

Figure 17.53 Network efficiency report, part 3

STUDY PERIOD DATE: 10/20/85

FORECAST PERIOD: 4P85

NETWORK AVERAGE CCS/TRK: 30.7

A SWITCH	Z SWITCH	ENG LSP	CCS/TRK
SACRAMENTO, CA	ST. LOUIS, MO	15	33.3
	WAYNE, PA	15	33.8
	WICHITA, KS	13	29.4
	WHITE PLAINS, NY	15	19.5
ST. LOUIS, MO	WAYNE, PA	5	34.3
	WICHITA, KS	3	32.7
	WHITE PLAINS, NY	3	31.2
WAYNE, PA	WICHITA, KS	2	28.3
	WHITE PLAINS, NY	3	16.4
WICHITA, KS	WHITE PLAINS, NY	3	33.0

PAGE 4 (LAST)

Figure 17.54 Network efficiency report, part 4

In investigating to find out why the CCS per trunk were so low, capacity managers found that Wichita was not marked in the forecast input files as a switch capable of screening 800 traffic, so in the forecast, 800 traffic was being routed from Wichita to Kansas City, because Kansas City is an 800-screening switch. In actuality, Wichita is also an 800-screening switch, and its capacity for 800 traffic was not being used. Hence, the evaluator function sometimes helps to identify problems.

17.5.2 Implementation of routing design

Capacity managers update the routing tables four times a year, at the beginning of each quarterly period, based on the forecast. Therefore, each quarterly period, the system screens the data and updates the database to establish new routing tables for the next quarterly period. The forecast-period parameter display (Figure 17.55) is used to set the effective date for the quarterly period, the hour to implement the routing, the forecast view that was used, and the quarterly period. This also is where the system threshold tables are set for the next quarterly period.

With this procedure, capacity managers can look at the quarterly period routing data before the new routing, together with the updated link capacity, is input for the next quarterly period. The route configuration display (Figure 17.56) is used to view the routing plan. In the example in Figure 17.56, the

```
FORECAST PERIOD PARAMETER                (OMN SGFPP)          PAGE 1 LAST
ACTION: I                                                     11/08/85 10:11:08
FPID# 4P85


IMPLEMENTATION REQUEST STATUS:    C

                  EFFECTIVE DATE:    10 05 85

                  EFFECTIVE HOUR:    05

                       IFS VIEW:    06 85

            IFS FORECAST PERIOD:    02

     SYSTEM BLOCKING THRESHOLD:     01 . 00 %


Z015 TRANSFER INQUIRY COMPLETED                 COMMAND:_____


1-CLEAR DATA FROM SCREEN  2-SGNDF - NODE DEFINITION  3-SGFPL - FP LIST
4-SGBTT - BLKING THRSH TBL  5-SGLSD - LSP DEFINITION    6-SGBRA - BLKING RESP ASGN
```

Figure 17.55 Forecast-period parameter display

```
ROUTE CONFIGURATION                  (OMN SRRTC)              PAGE 1 LAST
ACTION I                                                      11/08/95 09:35:30
EFFECTIVE DATE# 10 05 85              FINAL ROUTING            RESP CODE: AW10

A SWITCH# ANHMCA0211T    Z SWITCH# WHPLNY0504T    LSP# 03    DOM/INTL# D
```

PATH	VIA NODE	TYPE	SATITE	AV TRUNKS	AV TRUNKS	VZ TRUNKS
1		ED	N	144		
2	WAYNPALA42T	EV	N		60	720
3	DLLSTXTL34T	EV	N		300	144
4	WCHTKSBR24T	RT	N		84	36
5	SCRMCA0404T	RT	N		864	432
6	PHNXAZMA03T	RT	N		852	120
7	KSCYMO0904T	RT	N		240	156
8	DNVRCOZJ05T	RT	N		252	180
9	PITBPADG43T	RT	N		312	396
10	PTLDOR6203T	RT	N		264	48

```
                                                    COMMAND:_____


1-CLEAR DATA FROM SCREEN      2-SGNDF - NODE DEFINITION    3-SGBRA - BLKING RESP ASGN
4-SADBD - DAILY BLKNG DET     5-SADBS - DAILY BLKNG SUM    6-SGLSD - LSP DEFINITION
```

Figure 17.56 Route configuration display

Anaheim–White Plains switch pair has the direct path and two engineered via paths; the remaining routing table is filled up with real-time paths that meet the transmission constraints. This information is passed to the routing system, which in turn updates each switch in the network for the date and the hour that are specified in the forecast period parameter display to implement the routing changes for the next quarterly period.

In the route configuration display, capacity managers can edit the routing paths if they do not meet the transmission constraints or for some other reason, such as a via switch that does not meet transmission considerations. Within the capacity management system, the transmission constraint table, illustrated in Figure 17.57, identifies paths that satisfy the transmission standards and can be used as additional real-time routing paths to fill up the routing table.

The blocking threshold tables illustrated in Figures 17.58 and 17.59 are used to identify average business day and weekend blocking exceptions. If there is a blocking problem, the path-EFO model is run within the capacity management system and, subsequently, routing table changes are made for that load set period, capacity managers start the study-period average over for every hour in that load set period for that switch pair. If trunks are added to the network, the study-period average is started over for every load set period for every switch pair that used those trunks in the network. Thus, any time during the quarterly period, especially after there are up to four to five weeks of data, various switch pairs and load set periods have study-period averages starting and ending at different

```
TRANSMISSION CONSTRAINTS                  (OMN SGTMC)                    PAGE 1 LAST
ACTION:I                                                                11/08/85 09:52:40
FPID# 4P85


                                    FIBER OPTIC FACTOR:    1 . 5

             LONGTAIL ECHO CANCELER THRESHOLD (MILES):    0000

           ECHO CONTROL THRESHOLD DISTANCE (MILES):       1165

         DOMESTIC TWO LINK MAXIMUM DISTANCE (MILES):      2400

     INTERNATIONAL TWO LINK MAXIMUM DISTANCE (MILES):     2400

                   SHORTAIL ECHO MAXIMUM (MILES):         0452

                   LONGTAIL ECHO MAXIMUM (MILES):         1862

          INTERNATIONAL TWO TO ONE RATIO (MILES):         02 . 0


Z015 TRANSFER INQUIRY COMPLETED                           COMMAND:_____

1-CLEAR DATA FROM SCREEN      2-SGLSD - LSP DEFINITION     3-SGBTT - BLKNG THRSH TBL
4-SGFPL - FP LIST             5-SGBRA - BLKNG RESP ASGN    6-SGFPP - FP PARAMETERS
```

Figure 17.57 Transmission constraint table

```
BLOCKING THRESHOLD TABLE                    (OMN SGBTT)            PAGE 1 LAST
ACTION: I                                                    11/08/85 10:04:00
FPID# 4P85


LSP# 01                                 NUMBER OF VALID DAYS-BLOCKING

        LOAD - CCS                  01 06        07 10        11 14        15 20
     LOW            HIGH
>=        0    <=    000003         99 . 99      99 . 99      99 . 99      99 . 99
>=   000004    <=    000007         14 . 00      09 . 00      08 . 00      07 . 00
>=   000008    <=    000018         09 . 00      07 . 00      06 . 00      05 . 00
>=   000019    <=    000036         08 . 00      07 . 00      06 . 00      05 . 00
>=   000037    <=    000072         07 . 00      06 . 00      05 . 00      04 . 00
>=   000073    <=    999999         06 . 00      04 . 00      03 . 50      02 . 50


                                                     COMMAND:_____


1-CLEAR DATA FROM SCREEN     2-SGDST - DAYLITE SVNGS T      3-SGLSD - LSP DEFINITION
4-SGFPP - FP PARAMETERS      5-SRRTC - RTE CONFIGURATN      6-SGFPL - FP LIST
```

Figure 17.58 Blocking threshold table, example 1

```
BLOCKING THRESHOLD TABLE                    (OMN SGBTT)            PAGE 1 LAST
ACTION: I                                                    11/08/85 10:05:01
FPID# 4P85


LSP# 10                                 NUMBER OF VALID DAYS-BLOCKING

        LOAD - CCS                  01 06        07 10        11 14        15 20
     LOW            HIGH
>=        0    <=    000003         99 . 99      99 . 99      99 . 99      99 . 99
>=   000004    <=    000007         18 . 00      16 . 00      15 . 00      14 . 00
>=   000008    <=    000018         13 . 00      11 . 00      10 . 00      09 . 00
>=   000019    <=    000036         12 . 00      10 . 00      09 . 00      08 . 00
>=   000037    <=    000072         11 . 00      09 . 00      08 . 00      07 . 00
>=   000073    <=    999999         10 . 00      08 . 50      07 . 00      06 . 00


                                                     COMMAND:_____


1-CLEAR DATA FROM SCREEN     2-SGDST - DAYLITE SVNGS T      3-SGLSD - LSP DEFINITION
4-SGFPP - FP PARAMETERS      5-SRRTC - RTE CONFIGURATN      6-SGFPL - FP LIST
```

Figure 17.59 Blocking threshold table, example 2

times. However, whenever there is system processing to determine whether there is a blocking problem, the blocking threshold table is invoked and the number of valid days of data to set the blocking threshold is used to determine blocking exceptions. The load set period definition table (Figure 17.60) is used to define, for each of the 24 hours, the load set period associated with it. Capacity managers can also modify the retrial probability tables if the retrial probability assumptions need to be changed.

```
LOAD SET PERIOD DEFINITION                    (OMN SGLSD)              PAGE 1 LAST
ACTION: I                                                              11/08/85 09:52:09
FPID# 4P85


LOAD SET PERIOD DEFINITION TABLE


HOUR:  00  01  02  03  04  05  06  07  08  09  10  11  12  13  14  15  16  17  18  19  20  21  22  23
       —   —   —   —   —   —   —   —   —   —   —   —   —   —   —   —   —   —   —   —   —   —   —   —

 LSP:
 ABD   10  10  10  01  01  01  01  01  01  02  03  03  04  05  05  06  06  07  07  08  09  10  10  10

 SAT   15  15  15  11  11  11  11  11  11  11  11  12  12  12  13  13  13  14  14  14  15  15  15  15

 SUN   15  15  15  11  11  11  11  11  11  11  11  12  12  12  13  13  13  14  14  14  15  15  15  15

RETRIAL PROBABILITY %:

 LSP:  01  02  03  04  05  06  07  08  09  10  11  12  13  14  15
       86  86  86  86  86  86  86  86  86  86  86  86  86  86  86

                                                   COMMAND:_____

1-CLEAR DATA FROM SCREEN      2-SGTMC - TRANSMISSN CONS     3-SGBTT - BLKNG THRSH TBL
4-SGFPP - FP PARAMETERS       5-SGBRA - BLKNG RESP ASGN     6-SGNDF - NODE DEFINITION
```

Figure 17.60 Load set period definition table

17.5.3 Work center functions

We now discuss the workload of dynamic routing capacity managers in the fully deployed dynamic routing network. In 1987, all the 4ESS switches converted to the dynamic routing network, which involved roughly 700,000 circuits and about 4,200 links that capacity managers administer. Five to seven people do forecasting as well as capacity management for this entire network. On the daily blocking summary display (Figure 17.26), there is a responsibility code in the upper right-hand corner, which assigns individual switch pairs to each capacity manager. Certain sections of the network are assigned so that all capacity managers have an equal share of links that they are responsible for. Each capacity manager therefore deals primarily with one region. This is advantageous because dynamic routing capacity managers need to work with the hierarchical capacity managers in the region to coordinate the activity in the network that affects dynamic routing. Capacity managers also need to work with transport planners so that the transport planned for the links under the capacity manager's responsibility is available to the capacity manager. If, on a short-term basis, trunks have to be added to the network, capacity managers find out from the transport planner whether the transport capacity is available. If it is not available, they find out when it will be so that message trunks can be turned up.

The capacity management system is a highly automated system, and the time the capacity manager spends working with the system displays is small compared with other daily responsibilities. One of the most time-consuming work functions is following up on the circuit orders to determine status: Are they in the field? Does the field have them? Do they have the equipment working? If circuit orders are delayed, the capacity manager is responsible for making sure that the circuits are added to the network as soon as possible. With the normal amount of network activity going on, that is the most time-consuming part of the job.

17.5.4   Interfaces to other work centers

The capacity manager needs to work with the forecasters to learn of network activity that will affect the dynamic routing network. Of concern are new switches coming into the network circuit order and routing activity that affects the dynamic routing network. Capacity managers interact quite frequently with traffic managers to learn of network conditions such as cable cuts, floods, or disasters. Capacity managers detect such activities the next day in the data; the network problem stands out immediately. Before they exclude the data, however, capacity managers need to talk with the traffic managers to find out specifically what the problem was in the network. In some cases, capacity managers share information with them about something going on that they may not be aware of. For example, traffic managers investigated the MC3 condition in Anaheim, discussed earlier, to see what might have caused it to occur. Capacity managers can see these events in the data, and they share this type of information with the traffic managers. Other information capacity managers share with traffic managers relates to peak days. Capacity managers are able the next morning to give the traffic managers the actual reports and information of the load and blocking experienced in the network.

Capacity managers also work with the data collection work center. If they miss collecting data from a particular switch for a particular day, capacity managers discuss this with that work center to get the data into the capacity management system. In the capacity management system, capacity managers have until the time the study-week average is formed to get any data into the system that have been missed. So if data are missed one night on a particular switch, the switch is repolled to pull data into the system. This is a fairly common occurrence in the data collection work center.

Capacity managers frequently communicate with the routing work centers because there is so much activity going on with routing. For example, capacity mangers work with them to set up the standard numbering plans so that they can access new routing tables when they are entered into the network. Capacity managers also work with the people who are actually doing the circuit order activity on the links. Capacity managers try to raise the priority on trunk orders if there is a blocking condition, and often a single blocking condition causes multiple activities in the network. Normally, capacity managers succeed in getting a higher priority when the need is justified.

17.5.5   Comparison of dynamic routing
and hierarchical routing
capacity management

The hierarchical network has regional capacity managers, who use systems that have been in place for a number of years. For these systems, collection and movement of data take more time than in the capacity management system because of the various types of switching systems that they interface with. Because the traffic data collection system has a fully mechanized interface to the switch, the dynamic routing data are very reliable. The complexity of the dynamic routing network requires the capacity management system to automate the performance monitoring and short-term network adjustment functions, but the system is designed to handle that kind of complexity with a fast turnaround time. In the hierarchical network, it takes up to three weeks for the capacity manager to get data from the network. They make decisions to add trunks to the network based mostly on manual analysis. Some regions use mechanized tools on a day-to-day basis, but that varies from region to region. Some regions use sophisticated systems of their own; others rely on real-time traffic management; yet others use switch measurement reports.

For dynamic routing capacity management, the capacity management system is able to deliver data and run the process with an overnight turnaround to obtain the condition of the network. In that way, capacity managers are very current with the status of the network. They can also change and implement new routing tables on a weekly basis, and they have streamlined that process so that on Wednesday they run the path-EFO model to make routing changes, then pass these changes through the routing system to the switches, before the following Sunday, when the next week's activity starts. That gives a two- or three-day turnaround time for routing table updates. Capacity managers implement the forecast period routing changes within a five-day interval from delivery of the new routing tables through the routing system to the switches in time for the next quarterly period.

17.6   Network Management and Design Changes
with Real-Time Dynamic Routing

Real-time network routing (RTNR) replaced DNHR starting in 1991. With RTNR, several changes and improvements occurred in network management and design functions.

Real-time traffic management of real-time dynamic routing networks is similar to that of preplanned dynamic networks in many respects, but with some significant differences. Real-time dynamic routing networks, like preplanned dynamic networks, are best managed with centralized rather than decentralized real-time traffic management, as a single rather than distributed entity. Real-time traffic management caused NEMOS to be enhanced to accommodate new network displays and control requirements for real-time dynamic routing networks. Under preplanned dynamic routing, NEMOS automatically put in reroutes to solve blocking problems by looking everywhere in the network for additional available capacity and adding additional overflow paths to the existing preplanned engineered and real-time paths, on a five-minute basis. Real-time dynamic routing, on the other hand, replaces this automatic rerouting function. Because real-time dynamic routing examines all possible direct and two-link paths, the automatic reroute function provides no additional two-link choices that are not examined by real-time dynamic routing and hence is not required. This is yet another operational simplification introduced with real-time dynamic routing networks.

In addition to improving service quality, real-time dynamic routing networks also bring reduced network capital requirements. The real-time dynamic routing networks can carry the same traffic load as could preplanned dynamic networks, but with 5 to 8 percent less capacity in the network, based on the real-time dynamic routing network design models discussed in Chapter 13. With the introduction of real-time dynamic routing, it is important for network design to achieve the kind of efficiencies that are possible with real-time dynamic routing. A new network design model for real-time routing networks—which can be based on the fixed-point erlang flow optimization model, transport flow optimization model, or discrete event flow optimization model, as discussed in Chapter 13—achieves benefits within the forecasting and capacity management functions.

An important simplification introduced with the design of real-time dynamic networks is that routing tables need not be calculated by the design model, because these are computed in real time by the switch. This leads to simplifications in that the routing tables computed in preplanned dynamic networks are no longer needed. Real-time dynamic routing introduces simplifications into the administration of network routing. Under preplanned dynamic routing, routing tables must be periodically reoptimized and downloaded into switching systems via the automated routing administration system. Reoptimizing and changing the routing tables in the preplanned dynamic network represents an automated yet large administrative effort involving millions of records. This function is simplified by real-time dynamic routing because real-time dynamic routing does not have routing tables designed in the forecasting and capacity management systems, as does the preplanned dynamic routing network. Furthermore, the routing is generated in real time for each call and then discarded. Also, because real-time dynamic routing adapts to network conditions, less network churn and short-term capacity additions are required. This is one of the operational advantages of real-time dynamic routing networks—that is, to automatically adapt traffic routing so as to move the traffic load to where capacity is available in the network.

18

Dynamic Routing
Implementation Experience

18.1 Introduction

Dynamic routing in telecommunications networks has been the subject of worldwide study and interest. During the 1980s, dynamic nonhierarchical routing (DNHR) was fully deployed in the AT&T long-distance network. DNHR provided considerable benefits in improved performance quality and reduced costs, as discussed throughout the book, and has motivated the extension of dynamic routing to all classes-of-service. Real-time network routing (RTNR) is a real-time dynamic routing method that replaced DNHR starting in 1991. RTNR provides the platform for a class-of-service dynamic routing method that supports dynamic routing on an integrated transport network for all new and existing voice, data, and wideband services. RTNR also provides a multiple ingress/egress routing arrangement to ensure reliability and flexibility for international and access networks.

Usually the shortest distance between two points is a straight line. In a network, however, we are looking for the fastest way to complete a call, not the shortest.

With dynamic routing, the best and fastest way to complete a call from Atlanta to New York, for example, may well be through San Francisco. Although this explanation of dynamic routing may make it sound like it would take longer to get a call through, the route selection activity is actually handled in milliseconds, and the end result of dynamic routing is that more calls are completed more quickly and more accurately than before. Dynamic routing makes business sense, because it allows customers to use network capacity whenever available. That lets a network provider manage the network at greater efficiency levels, helps keep costs down, and helps ensure that calls go through on the first try. Dynamic routing also brings customers an additional benefit on some of the busiest and most congested days of the year—Mother's Day, for example. Because of dynamic routing, on Mother's Day the number of calls failing to complete on the first try is reduced by a factor of 10 or more.

In this chapter we review DNHR implementation and performance. During 1987, DNHR conversion was completed. DNHR is a computer-based system that uses several layers of network-based intelligence to move a call around possible congestion areas. DNHR identifies the uncluttered parts of the network that can be used to make sure telephone calls or data transmissions get through. In addition, traffic managers at the network operations center continually monitor dynamic routing activities to ensure that the connections are provided.

We illustrate the performance of RTNR in its first year of operation in the multiservice integrated network. This network provides connections for voice, data, and wideband services on a shared transport network. These connections are distinguished by resource requirements, traffic characteristics, and design performance objectives. Class-of-service routing allows independent control, by virtual network, of service-specific performance objectives, routing rules and constraints, and traffic data collection. Class-of-service routing allows the definition of virtual networks by class-of-service. Each virtual network is allotted a predetermined amount of the total network bandwidth. This bandwidth can be shared with other virtual networks but is protected under conditions of network overload or stress. With RTNR, the originating switch first tries the direct link and, if it is available, sets up the call on a direct trunk to the terminating switch. If a direct trunk is not available, the originating switch tries to find an available two-link path by first querying the terminating switch through the common-channel signaling (CCS) network for the busy/idle status of all links connected to the terminating switch. The originating switch then compares its own link busy/idle status information with that received from the terminating switch in order to find the least-loaded two-link path to route the call over.

18.2   DNHR Network Performance

18.2.1   Review of DNHR

Figure 1.23 illustrates the contrast between the hierarchical routing method and DNHR. In the hierarchical routing method, which began to be replaced in 1984 and had its beginnings back in the 1930s, a rigid structure is used to make route choices within the network. For example, a call in the hierarchical network starting from Phoenix, Arizona, destined for Wichita, Kansas, has only three other switches in the country that it could use (Denver, Colorado; St. Louis, Missouri; and Kansas City, Kansas), which have to be used in a prescribed order. In this case if all of the circuits from Phoenix to Wichita are busy, the call then routes to Kansas City. Once that call reaches Kansas City, if the link from Kansas City to Wichita, called a final link in the hierarchical routing network, is busy, the call is blocked. No other possible path in the structure could be accessed, which means that hierarchical routing has constrained alternate routes. Fixed routing indicates that the routing choices are not changed with time. In addition, hierarchical routing uses a progressive call-routing method; that is, whichever switch the call gets to in the network, its future is determined at that switch. If it cannot complete from that point forward, the call is blocked.

In DNHR, the same call from Phoenix to Wichita has many more possible path choices. In Figure 1.23 we have shown seven path choices—in reality, up to 21 path choices can be accessed. In addition, those path choices and the order in which they are chosen can change hour by hour, and even in real time, as we will see. They can be selected from any of the DNHR switches in the network. Finally, the originating call control feature means that if any switch in the network the call reaches is unable to complete the call, that call is returned to the originating switch, or "cranked back," and is then allowed to try all other potential routes through the network. Those features of DNHR give it a significant advantage over the hierarchical routing method.

Figure 18.1 illustrates how DNHR was introduced into the network over a $3\frac{1}{2}$ year period, in a phased introduction, due to the complexity of this task and

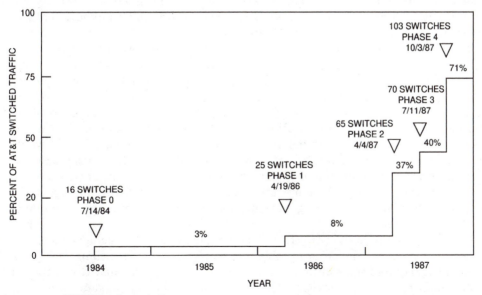

Figure 18.1 DNHR implementation

due to the fact that the network had to continue to run the hierarchical portions as DNHR was phased in. This called for a very careful introduction of DNHR into the network. Frank Blount, the vice president of the Network Operations Group during DNHR cutover, characterized this conversion activity as being akin to changing the wings on a 747 in midflight. Although perhaps not that dramatic, this is a real testament to the skill required to manage a very large conversion process—completely replacing old technology with a new technology—on an actively working network without disrupting any of the existing services.

18.2.2 DNHR systems engineering and development

Work on DNHR started in the late 1970s with an assessment of how ESS and CCS technology could be used to improve the efficiency and profitability of the network. Increasing the efficiency of the network positions a service provider more favorably in today's competitive environment: Through improvement in service quality and reducing costs, a service provider can increase profit margins. A key ingredient in making DNHR possible was the technology, and we now illustrate some of these technological trends that made DNHR possible.

Starting in the 1970s, switching in the network was converted to the 4ESS electronic switching system. That switching system replaced earlier electromechanical cross-bar systems and opened the door to more sophisticated routing in the network than was possible without stored program control. Another technological advance taking place at the same time was the conversion of network signaling from multifrequency, which allowed switching centers to communicate with one another over the actual trunk but was expensive and limited in the amount of information that could be sent, to CCS, which allowed much more information to be transmitted between switching systems much more rapidly. For example, it is CCS that makes the traveling class mark and crankback features in DNHR possible, as discussed in Chapter 16.

Another technological trend that made DNHR possible was the introduction of network management and design systems into the network. In the 1960s and 1970s the network saw a major introduction of mechanized network management and design systems to replace manual, paper-driven processes. The hierarchical network was designed, administered, maintained, and operated on a near-totally manual basis. DNHR is quite a different situation. Due to its complexity, and the amount of information that has to be put into the switches, DNHR would be impossible to administer and run on a manual basis. The introduction of these network management and design systems to mechanize the earlier manual techniques set the stage for more sophisticated real-time traffic management, control, performance monitoring, design, capacity management, and planning in the DNHR network.

Another key ingredient in DNHR is designing the network to achieve increased network efficiencies. The design problem is a very large-scale optimization problem, which requires the solution of a linear programming model. Given the set of demands between the switches in the network, the design determines the optimal set of routes and link sizes to accommodate those demands. In a network with more than 100 switches, the simplex linear programming algorithm on standard computing hardware requires hundreds of hours of processing time to solve a single problem, which would make DNHR design impractical. Therefore, a heuristic solution within the path-erlang flow optimization (path-EFO) model was developed, as described in Chapter 6. This is a simplified yet accurate and efficient technique to design the DNHR network—but orders of magnitude more rapidly than the simplex algorithm—which makes DNHR design realistic and possible.

DNHR was the first nationally deployed dynamic routing network in the world, and it provided excellent performance. Next, we illustrate that performance experience.

18.2.3 DNHR performance examples

Now we give some performance experiences with DNHR. Figure 1.3 illustrates two examples of peak-day performance: Thanksgiving and Christmas 1987. In both of these cases, we contrast the performance after the full conversion to DNHR late in 1987 with the performance a year earlier, when there was far less DNHR implementation. We can see from both of these results that the performance improvements are dramatic. In the case of Thanksgiving, average networking blocking, which in 1986 had run 34 percent, was down to 3 percent in 1987. In particular, during the midperiod of the day (the calling peak on Thanksgiving), blocking in 1986 had reached 60 percent, but blocking in 1987 was down in the 10 to 15 percent region. Similarly, on Christmas, which is a much heavier peak day, blocking in 1986 reached the 50–60 percent level, virtually throughout the entire day. This is in contrast to the 1987 Christmas performance when, although at the midday peak there was blocking on the order of 20 percent, during other portions of the day blocking was down to just a few percent.

A peak day such as Christmas, Mother's Day, or Thanksgiving is, in effect, one of the greatest tests for a routing method. These are the days on which the traffic loads on the network and the patterns of traffic deviate most severely from the normal business day and weekend loads and patterns for which the network is designed. Handling these kinds of deviations well suggests that the dynamic routing network can better handle most load patterns with equal or better quality. This is a particular advantage in a competitive environment in which the traffic loads can change rapidly, and dynamic routing gives the ability to respond to those changes quickly.

Figure 18.2 illustrates DNHR performance on nonpeak days by showing the performance of the DNHR network during the typical week of March 7 through March 13, 1988. By plotting the load variations in the network, we illustrate the normal behavior of the network. The peak load occurs on Monday morning, another peak occurs Monday afternoon, and another on Monday evening. Then, with slightly less intensity, that pattern is repeated for each of the five days of the week. On Saturday the loads are down, and on Sunday evening they are up, due primarily to residential calling. We also show the blocking performance of the network, but note the scale here. The blocking level in the DNHR network

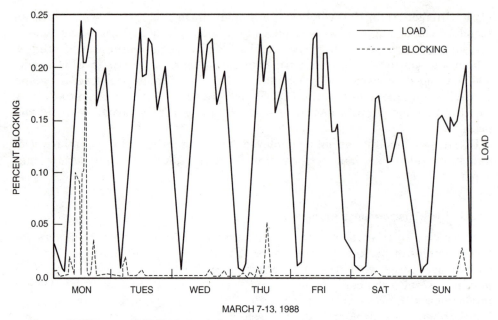

Figure 18.2 DNHR network blocking (typical week)

is fractions of a percent. Even in the worst periods, on the order of only one out of a thousand calls is blocked, and the average blocking is in the hundredths of a percent. So, the DNHR network under normal load conditions is virtually nonblocking. This is in contrast with the performance of the hierarchical network, where under similar normal business day loads, particularly on Monday mornings, blocking would frequently reach several percent, and in some cases, tens of percent. This represents, in fact, orders of magnitude improvement in the performance of the network, even under ordinary conditions, and means additional revenue, as those formerly blocked calls get through.

In addition to improving performance quality, DNHR also brings reduced network capital requirements—the DNHR network can carry the same load as the hierarchical network but uses about 15 percent less network capacity. That translates into a major capital savings. In addition, there are reduced network management and design costs, because with DNHR there is automated real-time traffic management, routing administration, and capacity management that in the hierarchical network had been done manually. In the operational areas, automation has been able to reduce the work of hundreds of people down to the work of a few people. The savings in network management and design costs have been large, as discussed in Chapters 16 and 17. Improved service translates to fewer defections of customers to competitive carriers and, thus, revenue retention.

Real-time traffic management is controlled from the network operations center in Bedminster, New Jersey, using the NEMOS system (Network Management Operations System). NEMOS gathers data from the network every five minutes and can change routes in the network and put in controls to deal with network

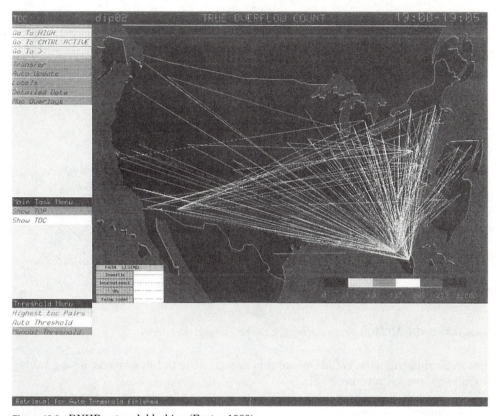

Figure 18.3   DNHR network blocking (Easter 1988)

events such as failures, overloads, and other problems that cannot be fully
foreseen or planned for in advance.

Figure 18.3 gives an example of NEMOS operation, showing details of network
performance for Easter 1988 at 7:00 P.M. Eastern Standard Time. The figure shows
the total calls that the DNHR network was not able to carry, where each line
connecting city pairs on the map represents some number of calls that are not
being carried on the network. Several things stand out from this kind of national
display. We can see that much of the problem is getting into cities in Florida.
This is probably not too surprising—a number of people have mothers, fathers,
and other relatives in Florida, and on a holiday such as Easter many people are
trying to call their families. As discussed in Chapter 8, NEMOS automatically
puts reroutes in place to solve those blocking problems. Of course, NEMOS is not
able to solve all of the blocking problems, but the reroute model in NEMOS is
able to look everywhere in the network for additional available capacity and add
up to 7 additional paths to the 14 existing ones used by DNHR at any given point
in time. Together these paths make up the 21 DNHR paths referred to earlier.
NEMOS, on a five-minute basis, can look at every one of the possible alternate
paths through the network and check them for available capacity. When it finds

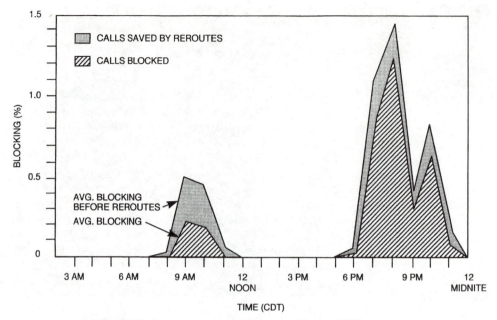

Figure 18.4 DNHR/NEMOS automatic reroute performance (Easter 1988)

that available capacity, NEMOS can add those paths to the switch routing tables within the five-minute interval.

The end result is illustrated in Figure 18.4, which shows the performance of the dynamic routing network on Easter Sunday 1988. As can be seen from the figure, the average blocking performance was indeed good. The figure also illustrates what the performance of the network would have been without the real-time traffic management reroute capability. The shaded areas on the top show the calls that were saved by the rerouting capability. We can see that in the morning rerouting was very effective, in that NEMOS was able to reduce blocking by about a factor of 2 through the use of the reroute capability. In the evening, as the network became more congested, there was less free capacity and less opportunity to find reroutes. But nonetheless, overall, reroutes were able to add a significant additional completion performance.

As discussed in Chapter 17, DNHR has a capacity management system that each day analyzes the latest traffic data and then each week reoptimizes the routing and sizing of the network to accommodate ongoing changes in demand from what was originally designed months earlier in the forecast design. Figure 9.6 gives one example of the operation of the capacity management system and quantifies the service between two cities in the DNHR network, Atlanta, Georgia, to Miami, Florida, during October 1987. On this particular route, despite the fact that the overall network performance was excellent, there was high blocking between this pair of cities. The data collected during those two weeks were transmitted to the capacity management system to process. The system then analyzed and solved the blocking problem, not by adding trunks to the network but by reoptimizing and then changing the routing tables in the network. After the new routing was downloaded to the network, the blocking problem was essentially eliminated. This illustrates one of the important operational advantages of dynamic routing. In contrast to the hierarchical network, where trunks are moved and rearranged to follow the demand, in the dynamic routing network there is an ability to move traffic around to where there is capacity in the network. This is simpler, less expensive, and faster. Dynamic routing teleprocessing moves both the traffic data and routing table information within the switched network. Those data are collected at the capacity management system over a packet-data network once a day. Every switch is polled and the traffic data are brought by the data collection system to the capacity management system. Each night the data are summarized, assembled, and analyzed. Each week the path-EFO model runs and finds a new routing design, which is then downloaded into the DNHR switches through the packet-data network. The routing data are formatted and transmitted to the DNHR switches in a single night's run, which involves sending millions of records to change all switch routing tables. Once that downloading has successfully taken place, the new routing design is then implemented across the network in one moment, early on Saturday morning, to put the new routing in operation.

18.3   RTNR Network Performance

18.3.1   Review of RTNR for multiservice integrated networks

Three motivations for developing the RTNR network to replace DNHR were as follows: (1) to support multiple classes-of-service with dynamic routing on a multiservice integrated network, (2) to provide a more robust routing method for self-healing networks, and (3) to simplify network management and design functions. These motivations are now briefly discussed.

Based on DNHR experience, it became clearly desirable to introduce all new services using dynamic routing. These services include 64-kbps, 384-kbps, and 1,536-kbps switched digital data services, international switched transit services, priority routing services, ISUP preferred service, international virtual private network services, and others. Such needs have led to the concept of multiservice integrated dynamic routing for telecommunications networks, in which individual voice and data services utilize dynamic routing and share network bandwidth. There clearly is benefit in developing a robust routing method for self-healing networks, as discussed in Chapter 14. A self-healing network is one that responds in near real time to a network failure and continues to provide connections to customers with essentially no perceived interruption of service. Here, dynamic routing shifts bandwidth rapidly among switch pairs and services and provides transport-diverse routing, as well as multiple ingress/egress routing, to help respond to link and switch failures. Transport-diverse routing allows selection of diverse traffic paths through the transport network, and multiple ingress/egress routing allows selection of diverse traffic paths into and out of the switching network.

There is a need to simplify the operational environment because operations costs are significant. There is also motivation to develop an efficient decentralized real-time routing method as a basic flexible routing method for international dynamic routing. These needs are met by the RTNR method, and the real-time internetwork routing (RINR) extension discussed in Chapter 12 for multiservice integrated networks.

In contrast to DNHR, switches using RTNR first select the direct link between the originating switch, denoted here as OS_j, and the terminating switch, denoted here as TS_k. When no direct trunks are available, the originating switch checks the availability and load conditions of all of the two-link paths to TS_k on a per-call basis. If any of these two-link paths are available, the call is set up over the least-loaded two-link path. Traffic loads are dynamically balanced across trunks throughout the network to maximize the call throughput of the network. As illustrated in Figure 1.27, an available two-link path from OS_1 to TS_2 goes through a via switch to which both OS_1 and TS_2 have idle trunks—that is, neither link is busy. An available two-link path is considered to be lightly loaded if the number of idle trunks on both links exceeds a threshold level. In order to determine all of the switches in the network that satisfy this criterion, the originating switch sends a message to TS_2 over the CCS network, requesting TS_2 to send a list of the switches to which it has lightly loaded links. Upon receiving this list of switches from TS_2, the originating switch compares this list with its own list of switches to which it has lightly loaded links. Any switch that appears in both lists currently has lightly loaded links to both OS_1 and TS_2 and therefore can be used as the via switch for a two-link connection for this call. In Figure 1.27 there are two lightly loaded paths found between OS_1 and TS_2.

The switch identifiers used in the switch list sent by TS_2 must be recognized by the originating switch. Each switch in the network is assigned a unique network switch number (NSN); these NSNs are used as switch identifiers. In the example depicted in Figure 1.27, there are five switches in a network that have been arbitrarily assigned NSNs. With these NSN assignments, a list of switches can be represented by a bit map that has a 1-bit entry for each NSN in the network. In Figure 1.27, a "1" entry is made in the bit map for each NSN having a lightly loaded link to TS_2. OS_1 also maintains its own bit map, listing each NSN having a lightly loaded link to OS_1. Using bit maps makes it very easy and efficient for the originating switch to find all lightly loaded two-link paths. The originating switch simply ANDs the bit map it receives from TS_2, which lists all of the lightly loaded links out of TS_2, with its own bit map to produce a new bit map that identifies all the via switches with lightly loaded links to both OS_1 and TS_2. Bit maps are also a very compact way to store a list of switches. This is an important consideration because the list is sent in a CCS message. In the simplest case, only 16 bytes of data are needed for a network with 128 switches.

Because some of the available two-link paths may not provide good voice transmission quality, network administrators can restrict path selection to the two-link paths that provide good transmission quality through use of another bit map, the allowed via switch list, which specifies the acceptable via switches from OS_1 to TS_2. As illustrated in Figure 1.27, ANDing this bit map with the bit map containing the via switches of all the available two-link paths removes the via switches of paths with unacceptable transmission quality. When two or more available paths have the same load status, as in Figure 1.27, where two lightly loaded paths are found, the originating switch randomly picks one of these paths to use for the call. To pick a path for this particular call to TS_2, the originating switch starts a circular search through the bit map list of paths with the same load status, beginning with the entry immediately following the via switch it last used for a call to TS_2.

Requesting trunk status on a per-call basis ensures that the current network conditions are known for selecting a two-link path for the call. However, sending a request message and waiting for the response adds to the setup time for the call. This additional call setup delay can be avoided. When the originating switch does not find any available direct trunks to TS_k, it requests the current list of switches to which TS_k has idle trunks. Rather than waiting for a response from TS_k to set up the call, the originating switch can select a two-link path using the most recently received status response from TS_k. When the new status response is received, the originating switch stores it away for use the next time it needs to select a two-link path to TS_k. Because the status of the idle trunks is not as current using this method, the two-link path selected has a greater probability of being blocked. When this happens, the via switch uses a CCS crankback message to return the call to the originating switch. The originating switch must then wait for the status response from TS_k and pick a new via switch for the call using this up-to-date information. With these methods, RTNR reduces call setup delay in comparison with preplanned dynamic routing methods, as discussed in Chapter 12. Further details of RTNR operation are given in Chapter 12, along with a description of the RINR extension, which is illustrated in Figure 1.29.

18.3.1.1   RTNR class-of-service routing. RTNR provides a platform to implement a multiservice integrated network and therefore is used for voice services, 64-kbps data services, 384-kbps data services, and 1,536-kbps data services. The simplicity of RTNR operation and the lack of routing data administration make RTNR an attractive choice for multiservice integrated networks. As illustrated in Figure 1.25, the various virtual networks share bandwidth on the network links. For this purpose the multiservice integrated network is designed to handle the combined forecasted call loads for many classes-of-service. The network design process allots a certain number of direct virtual trunks between OS_j and TS_k to virtual network i, which is referred to as VTeng_k^i. VTeng_k^i is the minimum guaranteed bandwidth for virtual network i, but if virtual network i is meeting its GOS^i blocking objective, other virtual networks are free to share the VTeng_k^i bandwidth allotted to virtual network i. The quantities VTeng_k^i are chosen in the network design process so that their sum over all virtual networks sharing the bandwidth of the direct link is less than or equal to the total bandwidth on the link, as illustrated by the three VTeng_k^i segments in Figure 1.25.

For each virtual network i, RTNR maintains switch-to-switch blocking rates NN_k^i, switch-to-switch traffic load estimates VTtraf_k^i, and current switch-to-switch calls-in-progress counts CIP_k^i. RTNR controls the use of direct and two-link paths based on parameters defining each virtual network for each class-of-service. A virtual network comprises the above traffic measurements, bandwidth allocation parameters (VTeng_k^i), routing priority, performance objective parameters, and voice/data transport capability. Through use of the virtual network parameters, RTNR implements routing that is able to meet different blocking objectives and load levels for different virtual networks. For example, virtual networks with different transport capabilities—such as voice and 384-kbps switched digital services—can have different blocking objectives, as can virtual networks that require the same transport capabilities, such as domestic and international voice services. In this way, RTNR maximizes the performance of a multiservice integrated network in meeting blocking and call throughput objectives for all classes-of-service. For purposes of call establishment, RTNR class-of-service routing executes the following steps:

1. At the originating switch, the class-of-service, virtual network, and terminating switch TS_k are identified.

2. The class-of-service, virtual network, and TS_k information are used to select the corresponding routing data, which include voice/data transport, data rate, performance objectives, dynamic reservation thresholds, routing priority, bandwidth allocation, and traffic data registers.

3. An appropriate path is selected through execution of dynamic path selection logic and possible exchange of network status bit maps, and the call is established on the selected path.

Switch OS_j uses the quantities VTtraf_k^i, CIP_k^i, NN_k^i, and VTeng_k^i to dynamically allocate link bandwidth to different virtual networks. Under normal nonblocking network conditions, all virtual networks fully share all available capacity. Because of this, the network has the flexibility to carry a call overload between two switches for one virtual network if the traffic loads for other virtual networks are sufficiently below their design levels. An extreme call overload between two switches for one virtual network may cause calls for other virtual networks to be blocked, in which case trunks are reserved to ensure that each virtual network gets the amount of bandwidth allotted by the network design process. This reservation during times of overload results in network performance that is analogous to having a number of trunks dedicated between the two switches for each virtual network. When blocking occurs for virtual network i, RTNR trunk reservation is enabled to prohibit alternate-routed traffic and traffic from other virtual networks from seizing direct link capacity designed for virtual network i. When the originating switch detects that the current blocking level NN_k^i for calls for virtual network i exceeds the GOS^i objective, trunk reservation is triggered for direct traffic for virtual network i, and the reserved bandwidth on the direct link for any virtual network is at most $\text{VTeng}_k^i - \text{CIP}_k^i$.

Sharing of bandwidth on the direct link is implemented by allowing calls for virtual network i to always seize a virtual trunk on the direct link if the number of calls in progress CIP_k^i is below the level VTeng_k^i. But if CIP_k^i is equal to or greater than VTeng_k^i, then calls for virtual network i can seize a virtual trunk on the direct link only when the idle bandwidth on the link is greater than the bandwidth reserved by other virtual networks that are not meeting their blocking objectives. That is, if

$$\mathrm{CIP}^i_k \geq \mathrm{VTeng}^i_k$$

and the idle link bandwidth ILBW_k is greater than a reserved bandwidth threshold, we select a virtual trunk on the direct link. Hence, traffic for other virtual networks is restricted from seizing direct link capacity that is reserved for virtual network i to meet its VTeng^i_k design call load level. In this manner each virtual network is ensured a minimal level of network throughput determined by the network design process. When the number of calls in progress CIP^i_k exceeds VTeng^i_k, the design capacity allotted for virtual network i is used up, and calls can then be routed to unreserved direct link capacity or to available via paths. Finally, when CIP^i_k exceeds VTtraf^i_k, reservation is no longer needed to meet the GOS^i objective for virtual network i, and virtual trunks are shared by all traffic, including alternate-routed traffic.

With the introduction of dynamic routing based on real-time network status, the network is more adaptive or robust in coping with traffic fluctuations and network failure situations. RTNR implements the concepts of transport-diverse routing, illustrated in Figure 18.5, and multiple ingress/egress routing, illustrated in Figure 1.28 and described in the following section. Transport-diverse routing provides immediate access to all surviving capacity after a network transport failure. RTNR achieves a maximally efficient transport-diverse routing method. This is because with fixed routing, switch pairs such as A–D in Figure 18.5 can be isolated by a single transport cable cut, whereas with RTNR all surviving capacity is immediately accessed, such as through switches E and F.

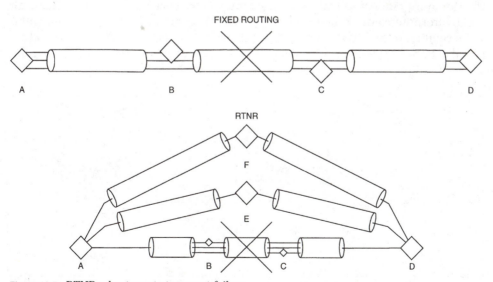

Figure 18.5 RTNR robustness to transport failure

18.3.1.2 RTNR multiple ingress/egress routing. Multiple ingress/egress routing is also incorporated in RTNR and, as illustrated in Figure 1.28, allows international networks, national intercity switches, end-offices, or customer private branch exchanges (PBXs) to be multiply connected to more than one switch in the network in order to provide flexible access to all available connecting capacity, as well as robust routing diversity and protection from switch failures, transport failures, or failures in the ingress/egress network.

Traditionally, the traffic to and from a particular geographical area goes to a particular switch nearest to it in order to carry the ingress and egress traffic. For example, in Figure 1.28, if all the traffic for the PBX location enters and exits the network through switch T_4, then if switch T_4 fails, the PBX location becomes isolated. To protect against this, the PBX location is connected to more than one switch, T_4 and T_3 in Figure 1.28. RTNR path selection might first be tried to switch T_4 to reach the PBX location, but if that is unsuccessful, for whatever reason, the call is returned to the originating switch with CCS crankback, and RTNR path selection is attempted to switch T_3 to reach the PBX location. In a similar manner, flexible international routing to other national networks is achieved with this multiple ingress/egress routing concept, together with the RINR capability illustrated in Figure 1.29 and described in Chapter 12. If a call from the originating switch in Figure 1.28 cannot reach the UK through switch T_3, perhaps because all international trunks from switch T_3 to the UK are busy, the call is returned to the originating switch via CCS crankback for RTNR path selection to switch T_4 and then to the UK. Full access to international capacity provides additional throughput flexibility and robustness to failure.

Figure 18.6 compares multiple ingress/egress routing with RTNR with the previous routing method with DNHR. Note that in DNHR, routing to the UK had a choke point. That is, traffic destined for the UK from Anaheim had to go through the Pittsburgh international switching center. Once such a call gets to Pittsburgh, it can alternate-route through the New York NY55 switch, but that is the point of vulnerability. If the Pittsburgh switch fails, or if there is a transport failure into or out of Pittsburgh, a large part of the incoming and outgoing international traffic

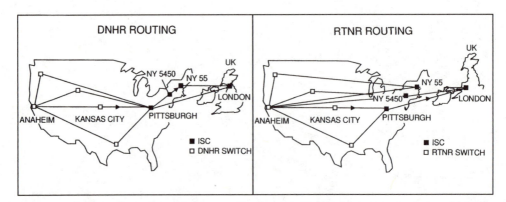

Figure 18.6 RTNR multiple ingress/egress routing robustness to switch failure

will be blocked. This vulnerability is eliminated by the multiple ingress/egress routing introduced by RTNR: a switch failure at Pittsburgh or a transport failure on that path can be circumvented by the additional path through the NY55 international switching center, leading to a significant improvement in network robustness for international traffic.

18.3.2 RTNR performance examples

RTNR was introduced into the network over the five-month period between March and July 1991, in a phased introduction, and over this time period it completely replaced DNHR in the network. This very large conversion process was made on an actively working network without affecting any of the existing services. After RTNR was fully cut over in July 1991, eight virtual networks were established across the classes-of-service listed in Table 18.1. These virtual networks share the transport capacity of the network but provide individual control and monitoring of traffic by class-of-service, as well as providing new priority-service capabilities. Multiple ingress/egress routing was implemented along with RTNR, and with that, international networks, as well as other applications discussed in Chapter 12, were provided the added flexibility of RTNR. RINR was implemented starting in 1995, replacing bilateral hierarchical routing between the U.S. national network and all other countries of the world.

RTNR allows a marked improvement in network reliability by providing for the selection of up to 134 paths between every pair of switches for every call. RTNR also enables the flexible ingress/egress routing arrangement, to ensure reliability and flexibility for international and access networks. RTNR is the first nationally deployed real-time dynamic routing network in the world, and has provided excellent performance. Next, we illustrate that performance experience.

18.3.2.1 Results for high-day traffic loads. Table 18.2 illustrates RTNR network performance for the highest-day load on the network to that date, December 2, 1991, the Monday after Thanksgiving, which is normally the highest business calling day of the year—and it was. We also show the blocking performance of the

**TABLE 18.1 Eight Virtual Networks Implemented
by RTNR Class-of-Service**

Virtual network number	Class-of-service
1	business voice
2	long-distance voice
3	international voice
4	64-kbps data
5	384-kbps data
6	1,536-kbps data
7	priority voice
8	priority 64-kbps data

TABLE 18.2   Network Blocking Performance for High-Day Traffic
Loads, Monday after Thanksgiving (12/2/91)

Virtual network number	Class-of-service	Blocking performance (% blocking)
1	business voice	0.00
2	long-distance voice	0.00
3	international voice	0.00
4	64-kbps data	0.00
5	384-kbps data	0.00
6	1,536-kbps data	0.00
7	priority voice	0.00
8	priority 64-kbps data	0.00

network. We see that for the Monday after Thanksgiving in 1991, which set a record for the number of calls on the network, 157.5 million, only 228 calls were blocked on intercity connections, providing a completion rate of 99.999 percent on the first try. This is in contrast to the Monday after Thanksgiving in 1990, when DNHR was fully implemented, when the number of calls on the network was 136.5 million, and 17,086 were blocked on intercity connections. Hence, with DNHR about 1 out of every 10,000 calls was blocked on the busiest day of the year, whereas with RTNR only 1 out of every 1,000,000 calls was blocked. This illustrates that the dynamic routing network under normal load conditions is virtually nonblocking, which means that real-time dynamic routing leads to additional service revenues as formerly blocked calls are completed. In addition, the results show that voice and data services integration leads to excellent network performance with independent traffic control for each voice or data virtual network, at the same time providing efficient sharing of transport network capacity.

18.3.2.2 Results for peak-day traffic loads. Peak days such as Christmas, Mother's Day, and Thanksgiving are among the greatest tests for dynamic routing, because the traffic loads and patterns of traffic deviate most severely from the normal business day traffic for which the network is designed. If the network performs well for such severe overload patterns, it can perform well for essentially any load pattern generated by customers with equal or better quality. This is an advantage in today's competitive environment, in which the traffic loads and the services that generate those loads can change rapidly. For example, sales promotions of discount calling programs, such as a campaign promising free calls all day Sunday, are frequently undertaken in the competitive environment and can lead to large traffic surges during the discount periods. Also, large business customers can move from one service provider to another, on short notice, bringing large amounts of traffic along with their move very suddenly. New services introduced to the network can take off faster than expected, again leading to traffic surges. Real-time routing methods such as RTNR give the ability to respond to those changes quickly.

Figure 1.3 illustrates two examples of peak-day performance: Thanksgiving and Christmas. In both of these cases, we contrast the performance after the conversion to RTNR in 1991 (with the performance of DNHR after its full deployment in 1987) with the performance of the hierarchical (HIER) network in 1986, when there was a far smaller DNHR implementation. The performance of these three networks is based on measured data. The network is not designed for peak days such as Christmas—it is designed for normal average business day and weekend traffic loads. On a peak day such as Christmas we try to complete as much traffic as possible, but it is often not possible to complete calls on the first try.

We can see from these results that the performance improvements of the DNHR network over the hierarchical network, and the RTNR network over the DNHR network, are dramatic. In the case of Thanksgiving, average network blocking, which in 1986 had run 34 percent, was down to 3 percent in 1987, and then down to 0.4 percent in 1991, which is nearly a factor of 100 improvement. In particular, during the midperiod of the day, the calling peak on Thanksgiving, blocking in 1986 had reached 60 percent, blocking in 1987 was down in the 10–15 percent region, and in 1991 down to 2–3 percent. Similarly, on Christmas, which is the second-heaviest peak calling day of the year (after Mother's Day), blocking in 1986 had run at the 50–60 percent level, virtually throughout the entire day. In contrast, the 1987 DNHR network yielded Christmas Day performance with blocking on the order of 20 percent at the midday peak, and the 1991 RTNR network had blocking on the order of 15 percent at the midday peak. During other portions of the day, blocking was down to just a few percent. The average blocking for Christmas Day was down from 58 percent with the hierarchical network to 12 percent with DNHR and to 6 percent with RTNR, which is almost a factor of 10 improvement. Also, in 1991 versus 1986, there were fewer trunks relative to demand because of the efficiencies of dynamic routing design, and there was more traffic load.

18.3.2.3 Results for unexpected network overloads. Table 18.3 illustrates the network blocking performance during the evening hours of August 19, 1991, when Hurricane Bob caused severe overloads in the northeastern United States. As is clear from the results in the table, the priority-service capability of the RTNR class-of-service network provides essentially zero blocking performance, even under such severe network conditions.

TABLE 18.3 Network Blocking Performance for Unexpected Traffic Overloads, Hurricane Bob (8/19/91)

Hour of day	Normal voice services (% blocking)	Priority voice services (% blocking)
6 to 7 P.M.	0.04	0.00
7 to 8 P.M.	0.99	0.00
8 to 9 P.M.	0.47	0.00

18.3.2.4 Results for network failure. Figures 12.24 and 12.25 give simulation results for an actual cable cut in Sayreville, New Jersey—a cut of approximately 33,000 trunks. The results demonstrate that RTNR not only improves network throughput (Figure 12.24) but also is better in preventing switch-pair-traffic isolation (Figure 12.25). Call attempts are also reduced, which means that customers require fewer attempts to get through. This self-healing property is very valuable for network reliability.

Figure 1.4 shows the performance of the network during a fiber cut that occurred at 4:26 P.M. E.S.T. on September 25, 1991. The fiber cut, which was caused by a bullet, was near Austin, Indiana, on a cable that runs between Columbus, Indiana, and Louisville, Kentucky. In all, more than 129,000 circuits were lost due to the cut, and within these circuits, 67,000 trunks were lost in the switched network. An automatic transport restoration system, FASTAR, restored the first 672 circuits (one DS3) in 10 minutes, and the next 87,000 circuits (130 DS3s) were restored within the next minute, which included 37,000 trunks in the switched network, leaving a total of 30,000 trunks still out of service in the switched network after the FASTAR restoration. Over the duration of this event, more than 12,000 calls were blocked in the switched network, almost all of them originating or terminating at the Nashville, Tennessee, switch.

The blocking at the Nashville switch is plotted at five-minute intervals in Figure 1.4, and reaches almost 25 percent blocking at its peak. It is noteworthy that the blocking in the network returned to zero after the 37,000 trunks were restored in the first 11 minutes, even though there were 30,000 trunks still out of service. RTNR was able to find paths on which to complete traffic even though there were still far fewer trunks than normal after the FASTAR restoration. Hence, RTNR in combination with FASTAR provides a significant self-healing capability for the network. It is also noteworthy that the remaining circuits were restored about two hours later. Without RTNR and FASTAR, degradation of service would have lasted far longer than it did. As discussed earlier, RTNR provides priority routing for selected customers and services, which permits priority calls to be routed in preference to other calls. As illustrated in Figure 1.4, blocking of the priority services was essentially zero throughout this entire network event.

18.3.2.5 Results for international traffic with multiple ingress/egress routing and real-time internetwork routing. Table 18.4 compares the performance of international traffic between the United States and China during the Chinese New Year Celebration for (1) bilateral hierarchical (HIER) international routing plus HIER domestic routing, without multiple ingress/egress routing (MIER), in 1991; (2) HIER plus RTNR/MIER in 1992; and (3) RINR plus RTNR/MIER in 1996. The results represent actual performance measurements over the period during which the international routing methods were upgraded to RTNR/MIER and then to RINR. Over this period, traffic levels generally increased. We see from these results the considerable improvement in the quality of service following the introduction of RTNR/MIER and then RINR for Chinese New Year traffic. It is also clear that key-service protection provides essentially zero blocking even

TABLE 18.4 Network Blocking Performance Comparison for Chinese New Year (%)

Virtual network	HIER (1991)	HIER + RTNR/MIER (1992)	RINR + RTNR/MIER (1996)
voice	73.2	31.6	0.7
key voice	—	—	0.0
average	73.2	31.6	0.7

under the most severe overload conditions in the network. Hence, from 1991 to 1996, as routing flexibility increased, quality of service substantially improved for the benefit of customers and increased profitability of the network.

18.4 Network Management and Design

18.4.1 DNHR network management and design

In order to manage and design a preplanned dynamic routing network such as the DNHR network, we use the family of network management and design systems illustrated in Figures 1.1, 1.33, and 1.37. The real-time traffic management function (Figure 1.33) is controlled from the network operations center in Bedminster, New Jersey, using NEMOS. NEMOS gathers data from the network every five minutes and can change routes in the network and put in controls to deal with unforeseen network events. To design network capacity we use a network forecasting system, as illustrated in Figure 1.37, to perform the design function of forecasting the link capacity of the network over the next several months and years in the future. We have a capacity management system, shown in the lower three blocks of Figure 1.37, that every week analyzes the latest traffic data and then reoptimizes the routing and sizing of the network to accommodate ongoing changes in demand from what was originally designed months earlier in the network forecasting system. The switch- and transport-planning systems determine the switch and transport capacity needed in the network as demand changes over time. Planning is done on a multimonth cycle.

An operational requirement of the network is transmission quality assurance. One of the features inherent in the hierarchical network is that the hierarchical restrictions themselves embodied the transmission constraints in that network. Calls could not go over too long a distance, calls could not encounter routes without proper echo control, and contrast between short routes and long routes was minimized through the rigid structure of the hierarchical routing methodology. In dynamic routing, where in essence any route becomes possible in the network, transmission quality becomes an important issue. In order to make sure that DNHR achieves the same high quality standards, and in fact higher quality standards than in the hierarchical network, the DNHR network design model implements a transmission quality assurance design. The transmission quality assurance design checks the transmission quality components, including noise and contrast. Contrast means that a call from New York to California, for example, is expected to have, time after time, the same transmission characteristics. We require that contrast be kept down to acceptable levels, and any transmission path that cannot achieve the contrast objective is not allowed. Similarly, a constraint is imposed on delay, in which echo cancelers have rigidly imposed control mileages to make sure that the so-called tail delay is not so long that the echo canceler cannot cancel all of the echo. Again, the transmission quality assurance design checks the mileage on all routes to ensure that the echo canceler constraint is met. DNHR accomplishes these design checks within the design systems, and in doing so achieves higher levels of transmission quality than in the hierarchical network. This is also the case because dynamic routing paths are limited to at most two links, whereas hierarchical routing paths could have up to seven links.

Real-time traffic management at the network operations center had to be completely revamped to handle network management for DNHR. In the hierarchical routing network, the network operations center primarily had the role of the coordination center between the regional centers, where due to the regional structure of the hierarchical network most of the real-time traffic management of the network was done. But with DNHR the concept of regions becomes meaningless. With DNHR there was now a national network, which needed to be managed as a single national entity. This meant completely changing the network management systems used at the network operations center, and thus the development of NEMOS was undertaken, which entailed completely rebuilding the displays and consoles used at the network operations center. In the hierarchical network operations center, the visual wall display mirrored the regional structure of the hierarchical network. With DNHR, the network operations center has evolved to video displays that are programmable, are national in scope, and can be changed through software much as other things can change in the DNHR network. This results in a much more flexible network management system and control center.

18.4.2 RTNR network management and design

The real-time traffic management of a real-time dynamic routing network, such as the RTNR network, is similar to DNHR in many respects but also has some significant differences. The RTNR network, like the DNHR network, is managed as a single national entity. With RTNR, some new network displays and controls were implemented and are used at the network operations center, where the video displays are changed through software updates to the NEMOS system. Under DNHR, NEMOS automatically put in reroutes to solve blocking problems. The reroute model in NEMOS was able to look everywhere in the network for additional available capacity and add up to 7 additional paths to the 14 existing paths used by DNHR, on a five-minute basis. RTNR replaces this automatic rerouting function that NEMOS performed under DNHR. That is, because RTNR examines *all* possible one- and two-link paths, the NEMOS reroute function provides no additional two-link choices not examined by RTNR and thus is not required. This is one operational simplification introduced under RTNR.

With the introduction of RTNR, new network design models, which include transport flow optimization models and a discrete event flow optimization model, were developed that capture the improved efficiencies possible with RTNR. In addition to improving service quality, RTNR brings reduced network capital requirements. That is, the RTNR network can carry the same traffic load as could the DNHR network, but with 5–8 percent or more reduction in network capacity, based on the real-time dynamic routing design models mentioned above. RTNR introduces simplifications into the capacity management of the network. Under DNHR, routing tables had to be periodically reoptimized in the capacity management system and downloaded into switching systems via the routing administration system. Reoptimizing and changing the routing tables in the network represented an automated but large administrative effort involving millions of records. This function is eliminated by RTNR because in RTNR the routing table is generated automatically in real time for each call and then discarded. Also, because RTNR adapts to network conditions, less short-term capacity adjustment is required. This is one of the operational advantages of RTNR: to automatically adapt the traffic routing so as to move the traffic load to where capacity is available in the network and to accommodate ongoing changes in demand from that designed in the network forecasting system.

18.5   Conclusion

We describe dynamic routing implementation for both preplanned and real-time dynamic routing, as illustrated by the DNHR and RTNR experiences, respectively, and the motivation for introducing these new dynamic routing methods. Dynamic routing brings benefits to customers in terms of new service flexibility and significantly higher service quality and reliability, all at reduced cost. Dynamic routing provides faster call setup by ensuring that calls go through on the first try. Dynamic routing, as illustrated by RTNR, implements a service-based multiservice integrated routing feature for extending dynamic routing to emerging services and provides a self-healing network capability to ensure a networkwide path selection and immediate adaptation to failure. We present network performance results across a spectrum of network conditions, which include peak-day loads on Thanksgiving and Christmas 1986, 1987, and 1991; high-day loads on the Monday after Thanksgiving 1991; a cable cut near Austin, Indiana, on September 25, 1991; a focused overload caused by Hurricane Bob on August 19, 1991; and the international network performance during the Chinese New Year holiday celebration after the introduction of multiple ingress/egress routing and real-time internetwork routing.

We conclude that dynamic routing, as illustrated by DNHR and RTNR experience, (1) improves network performance by lowering blocking, reducing call setup delay, and improving transmission performance; (2) allows significant reductions in network capacity requirements and capital cost savings; (3) simplifies the network management and design environment and lowers network management and design cost; (4) automates the routing functions in the network; (5) standardizes class-of-service routing data structures for new service introduction; (6) provides new routing services such as priority routing; and (7) allows sharing of voice and data network capacity in the deployment of ISDN. Telecommunications providers must differentiate their service quality and improve their network efficiencies to stay competitive, and dynamic routing is an important part of that competitive differentiation.

Bibliography

[ABS88] Ash, G. R., Blake, B. M., Schwartz, S. D., "Integrated Network Routing and Design," *Proceedings of the Twelfth International Teletraffic Congress,* Torino, Italy, June 1988.

[ACF91] Ash, G. R., Chen, J-S., Frey, A. E., Huang, B. D., "Real-Time Network Routing in a Dynamic Class-of-Service Network," *Proceedings of the Thirteenth International Teletraffic Congress,* Copenhagen, Denmark, June 1991.

[ACF92] Ash, G. R., Chen, J-S., Frey, A. E., Huang, B. D., Lee, C-K., McDonald, G., "Real-Time Network Routing in the AT&T Network—Improved Service Quality at Reduced Cost," *Proceedings of the IEEE Global Telecommunications Conference,* Orlando, Florida, December 1992.

[Ack79] Ackroyd, M. H., "Call Repacking in Connecting Networks," *IEEE Transactions on Communications,* Vol. 27, No. 3, March 1979.

[ACK89] Ash, G. R., Chemouil, P., Kashper, A. N., Katz, S. S., Yamazaki, K., Watanabe, Y., "Robust Design and Planning of a Worldwide Intelligent Network," *IEEE Journal on Selected Areas in Communications,* Vol. 7, No. 8, October 1989.

[ACL94] Ash, G. R., Chan, K. K., Labourdette, J. F., "Analysis and Design of Fully Shared Networks," *Proceedings of the 14th International Teletraffic Congress,* Antibes, France, June 1994.

[ACM81] Ash, G. R., Cardwell, R. H., Murray, R. P., "Design and Optimization of Networks with Dynamic Routing," *Bell System Technical Journal,* Vol. 60, No. 8, October 1981.

[ACM81] Ash, G. R., Cardwell, R. H., Murray, R. P., "Design and Optimization of Networks with Dynamic Routing," *Proceedings of the International Conference on Communications,* 1981.

[ACM91] Ash, G. R., Chang, F., Medhi, D., "Robust Traffic Design for Dynamic Routing Networks," *IEEE INFOCOM '91,* 1991.

[AKA94] Ash, G. R., Kashper, A. N., Alfred, J. A., "Dynamic Routing Interworking," Study Group 2 Delayed Contribution D.48, International Telecommunications Union—Telecommunications Standardization Sector, Geneva, Switzerland, March 1994.

[Aki83] Akinpelu, J. M., "The Overload Performance of Engineered Networks with Nonhierarchical and Hierarchical Routing," *Proceedings of the Tenth International Teletraffic Congress,* Montreal, Canada, June 1983.

[Aki84] Akinpelu, J. M., "The Overload Performance of Engineered Networks with Nonhierarchical and Hierarchical Routing," *Bell System Technical Journal,* Vol. 63, 1984.

[AKK81] Ash, G. R., Kafker, A. H., Krishnan, K. R., "Servicing and Real-Time Control of Networks with Dynamic Routing," *Bell System Technical Journal,* Vol. 60, No. 8, October 1981.

[AKK83] Ash, G. R., Kafker, A. H., Krishnan, K. R., "Intercity Dynamic Routing Architecture and Feasibility," *Proceedings of the Tenth International Teletraffic Congress,* Montreal, Canada, June 1983.

[AkN63] Akimara, H., Nishimura, T., "The Derivatives of Erlang's B Formula," *Review of the Electrical Communications Laboratories,* Vol. 11, No. 9–10, September–October 1963.

[AsC93] Ash, G. R., Chang, F., "Transport Network Design of Integrated Networks with Real-Time Dynamic Routing," *Journal of Network Systems and Management,* Vol. 1, No. 4, 1993.

[Ash85] Ash, G. R., "Use of a Trunk Status Map for Real-Time DNHR," *Proceedings of the Eleventh International Teletraffic Congress,* Kyoto, Japan, September 1985.

[Ash87] Ash, G. R., "Traffic Network Routing, Control, and Design for the ISDN Era," Fifth ITC Specialists Seminar: Traffic Engineering for ISDN Design and Planning, Lake Como, Italy, May 1987.

[Ash90] Ash, G. R., "Design and Control of Networks with Dynamic Nonhierarchical Routing," *IEEE Communications Magazine,* Vol. 28, No. 10, October 1990.

[Ash95] Ash, G. R., "Dynamic Network Evolution, with Examples from AT&T's Evolving Dynamic Network," *IEEE Communications Magazine,* Vol. 33, No. 7, July 1995.

[AsH93] Ash, G. R., Huang, B. D., "An Analytical Model for Adaptive Routing Networks," *IEEE Transactions on Communications,* Vol. 41, No. 11, November 1993.

[AsH94] Ash, G. R., Huang, B. D., "Comparative Evaluation of Dynamic Routing Strategies for a Worldwide Intelligent Network," *Proceedings of the 14th International Teletraffic Congress,* Antibes, France, June 1994.

[AsM84] Ash, G. R., Mummert, V. S., "AT&T Carves New Routes in Its Nationwide Network," *AT&T Bell Laboratories Record,* August 1984, pp. 18–22.

[AsO89] Ash, G. R., Oberer, E., "Dynamic Routing in the AT&T Network—Improved Service Quality at Lower Cost," *Proceedings of the IEEE Global Telecommunications Conference,* Dallas, Texas, November 1989.

[AsS89] Ash, G. R., Schwartz, S. D., "Network Routing Evolution," Network Management and Control Workshop, Tarrytown, New York, September 1989.

[AsS90] Ash, G. R., Schwartz, S. D., "Traffic Control Architectures for Integrated Broadband Networks," *International Journal of Digital and Analog Communication Systems,* Vol. 3, No. 2, April–June 1990.

[ATM96] ATM Forum Technical Committee, "Private Network–Network Interface Specification Version 1.0 (PNNI 1.0)," af-pnni-0055.000, March 1996.

[ATT77] AT&T, "Traffic Facilities Practices—Division G," AT&T, 1977.

[Bha97] Bhandari, R., "Optimal Physical Diversity Algorithms and Survivable Networks," *Proceedings of IEEE Symposium on Computers and Communications,* ISCC'97, Alexandria, Egypt, July 1997.

[BoM66] Boehm, B. W., Mobley, R. L., "Adaptive Routing Techniques for Distributed Communications Systems," *The RAND Corporation Memorandum* RM-4781-PR, February 1966.

[BNR86] BNR, Special Issue: Dynamic Network Controller Family, *Telesis Magazine,* Vol. 13, No. 1, 1986.

[Bro95] Brodie, D., "Dynamic Call Routing Methods," Network Management Development Group, Stockholm, Sweden, June 1995.

[BST77] "No. 4ESS Electronic Switching System," *Bell System Technical Journal,* Vol. 56, No. 7, September 1977.

[BST81] "No. 4ESS Electronic Switching System," *Bell System Technical Journal,* Vol. 60, No. 6, July–August 1981.

[BTL77] Bell Telephone Laboratories, "Engineering and Operations in the Bell System," Bell Telephone Laboratories, Inc., 1977.

[Bul75] Bulfer, A. F., "Blocking and Routing in Two-Stage Concentrators," National Telecommunication Conference, December 1–3, 1975, New Orleans, Louisiana.

[Bur61] Burke, P. J., "Blocking Probabilities Associated with Directional Reservation," unpublished memorandum, 1961.

[Cam81] Cameron, W. H., "Simulation of Dynamic Routing: Critical Path Selection Features for Service and Economy," *International Conference on Communications,* Denver, Colorado, 1981, pp. 55.5.1–55.5.6.

[Car88] Caron, F., "Results of the Telecom Canada High Performance Routing Trial," *Proceedings of the Twelfth International Teletraffic Congress,* Torino, Italy, June 1988.

[CDK83] Carroll, J. J., DiCarlo-Cottone, M. J., Kafker, A. H., "4ESS™ Switch Implementation of Dynamic Nonhierarchical Routing," *Proceedings of the IEEE Global Telecommunications Conference,* San Diego, California, December 1983.

[CED91] Chao, C-W., Eslambolchi, H., Dollard, P., Nguyen, L., Weythman, J., "FASTAR—A Robust System for Fast DS3 Restoration," *Proceedings of GLOBECOM 1991,* Phoenix, Arizona, December 1991, pp. 1396–1400.

[CFG86] Chemouil, P., Filipiak, J., Gauthier, P., "Analysis and Control of Traffic Routing in Circuit-Switched Networks," *Computer Networks and ISDN Systems,* Vol. 11, No. 3, March 1986.

[CGG80] Cameron, W. H., Galloy, P., Graham, W. J., "Report on the Toronto Advance Routing Concept Trial," *Proceedings of the Networks Conference,* Paris, France, 1980.

[CGH81] Chung, F.R.K., Graham, R. L., Hwang, F. K., "Efficient Realization Techniques for Network Flow Patterns," *Bell System Technical Journal,* Vol. 60, No. 8, October 1981.

[ChR90] Chung, S. P., Ross, K. W., "Reduced Load Approximations for Multirate Loss Networks," Technical Report, University of Pennsylvania, 1990.

[CKP91] Chaudhary, V. P., Krishnan, K. R., Pack, C. D., "Implementing Dynamic Routing in the Local Telephone Companies of USA," *Proceedings of the 13th International Teletraffic Congress,* Copenhagen, Denmark, June 1991.

[Coo72] Cooper, R. B., *Introduction to Queueing Theory,* New York: Macmillan; London: Collier-Macmillan Limited, 1972.

[CRG82] Cameron, W. H., Regnier, J., Galloy, P., Savoie, A. M., "Dynamic Routing for Intercity Telephone Networks," *Proceedings of the Tenth International Teletraffic Congress,* Montreal, Canada, June 1983.

[DaF83] David, A. J., Farber, N., "The Switch Planning System for the Dynamic Nonhierarchical Routing Network," *Proceedings of the Tenth International Teletraffic Congress,* Montreal, Canada, June 1983.

[Dij59] Dijkstra, E. W., "A Note on Two Problems in Connection with Graphs," *Numerical Mathematics,* Vol. 1, 1959, pp. 269–271.

[Dre95] Dressler, J., "LAW, the New German Load Dependent Routing," Network Management Development Group, Stockholm, Sweden, June 1995.

[Eis79] Eisenberg, M., "Engineering Traffic Networks for More than One Busy Hour," *Proceedings of the Ninth International Teletraffic Congress,* Melbourne, Australia, 1979.

[Els79] Elsner, W. B., "Dimensioning Trunk Groups for Digital Networks," Ninth International Teletraffic Congress, Torremolinos, Spain, 1979.

[Fel57] Feller, W., *An Introduction to Probability Theory and Its Applications,* New York: John Wiley & Sons, 1957.

[FGH78] Fischer, M. J., Garbin, D. A., Harris, T. C., Knepley, J. E., "Large Scale Communication Networks—Design and Analysis," *The International Journal of Management Science,* Vol. 6, No. 4, 1978.

[FGK73] Fratta, L., Gerla, M., Kleinrock, L., "The Flow Deviation Method: An Approach to Store-and-Forward Communication Network Design," *Networks,* Vol. 3, No. 3, 1973.

[FHH79] Franks, R. L., Heffes, H., Holtzman, J. M., Horing, S., Messerli, E. J., "A Model Relating Measurements and Forecast Errors to the Provisioning of Direct Final Trunk Groups," *Bell System Technical Journal,* Vol. 58, No. 2, February 1979.

[Fie83] Field, F. A., "The Benefits of Dynamic Nonhierarchical Routing in Metropolitan Traffic Networks," *Proceedings of the Tenth International Teletraffic Congress,* Montreal, Canada, June 1983.

[FiK73] Fisher, M. J., Knepley, J. E., "Circuit-Switched Network Performance Algorithm (MOD 1)," Defense Communication Agency, Technical Note TN 1-73, January 1973.

[FoF62] Ford, L. R., Jr., Fulkerson, D. R., *Flows in Networks,* Princeton, New Jersey: Princeton University Press, 1962.

[FrC71] Frank, H., Chou, W., "Routing in Computer Networks," *Networks,* Vol. 1, 1971.

[Fre89] Freeman, R. I., *Telecommunication System Engineering,* New York: John Wiley & Sons, 1989.

[FrR73] Franks, R. L., Rishel, R. W., "Overload Model of Telephone Operation," *Bell System Technical Journal,* Vol. 52, No. 9, November 1973.

[GaK81] Garbin, D. A., Knepley, J. E., "Marginal Cost Routing in Non-Hierarchical Networks," *Proceedings of the International Conference on Communications,* Denver, Colorado, 1981.

[Gal77] Gallagher, R. C., "A Minimum Delay Routing Algorithm Using Distributed Computation," *IEEE Transactions on Communications,* Vol. COM-25, No. 1, January 1977, pp. 73–85.

[Gar97] Garcia-Ayllón, F., "EDA Application in the Telefonica Network," *Proceedings of the IEEE Symposium on Computers and Communications, 1997,* Alexandria, Egypt, July 1997.

[GCK87] Gauthier, P., Chemouil, P., Klein, M., "STAR: A System to Test Adaptive Routing in France," *Proceedings of the IEEE Global Telecommunications Conference,* Tokyo, Japan, 1987.

[Gib86] Gibbens, R. J. "Some Aspects of Dynamic Routing in Circuit-Switched Telecommunications Networks," Statistics Laboratory, University of Cambridge, January 1986.

[Gir90] Girard, A., *Routing and Dimensioning in Circuit-Switched Networks,* Reading, Massachusetts: Addison-Wesley, 1990.

[GiW64] Gimpelson, L. A., Weber, J. H., "UNISM—A Simulation Program for Communication Networks," *Proceedings of Fall Joint Computer Conference,* 1964, pp. 233–249.

[GLC85] Garcia, J. M., Le Gall, F., Castel, C., Chemouil, P., Gauthier, P., Lechermeier, G., "Comparative Evaluation of Centralized/Decentralized Traffic Routing Policies in Telephone Networks," *Proceedings of the Eleventh International Teletraffic Congress,* Kyoto, Japan, September 1985.

[GoH64] Gomory, R. E., Hu, T. C., "Synthesis of a Communication Network," *Journal of the Society of Industrial and Applied Mathematics,* Vol. 12, No. 2, June 1964, pp. 348–369.

[Gra67] Grandjean, C., "Routing Strategies in Telecommunications Networks," *Proceedings of the Fifth International Teletraffic Congress,* New York, New York, 1967.

[GrP80] Grillo, D., Panaioli, "Symmetrical Alternate Routing for Enhanced Telephone Network Resilience," Telecommunications Networks Planning Conference, Paris, France, September 1980.

[HiN76] Hill, D. W., Neal, S. R., "The Traffic Capacity of a Probability Engineered Trunk Group," *Bell System Technical Journal,* Vol. 55, No. 7, September 1976.

[HKO83] Haenschke, D. G., Kettler, D. A., Oberer, E., "DNHR: A New SPC/CCIS Network Management Challenge," *Proceedings of the Tenth International Teletraffic Congress,* Montreal, Canada, June 1983.

[HSS87] Hurley, B. R., Seidl, C.J.R., Sewell, W. F., "A Survey of Dynamic Routing Methods for Circuit-Switched Traffic," *IEEE Communications Magazine,* Vol. 25, 1987.

[Hub85] Huberman, R., "Multihour Dimensioning for a Dynamically Routed Network," *Proceedings of the Eleventh International Teletraffic Congress,* Kyoto, Japan, September 1985.

[Jag74] Jagerman, D. L., "Some Properties of the Erlang Loss Function," *Bell System Technical Journal,* Vol. 53, No. 3, March 1974, pp. 525–551.

[KaI95] Kawashima, K., Inoue, A., "State- and Time-Dependent Routing in the NTT Network," *IEEE Communications Magazine,* Vol. 33, No. 7, July 1995.

[KaJ56] Kalaba, R. E., Juncosa, M. L., "Communication Networks I—Optimal Design and Utilization," *The RAND Corporation Memorandum* RM-1687, April 1956.

[KaJ56a] Kalaba, R. E., Juncosa, M. L., "Communication Networks II—Interoffice Trunking Problems," *The RAND Corporation Memorandum* RM-1688, November 1956.

[KaJ59] Kalaba, R. E., Juncosa, M. L., "Optimal Utilization and Extension of Interoffice Trunking Facilities," *AIEE Transactions,* Part 1: *Communications,* Vol. 77, No. 40, January 1959.

[KAK88] Katz, S. S., Ash, G. R., Kashper, A. N., "Robust Design of a Worldwide Intelligent Network in the Digital Era," *International Seminar on Teletraffic and Networking,* Beijing, China, September 1988.

[Kar84] Karmarkar, N. K., "A New Polynomial Time Algorithm for Linear Programming," *Combinatorica,* Vol. 4, No. 4, 1984, pp. 373–395.

[KaR88] Karmarkar, N. K., Ramakrishnan, K. G., "Implementation and Computational Results of the Karmarkar Algorithm for Linear Programming, Using an Iterative Method for Computing Projections," 13th International Symposium on Mathematical Programming, Tokyo, Japan, 1988.

[KaS91] Kashper, A. N., Schmitt, J. A., "Multiperiod Stochastic Design Model for International Networks," *Proceedings of the Thirteenth International Teletraffic Congress,* Copenhagen, Denmark, June 1991.

[Kat67] Katz, S. S., "Statistical Performance Analysis of a Switched Communication Network," *Proceedings of the Fifth International Teletraffic Congress,* New York, 1967.

[Kat72] Katz, S. S., "Alternate Routing for Nonhierarchical Communication Networks," National Telecommunications Conference Record, December 4–6, 1972.

[KaW95] Kashper, A. N., Watanabe, Y., "Dynamic Routing in the Multiple Carrier International Network," *IEEE Communications Magazine,* Vol. 33, No. 7, July 1995.

[KDP95] Krishnan, K. R., Doverspike, R. D., Pack, C. D., "Improved Survivability with Multi-Layer Dynamic Routing," *IEEE Communications Magazine,* Vol. 33, No. 7, July 1995.

[Kel86] Kelly, F. P., "Blocking Probabilities in Large Circuit-Switched Networks," *Advances in Applied Probability,* Vol. 18, 1986.

[KeW88] Key, P. B., Whitehead, M. J., "Cost-Effective Use of Networks Employing Dynamic Alternative Routing," *Proceedings of the Twelfth International Teletraffic Congress,* Torino, Italy, June 1988.

[Kha97] Khalil, A., "DCR Application in the MCI Network," *Proceedings of the IEEE Symposium on Computers and Communications, 1997,* Alexandria, Egypt, July 1997.

[Kle75] Kleinrock, L. I., *Queueing Systems, Volume I: Theory,* New York: John Wiley & Sons, 1975.

[Kle76] Kleinrock, L. I., *Queueing Systems, Volume II: Computer Applications,* New York: John Wiley & Sons, 1976.

[Kne73] Knepley, J. E., "Minimum Cost Design for Circuit Switched Networks," Technical Note Numbers 36–73, Defense Communications Engineering Center, System Engineering Facility, Reston, Virginia, July 1973.

[Kri82] Krishnan, K. R., "Routing of Telephone Traffic to Minimize Network Blocking," *Proceedings of the Control and Decision Conference,* 1982.

[KrO88] Krishnan, K. R., Ott, T. J., "State-Dependent Routing for Telephone Traffic: Theory and Results," *Proceedings of the 25th Conference on Decision and Control,* Athens, Greece, December 1988.

[KrO88a] Krishnan, K. R., Ott, T. J., "Forward-Looking Routing: A New State-Dependent Routing Scheme," *Proceedings of the 12th International Teletraffic Congress,* Torino, Italy, 1988.

[Kru37] Kruithof, J., "Telefoonverkeersrekening," *De Ingenieur,* Vol. 52, No. 8, February 1937.

[Kru79] Krupp, R. S., "Properties of Kruithof's Projection Method," *Bell System Technical Journal,* Vol. 58, No. 2, February 1979.

[Kru82] Krupp, R. S., "Stabilization of Alternate Routing Networks," *IEEE International Communications Conference,* Philadelphia, Pennsylvania, 1982.

[KuN79] Kumar, P.R.S., Narendra, K. S., "Learning Algorithm Model for Routing in Telephone Networks," Systems and Information Sciences Report No. 7903, Yale University, May 1979.

[LaR91] Langlois, F., Regnier, J., "Dynamic Congestion Control in Circuit-Switched Telecommunications Networks," *Proceedings of the Thirteenth International Teletraffic Congress,* Copenhagen, Denmark, June 1991.

[Lee55] Lee, C. Y., "Analysis of Switching Networks," *Bell System Technical Journal,* Vol. 34, No. 6, November 1955, pp. 1287–1315.

[LLS78] Lin, P. M., Leon, B. J., Stewart, C. R., "Analysis of Circuit-Switched Networks Employing Originating-Office Control with Spill-Forward," *IEEE Transactions on Communications,* Vol. 26, No. 6, June 1978.

[LyM88] Lynch, D. F., Moreland, J. P., "Economic Trunk Group Sizing for Stochastic Traffic Demands," *Proceedings of the Twelfth International Teletraffic Congress,* Torino, Italy, June 1988.

[McN78] McKenna, D. M., Narendra, K. S., "Simulation Study of Telephone Traffic Routing Using Learning Algorithms," Technical Report No. 7806, Yale University, 1978.

[Mcq77] McQuillan, J. M., "Routing Algorithms for Computer Networks—A Survey," *Proceedings of National Telecommunications Conference,* 1977.

[Med92] Medhi, D., "A Unified Framework for Survivable Telecommunications Network Design," *Proceedings of the International Conference on Communications,* Chicago, Illinois, 1992.

[Mee86] Mees, A. "Simple Is Best for Dynamic Routing in Circuit-Switched Telecommunications Networks," *Nature,* Vol. 323, September 11, 1986, p. 108.

[MGH91] Mitra, D., Gibbens, R. J., Huang, B. D., "Analysis and Optimal Design of Aggregated-Least-Busy-Alternative Routing on Symmetric Loss Networks with Trunk Reservation," *Proceedings of the Thirteenth International Teletraffic Congress,* Copenhagen, Denmark, June 1991.

[Min74] Mina, R. R., *Teletraffic Engineering,* Chicago, Illinois: Telephony Publishing Corporation, 1974.

[Min89] Minzer, S. E., "Broadband ISDN and Asynchronous Transfer Mode (ATM)," *IEEE Communications Magazine,* Vol. 27, No. 9, September 1989.

[MiS91] Mitra, D., Seery, J. B., "Comparative Evaluations of Randomized and Dynamic Routing Strategies for Circuit-Switched Networks," *IEEE Transactions on Communications,* Vol. 39, No. 1, January 1991.

[MRS80] McQuillan, J., Richter, I., Rosen, E., "The New Routing Algorithm for the ARPANET," *IEEE Transactions on Communications,* Vol. 28, No. 5, May 1980.

[Mum76] Mummert, V. S., "Network Management and Its Implementation on the No. 4ESS," International Switching Symposium, Japan, 1976.

[NaM73] Nakagome, Y., Mori, H., "Flexible Routing in the Global Communication Network," *Proceedings of the Seventh International Teletraffic Congress,* Stockholm, Sweden, 1973.

[NaT78] Narendra, K. S., Thathachar, M.A.L., "On the Behavior of a Learning Automaton in a Changing Environment with Application to Telephone Traffic Routing," Systems and Information Sciences Report No. 7803, Yale University, October 1978.

[NaT80] Narendra, K. S., Thathachar, M.A.L., "On the Behavior of a Learning Automaton in a Changing Environment with Application to Telephone Traffic Routing," *IEEE Transactions on Systems, Man, and Cybernetics,* Vol. SMC-10, No. 5, May 1980.

[Nea80] Neal, S. R., "Blocking Distributions for Trunk Network Administration," *Bell System Technical Journal,* Vol. 59, No. 6, July–August 1980.

[NWM77] Narendra, K. S., Wright, E. A., Mason, L. G., "Application of Learning Automata to Telephone Traffic Routing and Control," *IEEE Transactions on Systems, Man, and Cybernetics,* Vol. SMC-7, No. 11, November 1977.

[Osb30] Osborne, H. S., "A General Switching Plan for Telephone Toll Services," *Transactions of the AIEE,* Vol. 49, October 1930, pp. 1549–1557.

[OtK85] Ott, T. J., Krishnan, K. R., "State Dependent Routing of Telephone Traffic and the Use of Separable Routing Schemes," *Proceedings of the 11th International Teletraffic Congress,* Kyoto, Japan, 1985.

[PaW82] Pack, C. D., Whitaker, B. A., "Kalman Filter Models for Network Forecasting," *Bell System Technical Journal,* Vol. 61, No. 1, January 1982.

[Pil52] Pilliod, J. J., "Fundamental Plans for Toll Telephone Plant," *Bell System Technical Journal,* Vol. 31, No. 5, September 1952, pp. 832–850.

[Pio89] Pioro, M., "Design Methods for Non-Hierarchical Circuit Switched Networks with Advanced Routing," Warsaw University of Technology Publications, 1989.

[Pra67] Pratt, C. W., "The Concept of Marginal Overflow in Alternate Routing," *Proceedings of the Fifth International Teletraffic Congress,* New York, 1967.

[Rap62] Rapp, Y., "Planning of Multi-Exchange Networks, Part III," *Ericsson Review,* No. 4, 1962, pp. 102–104.

[RBC83] Regnier, J., Blondeau, P., Cameron W. H., "Grade of Service of a Dynamic Call-Routing System," *Proceedings of the Tenth International Teletraffic Congress,* Montreal, Canada, June 1983.

[RBC95] Regnier, J., Bedard, F., Choquette, J., Caron, A., "Dynamically Controlled Routing in Networks with Non DCR-Compliant Switches," *IEEE Communications Magazine,* Vol. 33, No. 7, July 1995.

[Rey83] Rey, R. F., "Engineering and Operations in the Bell System," Murray Hill, New Jersey, 1983, AT&T Bell Laboratories.

[Rud76] Rudin, H., "On Routing and Delta Routing: A Taxonomy and Performance Comparison of Techniques for Packet-Switched Networks," *IEEE Transactions on Communications,* Vol. 24, January 1976, pp. 43–57.

[Sch87] Schwartz, M. I., *Telecommunication Networks: Protocols, Modeling and Analysis,* Reading Massachusetts: Addison-Wesley, 1987.

[ScW94] Schmitt, J. A., Watanabe, Y., "Models and Results for Planning the Evolution to Worldwide Intelligent Network (WIN) Dynamic Routing," *Proceedings of the 14th International Teletraffic Congress,* Antibes, France, June 1994.

[Seg64] Segal, M., "Traffic Engineering of Communications Networks with a General Class of Routing Schemes," *Proceedings of the Fourth International Teletraffic Congress,* London, England, July 15–21, 1964.

[Seg77] Segall, A., "The Modeling of Adaptive Routing in Data-Communications Networks," *IEEE Transactions on Communications,* Vol. 25, January 1977, pp. 85–95.

[She75] Shearer, C. N., "Modified Forward Routing for CONUS/CSN AUTOVON," Technical Report Number 26-75, Defense Communications Engineering Center, Switched Networks Engineering Division (R500), Reston, Virginia, September 1975.

[StS87] Stacey, R. R., Songhurst, D. J., "Dynamic Alternative Routing in the British Telecom Trunk Network," *Proceedings of the International Switching Symposium,* Phoenix, Arizona, March 1987.

[Ste95] Steenstrup, M. E., *Routing in Communications Networks,* Englewood Cliffs, New Jersey: Prentice Hall, 1995.

[SzB79] Szybicki, E., Bean, A. E., "Advanced Traffic Routing in Local Telephone Networks: Performance of Proposed Call Routing Algorithms," *Proceedings of the Ninth International Teletraffic Congress,* Torremolinos, Spain, 1979.

[Sze80] Szelag, C. R., "Trunk Demand Servicing in the Presence of Measurement Uncertainty," *Bell System Technical Journal,* Vol. 59, No. 6, July–August 1980, pp. 845–860.

[Tan81] Tanenbaum, A. S., *Computer Networks,* Englewood Cliffs, New Jersey: Prentice-Hall, 1981.

[Top86] Topkis, D. M., "Reordering Heuristics for Routing in Communications Networks," *Journal of Applied Probability,* Vol. 23, 1986.

[Top88] Topkis, D. M., "A k Shortest Path Algorithm for Adaptive Routing in Communications Networks," *IEEE Transactions on Communications,* Vol. 36, No. 7, July 1988.

[Top89] Topkis, D. M., "All-to-All Broadcast by Flooding in Communications Networks," *IEEE Transactions on Computers,* Vol. 38, No. 9, September 1989.

[Tru54] Truitt, C. J., "Traffic Engineering Techniques for Determining Trunk Requirements in Alternate Routed Networks," *Bell System Technical Journal,* Vol. 31, No. 2, March 1954.

[Usr95] Usry, J., "DCR: Dynamically Controlled Routing," Network Management Development Group, Stockholm, Sweden, June 1995.

[WaD88] Wanamaker, D. M., Dorrance, D. A., "Dynamically Controlled Routing Field Trial Experience," *Proceedings of the Network Operations and Management Systems Conference,* New Orleans, Louisiana, March 1988.

[WaM87] Watanabe, Y., Mori, H., "Dynamic Routing Schemes for International ISDNs," *Proceedings of the Fifth ITC Specialists Seminar: Traffic Engineering for ISDN Design and Planning,* Lake Como, Italy, May 1987.

[Web62] Weber, J. H., "Some Traffic Characteristics of Communication Networks with Automatic Alternate Routing," *Bell System Technical Journal,* Vol. 41, 1962, pp. 769–796.

[Web64] Weber, J. H., "A Simulation Study of Routing and Control in Communications Networks," *Bell System Technical Journal,* Vol. 43, 1964, pp. 2639–2676.

[Wei63] Weintraub, S., *Tables of Cumulative Binomial Probability Distribution for Small Values of* p, London: Collier-Macmillan Limited, 1963.

[Whi85] Whitt, W., "Blocking When Service Is Required from Several Facilities Simultaneously," *AT&T Technical Journal,* Vol. 64, 1985.

[Wil56] Wilkinson, R. I., "Theories of Toll Traffic Engineering in the U.S.A.," *Bell System Technical Journal,* Vol. 35, No. 6, March 1956.

[Wil58] Wilkinson, R. I., "A Study of Load and Service Variations in Toll Alternate Route Systems," *Proceedings of the Second International Teletraffic Congress,* The Hague, Netherlands, July 1958, Document No. 29.

[Wil71] Wilkinson, R. I., "Some Comparisons of Load and Loss Data with Current Teletraffic Theory," *Bell System Technical Journal,* Vol. 50, October 1971, pp. 2807–2834.

[WMM85] Watanabe, Y., Matsumoto, J., Mori, H., "Design and Performance Evaluation of International Networks with Dynamic Routing," *Proceedings of the Eleventh International Teletraffic Congress,* Kyoto, Japan, September 1985.

[Wol90] Wolf, R. B., "Advanced Techniques for Managing Telecommunications Networks," *IEEE Communications Magazine,* Vol. 28, No. 10, October 1990.

[WoY90] Wong, E.W.M., Yum, T-S., "Selective Alternate Routing for Nonhierarchical Circuit-Switched Networks," Department of Electronics, The Chinese University of Hong Kong, Shatin, Hong Kong.

[WoY90] Wong, E.W.M., Yum, T-S., "Maximum Free Circuit Routing in Circuit-Switched Networks," *Proceedings of IEEE INFOCOM '90,* 1990.

[Yag71] Yaged, B., Jr., "Long Range Planning for Communications Networks," Polytechnic Institute of Brooklyn, Ph.D. Thesis, 1971.

[Yag73] Yaged, B., "Minimum Cost Design for Circuit Switched Networks," *Networks,* Vol. 3, 1973, pp. 193–224.

[YMI91] Yamamoto, H., Mase, K., Inoue, A., Suyama, M., "A State- and Time-Dependent Dynamic Routing Scheme for Telephone Networks," *Proceedings of the 13th International Teletraffic Congress,* Copenhagen, Denmark, June 1991.

Abbreviations Glossary

ABD	Average Business Day
ALBY	ALBanY, New York
ANHM	ANaHeiM, California
ARPANET	Advanced Research Projects Agency NETwork
AS	Access Switch
ATLN	ATLaNta, Georgia
ATM	Asynchronous Transfer Mode
ATR	Automatic Trunk Reservation
AUTOVON	AUTOmatic VOice Network
AVG	AVeraGe
BGL	BackGround Load
B-ISDN	Broadband Integrated Services Digital Network
BLK	BLocKed
BLKG	BLocKinG
BNR	Bell Northern Research
BRHM	BiRmingHaM, Alabama
BUS	BUSiness
CAMA	Centralized Automatic Message Accounting
CAP	CAPacity
Cavg	Completion rate average
CC	Cross-Connect
CCIS	Common-Channel Interoffice Signaling
CCITT	International Telegraph and Telephone Consultative Committee
CCS	Common-Channel Signaling *or* CCS hundred (Roman numeral C) Call Seconds
CGH	Chung-Graham-Hwang (routing)
CHG	CHanGe
CIP	Calls In Progress
CLLI	Common Language Location Identification
CLLT	Carried Load on Last Trunk
CMF	Common Master Files

CPU	Central Processing Unit
CRBK	CRankBacK
CS1	Congested Switch (level 1)
CT	CounT
DAR	Dynamic Alternative Routing
DCME	Digital Circuit Multiplexing Equipment
DCR	Dynamically Controlled Routing
DCS	Digital Cross-connect System
DDD	Direct Distance Dialing
DEFO	Discrete Event Flow Optimization
DLLS	DaLLaS, Texas
DNHR	Dynamic NonHierarchical Routing
DNST	Dialed Number Service Type
DNVR	DeNVeR, Colorado
DOC	Dynamic Overload Control
DOM	DOMestic
DR-5	Dynamic Routing—5 minutes
DTMF	Dual-Tone Multiple Frequency
DYN	DYNamic
ECCS	Economic hundred (Roman numeral C) Call Seconds
EFO	Erlang Flow Optimization
ENG	ENGineered
ENR	ENgineered Routing
EO	End-Office
EQUIV	EQUIValent
ERL	ERLang
ESS	Electronic Switching System
EXC	EXClusion
FASTAR	FAST Automatic Restoration
FHR	Fixed Hierarchical Routing
FP	Forecast Period
FRHD	FReeHolD, New Jersey
FTS	Federal Telecommunications System
GOS	Grade-Of-Service
GSDN	Global Software-Defined Network

HC	High Completion
HCthr	High-Completion threshold
HIER	HIERarchical (routing)
HL	Heavily Loaded
HR	HouR
HTR	Hard-To-Reach
HU	High-Usage (link)
IAM	Initial Address Message
IDDD	International Direct Distance Dialing
ILBW	Idle-Link BandWidth
ILDS	International Long-Distance Service
IND	INDicator
INTL	INTernationaL
INWATS	INward Wide Area Telecommunications Service
ISC	International Switching Center
ISDN	Integrated Services Digital Network
ISUP	ISDN User Part
ITU-T	International Telecommunications Union—Telecommunications (standardization sector)
kb	kilobits
kbps	kilobits per second
LAMA	Local Automatic Message Accounting
LBW	Link BandWidth
LC	Low Completion
LCthr	Low-Completion threshold
LDS	Long-Distance Service
LEC	Local Exchange Carrier
LIN	LINear
LL	Lightly Loaded
LLVT	Load on Last Virtual Trunk
LP	Linear Program
LRNG	LeaRNinG
LRR	Learning with Random Routing
LSP	Load Set Period
MAX	MAXimum

MAXFLOW	MAXimum FLOW
Mb	Megabits
mbps	megabits per second
MC3	Machine Congestion (level 3)
MCI	Microwave Communications Incorporated
MF	MultiFrequency
MGMT	ManaGeMenT
MIER	Multiple Ingress/Egress Routing
MIN	MINimum
MINCOST	MINimum COST
MPSX	Mathematical Programming System eXtended
MTGM	MonTGoMery, Alabama
MTS	Message Telecommunications Service
MXDR	MiXed Dynamic Routing
NANP	North American Numbering Plan
NAR	NonAlternate Routing
NC	No Circuit
NCP	Network Control Point
NEMOS	NEtwork Management Operations System
NET	NETwork
NFS	Network Forecasting System
NM	Network Management
NO	Number
NPA	Numbering Plan Area
NRFH	No Recent Failure History
NSN	Network Switch Number
NTC	National Trunk Congestion
NTT	Nippon Telegraph and Telephone (Company)
NWOR	NeW ORleans, Louisiana
OKLD	OaKLanD, California
OPT	OPTimization
ORIG	ORIGinating
OS	Originating Switch
OSO	Originating Screening Office
OTS	Originating Toll Switch

OUTWATS	OUTward Wide Area Telecommunications Service
OV	OVerflow (count)
OVFL	OVerFLow
PBX	Private Branch eXchange
PC	Peg Count
PCS	Packet Cross-connect System
PERF	PERFormance
PH	PatH
PHNX	PHoeNiX, Arizona
PNNI	Private Network–Network Interface
PROG	PROGram
PTYPE	Path TYPE
R	Reserved (state)
RBW	Reserved BandWidth
RECOM	RECOMmendation
RES	RESidential
RESLVL	REServation LeVeL
RESP	RESPonsibility
RESV	RESerVation
RINR	Real-time InterNetwork Routing
RNDM	RaNDoM
RT	Real-Time
RTG	RouTinG
RTNR	Real-Time Network Routing
SBND	South BeND, Indiana
SCRM	SaCRaMento, California
SDI	Switched Digital International (service)
SDN	Software-Defined Network (service)
SDOC	Selective Dynamic Overload Control
SDS	Switched Digital Service
SEQ	Sequential
SI	Service Identity
SLKC	Salt LaKe City, Utah
SNBO	SaN BernardinO, California
SNDG	SaN DieGo, California

SNFC	SaN FranCisco, California
SOLN	SOLutioN
SONET	Synchronous Optical NETwork
SPA	Sequential Projection Algorithm
SPC	Stored Program Control
SST	Signaling Service Type
STAR	System to Test Adaptive Routing
STP	Signal Transfer Point
STR	Selective Trunk Reservation *or* State- and Time-dependent Routing
STT	Success-To-the-Top (routing)
SW	SWitch
T1-IL	T1 Interswitch Link
TA	Terminal Adapter *or* Terminating Area
TAND	TANDem
TASI	Time Assignment Speech Interpolation
TAT9	TransATlantic (cable number 9)
TC	Transport Capability
TCM	Traveling Class Mark
TCSN	TuCSoN, Arizona
TD	Type of Destination
TERM	TERMinating *or* TERMination
TFO	Transport Flow Optimization
TK	TrunK
TKG	TrunKinG
TL	Traffic Level
TLBW	Total Link BandWidth
TO	Type of Origin
TOC	True-Overflow-Count
TOP	True-Overflow-Percent
TPC5	TransPacific Cable (number 5)
TPRL	TransPort Restoration Level
TRAF	TRAFfic
TRANS	TRANSport
TRK	TRunK
TRL	Traffic Restoration Level

TS	Terminating Switch *or* Toll Switch
TSB	Total Switch Blocking
TSL	Trunk SubLink
TSMR	Trunk Status Map Routing
TTS	Terminating Toll Switch
UK	United Kingdom
UNAS	UNASsigned
UPBD	UPper BounD
VAR	VARiation
VC	Virtual Circuit
VCI	Virtual Channel Identifier
VIRT	VIRTual
VN	Virtual Network
VP	Virtual Path
VPI	Virtual Path Identifier
VS	Via Switch
VT	Virtual Trunk
VTeng	Virtual Trunk engineered
VTSM	Virtual Trunk Status Map
WHPL	WHite PLains, New York
WIN	Worldwide International Network (routing)
WKE	WeeKEnd
ZPR	Zero-Profit Reroute

Index